Animal Behavior

ELEVENTH EDITION

Animal Behavior

ELEVENTH EDITION

Dustin R. Rubenstein • John Alcock
Columbia University Arizona State University

NEW YORK OXFORD
OXFORD UNIVERSITY PRESS

About the Cover

This male peacock spider (*Maratus elephans*) from Tamworth, NSW, Australia, unfurls from underneath his abdomen what looks like the tail of a male peacock. He also raises his legs as if to dance, something you can watch by following the QR code on the back cover. (Photograph by Michael Doe.)

Animal Behavior, Eleventh Edition

Oxford University Press is a department of the University of Oxford. It furthers the University's objective of excellence in research, scholarship, and education by publishing worldwide. Oxford is a registered trade mark of Oxford University Press in the UK and certain other countries.

Published in the United States of America by Oxford University Press
198 Madison Avenue, New York, NY 10016, United States of America

Sinauer Associates is an imprint of Oxford University Press.

For titles covered by Section 112 of the US Higher Education Opportunity Act, please visit www.oup.com/us/he for the latest information about pricing and alternate formats.

Address editorial correspondence to:
Sinauer Associates
23 Plumtree Road
Sunderland, MA 01375 USA
publish@sinauer.com

Address orders, sales, license, permissions, and translation inquiries to:
Oxford University Press USA
2001 Evans Road
Cary, NC 27513 USA
Orders: 1-800-445-9714

Library of Congress Cataloging-in-Publication Data

Names: Alcock, John, 1942- author. | Rubenstein, Dustin R., author.
Title: Animal behavior / Dustin R. Rubenstein, Columbia University, John Alcock, Arizona State University.
Description: Eleventh edition. | Sunderland, Massachusetts: Oxford University Press, [2019] | Revised edition of: Animal behavior: an evolutionary approach / John Alcock. 10th ed. c2013. | Includes bibliographical references.
Identifiers: LCCN 2018004580 | ISBN 9781605355481 (paperback) | ISBN 9781605357706 (ebook)
Subjects: LCSH: Animal behavior--Evolution.
Classification: LCC QL751 .A58 2019 | DDC 591.5--dc23
LC record available at https://lccn.loc.gov/2018004580

9 8 7 6 5 4 3 2 1
Printed in the United States of America

To my parents,
who inspired in me a love of nature
and animal behavior at an early age.
DRR

Contents in Brief

Contents

CHAPTER 3 The Developmental and Genetic Bases of Behavior 58

CHAPTER 4 The Neural Basis of Behavior 104

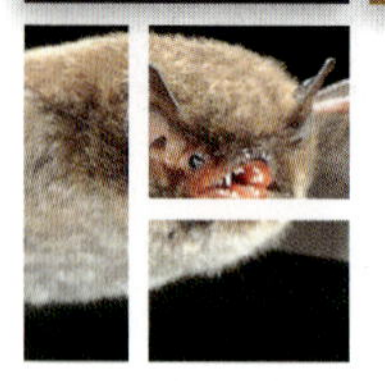

CHAPTER 5

The Physiological Basis of Behavior 144

CHAPTER 6

Avoiding Predators and Finding Food 184

CHAPTER 7 Territoriality and Migration 218

CHAPTER 8 Principles of Communication 256

CHAPTER 9

CHAPTER 10

CHAPTER 11 Parental Care 400

CHAPTER 12 Principles of Social Evolution 444

CHAPTER 13 Social Behavior and Sociality 476

CHAPTER 14 Human Behavior 512

Preface

For more than 40 years, *Animal Behavior* has been the leading textbook for introducing undergraduate students to the topic of animal behavior. John Alcock authored the first edition of this book in 1975, and after 9 subsequent versions, this eleventh edition brings on a new lead author, Dustin Rubenstein. The eleventh edition maintains its narrative tone as well as its focus on both evolutionary and mechanistic approaches to understanding how and why animals as different as insects and humans behave the way they do. In an effort to keep up with the rapidly evolving field of animal behavior, this new version also brings a more integrative approach to studying behavior, emphasizing the growing body of research linking behavior to the brain, genes, and hormones, as well as to the surrounding ecological and social environments. Topics like epigenetics and collective behaviors are highlighted for the first time. Additionally, the book covers the growing number of comparative phylogenetic studies in animal behavior that make use of ever-larger molecular phylogenies to generate and test new ideas in the evolution of animal behaviors. Ultimately, the book retains its primary goal of giving students a window into the various level of analysis that researchers use to explain why all living things—including humans—behave, often in complex ways.

New to the Eleventh Edition

In addition to a new lead author, the eleventh edition features several new approaches and features that support both student learning and instructor teaching. After extensive research concluded that most users prefer the classic organizational structure from early editions, rather than the changes that were made in the tenth edition, the book returns to its classic organizational structure, with proximate mechanisms introduced early in the book and before an extended discussion of the ultimate factors underlying behavior. Yet, each chapter attempts to highlight both proximate and ultimate explanations throughout, illustrating the integrative nature of the field today. This edition also includes a refined organizational structure within each chapter that makes material more accessible to students and instructors alike. Figures and photos throughout the book have been updated and revised as well, providing a fresher look that, along with the addition of error bars on most graphical figures, makes it easier for students to learn how to interpret data. The addition of new pedagogical features also makes it easier for students to learn complex concepts and apply behavioral and evolutionary thinking to thought-provoking questions and problems. Found in every chapter, these new boxes and tables emphasize hypothesis testing, data interpretation, and problem solving. Finally, since most students are drawn to animal behavior out of a fascination with something they have seen on TV or in nature, in each chapter we use QR codes offset in the margins to give students rapid access to high-definition video and audio clips of the behaviors discussed in the text.

Focus on Integration

This book provides a comparative and integrative overview of animal behavior, linking a diversity of behaviors and their adaptive functions to the brain, genes, and hormones, as well as to the surrounding ecological and social environments. Just as so many modern studies in animal behavior are taking advantage of new neurobiological or molecular approaches, this book introduces these and other cutting-edge techniques to its readers, all while maintaining a focus on the theoretical aspects of the field in an explicit hypothesis-testing framework. Ultimately, the book highlights both the evolutionary and mechanistic approaches to studying animal behavior, as well as the interdisciplinary approaches that emphasize the neural, genetic, and physiological mechanisms underlying adaptive behaviors. Because this new edition of the book is more integrative than previous versions, we have dropped the book's subtitle, *An Evolutionary Approach*, and now simply call it *Animal Behavior*. The concept of evolution by natural selection remains our guide as we expand in new and exciting directions.

In addition to the focus on integrative approaches, this edition features an update of specific topics such as how new technologies are revolutionizing the way we study animal movements and the genetic architecture of behavior, as well as new and expanded coverage of animal social behavior, animal communication, and human behavior, in addition to other topics. New empirical examples have been added to each chapter, some of which extend classic examples in new and exciting directions.

New Pedagogical Approach

Built upon the foundation of rich content and discussion topics from previous editions, we've applied more modern pedagogical tools and added new features in the eleventh edition. Found in every chapter, these new boxes and tables emphasize hypothesis testing, data interpretation, and problem solving. These features are designed to engage students, both inside and outside of the classroom, and aid instructors in developing their courses and delivering their lectures. New features include:

- **INTEGRATIVE APPROACHES:** Focusing attention on cutting-edge tools in the study of animal behavior, these boxes cover tools for studying birdsong, animal coloration, behavioral genetics, hormones, and how to conduct ethical studies of humans, among other topics.
- **HYPOTHESES TABLES:** In brief summaries, alternative or non-mutually exclusive hypotheses for specific animal behaviors are presented side-by-side.
- **EXPLORING BEHAVIOR BY INTERPRETING DATA:** In these boxes, we explore a concept in data analysis, such as analyzing spectrograms to show how song learning in birds can be adopted by other species, or drawing conclusions and generating new hypotheses from empirical data figures.

- **DARWINIAN PUZZLES:** Adapted from this popular feature in the tenth edition, these boxes deal specifically with unresolved "puzzles" in behavioral research, such as why females in some species have what looks like a penis but males don't.

In most boxes, students are challenged to "Think Outside the Box" by answering a series of thought-provoking questions related to the content and figures inside the box. These have been designed to foster in-class discussion between instructors and students. This feature complements Discussion Questions, which have now been consolidated online for easier student and instructor access.

QR Codes/Direct Web Links for Audio and Video Clips

Students taking a course on animal behavior want to see and hear animals behaving, and in this edition, we've integrated that directly into the text in an easily accessible manner. Using QR codes (and short URLs), students can immediately access audio and video directly related to the text in real-time, without having to wade through a large collection of resources on a companion website.

ABOUT THE TECHNOLOGY. On many iPhones and some Android phones, QR codes no longer require a QR reader app. (Any iPhone running iOS 11 or newer can read QR codes directly.) Simply open the camera, hold the phone over the QR code, and the audio/video link will appear. Other smartphones may require that you download a mobile app. For those who don't want to scan the QR codes, short URLs are provided for each individual audio and video clip (e.g., ab11e.sinauer.com/a2.1).

ABOUT THE AUDIO AND VIDEO LINKS. Most of the audio and video links in the book are provided by researchers whose examples appear in the text, or by the Macaulay Library at the Cornell Lab of Ornithology, drawing from their extensive collection of nearly 7,000,000 pictures, videos, and animal sounds.

Author Biographies

Dustin Rubenstein, an Associate Professor of Ecology, Evolution, and Environmental Biology at Columbia University, has studied animal behavior in birds, reptiles, mammals, insects, and crustaceans for nearly 20 years throughout Africa, Asia, Central and South America, and the Caribbean. As a leading expert in animal social behavior and evolution, his research has been published in top journals like *Science*, *Nature*, and *PNAS*, and he is co-editor of the book *Comparative Social Evolution*, published in 2017. In recognition of his research accomplishments, he has received young investigator awards from the Animal Behavior Society, the American Ornithologists' Union, the Society for Behavioral Neuroendocrinology, and the University of Michigan, and been made a Fellow of the American Ornithological Society. He was also recognized by the National Academy of Sciences as both a Kavli Fellow for his research accomplishments and an Education Fellow in the Sciences for his innovations in STEM teaching.

Throughout his education and training at Dartmouth College, Cornell University, and the University of California, Berkeley, Rubenstein has used *Animal Behavior* in his courses since its fifth edition. He seems to have learned a thing or two, having met his wife in his undergraduate animal behavior class. From his own work using stable isotopes to study avian migration, to studying stress hormones and breeding behavior in birds and lizards, to examining the genetic and epigenetic bases of reproductive behavior in insects and birds, Rubenstein approaches the study of animal behavior in an integrative and interdisciplinary manner. Over the past few years, he has helped lead a series of workshops and symposia about the integrative study of behavior, a topic of great emphasis in this new edition. His goal in taking over this classic textbook was to highlight the modern approach to animal behavior, while maintaining the book's timeless writing style and structure. Rubenstein thanks his wife Kate and two children, Renna and Ian, for their patience and support over the past few years of book writing. Many mentors helped teach and train Dustin in animal behavior over the years, but none were more important than his father, Daniel Rubenstein, a fellow behavioral biologist who has inspired him since his earliest days, together observing spiders in the backyard or zebras in Africa.

John Alcock retired in 2009 after having had an academic life full of rewards, such as trips to Australia on sabbaticals and during leaves of absence. His research focused on insect reproductive behavior and the many puzzles it provides for behavioral biologists, both in his home state of Arizona as well as in Costa Rica and Australia. Although retired, he continues to study insects, especially in Virginia, where he lives on the family farm in the spring and summer months. Alcock has written a number of natural history books for a general audience, such as *Sonoran Desert Spring* and his most recent effort along these lines, *After the Wildfire*. This book describes the account of the environmental recovery that occurred in the ten years that followed an intense wildfire in the mountains about an hour to the north of his home in Tempe,

Arizona. Now 75, Alcock enjoys being a grandfather, thanks to a son Nick and his wife Sara who have produced two grandchildren for him (and others). His other son Joe and his wife Satkirin are both doctors, and Joe is deeply interested in evolutionary medicine. This version of the textbook that Alcock first wrote long, long ago has been revised by Dustin Rubenstein, a good choice to carry the book forward as the readers of this text will shortly learn.

Acknowledgements

For over four decades, this book has greatly benefitted from the generosity of many colleagues who have contributed reviews, photographs, videos, and other means of support. We are especially grateful to the reviewers for this eleventh edition, whose suggestions improved the structure, content, and tone of many of the chapters. The reviewers for this edition include:

Patrick Abbot	Vanderbilt University
Andres Bendesky	Columbia University
Kirsten M. Bohn	Johns Hopkins University
Jennifer S. Borgo	Coker College
Patty L.R. Brennan	Mount Holyoke College
Molly Cummings	The University of Texas at Austin
Ben Dantzer	University of Michigan
Harold Gouzoules	Emory University
Sarah Guindre-Parker	University of Guelph
Kristina M. Hannam	State University of NY-Geneseo
Margret I. Hatch	Penn State Worthington Scranton
Mark E. Hauber	City University of New York
Hans Hofmann	University of Texas at Austin
Victoria Ingalls	Marist College
Roland Kays	North Carolina Museum of Natural Sciences
Darcy Kelley	Columbia University
Walt Koenig	Cornell University
Rafael Maia	Columbia University
Heather Masonjones	University of Tampa
Ronald L. Mumme	Allegheny College
Wiline M. Pangle	Central Michigan University
Kimberly N. Russell	Rutgers University
D. Kim Sawrey	University of North Carolina Wilmington
Joseph Sisneros	University of Washington
Nancy G. Solomon	Miami University
Elizabeth Tibbetts	University of Michigan
Sarah Woolley	Columbia University
Tim Wright	New Mexico State University
Ken Yasukawa	Beloit College

We are also indebted to our wonderful colleagues at Sinauer Associates. Without their dedication, patience, and careful oversight (and occasional prodding), this book—as well as the ten previous editions—would never have made it to print. The team that worked on this edition included:

Andrew D. Sinauer, Publisher
Rachel Meyers, Acquisitions Editor
Martha Lorantos, Production Editor
Chris Small, Production Manager
Beth Roberge Friedrichs, Production Specialist
Joanne Delphia, Book Designer
Mark Siddall, Photo Researcher
Elizabeth Pierson, Copy Editor

Media and Supplements

to accompany *Animal Behavior*, Eleventh Edition

eBook

(ISBN 9-781-60535-770-6)

Animal Behavior, Eleventh Edition, is available as an eBook, in several different formats, including RedShelf, VitalSource, and Chegg. All major mobile devices are supported.

For the Student

Companion Website (oup.com/us/rubenstein11e)

The *Animal Behavior*, Eleventh Edition, Companion Website includes a variety of study and review aids—all available at no cost to the student. The site includes the following:

- Chapter Summaries give the student a thorough review of each chapter's content.
- Audio and Video Links provide a set of online sites and resources relevant to each chapter.
- The Glossary provides definitions for all textbook bolded terms.

Student Lab Manual

This electronic lab manual walks students through the scientific process to improve their fluency with hypothesis development, observing and quantifying animal behavior, statistical analysis, and data presentation.

For the Instructor

Ancillary Resource Center (oup-arc.com)

The Ancillary Resource Center provides instructors using *Animal Behavior* with a wealth of resources for use in course planning, lecture development, and assessment. Contents include:

TEXTBOOK FIGURES AND TABLES

All the figures and tables from the textbook are provided in JPEG format, reformatted for optimal readability, with complex figures provided in both whole and split formats.

POWERPOINT RESOURCES

This presentation includes all of the figures and tables from the chapter, with titles on each slide and complete captions in the Notes field.

INSTRUCTOR'S MANUAL AND TEST BANK

These resources facilitate the preparation of lectures, quizzes, and exams. Contents include:

- Answers to the discussion questions in the textbook
- A list of films and documentaries about animal behavior, for use in the classroom
- A test bank, consisting of a range of multiple-choice and short answer questions drawn from textbook content, along with writing assignments to engage students in deeper thinking.

Learning the Skills of Research: Animal Behavior Exercises in the Laboratory and Field

Edited by Elizabeth M. Jakob and Margaret Hodge

Students learn best about the process of science by carrying out projects from start to finish. Animal behavior laboratory classes are particularly well suited for independent student research, as high-quality projects can be conducted with simple materials and in a variety of environments. The exercises in this electronic lab manual are geared to helping students learn about all stages of the scientific process: hypothesis development, observing and quantifying animal behavior, statistical analysis, and data presentation. Additional exercises allow the students to practice these skills, with topics ranging from habitat selection in isopods to human navigation. The manuals provide both student and instructor documentation. Data sheets and other supplementary material are offered in editable formats that instructors can modify as desired.

To learn more about any of these resources, or to get access, please contact your OUP representative.

Animal Behavior

ELEVENTH EDITION

ON
THE ORIGIN OF SPECIES
BY MEANS OF NATURAL SELECTION,
BY CHARLES DARWIN, M.A.,
FELLOW OF THE ROYAL, GEOLOGICAL, LINNÆAN, ETC., SOCIETIES;
AUTHOR OF 'JOURNAL OF RESEARCHES DURING H. M. S. BEAGLE'S VOYAGE
ROUND THE WORLD.'

CHAPTER 1

An Introduction to Animal Behavior

If you've ever been stung by a honey bee (*Apis mellifera*), you know that animals can behave in ways that seem illogical to humans. After all, just a few minutes after stinging you, that female honey bee will die. Why would an individual kill itself to temporarily hurt you? Although we will not fully answer this question until Chapter 12, the solution lies deep in the portion of the stinging bee's genetic code that it shares with its hive mates. As you will see in this chapter, behavioral theory gives us a framework within which to understand this and other seemingly paradoxical behaviors. As a student of animal behavior, you will use this book to begin exploring the natural world and examine how and why the diversity of animals that live in it behave in the ways they do.

Chapter 1 opening photograph by duncan1890/E+/Getty Images.

The discipline of animal behavior is vibrant and growing rapidly, thanks to thousands of behavioral biologists who are exploring everything from the genetics of bird song to why women find men with robust chins attractive. A major reason why the field is so active and broad ranging has to do with a book published more than 150 years ago, Charles Darwin's *On the Origin of Species* (Darwin 1859). In this book, Darwin (**FIGURE 1.1**) introduced the concept of **natural selection**, which argues that living species are the product of an unguided, unconscious process of reproductive competition among their ancestors. Natural selection theory provides the guiding principle for studying animal behavior, as well as most of biology more generally. Knowing that animal behavior, like every other aspect of living things, has a history guided by natural selection is hugely important. An understanding of evolutionary theory gives us a scientific starting point when we set out to determine why animals do the things they do—and why they have the genetic, developmental, sensory, neuronal, physiological, and hormonal mechanisms that make these behavioral abilities possible. To truly understand why a bee will kill itself to sting you, you must approach the study of animal behavior from an evolutionary perspective, from the perspective of Darwin himself. As the evolutionary biologist Theodosius Dobzhansky once said, "Nothing in biology makes sense except in the light of evolution" (Dobzhansky 1973). We hope to convince you that Dobzhansky was right when it comes to animal behavior as well. If we succeed, you will come to understand the appeal of the evolutionary approach to animal behavior, as it helps scientists identify interesting subjects worthy of explanation, steers them toward hypotheses suitable for testing, and ultimately produces robust conclusions about the validity of certain hypotheses.

Darwin's influence continues strongly to this day, which is why this book was titled *Animal Behavior: An Evolutionary Approach* for its first ten editions. Over the editions, we have explored the assortment of behaviors that animals exhibit and discuss not only how and why they evolved, but why they evolved in one species and not another, or in one sex and not the other. In this eleventh edition, we extend the evolutionary approach emphasized by the subtitle in the book's prior editions and take a more integrative approach. The concept of evolution by natural selection remains our guide, but we will also weave into our discussion topics such as epigenetics, development, and ecology, as well as tools such as game theory and molecular genetics. Indeed, as you will see in the following chapters, the study of animal behavior has itself evolved since Darwin made his observations about natural selection more than 150 years ago. In fact, the study of animal behavior has changed considerably since the first edition of this book was published more than 40 years ago. Therefore, in this edition, we go beyond an evolutionary approach to the study of animal behavior. We also consider an ecological approach. A mechanistic approach. A developmental approach. In short, we take an integrative approach to the study of animal behavior (Rubenstein and Hofmann 2015, Hofmann et al. 2016), discussing how the interaction among genes, development, and the environment shapes the evolution of animal behaviors. This means that students of animal behavior today may need to use the material learned in a biochemistry or molecular biology class just as much as they do the material learned in an ecology class. We illustrate the ways, for example, that genomics and neurobiology are changing how we study animal behavior in a manner that Darwin could not have imagined. And we demonstrate how new phylogenetic comparative methods are enabling us to use data collected from dozens to thousands of species to test existing, as well as generate novel, hypotheses.

FIGURE 1.1 Charles Darwin. Taken in 1881 by Herbert Rose Barraud (1845–1896), this is thought to be the last photograph of Darwin before his death. The original is in The Huntington Library in San Marino, California.

But first we introduce you to the fascinating field of animal behavior in this chapter. We begin with a detailed discussion of how natural selection governs the evolution of behavior. We introduce adaptationist thinking and the gene's eye view of behavioral trait evolution. We discuss the cost–benefit approach to behavioral biology, as well as the critical role that hypothesis testing plays in the scientific method. Ultimately, we argue that an integrative approach to studying animal behavior is essential for understanding how and why behaviors evolve, a topic we will explore in greater detail in Chapter 2 and then throughout the rest of the book.

Natural Selection and the Evolution of Behavior

When biologists ask questions about the behavior of animals, they are guided by Darwin's theory of evolution by natural selection. Darwin was puzzled by the fact that living organisms could increase their numbers geometrically, but that most didn't. Even in bacteria, which can reproduce rapidly and efficiently, some individuals replicate more than others. So which individuals reproduce more, and why? As Darwin came to realize after a lifetime of observing animals in their natural habitat, if in the past some individuals left more descendants than others, then these reproductively successful individuals would inevitably gradually reshape their species in their image. The logic of natural selection is such that evolutionary change is inevitable if just three conditions are met:

1. *Variation*, with members of a population differing in a particular characteristic
2. *Differential reproductive success*, with some individuals with particular characteristics having more offspring than others
3. *Heredity*, with parents able to pass on those characteristics to their offspring

If there is variation within a species (and there almost always is), if some of that variation is heritable and passed from parents to offspring, and if some of those individuals consistently reproduce more successfully than others, then the increased abundance of living descendants of the more successful types will gradually change the species. Over time, the "old" population—it is important to remember that natural selection acts on populations of individuals—evolves into one whose members possess the characteristics (or **traits**) that were associated with successful reproduction in the past. How successful an individual is at passing on its heritable traits to the next generation is referred to as **fitness**. As you will see throughout this book, fitness—which depends on both survival and reproduction—forms the foundation for understanding which traits are likely to evolve via natural selection and become more or less common in a population. After all, to reproduce and pass one's traits on to one's offspring, an individual has to survive long enough to breed.

Darwin not only laid out the logic of his theory clearly, he also provided abundant evidence that heritable variation in traits is common within species, and that high rates of mortality are also the rule. Indeed, the conditions necessary and sufficient for evolutionary change by natural selection to occur are present in all living things, a point that Darwin demonstrated by showing that people could cause dogs and pigeons to evolve by selectively breeding those individuals with traits that the breeders wanted in future generations of their domesticated animals. And although it was a heretical concept in Darwin's Victorian England, we now know that people, like all other organisms, also evolve by natural selection. For example, the ability of many humans to continue to digest lactose, the dominant sugar in milk, after childhood appeared around the time our ancestors domesticated livestock and has been hypothesized to have been strongly favored by natural selection because it allowed some individuals to survive on a milk diet.

We call the traits associated with successful survival and reproduction, and on which natural selection acts, **adaptations**. Figuring out exactly how a putative adaptation contributes to the **reproductive success** of individuals is perhaps the central goal for most behavioral biologists, some of whom are happy to be known as "adaptationists." Certainly Darwin would have agreed with the way adaptationists think about evolution, especially with respect to the hereditary foundation of evolutionary theory. Although Darwin himself knew nothing about **genes**, regions of DNA that encode traits, we now can reconfigure his argument to deal with selection at the level of the gene. Just as adaptations that increase the reproductive success of individuals will spread through populations over time, so too will the hereditary basis for these attributes. Because genes can be present in populations in different forms known as **alleles**, those alleles that contribute to traits linked to individual reproductive success will become more common over time, whereas those associated with reproductive failure will eventually disappear from the population and perhaps even the genome. In evolutionary biology, we refer to traits as **phenotypes** (aspects of an individual that arise from an interaction of the individual's genes with its environment), and to the set of alleles underlying the development of those traits as **genotypes**. The genetic basis of most complex phenotypes—including most behaviors—is rarely known in full, and when it is, it is typically only in model organisms. (Although as you will see in subsequent chapters, this is changing and scientists are now capable of identifying the hereditary basis of a trait in almost any organism.) Linking phenotypes to genotypes is not only an essential part of modern molecular and evolutionary biology, but with the powerful new tools and approaches from these disciplines being applied to non-model organisms in their natural habitats, it is also becoming crucial to behavioral biology as well.

It is critical to recognize three points about this so-called gene-centered, or gene's eye, view of evolution by natural selection:

1. Only genes replicate themselves; organisms do not. Instead, organisms—or groups of organisms—are vehicles within which replicators (genes) travel, a point we will discuss further in Chapter 12 (Dawkins 1989). Adaptive evolution therefore occurs through the differential survival of competing genes, which increases the frequency of those alleles whose phenotypic effects promote their own propagation.
2. Evolution is not natural selection. Evolution is gene frequency change within a population. Natural selection is one of several causes of evolution, as are mutation, migration, and genetic drift.
3. Natural selection is not guided by anything or anyone. Selection is not "trying" to do anything. Instead, it is the individuals that reproduce more that cause a population or species to evolve over time, a process that Darwin called **descent with modification**.

Notice also that the only kinds of heritable traits that will become more common in a species are those that promote individual reproductive success, which do not necessarily benefit the species as a whole. Although "for the good of the species" arguments were often made by biologists not so long ago, it is entirely possible for adaptations (and particular alleles) to spread through populations even if they do nothing to perpetuate the species (see Chapter 12). Indeed, traits and alleles that are harmful to group survival in the long run can still be favored by natural selection. Yet as we will discuss later in this book, there are also special cases when selection acts on traits that do benefit others, even seemingly at the expense of the individuals that possess those traits.

The Cost–Benefit Approach to Behavioral Biology

To help establish why a particular behavior has evolved, behavioral biologists must propose a **hypothesis**, or an explanation based on limited evidence. Of course, for most behaviors there may be multiple potential explanations, which requires posing multiple hypotheses. When only one of a series of competing hypotheses could explain a given behavior, we call these **alternative hypotheses**. In contrast, when multiple hypotheses could apply to a given behavior, we refer to them as **non-mutually exclusive hypotheses**. Throughout this book, we highlight hypotheses in blue in the text, and also define them in the page margins. When there are multiple hypotheses to explain a behavior, we group them together in a margin box and note whether they are alternative or non-mutually exclusive hypotheses.

Once a hypothesis has been generated, researchers can use it to make a **prediction**, or expectation that should follow if the hypothesis is true. Predictions allow for the discrimination among competing hypotheses. Generating, falsifying, and then generating new hypotheses and predictions is a natural part of the **scientific method**.

When behavioral biologists set out to test hypotheses and predictions related to a potentially adaptive behavior, they often consider the costs and benefits of that behavior; they take a cost–benefit approach to the study of animal behavior. When they speak of costs, behavioral biologists are talking about **fitness costs**, the negative effects of a trait on the number of surviving offspring produced by an individual or a reduction in the number of copies of its alleles that it contributes to the next generation. When they speak of **fitness benefits**, they are referring to the positive effects of a trait on reproductive (and genetic) success. Fitness costs and benefits are the units that behavioral biologists use to study adaptations and the process of evolution by natural selection. Most behavioral biologists study traits that they assume are adaptations, an assumption that they make in order to test specific hypotheses about the possible **adaptive value**, or the contribution to fitness, of the characteristic of interest. Recall that an adaptation is a heritable trait that has spread or is spreading by natural selection, and has replaced or is replacing any alternative traits in the population or species. Such an attribute has a better ratio of fitness benefits to costs than the alternative forms of this trait that have occurred in the species. Thus, studying why a behavioral adaptation has evolved depends on measuring the potential reproductive costs and benefits to the individual adopting that behavior.

Since fitness is an abstract term, behavioral biologists often have to settle for an indicator or correlate of reproductive success when they attempt to measure fitness. In the chapters that follow, the terms *fitness*, *reproductive success*, and *genetic success* are often used more or less interchangeably when referring to such indicators as offspring survival, the number of young that survive to fledging, the number of mates inseminated, or even more indirectly, the quantity of food ingested per unit of time, the ability to acquire a breeding territory, and so on. These proxies give us a currency with which to measure fitness costs and benefits and ultimately to test hypotheses underlying the evolution of a range of animal behaviors.

Using natural selection theory and the cost–benefit approach helps us identify why behaviors evolve, including seeming anomalies that require explanation such as why traits that appear to reduce rather than raise an individual's reproductive success persist or even spread through a population. As we discussed at the outset, a honey bee sting is a good example of a behavioral trait that at first seems counter-intuitive, or even maladaptive for the individual doing the stinging. (In Chapter 12

FIGURE 1.2 A male lion (*Panthera leo*) carrying a young lion cub that he has killed. Because infanticide by males occurs in several different species under similar social circumstances, the behavior is likely to be adaptive. (Photograph © Laura Romin and Larry Dalton/Alamy Stock Photo.)

we will explore why animals perform self-sacrificial acts such as this.) Throughout the book, we will refer to these challenges to evolutionary theory as **Darwinian puzzles**. Biologists deal with these puzzles by developing possible hypotheses based on natural selection theory for how the trait might actually help individuals reproduce and pass on their genes. As an illustration of this approach, we describe in **BOX 1.1** an example of a behavior that has long puzzled behavioral biologists: infanticide (**FIGURE 1.2**).

The Levels of Analysis

Now that we have discussed how traits—including behavioral ones—evolve via natural selection, let's consider what a behavioral trait is and how we study it. Although "behavior" seems like it should be an easy term to define, it actually means different things to different people, particularly those in different fields. You would think that behavioral biologists could at least agree on what a behavior is, but they don't. Daniel Levitis and colleagues posed a seemingly simple question—"What is animal behavior?"—to nearly 175 behavioral biologists at a series of scientific society meetings, and much to their surprise, there was no consensus on how behavioral biologists define animal behavior. Levitis and colleagues suggested defining behavior as the internally coordinated responses (actions or inactions) of whole living organisms (individuals or groups) to internal and/or external stimuli, excluding responses more easily understood as developmental changes (Levitis et al. 2009). This definition focuses largely on intrinsic factors. What if we gave you a different definition, something like this: behavior describes the way an individual acts or interacts with others or the environment? This definition has nothing to do with the internal processes of an individual. Instead, this definition emphasizes interactions between individuals or between an individual and its environment. Why do these definitions of behavior differ so much?

Part of the reason why behavioral biologists define behavior so differently is that animal behavior is a wide and varied discipline that has itself evolved over time. The field of animal behavior has inherited traditions from ethology and the

BOX 1.1

DARWINIAN PUZZLE

Natural selection and infanticide in primates

Hanuman langurs (*Semnopithecus entellus*) are monkeys that live in groups of several females and their offspring, accompanied by one or a few adult males (**FIGURE A**). In the course of a long-term research project in India, male Hanuman langurs were seen attacking and sometimes even killing the very young infants of females in their own group. The puzzle here is obvious: how can it be adaptive for a male langur to harm the offspring of females in his group, particularly since attacking males are sometimes injured by mothers defending their babies (**FIGURE B**)? Some primatologists have argued that the infanticidal behavior of these males was not adaptive but was instead the aberrant aggressive response by males to the overpopulation and crowding that occurred when langurs came together to be fed by Indian villagers. According to these observers, overcrowding caused abnormal aggressive behavior, and infanticide was simply a maladaptive result (Curtin and Dolhinow 1978).

(Continued)

(A)

FIGURE A Hanuman langur females and offspring. Males fight to monopolize sexual access to the females in groups such as this one. (Photograph © Heini Wehrle/AGE Fotostock.)

(B)

FIGURE B Male langurs commit infanticide. (Left) A nursing baby langur that has been paralyzed by a male langur's bite to the spine (note the open wound). This infant was attacked repeatedly over a period of weeks, losing an eye and finally its life at age 18 months. (Right) An infant-killing male langur flees from an aggressive protective female belonging to the band he is attempting to join. (Left, photograph by Carola Borries; right, photograph by Volker Sommer, from Sommer 1987.)

BOX 1.1 DARWINIAN PUZZLE *(continued)*

Behavioral biologist Sarah Hrdy used natural selection theory to try to solve the puzzle of infanticide in a different way, namely by asking whether the killer males were behaving in a reproductively advantageous manner (Hrdy 1977). She reasoned that by committing infanticide, the males might cause the babyless females to resume ovulating, which otherwise does not happen for several years in females that retain and nurse their infants. Once females began ovulating, the males could then mate with them. Hrdy tried to explain how infanticide might have spread through Hanuman langur populations in the past as a reproduction-enhancing tactic for individual males. This **infanticide hypothesis** leads to several expectations, of which the most important is the prediction that males will not kill their own progeny, but will instead focus their attacks on the offspring of other males. This prediction in turn generates the expectation that infanticide will be linked to the arrival of a new male or males into a group of females, with the associated ejection of the father or fathers of any baby langurs in the group. In cases of male turnover in groups, the new males could father offspring more quickly if they first killed the existing infants. Females that lose their infants do resume ovulating, which enables the new males in the group to father replacement offspring. Since these predictions have been shown to be correct for this species (Borries et al. 1999) as well as for some other primates (Beehner and Bergman 2008, Lyon et al. 2011) and for various carnivores, horses, rodents, and even a bat (Knörnschild et al. 2011), we can conclude that infanticide in Hanuman langurs is indeed an adaptation, the product of natural selection.

infanticide hypothesis Infanticide is a reproduction-enhancing tactic practiced by males.

Thinking Outside the Box

Given what you just read about the gene's eye view of evolution by natural selection, can you rephrase the infanticide hypothesis using this terminology? In other words, consider infanticide from a gene's perspective and discuss why it could be a beneficial behavior from the perspective of a gene that underlies that behavior. Would a gene for infanticidal behavior be more or less likely to be passed on than a gene that does not promote the killing of young?

observation of animals in the wild, as well as those from experimental psychology where experiments were generally conducted in the lab. As we said at the outset, animal behavior is becoming increasingly integrative and continues to evolve as a discipline to this day. Some scientists tend to take a more mechanistic or developmental approach, as our first definition implied. Still other scientists take an adaptive or evolutionary approach, as our second definition suggested. Indeed, these **ultimate causes** of behavior provide us with a perspective on the adaptive value and the evolutionary history of a trait of interest. However, there is much more to behavioral research than studies of adaptation and evolutionary history. As the behavioral biologist Nikolaas Tinbergen pointed out long ago, biologists also need to study how a behavior develops over the lifetime of an individual and how an animal's physiological systems make behaving possible if they are to really understand the evolution of a behavior (Tinbergen 1963). Because the mechanisms that underlie development and neurophysiology act within the life span of individuals, they are considered to be immediate and mechanistic, or what we call

TABLE 1.1 Levels of analysis in the study of animal behavior

Proximate Level	Ultimate Level
Development How genetic–developmental mechanisms influence the assembly of an animal and its internal components	**Evolutionary History** The evolutionary history of a behavioral trait as affected by descent with modification from ancestral species
Mechanism How neuronal–hormonal mechanisms that develop in an animal during its lifetime control what an animal can do behaviorally	**Adaptive Function** The adaptive value of a behavioral trait as affected by the process of evolution by natural selection

After Tinbergen 1963, and Holekamp and Sherman 1989.

proximate causes of behavior. Ultimately, every behavioral trait is the result of both proximate and ultimate factors; the self-sacrificing behavior of the honey bee, the infanticidal behavior of the langur, you name it—all are driven by immediate underlying mechanisms acting within the bodies of bees, monkeys, and every other animal, including humans.

To paint a complete picture of how and why an animal behaves, we need to explore and integrate both the proximate and the ultimate causes of the behavior, which are often referred to as **levels of analysis**. The two levels of analysis are clearly complementary, but one focuses on the underlying mechanistic and developmental features that give rise to behavioral variation, while the other deals with the role of evolutionary history in influencing behavioral variation and the fitness consequences of that behavioral variation. These two levels can be further subdivided into what are typically called **Tinbergen's four questions**, because in the 1960s, Niko Tinbergen argued that to best understand a behavior, one must conduct both a proximate and an ultimate analysis (**TABLE 1.1**). Within the proximate level of analysis fall questions relating to development (such as understanding how behaviors change ontogenetically over the course of an animal's lifetime) and to mechanism (such as determining the physiological, neurobiological, hormonal, or genetic correlates of behavior). Within the ultimate level of analysis fall questions relating to evolutionary history (such as understanding how a shared ancestry influences variation in behavior) and adaptive function (such as determining how behavior affects survival and reproduction—fitness).

The Integrative Study of Animal Behavior

Tinbergen's four questions are linked to one another because behaviors that have spread through a species due to their positive effects on fitness must have underlying proximate mechanisms that can be inherited. Let's illustrate the power of Tinbergen's integrative approach by examining the namesake behavior of the digger bee *Centris pallida*. Digger bees live in the deserts of the American Southwest and get their name because males dig holes in the desert floor (**FIGURE 1.3A**). Why do males do this? Not long before the first edition of this book was published, John Alcock became interested in understanding how and why this behavior evolved. He showed that males dig because they are searching for females that have completed their development in underground chambers that their mothers built and

FIGURE 1.3 A digger bee, *Centris pallida*, searching for a mate. (A) A male digging in the ground in search of a female. (B) A male copulating with a female that he had discovered before she emerged from the ground. (Photographs by John Alcock.)

provisioned with food for their female (and male) larval offspring. In other words, males dig to find a mate, and when a male reaches a female emerging from her chamber, he will immediately climb onto her and attempt to breed (**FIGURE 1.3B**). A digger bee male can find emerging females efficiently because this skill is adaptive, which is to say that it evolved and is now maintained by natural selection and passed on from parent to son.

But how does a male know where to dig to find a mate that is buried a centimeter or more beneath the desert floor? As Tinbergen first pointed out, because the immediate mechanisms of behavior have an evolutionary basis, the proximate and ultimate causes of behavior are related, and both are required for a full explanation of any behavioral trait (Alcock and Sherman 1994). Alcock hypothesized that digging males could somehow smell females that had burrowed up close to the surface (Alcock et al. 1976). This ability to smell hidden females—a proximate mechanism underlying the digging behavior—enabled males to reproduce more successfully than rivals without the same capacity. The ability to smell a mate might be related to specific genes, and identifying these genes or studying their expression and evolution might be useful. These genes have proximate effects when they, in conjunction with the cellular environment in which they operate and the ecological environment in which the bee lives, cause the bee to develop in a particular manner. The proximate developmental mechanisms within the egg, larva, prepupa, and pupa of digger bees have follow-on consequences, as they influence the production of the sensory and motor mechanisms with which adult male digger bees are endowed. The scent-detecting sensory cells in the bee's antennae in concert with olfactory neural networks in the bee's nervous system make it possible for males to find preemergent females in the desert soil. As a result, males are ready to dig up and pounce on sexually receptive females when they come out of their emergence tunnels.

There is also a history to the behavior, with adaptive changes such as extreme sensitivity to odors linked to buried females, layered onto traits that already had evolved previously in the history of the digger bee species. We can go even further and ask questions about not only how digging behavior varies among males and differentially influences their ability to reproduce, but also ask how this variation is proximately related to a male's age or developmental state. For example, are older, more experienced bees better at finding females that younger, less experienced ones? We can go back even further and wonder how such a small animal is able to dig so vigorously through the hard-packed desert soil. Since males of many bee species mate after finding receptive females at flowers, do digger bees have unique morphological adaptations, such as stronger leg muscles or specially shaped legs, that allow them to dig into the earth in search of mates instead of finding them elsewhere? Finally, we may wonder whether digger bees are unique, or if all bees that nest in the ground in deserts behave this way and have the same suite of morphological adaptations and even the same set of underlying genes. Ultimately, we can ask a range of questions across both levels of analysis to gain a richer and more integrated understanding of both how and why male digger bees dig into the desert floor, just as Tinbergen envisioned. And as we go forward in this book, we will endeavor to do just that when we introduce a new type of behavior. We will illustrate this integrative approach to studying animal behavior in Chapter 2, where we look at the proximate and ultimate explanations for bird song.

Approaches to Studying Behavior

To ask questions at different levels of analysis, behavioral biologists can use several complimentary approaches. The most basic approach is simply to watch animals behaving in nature or in the lab. Using this **observational approach**, behavioral biologists can define and record the behaviors of organisms, and then relate them to features of the animals' social or ecological environment, or their internal or developmental state. Observational approaches are inherently correlational, meaning one has to be very careful about drawing conclusions in relation to what is cause and what is effect. Instead of observing a relationship between traits and coming up with an explanation, a researcher might instead use an **experimental approach** where they can manipulate features of the animal or its environment to more directly establish a relationship among traits. Controlled experiments can be done in the lab, but also in the field if designed carefully to account for all of the natural variation in nature. Finally, some behavioral biologists may want to generate ideas that move beyond a single species by using a **comparative approach** for inferring the adaptive basis of a behavior utilizing many different species to look for general patterns of why behaviors evolved. A comparative approach provides information about the **evolutionary history** of a potentially adaptive trait by identifying the characteristics that extinct or ancestral species may have had and how those traits gave rise to the ones we see in the extant species today. This approach requires a resolved **phylogeny**, or evolutionary tree, that describes the historical relationships among species (**BOX 1.2**). The comparative approach can also inform us as to whether there were **evolutionary constraints**, or restrictions, limitations, or biases to the outcome of adaptive evolution, on the appearance of the trait.

The Adaptive Basis of Behavior: Mobbing in Gulls

All three of these approaches—observational, experimental, and comparative—can provide information about one level of analysis or another. To see how these different approaches can be useful for studying the adaptive evolution of a particular behavior, let's examine one behavior: mobbing in gulls. If you ever stumble upon a seabird colony during the nesting season, watch out! It would not be surprising for an adult bird to swoop about, yap loudly, dive at you, and even attempt to bomb you with

BOX 1.2 INTEGRATIVE APPROACHES

Phylogenies and the comparative method

The **FIGURE** in this box is a phylogenetic tree that represents the evolutionary history of three extant animal species (X, Y, and Z) and their links to two ancestral species (A and B). To create a phylogeny of this sort, it is necessary to determine which of the three extant species are more closely related and thus which are descended from the more recent common ancestor. Phylogenies can be drawn on the basis of anatomical, physiological, or even behavioral comparisons among species, but most often, molecular comparisons are used. The molecule DNA is very useful for this purpose because it contains so many "characters" on which such comparisons can be based, namely the specific sequences of nucleotide bases that are linked together to form an immensely long chain. Each of the two strands of that chain has a base sequence that can now be read by an automated DNA-sequencing instrument. Therefore, one can compare a cluster of species by extracting a specific segment of DNA from either the nuclei or mitochondria in cells from each species and identifying the base sequences of that particular segment.

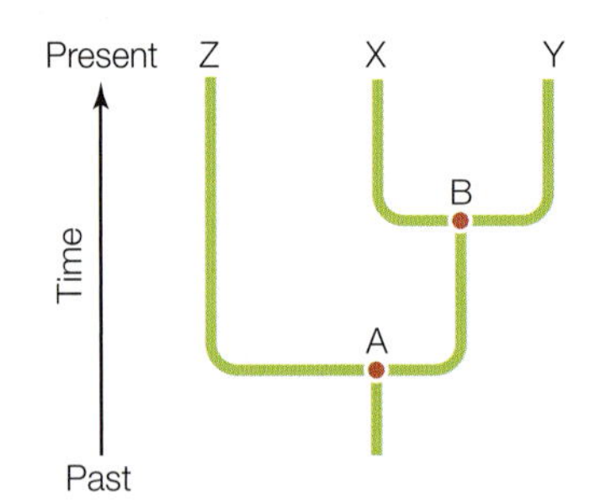

For the purposes of illustration, here are three made-up DNA base sequences that constitute part of a particular gene found in all three hypothetical extant species of animals:

Position	1	2	3	4	5	6	7	8	9	10	11	12	13	14	15
Species X	A	T	T	G	C	A	T	A	T	G	T	T	A	A	A
Species Y	A	T	T	G	C	A	T	A	T	G	G	T	A	A	A
Species Z	G	T	T	G	T	A	C	A	T	G	T	T	A	A	T

These data can be used to conclude that species X and Y are more closely related to each other than either is to species Z. The basis for this conclusion is that the base sequences of species X and Y are nearly identical (differing by a single change in position 11 of the chain), whereas species Z differs from species X and Y by four and five base changes, respectively. The shared genetic similarity between species X and Y can be explained in terms of their evolutionary history, which must have featured a recent common ancestor (species B in the phylogenetic tree). Species B must have split so recently into the two lineages leading to extant species X and Y that there has not been sufficient time for more than one mutation to become incorporated in this segment of DNA. The lesser, but still substantial, similarities among all three species can be explained in terms of their more ancient common ancestor, species A. The interval between the time when species A split into two lineages and the present has been long enough for several genetic changes to accumulate in the different lineages, with the result that species Z differs considerably from both species X and species Y.

Throughout this book, we will see (or discuss) several phylogenies that vary in size from a handful of taxa to a few thousand species. We will use these evolutionary trees—which are really just hypotheses about the ancestral relationships among species—to examine the evolution of various behavioral and morphological traits. These traits will be "mapped" onto the tree, which means they will be recorded for each branch tip (species or other taxonomic unit, which in this case are X, Y, and Z). Using the comparative method, the relationships among different traits—such as the relationship between animal social and mating systems (see Chapter 12)—will be examined and traced back through the most recent common ancestors to determine how various traits are

BOX 1.2 INTEGRATIVE APPROACHES *(continued)*

related to each other, as well as the sequence of events for how they might have co-evolved. We will provide another detailed example of this comparative approach in Chapter 2 when we consider the evolutionary history of bird song.

Thinking Outside the Box

Which two species are the least closely related to each other? Redraw the tree using the two additional species below and then determine which species are now the least and most closely related.

Position	1	2	3	4	5	6	7	8	9	10	11	12	13	14	15
Species Q	G	T	T	G	T	A	C	A	A	C	T	T	A	A	T
Species R	A	T	T	G	C	A	T	A	T	G	G	T	A	A	T

liquid excrement (**FIGURE 1.4**). It's not hard to see why birds become upset when you, a potential predator, get close to their nests—and their progeny. If the parents' assaults distract predators like you from their offspring, then mobbing birds may increase their own reproductive success, enabling them to pass on the hereditary basis for joining others in screaming at, defecating on, and even assaulting those who might eat their eggs or chicks. This possible explanation for such behavior led Hans Kruuk to investigate group mobbing in the black-headed gull (*Chroicocephalus ridibundus*), a colonial ground-nesting species that breeds across much of Europe and Asia (**FIGURE 1.5**) (**VIDEO 1.1**) (Kruuk 1964). Because Kruuk was interested in studying the adaptive basis of mobbing, he did what behavioral biologists do to solve problems of this sort—he considered the costs and benefits of the behavior. His

VIDEO 1.1 A colony of black-headed gulls **ab11e.com/v1.1**

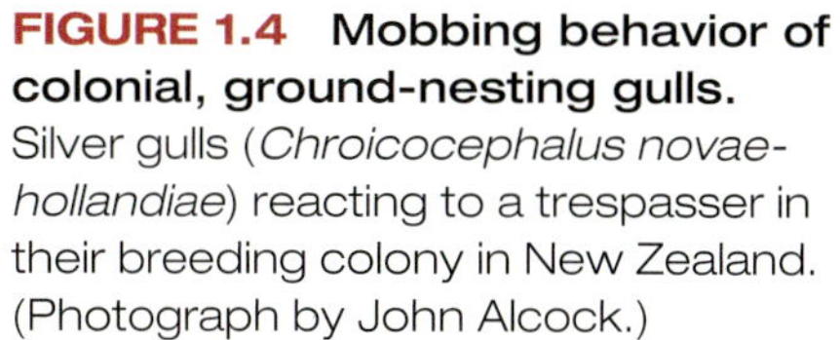

FIGURE 1.4 Mobbing behavior of colonial, ground-nesting gulls. Silver gulls (*Chroicocephalus novae-hollandiae*) reacting to a trespasser in their breeding colony in New Zealand. (Photograph by John Alcock.)

FIGURE 1.5 Colonial-nesting black-headed gulls. These birds build nests on the ground that are susceptible to predation. (Photograph © iStock.com/mauribo.)

working hypothesis was based on the notion that the costs to the mobbers (such as the time and energy expended by attacking potential predators, and the risk of getting injured or even killed) were outweighed by the fitness benefits to the birds from their social harassment of potential predators (such as increased offspring survival).

predator distraction hypothesis Mobbing potential predators distracts them from depredating nests.

To examine this **predator distraction hypothesis** and the idea that mobbing potential predators distracts them from depredating nests, Kruuk tested one of its central predictions, the idea that mobbing gulls should force distracted predators to expend more searching effort than they would otherwise. This prediction can be tested simply by watching gull–predator interactions using an observational approach (Kruuk 1964). Like most behavioral biologists studying animals in their natural habitat, Kruuk spent many hours watching the birds and observed that egg-hunting carrion crows (*Corvus corone*) must continually face gulls diving at them and so, while being mobbed, they cannot look around comfortably for their next meal (**FIGURE 1.6**). Because distracted crows are probably less likely to find their prey, Kruuk established that a probable benefit exists for mobbing. Moreover,

FIGURE 1.6 Why do gulls mob predators? Hans Kruuk proposed that the fitness costs to mobbers of harassing potential predators were outweighed by the fitness benefits. (Photograph by birdpix/Alamy Stock Photo.)

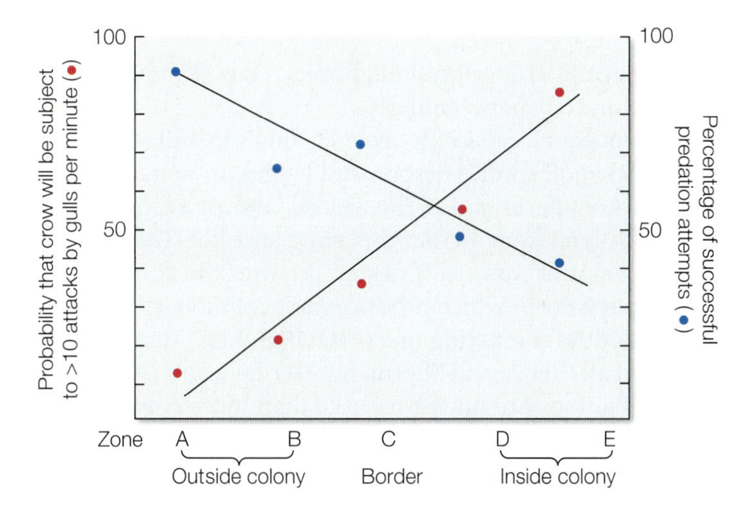

FIGURE 1.7 Does mobbing protect eggs? When chicken eggs were placed outside and inside a black-headed gull nesting colony, crows searching for the eggs within the colony were subject to more attacks by mobbing gulls (red circles), and as a result, they discovered fewer hen eggs (blue circles). (After Kruuk 1964.)

the benefit of mobbing crows plausibly exceeds the costs, given that crows do not attack or injure adult gulls.

Although Kruuk's observations were consistent with the predator distraction hypothesis, the hypothesis yields much more demanding predictions that require an experimental approach. Because adaptations are better than the traits they replace, we can predict that the benefit experienced by mobbing gulls in protecting their eggs should be directly proportional to the extent that predators are actually mobbed. To test this prediction, Kruuk designed an experiment that involved placing ten chicken eggs as stand-ins for gull eggs, one every 10 meters, along a line running from outside a black-headed gull nesting colony to inside it (Kruuk 1964). Eggs placed outside the colony, where mobbing pressure was low, were much more likely to be found and eaten by carrion crows and herring gulls (*Larus argentatus*) than were eggs placed inside the colony, where predators were communally harassed by the many parents whose offspring were threatened by the presence of a predator (**FIGURE 1.7**).

Kruuk assembled both observational and experimental evidence that was consistent with the predator distraction hypothesis and the idea that mobbing is an adaptation that helps adult black-headed gulls protect their eggs and young. Field experiments, along with correlational and observational field data, are critical for deciphering the adaptive value of animal behaviors. However, we could also use the comparative approach to test Kruuk's explanation for mobbing by black-headed gulls by testing the following prediction: If mobbing by ground-nesting black-headed gulls is an evolved response to predation on gull eggs and chicks, then other gull species whose eggs and young are at low risk of predation should not exhibit mobbing behavior. The rationale behind this prediction is as follows: The various fitness costs of mobbing (such as the increased risk of being killed by a predator) will be outweighed only if sufficient fitness benefits are derived from distracting predators. If predators have not posed a problem, then the odds are that the costs of mobbing (such as the energy spent dive-bombing a non-predator) would be greater than the benefits, and we would not expect the behavior to have evolved in those species.

There is good reason to believe that the ancestral gull was a ground-nesting species with many nest-hunting predators against which the mobbing defense would have been effective (**BOX 1.3**). Of the 50 or so species of extant gulls, most nest on the ground and mob enemies that hunt for their eggs and chicks (Tinbergen 1959). These behavioral similarities among gulls, which also share many other features, are believed to exist in part because all gulls are the descendants of a relatively recent common ancestor, from which they all inherited the genetic material that predisposes many extant gull species to develop similar traits. Thus, all other things being equal, we expect most gulls to behave similarly.

However, some things are not equal, especially when it comes to gull reproductive behavior. For example, a few gull species nest on cliff ledges or on trees rather than on the ground. Perhaps these species are all the descendants of a more recent cliff-nesting gull species that evolved from a ground-nesting ancestor. The alternative hypothesis, that the original gull was a cliff nester, requires the cliff-nesting trait to have been lost and then regained, which produces an evolutionary scenario that requires more changes than the competing one (**FIGURE 1.8**). Many evolutionary biologists, although not all (Reeve and Sherman 2001), believe that simpler scenarios involving fewer transitions are more probable than more complicated alternatives. This assumption is known as Occam's razor or the principle of **parsimony**, which holds that simpler explanations are more likely to be correct than complex ones—all other things being equal.

In any event, cliff-nesting gulls currently have relatively few nest predators because it is hard for small mammalian predators to scale cliffs in search of prey, and

BOX 1.3 EXPLORING BEHAVIOR BY INTERPRETING DATA

The benefit of high nest density for the arctic skua

The arctic skua (*Stercorarius parasiticus*), a close relative of gulls, also nests on the ground and mobs colony intruders, including another relative, the great skua (*Stercorarius skua*), a larger predator that eats many arctic skua eggs and chicks. In one population, hatching success and postfledging survival were greater for arctic skuas that nested in dense colonies than for those in low-density groups (**FIGURE**); the number of near neighbors was, however, negatively correlated with the growth rate of their chicks (Phillips et al. 1998).

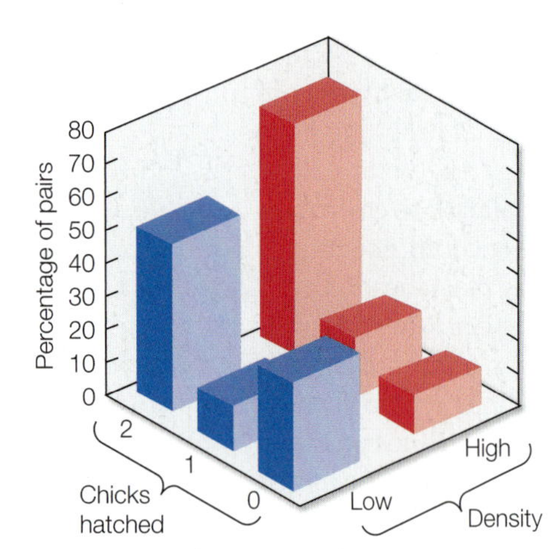

Arctic skua

Arctic skuas nesting with many nearby neighbors were more likely to rear two chicks than were individuals nesting in areas with a lower density of breeding pairs. (After Phillips et al. 1998; photograph © Andrew Astbury/Shutterstock.com.)

Thinking Outside the Box

Rephrase these findings in terms of the reproductive costs and benefits of communal nesting and mobbing by the arctic skua.

If adaptation meant a perfect trait, would communal mobbing by arctic skuas be labeled an "adaptation"?

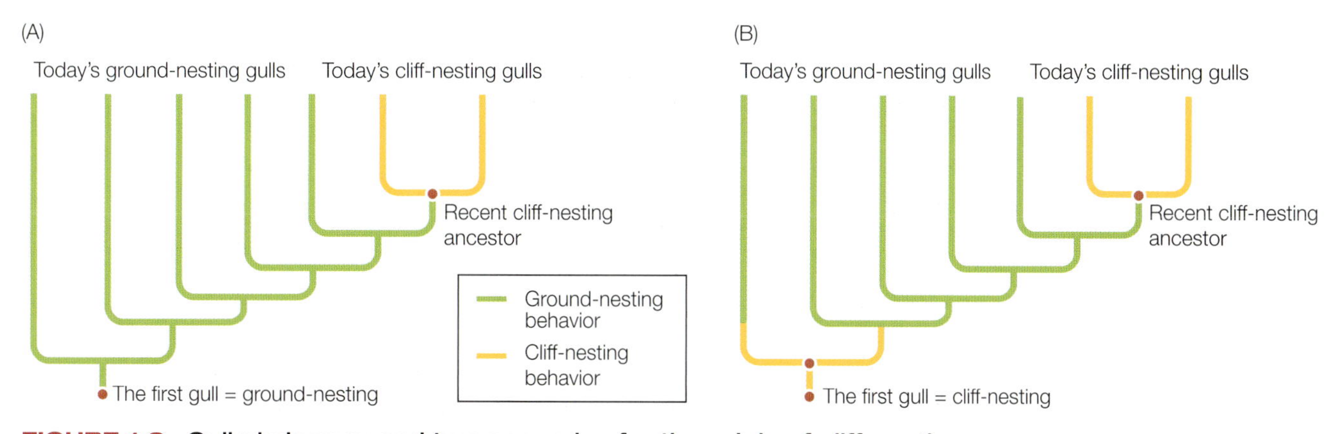

FIGURE 1.8 Gull phylogeny and two scenarios for the origin of cliff-nesting behavior. (A) Hypothesis A requires just one behavioral change, from ground nesting to cliff nesting. (B) Hypothesis B requires two behavioral changes, one from the ancestral cliff nester to ground nesting, and then another change back to cliff nesting.

because predatory birds have a difficult time maneuvering near cliffs in turbulent coastal winds. Thus, a change in nesting environment surely led to a reduction in predation pressure, which should have altered the cost–benefit equation for mobbing by these gulls. If so, cliff-nesting gulls are predicted to have modified or lost the ancestral mobbing behavior pattern. For example, the black-legged kittiwake (*Rissa tridactyla*) nests on nearly vertical coastal cliffs, where its eggs are relatively safe from predators (**FIGURE 1.9**) (Massaro et al. 2001). The relatively small size of the kittiwakes may also make the adults themselves more vulnerable to attack

(A)

(B)

FIGURE 1.9 Not all gulls nest on the ground. (A) Steep cliffs are used for nesting by kittiwake gulls, which appear in the lower half of this photograph. (Pictured above them are three thick-billed murres, *Uria lomvia*.) (B) Kittiwakes are able to nest on extremely narrow ledges; their young crouch in the nest, facing away from the cliff edge. (A, photograph by John Alcock; B, photograph © abi warner/123RF.)

by some nest predators, making the cost–benefit ratio for mobbing even less favorable. As predicted, groups of nesting adult kittiwakes do not mob their predators, despite sharing many other structural and behavioral features with black-headed gulls and other ground-nesting species. The kittiwake's distinctive behavior provides a case of **divergent evolution** and supports the hypothesis that mass mobbing by black-headed gulls evolved in response to predation pressure on the eggs and chicks of nesting adults (Cullen 1957). Take away the threat of predation, and over evolutionary time a gull will likely lose its mobbing behavior because it is no longer adaptive or useful to maintain.

Another use of the comparative approach for understanding the evolution of mobbing behavior is to test the idea that species from different evolutionary lineages that live in similar environments should experience similar selection pressures and thus can be predicted to evolve similar traits, resulting in **convergent evolution**. If this is true, these unrelated species will adopt the same adaptive solution to a particular environmental obstacle to reproductive success (**FIGURE 1.10**). All other things being equal, unrelated species should behave differently, unless they have been subjected to the same **selection pressures**, or drivers of differential survival and reproduction.

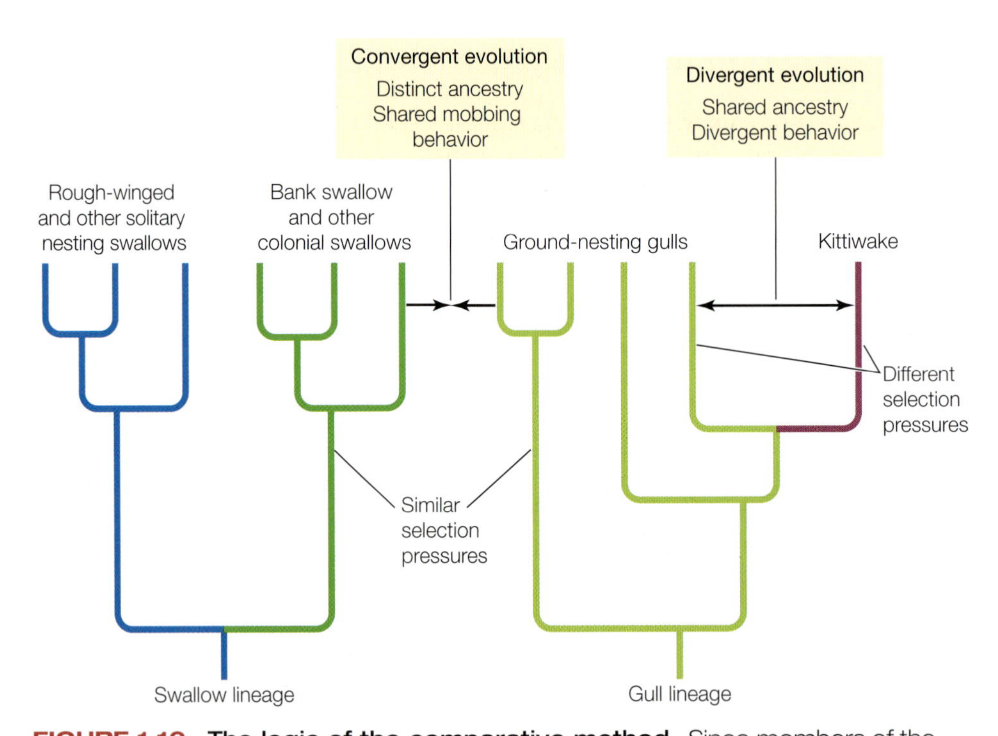

FIGURE 1.10 The logic of the comparative method. Since members of the same evolutionary lineage (such as gull species of the family Laridae) share a common ancestry and therefore share many of the same genes, they tend to have similar traits, such as mobbing behavior, which is widespread among ground-nesting gulls. But the effects of shared ancestry can be overridden by a novel selection pressure. A reduction in predation pressure has led to divergent evolution by the cliff-nesting kittiwake, which no longer mobs potential enemies. The other side of the coin is convergent evolution, which is illustrated by the mobbing behavior of some colonial swallows, even though gulls and swallows are not related, having come from different ancestors long ago. These colonial swallows and gulls have converged on a similar antipredator behavior in response to shared selection pressure from predators that have fairly easy access to their nesting colonies.

As predicted, mobbing behavior has evolved convergently in many other birds only distantly related to gulls, particularly those that nest colonially (Sordahl 2004). Among these species is the bank swallow (*Riparia riparia*), which also nests in colonies where predatory snakes and jays come to eat swallow eggs and nestlings (Hoogland and Sherman 1976). The common ancestor of bank swallows and gulls occurred long, long ago, with the result that the lineages of the two groups have evolved separately for millions of years—a fact recognized by the taxonomic placement of gulls and swallows in two different families, Laridae and Hirundinidae. Yet despite their evolutionary and genetic differences from gulls, bank swallows behave like gulls when they are nesting. As bank swallows swirl around and dive at their predators, they sometimes distract hunting jays or snakes that would otherwise kill their offspring.

Even some colonially nesting mammals have evolved mobbing behavior (Owings and Coss 1977). For example, adult California ground squirrels (*Otospermophilus beecheyi*), which live in groups and dig underground burrows, react to a hunting rattlesnake by gathering around it and shaking their infrared-emitting tails vigorously, a signal to the snake to encourage it to depart before the ground squirrels kick sand in its face (**FIGURE 1.11**) (**VIDEO 1.2**) (Rundus et al. 2007). Rattlers molested in this fashion cannot hunt leisurely for nest burrows to enter in search of vulnerable young ground squirrels. Thus, because mobbing behavior has evolved independently in several unrelated species whose adults can sometimes protect their vulnerable offspring by distracting predators, mobbing is almost certainly an antipredator adaptation.

VIDEO 1.2

California ground squirrel interacting with a rattlesnake
ab11e.com/v1.2

The Science of Animal Behavior

Throughout your reading of this book, it is important to remember that behavioral biology is a scientific discipline. By this, we mean that researchers studying animal behavior use a particular kind of logic to evaluate potential explanations for puzzling phenomena. As we detailed previously, this logic is grounded in natural selection

FIGURE 1.11 California ground squirrels signal to hunting rattlesnakes by shaking their tails. The tail of this rodent emits infrared to signal to potential predators to leave. (Illustration by R. G. Coss.)

theory, and as you will see in subsequent chapters, this theory can be expanded to encompass a range of seemingly paradoxical behaviors, such as why males use and evolve elaborate traits that decrease their own survival, why individuals in some group-living animals forego their own reproduction to help raise others' offspring, and why females in some species have what look like penises but males of those species do not. We must also remember that animal behaviors are grounded in, and sometimes constrained by, mechanism and evolutionary history. Therefore, integrative studies that cross levels of analysis can help explain how and why behaviors evolve. To ask questions at either level of analysis, behavioral biologists use the scientific method to develop hypotheses and then observationally, experimentally, or comparatively test predictions of those hypotheses using a variety of well-designed procedures. But at the heart of any good behavioral study remains the Darwinian logic to evaluate the puzzling phenomena in the world around us, the same logic we used to determine why nesting black-headed gulls mob predators or how digger bees locate mates.

One of our primary goals is to encourage our readers to realize how scientists evaluate hypotheses in ways that are generally considered fair and logical (at least by other scientists). Good science means that scientists approach a problem without preconceived answers. The scientific method allows researchers to distinguish among alternative explanations. The cases reviewed in the chapters ahead have been selected with this goal in mind. After a discussion in Chapter 2 of how the integrative study of animal behavior can help us better understand the evolution of song learning in birds, the book is organized to first present additional details about the proximate level of analysis, and then the ultimate level. Chapters 3, 4, and 5 emphasize the developmental and genetic, neural, and physiological bases of behavior, respectively. From there, we begin to explore the adaptive decisions that animals make, with Chapter 6 emphasizing survival decisions (including how to avoid predators and find food) and Chapter 7 examining settlement and movement decisions (including territoriality and migration). In Chapter 8 we consider the evolution of communication and how individuals transfer information in ways that influence their reproduction and survival. From there, we move to a discussion of sexual behavior, examining reproductive behavior in Chapter 9 and mating systems in Chapter 10. Chapter 11 introduces the topic of parental care and transitions us into a discussion of social behavior. We use eusocial insects to introduce the principles of social evolution in Chapter 12, followed by a broader examination of social behavior and group living in vertebrates in Chapter 13. Finally, Chapter 14 offers examples of how both proximate and ultimate questions contribute to an understanding of our own behavior. Let's get started.

SUMMARY

1. Evolutionary theory provides the foundation for the study of animal behavior.
2. Charles Darwin realized that evolutionary change would occur if "natural selection" took place. This process happens when individuals differ in their ability to reproduce successfully, as a result of their inherited attributes. If natural selection has shaped animal behavior, we expect that individuals will have evolved abilities that increase their chances of passing copies of their genes on to the next generation.
3. Most behavioral biologists now take a gene's eye view of evolution by natural selection. According to this view, adaptive evolution occurs through the differential survival of competing genes, which increases the frequency of those alleles whose phenotypic effects promote their own propagation.
4. Researchers interested in the adaptive value of behavioral traits use natural selection theory to develop particular hypotheses (tentative explanations) for how a specific behavior might enable individuals (not groups or species as a whole) to achieve higher reproductive success than individuals with alternative traits.
5. Behavioral traits have both ultimate (evolutionary) and proximate (immediate) causes that are complementary, not mutually exclusive. Questions about ultimate causes are those that focus on the possible adaptive value of a behavior as well as those that ask how an ancestral trait became modified over time, leading to a modern characteristic of interest. Questions about proximate causes can be categorized as those concerned with the genetic–developmental bases for behavior as well as those that deal with how physiological (neural and hormonal) systems provide the basis for behavior.
6. Adaptationist hypotheses can be tested in the standard manner of all scientific hypotheses by making predictions about what we must observe in nature in the outcome of an experiment or by comparatively exploring how traits evolved across a group of organisms over evolutionary time. Failure to verify these predictions constitutes grounds for rejecting the hypothesis; the discovery of evidence that supports the predictions means the hypothesis can be tentatively treated as true.
7. The beauty of science lies in the ability of scientists to use logic and evidence to evaluate the validity of competing theories and alternative hypotheses.

COMPANION WEBSITE

Go to **ab11e.com** for discussion questions and all of the audio and video clips.

CHAPTER 2

The Integrative Study of Behavior

As you learned in Chapter 1, animal behavior is a broad and varied discipline, and those who study it are becoming increasingly integrative. We used digger bees to illustrate how both proximate- and ultimate-level questions can be combined to create a richer understanding of bee behavior. In this chapter, we take that integration a step further. The integrative study of animal behavior is most often described as using tools (for example, molecular, endocrine, or neural, but also statistical and computational) from other disciplines to enrich the study of behavior, particularly in the wild. Indeed, since its inception, the field of behavioral biology has always drawn links to other fields—ethology and psychology in the earliest days, then ecology and evolution, and increasingly molecular biology and neuroscience.

Evolutionary game theory (see Chapter 6), for example, comes directly from the field of economics and has enriched the way behavioral biologists think about topics as diverse as foraging behavior, social evolution, and sex-ratio manipulation. Additionally, the application of tools to study hormones in blood samples or feces collected from free-living animals has changed the way behavioral biologists study aggression, reproduction, stress, and other topics (see Chapter 5). Similarly, the use of molecular genetic techniques to study parentage and kin structure in social groups has influenced the way scientists study mating behavior (see Chapter 10) and social evolution (see Chapter 12).

But is the integrative study of animal behavior more than simply designing studies that cross levels of analysis or incorporate tools from other disciplines? Like many behavioral biologists, we think it is. Indeed, true integration allows scientists to not only test long-standing hypotheses with new tools and techniques, but also to generate novel hypotheses that would otherwise not have come to mind without using these new approaches and ways of thinking (MacDougall-Shackleton 2011, Hofmann et al. 2016). Being integrative, then, means not just using cutting-edge tools from other disciplines, but also being comparative and studying multiple species or using phylogenetic methods to examine data from many species in the service of generating hypotheses, or explicitly interweaving insights from proximate and ultimate perspectives in ways that are mutually informative. Although we could choose from any number of behaviors to illustrate this integrative approach to studying behavior, this chapter serves as a case study focused on song learning in birds.

Bird song functions like many other animal sounds do, and can be analyzed in much the same way (**BOX 2.1**). Indeed, you are probably familiar with bird song, having heard birds sing early on spring mornings at some point during your life, but here we emphasize song learning—the ability to acquire vocalizations through imitation. We discuss not only how and why birds learn to sing, but we illustrate how understanding the way in which birds learn and produce song influences its function, and vice versa. We focus on both song production and song perception. Although subsequent chapters in this book will also highlight some of the reasons that birds and other animals vocalize, as well as the neural, genetic, developmental, and hormonal mechanisms that help them do so, here we briefly introduce all of these concepts in one place and demonstrate how by observing and recording singing birds in the field and lab, and by studying the brains, physiology, and genomes of birds during development and in adulthood, we can gain a richer and more integrative understanding of both the adaptive and mechanistic bases of vocal learning.

The Development of Song Learning

Our discussion begins with Peter Marler's classic work exploring why males in different populations of some bird species sing distinctive variants of their species' song, just as people in different parts of English-speaking countries speak different dialects. Marler became aware of bird dialects at a time when he was traveling from one British lake to another as part of a study on lake ecology. As he casually listened to the local songs of chaffinches (*Fringilla coelebs*), a common European songbird, he realized that each lake's population sang a somewhat different version of this species' standard song, which has been described as a rattling series of chips terminated by a descending flourish. It was these observations that changed Marler's scientific focus and set him along the path to studying bird song.

BOX 2.1 INTEGRATIVE APPROACHES

Characterizing sounds made by animals

Like all types of sounds, animal sounds are simply propagated disturbances in ambient pressure that travel in waves. Animal ears sense these changes in pressure and convert them to signals that are sent to the brain, much the way a microphone converts the same variations in pressure into electrical signals that are sent to a recording device. A **waveform** shows the changes in pressure over time that compose a sound. Imagine a waveform that looks like a repeating sine wave (**FIGURE A**). If the waveform changes in height (along the *y*-axis), we say that the **amplitude**—intensity—varies. If the waveform changes in width (along the *x*-axis), we say that the **frequency**—the rate at which amplitude increases and decreases—varies. Animals perceive changes in frequency as changes in pitch. The simplest way for behavioral biologists to analyze sounds is to use an **oscillogram**, a graph of amplitude as a function of time. However, since oscillograms do not reveal changes in frequency, researchers often use a **spectrogram** (sometimes referred to as a sonogram), a visual representation of sound frequencies over time. In a spectrogram, intensity is shown using color or grayscale. Simply by looking at a spectrogram, we can learn something about whether an animal produces high- or low-frequency sounds, the intensity of those sounds, and how the different sounds are structured (say, into a song) over some period of time.

To illustrate how to interpret a spectrogram, let's examine the song of the white-crowned sparrow, one of the most studied of all animal sounds. The male white-crowned sparrow (*Zonotrichia leucophrys*) song typically lasts 2 to 3 seconds and consists of several different elements, often referred to as **syllables**. Notice in the spectrogram that the shapes of these syllables differ greatly (**FIGURE B**) (Nelson et al. 2004). Now listen to some white-crowned sparrow songs (**AUDIO 2.1**). A typical song begins with a whistle, which sounds like one tone. The whistle is then followed by more complex sounds consisting of frequency sweeps, buzzy vibrato elements, and trills (Chilton et al. 1995). You might also have heard several other background noises in some of the recordings. From geese honking to cars driving to wind blowing, these sounds are not just distracting to us while listening—they can also influence how birds produce and hear song. As habitat is altered and humans encroach on more natural habitat, increasing anthropogenic noise may affect the way that many bird species produce and perceive song.

(Continued)

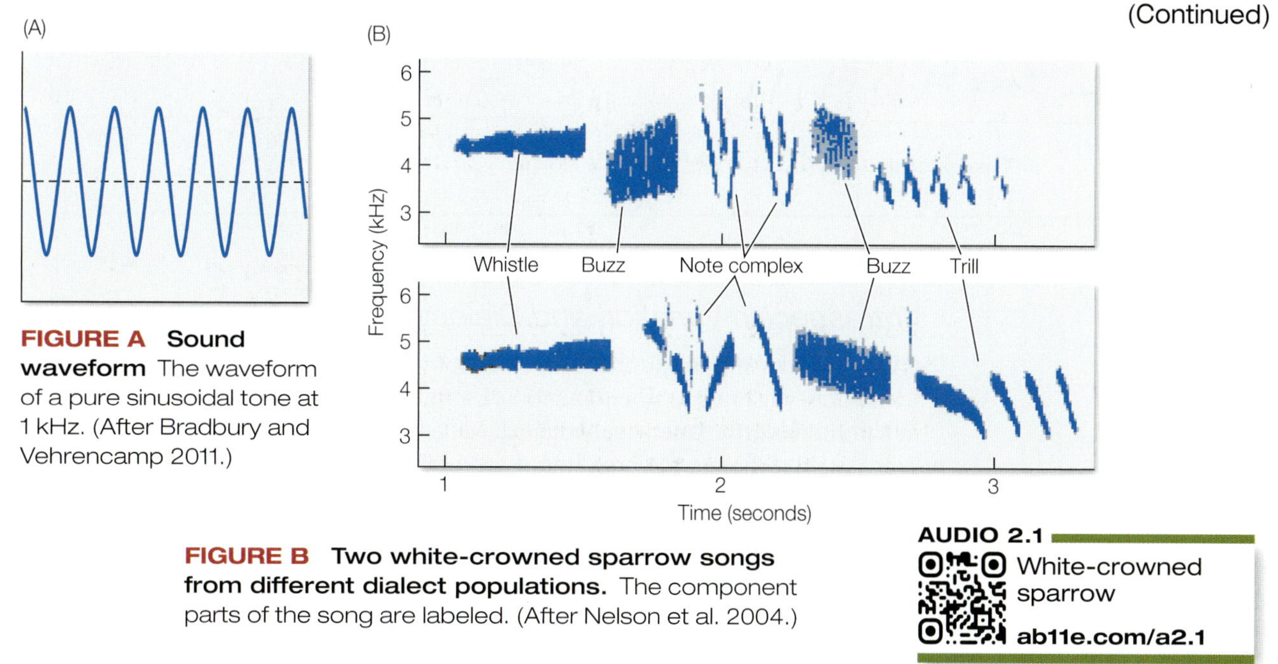

FIGURE A Sound waveform The waveform of a pure sinusoidal tone at 1 kHz. (After Bradbury and Vehrencamp 2011.)

FIGURE B Two white-crowned sparrow songs from different dialect populations. The component parts of the song are labeled. (After Nelson et al. 2004.)

AUDIO 2.1
White-crowned sparrow
ab11e.com/a2.1

BOX 2.1 INTEGRATIVE APPROACHES *(continued)*

Thinking Outside the Box

In some cases, birds have adapted to environmental noise by altering the songs that they produce. Several researchers have wondered whether bird songs have changed in cities in response to traffic noise, which comprises primarily low-frequency sounds. In one study, the authors documented the minimum sound frequency employed in three dialects of white-crowned sparrow song recorded over a 30-year period in San Francisco (Luther and Baptista 2010). As you can see from the graph in **FIGURE C**, frequency increased over time. To help understand how behavioral biologists collected these data, reconstruct the science behind this research from the question that stimulated the study. Then, think about the conclusion the authors must have reached by determining why the minimum song frequency might have increased in urban birds over the course of three decades.

FIGURE C Dialects of white-crowned sparrows in three parts of San Francisco. (i) The three dialects, which were recorded over several decades at three locations that exhibited increasing urbanization at the three locations, each represented by a different symbol. (ii) The minimum frequency contained in the songs in the three locations. The numbers above the vertical bars represent the number of birds recorded. (After Luther and Baptista 2010; photograph © Rolf Nussbaumer Photography/Alamy.)

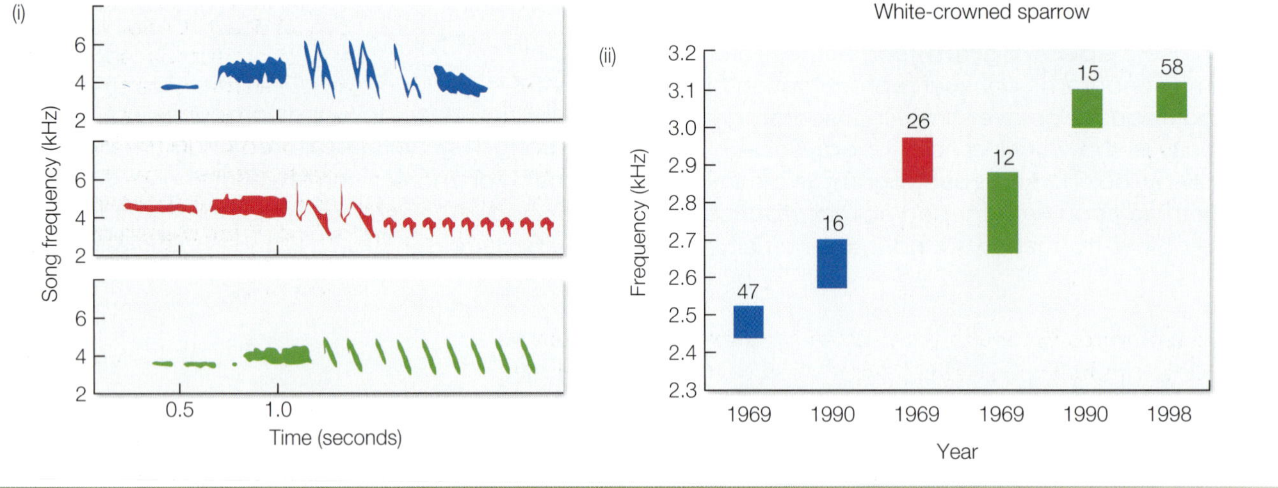

Intraspecific Variation and Dialects

When Marler moved to the University of California in the 1960s, he and his students investigated the phenomenon of song dialects in the white-crowned sparrow, a common North American songbird. Males of this species sing a complex whistled vocalization during the breeding season (**VIDEO 2.1**). Marler and his colleagues found that sparrows living in Marin, north of San Francisco, sing a song type that is easily distinguished from that produced by birds living in Berkeley to the east, or in Sunset Beach to the south (**FIGURE 2.1**) (Marler and Tamura 1964). For behavioral biologists tuned to bird song (or who that can at least read a spectrogram; see Box 2.1), distinguishing among dialects of different populations is as simple as distinguishing between someone from Boston and New Orleans after hearing them

VIDEO 2.1

White-crowned sparrow
ab11e.com/v2.1

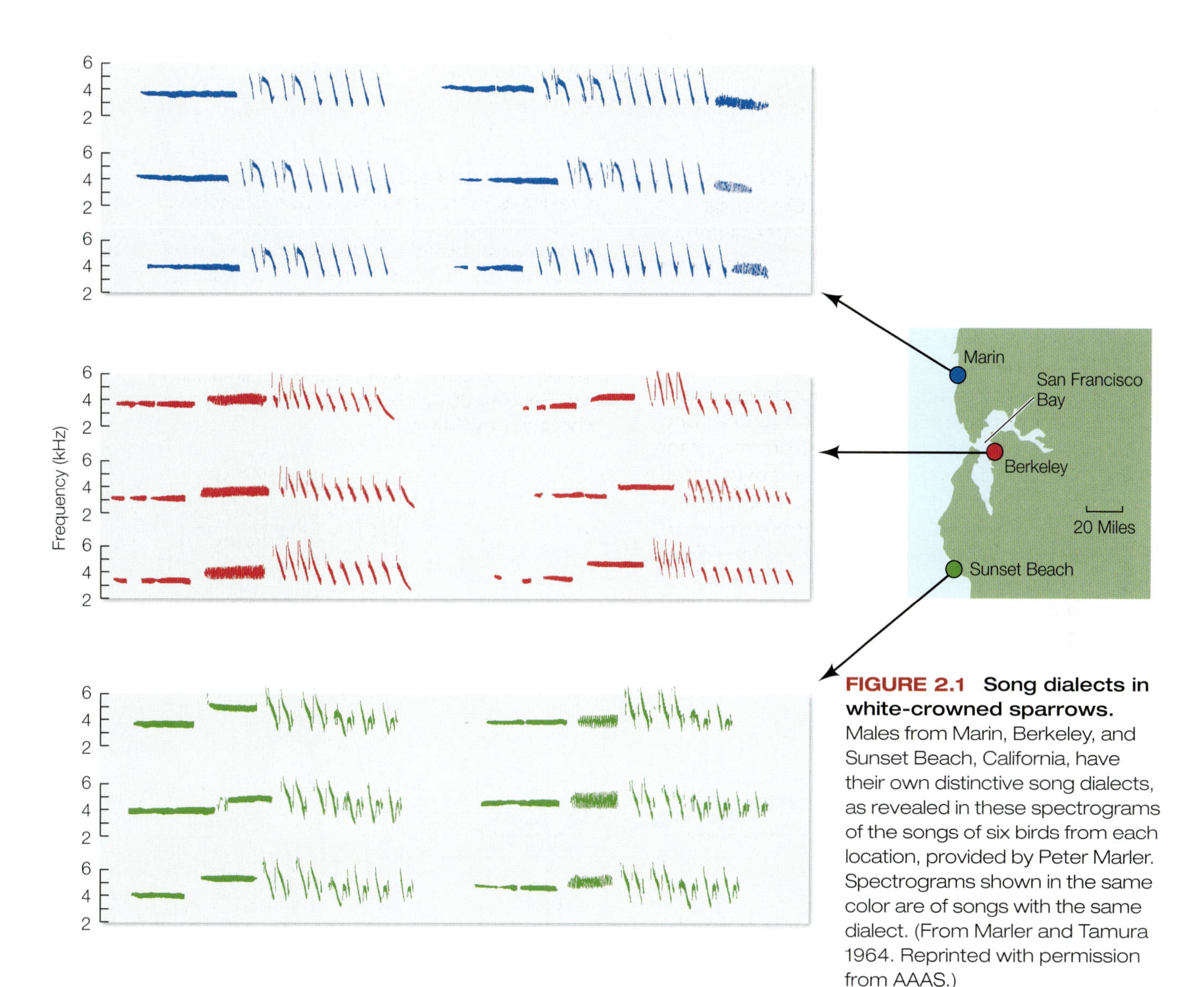

FIGURE 2.1 Song dialects in white-crowned sparrows. Males from Marin, Berkeley, and Sunset Beach, California, have their own distinctive song dialects, as revealed in these spectrograms of the songs of six birds from each location, provided by Peter Marler. Spectrograms shown in the same color are of songs with the same dialect. (From Marler and Tamura 1964. Reprinted with permission from AAAS.)

say just a few words. Although white-crowned sparrow dialects sometimes change gradually over time (Nelson et al. 2004), in at least some populations, local dialects tend to persist for decades with only modest changes (Harbison et al. 1999), showing that a bird's dialects can be relatively stable, just like human dialects. Many birds exhibit dialects, but they are particularly conspicuous and delineated in white-crowned sparrows, which has made this species a model system for the study of bird song (Toews 2017).

Why might birds sing distinct dialects in different geographic areas, especially when the areas are close to one another? To understand how birds can produce distinct dialects that resemble those of humans, Marler first explored the developmental factors that might be responsible for avian dialects. In other words, how do birds of the same species that live relatively close together develop such different dialects? There are several non-mutually exclusive hypotheses to explain how this happens (**HYPOTHESES 2.1**). Marler knew that one possible proximate explanation for the dialect differences was that Marin birds might differ genetically from those in Berkeley in ways that affected the construction of their nervous systems, with the result

HYPOTHESES 2.1

Non-mutually exclusive proximate hypotheses for the development of song and song dialects in birds

genetic differences hypothesis
Differences in song are the result of genetic differences.

acoustic stimulus hypothesis
Differences in song are the result of differences in a bird's acoustic environment.

social interaction hypothesis
Differences in song are the result of social interactions between a young bird and its tutor.

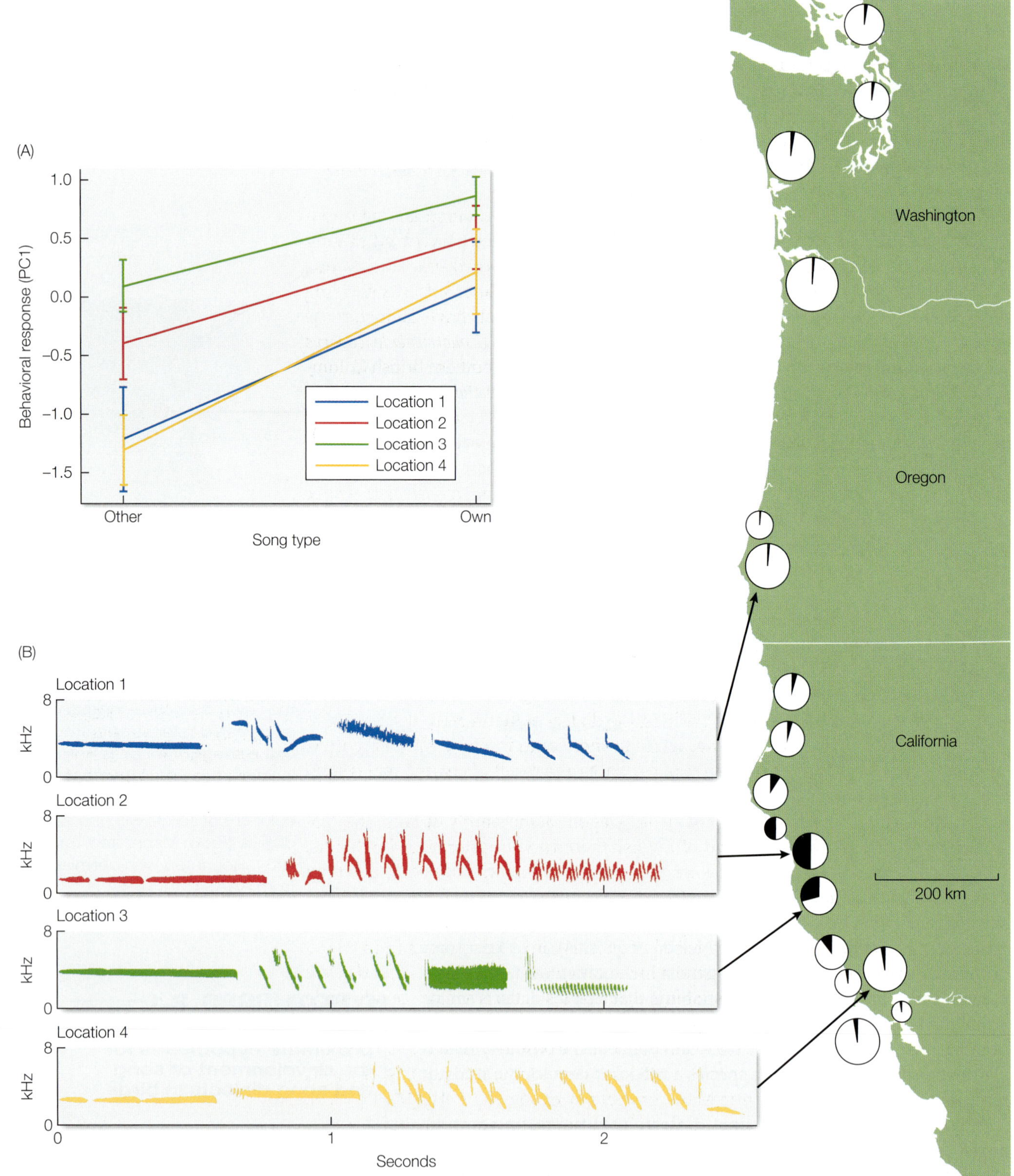

FIGURE 2.2 Sampling a transect in the western United States from the northern to the southern subspecies of white-crowned sparrow identified distinct genetic clusters. (A) Playback experiments comparing responses of males to their own subspecies song and to the song of the other subspecies demonstrated a stronger behavioral response to playback of their own subspecies song at all four localities. (B) Pie charts represent the proportions of genetic mixing between the two subspecies, with more genes from the northern subspecies in northern sites and more genes from the southern subspecies in southern sites. Circle size corresponds to the number of individuals selected for sequencing at each site. Black circle color indicates the proportion of the Nuttall's subspecies, whereas white circle color indicates the proportion of the Puget Sound subspecies. The spectrograms correspond to sites of playback experiments. (After Lipshutz et al. 2017.)

that birds in the two populations came to sing different songs. One way to test this **genetic differences hypothesis** is to check the prediction that groups of birds singing different dialects will be genetically distinct from one another. In one such series of studies, however, researchers found little genetic differentiation among six different dialect groups of white-crowned sparrows in the northern portions of their range (Soha et al. 2004, Poesel et al. 2017).

genetic differences hypothesis Differences in song are the result of genetic differences.

That is not to say, however, that there is no genetic differentiation among populations of white-crowned sparrows. It could be simply that genes involved in song divergence have not yet been surveyed. Moreover, white-crowned sparrows occupy a wide range, and in the western United States some populations do not just sing unique songs, they are also morphologically distinct. Scientists have divided these groups into different subspecies: the Puget Sound subspecies (*Zonotrichia leucophrys pugetensis*) is migratory and breeds from northern California to southern British Columbia, whereas the Nuttall's subspecies (*Zonotrichia leucophrys nuttalli*) is a year-round resident that breeds in coastal central and northern California. Early studies found few genetic differences between these two subspecies (Weckstein and Zink 2001), but using thousands of molecular markers scattered across the genome, Sara Lipshutz and colleagues found that the regional subspecies do in fact separate into distinct genetic clusters (**FIGURE 2.2**) (Lipshutz et al. 2017). These researchers also found that the two subspecies respond differently to each other's song. Thus, heritable differences exist at broad geographic scales (subspecies song differences), but these genetic differences do not seem to explain local song differences in dialect at the scale that Marler identified.

If genetic differences do not explain the song differences among local populations that Marler observed, then what does? Let's consider an alternative hypothesis for dialect differences, namely that these differences might be caused by differences in the birds' social environments. According to the **acoustic stimulus hypothesis**, perhaps young males in Marin *learn* to sing the dialect of that region by listening to what adult Marin males are singing, while farther south in Berkeley, young male white-crowns have different formative experiences as a result of hearing males singing the Berkeley dialect. After all, a young person growing up in Mobile, Alabama acquires a dialect different from someone in Bangor, Maine simply because the Alabaman child hears a different brand of English than the youngster reared in Maine. Moreover, experience also explains *which* language a person will speak; a person born in the United States but reared in China, for example, will speak fluent Chinese.

acoustic stimulus hypothesis Differences in song are the result of differences in a birds' acoustic environment.

Marler and his colleagues tested the hypothesis that differences in dialects may be the result of differences in the birds' social environments by taking eggs from white-crowned sparrow nests and hatching them in the laboratory, where the baby birds were hand-reared. Even when these young birds were kept isolated from the sounds made by singing birds and had no social guidance from adult sparrows, they still started singing when they were about 150 days old—but the best they could ever do was a twittering vocalization that never took on the rich character of the full song of a wild male white-crowned sparrow from Marin or Berkeley or anywhere else (Marler 1970).

This result suggested that something critical was missing from the hand-reared birds' environment. But what was missing? Perhaps it was the opportunity to hear the songs of adult male white-crowned sparrows. If this was the key factor, then a young male isolated in a soundproof chamber but exposed to recordings of white-crowned sparrow song should to be able to learn to sing a complete white-crowned song. Indeed, that is exactly what happened when 10- to 50-day-old sparrows were allowed to listen to recordings of white-crowned sparrow song. These birds also started singing on schedule when they were about 150 days old. At first their songs were incomplete, but by the age of 200 days, the isolated birds not only sang the species-typical form of their song, they closely mimicked the exact version they had heard on recordings. Play a Berkeley song to an isolated young male white-crowned

Male zebra finch

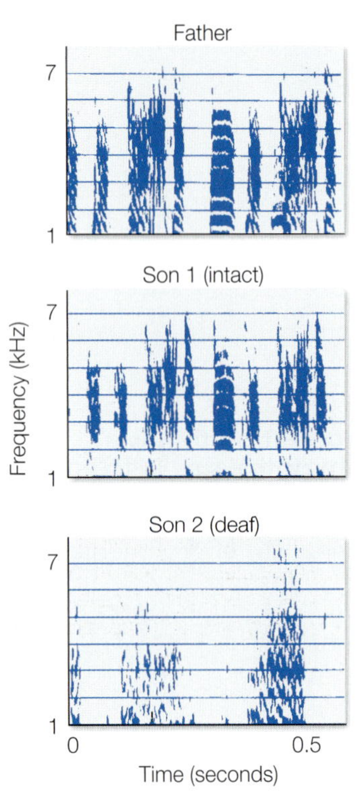

FIGURE 2.3 Hearing and song learning. In white-crowned sparrows and many other songbirds, young males have to hear themselves singing in order to produce an accurate copy of their species' song. A spectrogram of a male zebra finch's song is shown with those of two of his male offspring. The first son's hearing was intact, and he was able to copy his father's song. The second son was experimentally deafened early in life, and as a consequence, he never sang a typical zebra finch song, let alone one that resembled his father's song. (Spectrograms from Wilbrecht et al. 2002; photograph courtesy of Atsuko Takahashi.)

sparrow, and that male will come to sing the Berkeley dialect. Play a Marin song to another male, and he will eventually sing the Marin dialect. Thus, irrespective of where the bird was collected, it could readily be taught a new song.

These results offer powerful support for the acoustic stimulus hypothesis and the idea that the experience of hearing neighboring male sparrows sing affects the development of white-crowned sparrow dialects in young males. Birds that grow up in the vicinity of Marin hear only the Marin dialect as sung by older males in their neighborhood. Young male sparrows evidently store the acoustic information they acquire from their tutors and later match their own initially incomplete song versions against their memories of tutor song, gradually coming to duplicate a particular dialect. Along these lines, if a young hand-reared white-crowned sparrow is unable to hear itself sing (as a result of being deafened after hearing others sing but before beginning to vocalize itself), then it never produces anything like a normal song, let alone a duplicate copy of the one it heard earlier in life (Konishi 1965). Indeed, the ability to hear oneself sing appears to be critical for the development of a complete song in a host of songbird species including the zebra finch (*Taeniopygia guttata*) (**FIGURE 2.3**).

Marler and others did many more experiments designed to determine how song development takes place in white-crowned sparrows. For example, they wondered whether young males were more easily influenced by the stimuli provided by singing adults of their own species than by those of other species. In fact, young, isolated, hand-reared birds that hear only songs of another species almost never come to sing that kind of song (although they may incorporate notes from the other species' song into their vocalizations). If 10- to 50-day-old birds listen only to recordings of song sparrows (*Melospiza melodia*) instead of white-crowned sparrows, they develop aberrant songs similar to the "songs" produced by males that never hear any bird song at all. But if an experimental bird has the chance to listen to recordings of the white-crowned sparrow along with songs of another sparrow species, then by 200 days of age it will sing the white-crowned sparrow dialect that it heard earlier (Konishi 1985). The young bird's developmental system is such that listening to the songs of other sparrow species has little effect on its later singing behavior, indicating that this is a genetically encoded perceptual predisposition.

Although these studies demonstrate how dialects can be learned, they do not reveal how different dialects arise in the first place. Douglas Nelson proposed that white-crowned sparrow songs change gradually over time because of cultural evolution, whereby subtle, somewhat spontaneous changes to songs caused by imprecisions in imitative learning (analogous to mutations in DNA) are then learned by others and spread in the population (Nelson et al. 2004). Over time, the song a population sings will gradually change, resulting in a unique dialect. Nelson and colleagues demonstrated this process by examining white-crowned sparrow songs collected over a 30-year period, as did other researchers in a similar 31-year study of savannah sparrows (**AUDIO 2.2**) (*Passerculus sandwichensis*; Williams et al. 2013). In both studies, researchers traced the timing of specific changes to the song, observing when certain components were added or lost, and new dialects produced or altered.

AUDIO 2.2

Savannah sparrow

ab11e.com/a2.2

Social Experience and Song Development

The experiments with isolated white-crowned sparrows exposed to recorded songs in laboratory cages led Marler to summarize the path of song development in this species (**FIGURE 2.4**). At a very early age, the white-crowned sparrow's still immature brain is able to selectively store information about the sounds made by singing white-crowned sparrows while ignoring other species' songs. At this stage of life, it is as if the brain possesses a restricted computer file capable of recording only one kind of sound input. Then, when the bird begins to sing months later, it accesses the file that it stored earlier. By listening to its own plastic songs (incomplete versions of the more complex full song that it will eventually sing) and comparing those sounds against its memories of the full song it has heard, the maturing bird is able to shape its own songs to match its memory of the song it has on file. When it gets a good match, it then repeatedly practices this "right" song, and in so doing crystallizes a full song of its own, which it can then sing for the rest of its life.

The ability of male white-crowned sparrows to learn the songs of other males in their birthplace just by listening to them sing provides a plausible proximate explanation for how males come to sing a particular dialect of their species' full song. However, very occasionally, observers have heard wild white-crowned sparrows singing songs like those of other species. These rare exceptions led Luis Baptista to wonder whether some other factor, in addition to acoustic experience, might influence song development in white-crowned sparrows. One such factor might be social interactions, a variable excluded from Marler's famous experiments with isolated, hand-reared birds whose environments offered acoustic stimuli but not the opportunity to interact with living, breathing companions.

To test whether social stimuli can influence song learning in white-crowned sparrows (**social interaction hypothesis**), Baptista and his colleague Lewis Petrinovich placed young hand-reared birds in cages where they could see and hear living adult white-crowned sparrows or strawberry finches (*Amandava amandava*) in various treatment combinations (Baptista and Petrinovich 1984). Under these circumstances, white-crowned sparrows learned heterospecific song from strawberry finches, even when they could hear, but not see, adult male

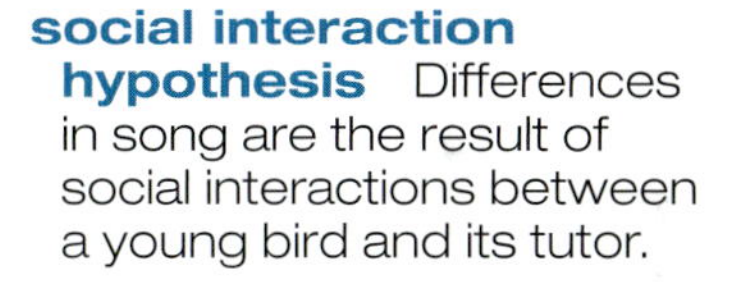

social interaction hypothesis Differences in song are the result of social interactions between a young bird and its tutor.

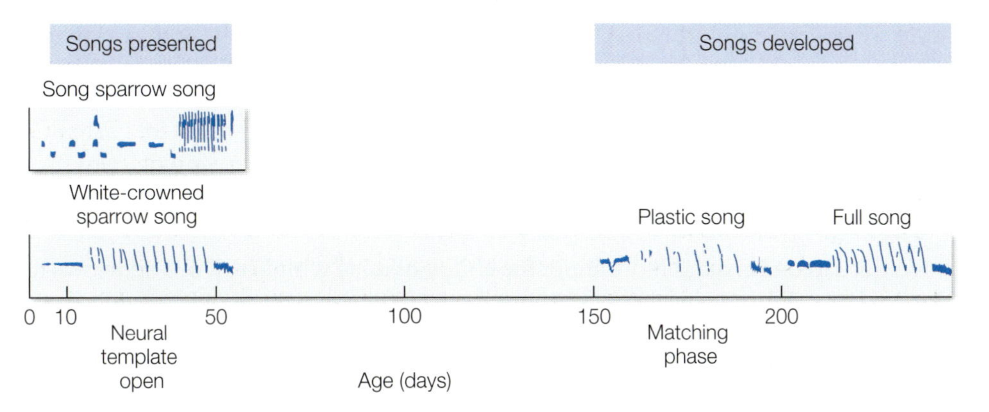

FIGURE 2.4 Song-learning hypothesis based on laboratory experiments with white-crowned sparrows. Young white-crowned sparrows have a special sensitive period 10 to 50 days after hatching, when their neural systems can acquire information from listening to their own species' song, but usually not to any other species' song. Later in life, the bird matches his own plastic song with his memory of the tutor's song and eventually imitates it perfectly—but does not sing elements of the song sparrow's song that he heard during his development. (After Marler 1970.)

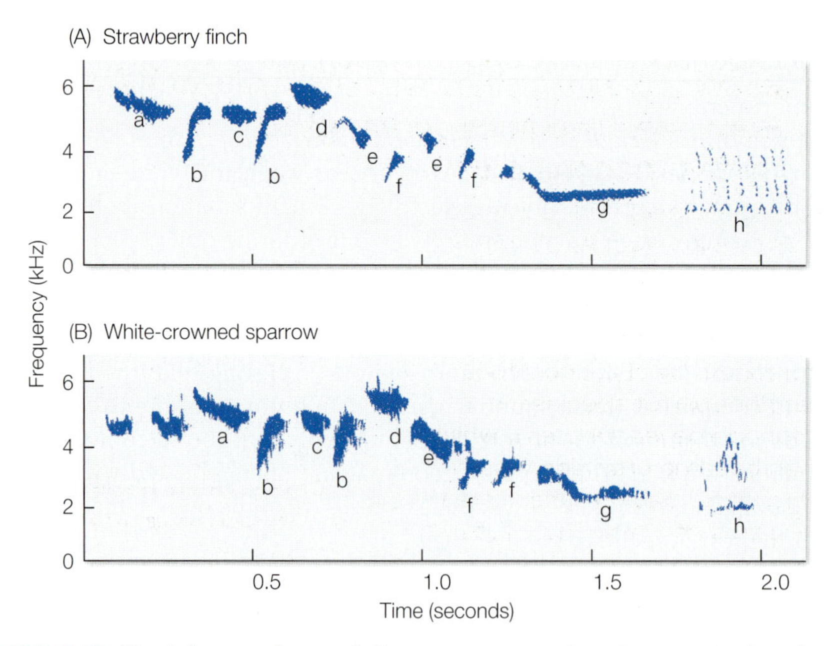

FIGURE 2.5 Social experience influences song development. A male white-crowned sparrow that has been caged next to a strawberry finch will learn the song of its social tutor but will not learn the song of a nearby, but unseen, white-crowned sparrow. (A) The song of a tutor strawberry finch. (B) The song of a white-crowned sparrow caged nearby. The letters beneath the spectrograms label the syllables of the finch song and their counterparts in the song learned by the sparrow. (From Baptista and Petrinovich 1984.)

white-crowned sparrows (**FIGURE 2.5**). Clearly, social acoustic experience can override purely acoustic stimulation during the development of white-crowned sparrow singing behavior, and this species is not unique in this regard. For example, researchers demonstrated in captive zebra finches that social interactions between young birds and their tutors dramatically enhance vocal learning and neural development in the brain compared with simply hearing song alone without any physical contact (**AUDIO 2.3**) (Chen et al. 2016). Such interspecific song learning also occurs occasionally in nature when one species tutors another (**BOX 2.2**). Thus, in white-crowned sparrows and other avian species, both social and acoustic cues influence song learning, highlighting how the acoustic stimulus and the social interaction hypotheses are not mutually exclusive.

AUDIO 2.3

Mechanisms of Song Learning

If we want to understand more about the ability of male white-crowned sparrows to learn a dialect of their species' song, we have to go beyond simply identifying those elements of the young bird's social acoustic environment that affect the development of its behavior at a later date. The process of development in a songbird results in the production of a fine-tuned brain. Understanding how a bird's brain works requires another level of analysis, a neurophysiological one, which we must investigate if we are to figure out how song learning occurs. For example, what drives a young male to attend to and memorize model songs? Where in a 1-month-old white-crowned sparrow's brain are memories stored of the songs it has listened to? What part of the brain controls the sounds that the bird will produce when it is 5 months old? And how does the young male come to match his song memories with his own initially simpler songs? These proximate questions require us to consider the internal devices

BOX 2.2 EXPLORING BEHAVIOR BY INTERPRETING DATA

Song learning in birds adopted by another species

A natural experiment occurred in Australian woodlands when galahs (*Eolophus roseicapilla*, a species of parrot) laid eggs in tree-hole nests that were stolen from them by pink cockatoos (another parrot species, *Lophochroa leadbeateri*), which then became unwitting foster parents for baby galahs. Listen to galahs and pink cockatoos calling (**AUDIO 2.4**). The young cockatoo-reared galahs produced begging and alarm calls that were identical to those produced by galahs cared for by their genetic parents. However, the adopted galahs eventually produced contact calls (used to promote flock cohesion) very much like those of their adoptive cockatoo parents, as you can see from the spectrograms in the **FIGURE** on the right (Rowley and Chapman 1986).

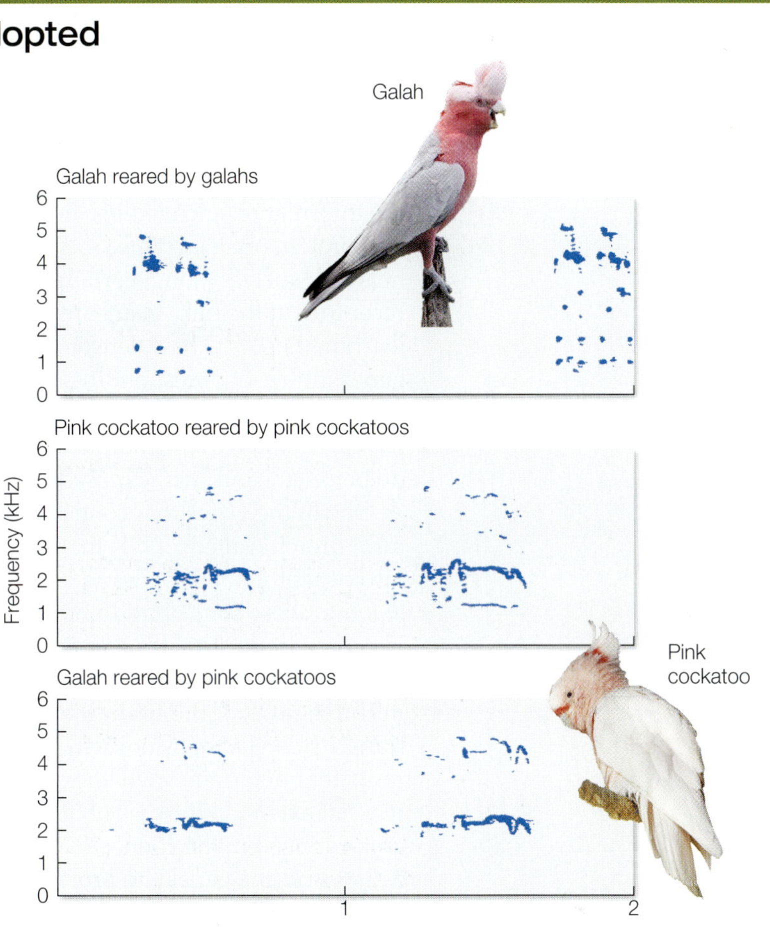

Spectrograms of contact calls of two parrot species. The top panel shows the contact call of a galah reared by galahs; the middle panel shows the call of a pink cockatoo reared by cockatoos; the bottom panel shows the call of a galah reared by pink cockatoo foster parents. (After Rowley and Chapman 1986. Top photograph © iStock.com/GCHaggisImages; bottom photograph © sdvonmb/123RF.)

Thinking Outside the Box

What can we make of these calls? If you were told that galah begging and alarm calls are genetically determined, whereas contact calls are environmentally determined, what would you think? Why? But it does make sense to claim that the alarm calls of the adopted galahs and their cockatoo foster parents are different as a result of genetic differences between them. Why? What other behavioral differences are the result of differences between the social environment of the adopted galahs and that of certain other individuals? Can you think of other cases in nature where this type of interspecific vocal learning could occur?

that the young male possesses that are capable of using social and acoustic inputs to steer his singing behavior along a particular developmental pathway. And as you will see, considering how, for example, these mechanisms of song learning are shared in different species—and even different taxonomic groups—can shed light on why birds and other animals have evolved this ability.

The Genetics of Song Learning

When a young bird is bombarded with sounds produced by singing adults of its own species, these sounds must activate special sensory signals that get relayed to particular parts of the brain. In response to these distinctive inputs, some cells in these locations must then alter their biochemistry to change the bird's behavior. If we trace the biochemical changes taking place over time in the brain cells of a sparrow, we will eventually find that these are linked to changes in gene expression, where the information encoded in a gene is actually used to produce a gene product, such as a protein or an enzyme. So, for example, when a young white-crowned sparrow hears its species' song, certain patterns of sensory signals generated by acoustic receptors in the bird's ears are relayed to song control centers in the brain where learning occurs. These sensory inputs are believed to alter the activity of certain genes in the set of responding cells, leading to new patterns of protein production and follow-on changes in cell biochemistry that reshape those cells. Once the cells have been altered, the modified song control system can do things that it could not do before the bird was exposed to the song of other males of its species.

One of the genes that contributes to these changes is called *ZENK*, which codes for a protein called ZENK that is expressed in particular parts of the brain after a bird hears the songs of its own species (Mello and Ribeiro 1998). For birds that only listen to these songs but do not sing in response, the protein can be detected only in certain brain structures associated with auditory processing. For those birds that respond vocally after hearing their species' song, the ZENK protein is also detected in the regions of the brain that control song production. The ZENK protein is a transcription factor (Moorman et al. 2011), meaning it is part of the regulatory apparatus that determines whether information in one or more genes is expressed and to what extent. In the case of bird song, these genes appear to affect which proteins are produced in the connections (synapses) between certain brain cells. Therefore, ZENK and the genes the protein regulates influence how one nerve cell communicates with another, and changes in these connections can alter a bird's behavior.

Studies of the relationship between genes and neural changes illustrate the intimate connection between the developmental and physiological levels of analysis of bird song. Changes in genetic activity in response to key environmental stimuli translate into changes in neurophysiological mechanisms that control the learning process, though how these mechanisms actually translate into memory remains an open question. Interestingly, these same mechanisms may be at play in female birds as well when it comes to learning song preferences (**BOX 2.3**). As you will see later in this chapter, an understanding of both developmental and neurophysiological systems is therefore necessary to give us a full account of the proximate causes of behavior.

Control of the Avian Song System

Having identified just a few of the many factors involved in the operation of the avian song-learning mechanism, let's now consider the parts of the brain that are essential for learning a song dialect. The brains of white-crowned sparrows and other songbirds feature many anatomically distinct clusters of neurons, or **nuclei**, as well as neural connections that link one nucleus to another. The various components of the brain are made up of cells (neurons) that communicate with one another via bioelectric messages (action potentials) that travel from one neuron to another (through synapses) via elongated extensions of the neurons (axons). Some components of the brain are deeply involved in the memory of songs, while others are necessary for the imitative production of memorized song patterns (Gobes and

BOX 2.3 **EXPLORING BEHAVIOR BY INTERPRETING DATA**

Proximate mechanisms underlying song preferences in females

Female European starlings possess the same *ZENK* gene that males do. In the female brain, the ventral caudomedial neostriatum, or NCMv, responds to signals sent to it from auditory neurons that fire when the bird is exposed to sounds, such as those made by singing male starlings. As you can see from the graph in the **FIGURE**, when captive female starlings are given a choice between perching next to a nest box where they can hear a long song versus perching next to another nest box where a shorter song is played, they spend more time at the long-song site.

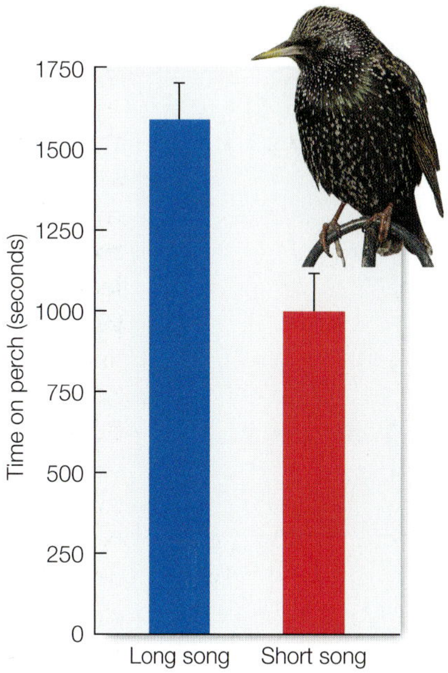

Song preferences of female European starlings. In this experiment, song preference was measured by the willingness of females to perch either near a nest box from which a long starling song (lasting about 50 seconds) was played or near one from which a shorter song (lasting about half as long) was played. Bars depict mean +/– SE. (After Gentner and Hulse 2000; photograph © Raymond Neil Farrimond/ShutterStock.)

Thinking Outside the Box

What proximate hypothesis could account for the song preferences of female starlings? What prediction can you make about the activity of *ZENK* in the cells that make up the NCMv of female starlings exposed to long versus short songs? How might you test your prediction? What would be the scientific point of collecting the data necessary to evaluate your prediction? Finally, begin to think about *why* females might choose males that sing long songs. For an ultimate explanation for the female preference for long songs, see work by Farrell and colleagues (Farrell et al. 2012).

Bolhuis 2007). Determining which anatomical unit does what in adult birds is a task for neurophysiologists more interested in the operational mechanics of the nervous system than in its development.

Neurophysiologists have long focused on a region of the brain of white-crowned sparrows and other songbirds called the high vocal center nucleus, or HVC. This dense collection of neurons connects to the robust nucleus of the arcopallium (mercifully shortened by anatomists to RA), which in turn is linked with the tracheosyringeal portion of the hypoglossal nucleus (whose less successful acronym is nXIIts). This bit of brain anatomy sends messages to the syrinx, the sound-producing structure of birds that is analogous to the larynx in humans. The fact that HVC and RA can communicate with nXIIts, which connects to the syrinx, suggests that these brain elements exert control over singing behavior (**FIGURE 2.6**). This hypothesis about the neural control of bird song has been well tested. For example, if neural messages from RA cause songs to be produced, then the experimental destruction of this center, or surgical cuts through the neural pathway leading from RA to nXIIts,

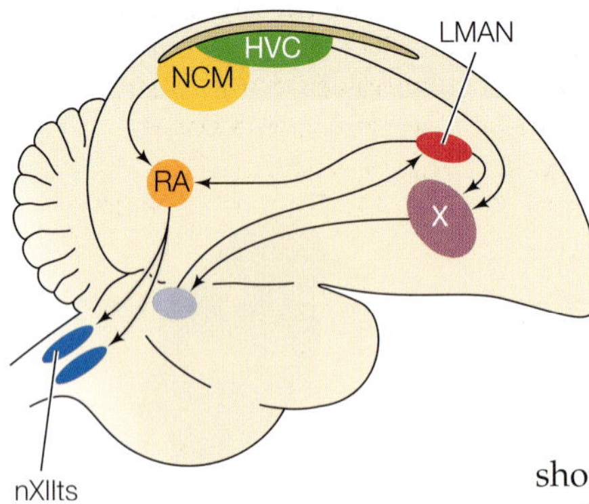

FIGURE 2.6 Song system of a typical songbird. The major components, or nuclei, involved in song production include the robust nucleus of the arcopallium (RA), high vocal center (HVC), the lateral magnocellular nucleus of the anterior nidopallium (LMAN), the caudomedial neostriatum (NCM), and area X (X). Neural pathways carry signals from HVC to the tracheosyringeal portion of the hypoglossal nucleus (nXIIts) to the muscles of the song-producing syrinx. Other pathways connect the nuclei, such as LMAN and area X, that are involved in song learning rather than song production. (After Brenowitz et al. 1997.)

should have devastating effects on a bird's ability to sing, which they do (Catchpole and Slater 2008). If RA plays an important role in controlling bird song, then in bird species such as the white-crowned sparrow, in which males sing and females do not, RA should be larger in male brains than in female brains, and it is (**FIGURE 2.7**) (Nottebohm and Arnold 1976, Baker et al. 1984, Nealen and Perkel 2000).

RA can also respond to the social environment of the bird. In European starlings (*Sturnus vulgaris*), for example, RA is the only song control nucleus that grows substantially in males that are exposed to high-quality (longer) songs of other males for a week (**FIGURE 2.8**) (Sockman et al. 2009). The suggestion here is that when some males are singing songs that are especially attractive to females, other eavesdropping males in the neighborhood benefit from modulating the motor region controlling their song production to sing more attractively themselves.

Other brain nuclei appear to be essential for song learning, rather than for song production. One such nucleus is the lateral magnocellular nucleus of the anterior nidopallium (LMAN), which projects to area X, another nucleus in the forebrain that is important for song learning. A team led by Arthur Arnold demonstrated that lesioning LMAN in juvenile male zebra finches disrupted song development (Bottjer et al. 1984). However, by the time these birds were adults, they produced normal songs, suggesting that LMAN lesions caused only temporary changes to song learning. Indeed, subsequent lesion studies in zebra finches demonstrated LMAN plays a role in circuit plasticity necessary for song learning (Scharff and Nottebohm 1991, Braindard and Douple 2000). More recently, Ölveczky and colleagues showed that using chemicals instead of lesions to inactivate LMAN in juvenile

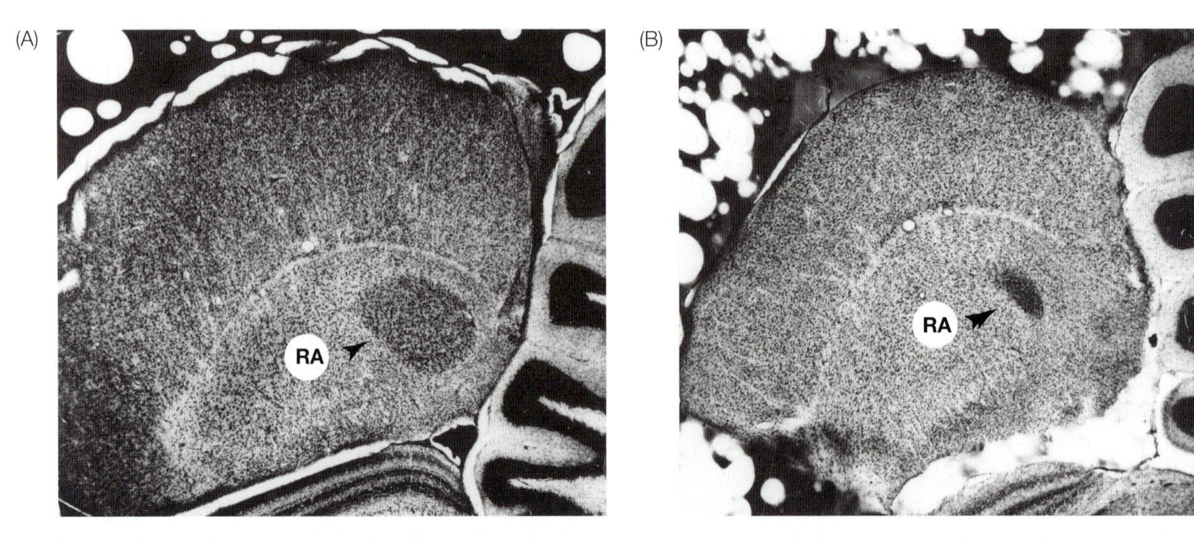

FIGURE 2.7 Difference in the size of one nucleus of the song system, the robust nucleus of the arcopallium (RA), in a male (A) and female (B) zebra finch. (Photographs courtesy of Fernando Nottebohm and Arthur Arnold.)

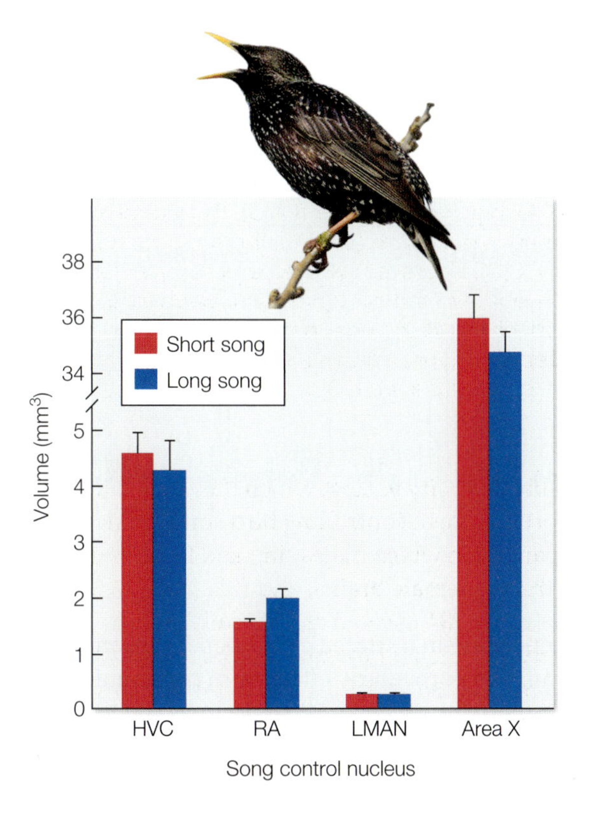

FIGURE 2.8 Song competition in the European starling changes the size of the robust nucleus of the arcopallium (RA). In males exposed to high-quality (long) songs of other starlings when they are adults, the RA increases in size, unlike some other song control nuclei. Bars depict mean +/– SE. (After Sockman et al. 2009; photograph © skapuka/Shutterstock.com.)

male zebra finches temporarily reduced song variability (Ölveczky et al. 2005). By measuring neuronal activity, the researchers further showed that LMAN was the source of this variability.

Relatedly, it has been proposed that large amounts of neural tissue are required if a male sparrow or other songbird is to learn a complex song or songs. Yet one could argue instead that the experience of singing a complex song or acquiring a large repertoire causes HVC to expand in response to stimulation of this region of the brain. If this were true, then males reared in isolation should have smaller HVCs than males that learn songs. This experiment has been done with sedge warblers (*Acrocephalus schoenobaenus*; **AUDIO 2.5**); isolated males had brains that were in no way different from those of males that had learned their songs by listening to the songs of other males (Leitner et al. 2002). Prior to this work, similar results came from a study in which some male marsh wrens (*Cistothorus palustris*; **AUDIO 2.6**) were given a chance to learn a mere handful of songs while another group listened to and learned up to 45 songs (Brenowitz et al. 1995). Thus, in both the warbler and the wren, the male brain develops largely independently of the learning experiences of its owner, suggesting that the production of a large HVC is required for learning, rather than the other way around.

AUDIO 2.5

Sedge warbler
ab11e.com/a2.5

AUDIO 2.6
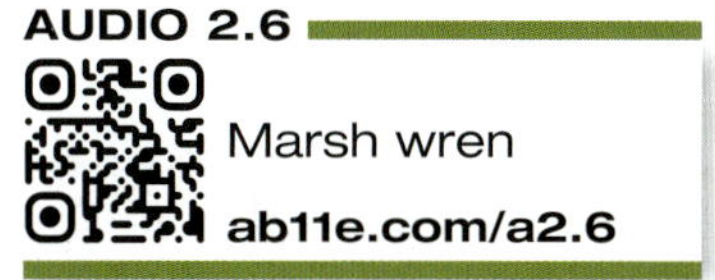
Marsh wren
ab11e.com/a2.6

Although the neurophysiology of song learning has been explored at the level of entire brain nuclei, neuroscientists could in theory study how given neurons contribute to communication between birds. In fact, this kind of research has been done by Richard Mooney and his coworkers with swamp sparrows (*Melospiza georgiana*) (Mooney et al. 2001). Males of this species sing two to five song types, each type consisting of a "syllable" of sound that is repeated over and over in a trill that lasts for a few seconds (**AUDIO 2.7**). If a young male swamp sparrow is to learn a set of song types, he must be able to discriminate among the types being sung by males around him, and later, he must be able to tell the difference between his own song types as he listens to himself sing. Mooney and colleagues' work demonstrated that the song types are presumably stored as discrete memories in different parts of the brain and that there is a clear link between learning and perceptual distinctiveness.

One mechanism that could help a young male control his song type output would be a set of specialized neurons in the HVC that respond selectively to a specific song type. Activity in these cells could contribute to a bird's ability to monitor what he is

AUDIO 2.7

Swamp sparrow
ab11e.com/a2.7

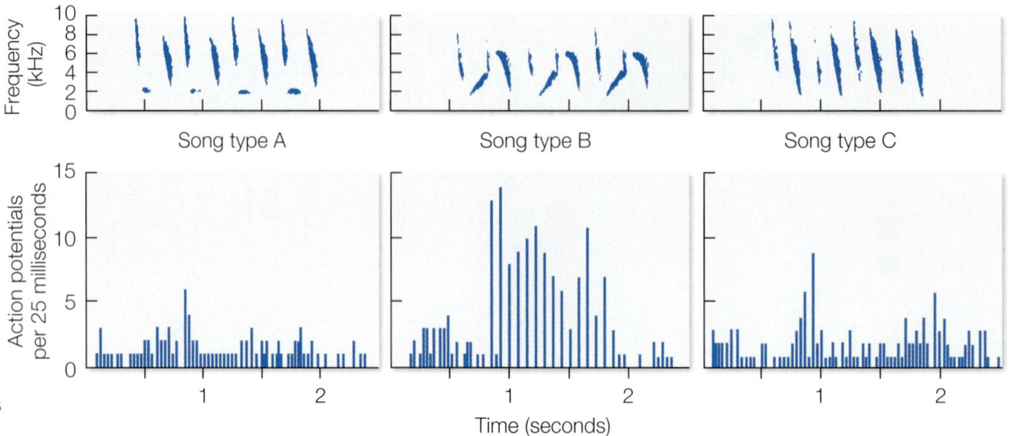

FIGURE 2.9 Single cells and song learning in the swamp sparrow. (Top) Spectrograms of three song types, A, B, and C. (Bottom) When the three songs are presented to the sparrow, one of its HVC relay neurons reacts substantially only to song type B. Other cells not shown here respond strongly only to song type A, while still others fire rapidly only when song type C is the stimulus. (After Mooney et al. 2001, © 2001 by the National Academy of Sciences, USA; photograph © Dennis Donohue/ShutterStock.)

singing so that he could adjust his repertoire in a strategic manner. For example, a male could select a song type that would be particularly effective in communicating with a neighboring male, which does happen, as you will see below. The technology exists to permit researchers to record the responses of single cells in a swamp sparrow's HVC to playbacks of that bird's own songs. In so doing, Mooney and his associates discovered several HVC relay neurons that generated intense volleys of action potentials when receiving neural signals from other cells upon exposure to one song type only (Mooney et al. 2001). Thus, one relay neuron, whose responses to three different song types are shown in **FIGURE 2.9**, produces large numbers of action potentials in a short period when the song stimulus is song type B. This same cell, however, is relatively unresponsive when the stimulus is song type A or C. So here we have a special kind of cell that could (in conjunction with many other neurons) help the sparrow identify which song type it is hearing, the better to select the best response to that signal.

The Evolution of Song Learning

Although a great deal has been learned about how the song control system of songbirds develops and operates, we still have much to learn about the underlying proximate mechanisms of singing behavior. But even if we had this information in hand, our understanding of singing by white-crowned sparrows would still be incomplete until we dealt with the ultimate causes of the behavior to understand why these mechanisms evolved. Since the complex and elaborate proximate mechanisms underlying bird song are unlikely to have materialized out of thin air, we can take a comparative approach to ask questions such as, when in the distant past did an ancestral bird species start learning its species-specific song, thereby setting in motion the changes that led eventually to dialect-learning abilities in birds such as the white-crowned sparrow?

An Evolutionary History of Bird Song

Evidence relevant to this historical question includes the finding that song learning occurs in members of just 3 of the 23 avian orders: the parrots, the hummingbirds, and the "songbirds," which belong to that portion of the Passeriformes—the oscines—that includes the sparrows and warblers, among other species (**FIGURE 2.10**) (Brenowitz 1991). The other group of Passeriformes is the suboscines, which comprise the non-learning singers. The suboscines and members of most of the remaining 20 orders of birds produce complex vocalizations, but they apparently

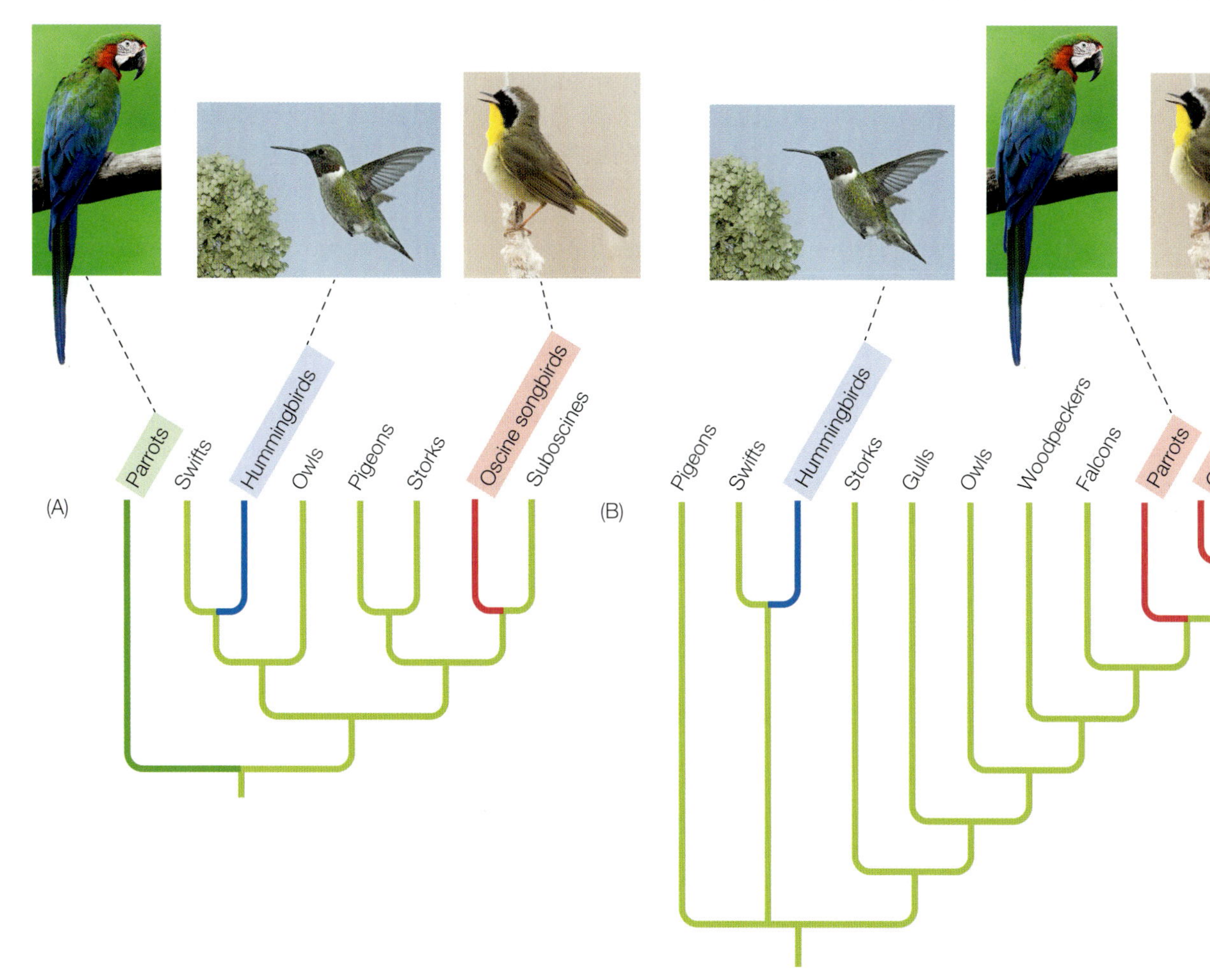

FIGURE 2.10 Two phylogenies of song learning in birds. (A) The bird phylogeny available in 1991 suggests that song learning evolved independently in three different lineages of modern birds: the oscine songbirds, the hummingbirds, and the parrots. (B) A newer phylogeny published in 2008 and confirmed in 2014 suggests that song learning may have evolved just twice, in the ancestor of the oscine songbirds and parrots, which are close relatives, and independently in the hummingbirds. Ancestral state reconstruction is only as good as the phylogeny being used. (A after Sibley et al. 1988; B after Hackett et al. 2008; parrot © Prin Pattawaro/123RF; hummingbird © Daniel Hebert/ShutterStock; songbird © iStock.com/pchoui.)

do not have to learn how to do so, as shown in some cases by experiments in which young birds that were never permitted to hear a song tutor, or were deafened early in life prior to the onset of song practice, nevertheless came to sing normally (Kroodsma and Konishi 1991).

A central question about the historical sequence underlying the evolution of song learning is, did the song-learning ability exhibited by the three groups of birds evolve independently or not? Answering this question requires extrapolation back in time from measured characters of extant species to their common ancestors, an approach called **ancestral state reconstruction**. As you can imagine, inferring the behavioral traits of extinct species is no easy task and requires several assumptions. Ancestral state reconstruction—as well as comparative analyses more generally—relies on two important factors: (1) an accurate phylogeny and (2) reliable character or trait data. It is important to remember that, as we discussed in Chapter 1, phylogenies are simply hypotheses about the evolutionary relationships among species, and as more genetic data become available, the shape of the trees—and therefore the evolutionary hypothesis itself—can change. Birds are one such taxonomic group where the phylogeny has been a source of great debate, which in turn has led to considerable discussion about the phylogenetic relationships among the three orders of song learners.

To illustrate both the power and the challenge associated with ancestral state reconstruction, let's consider the evolutionary history of song learning in birds. If we were to accept a more traditional phylogenetic hypothesis (see Figure 2.10A), we would conclude that the nearest living relatives of each of the three song-learning orders do not learn their songs. This conclusion suggests one of two scenarios: either song learning originated three different times in the approximately 65 million years since the first truly modern bird appeared, or song learning originated in the ancestor of all the lineages shown in the tree, but then was lost four times: (1) at the base of the evolutionary line leading to the swifts; (2) at the base of the owl lineage; (3) at the base of the pigeon and stork lineages; and (4) at the base of the suboscines. In other words, either there have been three independent origins of song learning, or one origin followed by four separate losses of the trait (Jarvis et al. 2000). Using the principle of parsimony—choosing the simplest scientific explanation that fits the data—we would conclude that song learning evolved three times independently in birds. However, newer phylogenies (see Figure 2.10B) indicate that, surprisingly, parrots are the closest relatives of oscine songbirds (Suh et al. 2011, Wang et al. 2012, Jarvis et al. 2014, Prum et al. 2015). What do these new phylogenies tell us about the evolution of song learning? Interestingly, they are consistent with either three independent gains of song learning, or two gains—in the hummingbirds and the common ancestor of parrots and oscine songbirds—followed by two losses within the passerines, the New Zealand wrens and the suboscines (Jarvis et al. 2014).

You can see from this example how the phylogenetic hypothesis greatly influences ancestral state reconstruction, and in our case, the conclusion about how many times song learning has been gained (or lost) in birds. However, the story is becoming even more complicated as we generate better character data by studying song learning in additional passerine species. That is, there is mounting evidence that several suboscine species can actually learn to sing (Kroodsma et al. 2013, Touchton et al. 2014), challenging the idea that the passerine clade can be divided cleanly into a group of song learners and a group of non-song learners. Using these newer character data and phylogenies, the ancestral state reconstruction would suggest that there could have been (at least) a fourth gain of song learning within the suboscines, or a regain after a loss of vocal learning in other suboscines (Jarvis et al. 2014). Thus, it is fair to say that ancestral state reconstruction in birds cannot (yet) determine how many times song learning evolved independently.

Mechanisms of Song Learning and the Comparative Approach

Excluding for a moment the issue of song-learning ability in the suboscines, how might we go about testing the two equally likely hypotheses generated from the ancestral state reconstruction that song learning evolved independently in songbirds, hummingbirds, and parrots, or was ancestral in all birds and subsequently lost a number of times? One approach might be to identify the brain structures that endow these groups of birds with their singing and song-learning abilities and look for similarities or differences. This could be done by capturing birds that have just been singing (or listening to others sing) and killing them immediately to examine their brains for regions in which a product of the *ZENK* gene can be found (Jarvis et al. 2000). Recall from earlier that this gene's activity is stimulated in particular parts of the brain when birds produce songs, and in different parts of the brain when birds hear songs. Other stimuli and activities do not turn the gene on, so its product is absent or very scarce when the birds are not communicating. By comparing how much ZENK protein appears in different parts of the brains of birds that were engaged in singing or listening to songs immediately before their deaths, one can effectively map those brain regions that are associated with bird song production and perception.

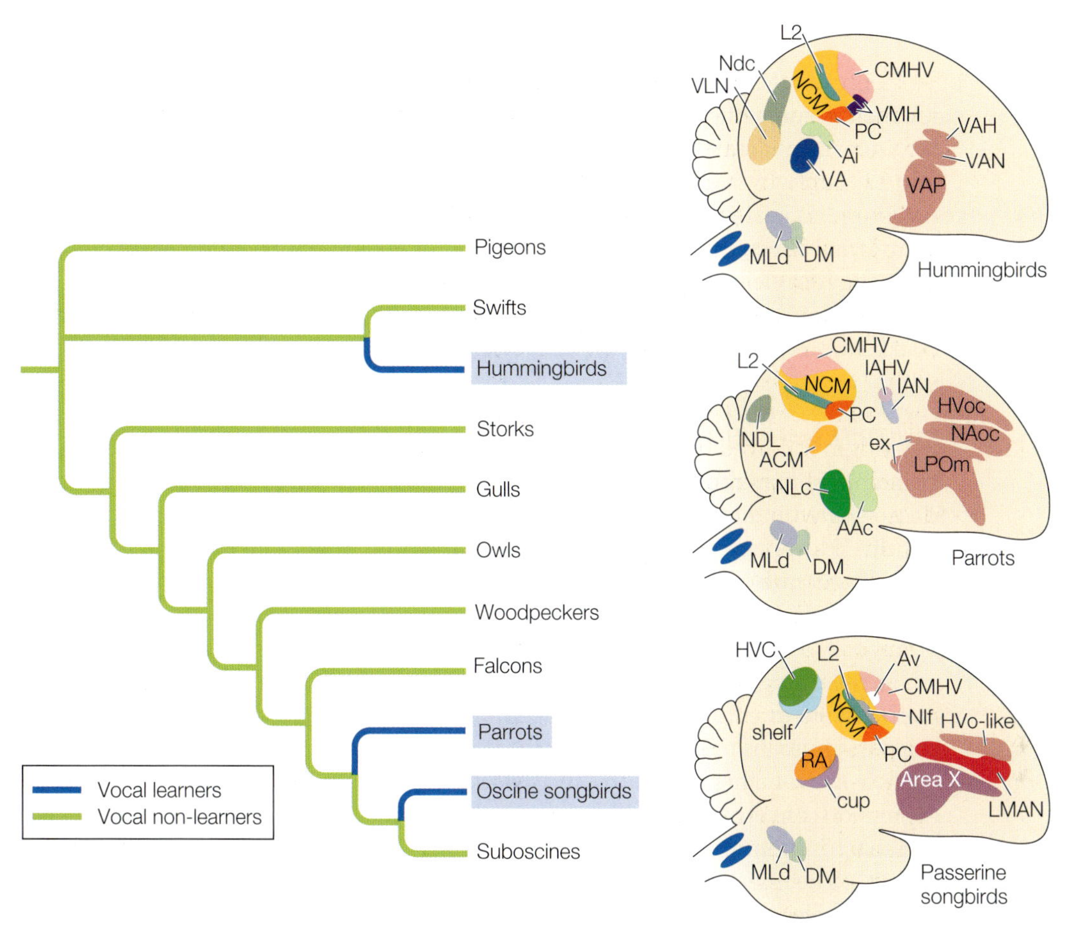

FIGURE 2.11 The song control systems of parrots, hummingbirds, and oscine songbirds are distributed throughout the brain in remarkably similar patterns. On the left is a diagram of the evolutionary relationships among some major groups of birds, including the three orders of vocal learners. On the right are diagrams of the brains of these groups, with the various equivalent components of the song control systems labeled (such as NCM). (After Jarvis et al. 2000.)

Research on the *ZENK* activity maps of the brains of selected parrots, hummingbirds, and songbirds reveals strong similarities in the number and organization of discrete centers devoted to song production and processing. In all three groups, for example, cells that form the caudomedial neostriatum (NCM) activate their *ZENK* genes when individuals are exposed to the songs of others. NCM is located in roughly the same part of the brain in all three groups, where it constitutes part of a larger aggregation of anatomically distinct elements that contribute to the processing of song stimuli. When parrots, hummingbirds, and songbirds vocalize, other brain centers, many in the anterior part of the forebrain, respond with heightened *ZENK* gene activity. Once again, there is considerable (but not perfect) correspondence in the locations of these distinctive song production centers in parrots, hummingbirds, and sparrows (**FIGURE 2.11**). The many similarities in brain anatomy among the three groups of vocal learners argue against the hypothesis that song-learning abilities evolved independently in the three groups of birds (or the two lineages). Moreover, like some parrots and songbirds, at least some species of hummingbirds are open-ended song learners that can continue to learn new songs throughout their lives and outside the critical window during development (Araya-Salas and Wright 2013), again suggesting a shared evolutionary history. Thus, we have to take seriously the possibility that the ancestor of all those birds in the orders sandwiched between the hummingbirds and the parrots–songbirds was a song learner, and that the mechanism for vocal learning possessed by that bird was then retained in some lineages while dropping out early on in others—a hypothesis that continues to generate debate (Jarvis et al. 2000, Farries 2001). This conclusion might also explain why we are discovering song learning in so many suboscine species. Of course, it is also possible that song learning evolved in both clades independently via convergence, though this is the least parsimonious hypothesis. But what about in vocal-learning species more distantly related to birds? Could convergent evolution explain, for example, why species as different as birds and humans learn vocalizations in surprisingly similar ways?

Human versus Avian Vocal Learning

If the mechanism of vocal learning was shared in unrelated lineages of songbirds, is it also shared in more much more distantly related species of animals that learn to vocalize? Although vocal communication is common in animals, vocal learning is exceedingly rare. Outside of birds, the ability to learn to produce vocalizations occurs in only a few groups of mammals, including some cetaceans (whales and dolphins), pinnipeds (seals and sea lions), bats, and of course humans. Indeed, the similarities in song-learning mechanisms between birds and many of these mammals—including humans—are quite high. Therefore, to test the hypothesis that song learning in birds and humans has evolved from shared genetic architectures, a team led by Erich Jarvis asked whether behavioral and anatomical convergence is associated with gene expression convergence in the brains of vocal-learning birds and humans, species that last shared a common ancestor between 310 and 68 million years ago (Pfenning et al. 2014). The researchers identified several specific song and speech brain regions of avian vocal learners and humans that shared patterns of gene expression. Specifically, they found that area X and RA in birds are most similar in gene expression to regions of the human brain associated with speech. Moreover, gene expression in bird motor and song-learning nuclei corresponds with expression in the human laryngeal motor cortex and parts of the brain that control speech production and learning, respectively. Thus, birds and humans exhibit convergent neural circuits for vocal learning that are accompanied by convergent molecular changes of multiple genes. In other words, the mechanisms have evolved along independent lines, but have adopted parallel functions in both taxa. For this reason, birds continue to be a model system for studying vocal learning in ways that ultimately may be applied to humans.

The Adaptive Value of Song Learning

Up until this point, we have discussed how birds learn new songs, both from a developmental and neurobiological perspective. We have also discussed the evolutionary history of bird song, comparatively exploring the phenomenon in the three lineages of birds that learn to sing, as well as comparing the process to that in mammals, including humans. But no analysis of song learning would be complete without discussing *why* vocal learning has evolved. Whatever the historical sequence of events underlying the evolutionary history of song learning in birds, we can still ask, why did the ability to learn spread through one or more ancestors of today's song learners? What are the fitness benefits of song learning for individual parrots, hummingbirds, and songbirds? To answer these and other questions about adaptive value, let's continue to focus on white-crowned sparrows and their learned dialects. The Darwinian puzzle here is that song-dialect learning has some obvious reproductive disadvantages—including the time, energy, and special neural mechanisms that this behavior requires—yet in many species of birds, males learn to sing, presumably giving them a reproductive advantage (**BOX 2.4**).

BOX 2.4 **DARWINIAN PUZZLE**

Why might song learning make males communicate more effectively with rivals or potential mates?

To answer this and other Darwinian Puzzles, we must first determine the costs and benefits associated with a given behavior. As we discussed in Chapter 1, for any behavior to be adaptive, its benefits must outweigh its costs. One potential cost of song learning in birds is captured in the **nutritional stress hypothesis**, which proposes that the brain structures underlying song learning (and production) develop during a period early in life when young birds are likely to experience developmental stress resulting from undernutrition (Nowicki et al. 1998). Similarly, perhaps not just developing large brains, but also maintaining them as adults is costly because brain tissue is among the most metabolically expensive tissues. Indeed, passerine songbirds and parrots—two of the three avian song learners—have unusually large brains highlighted by an expanded telencephalon with increased neuronal densities (Olkowicz et al. 2016). Many of these bird species actually have more neurons than mammalian brains, even those of primates with similar mass (**FIGURE**). Thus, producing the machinery necessary to learn song may be extremely costly in birds.

If song learning is costly because building or maintaining a complex brain is nutritionally and metabolically expensive, then the benefits of learning song must outweigh these energetic costs. What types of potential benefits might therefore explain why song learning might make males communicate more effectively with rivals and/or potential mates to increase their fitness? Behavioral biologists have explored a number of potential benefits of song learning in birds, including (1) improving reproductive success directly, either through male-male competition and female choice or assortative mating due to local adaptation to specific geographic areas or environments, and (2) increasing information flow, either by recognizing neighbors or groups mates or sharing information with them (Nowicki and Searcy 2014).

(Continued)

nutritional stress hypothesis Brain structures underlying song learning (and production) develop during a period early in life when young birds are likely to experience developmental and nutritional stress.

Neuronal density in the brains of song-learning birds and primates Brains of corvids (jay and raven), parrots (macaw), and primates (monkeys) are drawn at the same scale. Numbers under each brain represent the mass of the pallium (in grams). Circular graphs show the proportions of neurons contained in the pallium (green), cerebellum (red), and the rest of the brain (yellow); numbers in the circles are the total numbers of pallial/cortical neurons (in millions). Notice that the brains of these highly intelligent birds harbor absolute numbers of neurons that are comparable to, or even larger than, those of primates with much larger brains. (After Olkowicz et al. 2016.)

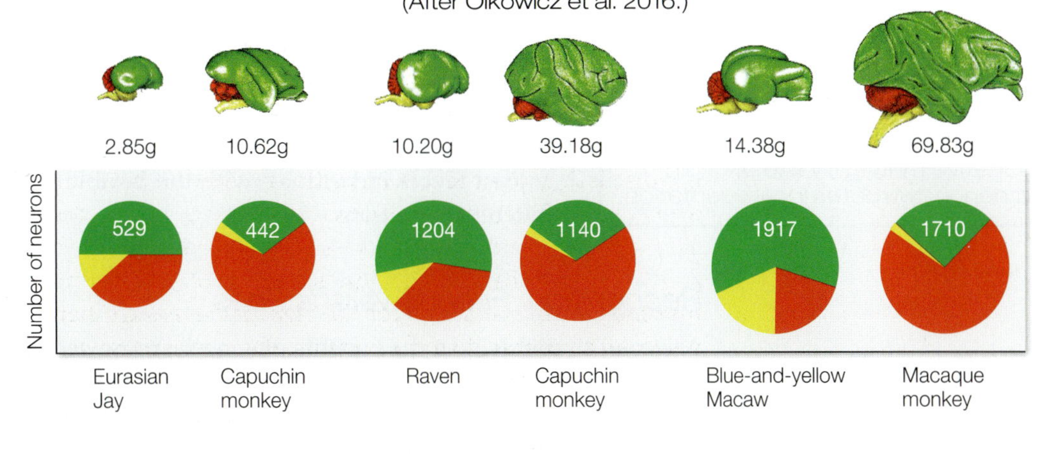

BOX 2.4 DARWINIAN PUZZLE *(continued)*

Thinking Outside the Box

We will discuss these benefits—as well as some of the costs—in greater detail in the text. But before we do that, consider how you might solve this song-learning puzzle. Start by listing as many potential costs and benefits as you can think of for why birds learn to sing, and then develop a test of each. For example, one approach for determining the costs of producing a big brain might be to manipulate nutrition during development and then measure the effects on brain size and singing ability. Similarly, one approach to determining the benefits of producing complex, learned songs might be to manipulate song learning in juvenile males and then subsequently determine how they fare in mate-choice trials as adults. As you will learn in the text, these types of experiments have been conducted in several bird species, allowing us to refine our hypotheses underlying avian song learning and gain a better understanding of the costs and benefits of this behavior. Try coming up with a few more experiments on your own.

HYPOTHESES 2.2

Ultimate hypotheses for the function of vocal learning in birds

environmental adaptation hypothesis Vocal learning promotes acoustic adaptation of vocal signals to the local habitat.

recognition hypothesis Vocal learning allows vocal signals to become more recognizable and thereby promotes identification of neighbors or social cohesion within groups (via group, kin, or individual recognition).

information-sharing hypothesis Vocal learning enables expansion of the vocal repertoire in systems where living with kin favors greater information sharing.

sexual selection hypothesis Vocal learning enables increases in the complexity of the vocal repertoire that is used in male–male competition or is favored by female preferences.

geographic matching hypothesis Vocal learning evolves to promote geographic variation in vocal signals, which in turn allows assortative mating by site of origin and hence promotes local adaptation.

Although there are numerous adaptive hypotheses to explain why white-crowned sparrows and other avian species have evolved vocal learning, they generally have to do with how song learning is shaped by both the ecological or and the social environment (Nowicki and Searcy 2014). We summarize the primary functional hypotheses for avian song learning in **HYPOTHESES 2.2**, and explore each in detail below.

Adapting to the Local Environment

One possible advantage of song learning might be the ability of a young male to fine-tune his song so that it acquires a dialect that can be transmitted unusually effectively in the particular habitat where he lives. According to this **environmental adaptation hypothesis**, a young male that learns his song from his older neighbors might then generate songs that travel farther and with less degradation than if he sang another dialect better suited to a different acoustic environment (Catchpole and Slater 2008). And in fact, male white-crowned sparrows in densely vegetated habitats sing songs with slower trills and lower frequencies than do birds in places with lower vegetation density (Derryberry 2007). This same pattern has been observed for other birds, such as the satin bowerbird (*Ptilonorhynchus violaceus*) and the great tit (*Parus major*), which employ lower frequencies in dense forests rather than in more open areas (**FIGURE 2.12**) (**AUDIO 2.8**) (Hunter and Krebs 1979, Nicholls and Goldizen 2006). In all of these cases, the environment either influences the evolution of communication signals or at least favors individuals with the flexibility to adjust their songs to different conditions.

AUDIO 2.8

Great tit

ab11e.com/a2.8

Recognition: Friends versus Foes

A second hypothesis on the benefits of song learning centers on the advantages a singer may gain by matching songs to their social environment (Beecher and Brenowitz 2005). This **recognition hypothesis** is based on the idea that birds need to recognize others, be it their neighbors as competitors, their group mates as allies, or sometimes even specific individuals. Perhaps males that are able to learn the local

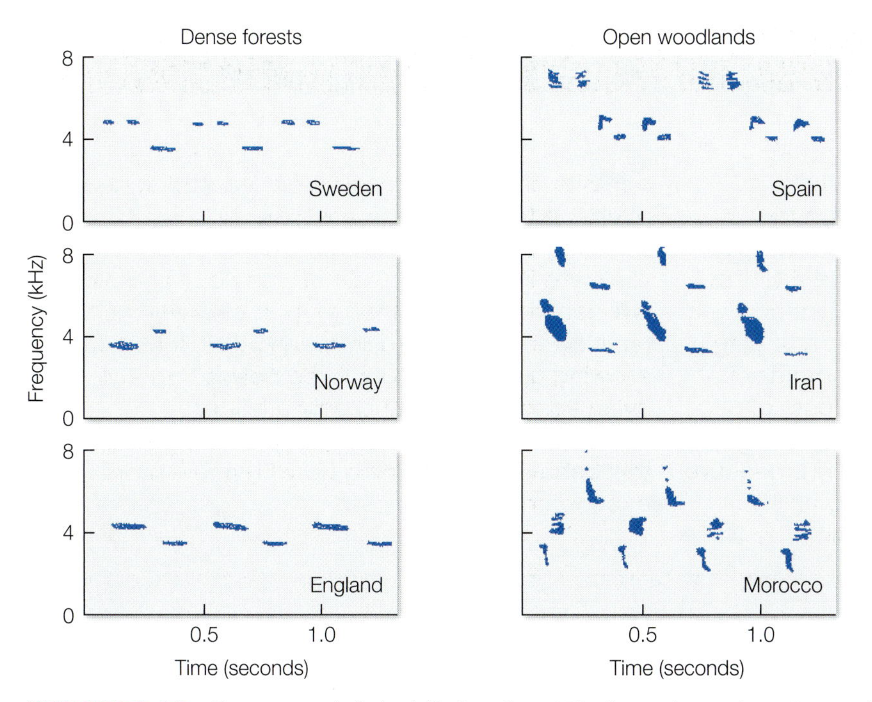

FIGURE 2.12 Songs match habitats. Great tits from dense forests produce pure whistles of relatively low frequency, whereas males of the same species that live in more open woodlands use more and higher sound frequencies in their more complex songs. (After Hunter and Krebs 1979; photograph © David Dohnal/ShutterStock.)

version of their species-specific song can communicate better with rivals that are also singing that particular learned song variant. If this is true, then we can predict that young males should learn directly from their territorial neighbors, using those individuals as social tutors. By adopting the song of a neighbor, a relative newcomer could signal his recognition of that male as an individual and demonstrate that he is not a novice competitor but old enough and experienced enough to have used the information provided by his neighbor's song. We might also predict that young males should alter their song selection to match that of older territorial neighbors, and that these older males should take males that match their songs more seriously as rivals as compared with young novice challengers with unmatched song types. Both of these predictions have been supported in white-crowned sparrows (Baptista and Morton 1988, Poesel and Nelson 2012).

The ability of males to produce the songs, or at least song elements, of rival neighbors is also a characteristic of the song sparrow, a species that has a repertoire of a dozen or more different, distinctive song types rather than a single vocalization of a particular dialect (**AUDIO 2.9**). Young song sparrows usually learn their songs from tutors that are their neighbors in their first breeding season, with their final repertoire tending to be similar to that of immediate neighbors (Nordby et al. 1999). In fact, Michael Beecher and his colleagues found through playback experiments that when a male heard a recording of a neighbor's song coming from the neighbor's territory, he tended to reply to that recording by singing a song from his own repertoire that matched one in the repertoire of that particular neighbor. This kind of type matching occurred when, for example, male BGMG heard the recorded version of male MBGB's song type A (**FIGURE 2.13**), which led him to answer with his own song type A (Beecher et al. 2000b).

AUDIO 2.9

Song sparrow
ab11e.com/a2.9

Recognition need not be only for identifying rival neighbors, but might also help in identifying or recognizing group mates. Indeed, complex vocal communication

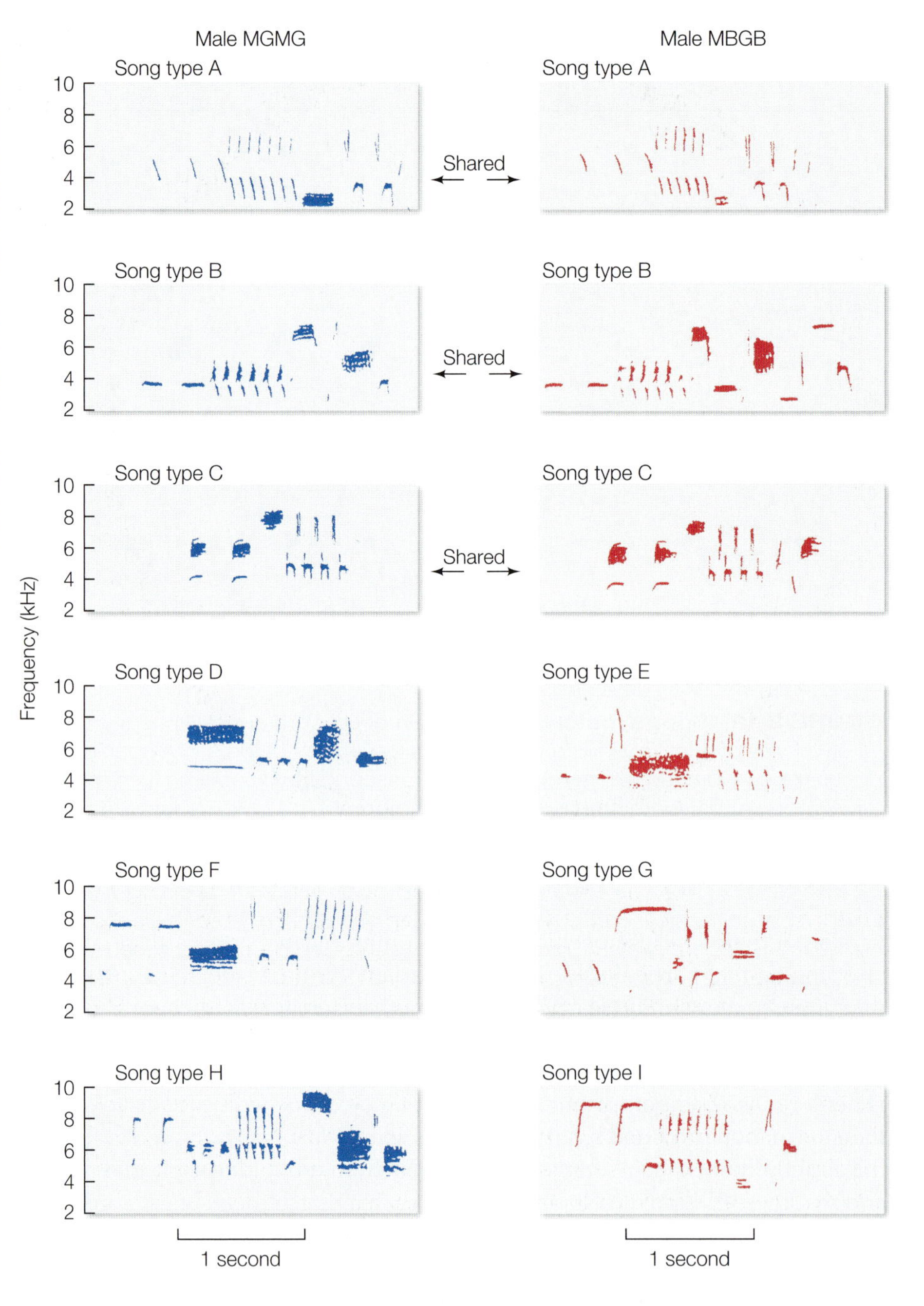

FIGURE 2.13 Song type matching in the song sparrow Males BGMG and MBGB occupy neighboring territories and share three song types (A, B, and C: the top three rows of spectrograms); six unshared song types (D, E, F, G, H, and I) appear on the bottom three rows. (After Beecher et al. 1996; photograph © Michael Woodruff/ShutterStock.)

systems have evolved in many cooperatively breeding bird species where individuals live in social groups to help rear young (see Chapter 13). Depending on the species, vocalizations in avian cooperative breeders can signal kinship, group membership, or individual identification. Sara Keen and colleagues hypothesized that in the most complex societies comprising a mix of related and unrelated individuals, signaling individuality and group association, rather than simply kinship (genetic relatedness), may be essential (Keen et al. 2013). For example, in noisy miners (*Manorina melanocephala*) (McDonald 2012), superb starlings (*Lamprotornis superbus*) (Keen et al. 2013), and apostlebirds (*Struthidea cinerea*) (Warrington et al. 2015)—all species that live in large and flexible social groups comprising both kin and non-kin—birds can differentiate among specific individuals using calls (**AUDIO 2.10**). In contrast, in species that live in less complex societies with smaller groups made up primarily of kin, calls appear to more often signal only kinship or group identity and not individual identity

AUDIO 2.10

Vocalization of cooperative breeders

ab11e.com/a2.10

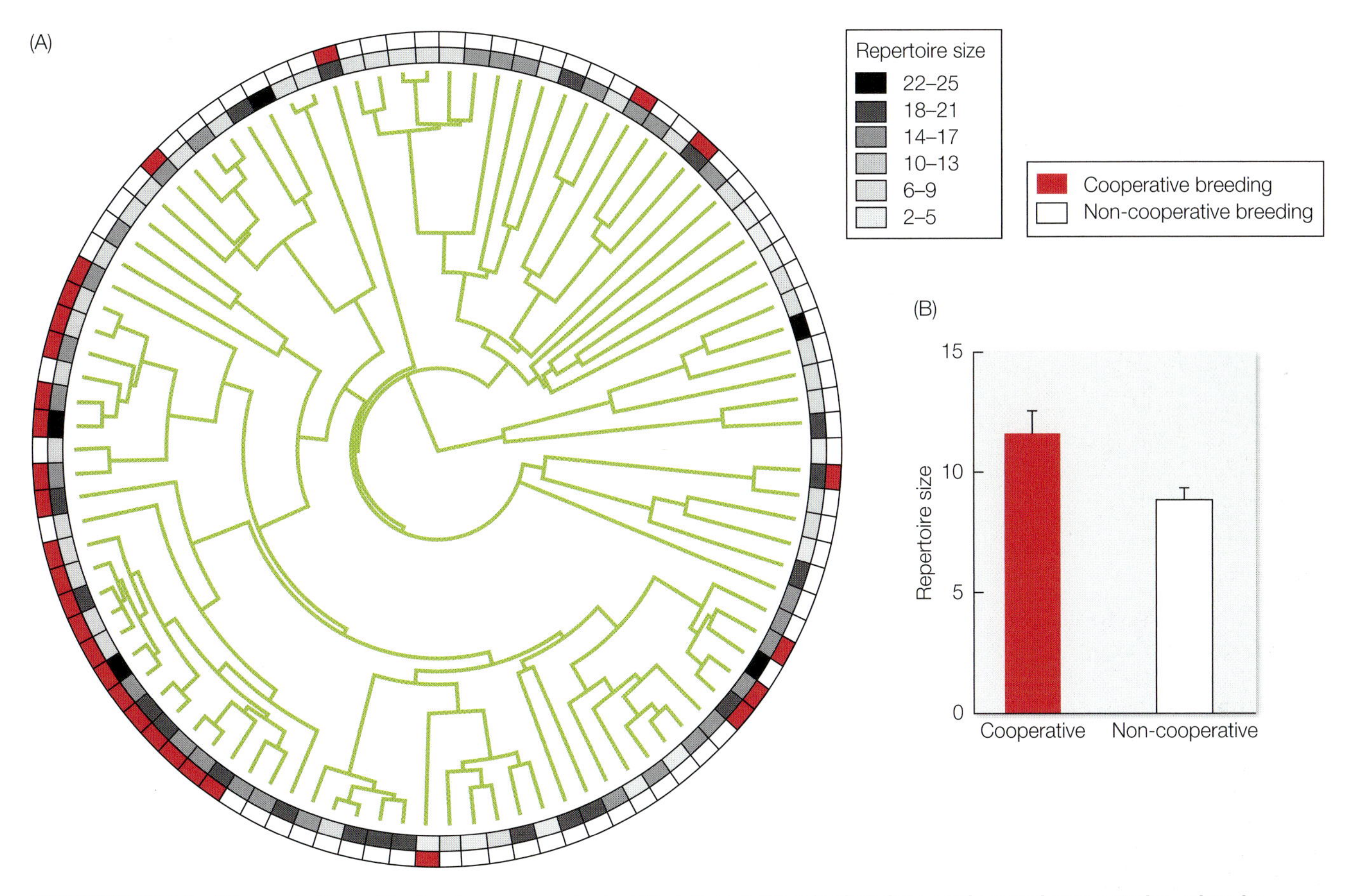

FIGURE 2.14 The relationship between cooperative breeding behavior and vocal repertoire size in birds (A) A phylogeny of 97 avian species with cooperative breeding behavior and vocal repertoire size mapped onto the tree. The interior circle represents different tiers of repertoire size (see key). The outer circle represents whether species are cooperative breeders or not. Species names have been removed from the tree. (B) Cooperatively breeding species have larger repertoire sizes than non-cooperatively breeding ones. Bars depict mean +/– SE. (After Leighton 2017.)

(Keen et al. 2013). Young birds learn these vocalizations in the group in which they were born, but since one sex (usually females in most bird species; see Chapter 7) immigrates into new social groups to breed, birds can apparently continue to learn additional group-specific vocalizations as adults.

As you will learn in more detail in Chapters 12 and 13 when we discuss kin selection, most animals that live in social groups live with kin with whom they share genes. Social groups can therefore often be considered family groups, and family members benefit by being able to share information not just about kinship or identity, but about resource availability, predation risk, and more. Researchers have argued that one of the earliest steps in human language was the acquisition of the ability to use a large class of symbols related to these issues (Jackendoff 1999). Stephen Nowicki and William Searcy have argued that vocal learning is needed for vocabulary expansion and that vocabulary expansion is favored in family-living species (Nowicki and Searcy 2014). This **information-sharing hypothesis** predicts that species that live in kin groups should evolve complex forms of communication and sophisticated vocal-learning mechanisms. Indeed, cooperatively breeding species have larger vocal repertoires—particularly alarm and contact calls—than do non-cooperatively breeding species (**FIGURE 2.14**) (Leighton 2017). Moreover, several cooperatively breeding birds have learned to rearrange and interpret combinations of sounds to produce "language" much in the same way as humans. For example, cooperatively breeding chestnut-crowned

information-sharing hypothesis Vocal learning enables expansion of the vocal repertoire in systems where living with kin favors greater information sharing.

AUDIO 2.11
"Language" producers
ab11e.com/a2.11

babblers (*Pomatostomus ruficeps*) can rearrange two simple song elements to create two functionally distinct elements (Engesser et al. 2015), and Japanese tits (*Parus minor*), which occasionally breed in cooperative groups, use a simple rule to discriminate and extract information from different orderings of novel call sequences (**AUDIO 2.11**) (Suzuki et al. 2017).

Sexual Selection: Male–Male Competition

Since we now know that males typically sing songs to defend territories, male–male competition in birds often involves disputes over territory ownership or boundaries. And if securing a territory is critical because males court females with their songs (often from their territories), males must also compete with other males for access to those potential mates. Another kind of response involves repertoire matching, in which a bird exposed to a song type from a neighbor responds not with the exact same type but with a song type drawn from their shared repertoire. For example, repertoire matching would occur if male song sparrow BGMG answered back with song type B or C when he heard song type A from male MBGB's territory (see Figure 2.13). A third option for a male song sparrow is to reply to a song from a neighbor with an unshared song. This kind of mismatched response would be illustrated by male BGMG singing song type D, F, or H (see Figure 2.13) upon hearing male MGBG's song type A.

The fact that song sparrows type-match and repertoire-match so often indicates that they not only recognize their neighbors and know what songs they sing, but that they use this information to shape their replies (Beecher et al. 2000b). But how do male song sparrows benefit from their selection of a song type to reply to a neighbor that is their competitor when it comes to attracting a mate? Since previous work established that song type matching can be highly aggressive, whereas nonmatching is more conciliatory, perhaps a male sparrow's choice of song enables him to send graded threat signals to his neighbors, with type matches telling the targeted receiver that the singer is highly aggressive. In contrast, replying with an unshared song could signal a desire to back off, while a repertoire match might signal an intermediate level of aggressiveness. According to this male–male competition version of the **sexual selection hypothesis**, playback experiments should therefore reveal differences in the responses of territorial birds to recorded songs that contain a type match, a repertoire match, or neither. When Michael Beecher and colleagues tested this prediction, they found that playback of a recording containing a type match did indeed elicit the most aggressive response from a listening neighbor, whereas a repertoire match song generated an intermediate reaction, and nonshared song types were treated less aggressively still (**FIGURE 2.15**) (Burt et al. 2001,

sexual selection hypothesis
Vocal learning enables increases in the complexity of the vocal repertoire that is used in male–male competition or is favored by female preferences.

FIGURE 2.15 Song matching and communication of aggressive intent in the song sparrow. Male song sparrows can control the level of conflict with a neighbor by their selection of songs. When a focal male sings a shared song at a rival, the neighbor has three options: one that will escalate the contest, one that will keep it at the same level, and one that will de-escalate it. Likewise, the initiator of the contest can use his ability to select either a matching song type, a repertoire match, or an unshared song type. The three different kinds of songs convey information about the readiness of the singer to escalate or defuse an aggressive encounter. (After Beecher and Campbell 2005.)

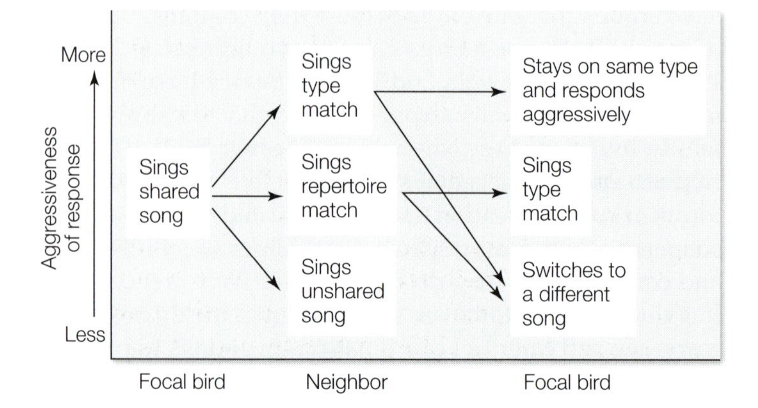

Beecher and Campbell 2005). This result supports the hypothesis that a male song sparrow that is able to learn songs from a neighbor can convey information about just how strongly he is prepared to challenge that individual.

If the ability to modulate song challenges is truly adaptive, then we can predict that male territorial success, as measured by the duration of territorial tenure, should be a function of the number of song types that a male shares with his neighbors. Song sparrows can hold onto their territories for up to 8 years, which tends to create a relatively stable community of neighbors. And in fact, the number of years that an individual male song sparrow holds his territory increases about fourfold as the number of song types shared with his neighbors goes from fewer than 5 to more than 20 (Beecher et al. 2000a). This finding strongly suggests that the potential for neighbors to form long-term associations in song sparrow communities selects for males able to advertise their aggressive intentions to particular individuals. By contrast, in the fluid social settings of certain other species in which territorial males come and go quickly (that is, in which the identity of neighbors is fluid), song learning and sharing of the sort exhibited by song sparrows should not occur. Instead, under these different ecological circumstances, males should acquire songs characteristic of their species as a whole to facilitate communication with any and all conspecifics, rather than developing a dialect or a set of shared song types characteristic of a small, stable population.

Donald Kroodsma and his colleagues have tested this proposition by taking advantage of the existence of two very different populations of the sedge wren (*Cistothorus platensis*) (**AUDIO 2.12**). In the Great Plains of North America, male sedge wrens are highly nomadic, moving from breeding site to breeding site throughout the summer. In this population, song matching and dialects are absent. Instead, the birds improvise variations on songs heard early in life as well as invent entirely novel songs of their own, albeit employing the general pattern characteristic of their species (Kroodsma et al. 1999a). In contrast, male sedge wrens living in Costa Rica and Brazil remain on their territories year-round. In these areas, dialects and song matching are predicted to occur, and these learning-proficient males do have dialects and do sing like their neighbors (Kroodsma et al. 1999b). The differences in song learning that occur within this one species indicate that learning and dialects do evolve when males gain by communicating specifically with other males that are their long-standing neighbors.

AUDIO 2.12
Sedge wren
ab11e.com/a2.12

Sexual Selection: Female Choice and Assortative Mating

Although our focus in this chapter has been largely on males, it is important to remember that females in many species of songbirds also sing. In fact, in a comparative study of nearly all of the songbirds, Karan Odom and colleagues showed that female song is present in more than 70 percent of species, and that in the common ancestor of modern songbirds, both males and females likely sang (Odom et al. 2014). Like males, many females sing in a competitive context with other females, sometimes even to attract males. Yet even in species where females do not sing, females still rely on male song to help choose a mate and therefore must be able to discriminate among the songs of different males, a process that often also requires learning. Therefore, another ultimate hypothesis for song learning in males is the female version of the sexual selection hypothesis, which takes aim at the social environment provided by females, typically the intended recipients of bird songs. Thus, to understand why males learn to sing, we must expand our focus on the role of song in male–male interactions to include a look at male–female communication.

When a female Cassin's finch (*Haemorhous cassinii*) disappears, leaving her partner without a mate, the number of songs he sings and the time he spends singing increase dramatically, either to lure his partner back or to attract a replacement (**AUDIO 2.13**). The evidence reviewed in the caption to **FIGURE 2.16** supports

AUDIO 2.13
Cassin's finch
ab11e.com/a2.13

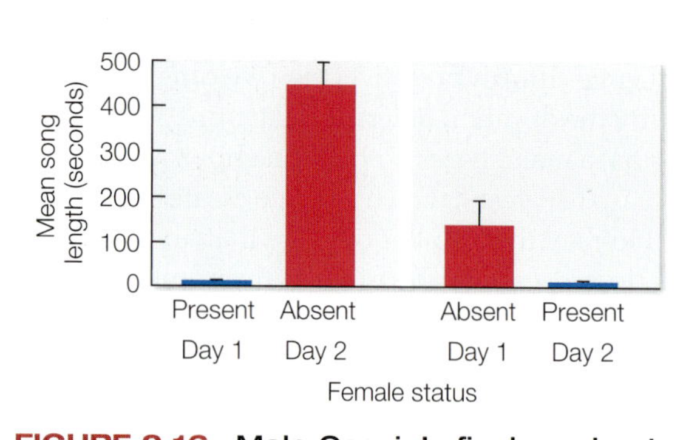

Cassin's finch

FIGURE 2.16 Male Cassin's finches sing to attract females. When a male is paired off with a female on day 1, he sings relatively little. But when the female is removed on day 2, the male invests considerable time in singing, perhaps to call her back. If the male is caged without a female on day 1, he allocates little time to singing, and he sings even less when a female is introduced into his cage, indicating that it is the loss of a potential mate that stimulates the male to sing. Bars depict mean +/– SE. (After Sockman et al. 2005; photograph courtesy of Dave Menke/U.S. Fish and Wildlife Service.)

geographic matching hypothesis Vocal learning evolves to promote geographic variation in vocal signals, which in turn allows assortative mating by site of origin and hence promotes local adaptation.

this hypothesis (Sockman et al. 2005). But could the fact that male Cassin's finches *learn* their songs contribute to a male's ability to attract females? Consider a species that is divided into stable subpopulations. In such a species, males in each of these groups are likely to have genes that have been passed down for generations by those males' successful ancestors. According to the **geographic matching hypothesis**, by learning to sing the dialect associated with their place of birth, males could announce their possession of traits (and underlying genes) well adapted for that particular area. Females hatched in that area might gain by having a preference for males that sing the local dialect, because they would endow their offspring with genetic information that would promote the development of locally adapted characteristics (Baker and Cunningham 1985). This hypothesis matches the discovery that male white-crowned sparrows that sang the local dialect around Tioga Pass, California, were less infected with a blood parasite than were non-dialect singers (**FIGURE 2.17**) (MacDougall-Shackleton et al. 2002). Therefore, female white-crowned sparrows in this population could potentially use song information to secure healthier mates. Accordingly, males with songs not matched to the local dialect fathered fewer offspring than local-dialect singers, a result consistent with the hypothesis that females prefer to mate with males that sing the local dialect.

If female preferences are designed to help them identify prospective mates from their natal area, then female white-crowned sparrows should prefer males with the dialect that the females heard while they were nestlings—namely, their father's dialect. In at least one Canadian population of white-crowned sparrows, they do not (Chilton et al. 1990). Furthermore, it turns out that young male white-crowned sparrows are not always locked into their natal dialects but may be able to change them later in life (DeWolfe et al. 1989) if they happen to move from one dialect zone to another. Therefore, female white-crowned sparrows cannot rely on a male's dialect to identify his birthplace with complete certainty. All of this casts doubt on the proposition that female song preferences enable females to endow their offspring with locally adapted gene complexes from locally hatched partners, as does the lack of genetic differentiation found among local-dialect groups, as we noted earlier (Soha et al. 2004, Poesel et al. 2017). However, experiments in other related species of sparrows are consistent with this hypothesis. For example, laboratory-reared

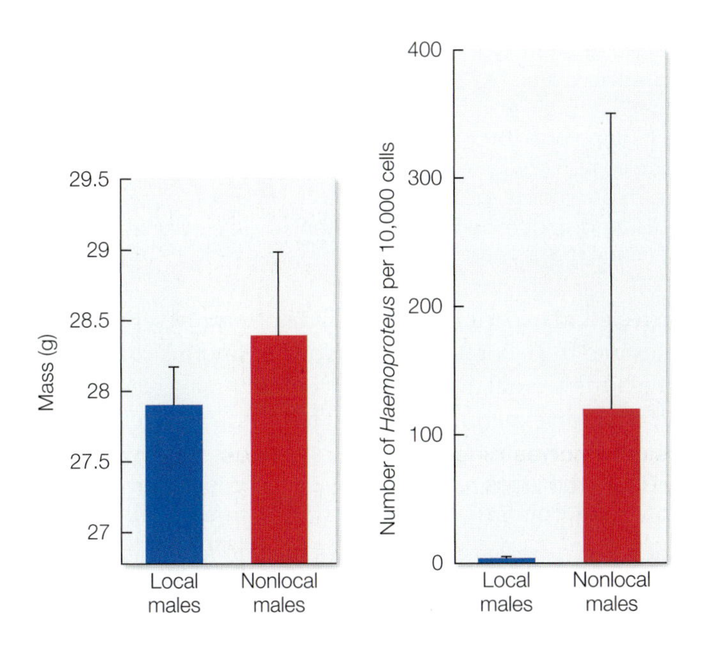

FIGURE 2.17 Differences in life history traits in male white-crowned sparrows singing local versus nonlocal dialect around Tioga Pass, California. Despite similar body masses in local and nonlocal males (left panel) and high variation in infection rates, nonlocal males had significantly higher *Haemoproteus* infections in their blood (right panel). Notice the large variation in the *Haemoproteus* infection among nonlocal males. Bars depict mean +/– SE. (After MacDougall-Shackleton et al. 2002.)

female swamp sparrows not only showed a preference for local song types with which they were tutored, but they also showed a preference for unfamiliar local songs over unfamiliar foreign songs (Anderson et al. 2014). Thus, there appear to be differences between sparrow species in the extent to which female song preferences enable females to choose mates born into their natal area.

Another version of the female choice hypothesis for male song learning builds on the possibility that females look to the learned songs of potential mates for information about their developmental history. If a female could tell just by listening to a male that he was unusually healthy or had a successful rearing history, she could acquire a mate whose genes had worked well during his development and so should be worth passing on to her offspring. In song sparrows, males with larger song repertoires have a larger-than-average HVC and are also in better condition as judged by their relatively large fat reserves and by their apparently more robust immune systems (Pfaff et al. 2007). Why might females prefer males of this sort? A healthier partner might be able to provide his offspring with above-average parental care, or alternatively, he might be able to endow his male offspring with genes that could give them a competitive edge (assuming there is a hereditary component to HVC size and body condition). In some European warblers, for example, females prefer to mate with males with large song repertoires (Nowicki et al. 2000). Katherine Buchanan and Clive Catchpole have shown that song-rich males of the sedge warbler bring more food to their offspring, which grow bigger—a result that almost certainly raises the reproductive success of females that find large song repertoires sexually appealing, which in turn selects for males that are able to sing in the favored manner (Buchanan and Catchpole 2000).

Yet the question still remains: why might female choice favor males that *learn* their complex songs, rather than males that simply manufacture an innately complex song with no learned component? One possibility is that the quality of vocal learning could give female songbirds a valuable clue about the quality of the singers as potential mates. The key point here is that song learning occurs when males are very young and growing rapidly. If rapid growth is difficult to sustain, then young males that are handicapped by genetic defects or nutritional stress should be unable to keep up, resulting in suboptimal brain development (see Box 2.4). According to the nutritional stress hypothesis, individuals with even slightly deficient brains may be less able to meet the demands of learning a complex species-specific song (Nowicki et al. 1998).

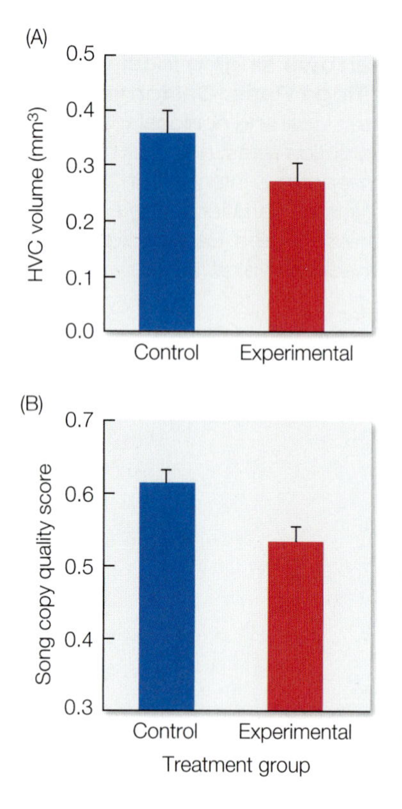

FIGURE 2.18 Nutritional stress early in life has large effects on song learning by male swamp sparrows. (A) Stress and brain development, as expressed by the volume of HVC, and (B) stress and song learning, as measured by the match between learned songs and tutor recordings. Nestling swamp sparrows were hand-reared and exposed to tutor recordings. The control group was fed all they could eat, whereas the experimental group received only 70 percent of that amount for about 2 weeks. Bars depict mean +/– SE. (After Nowicki et al. 2002a.)

These predictions have been tested in experiments with swamp sparrows. By bringing very young nestling males into the laboratory, Steve Nowicki and his coworkers were able to control what the birds were fed. A control group of males received as much food as they could eat, while an experimental group received 70 percent of the food volume consumed by the control males. During the first 2 weeks or so, when the sparrows were totally dependent on their handlers for their meals, the controls came to weigh about one-third more than the experimentals, a difference that was then gradually eliminated over the next 2 weeks as the sparrows came to feed themselves on abundant seeds and mealworms. Even though the period of nutritional stress was brief, this handicap had large effects on both brain development and song learning, such that the underfed birds showed poorer copying accuracy (**FIGURE 2.18**). Correspondingly, neural components of the song system in the food-deprived experimental group were significantly smaller than the equivalent regions in the control sparrows (Nowicki et al. 2002a).

Similarly, nestling song sparrows that were subjected to food shortages soon after hatching developed smaller HVCs than nestlings that were fed as much as they wanted. The effect manifested itself by the time of fledging, even before the young birds had begun to learn their songs (MacDonald et al. 2006). In the case of the swamp sparrows, the experimentally deprived birds came to sing poorer copies, compared with the controls, of the recorded song that both groups listened to during their early weeks of captivity (Nowicki et al. 2002a). Subsequent experiments like these have shown that both different song nuclei and different components of the male song are affected differently by stressors during development. For example, a team led by Scott and Elizabeth MacDougall-Shackleton examined in song sparrows the effects of early-life food restriction and corticosterone (a hormone related to stress) treatment on adult song production and neuroanatomy (Schmidt et al. 2013). The researchers found that their treatments influenced song complexity and song-learning accuracy, but not how much the birds sang. Moreover, while their treatments led to a reduction in the size of the birds' RA, there were no effects on HVC or area X. Thus, diverse early-life stressors may affect distinct regions of the brain and components of the song-learning process in different bird species.

Poor nutrition during a bird's early development could presumably affect more than a male's capacity to learn his species song. For example, when young zebra finches were divided into two groups and fed either a high-protein diet or a low-protein alternative, the birds with the better diet grew larger and "solved" a food-related trial-and-error task faster than did their less fortunate companions (Bonaparte et al. 2011). This result suggests that a superior diet enables birds to develop superior learning abilities, which could affect the ability of a male to forage efficiently, a factor that might come into play for paternal males that provide food for their offspring. If so, females that gained information about a male's developmental history could choose superior mates by preferring those that ate well during their early development.

But do female birds actually pay attention to the information about mate quality that appears to be encoded in male songs? Nowicki's team collected songs of song sparrows that had copied training songs from recordings with varying degrees of

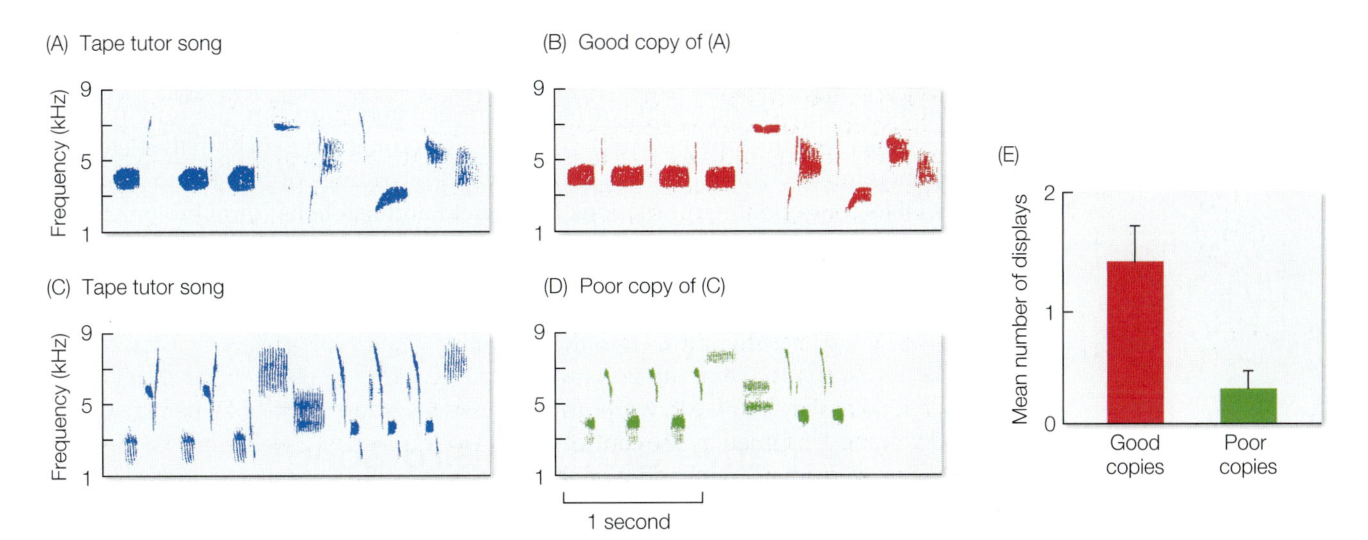

FIGURE 2.19 The mean number of precopulatory displays by female birds changes in response to differences in male song quality. Female song sparrows were tested with playback of the songs of males that had been able to copy their tutor songs very accurately versus males with lower copying accuracy. The top panel shows (A) a recorded tutor song and (B) a copy sung by a male able to copy a high proportion of the notes in the recorded song it listened to. The bottom panel shows (C) a recorded tutor song that was (D) poorly copied by another individual. (E) Female responses to good and poor song copies. Bars depict mean +/– SE. (A–E after Nowicki et al. 2002b; A–D after Nowicki and Searcy 2004.)

accuracy, and then the male songs were played to female sparrows in the laboratory (Nowicki et al. 2002b). Female sparrows can be hormonally primed to respond with a tail-up precopulatory display to male songs they find sexually stimulating. The more accurately copied songs elicited significantly more precopulatory displays from the females than did the songs that were copied less accurately (**FIGURE 2.19**). Somewhat similar results have been obtained by researchers working with zebra finches, who have found that stressed young males come to produce shorter and less complex songs than less handicapped males (Spencer et al. 2005).

Female songbirds may also prefer songs that are relatively difficult to produce (Woolley and Doupe 2008). For example, the faster a male swamp sparrow sings its trilled song, the harder it is to sing, especially if the syllables of the trill cover a relatively wide range of sound frequencies. Females are attracted to males that sing right up to the limits of performance (Ballentine et al. 2004). Likewise, female European serins (*Serinus serinus*), a species of small finch prefer males that sing at relatively high frequencies, another physiological challenge for males (**AUDIO 2.14**) (Cardoso et al. 2007). By basing their mate preferences on male performance, females are in effect favoring males likely to be healthy and in good condition, attributes that could make the male a superior donor of genes or a superior caregiver to offspring.

AUDIO 2.14
European serin
ab11e.com/a2.14

Do females learn these preferences for male songs the same way that males do? Evidence from zebra finches suggests that females indeed might learn these preferences early in development. Female zebra finches prefer songs that are both complex and contain high-performance phrases. Unlike in males, however, preferences for complex phrases are not disrupted by stressors early in development (Woodgate et al. 2011). In contrast, preferences for high-performance songs are disrupted when females do not hear them during development (Chen et al. 2017). Thus, at least some components of a female's preference for male song may be influenced during development, as different parts of the female auditory system exhibit different degrees of developmental plasticity in song learning.

As you can no doubt see by this point, there are a host of adaptive possibilities associated with learned bird song, and it is therefore not surprising that this topic continues to attract some highly inventive researchers. Birds are not only readily observable in nature, but they are amenable to experimentation in both the field and lab. Therefore, by combining observations and recordings of free-living animals, as well as controlled manipulations in the field and lab, behavioral biologists are gaining a better understanding of the adaptive consequences of learned song. This theme of marrying observation with experimentation to test explicit hypotheses—not only ultimate ones but proximate ones as well—is something we will continue to observe and highlight throughout this book.

The Integrative Study of Bird Song

We have reviewed just a small portion of what is known about the proximate and ultimate causes of song learning in birds. The wealth of competing hypotheses within any one level of analysis can be a bit overwhelming, as illustrated by the many hypotheses on why male birds of some species learned to sing (see Hypotheses 2.2). But a crucial point to which we return here is that the proximate and ultimate causes of behavior are not only complementary, they can also be mutually informative. In addition to the examples we already provided, another key example of how considering both levels of analysis can be mutually informative comes from thinking about whether males and females use the same brain regions for song learning. Since in many species the females have much smaller HVCs than the males, it seems unlikely that this brain region would be useful for song perception in either sex (MacDougall-Shackleton 2011). And it is not. Lesions in the HVC of female zebra finches did not alter song perception, but those in the caudomedial mesopallium (CMM) did (MacDougall-Shackleton et al. 1998). Thus, considering the functional reasons underlying song learning (such as the female preference version of the sexual selection hypothesis) allowed researchers to predict which brain regions might be involved and which might not.

As you will see throughout this book, considering multiple levels of analysis simultaneously when approaching a behavioral question can be important and enriching. We demonstrated that on the ultimate side, if we really want to understand why male white-crowned sparrows from Marin and Berkeley sing somewhat different songs, we must also understand how the song control mechanisms develop and how they operate. Males differ genetically from females, a fact that guarantees that different gene–environment interactions will take place in embryonic male and female birds. And yet both sexes show some similarities in how they learn to sing or develop preferences for song. The cascading effects of differences in gene activity and in the protein products that result from gene–environment interactions affect the assembly of all parts of the bird's body, including its brain and nervous system. The way in which a male white-crowned sparrow's brain, with its specialized subsystems and nuclei, can respond to experience is a function of the way (and environment) in which the male has developed. By the time a male sparrow has become an adult, he possesses a large, highly organized song control system whose physiological properties permit him to produce a very particular kind of song. The learned dialect he sings is a manifestation of both his developmental history and the operating rules of his brain, just as is true for the female preference for song.

Yet, because these proximate differences in white-crowned sparrows that generate differences in their dialects emerge through evolutionary processes, we must also consider the ultimate perspective. In the past, males in the lineage leading to today's males surely differed in their song control systems and thus in their ability to learn to sing a dialect accurately. Some of these differences were doubtless hereditary, and if the genetic differences among males translated into differences in acquiring a dialect and thereby their desirability as mates or their capacity to cope with rival males, selection would have resulted in the spread of those attributes associated with greater reproductive success. The males that made the greatest genetic contributions to the next generation would have supplied more of the genes available to participate in interactive developmental processes within members of that generation. The hereditary attributes of those individuals would in turn be measured against one another in terms of their ability to promote genetic success in a new round of selection. Indeed, studies demonstrating convergence in song learning between birds and humans suggest that some of the same genes that influence song learning in birds also influence our own ability to speak and learn languages. In other words, human speech is shaped by many of the same forces that shape bird song. Thus, the selective process that links the proximate bases of behavior with their ultimate causes results in a never-ending spiral through time. In subsequent chapters of this book, we will discuss these links and processes in greater detail.

SUMMARY

1. Different populations of birds often sing distinct forms of the same song, called dialects. To understand how birds learn different dialects, we must explore both proximate and ultimate hypotheses for avian song learning.
2. There appears to be little evidence of hereditary differences underlying dialects in several bird species. Instead, young birds appear to learn their distinct songs during a critical window during development. The experience of hearing other males is critical to a young male bird learning its proper song (and a young female learning the preference for that song). Moreover, social interactions in many avian species enhance vocal learning and neural development in the brain during this critical learning window.
3. During the period of song learning, distinct clusters of neurons in the brain called nuclei play a critical role. The high vocal center nucleus, or HVC, is particularly important for male song learning. HVC is bigger in males of species that learn songs than in those that do not, and in many species it is larger in males than in females.
4. Although most birds sing, only approximately 3 of 23 avian orders (hummingbirds, parrots, and oscine songbirds) include species that learn to sing. It is possible that these three orders of birds shared a common ancestor that was a vocal learner. What's more, humans and birds—which are clearly unrelated—exhibit convergent neural circuits for vocal learning that are accompanied by convergent molecular changes in multiple genes.
5. There are several hypotheses to explain why vocal learning is adaptive, all of which have to do with the acoustic, social, or ecological environment. These include acoustic adaptation to the local environment, recognition of neighbors or allies within a group, information sharing within groups, mate choice or male–male competition, and the promotion of local adaptation via distinct dialects.

COMPANION WEBSITE

Go to **ab11e.com** for discussion questions and all of the audio and video clips.

The Developmental and Genetic Bases of Behavior

Chapter 1 introduced the idea of proximate and ultimate causes of behavior, and Chapter 2 provided an example of how integrative studies of animal behavior can cross levels of analysis—from proximate to ultimate and back again. In the next few chapters, we will focus primarily on the proximate level of analysis, setting the stage for a more in-depth discussion of the ultimate level later on. Recall that we made the key point that one can categorize proximate causes into two complementary—but overlapping—concepts: the developmental component and the mechanistic component. We also made the point that an integrative approach to the study of animal behavior requires crossing levels of analysis and integrating across Tinbergen's four questions. The same type of integration is true at the proximate level; for example, the activity of genes

Chapter 3 opening photograph courtesy of Solvin Zankl/Minden Pictures.

underlying specific behaviors often changes over the course of development. This chapter expands on what is known about the developmental basis of behavior and discusses the role that genes play throughout ontogeny, as well as how their expression is influenced by the environment. It explores the developmental and genetic bases of a range of animal behaviors—from foraging to migration to social behavior—for which we will detail the ultimate explanations later in the book.

If you wake up early in the morning in most parts of the world, you will almost certainly hear birds singing. Although, as we discussed in Chapter 2, genes are clearly involved in the development of bird song, the developmental process is dependent on not only the genetic information a bird possesses but also on a host of environmental influences, ranging from the nongenetic materials in the egg yolk, to the hormones and other chemicals that certain of the bird's cells manufacture and transport to other cells, to the sensory signals generated when a baby sparrow hears its species' song, to say nothing of the neural activity that occurs when a young adult male interacts with a neighboring territorial male. Moreover, the time of year when a bird sings is dependent on environmental and ecological factors, which trigger the expression of genes and the release of specific gene products. In temperate regions, for example, the dawn chorus reaches its peak in the spring when many species of birds defend territories and begin to attract mates. A bird's song therefore highlights a central idea in behavioral biology, namely that development is an interactive process in which genetic information works together with changing internal and external environments. The process occurs because some of the genes in the organism's cells can be turned on or off by the appropriate signals that come from the developing animal's external environment (such as changes in day length or food availability). As genetic activity (gene expression) changes within an organism, the chemical reactions in its cells also change, building (or modifying) the proximate mechanisms that underlie an organism's characteristics and capacities, from the release of specific gene products such as hormones to the production of actual notes from the avian syrinx.

The role that genes play in this developmental process means that genetic differences among individuals can lead to differences in the way proximate mechanisms arise, which in turn can cause individuals to differ behaviorally. This individual variation in behavior not only influences differences in fitness, but behavioral variation among individuals can also affect their ability to pass on their genes. In other words, as the previous chapter also stressed, there are strong links between evolutionary processes and the proximate causes of behavior, a point that this chapter will reemphasize in more detail. Specifically, this chapter discusses how genes interact with the environment—both social and ecological—to influence the production of behavioral traits. In particular, we emphasize how the early life environment is critical to the development of behavior. We also examine how an evolutionary development (or "evo-devo") approach is being used to study ancient genetic mechanisms underlying behavior, as well as the development of learning and cognition, and the role that the environment plays in these processes. But first, we need to deal with the mistaken view that some traits are "genetically determined," as well as the equally erroneous claim that some traits are "environmentally determined."

Behavior Requires Genes and the Environment

Recall that in Chapter 1 we introduced the concept of natural selection and discussed the evolutionary requirement of heritable (genetic) variation underlying a behavioral trait as well as the importance of the environment in selecting for specific forms of that trait. Evolutionary analyses of behavioral development cannot proceed if we believe that some behaviors are purely "genetic" or "genetically determined" and so could evolve, while other behaviors are "learned" and so are "environmentally determined" and, by extension, are immune to natural selection. This major misconception is applied with particular fervor to humans, whose behavior is of course highly influenced by learning. Because so much of human behavior is learned, many people have concluded that nature (our genetic heritage) is trumped by nurture (growing up in a cultural environment where learning is of paramount importance). But as you saw in Chapter 2, many animals learn, and neither we nor they could learn a thing from our environments without our brains, which are ultimately a collection of cells that, like other cells in the body, encapsulate our DNA, our genes. Every cell in our body contains the same genetic material, yet not all cells perform the same function. The activity of our genes in conjunction with environmentally supplied inputs endows our cells with special properties. As a result, our brain cells help us respond to sensory information in our environment by learning things. So the idea of a purely environmentally determined behavior (one that would develop even in the absence of genetic information) is as nonsensical as the idea of a purely genetically determined behavior (one that would develop even in the absence of environmental inputs) (see chapter 2 in Dawkins 1982; also Robinson 2004). Instead, the **interactive theory of development** suggests that the development of behavioral traits—or any other traits for that matter—requires both genetic information and environmental inputs. It is this interaction between genes and the environment that leads to not just the production of a behavior, but also the evolution of that behavior.

The Interactive Theory of Development

One of the best-studied examples of behavioral ontogeny that illustrates the interactive theory of development is the set of changes that takes place in the behavioral roles adopted by a worker honey bee (*Apis mellifera*) over the course of her lifetime. After a worker bee metamorphoses into a young adult, her first job in the hive is a humble one, the cleaning of comb cells. As time passes, she then becomes a nurse bee that feeds honey to larvae in the brood comb before making the next transition to being a distributor of food to her fellow workers. The final phase of her life, which begins when she is about 3 weeks old, is spent foraging for pollen and nectar outside the hive (**FIGURE 3.1**) (Lindauer 1961).

What causes a worker bee to go through these different developmental stages? If behavioral development requires both genetic

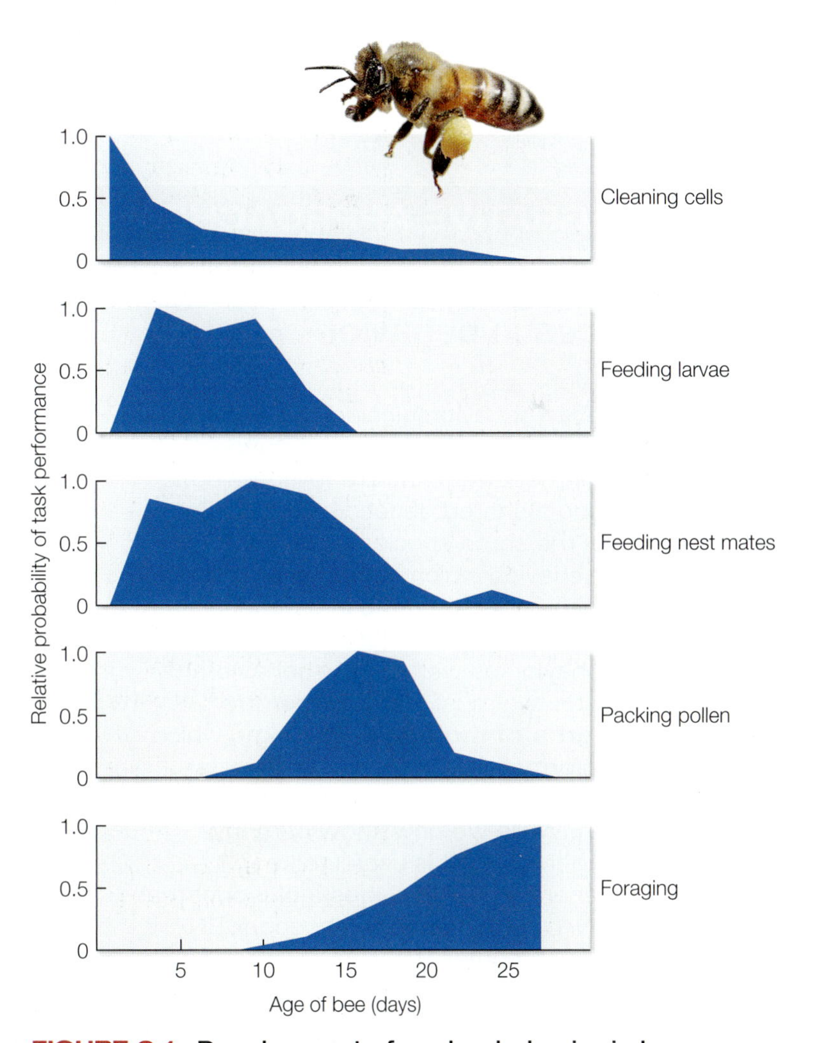

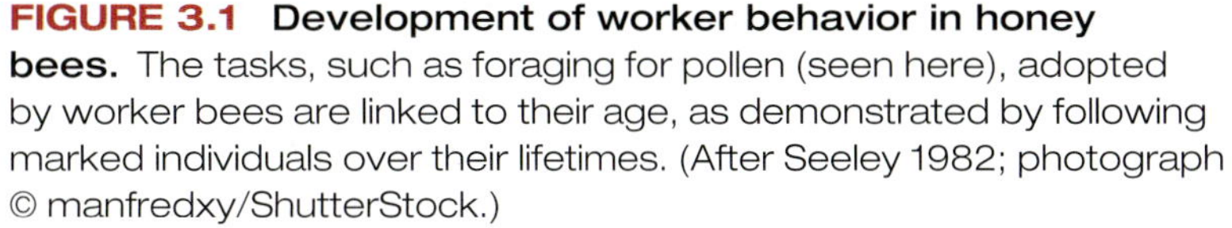

FIGURE 3.1 Development of worker behavior in honey bees. The tasks, such as foraging for pollen (seen here), adopted by worker bees are linked to their age, as demonstrated by following marked individuals over their lifetimes. (After Seeley 1982; photograph © manfredxy/ShutterStock.)

information and environmental inputs, the information in some of the bee's many thousands of genes (the bee's genotype) must respond to the environment in ways that influence the development of her phenotypic characteristics—traits such as the color of her body, the length of her antennae, or the quantity of juvenile hormone in her hemolymph (blood). In other words, if the interactive theory of development is correct, as a bee's behavioral phenotype changes from nurse to forager during her lifetime, there must also be changes in the interplay between genes and some aspect of the bee's environment. To begin to unravel how genes and the environment interact to influence behavior in honey bees—or any other organism—we first need a little refresher on how genes influence behavior, beginning with a review of some basic molecular biology (**BOX 3.1**).

Charles Whitfield and his coworkers examined the genetic bases of behavioral variation in honey bees by running brain extracts from nurses and foragers through microarrays to quantify gene expression. They were able to compare the activity of about 5500 genes (of the roughly 14,000 in the bee genome) for these two kinds of individuals (Whitfield et al. 2003). The researchers discovered that some of the genes that were turned on in nurse brains differed substantially and consistently from those that were active in foragers' brains, and vice versa. Indeed, about 40 percent of the surveyed genes showed different levels of product output (mRNA expression) in the two kinds of bees. These changes in genetic activity are correlated

BOX 3.1 INTEGRATIVE APPROACHES

Behavioral genetics: Identifying the genetic basis of differences in behavior

The genome of every species encodes for a blueprint of its behaviors (including innate ones), and delimits how they can be shaped by the environment (for example, by learning). Despite having a shared genetic blueprint, individuals of the same species always exhibit variation in behavior, which often has a heritable component. To understand the interaction between genetic information and environmental inputs on behavior, as well as how behavioral variation arises, we must first consider the **central dogma of molecular biology**, which posits that information encoded in DNA gets transcribed to RNA, which in turn is translated into a protein. While we now know of many exceptions to this rule, it is useful to keep in mind that generally DNA provides a blueprint that codes for RNA (in the form of messenger RNA, or mRNA), which carries those instructions to ribosomes for synthesis into proteins (sequences of amino acids that constitute molecules such as hormones or enzymes, or structures such as hair and antlers).

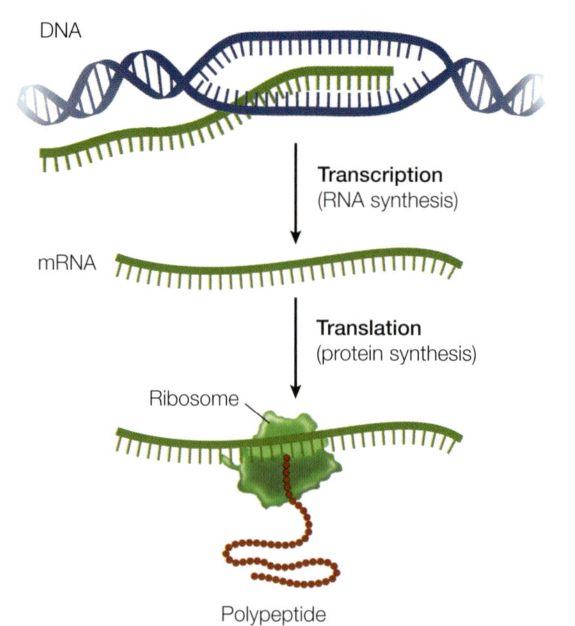

The central dogma of molecular biology. DNA is transcribed into mRNA, which is then translated at a ribosome into a protein. The protein is made of multiple other polypeptide chains.

BOX 3.1 INTEGRATIVE APPROACHES (*continued*)

Given this set of basic molecular processes, how does one actually identify the genetic basis of differences in behavior—a field known as behavioral genetics? In humans, for example, there are estimated to be nearly 20,000 genes that code for a variety of proteins, and many human behaviors are often linked to variation in more than one gene. Whether studying behavior or any other trait, biologists typically take one of two approaches to link genetic variation to phenotypic variation: (1) they examine how DNA leads to evolutionary change by comparing sequence variation among species or individuals that exhibit different traits (an approach known as **evolutionary genetics**); or (2) they compare gene or protein expression in individuals or species that exhibit different traits (an approach known as **functional genomics**). Evolutionary genetics often requires identifying genetic variation in all or part of the genome from large numbers of individuals to identify genetic differences that correlate with patterns of phenotypic differences. In contrast to evolutionary genetics, functional genomics examines only a subset of the genome—those genes being expressed as mRNA at the time of sample collection. Early functional genomic approaches (such as quantitative polymerase chain reactions, or qPCR) probed one gene at a time to examine short fragments of mRNA that had been reverse transcribed into complementary DNA (cDNA). Later, microarray technology made it possible to analyze the activity of a few thousand genes at a time in particular tissues (such as the brain), and RNA sequencing (RNA-seq) enabled researchers to analyze all of the mRNA being expressed in a given tissue at a given time point (referred to as the **transcriptome**).

Thinking Outside the Box

Although evolutionary and functional genomic approaches can both be used to study the genetic basis of behavior, some researchers would say that they address fundamentally different questions. Why? Evolutionary genetics, for example, leads to the identification of the genetic causes and evolutionary trajectories of behavior, whereas functional genomics can inform about intermediate steps (sometimes generating mechanistic hypotheses) between genetic variation and behavior. In other words, evolutionary genetics searches for signatures of natural selection that are evident in the genome, whereas functional genomics examines gene activity (expression), something that can change rapidly and can even vary among tissue types. In addition to these differences, when and why might a behavioral biologist want to take an evolutionary versus a functional approach to studying the genetic basis of behavior? Oftentimes, it might come down to practicality. For example, one benefit of evolutionary genetics is that researchers can use DNA from any tissue, including from museum specimens that may have been collected hundreds of years ago, or even fossils of organisms that may have lived tens of thousands of years ago. However, one of the downsides of this approach is that sequencing and assembling entire genomes is costly, complicated, and time-consuming. In contrast, a functional genomics approach works with the transcriptome rather than the entire genome, which means it is often simpler and cheaper to collect and analyze data. However, working with RNA typically requires fresh tissue, something that often necessitates storing samples in special buffers or at very cold temperatures soon after sampling so that RNA can be preserved (though technology is rapidly improving and even older tissues can now sometimes be used to sample RNA). Indeed, both approaches have their pros and cons when it comes to studying animal behavior, and often they can be paired to produce a richer understanding of how genes and the environment interact to influence a behavior. Think about some behaviors (and animal species) where one approach might be more useful than the other. As you come across other empirical examples of behavioral genetics throughout this book, remember to consider the approaches that were used to generate the studies' conclusions, and think about additional ways that researchers could address the question with other tools and techniques.

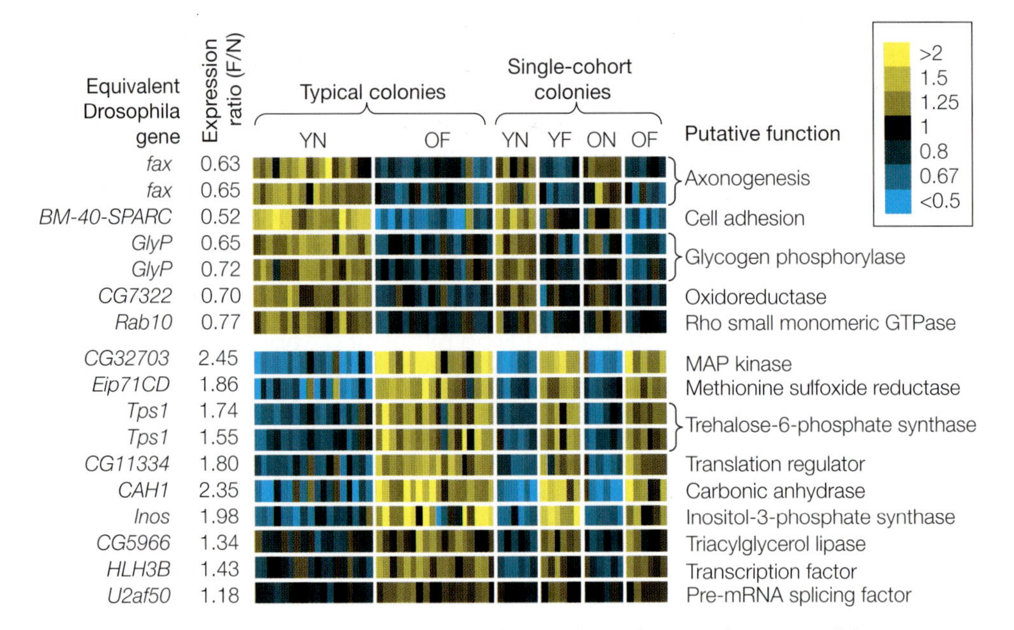

FIGURE 3.2 Gene activity varies in the brains of nurse bees and foragers. Each bar represents an individual bee's gene expression record for a particular gene from bees collected from typical and manipulated colonies. The 17 selected genes surveyed here were the ones that showed the largest difference in expression between the brains of nurses and foragers. In addition, these 17 genes are highly similar to genes found in fruit flies (*Drosophila*); the function of the corresponding gene in fruit flies has been established and is noted here. The activity level of a gene in forager, relative to nurse, bee brains is color coded from high (>2) to low (<0.5); see the reference code at the upper right. The left portion of the figure shows gene activity for young nurses (YN) and old foragers (OF) from unmanipulated colonies; the right portion shows those for young foragers (YF) paired with nurses of the same age and old nurses (ON) paired with foragers of the same age taken from colonies manipulated to contain either all young or all old workers. The expression ratio (F/N) is calculated by dividing the activity score of a given gene in foragers by that for nurses. (After Whitfield et al. 2003, © 2004 by the American Association for the Advancement of Science.)

with the typical age-related transformation of a nurse into a forager (**FIGURE 3.2**). However, these changes also occur when nurses are induced to make the transformation earlier in their adult lives than is normal, which can be done experimentally by creating colonies in which older foragers are absent. Under these experimental circumstances, some nurses adopt the forager role much sooner than they would otherwise. This study provides a key example of how the bee's genes are sensitive to the environment in which the developing individual finds itself, providing bees with the flexibility to switch between tasks in a way that meets the needs of the queen and her brood. In this case, the "environment" was actually the social environment, or the social makeup of the colony. As the social environment changes and the proportions of different types of worker bees vary, honey bees can utilize the plastic nature of their genomes to adopt the behavioral approach that is most needed at any given time.

This **forward genetic approach**, which is used to identify the genes responsible for a given phenotype, has been expanded to examine gene activity in the brain cells of hundreds of honey bees that were collected as they engaged in a more diverse array of activities in the wild. (In contrast, **reverse genetics** seeks to determine which phenotypes arise as a result of particular genetic sequences, often by experimentally manipulating the genetic code). Using a forward genetic approach, honey bee researchers have identified large numbers of genes whose expression differs depending on the behavioral state of the bee; the set of activated genes depends, for example, on whether the bees had been collected while defending the hive or had been sacrificed as they were foraging. An illustration of this approach comes from Chandrasekaran and colleagues, who found that large numbers of activated genes had related functions and were involved in a network, including, importantly, those that code for various transcription factors, proteins that regulate the activity of several target genes (Chandrasekaran et al. 2011). The different behaviors exhibited by the honey bees in the study could be linked to modules that generate a specific set of transcription factors, which in turn promote or suppress the activity of particular groups of target genes in the bee's brain (Chandrasekaran et al. 2011, Fernald 2011). Moreover, other transcription factors are involved in the control of neural genes with more general effects on the bee's behavior.

This work illustrates how a bee's social environment influences its gene expression and the transition among behavioral roles. A demonstration of how the ecological environment can also affect gene activity in honey bees comes from a study designed to determine whether individuals that differ in their perception of how far they have flown exhibit differences in which of the genes in their brains are active. When foragers return to the hive after drinking sugar water at a feeding station, they monitor the distance they have traveled by observing the flow of environmental images on their retina. These bees can be tricked into "thinking" they have flown a longer distance than they actually have, by having them fly through a tunnel with vertical-striped walls (Sen Sarma et al. 2010). Using this technique, a research team created two categories of bees, both of which had flown the same distance. However, one group had flown through a tunnel with vertical stripes and been made to think it flew farther than it did, while the other had flown through a tunnel without that pattern. The genes expressed in the two groups of bees differed greatly, demonstrating that the different perceptions of distance flown were based in part on a difference in genetic activity in particular parts of the bee's brain that differed as a result of flight experience.

These studies in honey bees illustrate how the environment—both social and ecological—is important for behavioral development and brain functioning. Indeed, environmental factors are critical for every element of gene expression within organisms, in part because the environment supplies the molecular building blocks that are essential if the information in DNA is to be used to make mRNAs, some of which are involved in gene regulation via transcription factors, others of which carry the information about making specific proteins from the nucleus of the cell to the protein assembly sites in the cytoplasm. In addition, the cellular environment must contain the precursors of these constituents of living things if they are to be produced. In the case of honey bees, these chemicals ultimately come from substances consumed by the queen prior to laying her eggs, as well as from the honey and pollen eaten by the larvae and adults that develop from those eggs. Some of the resultant developmental products may play a special role in changing the activity of one or more key genes in an individual, initiating a cascading series of biochemical changes that eventually alter the development of the brain, and thus the behavior of the bee.

One of these chemical products is royal jelly, a mixture of water, protein, sugars, and fatty acids that is produced in special glands in the workers and fed to developing larvae. Larvae destined to become queens are fed royal jelly for a longer period of time than are larvae that will become drones or workers. Royalactin is the key protein in royal jelly that causes a bee to develop into a queen. But how does a simple protein influence the morphological and behavioral development

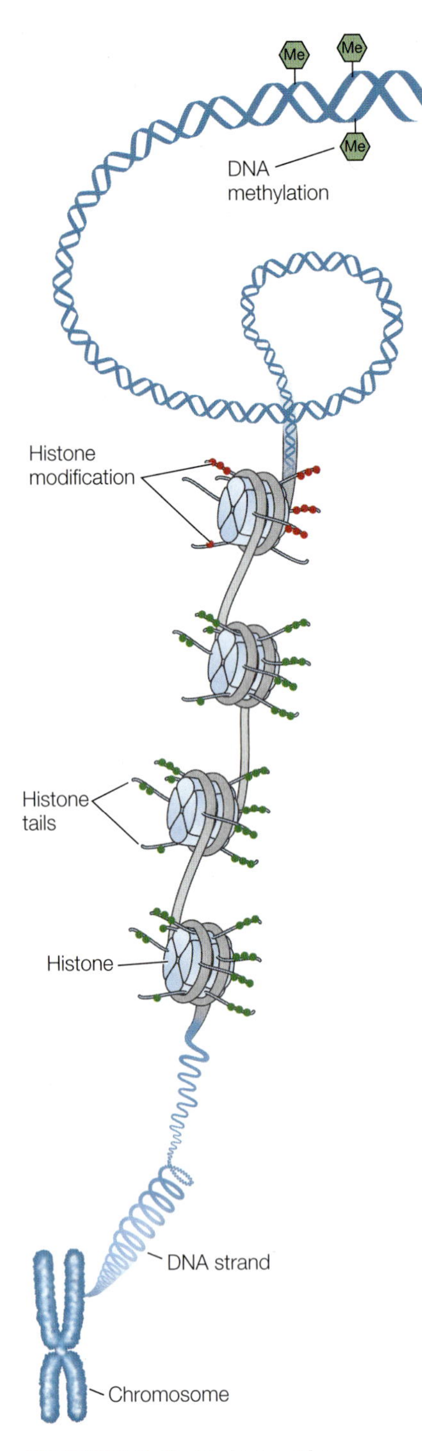

FIGURE 3.3 Epigenetic modifications. DNA is coiled tightly around histone proteins. Chemical or other changes to these proteins (histone modifications) or to the DNA itself (DNA methylation) can influence the expression of DNA.

of growing larvae through changes in expression of the genome? **Epigenetic modifications**, or alterations to the genome that do not change the DNA sequence (the sequence of nucleotides—A, T, C, and G—that make up DNA), are thought to play a role in this process. Epigenetic modifications include chemical or other changes to the histones (termed **histone modification**), or the structures on which DNA is wound, or chemical modification to the DNA bases themselves (a process referred to as **DNA methylation**) (**FIGURE 3.3**). Both histone modification and DNA methylation alter DNA accessibility, thereby regulating patterns of gene expression. In honey bees, there is evidence that DNA methylation (the addition of a chemical called a methyl group) at specific CpG sites—regions of the genome where a cytosine nucleotide is adjacent to a guanine nucleotide—influences behavioral development. Using RNA interference (RNAi), a process in which RNA molecules are used to experimentally silence gene expression, Kucharski and colleagues silenced an enzyme called DNA cytosine-5-methyltransferase (Dnmt3) that controls DNA methylation, demonstrating that doing so has similar effects on growing larvae as does royal jelly (Kucharski et al. 2008). The researchers' experiment showed that more than 70 percent of the emerging adults in which Dnmt3 expression was interfered with became queens, whereas nearly 80 percent of those in the control groups became workers (**FIGURE 3.4**).

This experimental manipulation of the enzyme controlling DNA methylation suggests that this epigenetic process is related to the differential production of queens and workers in honey bees. But which genes were likely to be differentially methylated in these two groups of bees? A subsequent study looking at DNA methylation across the entire honey bee genome found that more than 550 genes showed significant DNA methylation differences between queens and workers (Lyko et al. 2010). However, follow-up work was unable to find any DNA methylation differences in the genes from the brains of queens and workers at the time of emergence (Herb et al. 2012). Given these conflicting results about whether DNA methylation plays a role in the behavioral flexibility of honey bees, perhaps then DNA methylation is what drives the behavioral ontogeny and gene expression differences in workers? A study by Herb and colleagues tested this idea and not only found substantial variation in DNA methylation across the genome in bees transitioning from nurses to foragers, but that experimentally reverting foragers back to nurses reestablished the original patterns of DNA methylation (Herb et al. 2012). Together, these results suggest that behaviorally relevant DNA methylation is reversible over the course of a bee's lifetime and that epigenetic mechanisms may play a role in honey bee behavioral ontogeny.

Another particularly potent developmental product is a substance called juvenile hormone, which is found in low concentrations in the hemolymph of young nurse workers but in much higher concentrations in older foragers. Juvenile hormone is an important and ancient hormone in insects that regulates many aspects of development, physiology, reproduction, and behavior. As one might predict, if young bees are treated with juvenile hormone, they become precocious foragers (Robinson 1998), but if one removes a bee's corpora allata (the gland that produces juvenile hormone), the bee delays its transition to foraging. Moreover, bees without corpora allata that receive hormone treatment regain the normal timing of the switch to foraging (Sullivan et al. 2000). Thus, it appears that changes in juvenile hormone production have fundamental developmental effects in honey bees, a widespread feature of hormones in many other animals as well. For example, as we will discuss in Chapter 5, vertebrate reproductive hormones (such as testosterone and estrogen) are often intimately involved in the construction of neural mechanisms that are responsible for the differences between male and female mammals in their sexual behavior (Levine and Mullins 1966).

As you might expect from what we have already discussed about honey bees, hormone production in honey bees responds to the social environment. When researchers created experimental honey bee colonies with a worker force consisting of uniformly young bees of the same age, a division of labor still manifested itself, with some individuals remaining nurses much longer than usual, while others began foraging as much as 2 weeks earlier than average. What enabled the bees to make these developmental adjustments? Perhaps a deficit in social encounters with older foragers may have stimulated an early developmental transition from nurse to forager behavior. This possibility has been tested by adding groups of older foragers to experimental colonies made up of only young workers. The higher the proportion of added older bees, the lower the proportion of young nurse bees that undergo an early transformation into foragers (**FIGURE 3.5**) (Huang and Robinson 1992). As it turns out, the behavioral interactions between the young residents and the older transplants inhibit the development of foraging behavior because transplants of young bees have no such effect on young resident bees. The inhibiting agent has been traced to a fatty acid compound called ethyl oleate, which only foragers manufacture and store in a special chamber (the crop) off the digestive tract (Leoncini et al. 2004). When returning foragers pass nectar contained in the crop to nurses back at the hive, ethyl oleate is likely transferred as well. The more foragers there are in a hive, the more likely nurses are to receive quantities of this chemical, which slows their transition to foraging status.

All of this research from honey bees illustrates why it is incorrect to say that some behavioral phenotypes are genetic and others are not. Worker foraging behavior cannot be purely "genetically determined" because the behavior is the product of literally thousands of chemical interactions between the bee's genes and its social and ecological environment. As Gene Robinson puts it, "DNA is both inherited and environmentally responsive" (Robinson 2004). Environmental signals, such as those provided by royal jelly, juvenile hormone, and ethyl oleate, influence gene activity. When a gene is turned on or off by changes in the environment, the resulting changes in protein production can directly or indirectly alter the activity of other genes in affected cells. A multitude of precisely timed and well-integrated biochemical changes involving both the genes and the cellular environments that surround them are responsible for the construction of all traits. Thus, no trait, behavioral or otherwise, can be purely "genetic" or "environmentally determined," especially

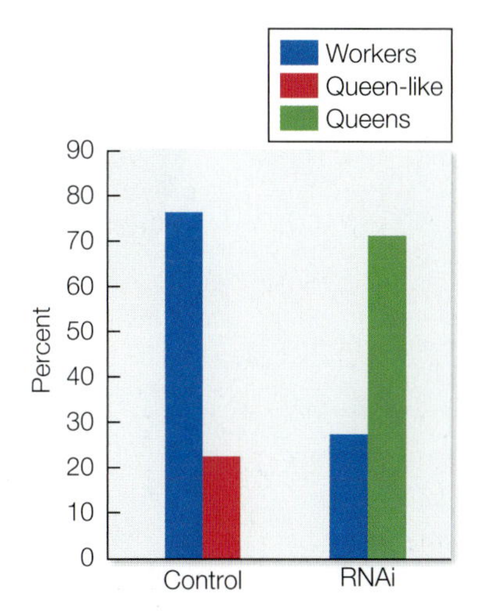

FIGURE 3.4 Silencing a gene that influences DNA methylation in honeybees influences caste development. Newly emerged honey bee larvae were injected with interfering RNA for either control or for the DNA cytosine-5-methyltransferase (*Dnmt3*) gene, an enzyme that controls DNA methylation. Most larvae in the control treatment developed into workers, whereas most of the larvae in which *Dnmt3* expression was interfered with became queens. (After Kucharski et al. 2008.)

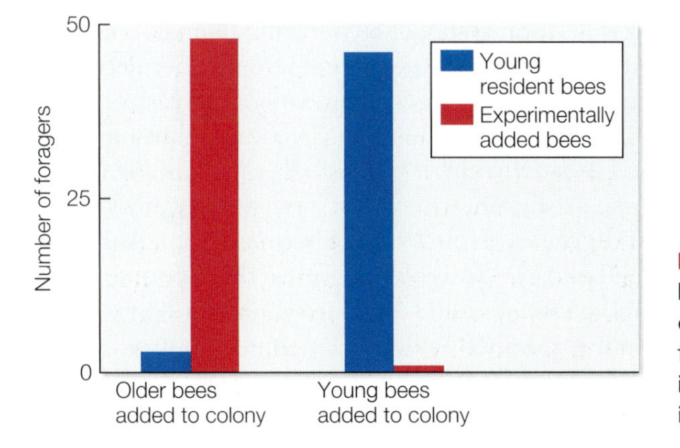

FIGURE 3.5 Social environment and task specialization by worker honey bees. In experimental colonies composed exclusively of young workers (residents), the young bees do not forage if older forager bees are added to their hive. However, if young bees are added instead, the young residents develop into foragers very rapidly. (After Huang and Robinson 1992.)

FIGURE 3.6 Kin discrimination in Belding's ground squirrels. Sisters reared apart display significantly less aggression toward each other than do other combinations of siblings reared apart, which are as aggressive toward each other when they meet in an experimental arena as are nonsiblings reared apart. Bars depict mean +/– SE. (After Holmes and Sherman 1982; photograph © istock.com/4FR.)

when we begin to consider the role of epigenetics in influencing behaviors.

Environmental Differences Can Cause Behavioral Differences

Although no behavior is either purely genetic or purely environmental, variation among individuals can arise as a result of developmental differences stemming from differences in either their genes or their environments. For example, the nonbreeding workers on a *Polistes* paper wasp nest typically react calmly toward other female workers that have been reared on that nest and have stayed on to become helpers. Much as in honey bees, there are genetic differences between those individuals that become queens and those that remain as workers, including in genes related to nutritional and reproductive pathways (Toth et al. 2007). But should a worker from another nest try to join the colony, the workers react very differently, by attacking these "foreigners." Workers appear to learn to recognize one another as nest mates in large part because they have acquired the special odor of the nest in which they were reared by other adult members of the colony (Gamboa 2004). And as you will see later in Chapter 8, some species of *Polistes* wasps can actually recognize individuals by the markings on their faces, a skill they might also use to distinguish foreigners from nest mates.

The ability to record information about the odor or appearance of one's nest mates requires genetic information, a point illustrated by a study of Belding's ground squirrels (*Urocitellus beldingi*) in which the newborn offspring of captive females were switched around at birth, creating four classes of individuals: (1) siblings reared apart; (2) siblings reared together; (3) nonsiblings reared apart; and (4) nonsiblings reared together. After they had been reared and weaned, the juvenile ground squirrels were placed in an arena in pairs and given a chance to interact. In most cases, animals that were reared together—whether actual siblings or not—tolerated each other, whereas animals that had been reared apart tended to react aggressively toward each other (Holmes and Sherman 1982). In this experiment, the young squirrels learned something about their nest mates thanks to brains primed to record certain information about the olfactory cues provided by their companions. Additionally, biological sisters reared apart engaged in fewer aggressive interactions than did nonsiblings reared apart (**FIGURE 3.6**). In other words, the squirrels had some way of recognizing their siblings that was not dependent on living with them as youngsters (Holmes and Sherman 1982, Holmes 1986). Instead, a different kind of learning was probably involved, one that goes informally by the indelicate label "the armpit effect" (Mateo and Johnston 2000). According to the **armpit effect hypothesis**, if individuals can learn what they themselves smell like, then they can use this information as a reference against which to compare the odors of other individuals (Holmes and Sherman 1983). And once animals can learn what they smell like, they might be able use this information to discriminate between relatives and nonrelatives (Hauber and Sherman 2001).

armpit effect hypothesis If individuals can learn their own smell, then they can use this information as a reference against which to compare the odors of other individuals, and discriminate between kin and non-kin.

The armpit effect hypothesis was examined in Belding's ground squirrels by Jill Mateo (Mateo 2002), who showed that individuals possess several scent-producing glands, including one around the mouth and another on the back (Mateo 2006).

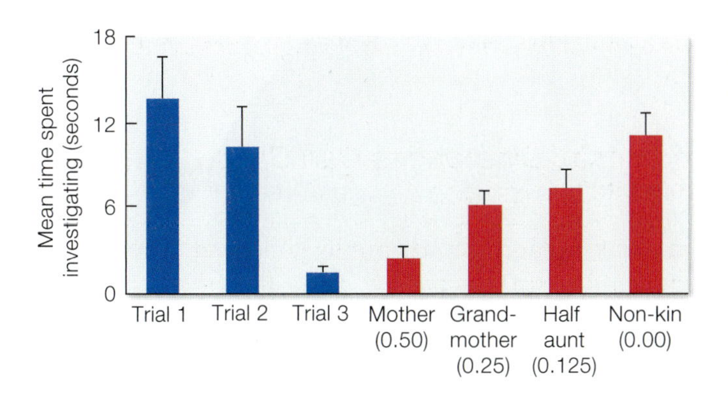

FIGURE 3.7 Belding's ground squirrels learn their own odor. Juvenile squirrels were first given three trials during which they could investigate their own odors applied to plastic cubes. Note the squirrels' decline in responsiveness to their own dorsal gland odors over these initial trials. Then the squirrels were provided with plastic cubes daubed with dorsal gland odors from four categories of individuals. Cubes with odors from close relatives received less attention than those with odors from distant relatives or non-kin. Numbers in parentheses show the degree of genetic relatedness (see Chapter 12) of the individual to the tested ground squirrel. Bars depict mean +/– SE. (After Mateo 2002.)

These squirrels regularly sniff the oral glands of other individuals, as if they are acquiring odor information that could conceivably be compared with the sniffers' own scents. By capturing pregnant ground squirrels and moving them to laboratory enclosures, Mateo was able to observe their juvenile offspring investigating objects (plastic cubes) that had been rubbed on the dorsal glands of other squirrels of varying degrees of genetic relatedness to the youngsters. Because the captive squirrels had been separated from some of their relatives, they had never met them and so had no prior experience with their odors. If, however, the test animals had learned what they themselves smelled like, and if close relatives produce odors more similar to their own than do distant relatives, an inexperienced youngster could in theory discriminate between unfamiliar relatives and nonrelatives on the basis of odor cues alone. As it turns out, the longer a Belding's ground squirrel sniffs an object, the greater its interest in that object. Cubes that have been rubbed on a fairly close relative, genetically speaking, receive only a cursory inspection. Items smeared with odors of a more distant relative are given a significantly longer sniff, and the inspection time increases again for cubes daubed with a nonrelative's odor. Belding's ground squirrels are odor analyzers, spending less time with scents similar to their own and increasingly more time with odors less like their own (**FIGURE 3.7**). They therefore have the capacity to treat individuals differently based solely on the basis of this highly specific learned label of relatedness (Mateo 2002).

Genetic Differences Can Also Cause Behavioral Differences

Although a great many differences in behavioral phenotypes have been traced to differences in the environment, others have been linked to genetic differences among individuals, which follows from the interactive theory of development. An example comes from a study of blackcap warblers (*Sylvia atricapilla*), some of which spend the winter in southern Great Britain while others migrate south to Africa (**FIGURE 3.8**) (Berthold 1991). If the differences between the two groups of blackcaps occurred because of genetic differences between them, the offspring of winter-in-Britain birds should differ in their migratory behavior from the offspring of blackcaps that winter elsewhere. Therefore, Peter Berthold and colleagues captured some wild blackcaps in Britain during the winter and took them to a laboratory in Germany, where the birds were kept indoors. Then, with the advent of spring, pairs of warblers were allowed to breed in outdoor aviaries, providing Berthold with a cohort of young that had never migrated (Berthold and Pulido 1994).

Once the young birds were several months old, Berthold's team placed some of them in special cages that had been electronically wired to record the number of times a bird hopped from one perch to another. The electronic data revealed that

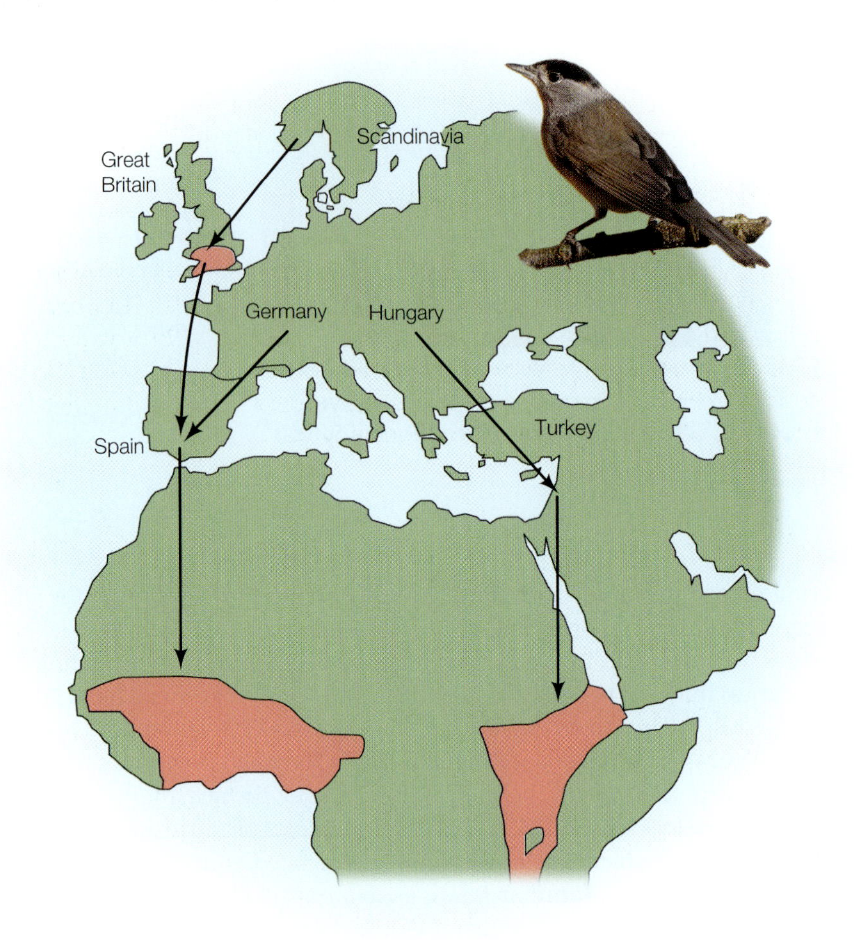

FIGURE 3.8 Fall migratory routes of the blackcap warbler. Blackcaps living in southern Germany and Scandinavia first go southwest to Spain before turning south to western Africa. Blackcaps living in eastern Europe migrate southeast before turning south to fly to eastern Africa. (Photograph © Rafa Irusta/ Shutterstock.)

when fall arrived, the young warblers became increasingly restless at night, exhibiting the kind of heightened activity characteristic of songbirds preparing to migrate (termed migratory restlessness; see **BOX 3.2**). The immature blackcaps' parents also became nocturnally restless when placed in the same kind of cage in the fall. The floors of these funnel-shaped cages were lined with paper that acquires marks when touched. Whenever a bird leapt up from the base of the funnel in an attempt to take off, it landed on the paper and left scratch marks, indicating the direction in which the bird was trying to go. Berthold's subjects, experienced adults and young novices alike, oriented due west, jumping up in that direction over and over. These data showed that the adults, which had been captured in wintery Britain, must have traveled there by flying west from Belgium or from central Germany, a point eventually confirmed by the discovery of some blackcaps in Britain that had been banded earlier in Germany. Young birds whose parents oriented in a southerly direction did the same in the fall, convincing Berthold that he had a case in which behavioral differences in migratory orientation were caused by genetic differences between the two populations of individuals (Berthold 1991).

But what were the genetic differences underlying the migratory behavior of the different blackcap populations? To answer this question, a team led by Bart Kempenaers explored a series of candidate genes that have been linked to migratory differences in other species of birds (Mueller et al. 2011). One gene that seemed to be of particular importance in blackcaps was the adenylate cyclase-activating polypeptide 1 gene (*adcyap1*), a neuropeptide that influences physiology and behavior in a variety of ways. In blackcaps, this gene was related to both individual and population variation in migratory behavior. At the individual level, autumn migratory

BOX 3.2 **EXPLORING BEHAVIOR** BY **INTERPRETING DATA**

Migratory restlessness

The black redstart (*Phoenicurus ochruros*) is a bird that migrates a relatively short distance from Germany to the Mediterranean region of Europe. In contrast, the closely related common redstart (*Phoenicurus phoenicurus*) travels as much as 5000 kilometers from Germany to central Africa. The scale in the **FIGURE** shows the duration of migratory restlessness in captive black redstarts, common redstarts, and hybrids between the two species, all of which had been hand-raised under identical conditions.

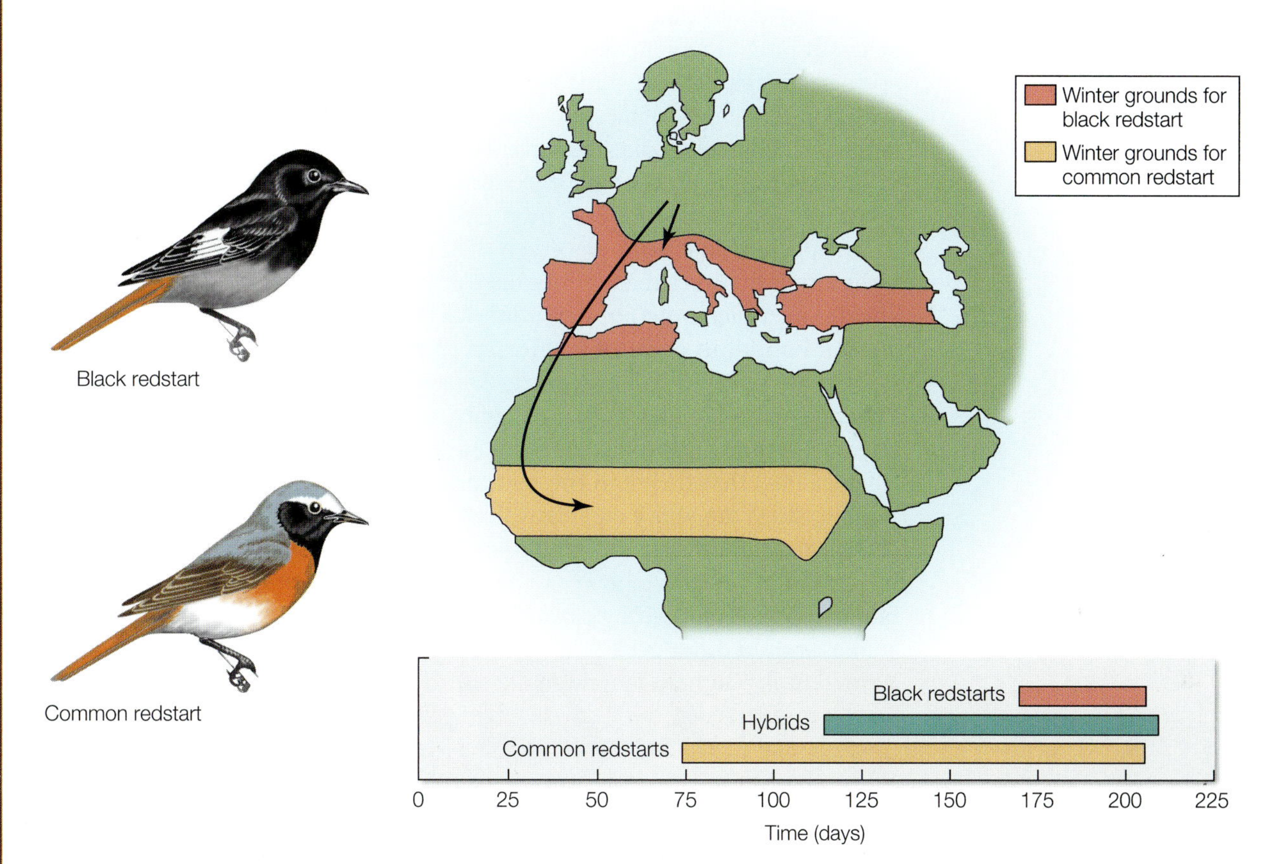

Differences in the migratory behavior of two closely related birds, the black redstart and the common redstart. The scale at the bottom shows the periods after hatching (day 0) when the young birds exhibit migratory restlessness at night. (After Berthold and Querner 1995.)

Thinking Outside the Box

Why do black redstarts exhibit migratory restlessness at night for fewer days than common redstarts? What are the potential costs and benefits for each species to migrating earlier or delaying migration longer? Design a study to examine the costs and benefits of migratory behavior in each species. Perhaps consider looking at intraspecific variation in migratory timing and where birds spend the winter. Finally, what does the behavior of the hybrids tell us about the role of genetic differences underlying the difference in the duration of migratory restlessness in the two parental species? Given what you read in the text about the genetics of migration in another European species, the blackcap warbler, think about how you would study the genetic basis of migration in redstarts. Can you make some predictions about what you would expect for a specific gene such as *adcyap1*?

restlessness behavior in populations from southern France and southern Germany was positively related to the allele length of a microsatellite in the *adcyap1* gene. Microsatellites are sequences of DNA base pairs that occur in repeats, such as ACACAC (a dinucleotide repeat) or CTGCTGCTG (a trinucleotide repeat). The repeat number often varies among individuals within a species. At the population level, there was a correlation between the among-population differentiation in migratory status of 14 populations and the genetic differentiation at the *adcyap1* locus. In a follow-up study in blackcaps, Mettler and colleagues explored the role that *adcyap1* plays in spring migration of blackcaps by determining if longer microsatellite alleles were associated with earlier spring arrival on the European breeding grounds (Mettler et al. 2015). Although the researchers did not find a relationship between allele length and spring arrival dates, they did find an interaction between wing shape and allele size, but only in female blackcaps. Specifically, allele size had a positive effect on spring arrival for female birds with rounded wings, but a negative effect on females with pointed wings. Wing shape has long been known to influence avian migratory behavior, and in blackcaps it seems to interact with migration-related genes to shape migratory timing and behavior in some birds.

Another example of a case in which the differences among individuals of the same species have been traced to differences in the genes possessed by these individuals involves the western terrestrial garter snake (*Thamnophis elegans*). This species occupies much of dry inland western North America, as well as foggy coastal habitat in parts of California (Arnold 1981). The diets of snakes in the two areas, referred to hereafter as inland and coastal snakes, differ markedly. Whereas inland snakes feed primarily on the fish and frogs found in lakes and streams in the arid West, coastal snakes regularly eat the banana slugs (*Ariolimax columbianus*) that thrive in the wet forests of coastal California (**FIGURE 3.9**). Banana slugs are slimy creatures that produce a thick mucus that contains toxic chemicals. Although many predators avoid the slugs because of their mucus, garter snakes have evolved ways to bypass the nasty chemicals. If the preference for banana slugs exhibited by coastal garter snakes has a hereditary basis, then these snakes should differ genetically from inland snakes. To test this hypothesis, Steve Arnold took pregnant female snakes from the two populations into the laboratory,

FIGURE 3.9 A coastal western terrestrial garter snake about to consume a banana slug. (Photograph by Robert Clay/Alamy Stock Photo.)

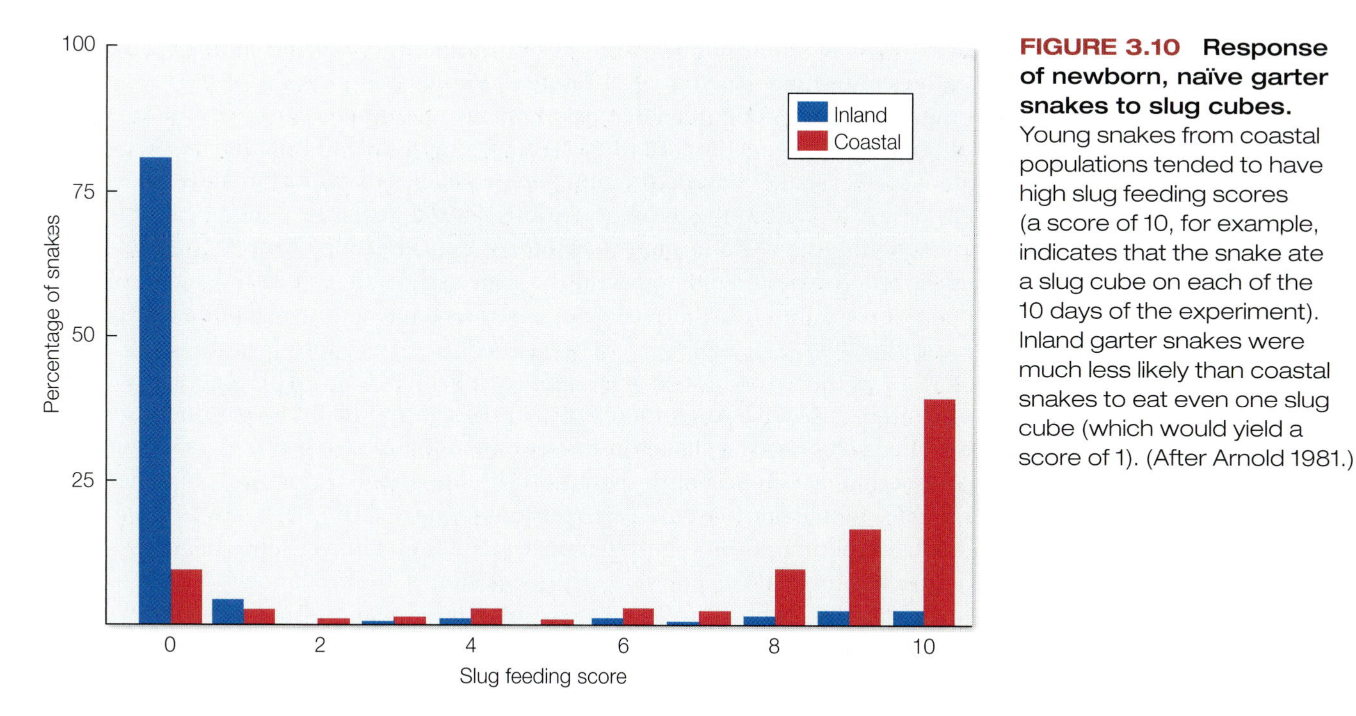

FIGURE 3.10 Response of newborn, naïve garter snakes to slug cubes. Young snakes from coastal populations tended to have high slug feeding scores (a score of 10, for example, indicates that the snake ate a slug cube on each of the 10 days of the experiment). Inland garter snakes were much less likely than coastal snakes to eat even one slug cube (which would yield a score of 1). (After Arnold 1981.)

where they were held under identical conditions. When the females gave birth to a litter (garter snakes produce live young rather than laying eggs), each baby snake was placed in a separate cage away from its litter mates and its mother to remove these possible (social) environmental influences on its behavior. Some days later, Arnold offered each baby snake a chance to eat a small chunk of freshly thawed banana slug by placing it on the floor of the young snake's cage. Most naïve young coastal snakes ate all the slugs they received, but most of the inland snakes did not (**FIGURE 3.10**). In both populations, slug-refusing snakes ignored the slug cube completely.

In a final experiment, Arnold took another group of isolated newborn snakes that had never fed on anything and offered them a chance to respond to the odors of different prey items. He took advantage of the readiness of newborn snakes to flick their tongues at, and even attack, cotton swabs that have been dipped in fluids from prey animals (**FIGURE 3.11**). Chemical scents are carried by the tongue to the vomeronasal organ in the roof of the snake's mouth, where the odor molecules are analyzed as part of the process of detecting prey. By counting the number of tongue flicks that hit the swab during a 1-minute trial, Arnold measured the relative responsiveness of inexperienced baby snakes to different odors. He found that populations of inland and coastal snakes reacted about the same to swabs dipped in toad tadpole solution (a prey of both groups), but they behaved very differently toward swabs daubed with slug scent. Almost all inland snakes ignored the slug odor, whereas almost all coastal snakes

FIGURE 3.11 A tongue-flicking newborn western terrestrial garter snake senses odors from a cotton swab that has been dipped in slug extract. (Photograph by Stevan Arnold.)

rapidly flicked their tongues at it. Because all the young snakes had been reared in the same environment, the differences in their willingness to eat slugs and to tongue-flick in reaction to slug odor appear to have been caused by their genetic differences. If the feeding differences between the two populations arise because most coastal snakes have a different allele or alleles than most inland snakes, then crossing adults from the two populations should generate a great deal of genetic and phenotypic variation in the resulting hybrid group. Arnold conducted the appropriate experiment and found the expected result, confirming again that the behavioral differences between populations have a strong genetic component (Arnold 1981).

What is the genetic basis of these intraspecific feeding differences in western terrestrial garter snakes? Although no one has yet studied this question in relation to banana slugs, garter snakes in some populations also feed on newts in the genus *Taricha* that produce a neurotoxin call tetrodotoxin (TTX), which can be lethal to humans, though garter snakes are resistant to the toxin. Work led by Edmund Brodie, Jr. and Edmund Brodie III demonstrated that toxin resistance in garter snakes occurs because of the expression of TTX-resistant sodium channels (Na_v) in skeletal muscle (Geffeney et al. 2002), which result from mutations in the gene underlying $Na_v1.4$ (Geffeney et al. 2005). By comparing individuals from different populations of California garter snakes that varied in their preference for newts, the Brodies' team found that TTX resistance correlates strongly with allelic variation in $Na_v1.4$, but not in so-called neutral genetic markers that are unrelated to toxin resistance (Feldman et al. 2010). It is likely that different populations of the same species that vary in their ability to eat newts—and probably banana slugs as well—differ in the genes underlying $Na_v1.4$. What's more, the genetic basis for TTX resistance is not unique to garter snakes, as many other snakes are resistant to the toxin. The Brodies and their colleagues have shown in snakes that TTX resistance has evolved independently at least five times over the past 12 million years in the same gene, but only after other mutations occurred in the genes for two other peripheral nerve channels, $Na_v1.6$ and $Na_v1.7$ (McGlothlin et al. 2016). Thus, for banana slug–eating populations of western terrestrial garter snakes, it is quite possible that the genes underlying $Na_v1.4$ (or something similar) differ from those of snakes that do not consume the slimy morsels.

Learning and Cognition

As we've seen in animals as diverse as ground squirrels and paper wasps, learning with whom to interact in these cases is central to the developmental process. In many species, learning begins almost from the moment an individual is born. For example, the first thing that many young animals see after birth is one or both of their parents. During this critical period, a young animal's early social interactions lead to it learning such essential issues as what constitutes an appropriate sexual partner. The development of the ability to learn is called **imprinting** and depends on specialized features of the brain, which in turn arise developmentally through the interplay between genes and key elements of the animal's social environment. In some of the earliest work in this area, behavioral biologist Konrad Lorenz studied imprinting in graylag geese (*Anser anser*) by letting himself be the first animal that newly hatched chicks observed. Rather than imprinting on a mother goose, the graylag goslings imprinted on Lorenz (**FIGURE 3.12**), forming both a learned attachment to him (Lorenz 1952) and, in the case of the male graylags when they reached adulthood, a sexual preference for humans as mates.

FIGURE 3.12 Imprinting in graylag geese. These goslings imprinted on the behavioral biologist Konrad Lorenz and followed him wherever he went. (Photograph © Nina Leen/Time Life Pictures/Getty Images.)

Learning Requires Both Genes and Environment

The experience of following a particular individual early in life must have somehow altered those regions of the male graylag goose's brain responsible for sexual recognition and courtship in the adult bird. The special effects of imprinting could not have occurred without a "prepared" brain, one whose genetically influenced development enabled it to respond in special ways to particular kinds of information available from its social environment. The fact that different species exhibit different imprinting tendencies provides further circumstantial evidence for the genetic contribution to learning. A group of Norwegian researchers provided this kind of evidence when they performed a cross-fostering experiment and switched broods from blue tit (*Cyanistes caeruleus*) nests into the nests of closely related great tits (*Parus major*), and vice versa. Some of the cross-fostered chicks grew up and survived to court and form pair-bonds with members of the opposite sex (**FIGURE 3.13A**). But were they able to find mates? Of the surviving fostered great tits, only 3 of 11 found

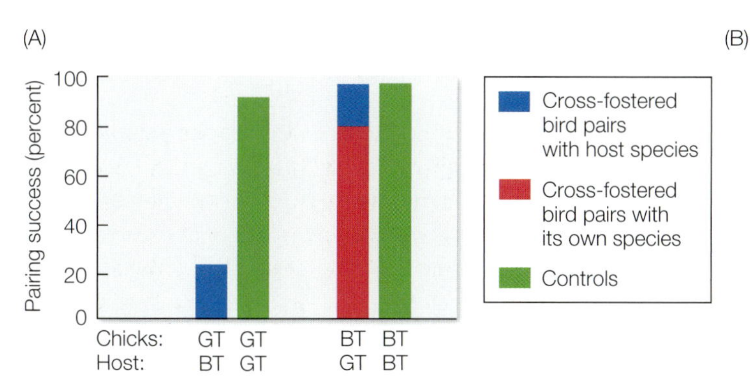

FIGURE 3.13 Cross-fostering has different imprinting effects in two related songbirds. (A) Males of the great tit (GT) that have been reared by blue tit (BT) foster parents try to pair with blue tit females, but only a small fraction succeeds. In contrast, cross-fostered blue tits always find mates, generally of their own species. Control birds, which were reared by their own species, always find mates and pair with members of their own species. (B) When blue tit females pair with great tit males, they also copulate with male blue tits. Here a female blue tit (far left) paired with a male great tit (upper left) reared a brood together that consisted entirely of blue tit nestlings. (After Slagsvold et al. 2002; photograph by Tore Slagsvold.)

mates—all of which were blue tit females that had been fostered by great tits. Of the surviving fostered blue tits, all 17 found mates, although 3 of these were females that socially mated with cross-fostered male great tits (Slagsvold et al. 2002). Although some individuals of both species became imprinted on another species as a result of their foster-care experiences, the degree to which individuals imprinted on their foster parents differed between the two species of songbirds. The researchers also found that none of the cross-fostered great tits mated with a member of their own species, whereas most of the cross-fostered blue tits did. Moreover, each blue tit female that had a great tit as a social partner must have mated with a blue tit male on the side, because all 33 offspring produced by those females were blue tits, not hybrids (**FIGURE 3.13B**). Thus, although mis-imprinting occurred in both species, the developmental effect of being reared by members of another species was far greater for great tits than for blue tits, an indication that the genetic basis of the imprinting mechanism was very different for these two closely related species.

Learning in Complex Environments

In addition to imprinting, bird species possess other specialized learning abilities, including the ability to remember where they have hidden food. The black-capped chickadee (*Poecile atricapillus*) is especially good at this task. This bird's spatial memory enables it to relocate large numbers of seeds or small insects that it has hidden in bark crevices or patches of moss scattered throughout its woodland environment, an ability that David Sherry investigated by providing captive chickadees with a chance to store food in holes drilled in small trees placed in an aviary (**FIGURE 3.14**). After the chickadees had placed sunflower seeds in 4 or 5 of 72 possible storage sites, they were shooed into a holding cage for 24 hours. During this time, Sherry removed the seeds and closed each of the 72 storage sites with a Velcro cover. When the birds were released back into the aviary, they spent much more time inspecting and pulling at the covers at their hoard sites than at sites where they had not stored food 24 hours

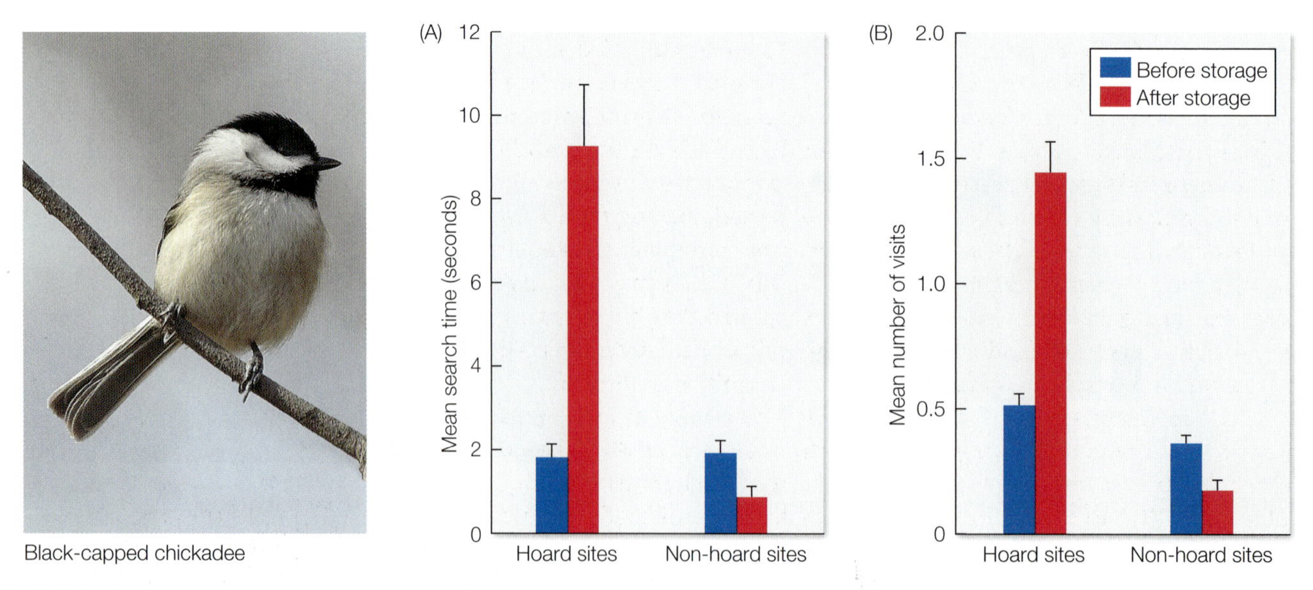

FIGURE 3.14 Spatial learning by black-capped chickadees. (A) Black-capped chickadees spent much more time at sites in an aviary where they had stored food 24 hours previously (hoard sites) than they had spent during their initial exposure to those sites, even though experimenters had removed the stored food. (B) The chickadees also made many more visits to hoard sites than to other sites, presumably because they remembered having stored food there. Bars depict mean +/– SE. (After Sherry 1984; photograph © Bryan Eastham/ShutterStock.)

earlier. Because the storage sites were all empty and covered, there were no olfactory or visual cues provided by stored food to guide the birds in their search. Instead, birds had to rely solely on their memories of where they had hidden seeds (Sherry 1984). In nature, these birds store only one food item per hiding spot and never use the same location twice, yet they can relocate their caches as much as 28 days later (Hitchcock and Sherry 1990).

Sherry's work suggests that chickadees have impressive cognitive abilities when it comes to remembering where they store food. The part of the chickadee brain used for this task is called the hippocampus. In general, food-caching birds such as chickadees tend to have a larger hippocampus than non-food-caching species. Moreover, hippocampal lesions in chickadees and other songbirds cause a decline in memory for location (Hampton and Shettleworth 1996). In black-capped chickadees, whose food-caching behavior varies seasonally—peaking in the fall before the onset of winter—there is mixed evidence that hippocampus size increases with a change of season. However, neurogenesis (the production of new neurons) in the hippocampus does seem to vary seasonally and increases at the time of year when food-storing behavior increases (Sherry and Hoshooley 2010). Thus, environmental inputs to the brain cause a change in neuronal activity that likely facilitates a bird's ability to learn and remember where it stored food.

In a related species, the mountain chickadee (*Poecile gambeli*), a team led by Vladimir Pravosudov explored how the harshness of the environment influences learning and cognitive ability more generally (Croston et al. 2017). To determine whether the degree of caching specialization was associated with differences in cognitive flexibility, the researchers used a spatial learning and memory reversal task on populations of chickadees from high and low elevations—sites that differed in their degree of winter climate severity. The learning task involved an array of radio-frequency identification (RFID) "smart" feeders that controlled birds' access to good food, in this case sunflower seeds (**VIDEO 3.1**). In "open" mode, all birds could feed, and the feeders recorded the identity of each bird that did so. In "feed all" mode, the feeder door was closed but would open for birds that had PIT (passive integrated transponder) tags. In "target" mode, the feeder door would open only for birds with unique RFID codes that were programmed to a specific feeder. During the initial learning phase of the experiment, birds were presented with "open"-mode feeders for 10 weeks, followed by a few weeks of feeders in "feed all" mode. Birds were then randomly assigned a "target"-mode feeder for a few weeks before learning was reversed by assigning each bird to a new "target" feeder. All chickadees exhibited the ability to learn in these tasks, but high-elevation birds visited the feeders more often and performed significantly worse during the reversal period, making more location errors than low-elevation birds. These results suggest that mountain chickadees are more reliant on food caches in harsher environments, where they may be initially better at spatial learning but at the expense of reduced cognitive flexibility. Thus, even in the same species of complex food-cachers, the environment greatly influences not only a bird's ability to learn, but also its ability to solve cognitive challenges.

Clark's nutcrackers (*Nucifraga columbiana*) may have an even more impressive memory than chickadees, as they scatter as many as 33,000 seeds in up to 5000 caches that may be up to 25 kilometers from the harvest site (**VIDEO 3.2**). Nutcrackers dig tiny holes in the earth for each store of seeds, then completely cover the cache. Each bird does this work in the fall and then relies on its stores through the winter and into the spring, recovering an estimated two-thirds of the hidden seeds, often months after hiding them (Balda 1980, Milius 2004). But how do nutcrackers remember exactly where to find their hidden bounties? It could be that nutcrackers do not really remember where each and every seed cache is, but instead rely on a simple rule of thumb, such as "look near little tufts of grass." Alternatively, they might remember only the general location where food

was stored and, once there, look around until they see disturbed soil or some other indicator of a cache. However, experiments similar to those performed with black-capped chickadees show that the nutcrackers do remember exactly where they hid their food. In one test, a nutcracker was given a chance to store seeds in a large outdoor aviary, after which it was moved to another cage. The observer, Russ Balda, mapped the location of each cache and then removed the buried seeds and swept the cage floor, removing any signs of cache making. No visual or olfactory cues were available to the bird when it was permitted to go back to the aviary a week later and hunt for the food. Balda mapped the locations where the nutcracker probed with its bill, searching for the nonexistent caches. The bird's spatial memory served it well, for it dug into as many as 80 percent of its ex-cache sites, while only very rarely digging in other places (Balda 1980). Other long-term experiments on nutcracker memory have demonstrated that birds can remember where they have hidden food for at least 6, and perhaps as long as 9, months (Balda and Kamil 1992). Indeed, when Balda tested one of his graduate students as if he were a food-storing bird, the student did only about half as well as a typical nutcracker when tested a month after making his caches (Milius 2004). The birds can even remember the size of the seeds they have hidden, as demonstrated by their tendency to spread their bills farther apart when probing the earth for large cached seeds as opposed to smaller ones. Because nutcrackers retrieve one seed at a time from their underground stores, they can secure and process them more efficiently by opening their beaks just the right distance to grasp and pluck a seed of a given size out of a cache (Möller et al. 2001).

The extraordinary ability of nutcrackers, chickadees, and other bird species to store spatial information in their hippocampus is surely related to the ability of certain brains to change biochemically and structurally in response to the kinds of sensory stimulation associated with hiding food. These changes in brain structure and size could not occur without the genes needed to construct the learning system and the genes that are responsive to key sensory stimuli relevant to the learning task. One logical stimulus that might drive changes in brain activity is photoperiodicity, or seasonal changes in day length. In the case of black-capped chickadees, experimental manipulation of photoperiod with captive birds did not appear to affect caching behavior (Shettleworth et al. 1995), nor did it influence hippocampus size (MacDougall-Shackleton et al. 2003), both important negative results suggesting that other, as yet undiscovered, mechanisms must be at play. Although clearly more work is needed to understand how the environment influences these types of behaviors, the general point is that even learned behaviors, which are obviously environmentally dependent, are gene-dependent as well.

The Adaptive Value of Learning

The previous examples clearly illustrate that animals that learn modify their behavior according to certain experiences (both social and ecological) that they have had. However, learning does not produce behavioral change just for the sake of change. Instead, selection favors investment in the mechanisms underlying learning only when there is environmental unpredictability that has reproductive relevance for individuals. Thus, we have another cost–benefit argument, one that presupposes that any proximate mechanism that enables individuals to learn comes with a price tag. We can test this assumption by predicting, for example, that the brains of male marsh wrens (*Cistothorus palustris*) living in the western United States should be larger than those of their East Coast counterparts because young West Coast wrens learn nearly 100 songs by listening to others, whereas East Coast wrens have much smaller learned repertoires of about 40 songs (Kroodsma and Canady 1985). When the birds' brains were examined, the song control systems of West Coast wrens weighed on average 25 percent more than the equivalent nuclei of East Coast wrens (see also Chapter 2).

At this point, you might think, wait a minute. There are many other differences between East and West Coast marsh wrens beyond song repertoire size that could influence brain size. Indeed, it is appropriate to be cautious, because as with all comparisons of just two populations or species, other factors such as environmental differences between East and West Coast birds could at least partially explain the differences in the wrens' brain sizes. To further examine the role of environmental variation on brain size, a group led by Daniel Sol used remotely sensed climatic data (measures such as temperature and rainfall) and the comparative approach to examine more than 1200 species of birds (Sayol et al. 2016). They found that larger brains (relative to body size) are more likely to occur in species that experience seasonal environments with greater environmental variation across their geographic ranges. Moreover, the researchers demonstrated that larger brains evolved when species invaded these more seasonal environments, perhaps allowing them to cognitively buffer these harsher environments through information gathering and learning, much like in the mountain chickadee we discussed earlier.

That spatial learning evolves in response to particular ecological pressures can also be seen by comparing the learning abilities of four bird species, all members of the crow family (Corvidae), that vary in their predisposition to store food—a task that puts a premium on spatial memory (**FIGURE 3.15**). As we have seen, Clark's

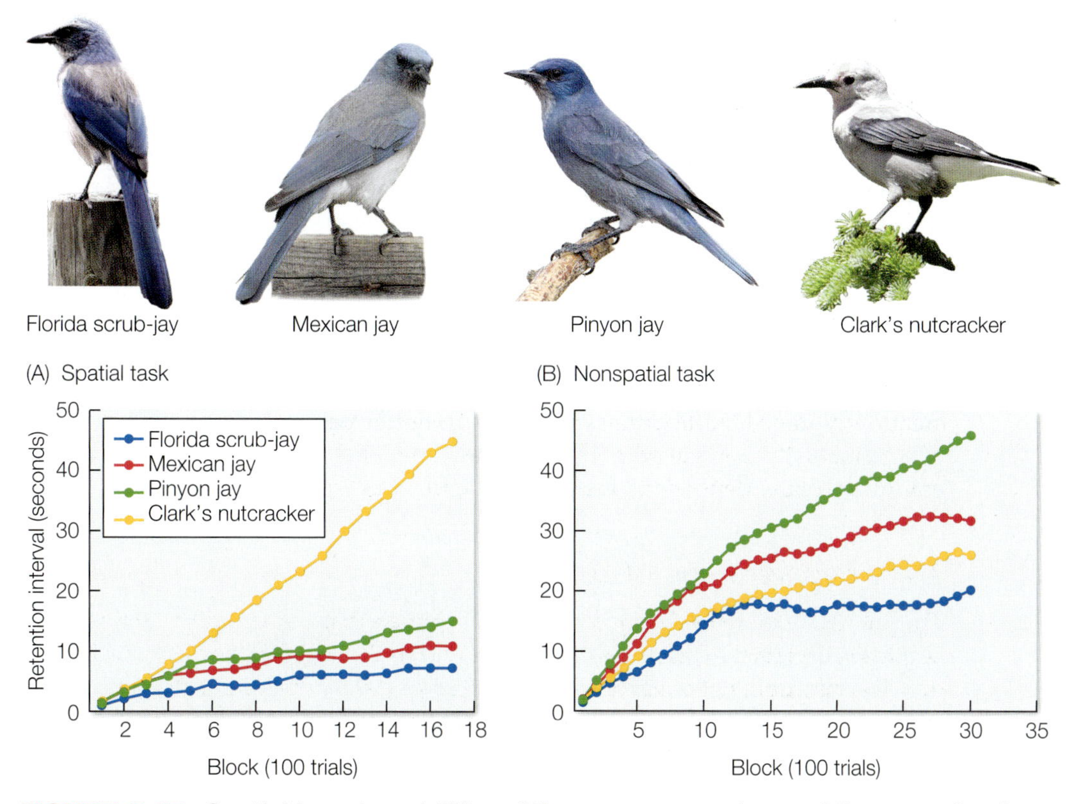

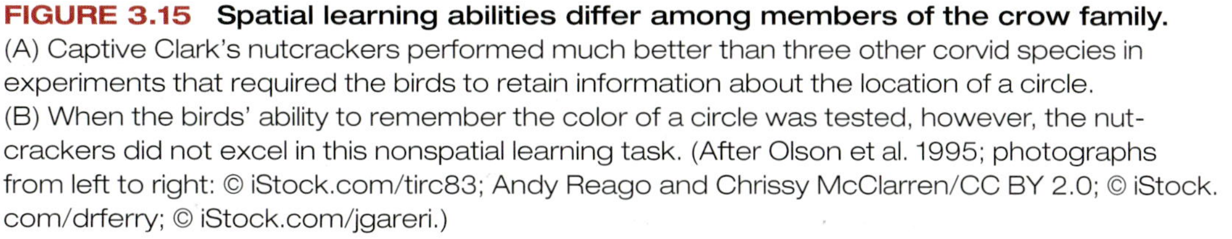

FIGURE 3.15 Spatial learning abilities differ among members of the crow family. (A) Captive Clark's nutcrackers performed much better than three other corvid species in experiments that required the birds to retain information about the location of a circle. (B) When the birds' ability to remember the color of a circle was tested, however, the nutcrackers did not excel in this nonspatial learning task. (After Olson et al. 1995; photographs from left to right: © iStock.com/tirc83; Andy Reago and Chrissy McClarren/CC BY 2.0; © iStock.com/drferry; © iStock.com/jgareri.)

nutcracker is a food-storing specialist, and it has a large pouch for the transport of pine seeds to storage sites. The pinyon jay (*Gymnorhinus cyanocephalus*) also has a special anatomical feature, an expandable esophagus, for carrying large quantities of seeds to hiding places. However, the Florida scrub-jay (*A. coerulescens*) and the Mexican jay (*Aphelocoma wollweberi*) lack special seed-transport devices and appear to hide substantially less food than their relatives. Individuals from these four species were tested on two different learning tasks in which they had to peck a computer screen to receive a reward. One task required the birds to remember the color of a circle on the screen (a nonspatial learning task), and the other required them to remember the location of a circle on the screen (a spatial task). When it came to the nonspatial learning test, pinyon jays and Mexican jays did substantially better than Florida scrub-jays and nutcrackers (Olson et al. 1995). But in the spatial learning experiment, the nutcracker went to the head of the class, followed by the pinyon jay, then the Mexican jay, and finally the Florida scrub-jay (but see de Kort and Clayton 2006, Pravosudov and de Kort 2006 for examples that do not fit this pattern and why they might occur). To the extent that nutcrackers and pinyon jays really are especially good at spatial learning only, we can conclude that birds have not evolved all-purpose learning abilities, but rather that their learning skills are designed to promote success in solving ecologically relevant problems (Olson et al. 1995).

The logic of an evolutionary approach to learning leads us to hypothesize that if males and females of the same species differ in the benefits derived from a particular learned task, then a sex difference in learning skills should evolve (**sex differences learning hypothesis**). The pinyon jay provides a case in point. As just noted, this jay hides large numbers of pinyon seeds when they are available, and then retrieves them up to 5 months later when food is scarce. But males are more likely than females to have to relocate old caches because they provide their mates and young with recovered food while their female partners spend their time instead at the nest, incubating their eggs and young. As predicted, males appear to have evolved better long-term memory than females. When captive birds of both sexes were tested during the nesting season, males made fewer errors than females (**FIGURE 3.16**) (Dunlap et al. 2006). The poorer performance of females suggests that the ability to learn has costs that require special benefits if the trait is to evolve.

sex differences learning hypothesis A sex difference in learning skills should evolve when males and females of the same species differ in the benefits derived from a particular learned task.

The sex differences learning hypothesis has also been tested by Steve Gaulin and Randall FitzGerald in their studies of spatial learning in voles (Gaulin and FitzGerald 1989). Males of the polygynous and wide-ranging meadow vole (*Microtus pennsylvanicus*) do better on maze-learning tests than

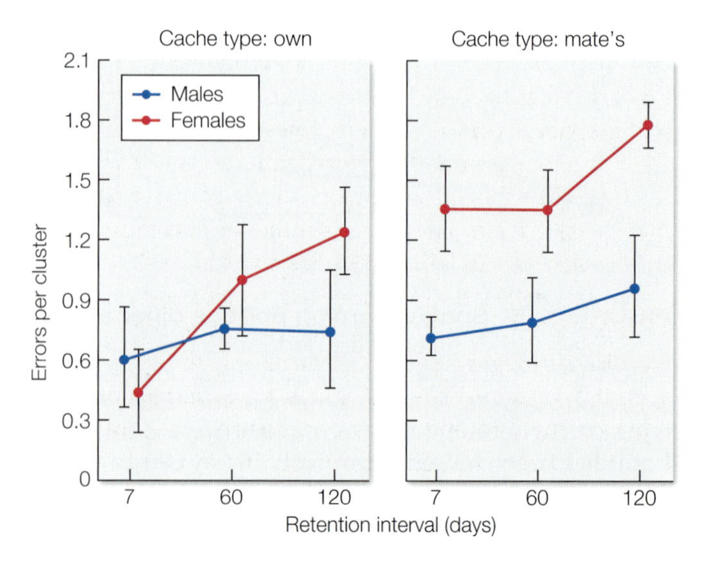

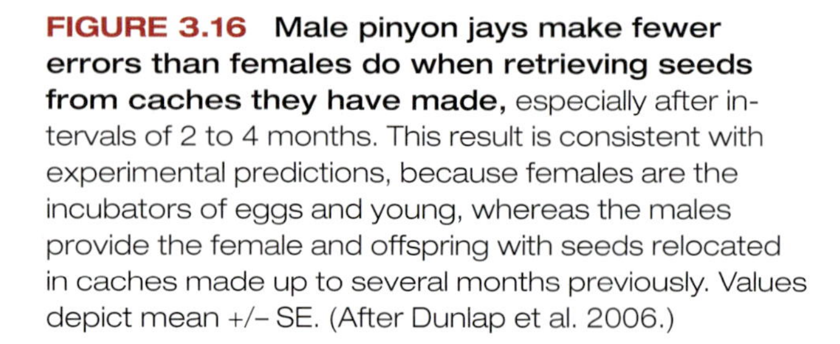
FIGURE 3.16 Male pinyon jays make fewer errors than females do when retrieving seeds from caches they have made, especially after intervals of 2 to 4 months. This result is consistent with experimental predictions, because females are the incubators of eggs and young, whereas the males provide the female and offspring with seeds relocated in caches made up to several months previously. Values depict mean +/– SE. (After Dunlap et al. 2006.)

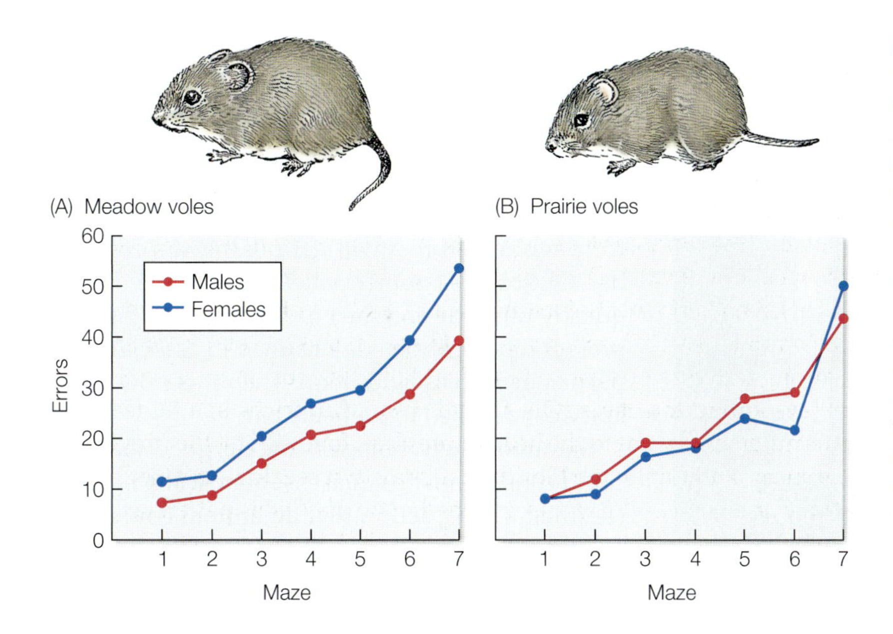

FIGURE 3.17 Sex differences in spatial learning ability are linked to home range size. Spatial learning in voles was tested by giving individuals opportunities to travel through seven different mazes of increasing complexity in the laboratory and then letting them run through each maze again. (A) Polygynous male meadow voles, which roam over wide areas in nature, consistently made fewer errors (wrong turns) on average than the more sedentary females of their species. (B) In contrast, in the monogamous prairie vole, a species in which males and females live together on the same territory, females matched males in performance. (After Gaulin and FitzGerald 1989.)

do females of this species. In the monogamous prairie vole (*M. ochrogaster*), in which males and females share the same living space, the two sexes did equally well on the same set of maze tests given to the meadow vole (**FIGURE 3.17**).

In a species in which females face greater spatial challenges than males, we would expect females to make larger investments in the neural foundations of spatial learning. The brown-headed cowbird (*Molothrus ater*) is such a species, because cowbirds are brood parasites that lay their eggs in other birds' nests (see Chapter 11). But it is only the female that must face the spatial challenge of locating an appropriate nest. That is, a female must not only search widely for nests to parasitize, but she must also remember where potential victims have started their nests in order to return to them up to several days later when the time is ripe for her to add one of her eggs to those already laid by the host bird. What's more, female cowbirds often parasitize the same nests year after year (because host species often build nests in the same location each year), suggesting that they must build a spatial map that lasts for long periods of time. In contrast, male cowbirds do not confront such difficult spatial problems. As predicted, the hippocampus (but no other brain structure) is considerably larger in female brown-headed cowbirds than in males. As we saw for chickadees, the hippocampus is thought to play an important role in spatial learning when it comes to nest finding, just as it does in food finding. As you might predict, in two nonparasitic relatives of the cowbird—the red-wing blackbird (*Agelaius phoeniceus*) and common grackle (*Quiscalus quiscula*)—the hippocampus is the same size in both sexes (**FIGURE 3.18**) (Sherry et al. 1993).

It is not just spatial learning that bears the clear imprint of natural selection. Consider **operant conditioning** (or trial-and-error learning), in which an animal learns to associate a voluntary action with the consequences that follow from that action (Skinner 1966). Operant conditioning does occur outside psychology laboratories, but it has been studied extensively in Skinner boxes, named after the psychologist B. F. Skinner. After a Norway rat (*Rattus norvegicus*) has been introduced into a Skinner box, it may accidentally press

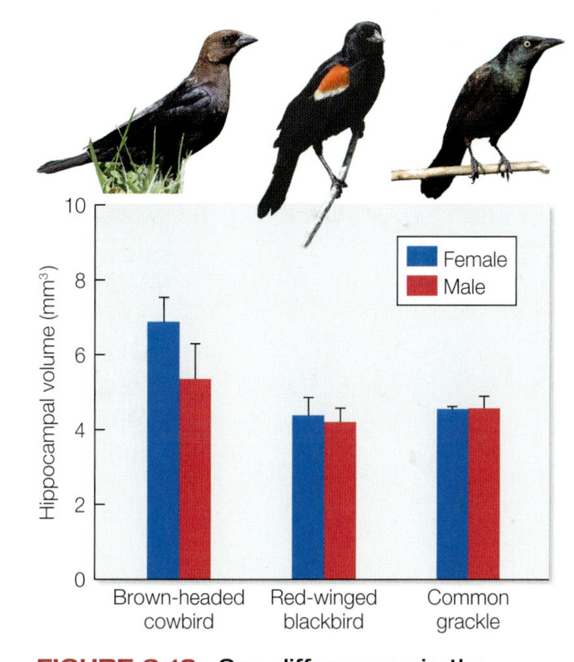

FIGURE 3.18 Sex differences in the avian hippocampus. Female brown-headed cowbirds have a larger hippocampus than males, as would be expected if this brain structure promotes spatial learning and if selection for spatial learning ability is greater on female than on male cowbirds. Red-winged blackbirds and common grackles do not exhibit this sex difference. (After Sherry et al. 1993; photographs from left to right: © iStock.com/SteveByland; Andy Reago and Chrissy McClarren/CC BY 2.0; © iStock.com/RT-Images.)

FIGURE 3.19 Operant conditioning exhibited by a Norway rat in a Skinner box. The rat approaches the bar and presses it. The animal awaits the arrival of a pellet of rat chow and consumes it, so the bar-pressing behavior is reinforced. (Photograph by Science History Images/Alamy Stock Photo.)

a bar on the wall of the cage (**FIGURE 3.19**), perhaps as it reaches up to look for a way out. When the bar is pressed down, a rat chow pellet pops into a food hopper. Some time may pass before the rat happens upon the pellet. After eating it, the rat may continue to explore its rather limited surroundings for a while before again happening to press the bar. Out comes another pellet. The rat may find it quickly this time and then turn back to the bar and press it repeatedly, having learned to associate this particular activity with food. The rat is now operantly conditioned to press the bar.

Skinnerian psychologists once argued that one could condition with equal ease almost any operant (defined as any action that an animal could perform). Indeed, the successes of operant conditioning are many; pigeons have been taught to play a kind of ping-pong, and blue jays (*Cyanocitta cristata*) have been conditioned to use computers. Rats also can be conditioned to do all sorts of things in the laboratory, such as avoid novel, distinctively flavored foods or fluids after being exposed to nausea-inducing X-ray radiation. However, John Garcia and colleagues found that the ability of these animals to learn to avoid certain punishing foods or liquids had some restrictions (Garcia and Ervin 1968, Garcia et al. 1974). The degree to which an irradiated rat rejects a food or fluid is proportional to (1) the intensity of the resulting illness, (2) the intensity of the taste of the substance, (3) the novelty of the substance, and (4) the shortness of the interval between consumption and illness (Garcia et al. 1974). But even if there is a long delay (up to 7 hours) between eating a distinctively flavored food and exposure to radiation and consequent illness, the rat still links the two events and uses the information to modify its behavior. In contrast, rats never learn that a distinctive sound (a click) is a signal that always precedes an event associated with nausea. Nor can rats easily make an association between a particular taste and shock punishment (**FIGURE 3.20**). If, after drinking a sweet-tasting fluid, the rat receives a shock on its feet, it often remains as fond of the fluid as it was before, as measured by the amount drunk per unit of time, no matter how often the rat is shocked after drinking this

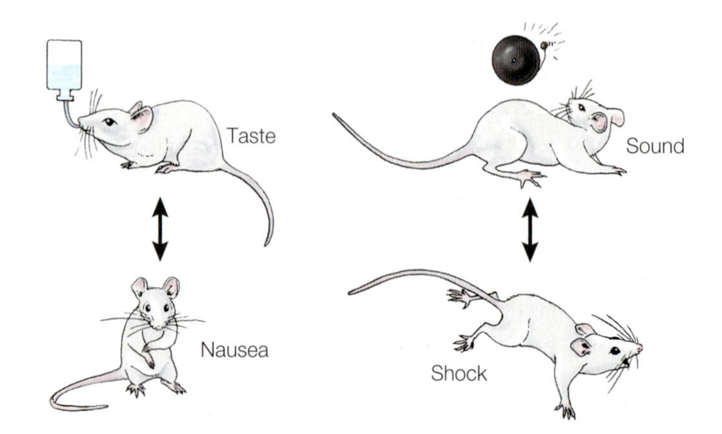

FIGURE 3.20 Biases in taste aversion learning. Although rats can easily learn that certain taste cues will be followed by sensations of nausea and that certain sounds will be followed by skin pain caused by shock, they have great difficulty forming learned associations between taste and consequent skin pain or between sound and subsequent nausea. (After Garcia et al. 1974.)

liquid. These failures surely relate to the fact that in nature, particular sounds are never associated with illness-inducing meals, any more than the consumption of certain fluids causes a rat's feet to hurt.

Understanding the natural environment from which rats came also helps explain why they are so adept at learning to avoid novel foods with distinctive tastes that are associated with illness, even hours after ingesting the food. Under natural conditions, a rat becomes completely familiar with the area around its burrow, foraging within that area for a wide variety of foods, plant and animal (Lore and Flannelly 1977). Some of these foods are edible and nutritious, but others are toxic and potentially lethal. A rat cannot clear its digestive system of toxic foods by vomiting. Therefore, the animal takes only a small bite of anything new. If it gets sick later, it avoids this food or liquid in the future, as it should, because eating large amounts might kill it (Garcia et al. 1974). This behavioral response suggests that even what appears to be a general, all-purpose form of learning is actually a specialized response to particular kinds of biologically significant associations that occur in nature.

If this argument is correct, then other mammals that are dietary generalists, which also run the risk of consuming dangerous, toxic items, should behave like the Norway rat, which is to say that they should also quickly form taste aversions to bad-tasting, illness-inducing items. And they do. Three bat species that feed on a range of foods behaved in the predicted manner: they rapidly formed taste aversions when fed a meal laced with an unfamiliar flavor, cinnamon or citric acid, before being injected with a chemical that made them vomit. When later offered a choice between food with and without cinnamon or citric acid, these three generalist consumers avoided foods spiced with the novel additives (Ratcliffe et al. 2003). In contrast, dietary specialists, which concentrate exclusively on one or a very few safe foods, should be unable to acquire taste aversions in this manner. The common vampire bat (*Desmodus rotundus*), a blood-feeding specialist, is quite incapable of learning that consumption of an unusual-tasting fluid will lead to gastrointestinal distress (**FIGURE 3.21**) (Ratcliffe et al. 2003).The difference between the specialist common vampire bat and its generalist relatives supports the idea that taste aversion learning is an evolved response to the risk of food or fluid poisoning. Just as is true for all aspects of behavioral development, the changes associated with learned behavior are worth the cost only if they confer a net fitness benefit on individuals capable of modifying their behavior in a particular way.

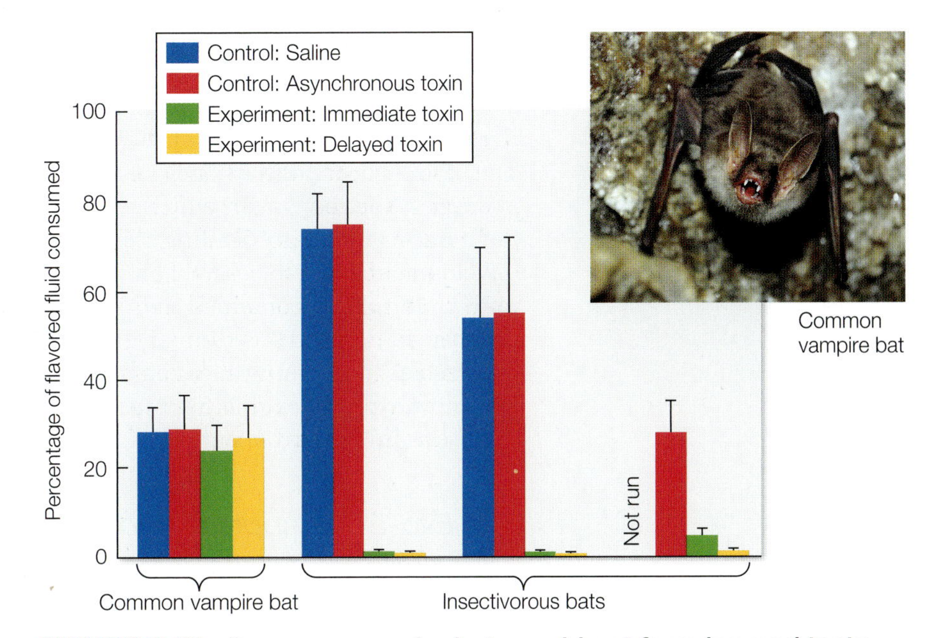

FIGURE 3.21 Common vampire bats could not form learned taste aversions. Instead, they continued to consume a flavored fluid even if, immediately after accepting this novel substance, they were injected with a toxin that caused gastrointestinal distress. In contrast, three insect-eating bat species completely rejected the novel dietary item when it was combined with injection of the toxin, no matter whether this was done immediately after feeding or after a delay. Two control groups were also used in the experiment, one in which the consumption of the novel food was paired with a harmless injection of saline solution, and another in which the toxin was injected but not in conjunction with feeding on the fluid. Bars depict mean +/– SE. (After Ratcliffe et al. 2003; photograph © kyslynskyyhal/Shutterstock.com.)

The Evolutionary Development of Behavior

As we showed in Chapter 2, proximate mechanisms have ultimate causes, and this ties the different levels of analysis together. Our discussion above of the adaptive value of learning illustrates how ultimate explanations can influence the development of behavior. Of course, developmental mechanisms also have an evolutionary history that can sometimes be described as a series of modifications of an ancestral pattern and its reconfiguration into a modern attribute. This kind of research is at the heart of what has been called the field of evolutionary development, or evo-devo (Carroll et al. 2005, Toth and Robinson 2007). Much of the early evo-devo work focused on the history behind the development of physical or structural features, but researchers are now beginning to apply this framework to behavioral traits as well. With respect to the often dramatic physical differences among species, even closely related ones, Darwin's theory of descent with modification has been used to identify the underlying genetic–developmental changes responsible for these differences. For example, a gene shared by honey bees and *Drosophila* fruit flies by virtue of their common ancestor contributes to the development of castes in bees by promoting the development of the relatively large size of queen bees as well as their large, fully functional ovaries (Kamakura 2011). In fruit flies, the same gene codes for the same protein (royalactin), and also promotes the development of large body size. This shared set of genes that developmentally lead to similar phenotypes forms the crux of the evo-devo approach.

Perhaps the most spectacular product of the evo-devo approach has been the discovery that creatures as different as *Drosophila* fruit flies and humans share a set of homeobox (or Hox) genes whose operation is critical for the developmental organization of their bodies. These genes, which must have originated in a very distant common ancestor, have been retained in flies, humans, and many other organisms as a result of their usefulness during the development of functional body structures. The base sequence of the genes has of course been altered somewhat from species to species, and the way in which their products influence the process of development differs markedly, leading to major differences in the structural products of development. Nonetheless, the imprint of history on the process can still be seen in the information contained within the homeobox genes, the "toolkit" that vast numbers of descendants from a very ancient ancestor use as they develop from a fertilized egg to a multi-celled organism with a body and limbs (Carroll et al. 2005). Let's now explore how this approach is being applied to the behavior of animals as diverse as insects and mammals.

The Evo-Devo Approach to Understanding Behavior

An example of the evo-devo approach to understanding animal behavior involves the *for* gene in *Drosophila*, a gene that also occurs in very similar form in the honey bee (Toth and Robinson 2007). The fruit fly *for* gene codes for a protein that, when produced, affects the operation of many other genes (Kent et al. 2009), leading to a broad range of chemical changes affecting the operation of the brain in both larval and adult fruit flies. These developmental changes control the larval flies' tendency to move about in a petri dish. In adults, the gene has a variety of additional metabolic and behavioral effects. The honey bee has inherited this same gene from a common ancestor of flies and bees. But over evolutionary time, the now modified gene has taken on a different but allied function in the honey bee by helping regulate the transition of a bee from a sedentary young adult that stays in the hive to a long-distance forager worker that collects food for the colony outside the hive. This transition is linked to an increase in the expression of the gene in the brains of the older workers (**BOX 3.3**) (Ben-Shahar et al. 2002). Furthermore, a microarray comparison of

BOX 3.3 **EXPLORING BEHAVIOR BY INTERPRETING DATA**

The genetics of foraging behavior in honey bees

Honey bees possess a gene related to foraging (called *for*) that contains information for the production of a particular enzyme called PKG. This gene is related to foraging behavior, and in *Drosophila*, allelic variation creates two phenotypes of foragers: "sitters," which show low locomotor and PKG activity when foraging, and "rovers," which show high locomotor and PKG activity when foraging. In honey bees, nurses and foragers also show differences in *for* expression, which presumably affects PKG activity and ultimately behavior (**FIGURE**).

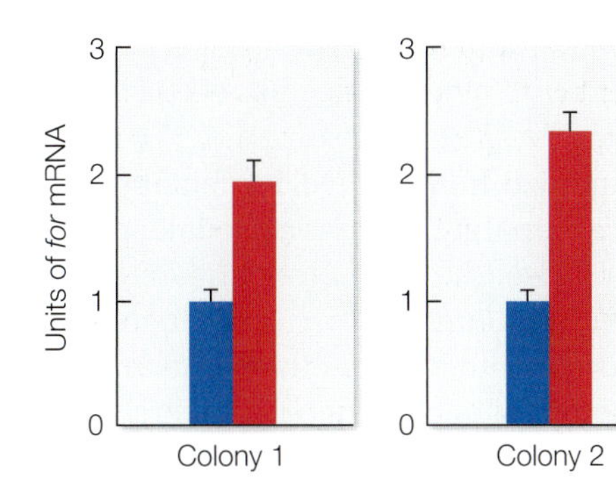

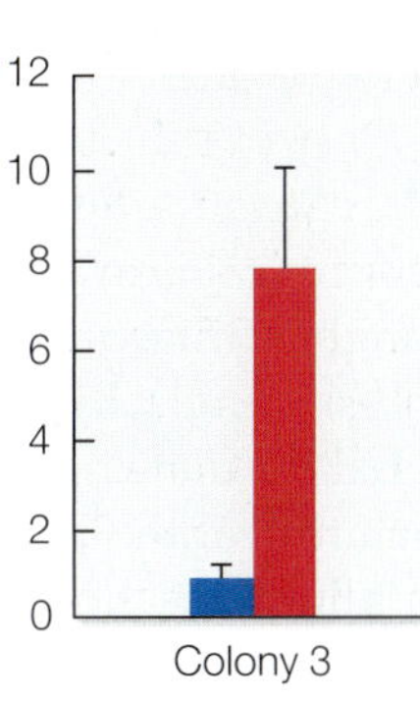

Levels of messenger RNA produced when the *for* gene is expressed in the brains of nurses and foragers in three typical honey bee colonies. Bars depict mean +/– SE. (After Ben-Shahar et al. 2002).

Thinking Outside the Box

If this genetic information is important (when expressed in brain cells) for foraging activity in worker honey bees, what prediction follows about the levels of the mRNA needed to produce the PKG enzyme present in foragers versus nonforagers taken from a typical bee colony? But since foragers are older than nurses in typical colonies, perhaps the greater activity of *for* is simply an age-related change that has nothing to do with foraging. What additional prediction and experiment are required to reach a solid conclusion about the causal role of *for* in regulating foraging in the honey bee? Suggestion: take advantage of the ability to create experimental colonies of same-age workers. Given that the *for* gene contributes to the onset of foraging by worker bees, how could this transition also be influenced by the workers' social environment? Generate a hypothesis that integrates the genetic and environmental contributions to the age-related change in bee behavior.

brain gene expression in the honey bee and paper wasp provided additional evidence of a shared genetic "toolkit," but only for some behaviors (Toth et al. 2010). Pathways known from the fruit fly to be related to lipid metabolism and heat and stress response were associated with behavior differences in wasps exhibiting different breeding roles. However, honey bees and paper wasps shared genes related to foraging behavior, but not to reproductive behavior. So while there may be shared toolkits in some genes underlying social behavior, novel genes with new functions may also be important in generating behavioral differences.

Perhaps one of the best examples of the conserved nature of behavioral development comes from studies of vasopressin, a neuropeptide hormone that plays an important role in pair-bonding in a variety of animals. As it turns out, all mammals, including our species, carry the gene for the production of this protein, which varies hardly at all from one mammalian species to another. In fact, molecules very similar to vasopressin have been found in everything from birds to insects to worms. Yet despite the similarities between the kinds of vasopressin produced in different

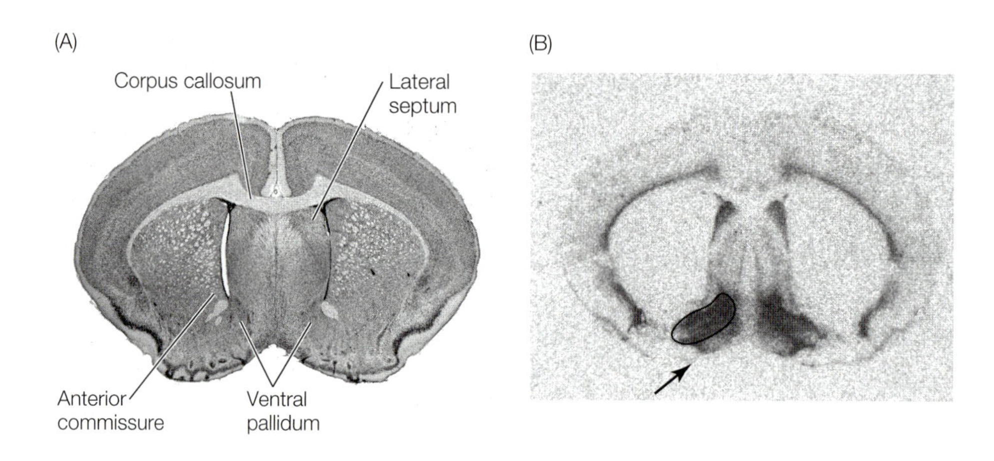

FIGURE 3.22 The brain of the prairie vole. (A) A cross section of the brain with just a few of its anatomically distinct regions labeled. The ventral pallidum contains many cells with receptor proteins that bind to the hormone vasopressin. (B) A brain section that has been treated so that regions with large numbers of vasopressin receptors appear black. The ventral pallidum occurs in both the left and right halves of the brain; the left-hand portion of the ventral pallidum is outlined in black (arrow). (After Lim et al. 2004a; photograph © 2004 by Wiley-Liss, Inc., a subsidiary of John Wiley & Sons, Inc.)

animals, the chemical is involved in many different developmental processes and behaviors, ranging from water balance to mating behavior (Donaldson and Young 2008). Vasopressin is produced and released into the bloodstream by brain cells when an animal copulates. Molecules of vasopressin are carried to the ventral pallidum, a structure near the brain's base in mammals and other vertebrates that helps provide rewarding sensations upon completion of certain behaviors.

In the case of voles, it is not just the hormone vasopressin that plays a role in social behavior, but also its receptor, the vasopressin 1a (V1a) receptor. When V1a receptor proteins in the ventral pallidum are stimulated by vasopressin, they trigger activity in the receptor-rich cells (**FIGURE 3.22**). A team led by Larry Young compared the distribution of V1a receptors in the brains of two species of voles that differ in their mating (or perhaps more precisely in the case of voles, social affiliative) behavior, the monogamous prairie vole and the polygynous mountain vole (*Microtus montanus*) (**FIGURE 3.23**) (Getz and Carter 1996). Although there is geographic variation in the degree to which monogamy occurs in the prairie vole (Streatfeild et al. 2011), males and females in captivity huddle together in their cages. In contrast, individuals in the polygynous mountain vole tend to avoid each other in captivity. Young believes that these hormonal rewards encourage the male prairie vole to remain in the company of his mate, forming a long-term social bond with her. In contrast, the reward system in the brain of the male polygynous mountain vole is different, in part because V1a receptors are less numerous in the ventral pallidum in this species. As a result, when these males copulate, their brains do not provide the kind of feedback that leads

(A)

(B)

FIGURE 3.23 Prairie voles are monogamous, whereas mountain voles are polygynous. (A) Male and female prairie voles in at least some populations form long-term relationships, pairing off as couples that live together, with both parents caring for offspring. (B) In contrast, male and female mountain voles typically do not form pairs, and only females care for young. (A photograph © Yva Momatiuk & John Eastcott/Minden Pictures; B photograph by wonderful-Earth.net/Alamy Stock Photo.)

to the formation of a durable social attachment between a male and his mate. Thus, polygynous males move on after copulating with a female, instead of staying put.

In addition to explaining male monogamy in terms of brain physiology, Young and his colleagues have also looked at a possible genetic basis for the monogamous mating system using a series of elegant experiments (Pitkow et al. 2001). Young's group knew that V1a receptor protein, which is so important in the prairie vole's vasopressin-based system of social bonding, is encoded by a specific gene, the *avpr1a* gene. By examining brain slices stained for the vasopressin receptor, the researchers found that the distribution of *avpr1a* is as divergent as the social behavior, with greater expression in the monogamous species than in the polygynous one (**FIGURE 3.24A**) (Young et al. 1999, Lim et al. 2004b). Experimentally increasing vasopressin in the brain only increased affiliative behavior of the monogamous species, which expressed more of the V1a receptor (**FIGURE 3.24B**). Additionally, Young's team used a viral vector to insert additional copies of *avpr1a* directly into cells in the ventral pallidum of monogamous prairie voles. Once in place, the "extra" genes enabled these particular cells to make even more V1a receptor proteins than they would have otherwise. The genetically modified, receptor-rich males did indeed form especially strong social bonds with female companions—even if they had not mated with them—when given a choice between a familiar female and one with which they had not previously interacted (**FIGURE 3.24C**). Therefore, the research team concluded that *avpr1a* contributes in some way to the

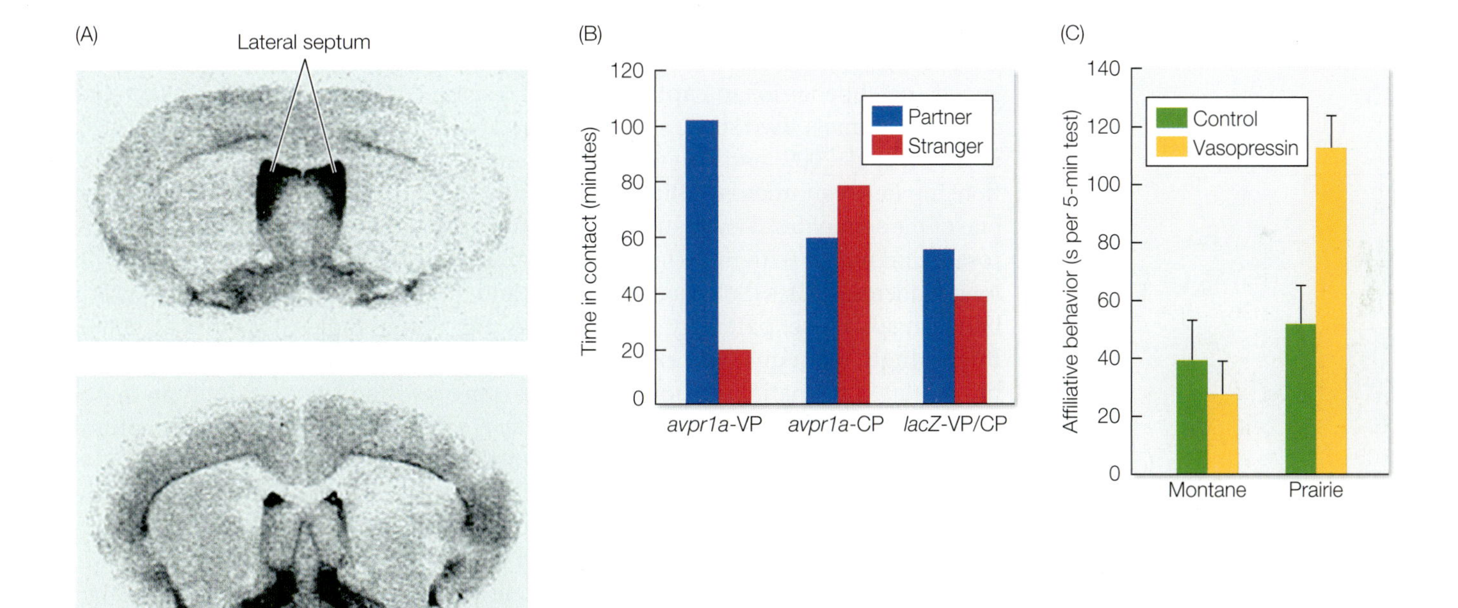

FIGURE 3.24 A gene that affects male pairing behavior in voles. (A) V1a-receptor binding patterns in the brains of the polygynous montane vole (top), and the monogamous prairie vole (bottom). Note the high intensity of binding in the lateral septum of the montane vole (but not the prairie vole), and in the diagonal band (DB) of the prairie vole (but not the montane vole). (B) Adding copies of the *avpr1a* gene affects male prairie vole pairing behavior in some brain regions but not others. Specifically, males that received added copies of the *avpr1a* gene in the ventral pallidum (VP) spent significantly more time with a familiar female versus an unfamiliar one, when given a choice over a 3-hour trial. No such effect occurred when males were given extra copies of the gene in the caudate putamen (CP) of the brain. Likewise, males given a different gene (*lacZ*) in the ventral pallidum or the caudate putamen as a control showed no special preference for a familiar female. (C) Male prairie voles, but not montane voles, exhibit elevated levels of affiliative behavior after vasopressin is administered directly into the brain. Bars depict mean +/– SE. (A, C from Young et al. 1999; B after Pitkow et al. 2001.)

development of monogamous behavior of male prairie voles (Young and Wang 2004). Indeed, inserting the monogamous prairie vole's receptor gene sequence into the mouse genome produced mice that expressed prairie vole–like receptor patterns and altered their mating behavior (Young et al. 1999).

This series of experiments provided convincing evidence that the *avpr1a* gene plays a role in shaping social behavior in these species. Yet, what was the genetic mechanism underlying the differences in receptor expression between the two species of voles? It turns out that just as in the migratory blackcap warbler that we discussed earlier, the genetic mechanism explaining differences in receptor expression in voles had to do with the length of a microsatellite repeat. The monogamous prairie vole contains a microsatellite repeat—located just upstream of the *avpr1a* gene, in its regulatory region—that the polygynous montane vole lacks (Young et al. 1999). Using the monogamous prairie voles to explore the microsatellite function further, Young's team demonstrated that differences in vasopressin receptor expression and mating/social affiliative behavior were directly related to the length of the microsatellite repeat (Hammock and Young 2005). Moreover, by selectively breeding animals for the length of the microsatellite allele, they found that males with long repeats were more inquisitive about social odors, novel juvenile individuals, and familiar females. What's more, this same mechanism may explain individual differences in pair-bonding and partner fidelity in humans. Like voles, humans have a microsatellite repeat in the regulatory region of our vasopressin receptor. Although the repeat is different from that of voles, there is an association between the human microsatellite and traits reflecting pair-bonding behavior in men, including partner bonding, perceived marital problems, and marital status (Walum et al. 2008).

Although microsatellite length in the regulatory region of the *avpr1a* gene influenced social behavior in captive voles, the same could not be said of wild and semiwild animals. Two studies in seminatural outdoor enclosures (Ophir et al. 2008, Solomon et al. 2009) and one with wild animals (Mabry et al. 2011) found no relationship between microsatellite length and social monogamy in males. However, one of the seminatural enclosure studies found that males with shorter *avpr1a* microsatellites mated with more females and sired more offspring than did males with longer microsatellites (Solomon et al. 2009), and the other study found a relationship between space use, paternity, and *avpr1a* expression in males (Ophir et al. 2008). Interestingly, subsequent work looking at vasopressin expression in regions of the brain related to spatial memory predicted space use and sexual fidelity in males (Okhovat et al. 2015). Genetic variation in *avpr1a* that appeared to be under natural selection influenced not only receptor expression but also patterns of DNA methylation in the gene. The authors suggest that DNA methylation in *avpr1a* of voles may shape heritable variation in environmental sensitivity, perhaps making individuals able to respond more rapidly to changes in their social or ecological environment.

Finally, although the vasopressin hormone (and its receptor) appears relevant to mating behavior, it also appears in voles and other species to be related to another form of social behavior, parental care. For example, experimental administration of vasopressin increases paternal behavior in the monogamous prairie vole, whereas administration of a vasopressin antagonist reduces parental care (Parker et al. 2001, Parker and Lee 2001). Andres Bendesky and colleagues examined the genetic basis of sex differences in parental care behavior in two species of deer mice that differ in mating behavior, the monogamous old-field mouse (*Peromyscus polionotus*) and the promiscuous deer mouse (*Peromyscus maniculatus*) (Bendesky et al. 2017). In the old-field mouse, both mothers and fathers provide parental care, whereas in the deer mouse only mothers provide care. Using a cross-fostering experiment in which they switched pups of one species to nests of the other, and vice versa, the researchers showed that paternal care behavior was heritable in these species. They then made hybrids of the two species by crossing individuals of each type and then backcrossing their offspring with the parental type. After screening the entire genome of these hybrids for genes that were associated with one form of parental care (nest-building behavior) and that differed between the two species, the researchers identified vasopressin as the top gene candidate for variation in nest-building behavior. They found that it is highly but differentially expressed in the brains of both mice species. To confirm the inverse relationship between vasopressin and nest building they observed, the researchers administered vasopressin to parents of the high-nest-building monogamous species and found that it indeed inhibited nest building but did not affect any other parental care behaviors. Moreover, the administration of a closely related hormone, oxytocin, had no effect on nest-building behavior. Thus, vasopressin and its receptor seem to be key modulators of mammalian social and reproductive behavior, including both mating behavior and parental care.

Perhaps the major conclusion of evo-devo is that although changes in the base sequences of genes are clearly important in the evolution of certain differences among species, another category of change may often be even more important. We speak of changes in the mechanisms regulating the expression of other genes, namely mutations or epigenetic modifications in those genes whose transcriptional products exert their effect by influencing the activity of other genes, often many other genes (Carroll 2005, Brakefield 2011). Altering the regulatory functions of the genome will affect how development proceeds because of the multiplicative effect that comes from a mutation in one gene that changes the way in which the regulated genes work. As Sean Carroll has said, the process is analogous to teaching old genes new tricks rather than having each species evolve its own particular novel major genes for any given developmental task (Carroll 2005). In many ways, animal behavior represents the frontier for these types of studies because it is the currency of selection; natural selection does not work directly on genes, proteins, hormones, or ecosystems, but instead on the behavioral phenotype itself (Rubenstein and Hofmann 2015). Behavior is a complex trait that often involves multiple genes and their interaction with the environment. But as you can see, we now have the tools to apply an eco-evo-devo approach to dissect the developmental, genetic, and even epigenetic bases of animal behaviors, not just in animals in the lab, but also in those that live and behave naturally in the wild.

Early Life Developmental Conditions

In addition to the historical level of analysis and the evo-devo approach, evolutionary biologists can examine all traits—including those related to development—in terms of their possible adaptive value, something we saw when we considered learning above and in Chapter 2. One of the features of developmental mechanisms that can be approached in this way is the ability of organisms to develop more or less normally even when endowed with genetic mutations that could be injurious. Despite the potential for genetic–developmental problems caused by the widespread distribution of novel mutations, most animals look and behave reasonably normally, a helpful outcome in terms of individual reproductive success. In fact, although gene knockout experiments (a reverse genetics approach in which one of an organism's genes is made inoperative) sometimes do have dramatic phenotypic effects, in many cases blocking the activity of a particular gene has little or no developmental effect. These findings have led some geneticists to conclude that genomes exhibit considerable information redundancy, which would explain why the loss of one protein linked to a particular gene is not fatal to the acquisition of certain traits of importance to the individual (Kerverne 1997, Picciotto 1999). What, then, are the factors that generate behavioral variation among individuals during development?

The Role of the Social Environment

We know that many animals overcome what you might think would constitute great environmental obstacles to normal development during early life. For example, some young birds lack the opportunity to interact with their parents and so cannot acquire the information that in other species is essential for normal social and sexual development via imprinting. When chicks of the Australian brushturkey (*Alectura lathami*) hatch from eggs placed deep within the immense compost heap of a typical nest, they dig their way out and walk away, often without ever seeing a parent or sibling (**FIGURE 3.25**). So how do they manage to recognize other members of their species? Ann Göth and Christopher Evans studied captive young brush turkeys in an aviary in which they were exposed to feathered robots that looked like other young. All that was required to elicit an approach from a naïve chick was a peck or two at the ground by the robot. Thus, young brush turkeys do not require extensive social experience to develop their basic social behaviors (Göth and Evans 2004). As adults, the birds are completely capable of normal sexual behavior despite having lived primarily by themselves beforehand.

Other researchers have created genuinely weird rearing environments, only to find that various forms of deprivation have little or no effect on the development of

FIGURE 3.25 Australian brushturkey mound. Male brushturkeys build mounds of earth, leaves, and other material that can reach 13 feet high. The mounds are visited by a succession of females who lay their eggs deep within the structure. The mound helps regulate the temperature for the developing eggs. Males monitor the temperature by using their beaks to probe the interior and then adjust as necessary by adding or removing nesting material. Maintaining the correct temperature is important for brushturkey development, as an equal number of male and female young are produced at an incubation temperatures of 34°C. However, more males are produced at cooler temperatures and more females at warmer temperatures. (Photograph © Jurgen Freund/Minden Pictures.)

normal behavior. Rear baby Belding's ground squirrels without their mothers, and they still stop what they are doing to look around when they hear a tape of the alarm call of their species (Mateo and Holmes 1997). Male crickets that live in complete isolation sing a normal species-specific song despite their severely restricted social and acoustic environment (Bentley and Hoy 1974). Captive hand-reared female cowbirds that have never heard a male cowbird sing nevertheless adopt the appropriate precopulatory pose when they hear cowbird song for the first time, if they have mature eggs to be fertilized (King and West 1983). One challenge for these and other species in atypical rearing environments is how to learn to recognize their species-specific traits. How, for example, do cowbirds learn to identify vocalizations and individuals of their species? Work by Mark Hauber and colleagues uncovered the species-specific signal, or "password," that triggers how brown-headed cowbirds learn about their own phenotype (Hauber et al. 2001). By presenting 6-day-old nestlings with a variety of avian sounds, the authors demonstrated that baby cowbirds respond to "chatters"—calls made by adult cowbirds of both sexes during social interactions during the breeding season—more strongly than to any other type of vocalization. The researchers proposed that passwords such as this one in cowbirds may be involved in the ontogeny of species recognition in brood parasites more generally (**password hypothesis**). But it is not just the brood parasites that have passwords. The host species may also have their own unique passwords, used to distinguish between parasitic chicks and their own offspring. Superb fairy-wrens (*Malurus cyaneus*) in Australia are parasitized by Horsfield's bronze-cuckoos (*Chrysococcyx basalis*). Mother fairy-wrens sing a special "incubation call" to their eggs that acts like a familial password (Colombelli-Negrel et al. 2012). Even though the chicks begin hearing the call while they are still embryos, they incorporate the notes into their begging calls after hatching. Bronze-cuckoos lay their eggs too late in the nesting cycle for their chicks to hear or learn the calls, so only the fairy-wren chicks learn the password.

password hypothesis Passwords—species-specific vocal signals—may be involved in the ontogeny of species recognition in brood parasites.

Developmental Homeostasis versus Developmental Constraint

The ability of many animals to develop more or less normally, despite defective genes or deficient rearing environments, has been attributed to a process called **developmental homeostasis**. This property of developmental systems reduces the variation around a mean value for a phenotype and reflects the ability of developmental processes to produce an adaptive phenotype quite reliably. A clear demonstration of this ability comes from a classic experiment on the development of social behavior in young rhesus monkeys (*Macaca mulatta*) deprived of contact with others of their species, conducted by Margaret and Harry Harlow more than 4 decades ago when animal rights were not the issue they are today (Harlow and Harlow 1962, Harlow et al. 1971). In one study, the Harlows separated a young rhesus from its mother shortly after birth and placed it in a cage with an artificial surrogate mother—a wire cylinder or a terry-cloth figure with a nursing bottle (**FIGURE 3.26A**). The baby rhesus gained weight normally and developed physically in the same way that non-isolated rhesus infants do. However, it soon began to spend its days crouched in a corner, rocking back and forth, biting itself. If confronted with a strange object or another monkey, the isolated baby withdrew in apparent terror.

The isolation experiment demonstrated that a young rhesus needs social experience to develop normal social behavior. But what kind of social experience—and how much—is necessary for normal social development? Interactions with a mother are insufficient since infants reared only with their mothers fail to develop truly normal sexual, play, and aggressive behavior. Perhaps standard social development in rhesus monkeys requires the young animals to interact with one another? To test

FIGURE 3.26 Young rhesus monkeys have been reared in isolation to study social deprivation during development. (A) Surrogate mothers used in social deprivation experiments. This isolated rhesus infant was reared with wire cylinder and terry-cloth dummies as substitutes for its mother. (B) Socially isolated rhesus infants that are permitted to interact with one another for short periods each day at first cling to each other during the contact period. (Photographs © Nina Leen/Time Life Pictures/Getty Images.)

social experience hypothesis Young animals need social experience to develop normal social behavior.

this **social experience hypothesis**, the Harlows isolated some infants from their mothers, but gave these infants a chance to be with three other such infants for just 15 minutes each day (Harlow et al. 1971). At first, the young rhesus monkeys simply clung to one another (**FIGURE 3.26B**), but later they began to play. In their natural habitat, rhesus babies start to play when they are about 1 month old, and by 6 months they spend practically every waking moment in the company of their peers. Even so, the 15-minute-play group developed nearly normal social behavior. As adolescents and adults, they were capable of interacting sexually and socially with other rhesus monkeys without exhibiting the intense aggression or withdrawal of individuals that had been completely isolated from other young.

Naturally, one wonders about the relevance of these studies for another primate species, *Homo sapiens*, whose intellectual development is often said to be dependent on the early experiences that children have with their parents and peers. But is this true? We cannot, of course, do social isolation experiments with human babies, but we can examine evidence of another sort regarding the resilience of intellectual development in the face of nutritional deprivation. Consider, for example, the results of a study of young Dutch men who were born or conceived during the Nazi transport embargo during the winter of 1944–1945, which caused many deaths from starvation in the larger Dutch cities (Susser and Stein 1994). For most of the winter, the average caloric intake of city people was only about 750 calories per day (today, the recommended daily caloric intake to maintain weight is about 2500 calories for men and 2000 for women). As a result, urban women living under famine conditions produced babies of very low birth weights. In contrast, rural women were less dependent on food transported to them, and the babies they had that were conceived at the same time were born at more or less normal birth weights.

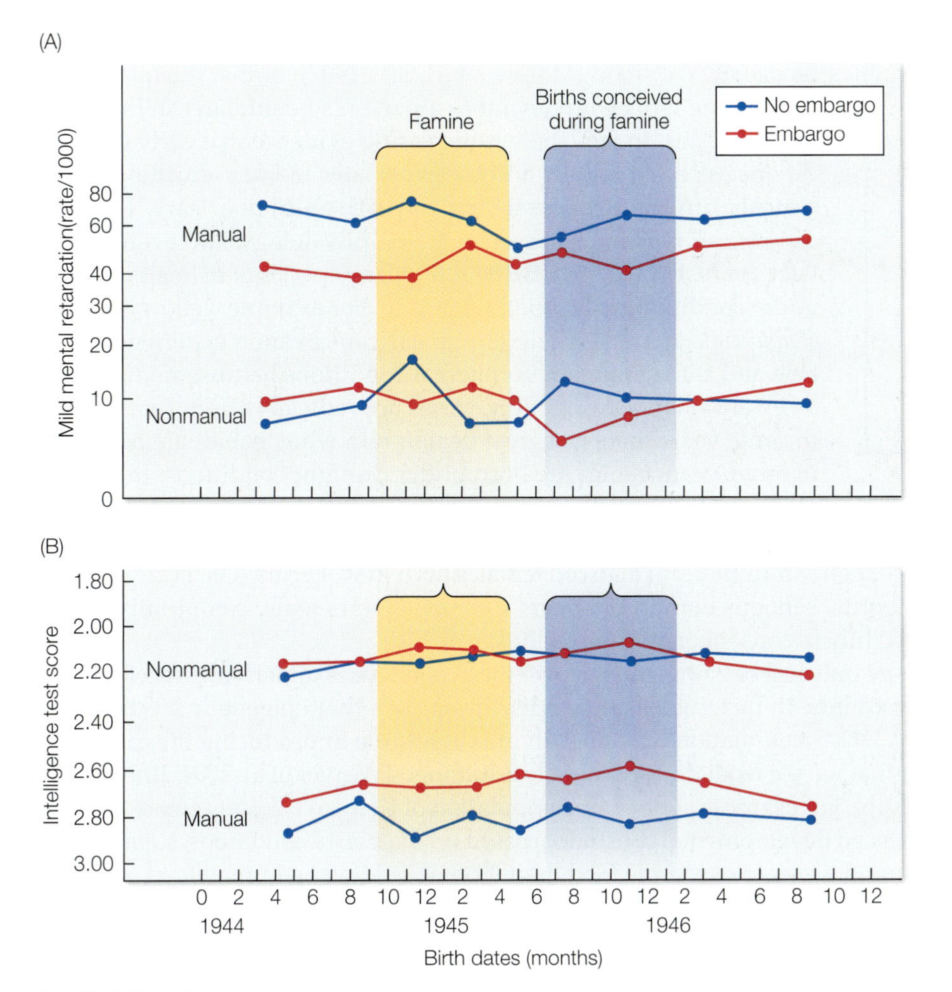

FIGURE 3.27 Developmental homeostasis in humans. Maternal starvation has surprisingly few effects on intellectual development in humans, judging from (A) rates of mild mental retardation and (B) intelligence test scores among 19-year-old Dutch men whose mothers lived under Nazi occupation while pregnant. (In this case, the lower the intelligence test score, the higher the intelligence of the individual.) The men were grouped according to the occupations of their fathers (manual or nonmanual) and whether their mothers lived or gave birth to them in a city subjected to food embargo by the Nazis (embargo) or in a rural area unaffected by the embargo (no embargo). Those men who were conceived or born under famine conditions exhibited the same rates of mental retardation and the same levels of intelligence test scores as men conceived by or born to unstarved rural women. (After Stein et al. 1972.)

One would think that full brain development depends on adequate nutrition during pregnancy, when much of brain growth occurs. However, Dutch men who were born in urban famine areas did not exhibit a higher incidence of mental retardation at age 19 than did rural boys whose early nutrition was far superior (**FIGURE 3.27A**). Nor did those boys born to food-deprived mothers score more poorly than their relatively well nourished rural counterparts when they took the Dutch intelligence test administered to draft-age men (**FIGURE 3.27B**) (Susser and Stein 1994). These results are buttressed by the discovery that Finnish adults who experienced severe nutritional shortfalls in utero (during a nineteenth-century famine) lived just as long on average as those who were born after the famine was over (Kannisto et al. 1997).

HYPOTHESES 3.1

Summary of alternative proximate hypotheses for how early life conditions influence fitness later in life

developmental constraint hypothesis Individuals born in low quality environments experience reduced fitness later in life.

predictive adaptive response hypothesis Individuals adjust their phenotype during development to match conditions later in life.

No one believes that pregnant women or young children should be deprived of food (Morley and Lucas 1997), or that the nutritional state of the fetus is irrelevant to a person's health later in life (Rasmussen 2001). In fact, there is increasing evidence that early life conditions *can* affect fitness and behavior later in life. According to the **developmental constraint hypothesis**, poor early life conditions can negatively affect behavior, physiology, and even fitness later in life (**HYPOTHESES 3.1**). Many species of animals are born under conditions when food is limiting. For example, yellow baboons (*Papio cynocephalus*) living on the African savanna experience variable and fluctuating environmental conditions because rainfall, and by extension food availability, varies unpredictably from year to year. In some years there is a great deal of rain when babies are born, but in other years babies are born under drought conditions. In a more than 4-decade-long study of yellow baboons in Kenya, Amanda Lea and colleagues found that females born in dry years showed greater decreases in fertility as adults in subsequent drought years than did females born in wet years (Lea et al. 2015). In other words, baboons born in dry years were developmentally compromised and suffered fitness consequences later in life.

We are only just beginning to uncover the genetic bases underlying developmental constraints. In fact, increasing evidence suggests that epigenetic mechanisms such as DNA methylation may play an important role in producing the carryover effects that we see from development to adulthood (Weaver et al. 2004, Rubenstein et al. 2016). For example, in the same population of Kenyan baboons where females experienced developmental constraints based on their birth conditions, some troops have found a way to overcome food limitation by feeding on human food scraps at tourist lodges. By comparing whole genome DNA methylation in individuals from troops that fed only on natural food with those that supplemented their diets with human refuse, Lea and colleagues showed that resource availability shapes patterns of DNA methylation across the baboon genome (Lea et al. 2016). Although many regions of the genome showed differences in DNA methylation between the natural and refuse-feeding groups of baboons, the region that showed the greatest differences contained the phosphofructokinase gene (*PFKP*), which has been implicated in obesity-related traits in several primate species. Experimental manipulation of DNA methylation at this gene in bacterial cells in the lab showed that it did indeed directly influence gene expression at this locus.

But are all individuals born in low-quality environments destined for a life filled with problems? According to the **predictive adaptive response hypothesis**, organisms are able to adapt their physiology and behavior to overcome any potential developmental constraints that occur early in life (see Hypotheses 3.1). Predictive adaptive responses can take place over the lifetime of an animal or across generations via **maternal effects,** whereby the phenotype of an organism is determined not just by the environment it experiences and its genotype, but also by the environment and genotype of its mother. Moreover, predictive adaptive responses are most likely to occur in species in which early and later life environments are similar. In African savannas, conditions during development are rarely predictive of conditions in adulthood, but in temperate environments they often are. A good example of this environmental matching occurs in Arctic populations of snowshoe hares (*Lepus americanus*), which cycle in a predictable manner because of changes in predation pressure, primarily from Canada lynx (*Lynx canadensis*) (Krebs et al. 1995). The risk of predation by lynx influences maternal body condition and levels of glucocorticoid stress hormones in the hares (Boonstra et al. 1998), which in turn can affect offspring survival. For example, in an experiment in which some wild-caught pregnant hares were exposed to a dog to simulate a predator, mothers exposed to predators had high levels of glucocorticoids (a class of hormone that is often released after a stressful event; see Chapter 5) and produced lighter young than control mothers (Sheriff et al. 2009). When the predator-rich environment that mothers experience is the same as their offspring experience when they are born, producing young with a heightened stress response may be adaptive and enable young to survive better (Sheriff and Love 2012, Dantzer et al. 2013).

Thus, depending on the species and environment in which it lives, early life conditions can have positive, negative, or no effects on behavioral development and fitness. At the genetic level, there is increasing evidence from a variety of animals that epigenetic modifications may play a role in regulating these adaptive or maladaptive responses. And the same mechanisms may occur in humans, even in cases where the fitness effects may not be apparent. For example, the Dutch men whose mothers had been exposed to famine during the Nazi embargo showed genome-wide patterns of DNA methylation that differed from that of their siblings born later during normal conditions (Tobi et al. 2014). In particular, some of the most differentially methylated genes were those related to growth, energy metabolism, and lipid metabolism (Tobi et al. 2018), which has also been shown in rats that are experimentally food deprived (Boque et al. 2013). Since studying the fitness effects of food limitation and other stressors is arguably easier to do in animals other than humans because of their shorter lifespans and generation times, as well as the ability to do manipulative studies and to study DNA methylation directly in the brain, behavioral studies such as these will continue to be important for predicting the effects of malnourishment on survival and reproduction.

Developmental Switch Mechanisms

An adaptationist approach to development has led to an appreciation of how individuals in diverse environments acquire traits that are essential for individual reproductive success. But there are many species in which two or three quite distinct alternative phenotypes coexist comfortably, a topic we will explore in greater detail in Chapter 9 as it relates to reproductive behavior. These differences arise proximately as a result of environmental differences among the individuals in question through a process called phenotypic or **developmental plasticity** (**FIGURE 3.28**) (West-Eberhard 2003, Simpson et al. 2011). A challenge associated with such **polyphenisms** (a special type of phenotypic plasticity where discrete phenotypes arise from a single genotype) is to identify the proximate environmental cues that activate the mechanisms underlying the development of either one or another distinct phenotype, not an intermediate form of some sort. We have spoken at length about one such example of a behavioral polyphenism already in this chapter: the distinct worker and queen castes in honey bees, which arise from the same genotype as a result of differences in larvae provisioning during development.

Another example of a behavioral polyphenism in insects is the solitary and gregarious forms of the migratory desert locust (*Schistocerca gregaria*), a major consumer of crops in Africa and other parts of the world. Locusts develop into the swarming gregarious form when the nymphs see and smell other locusts and when grasshopper densities are high enough so that the insects' hindlegs are touched repeatedly (Simpson et al. 2001). Under these conditions, mechanoreceptors on the hindlegs relay their tactile messages to the central nervous system, where the neural messages trigger genetic changes in target cells. These changes translate into the production of new brain chemicals, notably serotonin, which provides the apparent critical first step toward the developmental changes associated with the shift from the green solitary form to the black-and-yellow migratory form (Anstey et al. 2009).

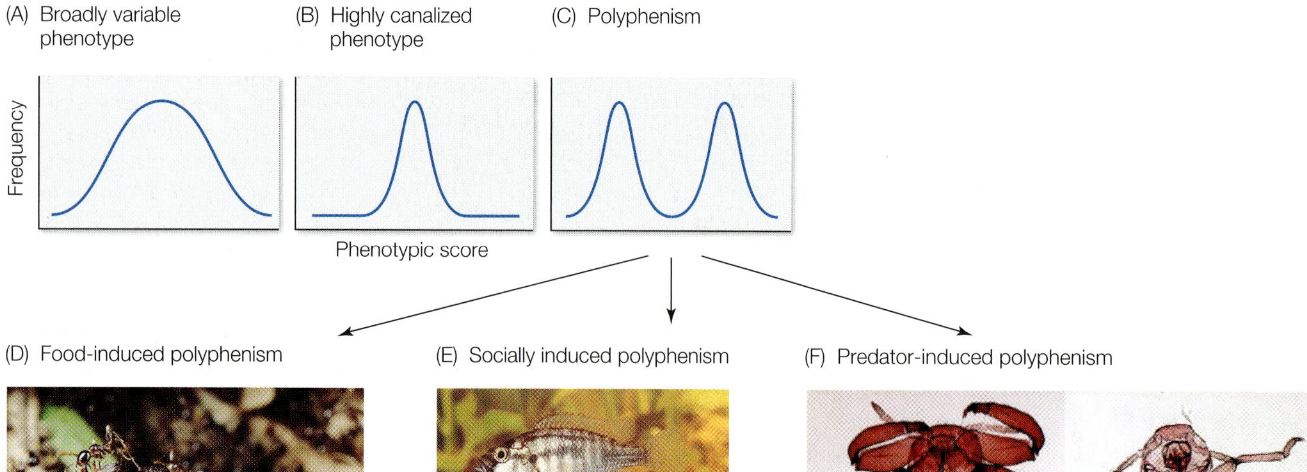

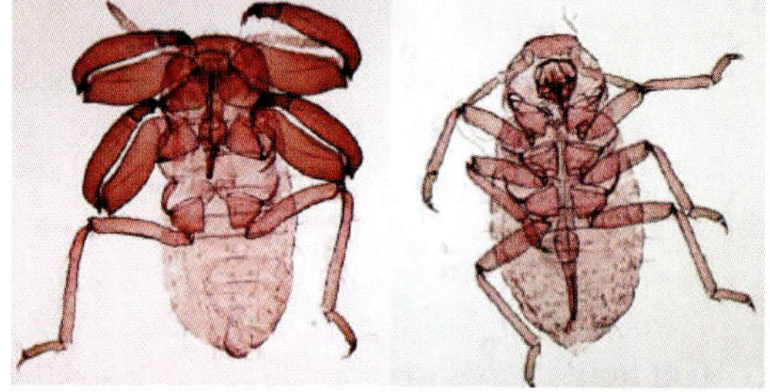

FIGURE 3.28 Developmental switch mechanisms can produce polyphenisms within the same species. Different phenotypes can arise when developmental switch mechanisms are activated in response to critical environmental cues. Top panel: Phenotypic variation within a species can range from (A) continuous, broad variation about a single mean value to (B) continuous but narrow variation about a single mean value to (C) discontinuous variation that generates several distinct peaks, each representing a different phenotype. Bottom panel: (D) In some cases, the amount or nature of the food eaten contributes to the production of certain polyphenisms, as in the castes of ants and other social insects. (E) In other cases, social interactions play a key role in switching phenotypes, as in the territorial and nonterritorial forms of the cichlid *Astatotilapia burtoni*. (F) In still other instances, the presence or activity of predators contributes to the development of an antipredator phenotype, as in the soldier caste (left) of some aphids, which possess more powerful grasping legs and a larger, stabbing proboscis than nonsoldier forms (right). (D, photograph by Mark Moffett; E, photograph by Russell Fernald; F, photographs courtesy of Takema Fukatsu.)

At the ultimate level of analysis, we can ask what advantage individual locusts gain from having the developmental flexibility to be migratory or sedentary. The migratory form is the one that does so much damage when the locusts gather together in areas where they eat everything in sight before flying away to new locations where there may still be plants to consume. You can imagine how well the sedentary and semi-migratory forms would do in competition with a truly migratory form when there are millions of grasshoppers competing for the palatable plants that they must eat to survive and reproduce.

Polyphenisms are common in vertebrates as well. For example, there are two very different tiger salamander (*Ambystoma tigrinum*) phenotypes: (1) a typical aquatic immature form, which eats small pond invertebrates such as dragonfly nymphs; and (2) a cannibal type, which grows much larger, has more powerful teeth, and feeds on other tiger salamander larvae unfortunate enough to live in its pond (**FIGURE 3.29**). The development of the cannibal type, with its distinctive form and behavior, depends proximately on the salamanders' social environment. For example, cannibals develop only when many salamander larvae live together (Collins and Cheek 1983). Moreover, they appear more often when the larvae in a pond (or aquarium) differ greatly in size, with the largest individual much more likely to become a cannibal than its smaller companions (Maret and Collins 1994). In addition, the cannibal form is more likely to develop when the population consists

FIGURE 3.29 Tiger salamanders occur in two forms: cannibal and non-cannibal. The typical non-cannibal form (being eaten here) feeds on small invertebrates and grows more slowly than the cannibal form (which is doing the eating). Cannibals have broader heads and larger teeth than their insect-eating victims. (Photograph by Tim Maret, courtesy of James Collins.)

largely of unrelated individuals than when many siblings live together (Pfennig and Collins 1993). If a larger-than-average salamander larva occupies a pond with many other young salamanders that do not smell like its close relatives, its development may well be switched from the typical track to the one that produces a giant, fierce-toothed cannibal (Pfennig et al. 1994).

What selective advantages do tiger salamanders derive from having two possible developmental pathways and a switch mechanism that enables them pursue one form or the other? Individuals with some developmental flexibility may be able to exploit a particular resource niche better than individuals stuck with a one-size-fits-all phenotype. Larval salamanders can access two different sources of potential nutrients, insect prey and their fellow salamanders. If numerous salamander larvae occupy a pond, and if most are smaller than the individual that becomes a cannibal, then shifting to the cannibal phenotype gives that individual an abundant food source that is not being exploited by its neighbors, so it can grow quickly. But a relatively small individual that was locked into becoming a large-jawed cannibal form would surely starve to death in a pond that lacked numerous potential victims of appropriate size. Because salamanders have no way of knowing in advance which of two different food sources will be more available in the place where they happen to be developing, selection appears to have favored individuals with the ability to develop in either one of two ways depending on the properties of their environments.

A similar phenomenon of developmental switch points can be observed in dung beetles. Males in many of the more than 5000 species of dung beetles produce horns, which are rigid outgrowths of the exoskeleton that are used in male–male competition for resources and access to females (**FIGURE 3.30A**). In most species, the resources in question are balls of dung that females use to lay their eggs. In the genus *Onthophagus*, females dig underground tunnels within which the dung balls are buried. Males then fight for and defend these tunnels to gain exclusive access to females (**FIGURE 3.30B**) (Emlen 2000). But not all males have horns and are able to guard tunnels. Small males that lack horns try to sneak by the large, heavily armored guards to reach the females buried deep inside the tunnels. What determines these different mating approaches of large and small males? It turns out that the nutritional conditions that developing larvae experience influence final adult body size and horn length. Male larvae that develop with lots of food grow large and produce horns, whereas those with little food grow to be small and lack horns.

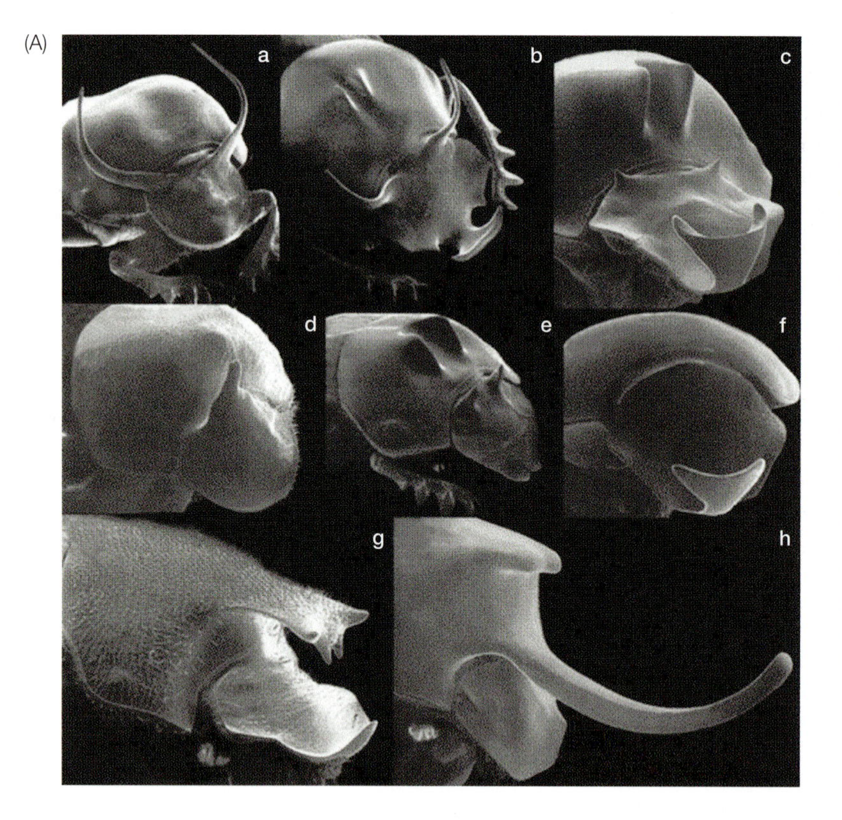

FIGURE 3.30 Beetle horns come in all shapes and sizes. (A) Dung beetles in genus *Onthophagus* produce horns that vary in shape, size, and location on the head or thorax. Depending on the species, horns can extend from the front of the head, the back of the head, or the thorax. All photographs are of male beetles unless otherwise indicated. (a) *Onthophagus taurus*. (b) *Onthophagus clypeatus*. (c) *Onthophagus praecellens*. (d) *Onthophagus nuchicornis*. (e) *Onthophagus clypeatus* (female). (f) *Onthophagus sharpi*. (g) *Onthophagus hecate*. (h) *Onthophagus nigriventris*. (B) In many *Onthophagus* species, males with large horns fight for and defend female-dug tunnels that contain balls of dung that are used by females during egg-laying. Small males that lack horns dig side tunnels in an attempt to sneak by the horned guards and mate with females inside the tunnels. (From Emlen 2000.)

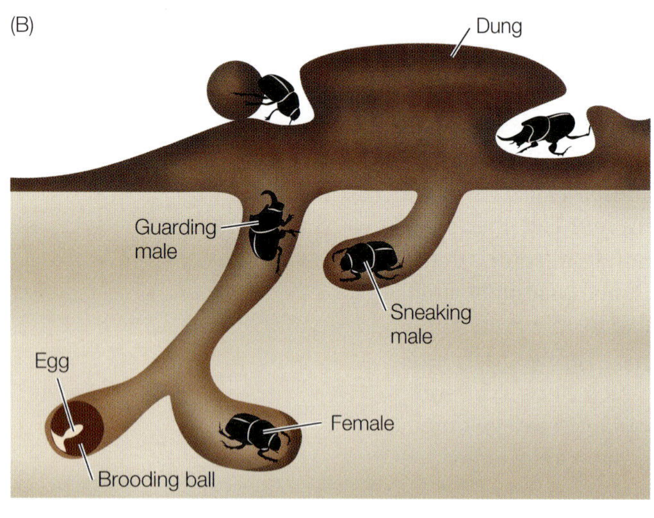

What is the genetic basis of horn development and different behavioral phenotypes in dung beetles? A network of genes related to insulin signaling links larval nutrition and horn growth in beetles (Emlen et al. 2006). Females and small males have reduced expression of the insulin receptor gene (*InR*), and knocking out this gene in males leads to reduced horn growth (Emlen et al. 2012). This same developmental and signaling pathway is shared with other insects, including fruit flies. But there may be an even more conserved developmental mechanism underlying beetle horns. Much like homeobox genes, which also play a role in beetle horn development, the gene hedgehog (*Hh*) is part of an ancient signaling pathway. In dung beetles, this pathway is directly linked to the nutrition-related "switch point" for horn growth (the development time point at which beetles either begin to grow horns or not) in developing larvae (Kijimoto and Moczek 2016).

More generally, when a species faces discrete ecological problems that require different developmental solutions, the stage may be set for the evolution by natural selection of sophisticated developmental switch mechanisms that enable individuals to develop the phenotype best suited for their particular circumstances. The existence of two nonoverlapping categories of food (large amounts versus small amounts) or two levels of risk (predators present versus predators absent) may select for the kind of developmental mechanism that can produce a limited number of very different specialist phenotypes, rather than a mechanism that generates a full range of intermediate forms (Pfennig et al. 2007). Thus, when developmental conditions are likely to vary for young, developmental plasticity appears to be an important mechanism for animals to adapt to these variable environments.

Not all polyphenisms are restricted to early life when organisms are going through development. Some animals can switch behavioral roles throughout their lifetime. A good example of this occurs in the cichlid fish *Astatotilapia burtoni*, in which males are either competitively superior or socially inferior to other males. This variation in social status among cichlid males helps explain why they have the capacity to shift between two different phenotypes (see Figure 3.28E). In this fish, males compete for a mate-attracting territory, with winners holding and defending sites until ousted by stronger intruders. In such an either–or social environment, it pays to be either aggressively territorial (and to signal that state with bright colors) or nonaggressive (and to signal that state with dull colors) (Fernald 1993, Francis et al. 1993). Fish that behave in some intermediate fashion will almost certainly fail to hold a territory against motivated rivals, but they will also fail to conserve their energy, which they can do only by temporarily dropping out from the competition to secure a territory. To this end, the fish respond to changes in their social status with changes in gene activity in specific brain cells (White et al. 2002). Indeed, when a subordinate male is experimentally given a chance to become dominant (through the removal of a rival), gonadotropin-releasing hormone (GnRH) nerve cells in the anterior parvocellular preoptic nucleus quickly begin to ramp up the activity of the early growth response protein 1 gene (*egr-1*) that codes for GnRH, which in turn regulates reproduction through the activation of other genes and the production of additional hormones. In socially ascendant males, *egr-1* is expressed twice as much in the target cells as it is in males that have been stable subordinates or longtime dominant territorial individuals (**FIGURE 3.31**). By a week later, the males have been transformed not only in terms of their appearance and behavior but also in the size of their GnRH neurons and the size of their testes. The initial rapid rise in *egr-1* expression in the brains of previously subordinate males in response to the chance to become dominant appears to act as a trigger for an entire cascade of associated genetic, hormonal, and developmental changes. These changes enable an ascendant subordinate to take advantage of his good fortune and become reproductively active while suppressing reproduction in the other males in his neighborhood (Burmeister et al. 2005).

Although polyphenisms are common, they are far from universal, perhaps because many environmental features vary continuously rather than discontinuously. Under

FIGURE 3.31 Subordinate males of the fish *Astatotilapia burtoni* react very quickly to the absence of a dominant rival. (A) Within minutes of the removal of the dominant male, a subordinate may begin to behave more aggressively than before. (B) This change in behavior is correlated with a surge in the activity of the *egr-1* gene in the preoptic region of the fish's brain. This gene may initiate a sequence of other genetic changes that provide the male with the physiological foundation for dominance behaviors. Note that *egr-1* activity ramps up during the transition from subordinate to dominant status but then falls back once the male has become truly dominant. Bars depict mean +/– SE. (After Burmeister et al. 2005; photograph courtesy of Russell Fernald.)

(A)

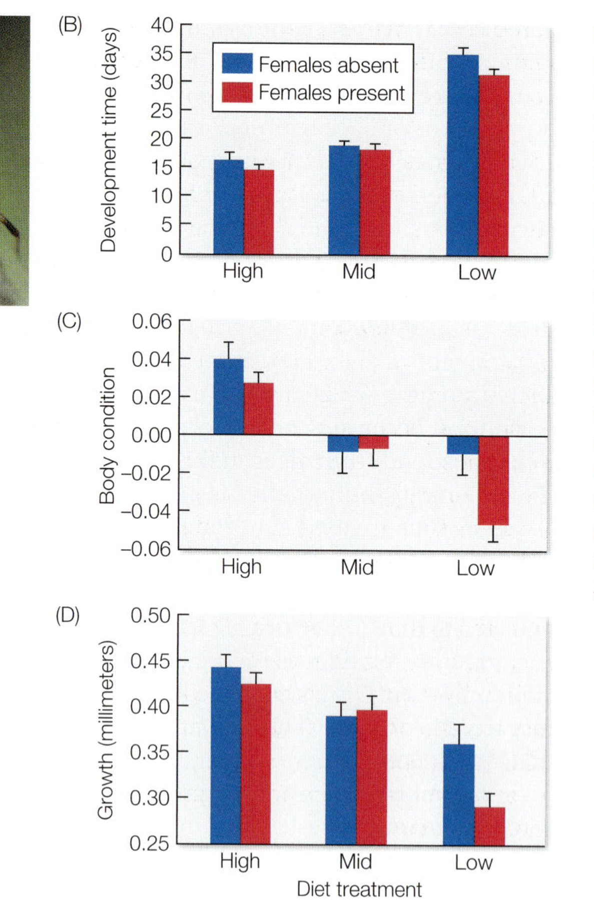

FIGURE 3.32 Developmental flexibility in redback spiders. (A) An adult male redback. (B–D) Immature males that grow to adulthood in the presence of females (red bars) develop more rapidly and so reach a smaller adult size in poorer physical body condition than do immatures that develop in the absence of females (blue bars). Adult males that develop in places without females are likely to encounter male competitors when they reach a web with a female somewhere else. (After Kasumovic and Andrade 2006; photograph by Ken Jones, © M. C. B. Andrade.)

these conditions, individuals may not benefit from developmental systems that produce a particular phenotype targeted at a narrow part of the entire range of environmental variation. Instead, selection may favor the ability to shift the phenotype by degrees in such a way as to generate a broad distribution of phenotypes, each one representing an adaptive response to one or more variable factors. So, for example, the final body size of a male redback spider (*Latrodectus hasselti*) varies considerably in response to variation in available food and the cues associated with virgin female redbacks (potential mates) and male redbacks (rivals for females) (Kasumovic and Andrade 2006). When males are reared in the presence of virgin females, they develop more quickly and achieve adulthood at a smaller size compared with males reared on the same diet but in the absence of the odor cues associated with virgin females (**FIGURE 3.32**). The developmental effects of growing up in the presence of males are exactly the opposite, a fine example of how evolutionary pressures can favor developmental plasticity that produces a wide range of phenotypes within one species.

Supergenes and Behavioral Polymorphisms

Not all animals are capable of plastically regulating behavior phenotypes, either during development or later in life. In some cases, species lose the ability to plastically regulate behavioral roles according to environmental conditions. Instead, behavioral phenotypes become relatively more influenced by genetic factors than environmental ones, a phenomenon referred to as a **behavioral polymorphism**. Although we are just beginning to scratch the surface on the genetic architecture of behavioral polymorphisms, a shared molecular mechanism keeps surfacing among many different animal taxa: a chromosomal inversion that leads to the formation of a **supergene**, a region of DNA containing many linked genes that influence a behavioral phenotype.

Perhaps the best-known example of behaviorally relevant supergenes comes from ants, where all of the more than 12,000 species live in eusocial colonies with a reproductive division of labor and a worker caste (see Chapter 12). Yet the structure of ant societies can vary greatly, as the number of queens in a colony is quite variable. Some ant species have populations that are **monogynous** and live in colonies with only a single queen, whereas other populations are **polygynous** and live in societies with multiple queens. Variation in social organization in the red imported fire ant (*Solenopsis invicta*) appears to be controlled by a supergene containing more than 500 linked genes (Wang et al. 2013). This supergene arose because of a chromosomal inversion, something that occurs when part of a chromosome breaks off, rotates 180 degrees, and reinserts in the opposite orientation. Evolutionary genetic analyses suggest that nearly all of the genetic differences between queens from monogynous and polygynous red imported fire ant colonies occur on this supergene. Similarly, differences between monogynous and polygynous colonies of the alpine silver ant (*Formica selysi*) are also controlled by another supergene that seems to differ in location and to have evolved independently from that in the red imported fire ant (Purcell et al. 2014). Once a portion of the red imported fire ant or alpine silver ant chromosome became inverted, it could no longer recombine during the process of meiosis. Since recombination (crossing over of parts of the chromosome) allows genes to reassemble in different ways, the inability of these ant chromosomes to recombine means that the genes underlying specific social traits in ants are permanently linked.

These two examples of genetic polymorphism in social behavior from ants do not, however, mean that the environment plays an insignificant role in defining these group-level phenotypes of monogyny versus polygyny. In fact, the social makeup of the colony strongly influences which form of social organization the colony will exhibit; the proportion of polygyny-specific alleles that the workers express influences whether a colony will exhibit polygyny or not. In red imported fire ants, the gene responsible for social organization is called *Gp-9*. It has two alleles, the *B* allele, which is the only allele found in monogynous colonies, and the *b* allele, which occurs with the *B* allele in polygynous colonies. Two *b* alleles never occur together because they are lethal in this arrangement (because of the chromosomal inversion that we described above) (Ross and Keller 1998). By manipulating the genotypes of the workers in a colony, Ross and Keller found that colonies with less than 5 percent of workers carrying the *b* allele behaved like monogynous colonies, whereas colonies with more than 10 percent of such workers behaved like polygynous colonies (Ross and Keller 2002). In other words, there is a genotype frequency threshold for a transition in colony social organization. So even though social organization in red imported fire ants is determined at a single genetic locus that results from a chromosomal inversion that hosts more than 500 socially related genes, the social environment of the colony actually influences whether that trait is expressed.

Supergenes also appear to explain behavioral differences in several bird species, including the ruff (*Philomachus pugnax*), a midsize sandpiper in which males exhibit one of three reproductive types. Up to 95 percent of "independent" males are territorial and have elaborately ornamented black and chestnut ruffs (curled chest feathers that give this species its name), which they use to court females in elaborate displays. Approximately 5 to 20 percent of males are "satellites" that have white ruffs and do not guard territories; instead, they enter the territories of independent males and attempt to mate with visiting females. Finally, less than 1 percent of males are "faeders," or female mimics, which despite dull plumage have much bigger testes than their ornamented counterparts. The three ruff types are controlled by two different non-recombining supergenes that evolved independently, one in satellites and another in faeders (**FIGURE 3.33**) (Küpper et al. 2015, Lamichhaney et al. 2015). Much as in the ants, these supergenes were the result of two independent chromosomal inversion events that resulted in chromosomes that can no longer recombine. A variety of genes relating to reproductive hormones and feather color have accumulated in these supergenes and underlie many of the phenotypic differences in the three types of ruffs.

Supergenes are beginning to show up more frequently in evolutionary genomic studies of behavioral polymorphisms. Indeed, improved sequencing technologies are allowing behavioral biologists to study the genomic bases of different behavioral morphs in several species. As we saw in red imported fire ants, however, even these large accumulations of behaviorally relevant loci illustrate how genes *and* the environment—the social environment in the case of the ants—influence behavior. After all, the genome of each species not only encodes a blueprint of its behavior, but also delimits how the environment—either social or ecological—can shape that behavior.

In the next two chapters we will explore the process of behavioral development further by discussing how the nervous system and physiological mediators such as hormones translate environmental inputs into behavior throughout ontogeny.

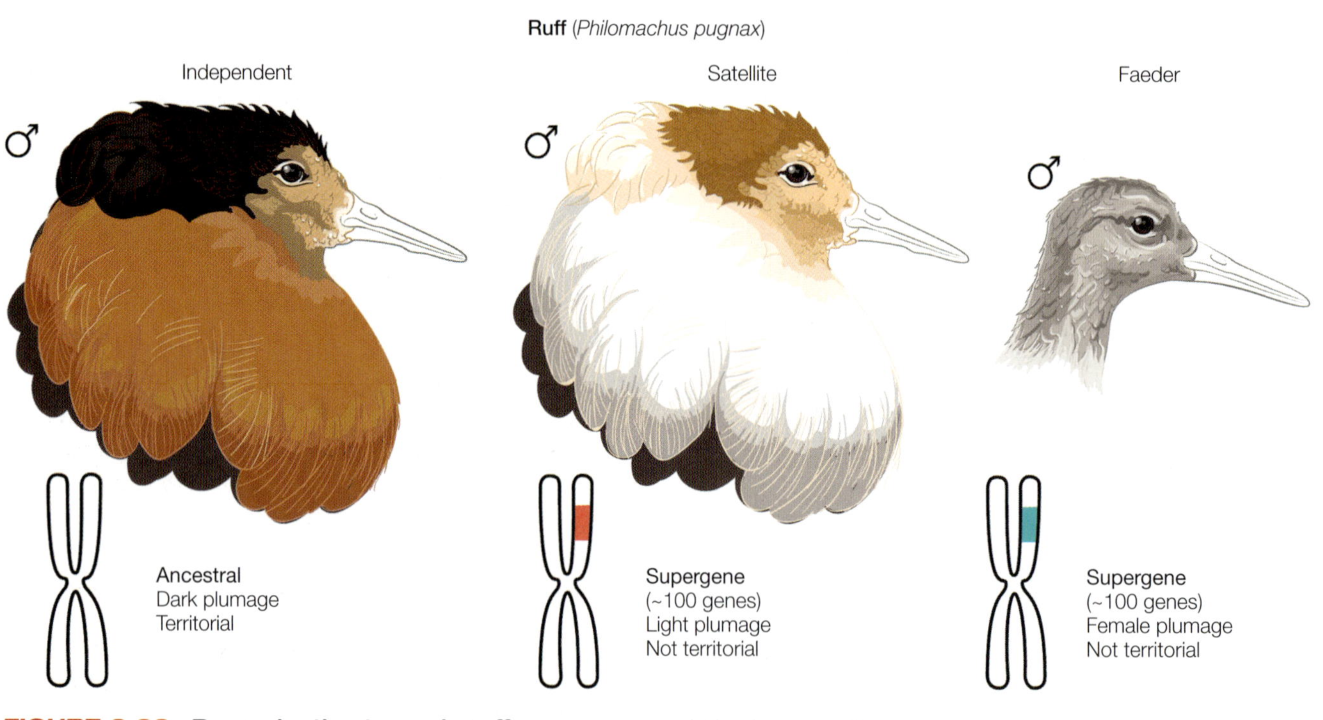

FIGURE 3.33 Reproductive types in ruffs are influenced by supergenes. Male ruffs occur as one of three types, each with a distinct reproductive strategy and appearance. Independent males defend territories and have dark, ornamented plumage. Satellite males are nonterritorial and have light, ornamented plumage. Faeder males are also nonterritorial but resemble females with drab plumage. The presence of inversion-generated supergenes determines morphological and behavioral traits in satellite and faeder ruffs. Each of their two independently-evolved supergenes contains about 100 genes, including those related to reproductive physiology and plumage coloration. (From Taylor and Campagna 2016.)

SUMMARY

1. The development of any trait, including a behavioral trait, is the result of an interaction between the genotype of a developing organism and its environment, which consists of not only the food it receives and the metabolic products produced by its cells but also its sensory experiences. Genes can respond to signals from both the social and ecological environments by altering their activity, leading to changes in the gene products available to the developing organism.

2. Despite the nature–nurture misconception (the idea that traits are influenced by genes *or* the environment), no measurable product of development (a phenotype) can be genetically determined. The statement "in garter snakes there is a gene for eating banana slugs" really means "a particular allele in a garter snake's genotype codes for a distinctive protein; if the protein is actually made, which requires an interaction between the gene and its environment, the protein may influence the development or operation of specific physiological mechanisms underlying the snake's ability to recognize slugs as food."

3. By the same token, the interactive nature of development means that no phenotype can be purely environmentally determined. The statement "a sparrow's dialect is environmentally determined" really means "acoustic experiences early in the sparrow's life led to chemical changes in the bird's brain, which altered the pattern of genetic activity in some parts of its brain. These changes set in motion subsequent genetic and neural changes in the physiological systems that an adult sparrow uses when singing a version of its species' song."

4. Because development is interactive, changes in either the genetic information or the environmental inputs available to an individual can potentially alter the course of its development by changing the interplay between genes and environment within that individual. Therefore, the behavioral *differences* between two individuals can be largely genetically determined or environmentally determined, or both.

5. The genetic mechanisms that affect development can be analyzed evolutionarily. On the one hand, these mechanisms have an evolutionary history. Thanks to the discipline of evo-devo, we now know of many examples of genes that have been retained in modern species from a distant ancestor. These ancient genes often have regulatory effects on the development of important traits. Modest modifications in these genes, or in the target genes whose expression they affect, can result over time in the evolution of major developmental differences, even among closely related species.

6. Because some differences between individuals are hereditary, populations have the potential to evolve by natural selection. The adaptationist level of analysis, which deals with the possible adaptive value of a trait, has revealed that behavioral development has adaptive features, such as developmental homeostasis, the capacity of the developmental process to ignore or overcome certain environmental or genetic shortfalls that might prevent animals from acquiring fitness-enhancing traits.

7. Likewise, developmental switch mechanisms are adaptive because they control alternative developmental pathways leading to alternative phenotypes. Each of the different traits helps individuals succeed within a particular part of the larger environment. Learning mechanisms provide another form of developmental flexibility, enabling individuals to use their experiences to make adaptive adjustments in behavior that help individuals cope with a variable environment.

COMPANION WEBSITE

Go to **ab11e.com** for discussion questions and all of the audio and video clips.

The Neural Basis of Behavior

In Chapter 3, we focused largely on the developmental basis of animal behavior, one of the two major proximate levels of analysis that define the research of behavioral biologists. We now begin to discuss the other proximate level of analysis, mechanism, starting with an exploration of the neural basis of behavior in this chapter and finishing with an examination of the physiological basis of behavior in Chapter 5. As multicellular animals develop, they acquire a nervous system, the properties of which are crucially important to the proximate control of behavior. In vertebrates, the nervous system consists of the central nervous system (the brain and spinal cord) and the peripheral nervous system (nerves). Because nervous systems represent one way in which information from the environment is translated into behavior, biologists have been eager to learn how these systems work. Although our current understanding of nervous systems has been greatly helped by the application of sophisticated technologies, even simple observations of animals in action can sometimes provide considerable information about the proximate properties of their neural mechanisms.

We begin this chapter with a basic description of behaviors that appear to be innate or instinctive, discussing the stimuli that elicit such behavioral responses. We then begin a foray into the nervous system, using the struggle between bats (predators) and moths (prey) to illustrate the neurobiological basis of how animals produce and interpret stimuli, which in the case of this system is based on ultrasound, sound frequencies that humans cannot hear. With the introduction of other animal systems as varied as chirping crickets, migrating butterflies, singing fishes, and subterranean moles, we begin to examine the attributes of individual neurons and neural clusters that are involved in the detection of certain kinds of sensory information, the relaying of those messages to other cells in the nervous system, and the control of motor commands that are sent to muscles that ultimately lead to the production of behaviors. Students interested in animal behavior are often surprised, and sometimes even intimidated, by the central role that neurons and **neural circuits** (groups of interconnected **neurons** that are able to regulate their own activity using a feedback loop) play in shaping behaviors, even the simplest ones. As you saw in Chapter 3, to understand the adaptive significance of animal behaviors, we also need to probe their mechanistic bases, from the developmental and genetic bases we discussed previously, to the neural basis we introduce here, and also the neuroendocrine basis, which we will introduce in Chapter 5. We continue our process of integration here, digging a bit deeper at the proximate level, but always relating what we discuss about the nervous system to the ultimate level. Natural selection continues to frame our understanding even as we begin to uncover the neural basis of behavior.

Responding to Stimuli

If given the opportunity, males of the digger bee *Centris pallida* will attempt to mate with your finger, despite the fact that a human finger does not appear to resemble a female *C. pallida* (**FIGURE 4.1**). When a sexually motivated male *C. pallida* grasps an object approximately the size of a female of his species, the sensory signals generated by his touch receptors travel to other parts of his brain, where messages are produced

FIGURE 4.1 A complex response to simple stimuli. (A) A male *Centris pallida* copulating with a female of his species. (B) A male of this digger bee species attempting to copulate with a human thumb. (Photographs by John Alcock.)

FIGURE 4.2 A male of the solitary bee *Colletes hederae* attracted to a cluster of larval blister beetles. After attempting to mate with the mass of beetles (some of which are indicated by the arrow), the bee becomes covered by the tiny larvae, which climb aboard for transport to a female bee, should the male succeed in finding a mate of the appropriate species. (Photograph by Nicolas Vereecken.)

that eventually translate into a complex series of muscle commands. The behavioral result is the sequence of movements that passes for courtship in *C. pallida*. The fact that these activities can be stimulated by a thumb instead of a female bee indicates that the nervous system of a male *C. pallida* is not particularly discriminating. Nor is this species unusual in this respect, given that males of the ivy bee (*Colletes hederae*) will attempt to mate, not with a thumb, but with a mass of tiny blister beetle larvae (**FIGURE 4.2**) (Vereecken and Mahé 2007; see also Saul-Gershenz and Millar 2006). Observations of bees trying to mate with objects other than females of their species are puzzling. How can the underlying neural mechanisms be adaptive if they generate such clearly maladaptive responses? These and other Darwinian puzzles have been solved in part by studying how complex responses to simple stimuli are triggered by animal nervous systems.

VIDEO 4.1

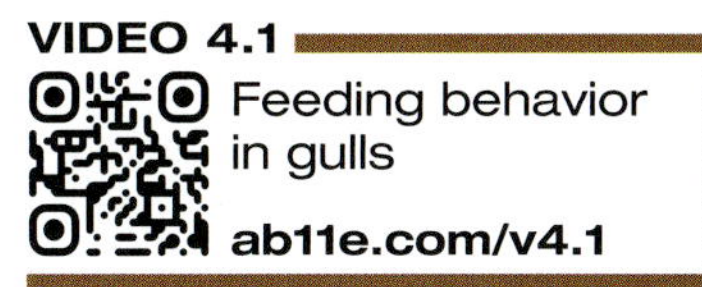

Feeding behavior in gulls

ab11e.com/v4.1

Complex Responses to Simple Stimuli

Let's consider Niko Tinbergen's classic work on begging behavior in herring gull (*Larus argentatus*) chicks that helped lead to a Nobel Prize (Tinbergen and Perdeck 1950). Recently hatched baby gulls will peck at the tip of their parent's bill (**FIGURE 4.3**), an action that induces the adult gull to regurgitate a half-digested fragment of fish or other delicacy, which the chick enthusiastically consumes (**VIDEO 4.1**). You might think a baby gull would find the bill of a living three-dimensional gull more stimulating than painted sticks and two-dimensional cardboard cutouts, but Tinbergen showed that sticks and cutouts work fine in eliciting the pecking response from very young chicks (**FIGURE 4.4**). Experiments with these and other models have revealed that recently hatched gull chicks ignore almost everything except the shape of the "bill" and the red dot at the end of it. Tinbergen proposed that when a young gull sees a pointed object with a contrasting dot at the tip, the resulting sensory signals reach the brain, where other neurons eventually generate the motor commands that cause the chick to peck at the **stimulus**—whether it is located on its parent's bill or a piece of cardboard or the end of a stick.

FIGURE 4.3 Begging behavior by a gull chick. A silver gull (*Chroicocephalus novaehollandiae*) chick is being fed regurgitated food by its parent after pecking at the adult's bill. (Photograph by John Alcock.)

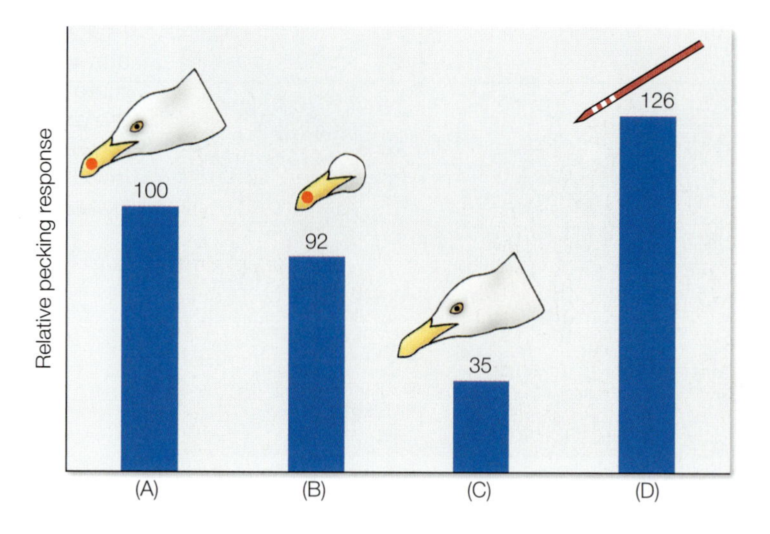

FIGURE 4.4 Effectiveness of different visual stimuli in triggering the begging behavior of young herring gull chicks. (A) A two-dimensional cardboard cutout of the head of an adult gull with a red dot on its bill is not much more effective in eliciting begging behavior in a gull chick than is a model of the bill alone (B), provided the red dot is present. Moreover, a model of a gull head without the red dot (C) is a far less effective stimulus than is an unrealistically long "bill" with contrasting bars at the end (D). (After Tinbergen and Perdeck 1950.)

Tinbergen and his friend Konrad Lorenz collaborated on another famous experiment that also identified a simple stimulus capable of triggering a complex behavior. The researchers found that if they removed an egg from under an incubating graylag goose (*Anser anser*) and put it a half meter away, the goose would retrieve the egg by stretching its neck forward, tucking the egg under its lower bill, and rolling the egg carefully back into its nest. If they replaced the goose egg with nearly any object that was egg-shaped, the goose would invariably run through its egg-retrieval routine. And if the researchers removed the object as it was being retrieved, the bird would continue pulling its head back just as if an egg were still balanced against the underside of its bill (Tinbergen 1951). From these results, Tinbergen and Lorenz concluded that the goose must have a perceptual mechanism that is highly sensitive to visual cues provided by egg-shaped objects. Moreover, that sensory mechanism must send its information to neurons in the brain that automatically activate a more or less invariant motor program for egg retrieval.

The gull chick's pecking response and the graylag goose's retrieval behavior are only two of many innate behaviors that Tinbergen and Lorenz studied. These founders of **ethology** (along with Karl von Frisch, who studied honey bees; see Chapter 8), the first discipline dedicated to the study of both the proximate and ultimate causes of animal behavior, were especially interested in the instincts exhibited by wild animals living under natural conditions (Burkhardt 2004, Kruuk 2004). An **instinct**, or innate behavior, can be defined as a behavior pattern that appears in fully functional form the first time it is performed, even though the animal may have had no previous experience with the cues that elicit the behavior. It is important to reemphasize what you learned in Chapter 3 and make the point that instincts are not "genetically determined." As with all behaviors, instinctive behaviors, like learned ones, are dependent on the gene–environment interactions that took place during development. In the case of a gull chick, these interactions led to the construction of a nervous system that contains a network that enables the young gull to respond to the red dot at the end of an adult gull's bill. The neural network responsible for detecting the simple cue (referred to as a **sign stimulus**, or **releaser**) and activating the instinct, or **fixed action pattern** (**FAP**), was called the **innate releasing mechanism** by Tinbergen and Lorenz (**FIGURE 4.5**) (Tinbergen 1951).

The simple relationship between an innate releasing mechanism, sign stimulus, and FAP is highlighted by the ability of some species to exploit the FAPs of other

FIGURE 4.5 The architects of instinct theory. (A) Niko Tinbergen and Konrad Lorenz (see Figure 3.12) proposed that simple stimuli, such as the red dot on a parent gull's bill (B), can activate or release complex behaviors, such as a gull chick's begging behavior. This effect is achieved, according to these ethologists, because certain sensory messages from the releaser are processed by an innate releasing mechanism (neuronal clusters) higher in the nervous system, leading to motor commands that control a fixed action pattern, a preprogrammed series of movements that constitute an adaptive reaction to the releasing stimulus. (Photograph by Keystone Pictures USA/Alamy Stock Photo.)

species, a tactic sometimes referred to as code breaking (Wickler 1968). For example, when a male ivy bee pounces on a ball of blister beetle larvae, he gets covered with little parasites that may later be transferred to a female bee, if the male is fortunate enough to find a real sexual partner (see Figure 4.2). After moving onto a female, the larvae will eventually be transported to the underground nest of the bee, where they can drop off and make their way into the food-containing brood cells. There they consume the provisions the mother bee collected and stored for her own offspring (Vereecken and Mahé 2007). The beetles have evolved to exploit the male bee's tendency to try to mate with anything that remotely resembles a female, thereby finding a way to maximize their own chances of developing in a food-rich environment.

In many species of estrildid finches, chicks have elaborate mouth markings, with complicated patterns of spots, stripes, and even brightly colored modules (Payne 2005). What function do these decorations serve? The most commonly studied estrildid finch is the zebra finch (*Taeniopygia guttata*), often referred to as the lab rat of the avian world. By comparing a natural strain of zebra finches that had mouth markings with a mutant strain that lacked markings, Immelmann and colleagues showed that the mouth markings are sign stimuli that parents use to feed chicks (Immelmann et al. 1977). Much like the blister beetles that have learned to exploit the bee's behavior, other species of birds have learned to exploit the estrildid mouth markings. Recall that in Chapter 3 we introduced avian brood parasites, which lay their eggs in the nests of other bird species. In species such as the common cuckoo (*Cuculus canorus*) and brown-headed cowbird (*Molothrus ater*), the nestlings must be able to deceive their hosts by looking and behaving in ways that stimulate their adult hosts to feed them (Davies 2000). This type of deception is taken to an extreme in the African vidua finches (*Vidua* spp.), a group that parasitizes estrildid finches. Vidua finch chicks also possess mouth markings, which are thought to have evolved as stimuli to look attractive to host parents so that they will feed the

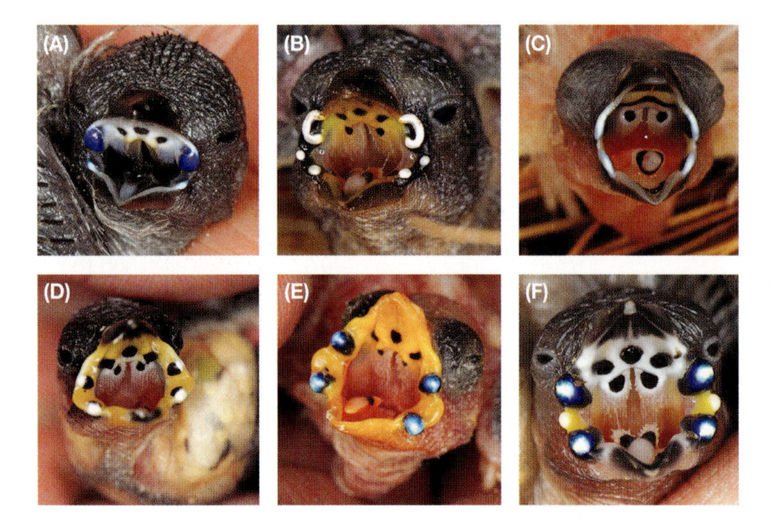

FIGURE 4.6 Estrildid finch mouth markings. Estrilidid finches are found across Australia, Asia, and Africa. The nestlings of many estrildid species have elaborate mouth markings, which may help parents differentiate their own offspring from those of brood parasitic vidua finches, which lay their eggs in estrildid nests. (A) Common grenadier (*Uraeginthus granatina*); (B) orange-cheeked waxbill (*Estrilda melpoda*); (C) plum-headed finch (*Aidemosyne modesta*); (D) pin-tailed parrot-finch (*Erythrura prasina*); (E) red-eared parrotfinch (*Erythrura coloria*); (F) Gouldian finch (*Erythrura gouldiae*). (Photographs © Hofmann-photography.de.)

parasites (**FIGURE 4.6**) (Schuetz 2005). In other words, brood parasite chicks exploit the sensory systems of their hosts to ensure that they get fed, even when the markings fail to match those of their host species' chicks perfectly.

Like brood parasite chicks or blister beetle larvae, humans have been known to exploit the sensory systems of other species. The exploitation of the proximate neural mechanisms of certain animals has even enabled some humans to make a living collecting earthworms to sell to fishermen. "Worm grunters" in Florida have learned that native earthworms (*Diplocardia mississippiensis*) will come rushing to the surface to crawl away when a stake is driven into the earth and vibrated, making the collection of the worms easy. The mystery of why worms behave this way has been solved by the discovery that worm-eating eastern moles (*Scalopus aquaticus*) also generate vibrations in the soil as they dig underground (Catania 2008). By using soil vibrations to activate escape behavior, worms generally leave places before they are captured by moles. Since the 1960s, humans have come to take advantage of this innate response of earthworms, indicating that the current maladaptive behavior of the worms can be explained as a response to recent novel environmental pressures.

How Moths Avoid Bats

Natural selection has endowed gull chicks with neural circuitry that usually enables these creatures to make wise choices about how to get fed. The role of natural selection in shaping individual nerve cells so that animals behave adaptively was the subject of classic research by Kenneth Roeder on how night-flying moths manage to escape from night-hunting bats. Roeder told his readers that they too could watch moths fleeing from bats by spending some time outdoors on summer nights, armed with "a minimum amount of illumination, perhaps a 100-watt bulb with a reflector, and a fair amount of patience and mosquito repellent" (Roeder 1963). Were you to follow Roeder's suggestion, you might see a moth come to the light only to turn abruptly away just before a bat swooped into view. Or you might observe a moth dive straight down just as a bat appeared.

But how does the moth sense the bat and know when it is near? If you took your keys out of your pocket and jangled them, you might also see moths turn away or dive down, suggesting that an acoustic cue provides the trigger for certain moth behaviors, just as simple visual cues are sufficient to activate begging behavior in a gull chick or feeding behavior in an adult finch. As it turns out, the idea that acoustic stimuli trigger the turning or diving behavior of moths is correct, but the sounds that the moth hears when a bat approaches or keys are rattled together are not the sounds that we hear. Instead, the moth detects the very high frequency sounds produced by bats or jangling keys, which makes sense when you consider that most night-hunting bats vocalize ultrasonically using ultrasound frequencies between 20 and 80 kilohertz (kHz)—well above the hearing range of humans, but not of moths.

Before we can understand how moths are able to avoid bats, we first need to learn a bit about how bats hunt moths. Bats use ultrasonic calls or echolocation to locate food and navigate at night—something that was not suspected until the 1930s, when researchers with ultrasound detectors were able to eavesdrop on bats. At that time, Donald Griffin proposed the proximate **echolocation hypothesis** and suggested that night-flying bats use high-frequency cries to listen for weak ultrasonic echoes reflected back from objects in their flight paths (Griffin 1958). When Griffin placed captive little brown bats (*Myotis lucifugus*) in a dark room filled with fruit flies and wires strung from ceiling to floor, his subjects had no trouble catching the insects while negotiating the obstacle course. However, when Griffin turned on a machine that filled the room with high-frequency sound, the bats began to collide with the wires and crash to the floor, where they remained until Griffin turned off the jamming device. In contrast, loud sounds with frequencies between 1 and 15 kHz (which humans can easily hear) had no effect on the ultrasound-using bats because these stimuli did not mask the high-frequency echoes bouncing back from objects in the room. Griffin rightly concluded that the little brown bat employs a sonar system to avoid obstacles and detect prey at night. We now know that the bat sonar system is capable of creating rich and detailed three-dimensional images of the world, even in complete darkness. Bats are able to estimate not only the distance to an object of interest by measuring the time delay from the emitted pulse and the returning echo, but also an object's shape and texture by the degrees of amplitude modulation of the returning echo. Moreover, bats can also accurately compute the target direction with a resolution of only 2 to 3 degrees by comparing the sounds arriving at each of their ears. Some species of bats are even able to identify individual rocks using the structure of the returning echoes (Geva-Sagiv et al. 2015).

echolocation hypothesis Ultrasonic calls or echolocation are used by bats to locate food and navigate at night.

AUDITORY RECEPTION IN MOTHS As Roeder watched moths evading echolocating bats, he hypothesized that the insects were able to hear pulses of bat ultrasound and use them to avoid detection. He began to test this **ultrasound evasion hypothesis** by finding the ears of a noctuid moth, one on each side of the thorax (**FIGURE 4.7**). Although the ears in some moth species occur in places other than the thorax (including the mouth, abdomen, and even on the wings), all moth ears have a similar structure (ter Hofstede and Ratcliffe 2016). Each ear consists of a thin, flexible sheet of cuticle—the tympanic membrane, or tympanum—lying over a chamber on the side of the thorax. Attached to the tympanum are two neurons, the A1 and A2 auditory receptors. These receptor cells are deformed when

ultrasound evasion hypothesis Insects are able to hear pulses of bat ultrasound.

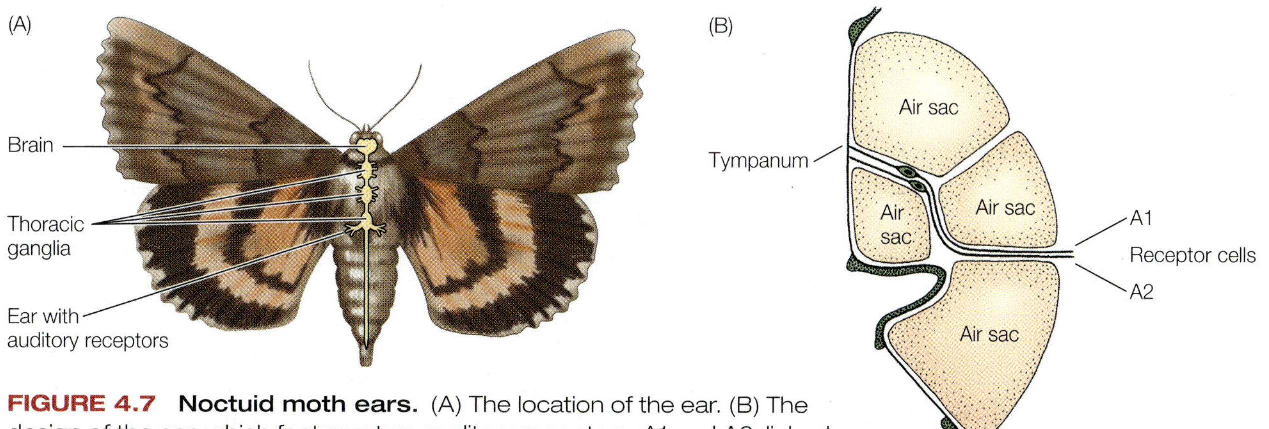

FIGURE 4.7 Noctuid moth ears. (A) The location of the ear. (B) The design of the ear, which features two auditory receptors, A1 and A2, linked to a tympanum that vibrates when exposed to sounds. (After Roeder 1963.)

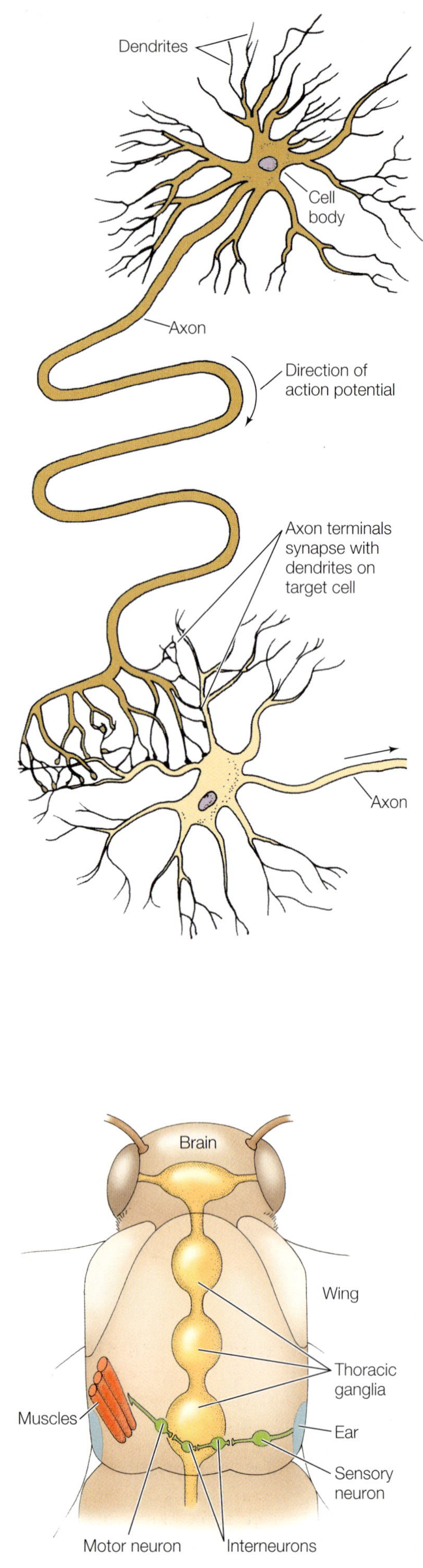

FIGURE 4.8 Neurons and their operation. Illustration of the structure of a generalized neuron with its dendrites, cell body, axon, and synapses. Electrical activity in a neuron originates with the effects of certain stimuli on the dendrites. If the electrical changes in a dendrite's cell membrane are sufficiently great, they can trigger an action potential, which begins near the cell body and travels along the axon toward the next cell in the network.

the tympanum vibrates, which it does when intense sound pressure waves sweep over the moth's body.

The moth's A1 and A2 receptor cells work much like other neurons: They respond to the energy contained in selected stimuli by changing the permeability of their cell membranes to positively charged ions. The effective stimuli for a moth auditory receptor appear to be provided by the movement of the tympanum, which mechanically stimulates the receptor cell, opening stretch-sensitive channels in the cell membrane. As positively charged ions flow in, they cause the normally negative inside of the cell to become more positive. If the inward movement of ions is sufficiently great, a substantial, abrupt, local change in the electrical charge difference across the membrane may occur and spread to neighboring portions of the membrane, sweeping around the cell body and down the axon—the "transmission line" of the cell (**FIGURE 4.8**). This brief, all-or-nothing change in electrical charge, called an **action potential**, is the signal that one neuron uses to communicate with another.

When an action potential arrives at the end of an axon, it may cause the release of a **neurotransmitter**, a chemical signal that diffuses across the narrow gap, or **synapse**, separating the axon tip of one cell from the surface of the next cell in the network. Neurotransmitters can affect the membrane permeability of the next cell in a chain of cells in ways that increase (or decrease) the probability that this next neuron will produce its own action potential. If a neuron generates an action potential in response to stimulation provided by the preceding cell in the network, the message may be relayed on to the next cell, and on and on. Volleys of action potentials initiated by distant receptors may have excitatory (or inhibitory) effects that reach deep into the nervous system, eventually resulting in action potential outputs that reach the animal's muscles and cause them to contract.

In the case of the noctuid moth studied by Roeder, the A1 and A2 receptor cells are linked to relay cells called **interneurons**, whose action potentials can change the activity of other cells in one or more of the insect's thoracic ganglia (a ganglion is a neural structure composed of a highly organized mass of neurons) that relay messages on to the moth's brain (**FIGURE 4.9**). As messages flow through these parts of the nervous system, certain patterns of action potentials produced by cells in the thoracic ganglia trigger other interneurons, whose action potentials in turn reach motor neurons that are connected to the wing muscles of the moth. When a motor neuron fires, the neurotransmitter it releases at the synapse with a muscle

FIGURE 4.9 Neural network of a moth. Receptors in the ear relay information to interneurons in the thoracic ganglia, which communicate with motor neurons that control the wing muscles.

fiber changes the membrane permeability of the muscle cell. These changes initiate the contraction or relaxation of muscles, and this drives the wings and thereby affects the moth's movements. Thus, the moth's behavior, like that of any animal, is the product of an integrated series of chemical and biophysical changes in a network of cells. Because these changes occur with remarkable rapidity, a moth can react to certain acoustic stimuli in fractions of a second, which helps the moth avoid bats zooming in for the kill.

AUDITORY RECEPTORS IN THE MOTH EAR Clearly, the auditory receptors of noctuid moths are highly specialized for the detection of ultrasonic stimuli. Roeder demonstrated this point by attaching recording electrodes to the A1 and A2 receptors of living, but restrained, moths (**BOX 4.1**) (Roeder 1963). Although the moth's ears have just two receptors each, they are able to provide an impressive amount of information to the moth's nervous system about echolocating bats. The key property of the A1 receptor is its great sensitivity to pulses of ultrasound, which enables it to begin generating action potentials in response to the faint cries of a little brown bat up to 30 meters away, long before the bat can detect the moth. In addition, because the rate of firing in the A1 cell is proportional to the loudness of the sound, the insect has the ability to determine whether a bat is getting closer or going farther away. The moth's ears also gather information that could presumably be used to locate the bat in space. For example, if a hunting bat is on the moth's left, the A1 receptor in its left ear is stimulated a fraction of a second earlier and somewhat more strongly than the A1 receptor in its right ear, which is shielded from the sound by the moth's body. As a result, the left receptor fires sooner and more often than the right receptor (**FIGURE 4.10A**). The moth's nervous system could also detect whether a bat is above it or below it. If the predator is higher than the moth, then with every up-and-down movement of the insect's wings, there is a corresponding fluctuation in the firing rate of the A1 receptors as they are exposed to bat cries and then shielded from them by the wings (**FIGURE 4.10B**). If the bat is directly behind the moth, there is no such fluctuation in neural activity (**FIGURE 4.10C**).

As neural signals initiated by the receptors race through the moth's nervous system, they may ultimately generate motor messages that cause the moth to turn and fly directly away from the source of ultrasonic stimuli (Roeder and Treat 1961). When a moth is moving away from a bat, it exposes less echo-reflecting area than if it were presenting the full surface of its wings at right angles to the bat's vocalizations. If a bat receives no insect-related echoes from its calls, it cannot detect a meal. Bats rarely fly in a straight line for long, and therefore the odds are good that a moth will remain undetected

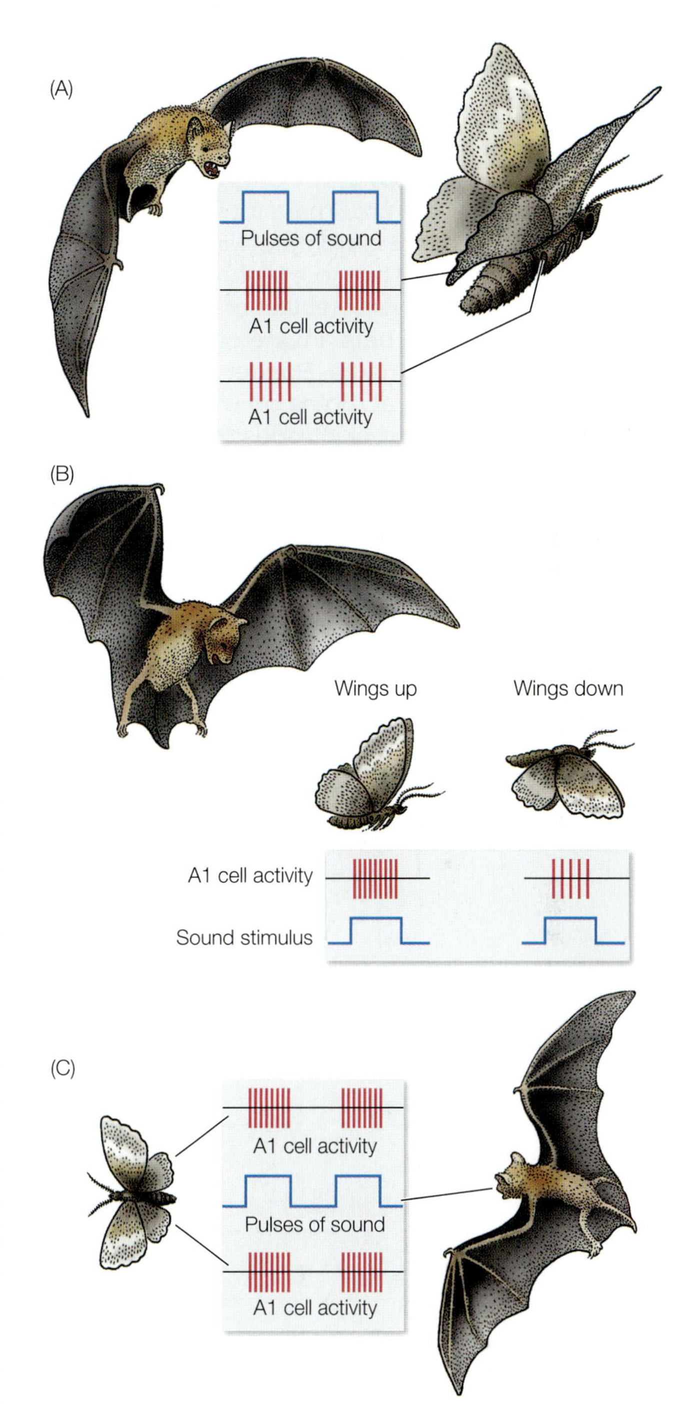

FIGURE 4.10 How moths might locate bats in space. (A) When a bat is to one side of the moth, the A1 receptor on the side closer to the predator fires sooner and more often than the shielded A1 receptor in the other ear. (B) When a bat is above the moth, activity in the A1 receptors fluctuates in synchrony with the moth's wingbeats. (C) When a bat is directly behind the moth, both A1 receptors fire at the same rate and time. (Figures are not drawn to scale.)

BOX 4.1 EXPLORING BEHAVIOR BY INTERPRETING DATA

Ultrasound detection in the moth ear

In many ways, Kenneth Roeder was a pioneer in integrating the study of proximate and ultimate aspects of behavior. Knowing that moths could somehow avoid predatory bats, he searched for and found the proximate foundation for that behavior. Combining ultimate with proximate research enabled him to eventually find the neural basis for the detection of ultrasound and avoidance of bats. In one of his classic proximate experiments, Roeder projected a variety of sounds at moths on which he had attached recording electrodes to the A1 and A2 receptors. The electrical responses of the receptors were relayed to an oscilloscope, which produced a visible record (Roeder 1963). He found that sounds of low or moderate intensity do not generate action potentials in the A2 receptor, and that the A1 receptor fires sooner and more often as sound intensity increases (**FIGURE A**). However, the A1 receptor initially reacts strongly to pulses of ultrasound but then reduces its rate of firing if the stimulus is a constant sound (**FIGURE B**).

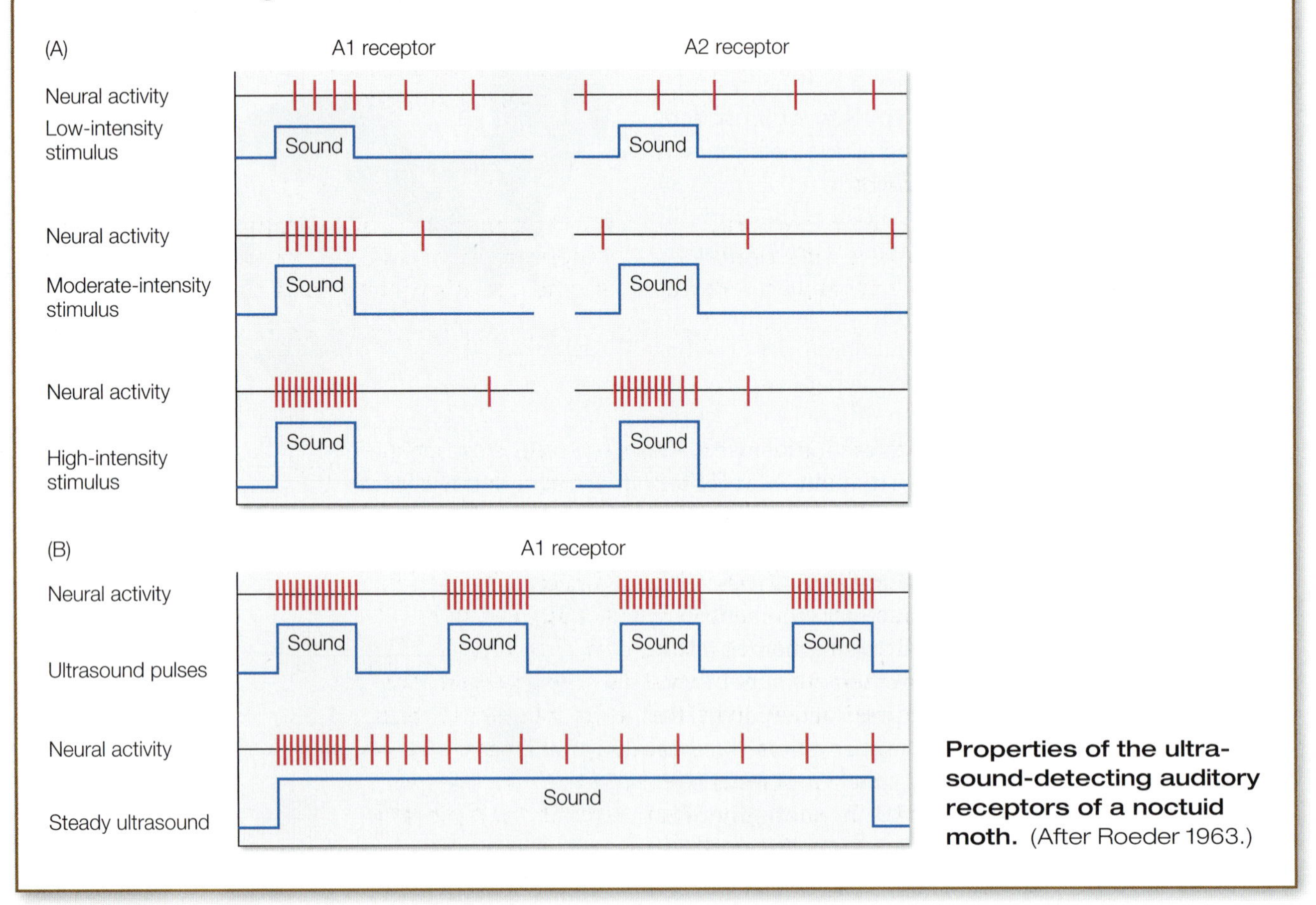

Properties of the ultrasound-detecting auditory receptors of a noctuid moth. (After Roeder 1963.)

if it can stay out of range for just a few seconds. By then the bat will have found something else within its 3-meter moth-detection range and will have veered off to pursue another meal. To employ its antidetection response most effectively, however, a moth must orient itself so as to synchronize the activity of the A1 receptors in each ear. Differences in the rate of firing by the A1 receptors in the two ears are monitored by the brain, which relays neural messages to the wing muscles via the thoracic ganglia and allied motor neurons. The resulting changes in muscular action steer the moth away from the side of its body with the ear that is more strongly

BOX 4.1 EXPLORING BEHAVIOR BY INTERPRETING DATA *(continued)*

These recordings revealed a number of special features of the ears of a noctuid moth:

1. The A1 receptor is sensitive to ultrasounds of low to moderate intensity, whereas the A2 receptor begins to produce action potentials only when an ultrasound is relatively loud.
2. As a sound increases in intensity, the A1 receptor fires more often and with a shorter delay between the arrival of the stimulus at the tympanum and the onset of the first action potential.
3. The A1 receptor fires much more frequently in response to pulses of sound than to steady, uninterrupted sounds.
4. Neither receptor responds differently to sounds of different frequencies over a broad ultrasonic range. Thus, a burst of 30-kHz sound elicits much the same pattern of firing as an equally intense sound of 50 kHz. In other words, because the shapes of the A1 and A2 frequency tuning curves are so similar, moths (but not all insects) are essentially tone-deaf and cannot discriminate between different frequencies (reviewed in Miller and Surlykke 2001).
5. The receptor cells do not respond at all to low-frequency sounds, which means that moths are deaf to stimuli that we can easily hear, such as the chirping calls and trills of night-singing crickets. Of course, we are deaf to sounds that nocturnal moths have no trouble hearing.

Thinking Outside the Box

Why do you think the A1 and A2 receptors function differently? Given these differences, how do they work together to enable a moth to evade a potential bat predator? Why are these receptors so specialized for ultrasound and not for the sounds that humans can hear?

stimulated. As the moth turns away from the relatively intense ultrasound reaching one side of its body, it reaches a point at which both A1 cells are equally active; at this moment, it is facing in the opposite direction from the bat and heading away from danger (see Figure 4.10C).

Although this reaction is effective if the moth has not been detected, it is useless if a speedy bat has come within 3 meters of the moth. At this point, a moth has at most a second before it and the predator will collide (Kalko 1995). Therefore, moths in this situation do not try to outrun their enemies but instead employ drastic evasive maneuvers, such as wild loops and power dives, that make it harder for bats to intercept them. A moth that executes a successful power dive and reaches a bush or grassy spot is safe from further attack because echoes from the leaves or grass at the moth's crash-landing site mask those coming from the moth itself (Roeder and Treat 1961).

How do moths perform these acrobatic escape responses? Roeder speculated that the physiological basis for this erratic escape flight lies in the neural circuitry leading from the A2 receptors to the brain and then back to the thoracic ganglia (Roeder 1970). When a bat is about to catch a moth, the intensity of the sound waves reaching the insect's ears is high. It is under these conditions that the A2 cells fire. Roeder believed that the A2 signals, once relayed to the brain, might shut down the central steering mechanism that regulates the activity of the flight motor neurons. If the steering mechanism were inhibited, the moth's wings would begin beating out of synchrony, irregularly, or not at all. As a result, the insect might not know where it was going—but neither would the pursuing bat, whose inability to plot the path of its prey could permit the insect to escape.

Although Roeder's ideas about the functions of the A1 and A2 cells were plausible and supported by considerable evidence, other scientists continued to look at how these cells control the response of moths to bats. As a result, we now know that notodontid moths, which are similar to noctuid moths but have just one auditory receptor per ear, still appear to exhibit a two-part response to approaching bats: the turning away from distant hunters and then the last-second erratic flight pattern when death is at hand. Thus, two cells may not be necessary for the double-barreled response of moths to their hunters (Surlykke 1984). Even in moths with two receptors per ear, the A1 cell's activity changes greatly as a bat comes sailing in toward them, since the bat's ultrasonic cries speed up and become much more intense (**FIGURE 4.11**) (Fullard et al. 2003). Presumably, higher-order neurons up the chain of

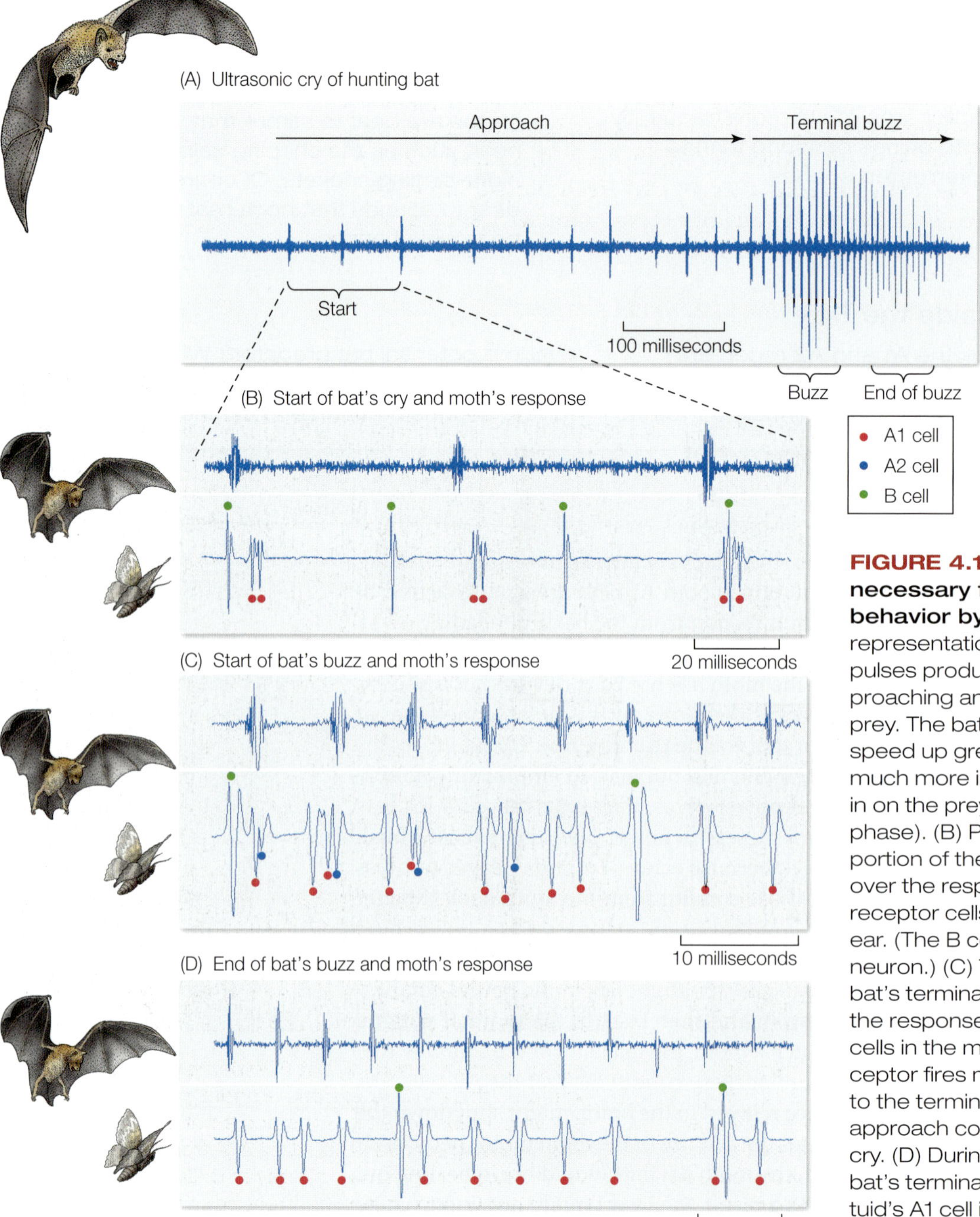

FIGURE 4.11 Is the A2 cell necessary for anti-interception behavior by moths? (A) A visual representation of the ultrasonic pulses produced by a bat approaching and then attacking a prey. The bat's ultrasonic cries speed up greatly and become much more intense as it closes in on the prey (the terminal buzz phase). (B) Part of the approach portion of the bat's cry is shown over the response of the A1 and B receptor cells in a noctuid moth's ear. (The B cell is a nonauditory neuron.) (C) The initial part of the bat's terminal buzz is shown over the response of the A1, A2, and B cells in the moth's ear. The A1 receptor fires more often in response to the terminal buzz than to the approach component of the bat's cry. (D) During the latter part of the bat's terminal buzz, only the noctuid's A1 cell is active, not the A2 receptor. (After Fullard et al. 2003.)

command could analyze the correlated changes in the activity of the A1 receptor alone and make adaptive adjustments accordingly without involvement of the A2 receptor.

More doubts that the A2 cell is necessary to trigger erratic evasive behavior come from the finding that in some noctuid moths, both the A1 and A2 cells may more or less stop firing during the terminal buzz phase—the last 150 milliseconds—of a bat attack. One would think that these cells would keep signaling if either one was truly important in controlling the last-gasp evasive maneuvers of moths under attack. James Fullard and colleagues suggest that perhaps these cells fail to signal at this late stage simply because the extremely loud and rapid attack vocalizations of a nearby bat incapacitate the cells (Fullard et al. 2003). Thus, we have reason to question whether a connection exists between A2 cell activity and the moth's response to onrushing bats.

Ultrasonic Hearing in Other Insects

Like moths, many other nocturnal insects have independently evolved the capacity to sense ultrasound, and they also take evasive action when bats approach them (**FIGURE 4.12**) (Yager and May 1990, Miller and Surlykke 2001, Yack et al. 2007). Bat-detecting ears evolved from ultrasound organs that are dispersed throughout the body in insects and even some crustaceans, likely explaining why ears occur in different parts of the body in moths as well as other insects. However, some species

(A)

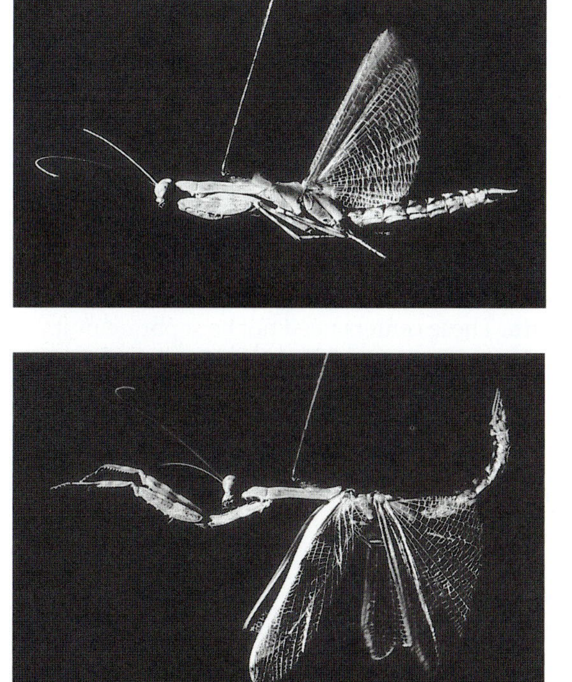

(B)

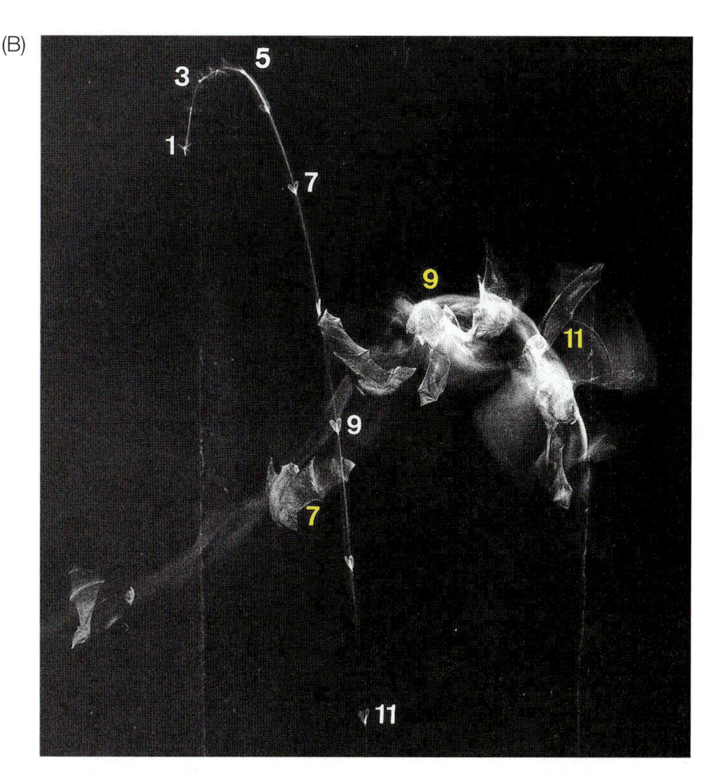

FIGURE 4.12 Evasive behavior of insects attacked by bats. (A) In normal flight, a budwing mantis (*Parasphendale agrionina*) holds its forelegs close to its body (top), but when it detects ultrasound, it rapidly extends its forelegs (bottom), which causes the insect to loop and dive erratically downward. (B) The anti-interception power dive of a lacewing when approached by a hunting bat. The numbers superimposed on this multiple-exposure photograph show the relative positions of a lacewing and a bat over time. (The lacewing survived the attack.) (A from Yager and May 1990; photographs courtesy of D. D. Yager and M. L. May; B, photograph courtesy of Lee Miller.)

of insects with ears, such as mantids, originated before bats (Yager and Svenson 2008), suggesting that not all insect species evolved hearing to escape bat predation (ter Hofstede and Ratcliffe 2016). In moths, however, the evidence is quite good that hearing is fine-tuned to reflect the diversity of echolocation call frequencies across the bat community in which a given species lives (reviewed in Fullard 1998, Ratcliffe 2009, ter Hofstede et al. 2013). For example, in tropical communities with high bat diversity, moths can hear a wider range of frequencies than can those in temperate communities where bat diversity is lower (Fullard 1982).

Neural Command and Control

Walk down the street of any major city and you will be bombarded by an overload of sensory stimuli. Horns honking. Pedestrians conversing. Sirens blaring. And if you're lucky, maybe even some birds singing. Since most animals are exposed to a continuous stream of sensory information, a widespread challenge facing the members of many species is how to prioritize stimuli and select which of several behaviors to perform when competing responses are possible. Natural selection appears to have favored individuals whose proximate systems not only activate useful responses to key stimuli, but also organize an animal's repertoire of behaviors in ways that establish an adaptive set of priorities among these options. For example, the neurons in a noctuid moth help it sense and respond to an approaching source of ultrasound generated by a hunting bat. But the moth can also detect and track down a distant source of sex pheromone produced by a female of its own species. As a male noctuid moth flies enthusiastically toward a distant female, he may encounter a hunting bat along the way. What should he do? Fortunately for the moth, his nervous system has been shaped by natural selection to respond to the threat. If the moth hears very loud ultrasonic pulses, he generally aborts his scent-tracking activity and dives for cover, an adaptive decision that enables the male to live to copulate on another day (Acharya and McNeil 1998, Skals et al. 2005).

But how does the moth override the urge to find a mate and take evasive action to avoid certain death? One way in which nervous systems might be organized to avoid maladaptive conflicts between competing stimuli involves **command centers** that communicate with one another, an idea formalized in the proximate **command center hypothesis**. These centers need not be separate units in the brain, but might instead consist of an assortment of cells capable of unified decision making, as might occur in some innate releasing mechanisms. If each command center were primarily responsible for activating a particular response but could inhibit or be inhibited by signals from other centers, an individual might avoid incapacitating or dangerous indecision or conflict.

command center hypothesis Nervous systems are organized to avoid maladaptive conflicts between competing stimuli via command centers in the brain.

Decision Making in the Brain

Kenneth Roeder used the command center hypothesis to examine decision making in the European mantis (*Mantis religiosa*) (Roeder 1963). A mantis can do many things: search for mates, sunbathe, copulate, fly, dive away from bats, and so on. Most of the time, however, the typical mantis remains motionless on a leaf until an unsuspecting bug wanders within striking distance. When the mantis's visual receptors alert it to the presence of the prey, the mantis makes very rapid, accurate, and powerful grasping movements with its front pair of legs. Roeder proposed that the mantis's nervous system sorts out its options thanks to inhibitory relationships between an assortment of command centers. The design of the mantis's nervous system suggested to Roeder that the control of muscles in each of the insect's segments is the responsibility of that segment's ganglion. Roeder tested this possibility by cutting one segmental ganglion's connections with the rest of the nervous system. Not surprisingly, the muscles in the neurally isolated segment subsequently failed to react when the mantis's nervous system became active elsewhere. However, when the isolated ganglion was stimulated electrically, the muscles and any limbs in that segment made vigorous, complete movements.

If the segmental ganglia are indeed responsible for telling the muscles in particular segments to carry out given movements, what is the mantis's brain doing? Roeder suspected that certain brain cells are responsible for inhibiting (or blocking) neural activity in the segmental ganglia, keeping cells in a ganglion quiet until they are specifically ordered into action by an excitatory command center in the brain. If this is correct, then cutting the connection between the inhibitory brain cells and the segmental ganglia should have the effect of removing this inhibition and inducing inappropriate, conflicting responses. When Roeder severed the connections between the protocerebral ganglion (the mantis's brain) and the rest of its nervous system, he produced an insect that walked and grasped simultaneously—not a particularly adaptive thing to do. The protocerebral ganglion apparently makes certain that an intact mantis either walks or grasps but does not do both at the same time.

What would happen if one removed both the subesophageal and protocerebral ganglia? Roeder conducted this experiment by removing the entire head, finding that when he did so, the mantis became immobile (Roeder 1963). Moreover, Roeder could induce single, irrelevant movements, but nothing more, by poking the mantid sharply. These results suggest that the protocerebral ganglion of an intact mantis typically sends out a stream of inhibitory messages to the subesophageal ganglion, preventing these neurons from communicating with the other ganglia. When certain sensory signals reach the protocerebral ganglion, however, neurons there stop inhibiting certain modules in the subesophageal ganglion. Freed from suppression, these subesophageal neurons send excitatory messages to various segmental ganglia, where new signals are generated that order muscles to take specific actions. Depending on what sections of the subesophageal ganglion are no longer inhibited, the mantis either walks forward, or strikes out with its forelegs, or flies, or does something else.

Interestingly, mature male mantises do not always obey the "rule" that complete removal of the head eliminates behavior. Occasionally, a headless adult male performs a series of rotary movements that swing its body sideways in a circle. While this is happening, the mantis's abdomen is twisting around and down, movements that are normally blocked by signals coming from the protocerebral ganglion. This odd response to decapitation begins to make sense when you consider that a male mantis sometimes literally loses his head over a female, when she grabs him and consumes him, head first (see Chapter 9). Even under these difficult circumstances, the male can still copulate with his cannibalistic partner (**FIGURE 4.13**), thanks to the nature of the control system regulating his mating behavior. Although he is headless, his legs carry what is left of him in a circular path until his body

FIGURE 4.13 Female mantis (top) decapitating a male. (Photograph © Jose B. Ruiz/Minden Pictures.)

touches the female's, at which point he climbs onto her back and twists his abdomen down to copulate competently (Roeder 1963).

The moth–bat story tells us that nervous systems have been shaped by natural selection to deal with the real-world challenges confronting an animal. Kenneth Roeder knew that noctuid moths live in a world filled with echolocating moth killers. Using the scientific method, he searched for, and found, a specialized proximate mechanism that helps some moths hear sounds that humans cannot sense, the better to deal with hunting bats. Many other researchers have continued to explore the proximate and ultimate aspects of moth–bat interactions and, in so doing, have produced a vast literature on how these creatures have influenced each other's evolution (Conner and Corcoran 2012). Among the many fascinating discoveries made in the post-Roeder era is that some species of the tiger moth *Bertholdia trigona* respond to big brown bat (*Eptesicus fuscus*) ultrasound by generating ultrasonic clicks of their own. At first glance, this reaction would seem to be suicidal, as the moths appear to be offering signals that their predators could use to track them down. However, the moth clicks are in fact loud jamming signals, which interfere with a bat's ability to zero in on a prey, causing the predator to misjudge the location of the moth (Corcoran et al. 2011). In addition, since some species of tiger moths are chemically defended and inedible to bats, their clicks may also signal bats to avoid the poor-tasting prey (Dowdy and Conner 2016). In an apparent case of acoustic aposematism (warning signaling and mimicry; see Chapter 6), when captive bats were presented with two species of sound-producing, noxious tiger moths, they quickly learned not only to avoid the tiger moths, but also sound-producing, non-chemically defended noctuid moths presented as controls (Barber and Conner 2007). Thus, the moth–bat story is not only a complicated one of coevolution between predator and prey, but one that illustrates how natural selection shapes the evolution of specialized neurophysiological mechanisms that have clear fitness benefits for both bats and moths.

From Ultrasound to Ultraviolet Radiation

Both the noctuid moths studied by Roeder and the tiger moths that jam bat sonar use their evolved acoustic mechanisms to detect sounds outside the range of human hearing. Yet these moths illustrate how animals differ not only in their behavioral reaction but also in the underlying neurophysiology, even in response to the same selective pressure: the risk of predation by bats. From an ultimate perspective, animals differ in their neuroanatomy and neurophysiology because their proximate mechanisms have been shaped by different selection pressures. We can illustrate this using another sensory modality, ultraviolet (UV) radiation, which comes from many objects in nature and yet, much like ultrasound, we humans cannot detect.

Many animals use UV radiation not to escape predators, but to find food, attract mates, or navigate great distances. For example, many bees use UV-reflecting bee guides on flowers to locate the nectar source more readily (**FIGURE 4.14**). Ultraviolet-reflecting patches on the wings of certain male butterflies are seen by and are sexually attractive to females of these butterflies (Rutowski 1998, Kemp 2008). The same is true for female three-spined stickleback (*Gasterosteus aculeatus*), which respond positively to the carotenoid pigments on male bodies (Boulcott et al. 2005). When we look at carotenoids, we see reds, oranges, and yellows, but not the UV signals that female sticklebacks apparently use to determine the quantity of carotenoids incorporated in a male's skin (Pike et al. 2011). Female preferences based on these signals help the females choose males capable of surviving the breeding season, during which time the males provide parental care for their choosy mates'

FIGURE 4.14 Ultraviolet-reflecting patterns are detected by many animals. In both sets of photographs, the image on the left shows the organism as it appears to humans, while the image on the right shows the organism's UV-reflecting (pale) surfaces. (A) The ultraviolet pattern on this daisy advertises the central location of food for insect pollinators. (B) Only males (top specimens) of this sulphur butterfly species have UV-reflecting patches on their wings, which helps signal their sex to other individuals of their species. (A, photographs by Tom Eisner; B, photographs by Randi Papke and Ron Rutowski.)

offspring (Pike et al. 2007). Signals containing UV wavelengths also play a role in certain aggressive interactions. For example, adult male collared lizards (*Crotaphytus collaris*) aggressively display to one another by opening their mouths wide, and when they do so, whitish UV-reflecting patches at the corners of their mouths become visible to conspecifics (Lappin et al. 2006). In another lizard species, males with experimentally reduced UV-reflecting throat patches were subject to more attacks than control males, an indication that the size of the UV patch is a signal of male fighting ability in this lizard as well (Stapley and Whiting 2005).

The ability to detect UV radiation also plays a role in the extraordinary navigational abilities of migratory monarch butterflies (*Danaus plexippus*), an idea referred to as the **UV navigation hypothesis**. Migrant monarchs fly south from Canada toward a select group of forest groves high in the mountains of central Mexico, a trip of as much as 3600 kilometers (**FIGURE 4.15**). Once in Mexico, the butterflies congregate by the millions in stands of oyamel firs, where they mostly sit and wait, day after day, through the cool winter season until spring comes. When the weather begins to warm with the arrival of spring, the monarchs rouse themselves and begin a return trip to the Gulf Coast, arriving in the southern United States in time to lay their eggs on milkweed plants growing there, which their caterpillar progeny will consume.

UV navigation hypothesis Some migratory animals use UV radiation to navigate.

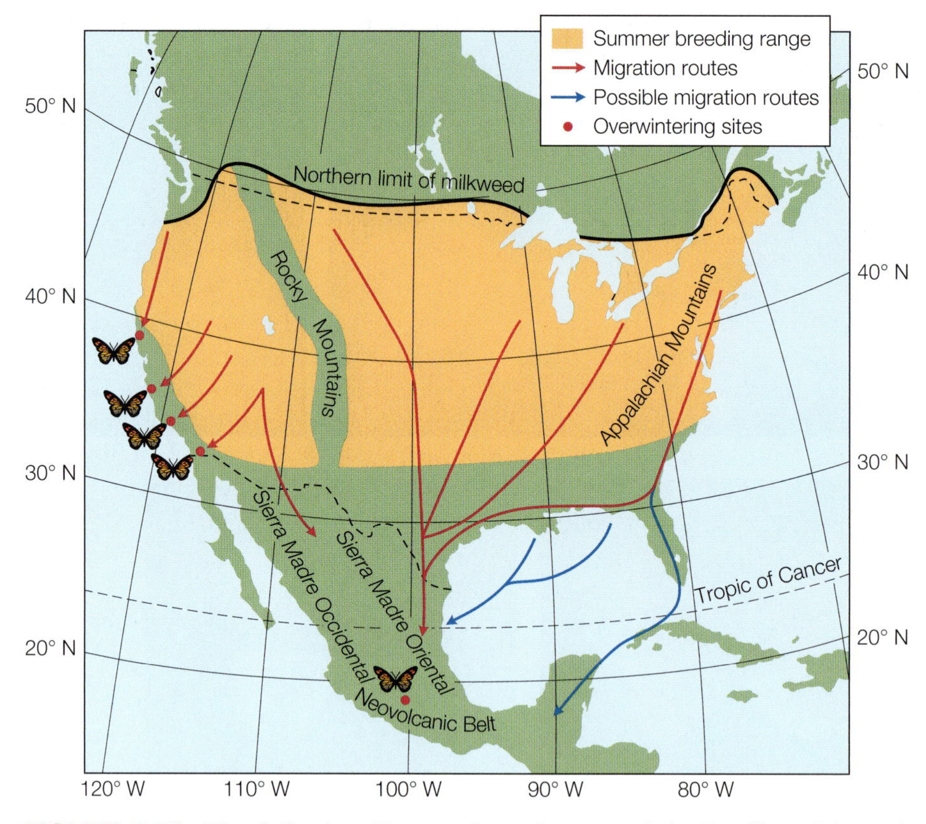

FIGURE 4.15 The fall migration routes of monarch butterflies. Monarchs that reach southern Canada then turn around to travel to Mexico, where they cluster in a few forest patches high in the central Mexican mountains. Monarchs in the western United States often migrate to coastal wintering sites. (After Brower 1995.)

Migrating monarchs fly during the daytime, which suggests that they might use the sun as a compass to guide them in a southwesterly direction during the fall migration (and in a northeasterly direction during the spring return flight). Sunlight is composed of many different wavelengths of light, and so the question arises, which wavelengths are critical for monarch navigation? To test the UV navigation hypothesis, researchers captured migrating butterflies in the fall and then permitted tethered individuals to start flying in a flight cage where they could orient in a direction of their choosing. When the team then covered the flight cage with a UV-interference filter that screened out this component of sunlight, the monarchs quickly became confused, some so much so that they stopped flying altogether. However, most individuals (11 out of 13) resumed flight as soon as the filter was removed, evidence that the sun's UV radiation is indeed essential for monarch navigation (Froy et al. 2003).

UV light may help monarchs get started flying in the right direction. But once airborne, the butterflies need to maintain their compass orientation, a task that they achieve with assistance from the cues provided by the polarized skylight pattern. Polarized light—another form of light that humans cannot see—is produced when sunlight enters Earth's atmosphere and the light scatters so that some light waves are vibrating perpendicular to the direction of actual sun rays. Because the three-dimensional pattern of polarized light in the sky created in this fashion depends on the position of the sun relative to Earth, this pattern changes as the sun moves across the sky. Therefore, if an animal on the ground is able to perceive the pattern of polarized light in the sky, this stimulus can indicate the position of the sun at any given time. As a result, creatures with a **circadian clock** (an internal oscillator modulated by external cues such as sunlight or temperature that regulates physiological processes) and the capacity to see polarized light can use the information in skylight as a compass in the same way that humans can use the sun's position in the sky as a compass. Being able to make use of the directional information in polarized light has real advantages for monarchs and some other animals, such as migrating salmon (Hawryshyn et al. 2010), because this cue is available even when the sun is hidden behind clouds or mountains.

A team of researchers established that monarchs can orient to polarized light information by tethering butterflies captured on their fall migration in a small walled arena where the insects could not see the sun but could look at the sky overhead. Under these conditions, when the butterflies flew, they were able to orient consistently to the southwest. Flying tethered monarchs retained this ability when a light filter was placed over the apparatus and aligned so as to permit the entry of linear polarized light waves in the same plane as occurs in the sky at the zenith (the highest point in the sky). However, when the filter was turned 90 degrees from its original orientation, changing the angle of entry of the polarized light visible to the monarchs at the zenith, they altered their flight orientation by 90 degrees as well, demonstrating their reliance on a polarized-light compass (**FIGURE 4.16**) (Reppert et al. 2004).

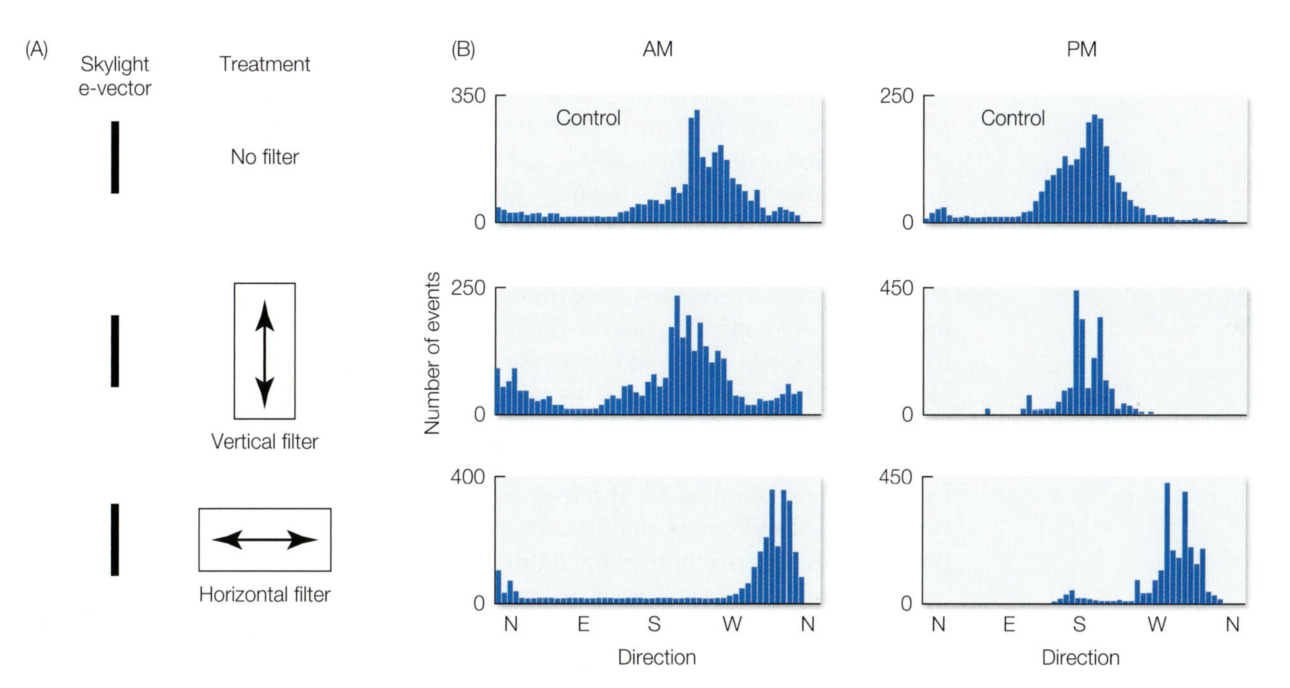

FIGURE 4.16 Polarized light affects the orientation of monarch butterflies. (A) Monarchs tethered in a flight cage received three treatments with respect to the angle of polarized light reaching them from the sky (the skylight e-vector): (1) no filter, which meant they could see the natural pattern of polarized light in the sky; (2) vertical filter, which did not interfere with the pattern of polarized light visible from the flight cage; and (3) horizontal filter, which shifted the angle of polarized light arriving from the sky by 90 degrees. (B) Orientation choices made by two flying monarchs in the flight cage, one tested in the morning (AM) and the other in the afternoon (PM). During the period of flight, a computer automatically recorded the orientation of the butterfly every 200 milliseconds, yielding a record composed of many "events" during the minutes when the monarch was flying. The data show that when the insects could observe polarized light in the natural skylight pattern, they tended to fly to the southwest, but when the filter shifted the angle of polarized light, altering the skylight pattern by 90 degrees, the monarchs shifted their flight orientation accordingly. (After Reppert et al. 2004.)

Selective Relaying of Sensory Inputs

In monarch butterflies, three-spined sticklebacks, collared lizards, and many other species, sensory receptors collect information about ultraviolet radiation. But how do light stimuli actually influence mating or migratory behavior? To act on that sensory information, an animal's receptors must forward messages to those parts of its brain that can process selected sensory inputs and order appropriate responses based on that information. We are beginning to understand how this works in monarchs, in which individuals respond to multiple light stimuli. Their light-sensitive clock mechanism is located in the antennae, where proteins produced from a series of circadian genes (called *period*, *timeless*, and *cry2*) that are shared with many other animals, including *Drosophila* and mice, help regulate the timing of migration. Zhu and colleagues demonstrated using microarray analysis that these genes are differentially upregulated (that is, turned on) in the brains of migratory versus non-migratory butterflies (Zhu et al. 2008). Additionally, the location of the sun is detected by the eyes, where light triggers neuronal signals in photoreceptors. These signals are then relayed directly to the lateral accessory lobes of the brain, which connect to descending motor pathways that control behavior. In other words, both time of day and the sun's position appear to be encoded independently in different structures by neuronal firing rates, which are carried to the brain to influence the expression of circadian genes and motor pathways (Shlizerman et al. 2016). In a model proposed by a team led by Steven Reppert, these structures are linked in a simple neural circuit that ultimately controls precise angular position and allows for the correction of flight direction (Shlizerman et al. 2016).

We can further illustrate how neural mechanisms contribute to the selective transfer of data by returning to an insect that hears ultrasound, the cricket *Teleogryllus oceanicus* (Moiseff et al. 1978). Like noctuid moths, the cricket uses this ability to avoid predatory bats. The process begins with the firing of certain ultrasound-sensitive auditory receptors in the ears, which are found on the cricket's forelegs. Sensory messages from these receptors travel to other cells in the cricket's central nervous system. Among the receivers of these messages is a pair of sensory interneurons called int-1, also known as AN2, one of which is located on each side of the insect's body. By playing sounds of different frequencies to a cricket and recording the resulting neural activity, Ronald Hoy and his coworkers established that int-1 plays a key role in the perception of ultrasound (Moiseff et al. 1978). The researchers found that these cells became highly excited when the cricket's ears were bathed in ultrasound. The more intense a sound was in the 40- to 50-kHz range, the more action potentials the cells produced and the shorter the latency was between stimulus and response—two properties that match those of the A1 receptor in noctuid moths.

The int-1 cells seem to be part of a neural circuit (which also consists of at least 20 ultrasound-sensitive neurons in the brain; Brodfuehrer and Hoy 1990) that helps the cricket respond to ultrasound. If this is true, then it follows that if one could experimentally inactivate int-1, ultrasonic stimulation should not generate the typical reaction of a tethered cricket suspended in midair, which is to turn away from the source of the sound by bending its abdomen (**FIGURE 4.17**). As predicted, crickets with temporarily inactivated int-1 cells do not attempt to steer away from ultrasound, even though their auditory receptors are firing. Thus, int-1 is necessary for the steering response. The corollary prediction is that if one could activate int-1 in a flying, tethered cricket (and one can, with the appropriate stimulating electrode), the cricket should change its body orientation as if it were being exposed to ultrasound, even when it is not. Experimental activation of int-1 is sufficient to cause the cricket to bend its abdomen (Nolen and Hoy 1984). These experiments establish that int-1 activity is both necessary and sufficient for the apparent bat-evasion response of flying crickets. Therefore, these interneurons are a key part of the relay apparatus

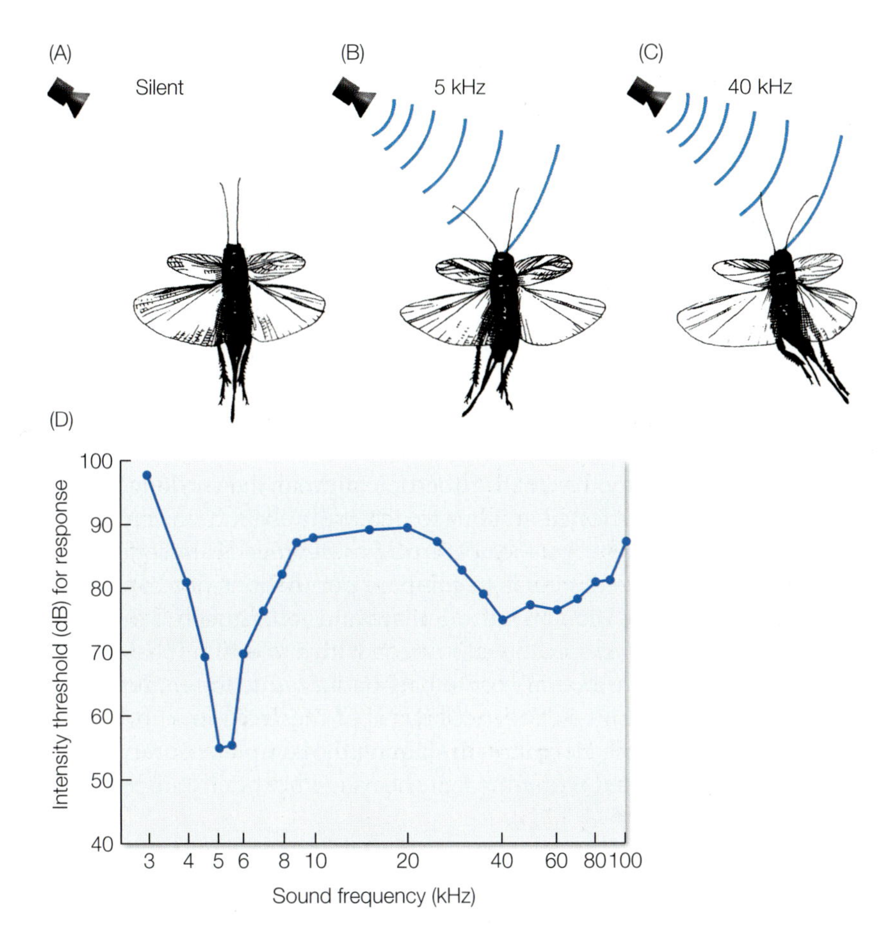

FIGURE 4.17 Avoidance of and attraction to different sound frequencies by crickets. (A) In the absence of sound, a flying tethered cricket holds its abdomen straight. (B) If it hears low-frequency sound, the cricket turns toward the source of the sound. (C) If it hears high-frequency sound, it turns away. Males of this species (*Teleogryllus oceanicus*) produce sounds in the 5-kHz range; some predatory bats produce high-frequency calls of about 40 kHz. (D) The tuning curve of the cricket's int-1 interneuron. Sounds in the 5- to 6-kHz range need only be about 55 decibels (dB) loud in order to trigger firing. The intensity threshold also dips in the neighborhood of 40 kHz, which is in the ultrasonic range commonly produced by bats. (After Moiseff et al. 1978.)

between the receptors and the neurons in the brain that enables the cricket to react adaptively to an ultrasonic stimulus.

How does a flying cricket carry out orders from its brain to steer away from a source of ultrasound? This question attracted the attention of Mike May, who began his study not by conducting a carefully designed experiment but instead by "toying with the ultrasound stimulus and watching the responses of a tethered cricket" (May 1991). As May zapped the cricket with bursts of ultrasound, he noticed that the beating of one hindwing seemed to slow down with each application of the stimulus. Crickets have four wings, but only the two hindwings are directly involved in flight. If the hindwing opposite the source of ultrasound really did slow down, that would reduce power or thrust on one side of the cricket's body, with a corresponding turning (or yawing) of the cricket away from the stimulus.

On the basis of his informal observations, May proposed that the flight path of a cricket was controlled by the position of the insect's hindleg, which when lifted into a hindwing, altered the beat of that wing and thereby changed the cricket's position in space (**FIGURE 4.18**). May went on to take several high-speed photographs of crickets with and without hindlegs. Without the appropriate hindleg

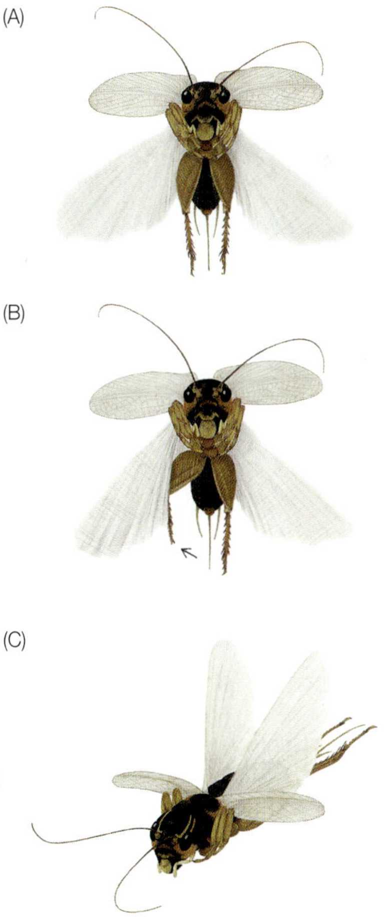

FIGURE 4.18 How to turn away from a bat—quickly. (A) A flying cricket typically holds its hindlegs so as not to interfere with its beating wings. (B) Ultrasound coming from the cricket's left side causes its right hindleg to be lifted up into its right wing. (C) As a result, the beating of the right wing slows, and the cricket turns right, away from the source of ultrasound, as it dives to the ground. (Drawings by Virge Kask, from May 1991.)

to act as a brake, both hindwings continued beating unimpeded when the cricket was exposed to ultrasound. As a result, crickets without hindlegs required about 140 milliseconds to begin to turn, whereas intact crickets started their turns in about 100 milliseconds. These findings led May to assert that neurons in the ultrasound detection network signal the appropriate motor neurons to induce muscle contractions in the opposite-side hindleg of the cricket. As these muscles contract, they lift the leg into the wing, interfering with its beating movement and thereby causing the cricket to veer rapidly away from an ultrasound-producing bat (May 1991).

Finally, building a sophisticated sensory detection and processing system to avoid bats is likely to be costly in terms of both energy expenditure and the diversion of time from other activities such as foraging and mating (Yager 2012). Given the potential costs of producing and using predator detection systems, one would predict that such systems would be lost if they are no longer needed. That is exactly what has happened to the few species of moths that have become diurnal; they no longer have ultrasonic hearing or the sophisticated systems to detect ultrasound. In contrast, a few species of butterflies that have become nocturnal have evolved ultrasonic hearing. Another way to test this prediction is to compare populations of ultrasound-hearing species that vary in predation pressure. Fullard and colleagues did this with populations of *Teleogryllus oceanicus* that lived in areas with and without bat predators and found that populations not subject to bats had a reduction in defensive behavior and in the activity of int-1 (Yager 2012). All of these pieces of evidence illustrate the role that natural selection plays in shaping the complex neurophysiology and neuroanatomy associated with ultrasonic hearing in bat-detecting insects.

Responding to Relayed Messages

We have seen how sensory receptors detect key stimuli before relaying sensory signals via interneurons to other neurons in the central nervous system. When these central cells respond, they can generate signals that turn on a set of motor neurons. The escape dives of moths, the evasive swerves of flying crickets pursued by bats, and even the more subtle changes in direction by migratory butterflies are effective one-step responses triggered by simple releasing stimuli in these animals' environments. Most behaviors, however, involve a coordinated series of muscular responses, which cannot result from a single command from a neuron or neural network. Consider, for example, the escape behavior of the sea slug *Tritonia diomedea*, which is activated when the slug comes in contact with a predatory sea star (**FIGURE 4.19**). Stimuli associated with this event cause the slug to swim in the ungainly fashion of sea slugs, by bending its body up and down (Willows 1971). If all goes well, the slug will move far enough away from the sea star to live another day.

FIGURE 4.19 Escape behavior by a sea slug. The sea slug (*Tritonia diomedea*) in this photograph has just begun to swim away from its deadly enemy, a predatory sea star. The slug's dorsal muscles are maximally contracted, drawing the slug's head and tail together. Soon the ventral muscles will contract and the slug will begin to thrash away to safety. (Photograph by William Frost.)

How does *Tritonia diomedea* manage its multistep swimming response, which requires from 2 to 20 alternating bends, each involving the contraction of a sheet of muscles on the slug's back followed by a contraction of the muscles on its belly? As it turns out, the dorsal and ventral muscles are under the

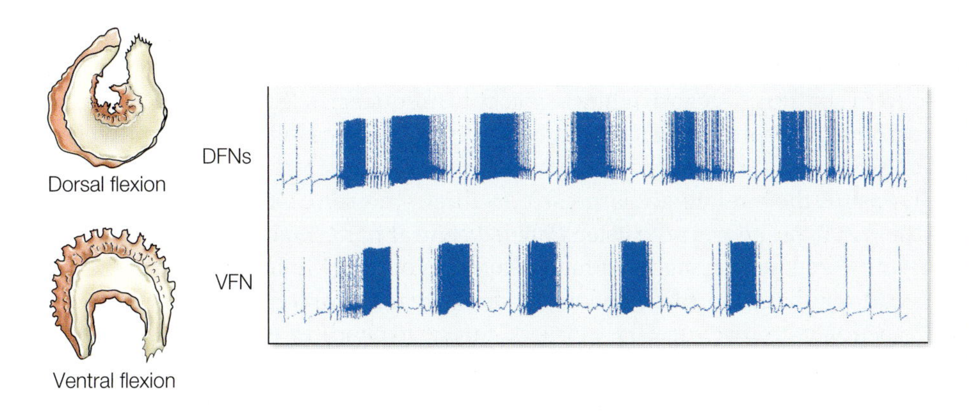

FIGURE 4.20 Neural control of escape behavior in *Tritonia diomedea*. The dorsal and ventral muscles of this sea slug are under the control of two dorsal flexion neurons (DFNs) and a ventral flexion neuron (VFN). The alternating pattern of activity in these two categories of motor neurons translates into alternating bouts of dorsal and ventral bending—the movements that cause this animal to swim. (After Willows 1971.)

control of a small number of motor neurons. The dorsal flexion neurons (DFNs) are active when the animal is being bent into a U, and the ventral flexion neuron (VFN) produces a pulse of action potentials that turn the slug into an inverted U (**FIGURE 4.20**). But what controls the alternating pattern of DFN and VFN activity?

The escape reaction begins when sensory receptor cells in the skin of *Tritonia diomedea* detect certain chemicals on the tube feet of its sea star enemy (**FIGURE 4.21**). The receptors then relay messages to interneurons, among them the dorsal ramp interneurons (DRIs), which, upon receipt of sufficiently strong stimulation, begin to fire steadily. This category of interneurons sends a stream of excitatory signals to several interneurons (the dorsal swim interneurons, or DSIs), which in turn are part of an assembly of interconnected cells, among them the ventral swim interneurons (VSIs) and cerebral neuron 2 (C2), as well as the flexion neurons mentioned already (Getting 1983, 1989). A web of excitatory and inhibitory relations exists within this cluster of interneurons such that, for example, activity in the DSIs turns on C2, which leads to excitation of the DFNs and contraction of the dorsal flexion muscles. After a short period of excitation, however, C2 begins to block the DFNs while sending excitatory messages to the VSIs, leading to activation of the VFN and contraction of the ventral flexion muscles. The situation then reverses. Alternating bouts of activity in the interneurons regulating the DFNs and VFN lead to alternating bouts of DFN and VFN firing, and thus alternation of dorsal and ventral bending (Frost et al. 2001).

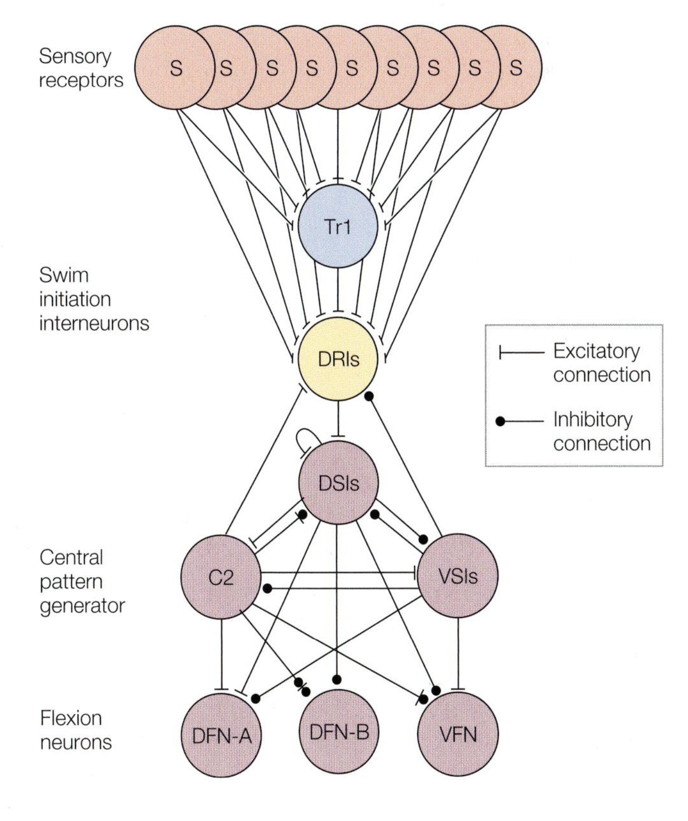

FIGURE 4.21 The central pattern generator of *Tritonia diomedea* in relation to the dorsal ramp interneurons (DRIs) that maintain activity in the cells that generate the sequence of signals necessary for the sea slug to swim to safety. These interneurons receive excitatory input from sensory receptor cells (S) and from another interneuron, Tr1. DRI cells in turn interact with three other categories of neurons—dorsal swim interneurons (DSIs), cerebral neuron 2 (C2), and ventral swim interneurons (VSIs), which are the cells that send messages to the flexion neurons. There are two kinds of dorsal flexion neurons (DFN-A and DFN-B) with somewhat different properties, and one kind of ventral flexion neuron (VFN). (After Frost et al. 2001.)

The capacity of the simple neural network headed by the DRIs to impose order on the activity of the motor neurons that control the dorsal and ventral flexion muscles in sea slugs means that this mechanism qualifies as a **central pattern generator**. Systems of this sort have been particularly well studied in invertebrates, especially with regard to locomotion, because the small number of neurons involved and their relatively large size facilitate their investigation (Clarac and Pearlstein 2007). The neural clusters labeled central pattern generators play a preprogrammed set of messages—a motor tape, if you will—that helps organize the motor output underlying movements of the sort that Tinbergen would have labeled a fixed action pattern.

Central pattern generators are also found in vertebrates, as we can illustrate by looking at the plainfin midshipman (*Porichthys notatus*), a fish in which males "sing" to attract mates by contracting and relaxing certain muscles in a highly coordinated fashion (Bass 1996, Feng and Bass 2017). Only the large type I males of this rather grotesque fish sing "humming" songs, which last more than a minute and sometimes for an hour or more (**VIDEO 4.2**). These large, territorial males sing only at night during spring and summer while guarding certain rocks. Male songs are so loud that they can annoy houseboat owners in the Pacific Northwest. Type I males sing to attract females to defended rocky nest sites where a male guards the eggs that females lay in his territory. In contrast to these larger, singing males, smaller type II males can produce grunts, but they do not sing to attract females as the larger males do. Instead, type II males mimic females and try to sneak onto type I male territories, where they spawn in an attempt to fertilize some of the eggs laid by females. If the sneaker type II males get caught, the territorial type I males chase them away and make a "growl" or series of "grunt" sounds. The type I and II midshipman males are another example of alternative behavioral polymorphisms, which you learned about in Chapter 3.

VIDEO 4.2A-C
Type I male midshipman songs
ab11e.com/v4.2

How do type I male midshipman produce their songs? When Andrew Bass and his coworkers inspected the anatomy of the fish's abdomen, they found a large, air-filled swim bladder with large attached sonic muscles (**FIGURE 4.22**). The bladder serves as a drum; rhythmic contractions of the muscles "beat" the drum, generating vibrations that other fish can hear. Muscle contractions require signals from motor neurons, which Bass found connected to the sonic muscles. He applied a cellular dye called biocytin to the cut ends of these motor neurons, which absorbed the material and stained the cells brown. And the stain kept moving, crossing the synapses between the first cells to receive it and the next ones in the circuit, and so on, through the whole network of cells connected to the sonic muscles. By cutting the brain into fine sections and searching for cells stained brown by biocytin, Bass and his colleagues mapped the fish's sonic control system. In so doing, they discovered three discrete collections of interrelated neurons that generate the signals controlling the coordinated muscle contractions required for midshipman humming. These three clusters in the vocal central

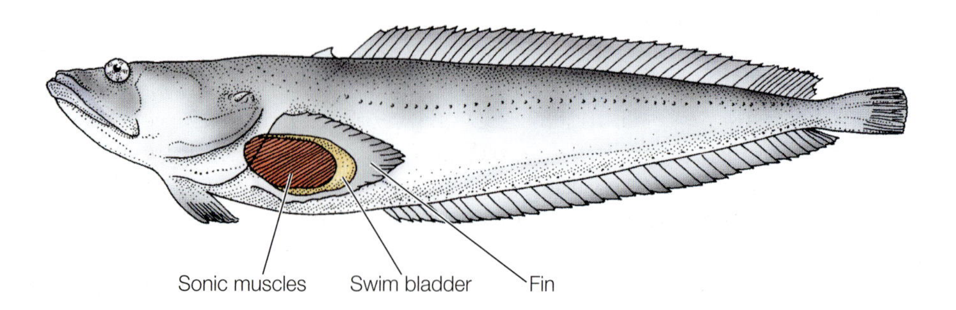

FIGURE 4.22 Song-producing apparatus of the type I male plainfin midshipman fish. The sonic muscles control the movement of the swim bladder, thereby controlling the fish's ability to sing. (After an illustration by Margaret Nelson in Bass 1996.)

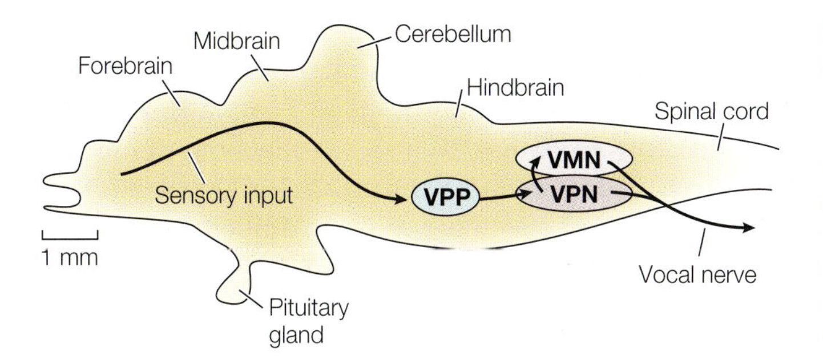

FIGURE 4.23 Neural control of the sonic muscles in the plainfin midshipman fish brain. Signals from the forebrain and midbrain travel to three anatomically and physiologically distinct neuronal populations in the hindbrain: vocal prepacemaker nucleus (VPP), vocal pacemaker nucleus (VPN), and vocal motor nucleus (VMN). Axons from the VMN coalesce into a vocal nerve that exits the brain and connects to vocal muscles in the swim bladder. (After an illustration by Margaret Nelson in Bass 1996.)

pattern generator—vocal motor nucleus (VMN), vocal pacemaker nucleus (VPN), and vocal prepacemaker nucleus (VPP)—are located in the upper part of the spinal cord near the base of the brain (**FIGURE 4.23**). The long axons of their neurons travel out from the brain, fusing together to form a vocal nerve, which reaches the sonic muscles. VMN is the final and farthest downstream node in the circuit that relays outputs of the central pattern generator to vocal muscles that ultimately help produce male song. VMN is regulated by input from VPN in the form of pacemaker-like action potentials. Upstream of VPN lies VPP, which generates sustained neuronal depolarizations, the duration of which regulates and correlates with the duration of VPN–VMN activity (Feng and Bass 2017).

As one might expect, there are differences in the size and developmental patterns of the vocal circuits in type I and type II male plainfin midshipman (Feng and Bass 2017). For example, type I males have larger motor neurons (and a larger volume VMN) than type II males, although there is no difference in the number of motor neurons between the two types of males. In contrast, type II males reach sexual maturity before type I males, suggesting there is a life history trade-off between investment in the strategy of type I males (larger body size and vocal motor system) and type II males (earlier reproduction and gonad development). And as you will see in Chapter 5, much of the vocal behavior that these male midshipman and other vertebrate species produce is actually influenced by steroid hormones, which play a central role in translating environmental stimuli into behavioral responses.

The Proximate Basis of Stimulus Filtering

We have now looked at the attributes of individual neurons, neural clusters called nuclei, and neural circuits that are involved in the detection of certain kinds of sensory information, the relaying of messages to other cells in the nervous system, and the control of motor commands that are sent to muscles. The effective performance of these basic functions is promoted by **stimulus filtering**, the ability of neurons and neural circuits to ignore—to filter out—vast amounts of potential information in order to focus on biologically relevant elements within the diverse stimuli bombarding an animal.

The noctuid moth's auditory system offers a lesson on the operation and utility of stimulus filtering. First, the A1 receptors are activated only by acoustic stimuli, not by other forms of stimulation. Moreover, these cells completely ignore sounds of relatively low frequencies, which means that moths are not sensitive to the stimuli produced by chirping crickets or croaking frogs—sounds they can safely ignore. Finally, even when the A1 receptors do fire in response to ultrasound, they do little to discriminate between different ultrasound frequencies.

Not all systems work this way, however. In humans, for example, auditory receptors produce distinct messages in response to sounds of different frequencies, which is why we can tell the difference between C and C-sharp. But the noctuid moth's sensory apparatus appears to have just one task of paramount importance: detecting cues associated with echolocating predators. To this end, the moth's auditory capabilities are tuned to pulsed ultrasound at the expense of all else. Upon detection of these critical inputs, the moth can take effective action. This fine-tuning of the moth's auditory system makes perfect sense in the context of natural selection. If we make a mistake differentiating between a C and a C-sharp, nothing much will happen, particularly in terms of our fitness (though the ability to discriminate between similar sounds plays an important role in speech discrimination in our species). But if a moth fails to detect a bat's ultrasound, mistaking it for some other type of sound, the moth will likely be eaten, and unable to pass its genes on to the next generation. In other words, the moth's auditory system is under very strong natural selection, and stimulus filtering plays a central role in this process.

The relationship between stimulus filtering and a species' reproductive success is evident in every animal whose sensory systems have been carefully examined. Consider the male midshipman fish that listens to the underwater grunts, growls, and hums produced by others of his species. These signals are dominated by sounds in the 60- to 120-hertz (Hz) range. The auditory receptor cells in the hearing organs of these fish are most sensitive to sounds in exactly that range (Sisneros 2007). In summer, however, when the humming "songs" of territorial males incorporate higher-frequency sounds, the female's auditory system changes, providing females in search of spawning partners with sensitivity to sounds that range up to 400 Hz (Sisneros and Bass 2003). In this case, the female auditory system is the one using stimulus filtering, and it is doing so on a seasonal basis as females listen to sounds in their underwater world.

The screening of acoustic stimuli also occurs in certain parasitoid flies, which use their hearing to locate singing male crickets (or other insects; **BOX 4.2**), the better to place their larvae on these insects. The little maggots burrow into the unlucky crickets and proceed to devour them from the inside out. For example, larvae-laden female *Ormia ochracea* flies can find food for their offspring because they have ears tuned to cricket calls, as researchers discovered when they found these flies coming to loudspeakers that were playing tapes of cricket song at night. The unique ears of the female fly consist of two air-filled structures with tympanic membranes and associated auditory receptors on the front of the thorax. Vibration of the fly's tympanic "eardrums" activates the receptors, just as in noctuid moth ears, and thus provides the fly with information about (certain) sounds in its environment. Moreover, even with their tiny ears, these flies are able to localize sound with a precision equal to that of humans (Mason et al. 2001). As predicted by a team led by Ronald Hoy, the female fly's auditory system is most sensitive to the dominant frequencies in cricket songs (**FIGURE 4.24**). That is, the female fly can hear sounds of 4 to 5 kHz (the sort produced by crickets) more easily than sounds of 7 to 10 kHz, which have to be much louder if they are to generate

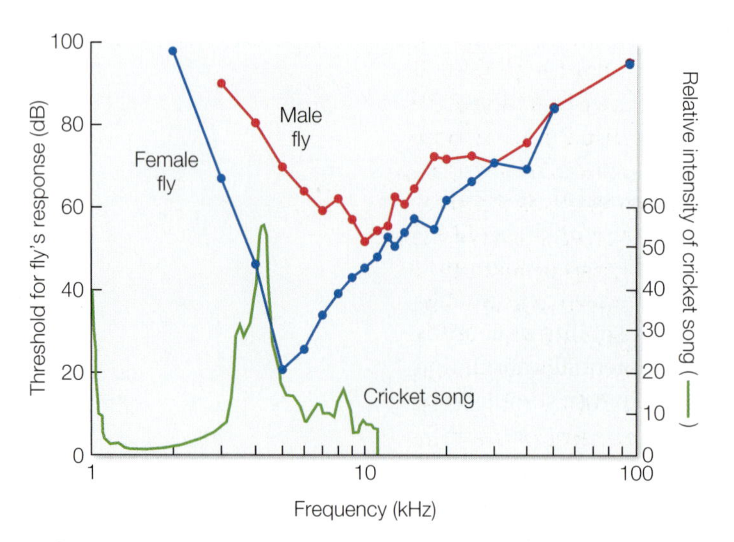

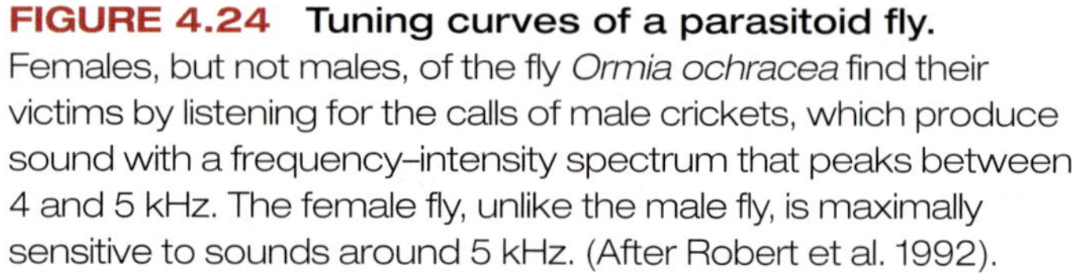
FIGURE 4.24 Tuning curves of a parasitoid fly. Females, but not males, of the fly *Ormia ochracea* find their victims by listening for the calls of male crickets, which produce sound with a frequency–intensity spectrum that peaks between 4 and 5 kHz. The female fly, unlike the male fly, is maximally sensitive to sounds around 5 kHz. (After Robert et al. 1992).

(A)

(B)

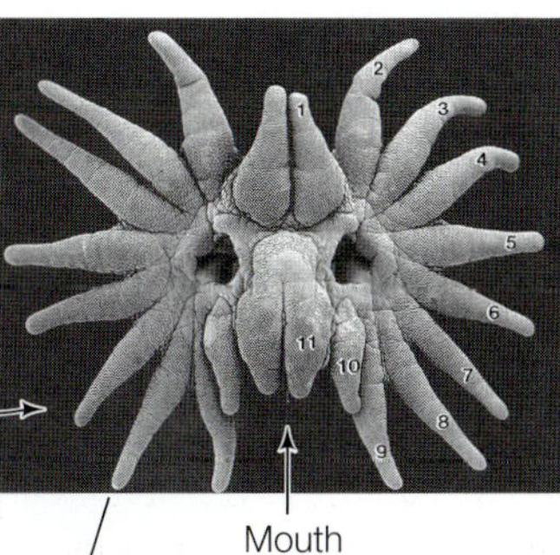

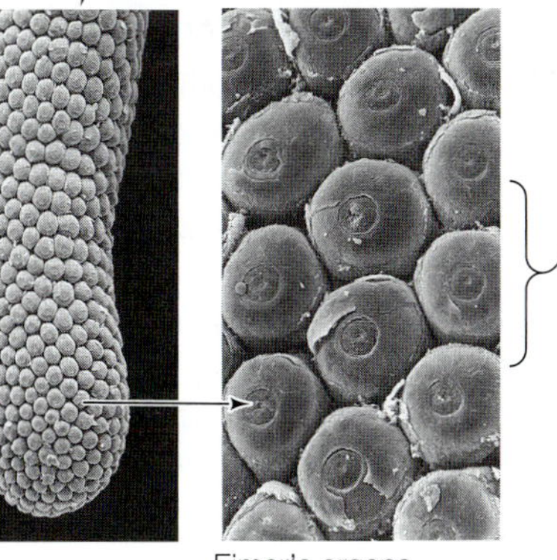

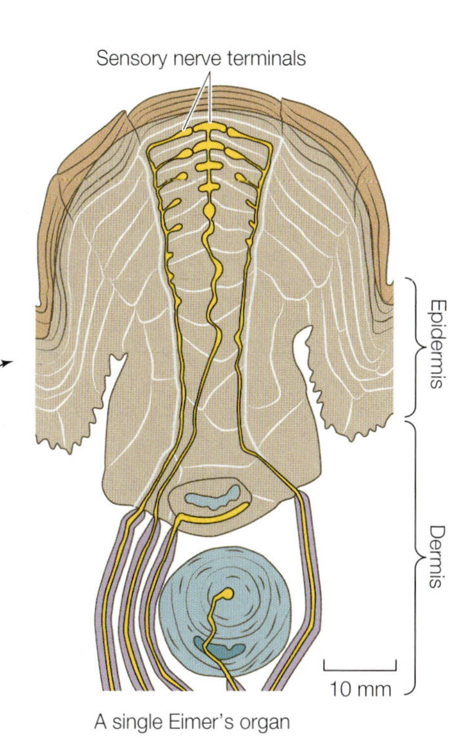

FIGURE 4.25 A special tactile apparatus. (A) The star-nosed mole has a remarkable pink nose. (B) The 22 appendages of the mole's nose are covered with thousands of Eimer's organs. Each organ contains a variety of specialized sensory cells that respond to mechanical deformation of the skin above them. (A, photograph © Rod Planck/Science Source; B, photograph courtesy of Kenneth C. Catania; drawing after Catania and Kaas 1996.)

any response (Robert et al. 1992). In contrast, male *Ormia ochracea* are not especially sensitive to sounds of 4 to 5 kHz, which makes ecological sense since males do not track down singing male crickets.

Stimulus filtering takes place not only at the level of sensory receptors and sensory interneurons, but also at the level of the brain, as we can see by looking at the star-nosed mole (*Condylura cristata*). This odd-looking mammal lives in wet, marshy soil, where it burrows in search of earthworms and other prey (**FIGURE 4.25A**). Because the mole inhabits dark tunnels where earthworms cannot be seen, its eyes are greatly reduced in size and it largely ignores visual information, even when light is available. Instead, the mole relies heavily on touch to find its food, using its wonderfully strange nose to sweep the tunnel walls as it moves forward. Its two nostrils are ringed by 22 fleshy appendages, 11 on each side of the nose (**FIGURE 4.25B**). Although these appendages cannot grasp or hold, they are covered with thousands of tiny sensory devices called Eimer's organs. The typical star-nosed mole has approximately 25,000 Eimer's organs (Catania 2011), each containing several different kinds of sensory cells that appear to be dedicated to detecting objects touching the nose (Marasco and Catania 2007). With these mechanoreceptors, the animal can collect extremely complex patterns of information about the things it encounters underground, enabling it to identify prey items in the darkness with great rapidity (Catania and Kaas 1996, Catania and Remple 2005). In fact, the star-nosed mole touches its nose to different areas of the environment 13 times per second, which means it could potentially investigate 46,000 square centimeters of surface area per hour if left uninterrupted (**VIDEO 4.3**) (Catania and Remple 2004).

Whenever the mole brushes an earthworm with, say, appendage 5, it instantly sweeps its nose over the prey so that the two projections closest to the mouth,

BOX 4.2 **EXPLORING BEHAVIOR** BY **INTERPRETING DATA**

Determining how female parasitoid wasps choose their singing male bush-cricket hosts

Females of the parasitoid fly *Therobia leonidei* search for singing male bush-crickets (commonly referred to as katydids in North America). The male Veluchi bright bush-cricket (*Poecilimon veluchianus*; **FIGURE A**) produces an ultrasonic stridulatory mate-attracting call that falls largely in the 20- to 30-kHz range (**AUDIO 4.1**). In Greece, *Therobia leonidei* parasitizes several species, including both *P. veluchianus* and *P. mariannae,* whose songs differ in structure but whose activity patterns, body sizes, and population sizes are similar. The primary difference in song between these two species is the number of syllables per chirp, which ranges from 1 in *P. veluchianus* to 5 to 11 in *P. mariannae.* Gerlind Lehmann and Klaus-Gerhard Heller tested the prediction that the species with the polysyllabic song should be parasitized more than the species with the monosyllabic song (Lehmann and Heller 1998).

(A)

FIGURE A Song in male bush-crickets. A male Veluchi bright bush-cricket. (Photograph © Wolfgang Wagner.)

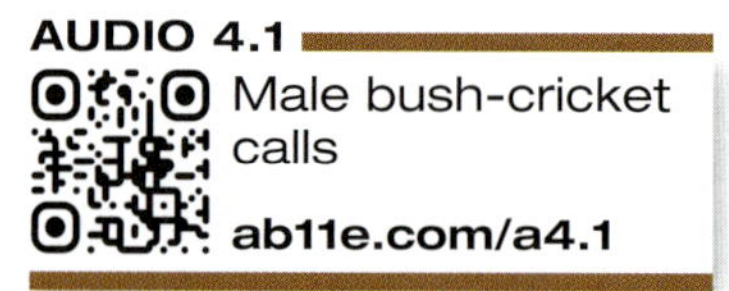

Thinking Outside the Box

Examine **FIGURE B**, which shows how the parasitoid fly's ears are tuned to different frequencies. What sound frequencies should elicit maximal response in the ears of this bush-cricket-hunting parasitoid fly if stimulus filtering enables the animal to achieve biologically relevant goals? Why is there a difference between the tuning curves of male and female flies? Now look at **FIGURE C**. Are these data consistent with Lehmann and Heller's hypothesis? Why or why not? What can you conclude from these experiments about how female *Therobia leonidei* flies choose hosts?

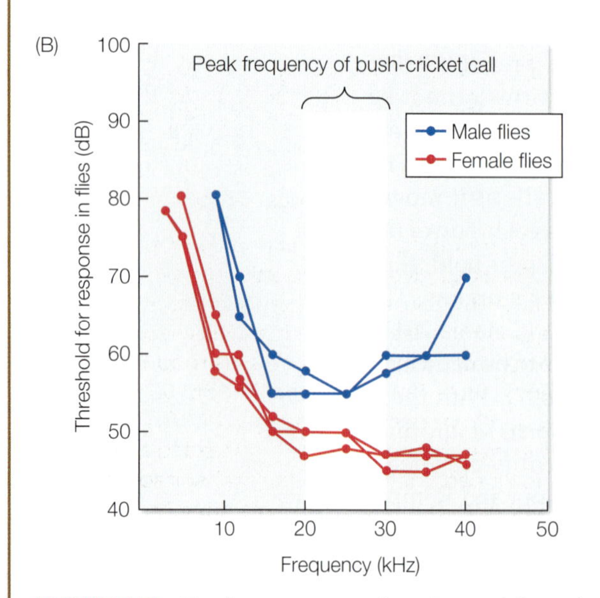

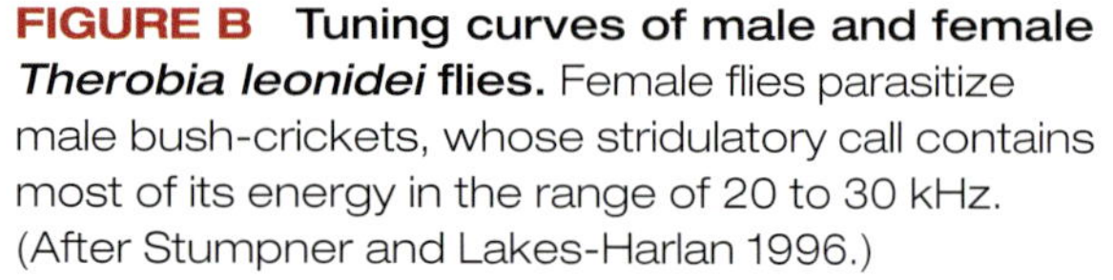

FIGURE B Tuning curves of male and female *Therobia leonidei* flies. Female flies parasitize male bush-crickets, whose stridulatory call contains most of its energy in the range of 20 to 30 kHz. (After Stumpner and Lakes-Harlan 1996.)

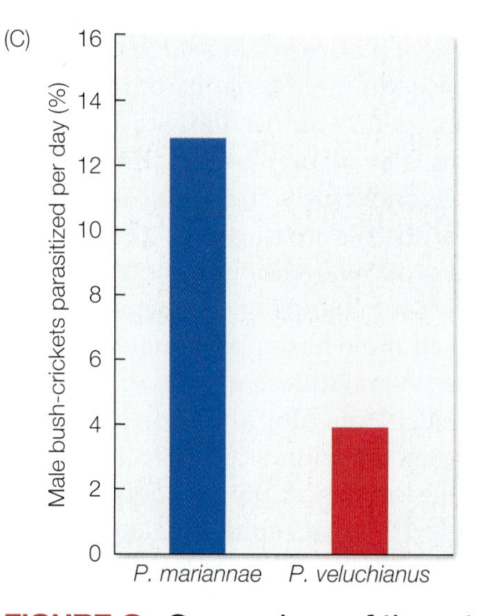

FIGURE C Comparison of the rates of parasitism by *P. mariannae* (polysyllabic songs) and *P. veluchianus* (monosyllabic songs) as the percentage of male bush-crickets parasitized per day. (After Lehmann and Heller 1998.)

labeled appendage 11 (**FIGURE 4.26A**), come in contact with the object of interest. The tactile receptors on each appendage 11 generate a volley of signals that are carried by nerves to the brain. Although these two nose "fingers" contain only about 7 percent of the Eimer's organs on the star nose, more than 10 percent of all the nerve fibers relaying information from the nose's touch receptors to the brain come from these two appendages. In other words, the mole uses relatively more neurons to relay information from appendage 11 than from any other appendage. For this reason, appendage 11 has been referred to as a tactile fovea (after the fovea in the eye, a small depression in the retina where visual acuity is highest).

Not only is the mole's relay system biased toward inputs from appendage 11, but the animal's brain also gives extra weight to signals from this part of the nose. The information from the nose travels through nerves to the somatosensory cortex, the part of the brain that receives and decodes sensory signals from touch receptors all over the animal's body (**FIGURE 4.26B**). Of the portion of the somatosensory cortex that is dedicated to decoding inputs from the 22 nose appendages, about 25 percent deal exclusively with messages from the two appendage 11s (**FIGURE 4.26C**) (Catania and Kaas 1997). This discovery was made by Kenneth Catania and Jon Kaas when they recorded the responses of cortical neurons as they touched different parts of an anesthetized mole's nose. Perhaps the mole's brain is "more interested in" information from appendage 11 because of its location right above the mouth; should signals from this appendage activate a cortical order to capture a worm and consume it, the animal is in the right position to carry out the action immediately (Catania and Kaas 1996).

(A)

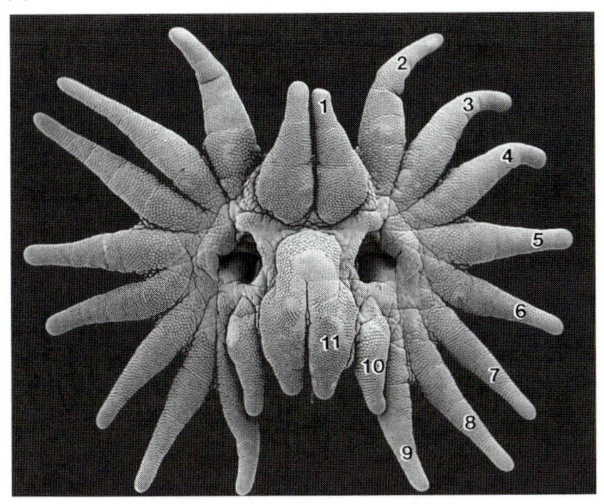

(B)

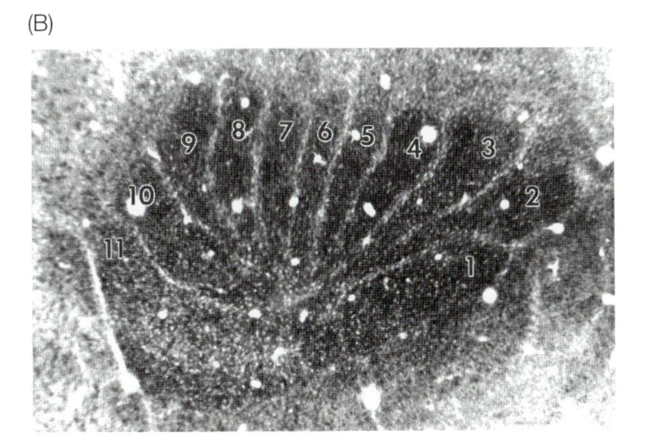

(C)

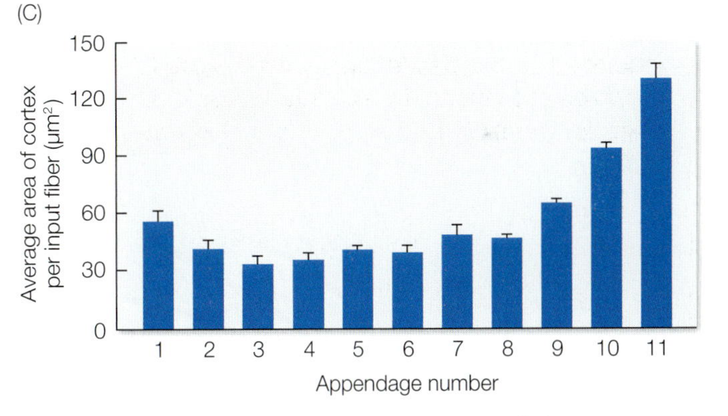

FIGURE 4.26 The cortical sensory map of the star-nosed mole. The map gives disproportionate weight to appendage 11 of the mole's nose. (A) The nose of the mole, with each appendage numbered on one side. (B) A section through the area of the somatosensory cortex that is responsible for analyzing sensory inputs from the nose. The cortical areas that receive information from each appendage are numbered. (C) The amount of somatosensory cortex devoted to the analysis of information from each nerve fiber carrying sensory signals from the different nose appendages. Bars depict mean +/– SE. (A after Catania 1995; B, C after Catania and Kaas 1995.)

In the star-nosed mole, the disproportionate investment in brain tissue to decode tactile signals from one part of the nose is mirrored on a larger scale by the biases evident in the somatosensory cortex as a whole, which focuses on signals from the mole's forelimbs and nose at the expense of other parts of the body. This pattern of cortical magnification—which occurs in all mammals (**BOX 4.3**)—makes adaptive sense for this species because of the biological importance of the mole's forelimbs for burrowing and of its nose for locating prey (**FIGURE 4.27A**). If this argument is correct, then we can predict that the allocation of cortical tissue to somatosensory inputs will differ from species to species in ways that make sense given the particular environmental problems each species has to solve. And it is true that other insectivores, even those related to the star-nosed mole, exhibit their own adaptively different patterns of cortical magnification whereby inputs come from completely different organs, such as vibrissae (whiskers) (**FIGURE 4.27B–D**).

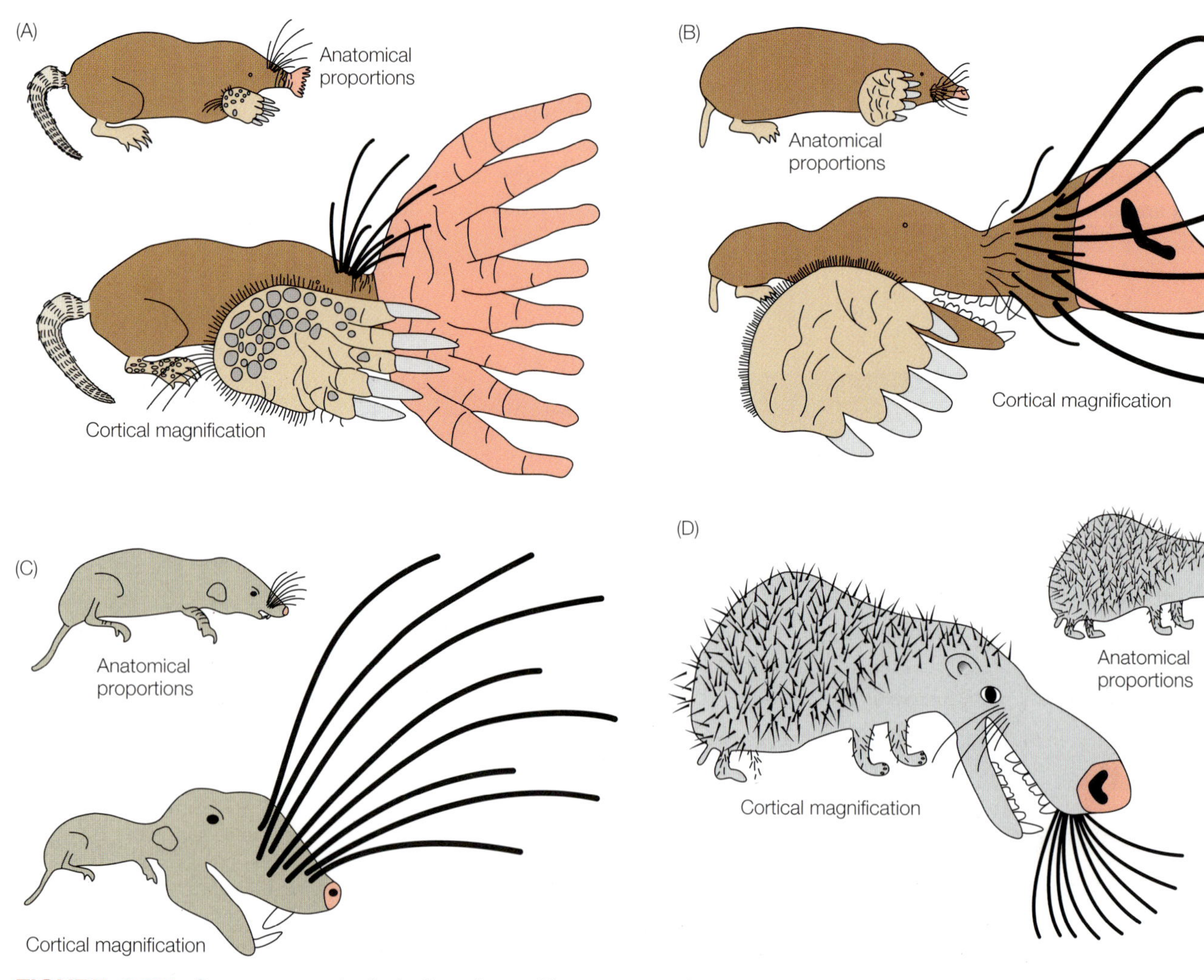

FIGURE 4.27 Sensory analysis in four insectivores. In each case, the smaller drawing shows the actual anatomical proportions of the animal; the larger drawing shows how the body is proportionally represented in the somatosensory cortex of the animal's brain. (A) The star-nosed mole devotes much more somatosensory cortex to processing inputs from its nose and forelimbs than it does to processing those coming from receptors in other parts of its body. (B) Cortical magnification in the eastern mole (*Scalopus aquaticus*) focuses on sensory inputs from the forelimbs, nose, and sensory hairs, or vibrissae, around the nose. (C) Cortical magnification in the masked shrew (*Sorex cinereus*) also reveals the importance of the vibrissae. (D) Sensory signals from the vibrissae are magnified cortically to a lesser degree in the East African hedgehog (*Atelerix albiventris*). (After Catania and Kaas 1996, Catania 2000.)

BOX 4.3 EXPLORING BEHAVIOR BY INTERPRETING DATA

Cortical magnification in mammals

Brains evolve in response to selection pressures associated with particular physical and social environments. The drawings in the **FIGURE** are based on the amount of brain tissue devoted to tactile signals coming from different parts of the bodies of human beings and of the naked mole rat (*Heterocephalus glaber*), which lives in groups in underground tunnels that the animals dig. For each species, the drawing on the left shows the actual anatomical proportions; the drawing on the right shows how the body is proportionally represented in the somatosensory cortex.

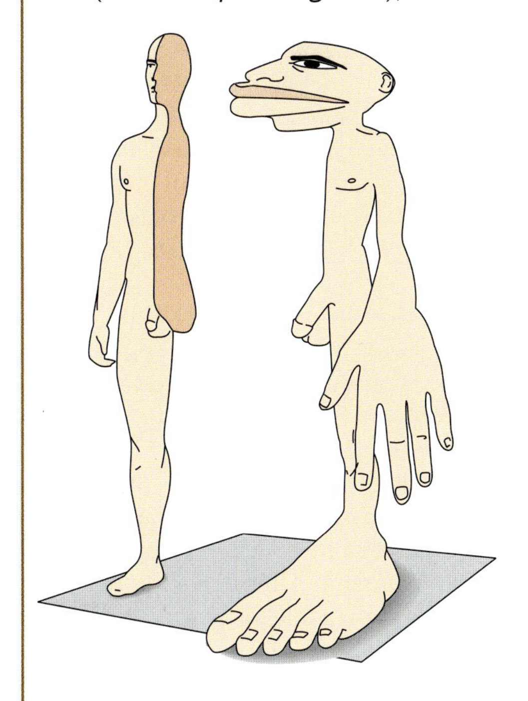

Cortical maps of humans and naked mole rats. (Cortical map of human male based on data from Kell et al. 2005; cortical map of naked mole rat drawn by Lana Finch in Catania and Remple 2002.)

Thinking Outside the Box

In what ways do these two maps support the argument that animal brains exhibit adaptive sensory biases? Think about the similarities and differences between humans and mole rats in terms of the environments in which they live and which sensory processes they rely on most.

The Evolution of Cognitive Skills

Up until now, this chapter has focused primarily on the nervous system and how attributes of individual neurons influence the detection of sensory information, which is passed to other cells in the nervous systems, oftentimes all the way to the brain where it ultimately leads to the production of behavior through changes in gene expression or motor control. The brain is a complex organ, and studying it in great detail is beyond the scope of this book. However, it is important to think about one of the main functions of the brain: **cognition**, the mental process of acquiring knowledge and understanding through thought, experience, and the senses. As a starting point, think back to Chapter 3, where we introduced the idea of developmental learning. Because of our considerable intelligence, we humans tend to be interested in the evolution of learning, thinking, and perception. If, as is generally assumed, our capacity to solve problems, to employ rational thought, and to think logically is related to our very large cerebral cortex, then we have a Darwinian puzzle to solve (**BOX 4.4**).

Although our cognitive skills appear more varied and sophisticated than those of honey bees and creatures with much smaller brains, consider what the New Caledonian crow (*Corvus moneduloides*) can do with a brain only a small fraction of the size of ours. This bird routinely modifies a wide variety of objects, such as pandanus palm leaves, sticks, grass stems, and the like, and uses them to extract beetle grubs and other food items from their hiding places (**FIGURE 4.28**). In the lab, captive New Caledonian crows have proven to be ingenious as they solve novel problems of food extraction that, for example, require a crow to use one short stick to secure a longer stick, which the bird then uses to secure some meat that is otherwise out of reach. In an experiment with this species, Taylor and colleagues found that four of seven birds tested made use of the two tools successfully on the first attempt (Taylor et al. 2010). Just why this crow, an inhabitant of a remote Pacific island, should have evolved such impressive problem-solving abilities is a bit of a mystery. Perhaps it is something unique about its island environment? For example, there are no woodpeckers on New Caledonia, possibly creating an open niche for a tool-using woodpecker substitute. Moreover, related species such as the broadly distributed common raven (*Corvus corax*) are widely recognized to be highly intelligent animals (Heinrich 1999).

social brain hypothesis Large brains evolved as a means of surviving and reproducing in large and complex social groups.

The rather asocial New Caledonian crow (Holzhaider et al. 2011) does not really fit the mold for a widely discussed hypothesis for the evolution of cognitive abilities called the **social brain hypothesis**, which proposes that advanced problem solving and the like evolved in the context of dealing with the obstacles to reproductive success posed by interacting socially with members of one's own species (Barton and Dunbar 1997, Dunbar 2003). The logic of this explanation is that an increase in intelligence in some individuals could favor an increase in (social) intelligence in others, setting up a positive feedback loop leading to ever greater species-wide cognitive abilities up to a point. Although early derivations of the social brain hypothesis emphasized the brain's role in sensory and technical

FIGURE 4.28 A New Caledonian crow using a tool. This bird is unusual in its ability to modify objects in its environment in ways that improve their usefulness as tools for extracting out-of-reach beetle larvae from their burrows. (Photograph © Nicolas-Alain Petit/Science Source.)

BOX 4.4 **DARWINIAN PUZZLE**

Do energetic demands explain why humans have such large brains?

Humans have bigger brains—up to three times larger—than other great apes. To understand why we have evolved such large brains, we must first examine another paradox in humans. In most animals, species that reproduce quickly also tend to die quickly. However, humans not only reproduce more often and have larger young than the other great apes, we also live the longest (**FIGURE A**). One hypothesis to explain our unique ability to reproduce quickly and live a long time is that humans have evolved an accelerated metabolic rate and larger energy budget, which accommodates not only greater reproductive output and longer life spans without the expected energetic trade-offs, but also larger brains (Pontzer et al. 2016). In other words, according to this **metabolic rate hypothesis**, both our unique life history traits and our large brains might be the result of our ability to collect and use more energy than our closest relatives.

We've long known that the human brain is an expensive organ, especially for a developing infant and the infant's mother (Roth and Dicke 2005). After all, very small brains are capable of remarkable feats of learning; witness the honey bee's ability to learn where and when flowers will be available to provide the bee and its recruits with pollen and nectar (see Chapter 8). Honey bees (*Apis mellifera*) have only about 960,000 neurons in their grass-seed-sized brains (Menzel and Giurfa 2001), whereas humans have 10,000 times as many neurons packed in a much larger organ.

To test the hypothesis that the human lineage has experienced an acceleration in metabolic rate, providing energy for larger brains and faster reproduction without sacrificing maintenance and longevity, a team of researchers led by Herman Pontzer used doubly labeled water measurements of total energy expenditure (TEE) in humans, chimpanzees and bonobos, gorillas, and orangutans from zoos all over the world (Pontzer et al. 2016). In support of this metabolic rate hypothesis, the researchers found that human TEE exceeded that of the other apes (**FIGURE B**), readily accommodating the cost of humans' greater brain size and reproductive output. Much of the increase in TEE was attributable to humans' greater basal metabolic rate, indicating increased organ metabolic activity. Moreover, humans also had the greatest body fat percentage. Thus, an increased metabolic rate, along with changes in energy allocation, was crucial in the evolution of human brain size and life history.

(Continued)

metabolic rate hypothesis Humans have experienced an acceleration in metabolic rate, providing energy for larger brains and faster reproduction without sacrificing maintenance and longevity.

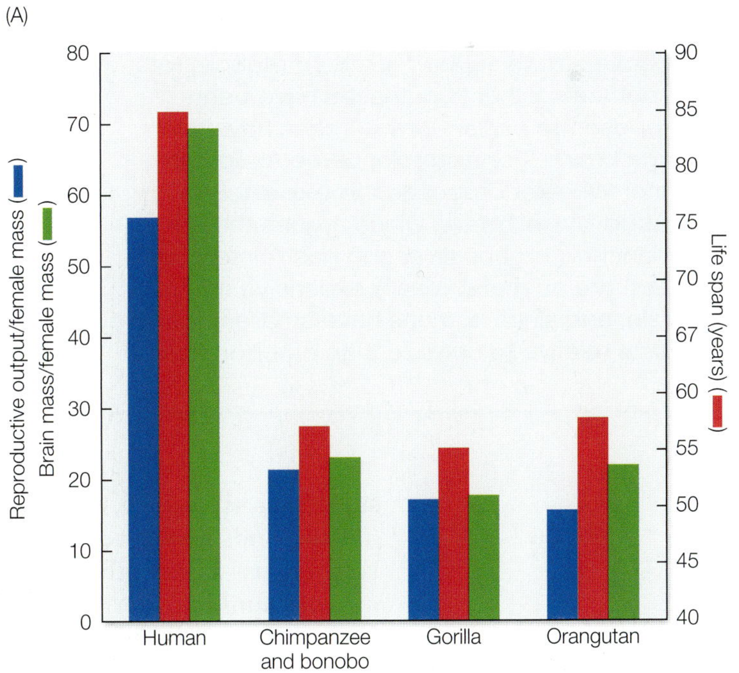

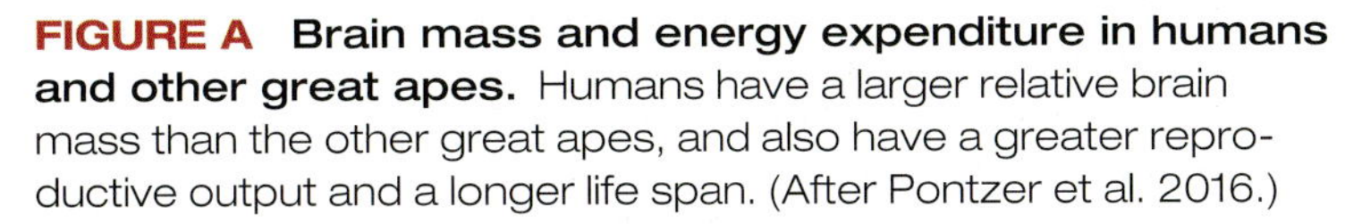

FIGURE A Brain mass and energy expenditure in humans and other great apes. Humans have a larger relative brain mass than the other great apes, and also have a greater reproductive output and a longer life span. (After Pontzer et al. 2016.)

BOX 4.4 DARWINIAN PUZZLE (*continued*)

FIGURE B Humans have a greater total energy expenditure (TEE) than the other great apes, here shown relative to fat-free mass. Lines and shaded regions indicate least squares regressions and 95 percent confidence intervals for each species. (After Pontzer et al. 2016.)

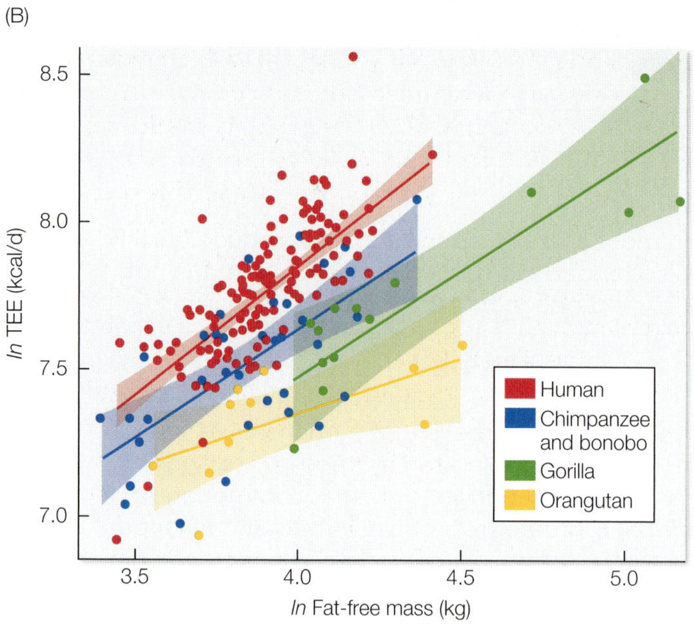

Thinking Outside the Box

Our ability to find and use food better than our ancestors sustains our reproductive output and our energetically expensive brain tissue. What benefits could we derive from a big brain that might also compensate for the metabolic and developmental costs of all those neurons? Despite strong support for the metabolic rate hypothesis, other hypotheses have been proposed to explain why humans have such large brains. For example, Leslie Aiello and Peter Wheeler proposed the **expensive-tissue hypothesis**, which argues that the splanchnic organs (liver and gastrointestinal tract) are as metabolically expensive as the brain, and since humans have smaller digestive tracts relative to our size than other primates, we escape paying those costs for the brain by reducing the size of these other organs (Aiello and Wheeler 1995). In contrast, Ana Navarrete and colleagues argued that the human ability of bipedal locomotion could have led to major reductions in energy expenditure (Navarrete et al. 2011). How might these hypotheses explain the evolution of large brain size? Do you think these are really alternative hypotheses, or might they also be related to energy demand and metabolic rate?

expensive-tissue hypothesis The high metabolic requirements of human brains are offset by a corresponding reduction of their splanchnic organs (liver and gastrointestinal tract), which are as metabolically expensive.

skills (such as foraging, innovation, and navigation), it later focused on the cognitive demands associated with living in large, complex societies that selected for large brains—tasks such as maintaining group cohesion during conflict, enhancing communication, or improving recognition skills (Dunbar and Shultz 2007). Later still, this hypothesis came to encompass the idea that it may have been the particular demands of the more intense forms of pair-bonding found in many social species that selected for a larger brain. Although the New Caledonian crow is not particularly social, it is a bird with long-lasting pair-bonds and biparental care, social features of birds known to have relatively large brains (Shultz and Dunbar 2010). Humans also form pair-bonds, of course, and they and many other species live in groups in which one's reproductive success may be linked to one's social standing and ability to persuade, coerce, or manipulate others to engage in tasks that require cooperation (such as defending the group's territory against intruders from other bands).

VIDEO 4.4 Interacting with puzzle box ab11e.com/v4.4

If the social brain hypothesis is correct, we would predict that social mammals would have particularly large brains for their body size. A case in point is the spotted hyena (*Crocuta crocuta*), which lives in highly stratified clans whose members cooperatively hunt prey and work together to battle other clans and other predators over control of food and good hunting areas. This species does, as predicted from the social brain hypothesis, have a larger brain relative to body volume than other closely related but solitary hyena species (Sakai et al. 2011). However, despite evidence from single species such as the spotted hyena and New Caledonian crow that is consistent with the social brain hypothesis, comparative evidence is more mixed. For example, Perez-Barberia and colleagues found an association between brain size and sociality in three orders of mammals (ungulates, carnivores, and primates) (Pérez-Barberia et al. 2007), but later work in carnivores (Finarelli and Flynn 2009) and primates (DeCasien et al. 2017) showed no support for the social brain hypothesis. In particular, the study in carnivores used a larger set of 289 species (including 125 fossil taxa) than the previous mammal study, and the study in primates found that ecological factors (diet) were a better predictor of brain size than were social factors.

Perhaps a better way to test the social brain hypothesis than looking at some relatively crude estimate of sociality is to look more directly at the relationship between brain size and problem solving. Brain size has been shown to be related to complex tasks in several species. For example, male satin bowerbirds (*Ptilonorhynchus violaceus*) build complicated structures called bowers to attract females, and species that build more complicated bowers have larger brains (Madden 2001) (see Chapter 9).

Christine Drea and Allisa Carter used knowledge about the socio-ecological factors affecting spotted hyenas to predict that this animal would use its large brain to cooperatively solve ecologically relevant problems (Drea and Carter 2009). To test this possibility, they devised a task for captive hyenas that required the coordinated activity of two individuals, each of which had to pull on a separate rope at the same time to open a trap door and spill food onto the floor of the cage (**FIGURE 4.29**). Pairs of hyenas spontaneously and quickly solved the problem, coordinating their rope pulling so as to gain access to the food—using much the same ability that helps wild hyenas capture big game animals. Using a similar approach, in a study with 39 species of carnivores housed in North American zoos, Sarah Benson-Amrama and colleagues presented individuals with a puzzle box baited with food (Benson-Amrama et al. 2016). To open the box, which was scaled to the body size of the test subject, individuals needed to solve a technical problem: sliding a latch (**VIDEO 4.4**). The researchers found that species with larger brains were more successful at opening the boxes. However, they also found that measures of the species' social complexity failed to predict opening success, again showing little support for predictions from the social brain hypothesis.

FIGURE 4.29 An experimental demonstration of hyena cooperation. Two spotted hyenas have spontaneously figured out that they both have to pull ropes at the same time in order to secure food that is out of reach. (Photograph by Christine Drea.)

If sociality or social complexity does not influence brain size or cognitive ability, why are so many social species considered

domestication hypothesis Domesticated animals evolved an inherent sensitivity to human gestures but non-domesticated animals did not.

to be "intelligent"? This question has been addressed in studies of how the intelligence of wolves differs from that of dogs, both of which are highly social species. Dogs are descended from wolves but are domesticated and have therefore interacted and evolved with humans for thousands of years. According to the **domestication hypothesis**, during domestication, dogs evolved an inherent sensitivity to human gestures that non-domesticated wolves do not share. Brian Hare and others predicted that dogs should be much more capable than wolves of detecting and responding to cues about the location of food provided by humans (Hare et al. 2002). Hare and colleagues have found that dogs—including puppies—are indeed far more attentive to the actions of humans than are wolves, even those that have been raised by people. For example, when dogs are presented with two dishes, one covering a food item and the other not, they respond to a person pointing to or even gazing at the dish with food (**FIGURE 4.30**), whereas wolves are significantly less likely to do so. The authors' conclusions, however, have been met with skepticism by some researchers, as wolves socialized by humans can match or even exceed the performance of dogs in some socio-cognitive tasks (Udell et al. 2008, 2010).

The eagerness with which dogs attend to human handlers is reflected in the results of another experiment in which a researcher hid a toy several times at a particular location, then repeated the experiment after hiding the toy in another place. The dogs typically continued to look for the toy at the first hiding place, just as human infants do, but socialized wolves were less likely to keep looking in the original spot (Topál et al. 2009). These findings suggest that dogs are like 10-month-old babies in that they pay more attention to the instructions they receive from a human companion than to their own observations. According to some researchers, dogs and humans are creatures with evolved brains that reflect the effect of a particular social environment associated with domestication.

Studies of this sort demonstrate that the kinds of problem-solving behaviors exhibited by animal species differ in ways related to the environments in which those species evolved. A final demonstration of this point comes from studies of *Polistes* paper wasps in which Elizabeth Tibbetts and her coworkers have shown that *P. fuscatus* does an excellent job of recognizing individual wasps by their facial color patterns (Sheehan and Tibbetts 2010, 2011). These wasps demonstrate they can tell the difference between faces when they are put into a T-maze with one non-electrified arm and another where a wandering wasp will receive a mild shock. The two

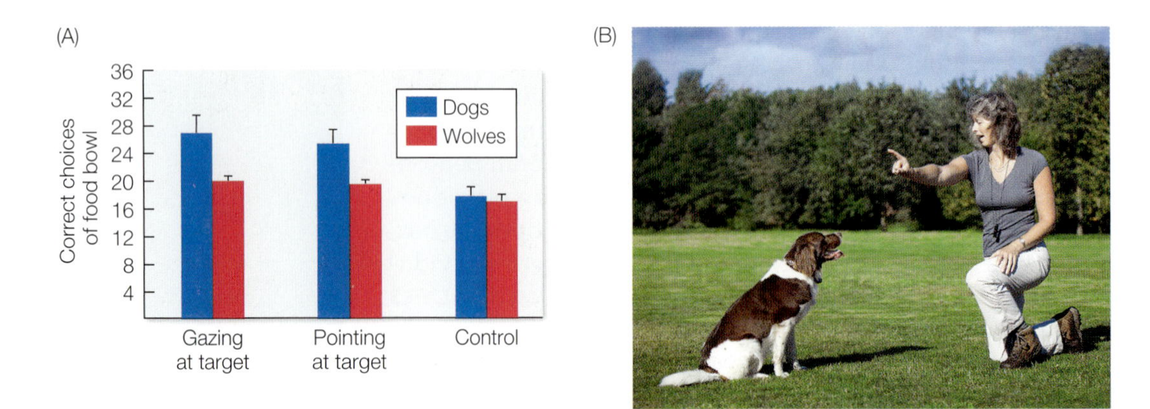

FIGURE 4.30 Dogs have evolved the ability to pay attention to humans. (A) Dogs do better than wolves reared in the same way when it comes to using human-provided cues (such as gazing at a food bowl or pointing at it) to go to the correct bowl with food. Bars depict mean +/– SE. (B) This dog is staring intently at its master, the better to attend to the behavior of its human companion. (A after Hare et al. 2002; photograph © iStock.com/Peter-verreussel.)

arms of the maze are associated with images of two different wasp faces. Members of *P. fuscatus* colonies quickly learn to avoid the image of a wasp face linked with an electric shock, whereas individuals of another member of the genus, *P. metricus*, perform essentially at random in tests of this sort (**FIGURE 4.31**).

Why the difference in face learning between the two species? Females of *P. fuscatus* regularly join others in founding colonies where individuals compete with their companions for dominance, and thus the right to reproduce. Knowing the identity of others in the dominance hierarchy helps females know when to

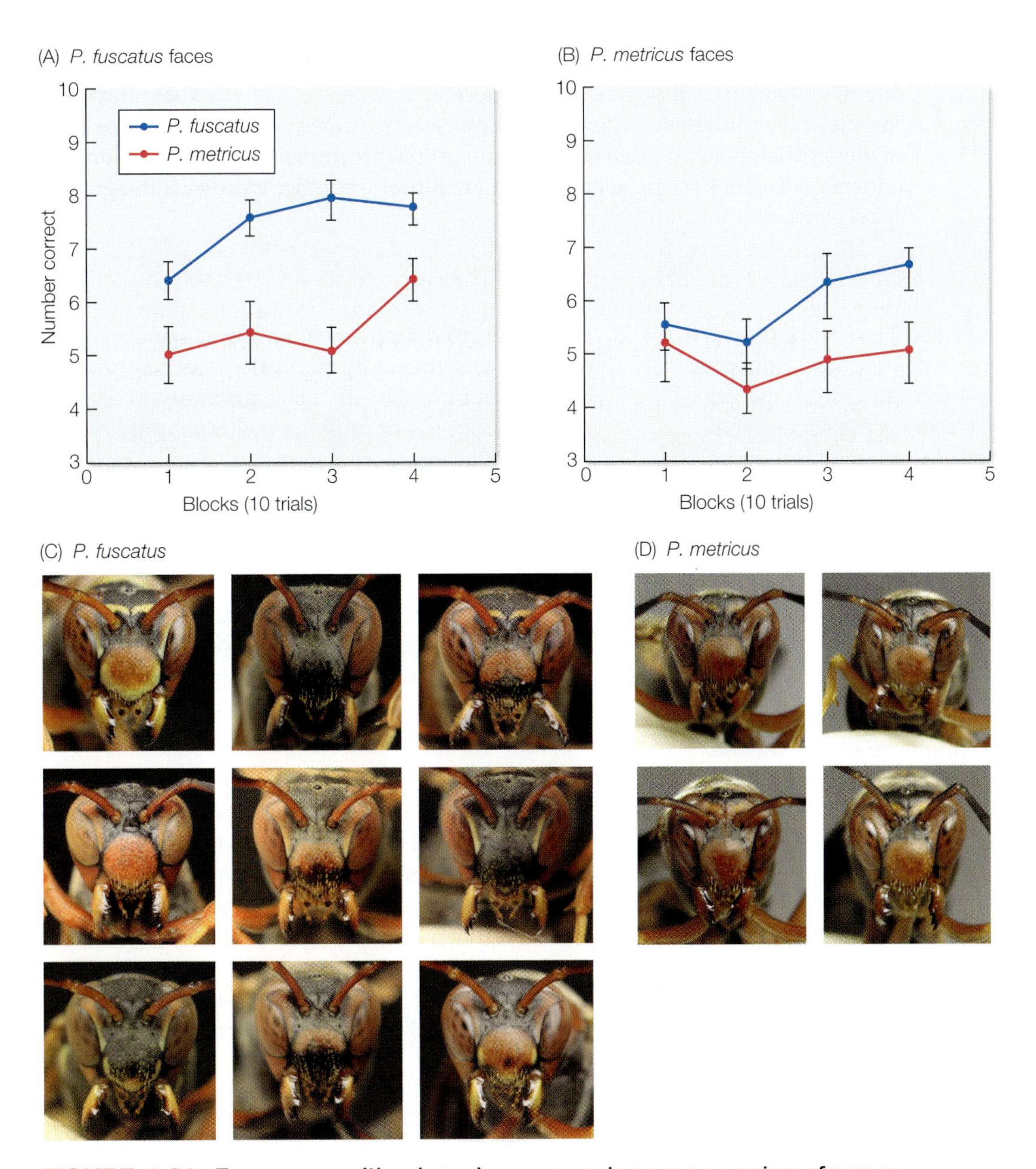

FIGURE 4.31 Face recognition learning occurs in some species of paper wasps, but not others. (A) Females of *Polistes fuscatus* learn to associate a particular face of a female of their species with a safe refuge in a T-maze. (B) Females of *P. metricus* perform no better when required to recognize faces of their own species than they do when required to recognize faces of *P. fuscatus*. Values depict mean +/– SE. (C) The faces of female *P. fuscatus* differ among individuals, whereas (D) the faces of *P. metricus* are similar, indicating that selection has acted to facilitate face recognition in one species but not the other. (From Sheehan and Tibbetts 2010, 2011.)

compete for higher status and when to accept a position subordinate to particular rivals. In other words, *P. fuscatus* is a social species. In contrast, females of *P. metricus* are solitary creatures that found nests by themselves. They gain no advantage by recognizing other wasps as individuals, and indeed, the faces of females of this species are far more uniform in appearance than the faces of *P. fuscatus* (Sheehan and Tibbetts 2011).

What paper wasps, dogs, and our own species appear to tell us is that cognitive skills evolve in response to particular selection pressures associated with particular environments—both social and ecological. The challenge, especially when studying species with brains as complex as ours, is to link these ultimate factors that select for learning and cognition to the neurons and circuits in the brain. What exactly is it about human brains that contributes to our superior intelligence? What are the differences in the brains of social and asocial paper wasps that allow social species to recognize individuals better than asocial species do? Answering questions such as these will not only require tools from neuroscience, but also an appreciation and understanding of natural selection and evolution—in other words, an integrative approach to studying animal behavior.

SUMMARY

1. The operating rules of neural mechanisms constitute proximate causes of behavior. Receptor cells acquire sensory information from the environment, interneurons relay and process that information through neural circuits, and other nerve cells in the central nervous system—often located in the brain—order appropriate motor responses to the events an animal can detect. Different species have distinct neural mechanisms and therefore perform these tasks differently, providing proximate reasons for why species differ in their behavior.

2. At the ultimate level, animals differ in their neurophysiology because their proximate mechanisms have been shaped by different selection pressures, as can be seen in the highly specialized sensory cells possessed by different species. For example, because of their auditory receptors, moths and other ultrasound-detecting species can hear sounds humans cannot sense, while certain bees, birds, fishes, and lizards, among others, easily detect ultraviolet radiation that is invisible to us.

3. In addition to adaptively specialized sensory receptors, stimulus filtering (the selective processing of potential stimuli) is apparent in all the components of nervous systems. Sensory receptors ignore some stimuli in favor of others, while interneurons relay some, but not all, of the messages they receive from receptor cells. Within the central nervous system, many cells and circuits are devoted to analyzing certain categories of information, although this means that other inputs are discarded. As a result, animals focus on the biologically relevant stimuli in their environments, increasing the odds of a prompt and effective reaction to the items that really matter.

4. The cognitive abilities of humans and other animals are also based on the evolved properties of nerve cells and nervous systems. Species that live with others in a competitive social environment may possess special problem-solving skills associated with this environment—another demonstration of the power of natural selection in shaping the proximate neural mechanisms of behavior.

COMPANION WEBSITE

Go to **ab11e.com** for discussion questions and all of the audio and video clips.

The Physiological Basis of Behavior

In this chapter, we continue to examine the relationship between the proximate mechanisms of behavior and their adaptive outcomes. In Chapter 4 you learned how neurons are involved in the detection of sensory information from the environment, and how those messages are relayed to other cells in the nervous system, where they ultimately lead to the production of behaviors. Here, we go one step further and explore not only how the nervous system is organized to anticipate both predictable and unpredictable environmental changes, but how an animal's physiology allows it to rapidly respond to those changes. First, we explore the role that endogenous clock mechanisms play in organizing behavior and physiology to adapt to environment cycles. We discuss circadian, or daily, cycles, but also other rhythms

Chapter 5 opening photograph by harum.koh/CC BY-SA 2.0.

of increasing length, including tidal cycles that influence mating and foraging behavior, as well as circannual cycles that help regulate migratory and hibernation behavior. Our discussion of endogenous clocks integrates not only the neural basis of these timing mechanisms, but also their genetic and endocrine bases. From there, we examine how the endocrine system allows animals to respond not only to predictable environmental factors such as the changing of the seasons, but also to unpredictable environmental challenges—both social and ecological—such as aggressive interactions, territorial intrusions, and severe weather events. We consider a variety of different hormones, or signaling molecules, that play central roles in mediating a range of animal behaviors. We emphasize the role that hormones play in reproductive behavior, as well as how they enable vertebrates to respond behaviorally to social and ecological stressors. By the end of this chapter, you will have a better understanding not only of the physiological mediators that organisms use to respond to environmental cues, but of how natural selection has shaped those responses.

Endogenous Rhythms and Changing Behavioral Priorities

The ability of neurons to communicate with and inhibit one another helps set an animal's behavioral priorities. But the relationship between different components of the brain can change adaptively over time. For example, female *Teleogryllus* crickets usually hide in burrows or under leaf litter during the day and move about only after dusk when it is relatively safe to search for mates (Loher 1979). Unsurprisingly, male crickets wait for the evening to start calling for mates (Loher 1972). These observations suggest that the inhibitory relationships between the calling center in a male cricket's brain and other neural elements responsible for other behaviors must change cyclically over a 24-hour period. Since these same changes are likely to occur day after day for most of the organism's adult life, a clock mechanism could be useful for the control of singing behavior, just as you learned in Chapter 4 that the biological clock of the monarch butterfly (*Danaus plexippus*) enables it to adjust its sun compass as the hours pass during the daytime (**FIGURE 5.1**) (Zhu et al. 2008). As you will see below, clock mechanisms are important for regulating behavior and changing behavioral priorities.

Mechanisms of Changing Behavioral Priorities

Two competing but non-mutually exclusive proximate hypotheses have been proposed to explain how an animal's priorities can change over time (**HYPOTHESES 5.1**). First, individuals might manage their priorities in response to a biological clock, an endogenous timing mechanism with a built-in schedule that acts independently of any cues from the animal's surroundings (**endogenous clock hypothesis**). The idea that at least partially environment-independent timing mechanisms might exist should be plausible to anyone who has flown across several time zones and then tried to adjust immediately to local conditions. Your internal state—when you feel hungry or tired, for example—seems to respond to the environment you just left and not to the one in which you just landed. Eventually, your body adjusts to the local conditions, but this takes time. The second hypothesis suggests that the relationships between command centers in an animal's nervous systems are altered largely on the basis of feedback information gathered by mechanisms that monitor the surrounding environment

HYPOTHESES 5.1

Non-mutually exclusive proximate hypotheses for how animals manage their priorities over time

endogenous clock hypothesis An endogenous timing mechanism with a built-in schedule acts independently of any cues from the animal's surroundings to control how animals change priorities over time.

environmental stimulus hypothesis Animals use feedback information from the surrounding environment to change priorities over time.

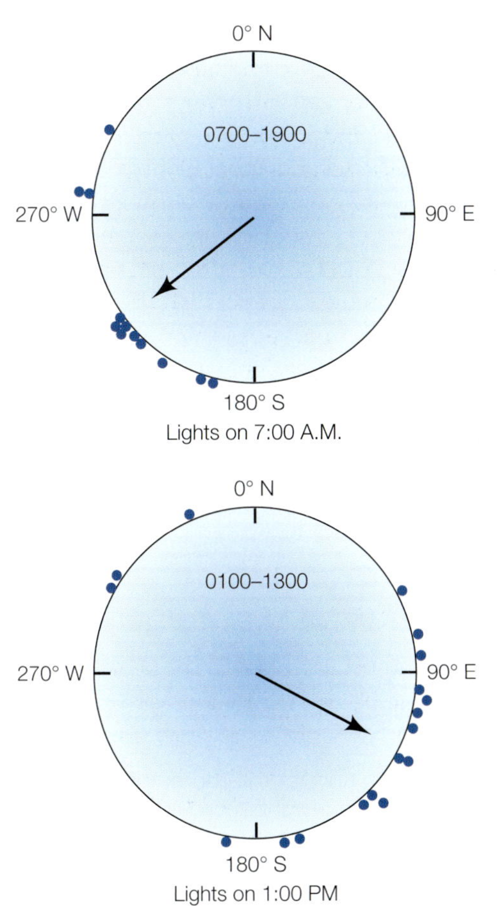

Monarch butterfly

FIGURE 5.1 The monarch butterfly possesses a biological clock. Experimental manipulation of the clock changes the orientation of migrating monarchs. Individuals were tested in an outdoor flight cage after one group had been held indoors under an artificial light–dark cycle with lights on at 7:00 AM and lights off at 7:00 PM. This group tried to fly in a southwesterly direction. A second group of butterflies that had been held with lights on at 1:00 AM and lights off at 1:00 PM flew in a southeasterly direction, evidence of the importance of a clock and sun compass in helping monarchs stay on course. (After Froy et al. 2003; photograph © Nancy Bauer/Shutterstock.com.)

(**environmental stimulus hypothesis**). Such devices would enable individuals to modulate their behavior in response to certain changes in the world around them, such as a decrease in light intensity as evening comes on.

Let's consider these two hypotheses in the context of the calling cycle of male *Teleogryllus* crickets. Each day's calling bout might begin at much the same time because the crickets possess an internal timer that measures how long it has been since the last bout began. Individuals might use this environment-independent system to activate the onset of a new round of chirping each evening at dusk. Alternatively, the insect's neural mechanisms might be designed to initiate calling when light intensity falls below a particular level. If this second hypothesis is correct, then crickets held under constant bright light should not call. But in fact, laboratory crickets held in rooms in which the temperature stays the same and the lights are on 24 hours a day still continue to call regularly for several hours each day. Under conditions of constant light, calling starts about 25 to 26 hours after it did the previous day (**FIGURE 5.2**). A cycle of activity like this one in crickets that is not matched to environmental cues is called a **free-running cycle**. Because the length, or period, of the free-running cycle of cricket calling deviates from the many 24-hour environmental cycles caused by Earth's daily rotation around its axis, we can conclude that, consistent with the endogenous clock hypothesis, the cyclical pattern of cricket calling is caused in part by an environment-independent internal **circadian rhythm** (circadian means "about a day").

What happens if the lighting conditions are altered by placing a cricket in a regime of 12 hours of light and 12 hours of darkness? The switch from light to darkness offers an external environmental cue that the crickets may use to adjust their

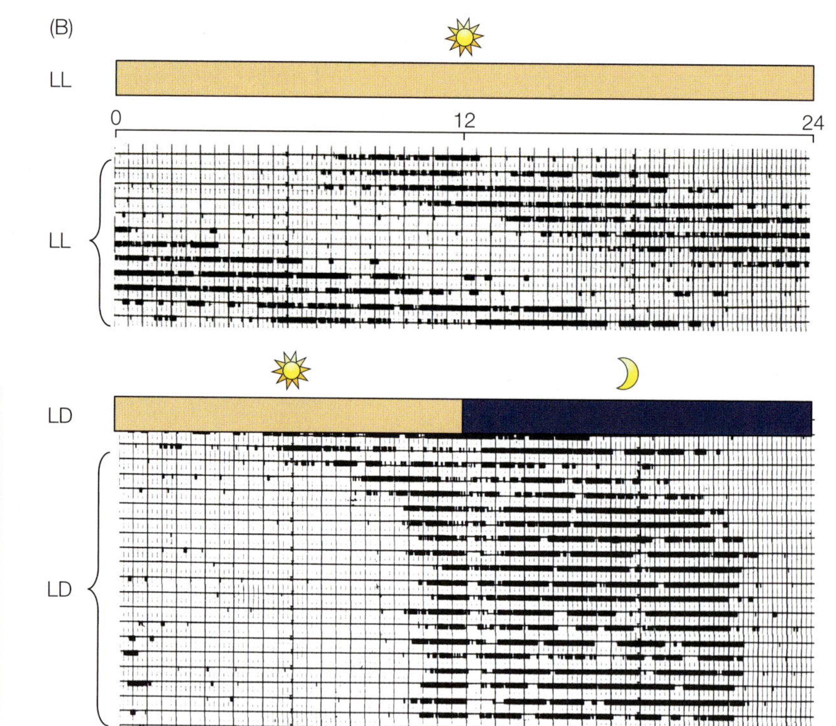

FIGURE 5.2 Circadian rhythms in cricket calling behavior. (A) A female *Teleogryllus* cricket (right) has tracked down a male (left) by his calls. (B) Calling occurs on a daily schedule. Each horizontal line on the grid represents 1 day; each vertical line represents a half hour on a 24-hour timescale. Dark marks indicate periods of activity—in this case, calling. The bars at the top and middle of the figure represent the lighting conditions. Thus, for the first 12 days of this experiment, male crickets are kept in constant light (LL), and for the remainder, they are subjected to 12-hour cycles of light and dark (LD). Male crickets held under constant light exhibit a daily cycle of calling and noncalling, but the calling starts later each day. The onset of "nightfall" on day 13 acts as a cue that resets the calling rhythm, which soon stops shifting and eventually begins an hour or two before the lights are turned off each day. (A, photograph by Denis Crawford/Alamy Stock Photo; B after Loher 1972.)

timing mechanism. In fact, they do, just as other species such as monarch butterflies and pigeons can reset their clocks when moved into a laboratory with artificial lighting, or as you do when you fly to a new time zone. After a few days, the male crickets start calling about 2 hours before the lights go off, accurately anticipating "nightfall," and they continue until about 2.5 hours before the lights go on again in the "morning" (see Figure 5.2). Consistent with the environmental stimulus hypothesis, this cycle of calling matches the natural one, which is synchronized with dusk. However, unlike the free-running cycle, this cycle does not drift out of phase with the 24-hour day but is reset, or **entrained**, each day so that it begins at the same time in relation to lights-out (Loher 1972). From these results, we can conclude that the complete control system for cricket calling has both an environment-independent timer, or biological clock, set on a cycle that is not exactly 24 hours long, but also an environment-activated entrainment device that synchronizes the clock with local light conditions. In other words, both the endogenous clock and the environmental stimulus hypotheses are supported in crickets, as they have a biological clock mechanism that can be adjusted by environmental cues.

What about a species that exhibits a polyphenism (when multiple phenotypes are produced from the same genotype) in the timing of its activity, such as the sand cricket (*Gryllus firmus*), a species that comes in two different forms: a long-winged, flight-capable, nocturnally active morph and a short-winged, flightless morph that

FIGURE 5.3 Hormonal differences are correlated with flight behavior differences in crickets. In the early part of the night, the long-winged, flight-capable form of the cricket *Gryllus firmus* (on the right in the photo) has higher concentrations of juvenile hormone (JH) in its blood than the short-winged form of the cricket (on the left). (After Zera et al. 2007; photograph courtesy of Derek Roff.)

is more likely to be active during the day. The long-winged form flies primarily at night, a safe time for an insect to leave its burrow and search for mates, whereas the short-winged form cannot fly and so stays in the safety of concealed vegetation on the ground during the day (Roff and Fairbairn 2007). To compare the circadian rhythms of the two types of sand crickets, Anthony Zera and colleagues brought recently collected individuals of both types to the laboratory and then took blood samples at intervals over 24-hour periods. The researchers focused on a molecule called juvenile hormone (JH), which is known to regulate many components of insect physiology, including reproduction and polyphenisms. When the team assayed JH concentrations in the blood of short-winged crickets, they found no significant changes in relation to time of day. However, the blood of flight-capable crickets revealed a very different story, with JH concentrations rising sharply in the late afternoon or evening from baseline levels comparable to those in the flightless crickets (**FIGURE 5.3**) (Zera et al. 2007). The different daily patterns of circulating JH in the two kinds of crickets are correlated with the different circadian rhythms of the two forms. It seems likely that the nicely timed surge in JH in long-winged crickets in some way helps prepare these individuals for a round of nocturnal flight. As we discuss in greater detail later in this chapter, hormones such as JH play a critical role not only in activating animal behavior, but also in regulating circadian control of behavior.

The Neurobiology of Circadian Timing

In addition to studying the role that hormones play in regulating circadian rhythms, researchers have examined other proximate mechanisms of circadian behavior, including the interaction between neural circuits and genes. For example, although the daily schedule of behavioral activity in long-winged *Gryllus firmus* appears to be linked to the cyclical release of JH, other aspects of the circadian rhythm of these insects involve the optic lobes of their brain. If one cuts the nerves carrying sensory information from the eyes of a male cricket to its optic lobes (**FIGURE 5.4**), the insect enters a free-running cycle. Visual signals of some sort are evidently needed to entrain the daily rhythm to local conditions, but a rhythm persists in the absence of this information. If, however, one cuts the connections between both optic lobes and the rest of the brain, the calling cycle breaks down completely and all hours

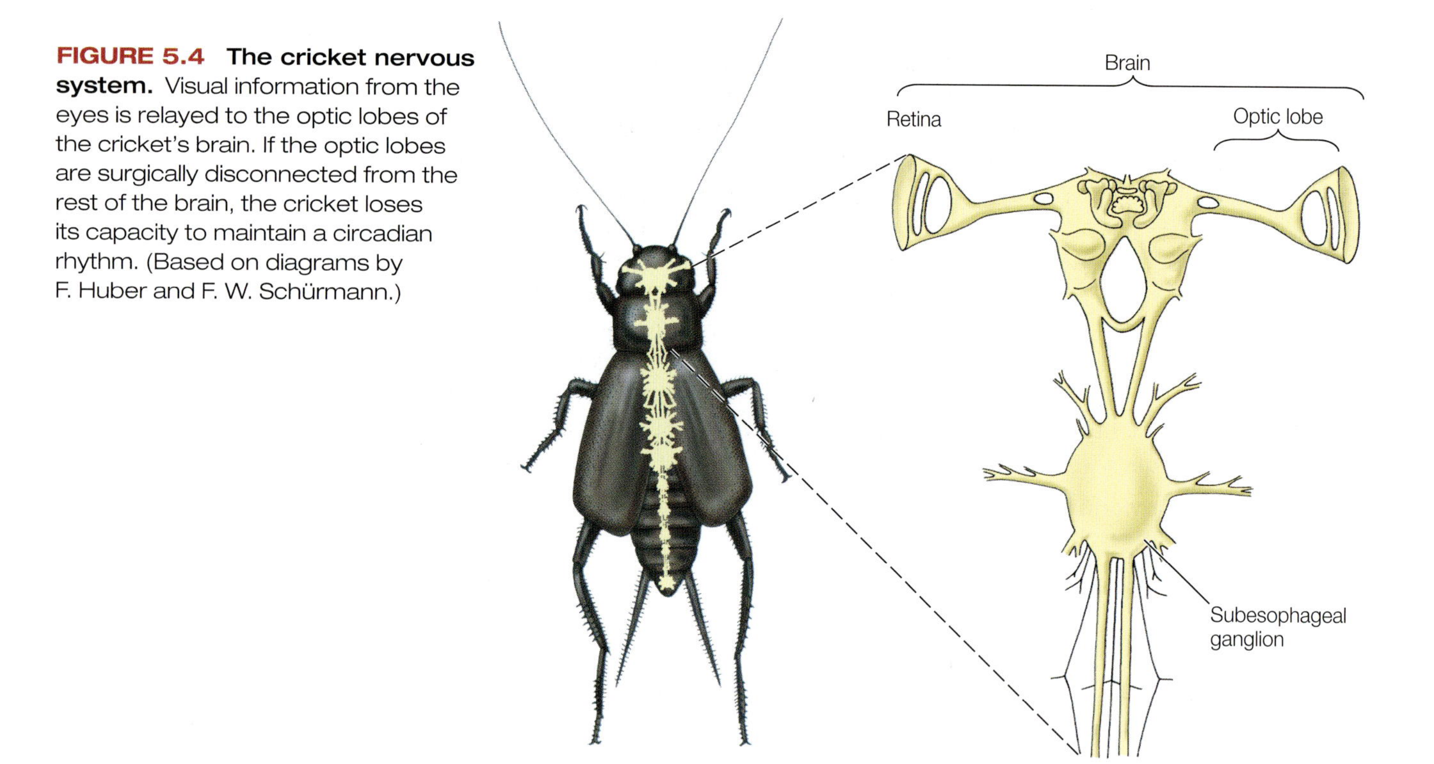

FIGURE 5.4 The cricket nervous system. Visual information from the eyes is relayed to the optic lobes of the cricket's brain. If the optic lobes are surgically disconnected from the rest of the brain, the cricket loses its capacity to maintain a circadian rhythm. (Based on diagrams by F. Huber and F. W. Schürmann.)

are now equally probable times for cricket calling. These results are consistent with the hypothesis that a master clock mechanism (**FIGURE 5.5**) resides within the optic lobes, sending messages to other regions of the nervous system (Page 1985, Johnson and Hasting 1986) as well as receiving hormonal signals generated by the animal's endocrine system.

As we mentioned earlier, circadian rhythms occur not just in insects, but also in a variety of vertebrates, including humans. Biologists interested in the control of circadian rhythms in mammals have focused on the hypothalamus of the brain, with special emphasis on the suprachiasmatic nucleus (SCN), a pair of hypothalamic neural clusters that receive inputs from nerves originating in the retina. The SCN is therefore a likely element of the mechanism that secures information (at least partly from the eyes) about day and night length—information that could be used to adjust a master biological clock. If the SCN contains a master clock or pacemaker that is critical for maintaining circadian rhythms, then neurons in the putative clock should change their activity in a regular fashion over the course of a 24-hour period.

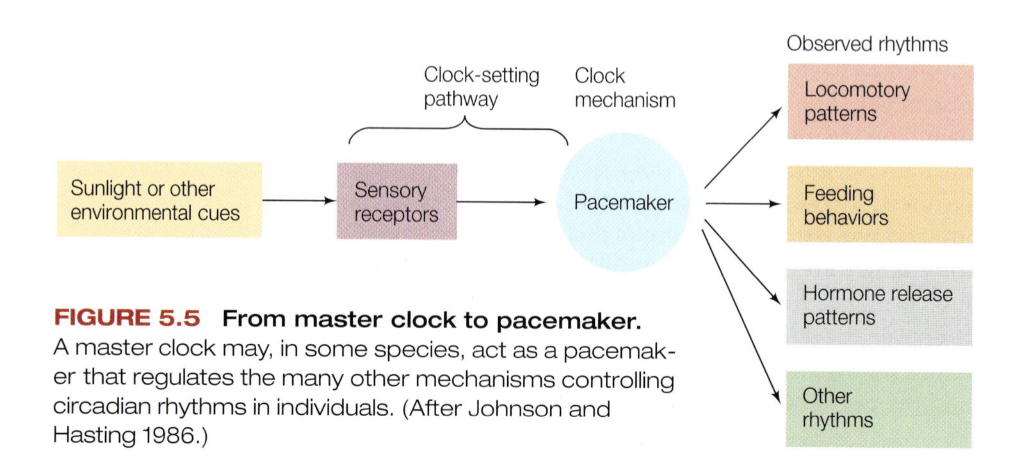

FIGURE 5.5 From master clock to pacemaker. A master clock may, in some species, act as a pacemaker that regulates the many other mechanisms controlling circadian rhythms in individuals. (After Johnson and Hasting 1986.)

Indeed they do, with some groups of cells in the SCN being electrically silent at night but becoming active at dawn and continuing to fire during the day (Colwell 2012). Another prediction of this hypothesis is that damage to the SCN should interfere with circadian rhythms. In experiments in which SCN neurons have been destroyed in the brains of hamsters and rats, the animals have subsequently exhibited arrhythmic patterns of hormone secretion, locomotion, and feeding (Zucker 1983). If arrhythmic hamsters receive transplants of SCN tissue from fetal hamsters, they sometimes regain their circadian rhythms, but this does not happen if the tissue transplants come from other parts of the fetal hamster brain (DeCoursey and Buggy 1989). Moreover, if an arrhythmic hamster gets a SCN transplant from a mutant hamster with a circadian period that is much shorter than the standard one of approximately 24 hours, the experimental subject adopts the circadian rhythm of the donor hamster, further evidence in support of the hypothesis that the SCN controls this aspect of hamster behavior (Ralph et al. 1990).

The Genetics of Circadian Timing

Perhaps the SCN clock of hamsters, rats, and other mammals operates via rhythmic changes in gene activity. A key candidate gene related to these changing rhythms is the *period* (*per*) gene, which codes for a protein (PER) whose production varies over a 24-hour schedule in concert with that of the product of another mammalian gene, called *tau* (**FIGURE 5.6**). The product of *tau* is an enzyme whose production is

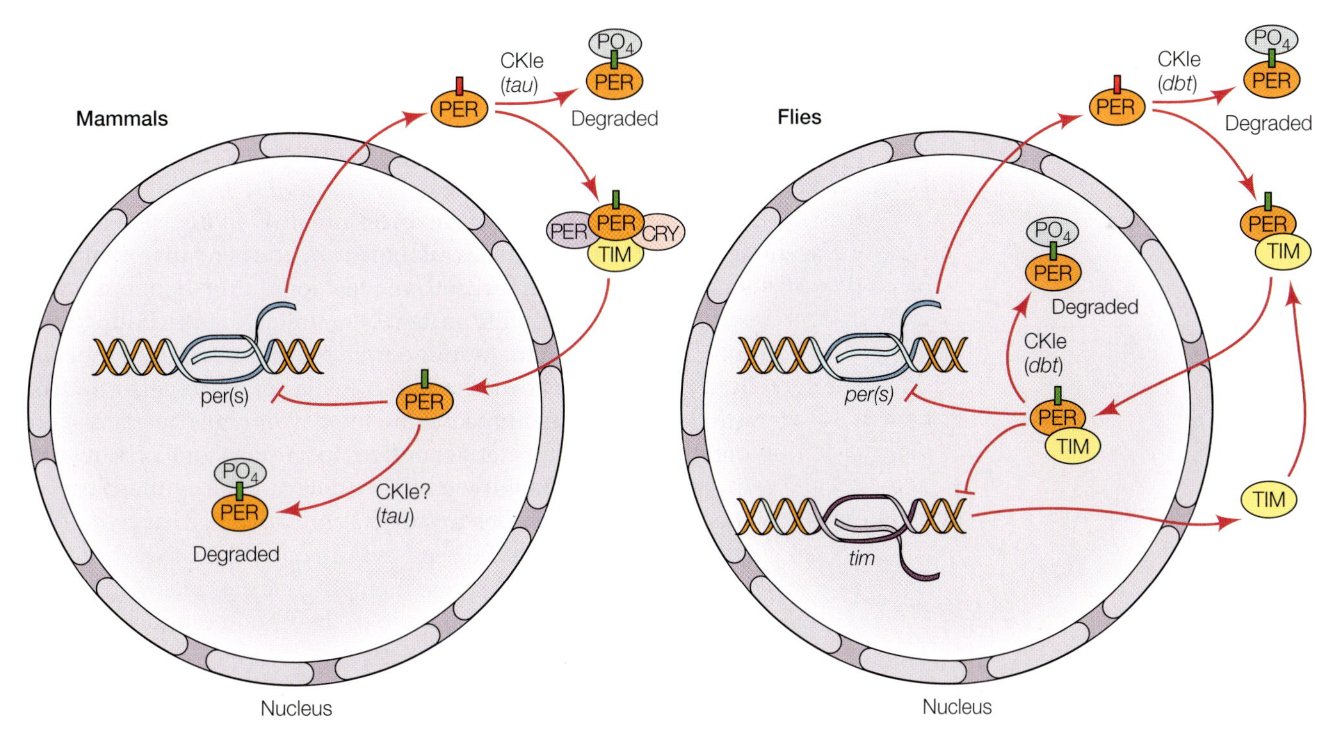

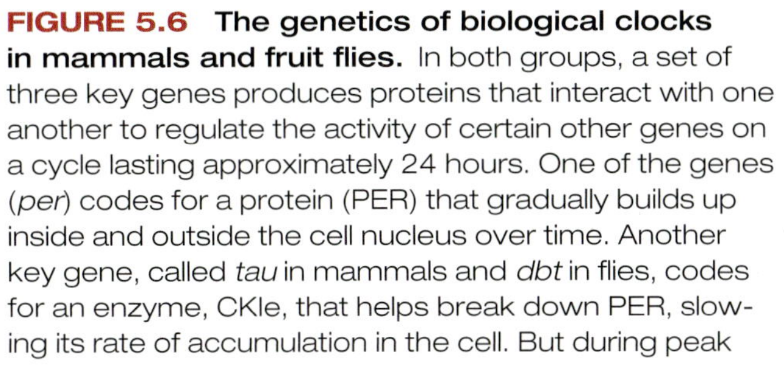

FIGURE 5.6 The genetics of biological clocks in mammals and fruit flies. In both groups, a set of three key genes produces proteins that interact with one another to regulate the activity of certain other genes on a cycle lasting approximately 24 hours. One of the genes (*per*) codes for a protein (PER) that gradually builds up inside and outside the cell nucleus over time. Another key gene, called *tau* in mammals and *dbt* in flies, codes for an enzyme, CKIe, that helps break down PER, slowing its rate of accumulation in the cell. But during peak periods of PER production, more PER is available to bind with another protein (TIM), coded for by a third gene (*tim*). When the PER protein is bound in complexes with TIM (and another protein, CRY, in the case of mammals), it cannot be broken down as quickly by CKIe. Therefore, more intact PER is carried back into the nucleus, where it blocks the activity of the very gene that produces it, though only temporarily. Then a new cycle of *per* gene activity and PER protein production begins. (After Young 2000.)

turned on when PER is at peak abundance in the cell. This enzyme degrades PER, contributing to a 24-hour cycle in which PER first increases in abundance and then falls (Young 2000). A mutation in *tau*, first discovered in the golden hamster (*Mesocricetus auratus*), effectively accelerates the cellular dynamics of *per* and speeds up the circadian cycle. Animals lacking the mutation have a regular 24-hour circadian cycle, but heterozygous individuals with one copy of the mutation have periods of 22 hours, and homozygous individuals with two copies of the mutation have periods of only 20 hours. Animals carrying the *tau* mutation—particularly two copies of it—are unable to entrain on a natural light–dark cycle (Ralph and Menaker 1988). Spoelstra and colleagues hypothesized that individuals with mutated *tau* genes would have reduced fitness compared with individuals lacking the mutation (Spoelstra et al. 2016). Consistent with their hypothesis, the researchers showed that in seminatural enclosures, mice with normal circadian rhythms survived longer and produced more offspring than those with shortened circadian rhythms.

A striking feature of this system, whose complexity is daunting, is that the key clock genes regulating cellular circadian rhythms in mammals are also present in a variety of other organisms, including many insects, a legacy of a very ancient shared ancestor. Nearly a dozen clock genes have been identified in insects (Tomioka and Matsumoto 2015). For example, *Drosophila* fruit flies, honey bees, some butterflies, beetles, grasshoppers, ants, and other insects also have the *per* gene, a chain of DNA composed of more than 3500 base pairs and that provides the information needed to produce the PER protein chain of nearly 1200 amino acids. Alterations in the base sequence involving as little as a single base-pair substitution can result in dramatically different circadian rhythms in fruit flies (**FIGURE 5.7**), as well as in humans. Interestingly, human carriers of one *per* mutation typically fall asleep around 7:30 in the evening and arise at about 4:30 in the morning (Toh et al. 2001).

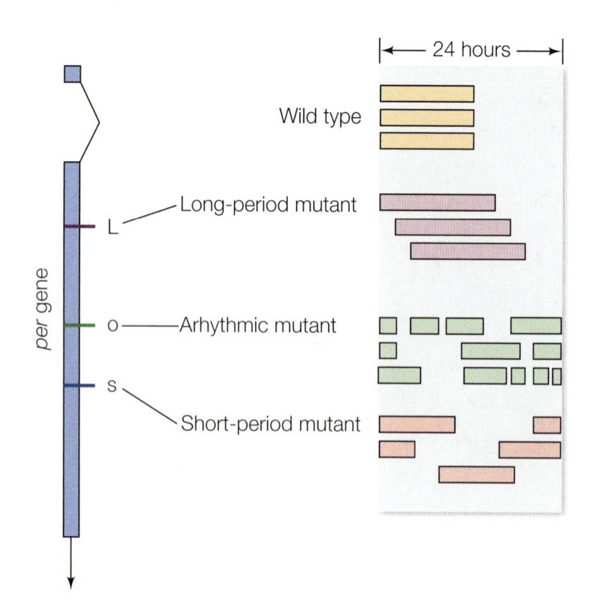

FIGURE 5.7 Mutations of the *per* gene affect the circadian rhythms of fruit flies. On the left is a diagram of the DNA sequence that constitutes the *per* gene. The locations of the base substitutions present in three mutant alleles found in fruit flies are indicated on the diagram. The activity patterns of wild-type flies and of those associated with each mutation are shown on the right. (After Baylies et al. 1987.)

These results from hamsters, mice, flies, and humans strongly suggest that circadian genes play a critical role in enabling circadian rhythms. If, however, expression of *per* influences circadian timing, then animals in which the *per* gene is relatively inactive should behave in an arrhythmic fashion. In the honey bee *Apis mellifera*, very little PER protein is manufactured in young bees, which generally remain in the hive to care for eggs and larvae. And young honey bees are in fact just as likely to perform these nursing tasks at any time during the day or night over a 24-hour period. In contrast, older honey bees that forage for food only during the daytime exhibit well-defined circadian rhythms, leaving the hive to collect pollen and nectar only during that part of the day when the flowers the bees seek are most likely to be resource-rich. Because foragers express their *per* gene vigorously, they have three times as much PER protein in their brain cells as do young nurse bees (Toma et al. 2000).

The Physiology of Circadian Timing

Because the fruit fly, the hamster, the honey bee, and humans all have the same gene serving much the same clock function, evolutionary biologists believe that we inherited this gene, as well as some others involved in the regulation of activity patterns, from a creature that lived perhaps 550 million years ago (Young 2000). In mammals and some other vertebrates, the *per* gene and certain other clock genes are expressed in the neurons of the SCN, which is usually viewed as the home of a master clock or circadian pacemaker that regulates many other tissues, thereby keeping many different behaviors on a daily schedule. However, other parts of the brain may also have their own biological clocks. For example, the olfactory bulb in mice exhibits environment-independent cyclical changes in genetic activity, with the result that mice are more sensitive to odors at night than during the day, an adaptive effect of a timing mechanism for a nocturnally active creature (Granados-Fuentes et al. 2006).

Nonetheless, the SCN clearly contains a major biological clock, which must send chemical signals to the target systems it controls. If so, we can make three predictions: (1) the molecule that relays the clock's information should be secreted by the SCN in a manner regulated by the clock's genes; (2) there should be receptor proteins for that chemical messenger in cells of the target tissues; and (3) experimental administration of the chemical messenger should disrupt the normal timing of an animal's behavior.

What is the chemical signal that might regulate the mammalian clock? As was the case in sand crickets, the critical messenger in mammals is another hormone, called melatonin. Melatonin, which is produced in the pineal gland in the brain, has been the subject of much research demonstrating its involvement in the regulation of animal circadian rhythms (Turek et al. 1976). Melatonin is produced at night by the SCN and modulates both circadian rhythms and sleep. It can "reset" the sleep–wake cycle, which is why people often take it to help with jet lag (Johnston and Skene 2015). In addition to playing a role in circadian timing, melatonin also influences seasonal reproduction in several mammals. Seasonal changes in day length affect the secretion of melatonin, which in turn interacts with other reproductive hormones to influence the timing of breeding.

Melatonin is not the only chemical that plays a critical role in circadian rhythms. More recently, researchers have demonstrated that a protein called prokineticin 2 (PK2), which is coded for by a gene of the same name, has the key properties expected of a clock messenger (Cheng et al. 2002). In normal lab mice (*Mus musculus*) living under a cyclical regime of 12 hours of light and 12 hours of darkness, PK2 is produced in a strongly circadian pattern by the SCN (**FIGURE 5.8**). Moreover, mice with certain mutations in key clock genes lack the circadian rhythm of PK2 production. Furthermore, only certain structures in the brain of the mouse produce a receptor protein that binds with PK2. These structures are linked to the SCN by neural pathways and are thought to contain command centers that control various behavioral activities in a circadian fashion. Finally, if one injects PK2 into the brains of rats (*Rattus norvegicus*) during the night, when the animals are normally active, the behavior of these animals changes dramatically. Instead of running in

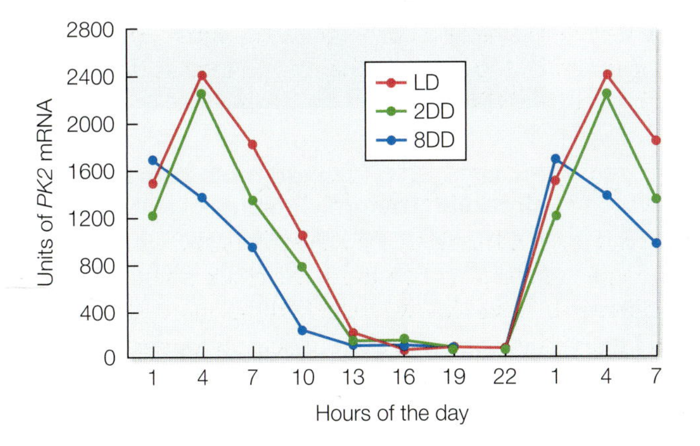

FIGURE 5.8 Expression of the gene that codes for PK2 in the SCN. Mice were exposed to a standard light-dark cycle of 12 hours of light and 12 hours of darkness (LD) prior to the start of the experiment. The graph shows hourly changes in the production of the messenger RNA encoded by the *PK2* gene. The gene was expressed in a circadian rhythm whether the animals were held under the standard light–dark cycle, kept in complete darkness for 2 days (2DD), or held in darkness for 8 days (8DD). (After Cheng et al. 2002.)

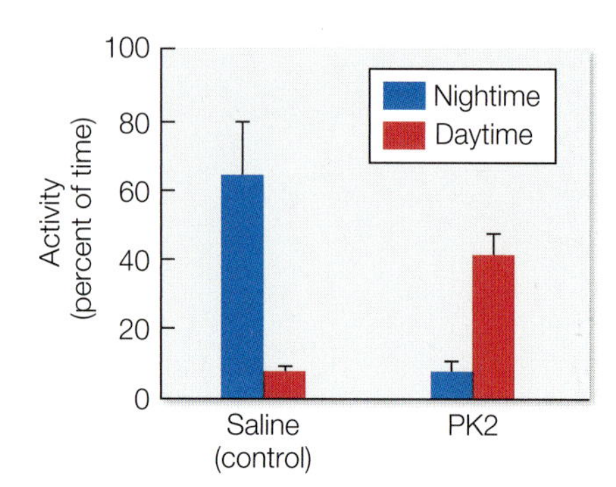

FIGURE 5.9 Locomotion and circadian rhythms in rats. Circadian control of wheel running by rats changes when the rats' brains are injected with PK2. The rats whose brains are injected with PK2 become active primarily during daylight hours, whereas control rats injected with saline exhibit the standard preference for nighttime activity. Bars depict mean +/– SE. (After Cheng et al. 2002.)

their running wheel, they sleep, shifting to daytime activity instead (**FIGURE 5.9**) (Cheng et al. 2002).

All of these lines of evidence point to PK2 as the chemical messenger that the mammalian SCN uses to communicate with and regulate target centers in the brain. The SCN, in turn, receives information from the retina about the light–dark cycle of the animal's environment so that it can fine-tune its autoregulated pattern of gene expression. The adaptive value of the environment-independent component of the clock system may be that a clock of this sort enables individuals to alter the timing of their behavioral and physiological cycles without having to constantly check the environment to see what time it is. The presence of an environment-dependent element, however, permits individuals to adjust their cycles in keeping with local conditions. As a result, a typical nocturnal mammal automatically becomes active at about the right time each night, while retaining the capacity to shift its activity period gradually to accommodate the changes in day length that occur as spring becomes summer, or summer becomes fall.

One way to determine the adaptive value of circadian rhythms would be to test the hypothesis that if there were animals for which the day–night cycle was biologically irrelevant, then these species should lack a circadian pattern of activity. As predicted for nurse honey bees that never venture outside the hive or see light, their brains exhibit no circadian rhythmicity in the expression of key genes associated with the honey bee's circadian clock. Foragers, by contrast, go outside the hive to collect pollen and nectar during the day. In contrast to nurse bee brains, forager brains exhibit regular oscillations in the activity of many genes believed to be linked to the biological clock system of the bee (Rodriguez-Zas et al. 2012).

The naked mole rat (*Heterocephalus glaber*) is a similar species in that these animals live underground in total darkness and almost never come to the surface. As predicted, naked mole rats lack a circadian rhythm. Instead, individuals engage in generally brief episodes of activity within longer periods of inactivity, with the pattern changing irregularly from day to day (**FIGURE 5.10**) (Davis-Walton and Sherman 1994).

Time
Individuals
12:30 00:30 12:30
K-3 (Queen) Day 1 Day 2 Day 3
K-17 Day 1 Day 2 Day 3
K-F Day 1 Day 2 Day 3
12:30 00:30 12:30
TT-+ (Queen) Day 1 Day 2 Day 3
TT-Z Day 1 Day 2 Day 3
TT-17 Day 1 Day 2 Day 3

Naked mole rats

FIGURE 5.10 Naked mole rats lack a circadian rhythm. Patterns of activity are shown for six individuals from two captive colonies held under constant low light. Dark bars indicate periods when the individual was awake and active. (After Davis-Walton and Sherman 1994; photograph © Raymond Mendez/Animals/age fotostock.)

Seasonal and Annual Cycles of Behavior

Because of their unusual lifestyle, naked mole rats do not have to deal with cyclically changing environments, and they have apparently lost their circadian rhythm as a result. In contrast, almost all other creatures confront not only daily changes in food availability or risk of predation, but also changes that cover periods longer than 24 hours, especially the seasonal changes that occur in many parts of the world. If circadian rhythms enable animals to prepare physiologically and behaviorally for certain predictable daily changes in the environment, might not some animals possess a **circannual rhythm** that runs on an approximately 365-day cycle (Gwinner 1996)? A circannual clock mechanism could be similar to the circadian master clock, with an environment-independent timer capable of generating a circannual rhythm in conjunction with a mechanism that keeps the clock entrained to local conditions.

Testing the proximate hypothesis that an animal has a circannual rhythm is technically difficult because individuals must be maintained under constant conditions for at least 2 years after being removed from their natural environments. One successful study of this sort involved the golden-mantled ground squirrel (*Callospermophilus lateralis*) of North America, which in the wild spends the late fall and frigid winter hibernating in an underground chamber. Five members of this species were born in captivity, then blinded and held thereafter in constant darkness and at a constant temperature while supplied with an abundance of food (Pengelley and Asmundson 1974). Year after year, these individuals entered hibernation at about the same time as other golden-mantled ground squirrels living in the wild (**FIGURE 5.11**).

Circannual rhythms have also been examined in birds. For example, several nestling European stonechats (*Saxicola rubicola*) were taken from Kenya to Germany to be reared in laboratory chambers in which the temperature and **photoperiod** (number of hours of light in a 24-hour period) were always the same. Needless to say, these birds and their offspring never had a chance to encounter the spring rainy season in Kenya, which heralds a period of insect abundance and is the time when stonechats must reproduce if they are to find sufficient food for their nestlings. Thus, wild stonechats exhibit an annual cycle of reproductive physiology and

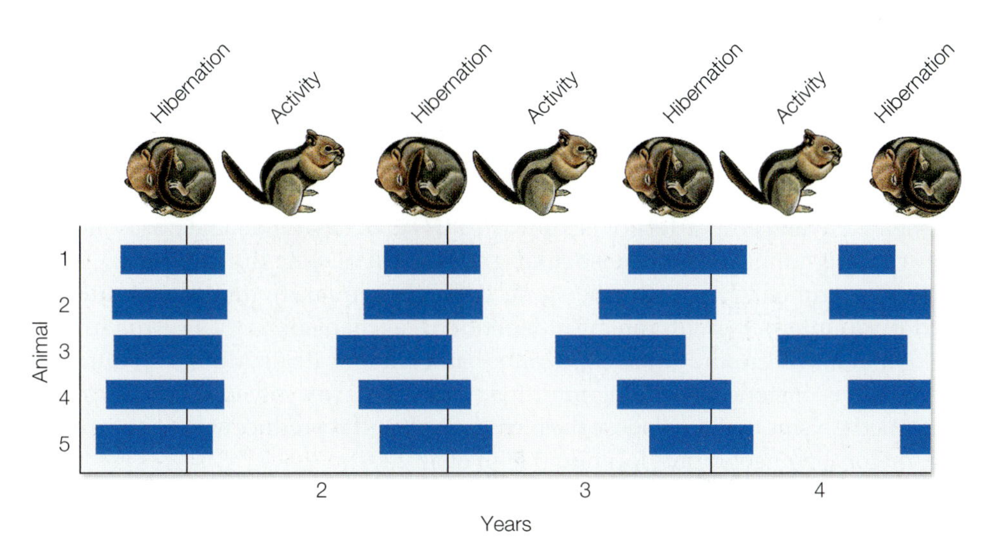

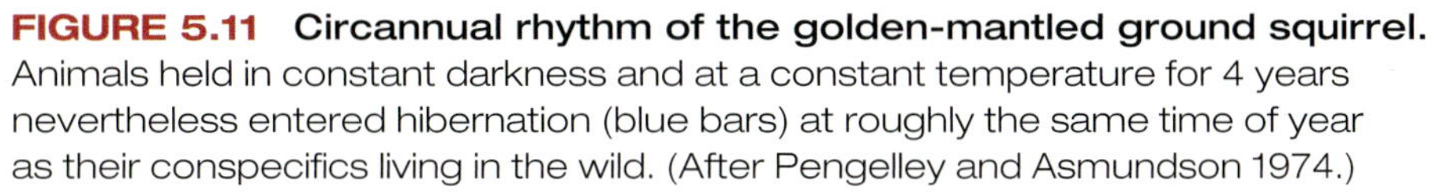

FIGURE 5.11 Circannual rhythm of the golden-mantled ground squirrel. Animals held in constant darkness and at a constant temperature for 4 years nevertheless entered hibernation (blue bars) at roughly the same time of year as their conspecifics living in the wild. (After Pengelley and Asmundson 1974.)

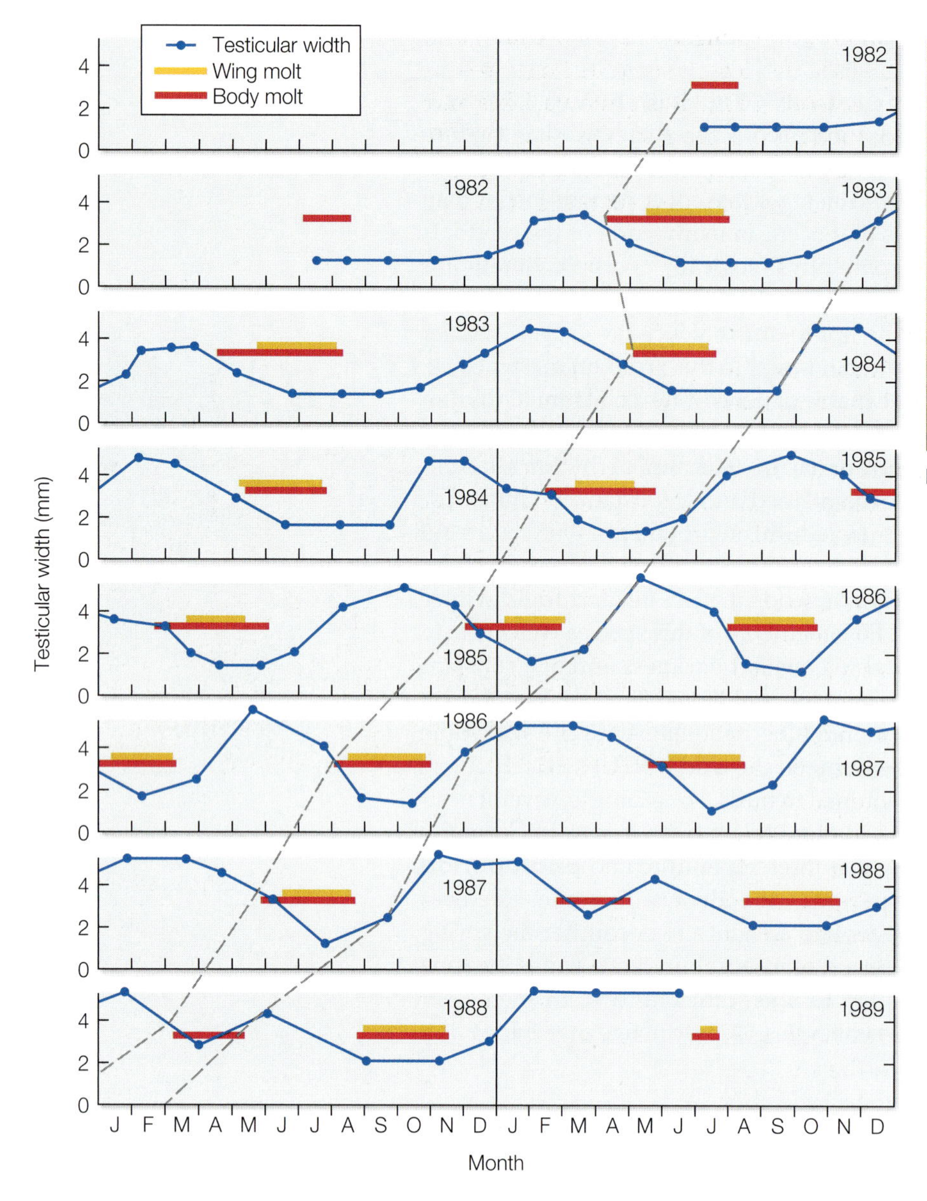

European stonechat

FIGURE 5.12 Circannual rhythm in a European stonechat. When transferred from Kenya to Germany and held under constant conditions, this male stonechat still underwent a regular long-term cycle of testicular growth and decline (blue lines), as well as regular feather molts (the two bars, which refer to wing and body molt). The cycle was not exactly 12 months long, however, so the timing of molt and testicular growth gradually shifted over the years (see the dashed lines that angle downward from right to left). (After Gwinner and Dittami 1990; photograph © M. Rose/Shutterstock.com.)

behavior. The transplanted stonechats, despite their constant environment, also exhibited an annual reproductive cycle, but one that shifted out of phase with that of their Kenyan compatriots over time (**FIGURE 5.12**). One male, for example, went through nine cycles of testicular growth and decline during the 7.5 years of the experiment. Evidently, the stonechat's circannual rhythm is generated in part by an internal, environment-independent mechanism, just as is true for the golden-mantled ground squirrel (Gwinner and Dittami 1990). Moreover, different populations of stonechats living in different places have evolved different circannual mechanisms that predispose the birds to molt and reproduce at the appropriate season for their geographic location (Helm et al. 2009).

Cues That Entrain Cycles of Behavior

The differences in circannual rhythms among European stonechat populations, as well as between stonechats and ground squirrels, are perhaps not surprising when we think about the environments in which these animals live. The seasonal temperature changes that North American ground squirrels experience are quite predictable from year to year, and the changes in day length are quite large. In contrast, the transition from the dry to rainy season in Kenya is quite variable and unpredictable, not only from location to location but also from year to year. And for these equatorial-dwelling birds, changes in day length are minimal throughout the year. Thus, ground squirrels and stonechats might rely on different types of cues to entrain their cycles of behavior.

Predictable Environmental Cues

In nature, environmental cues reset circadian and circannual clocks and thus produce behavioral rhythms that match the particular features of the animal's environment, such as the times of sunrise and sunset, or the onset of the rainy season in a given year, or the increasing day lengths associated with spring. This fine-tuning of behavioral cycles involves mechanisms of great diversity that respond to a full spectrum of environmental influences such as tidal cycles, lunar cycles, and even the behavioral patterns of other species, all of which can vary from species to species according to their ecological circumstances.

For many species living near the ocean, tidal cycles resulting from the gravitational forces exerted by the sun and moon can affect behavior. Take, for example, Atlantic marsh fiddler crabs (*Uca pugnax*), which plug their burrows with mud at high tide, emerging at low tide to forage and look for mates. Because the East Coast of North America, where these crabs live, experiences two periods of low and high tides daily, fiddler crabs have two distinct periods of peak activity each day. However, the timing of low tide advances by about 1 hour each day, because Earth's position relative to the sun and moon also changes daily. By observing crabs in the lab under constant conditions, Bennett and colleagues demonstrated that fiddler crabs maintain precise behavioral cycles that coincide with the daily progression of the tides for more than a week (Bennett et al. 1957). But crabs and other organisms that rely on tidal cycles must also be able to fine-tune their biological rhythms to local tidal variation. Studies of the Pacific fiddler crab (*Uca princeps*) have demonstrated that daily rhythms controlled by the light–dark cycle help individuals adjust behavior on a shorter time frame (Stillman and Barnwell 2004). Researchers found that crabs were most active on days when low tide occurred in midmorning, and that daily activity decreased as low tide shifted into the afternoon.

This interaction between daily and tidal cycles has also been observed to regulate foraging behavior in Galápagos marine iguanas (*Amblyrhynchus cristatus*), which feed on algae both intertidally at the water's edge and subtidally under the water's surface (**FIGURE 5.13A**). Dustin Rubenstein and Martin Wikelski showed that marine iguanas on the island of Santa Fe foraged primarily at low tide on intertidal algae exposed

(A)

(B)

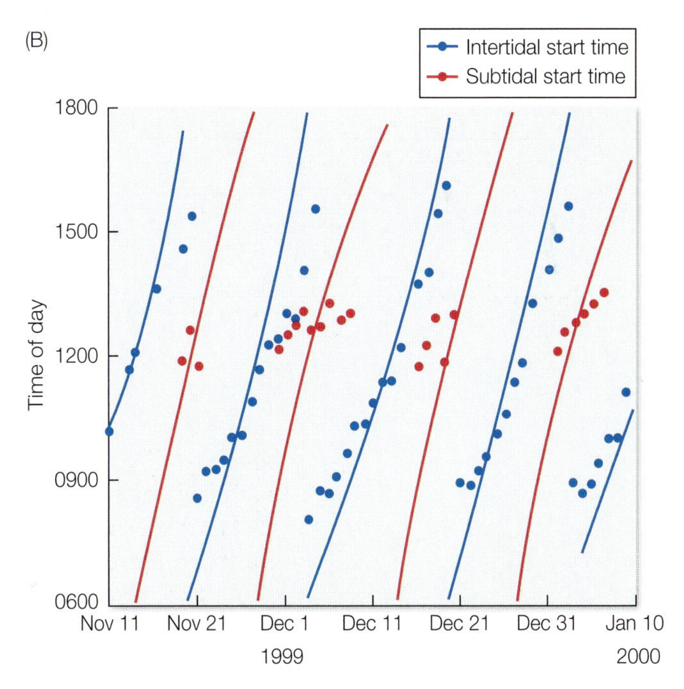

FIGURE 5.13 Foraging behavior of Galápagos marine iguanas is closely tied to the tidal cycle. (A) A marine iguana feeding underwater. (B) Mean intertidal foraging start times and mean subtidal foraging start times are shown. Low tide (blue line) and high tide (red line) cycles are also depicted. Iguanas spend most of their time foraging intertidally on exposed algae. Only when high tide occurs near midday do some of the larger individuals switch and feed subtidally on submerged algae. (A, photograph by wildestanimal/Moment/Getty Images; B after Rubenstein and Wikelski 2003.)

on rocks (Rubenstein and Wikelski 2003). However, when high tide took place around midday, something that occurred about once every 2 weeks, the iguanas switched to feeding subtidally on underwater algae (**FIGURE 5.13B**). Although nutritionally rewarding, feeding subtidally has been shown to be energetically costly because these ectothermic reptiles rapidly lose body heat in the cold waters of the Galápagos. Iguanas tend to forage subtidally only at that time of day when the sun is hottest, so that they can rapidly warm themselves on the rocks when they return from feeding (Vitousek et al. 2007). These high energetic demands also mean that only the largest animals tend to forage subtidally, as the energetic and thermal costs associated with subtidal foraging outweigh the benefits for smaller animals.

An example of how lunar cycles influence behavioral rhythms comes from studies by Robert Lockard and Donald Owings on banner-tailed kangaroo rats (*Dipodomys spectabilis*), which are more likely to stay in their underground retreats when moonlight is available to aid visually hunting nocturnal predators, such as coyotes (*Canis latrans*) and great horned owls (*Bubo virginianus*) (Lockard and Owings 1974, Lockard 1978). To measure the daily activity of kangaroo rats at a site in southeastern Arizona, Lockard invented an ingenious food dispenser that released very small quantities of millet seed at hourly intervals. To retrieve the seeds, an animal had to walk through the dispenser, depressing a treadle in the process. The moving treadle caused a pen to make a mark on a paper disc that turned slowly throughout the night, driven by a clock mechanism. When the paper disc was collected in the morning, it carried a temporal record of all nocturnal visits to the dispenser.

Lockard's data showed that in the fall, when the kangaroo rats had accumulated a large cache of seeds, they were selective about the timing of foraging, usually coming out of their underground burrows only at night when the moon was not shining (**FIGURE 5.14**). Because the predators of kangaroo rats (coyotes and owls) can see their prey more easily in moonlight, the rats are probably safer when they forage in complete darkness. For this reason, the kangaroo rats apparently possess a mechanism that enables them

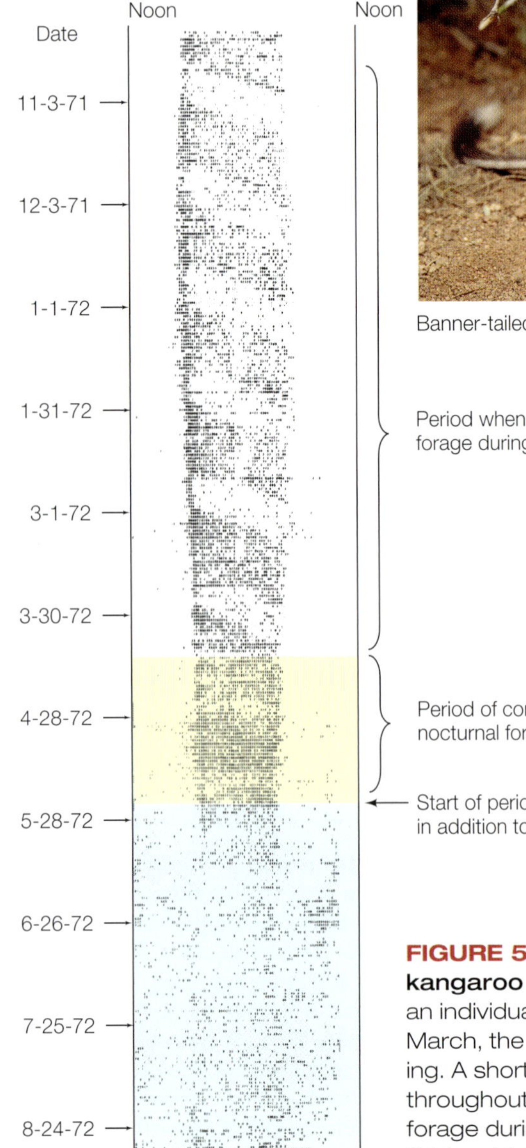

FIGURE 5.14 The lunar cycle and foraging by banner-tailed kangaroo rats. Each thin black mark represents a visit made by an individual to a feeding device with a timer. From November to March, the rats were active at night only when the moon was not shining. A shortage of seeds later in the year caused the animals to feed throughout the night, even when the moon was up, and later still, to forage during all hours of the day. (After Lockard 1978; photograph © Mary McDonald/Minden Pictures.)

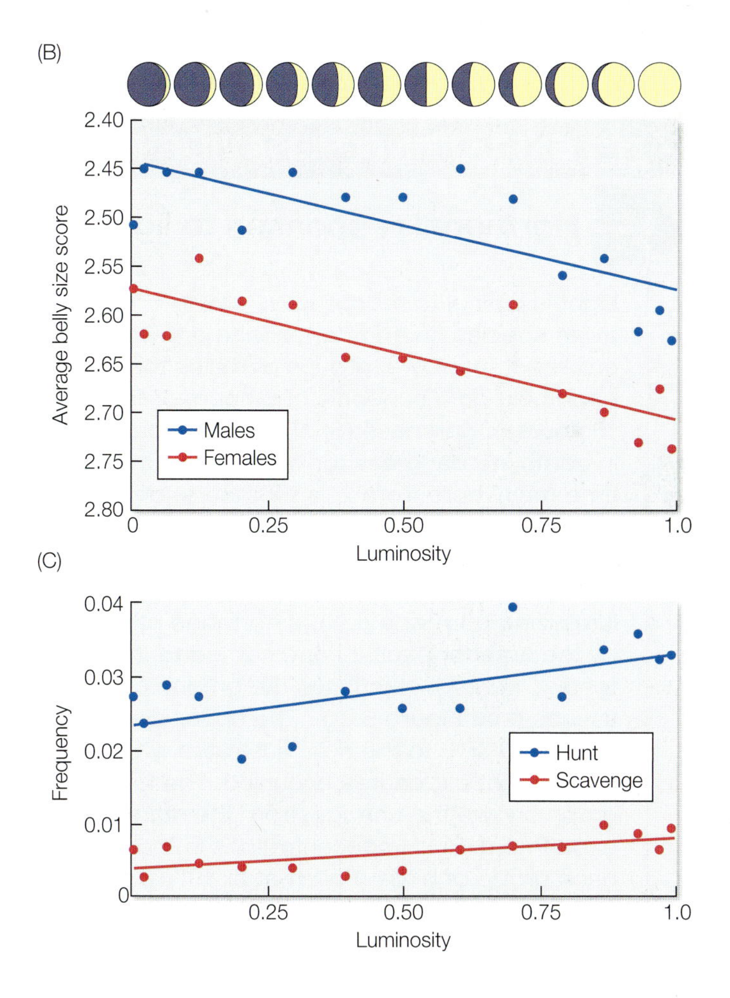

FIGURE 5.15 The lunar cycle affects lion hunting success. (A) A pride of African lions killed their prey at night, and this male continued to feast into the day. (B) As nighttime luminosity (brightness) increases, the bellies of male and female lions decrease in fullness, indicating greater difficulty in making kills under brighter conditions. (C) As the moon becomes full, lions are more likely to have to hunt during the day and to scavenge from others. (A, photograph courtesy of Dustin Rubenstein; B, C after Packer 1994, CC BY 4.0.)

to shift their foraging schedule in keeping with nightly moonlight conditions.

Of course, if more kangaroo rats are out and about when it is completely dark or nearly so, perhaps their predators also respond to the lunar cycle similarly. If so, the resulting overlap in their activity periods in relation to nighttime darkness would be an example of a coevolutionary arms race between predator and prey. Although it is not yet known how coyotes deal with moonlight, Craig Packer and his colleagues have established that African lions (*Panthera leo*) secure more food on dark nights near the new moon, when it is difficult for their prey to detect them (**FIGURE 5.15**). In contrast, when the moon is full, lions are more likely to hunt or scavenge during the day, a pattern indicative of a shift in foraging behavior linked to the lunar cycle. Because lions prefer to hunt when it is dark, people as well as antelopes and the like are more likely to fall prey to lions at this time (Packer et al. 2011). Given that *Homo sapiens* evolved in areas with lions and other large nocturnal predators, perhaps we can understand why we tend to be inactive at night, a decision fueled by the fear of darkness, a common feature of human behavior.

Whereas desert kangaroo rats and lions forage at night in accordance with the lunar cycle, and stonechats employ a circannual rhythm to breed at the right time in their tropical environment, temperate-zone birds such as white-crowned sparrows (*Zonotrichia leucophrys*) have their own behavioral control system that is well suited for coping with the dramatic seasonal changes that occur where they live. In the spring, males fly from their wintering grounds in Mexico and the southwestern United States to their distant summer headquarters in the northern United States, Canada, or Alaska. There they establish breeding territories, fight with rivals, and court sexually receptive females. In concert with these striking behavioral changes, the gonads of these birds—and of many other bird species—grow with dramatic rapidity, regaining all of the weight lost during the fall, when they shrink to 1 percent of their breeding-season weight. To restore their gonads in time for the onset of reproduction, the birds must somehow anticipate the spring breeding season. How do they manage this feat?

The sparrows' ability to change their physiology and behavior depends on their capacity to detect changes in the photoperiod, which grows longer as spring

photosensitivity hypothesis The circadian clock mechanism in animals exhibits a daily change in sensitivity to light, with a cycle that is reset each morning at dawn.

advances in temperate North America (Farner and Lewis 1971). According to the **photosensitivity hypothesis**, the clock mechanism of white-crowned sparrows and other birds exhibits a daily change in sensitivity to light, with a cycle that is reset each morning at dawn (**BOX 5.1**). During the initial 12 hours or so after the clock is reset, this mechanism is highly insensitive to light, an insensitivity that gives way to increasing sensitivity that reaches a peak 16 to 20 hours after the starting point in the cycle. Photosensitivity then fades very rapidly to a low point 24 hours after the starting point, at the start of a new day and a new cycle. Therefore, if the days have fewer than 12 hours of light, the system never becomes activated because no light is present during the photosensitive phase of the cycle. However, if the photoperiod is longer than 14 or 15 hours, light does reach the bird's brain during the photosensitive phase, initiating a series of hormonal changes that lead to the development of the bird's reproductive equipment and the drive to reproduce.

Photoperiodic systems work well for birds living in the temperate zone where day length varies greatly from winter to summer, but what about species that live near the equator where day length varies little from season to season? Michaela Hau and colleagues studied spotted antbirds (*Hylophylax naevioides*) in Panama,

BOX 5.1 **EXPLORING BEHAVIOR** BY **INTERPRETING DATA**

Hormonal responses to light in birds

Light is critical to reproductive timing in many avian species. In another experiment with white-crowned sparrows, groups of males that had been held on a schedule of 8 hours of light and 16 hours of darkness (8L:16D) were housed in complete darkness for variable lengths of time (anywhere from 2 to 100 hours) before being exposed to an 8-hour period of light. A few hours later, the researchers measured the concentrations in the birds' blood of luteinizing hormone (LH), a hormone released by the anterior pituitary and carried to the testes, where it stimulates the growth of these tissues (see Figure 5.30). The upper diagram in the **FIGURE** to the right illustrates when the 8-hour light exposures occurred. The lower graph shows the change, from the start of the experiment, in LH concentrations in the blood in each group of birds (Follett et al. 1974).

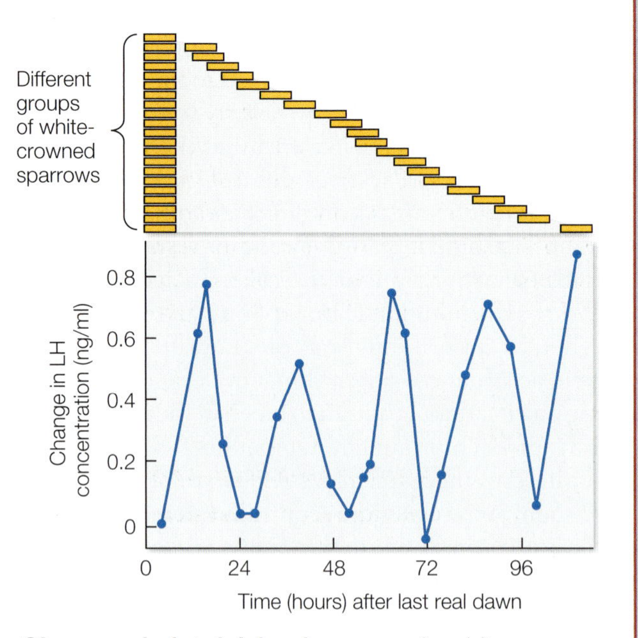

Changes in luteinizing hormone in white-crowned sparrows exposed to light after variably lengthened periods of darkness. (After Follett et al. 1974.)

Thinking Outside the Box

Do the data shown in the figure provide a test for the photosensitivity hypothesis? Can you design another experiment that might allow you to test this hypothesis in another way in this or a different species of animal?

where day length varies by less than 1 hour between the longest (June) and shortest (December) days of the year (Hau et al. 1998). The team found that wild birds had regressed (or shrunken) gonads in December, but increased gonad size ahead of the breeding season in May. To determine if slight changes in photoperiod could explain the gonadal development, the researchers housed birds in captivity under different light regimes. Birds exposed to "long" photoperiods of just 13 hours light (13L:11D) increased gonadal size eightfold and song activity sixfold compared with control birds exposed to "short" periods of 12 hours of light (12L:12D). In fact, the researchers found that an additional 17 minutes of light was all that was needed to trigger gonadal growth in this antbird. Subsequently, other tropical birds have also been shown to be similarly sensitive to small changes in photoperiod (Goymann and Helm 2014), suggesting that even the most subtle photoperiodic cues can trigger behavioral rhythms in some animals.

If this model of the photoperiod-measuring system is correct, it should be possible to deceive the system. William Hamner, working with house finches (*Haemorhous mexicanus*) (Hamner 1964), and Donald Farner, in similar studies with white-crowned sparrows (Farner 1964), stimulated testicular growth by exposing captive birds to light during the hypothesized photosensitive phase of their circadian rhythms. In Farner's experiment, birds that had been on a regular schedule of 8 hours of light and 16 hours of darkness (8L:16D) were shifted to an 8L:28D schedule. Because the light periods were now out of phase with a 24-hour cycle, these birds sometimes experienced light during the time when their brains were predicted to be highly photosensitive. The male birds' testes grew under these conditions, even though there was a lower ratio of light to dark hours than under the 8L:16D cycle, which did not stimulate testicular growth (**FIGURE 5.16**) (Farner 1964).

Photoperiodic changes are useful guides for breeding activity in birds that live in seasonal environments with predictable changes in food resources (even Panama has seasons, a dry season and a rainy season). But what about the rufous-winged sparrow (*Peucaea carpalis*), a bird that lives in the drought-prone Sonoran Desert, where the arrival of spring does not guarantee abundant supplies of the insects needed to rear a batch of offspring? Instead, food for this sparrow's brood may not become available until summer rains have fallen, in which case the bird waits to reproduce until after the onset of the monsoon, which can begin anywhere from early July to mid-August, when the photoperiod is already declining (Small et al.

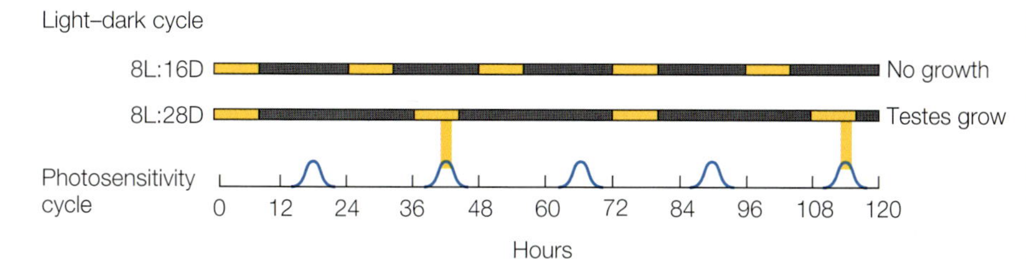

Male white-crowned sparrow

FIGURE 5.16 A cycle of photosensitivity. An experiment with white-crowned sparrows tested the hypothesis that these birds possess a clock mechanism that is especially sensitive to light between hours 17 and 19 of each day. The lower line represents these hypothetical periods of photosensitivity. The yellow and black sections of the two upper horizontal bars show the light and dark periods of two different light-dark regimes. Only sparrows under the 8L:28D experimental regime were exposed to light during the supposed photosensitive phase of the cycle, and only they responded with testicular growth. (After Farner 1964; photograph © Chad A. Ivany/ShutterStock.)

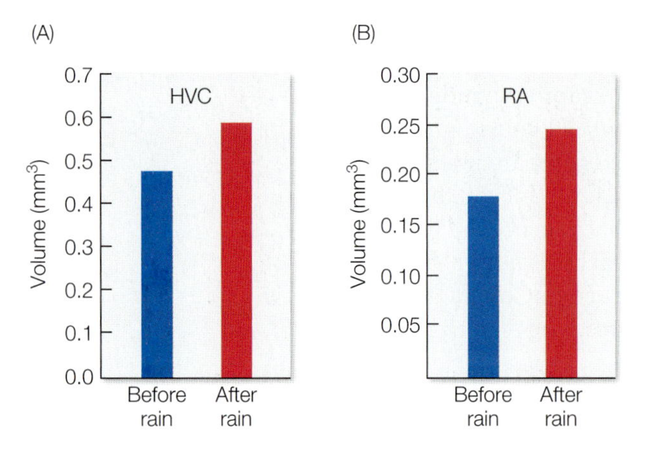

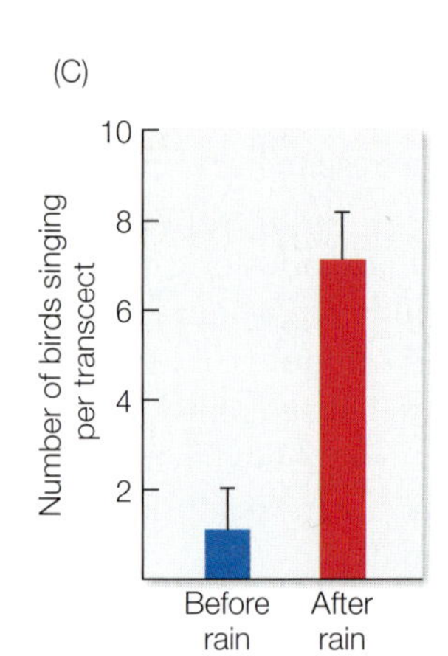

Male rufous-winged sparrow

FIGURE 5.17 Seasonal changes in the song control regions of the rufous-winged sparrow's brain. Summer rainfall triggers these changes, which lead to an increase in singing behavior. The sizes of the (A) high vocal center (HVC) and (B) robust nucleus of the arcopallium (RA) in the sparrow increase following monsoon thunderstorms in southern Arizona. (C) The neural changes are linked to an increase in singing after the monsoon has begun. Blue bars represent data collected before the first thunderstorms. Red bars represent data collected after the start of the monsoon. Bars depict mean +/– SE. (After Strand et al. 2007; photograph © Tim Zurowski/Shutterstock.com.)

2007). Summer rainfall after a dry spring appears to stimulate the production of luteinizing hormone, or LH, which in turn may cause the testes to produce testosterone, priming the male to establish a territory, start singing, and if all goes well, reproduce (**FIGURE 5.17**) (Strand et al. 2007).

Unpredictable Environmental Cues

Rufous-winged sparrows have evolved proximate mechanisms that respond to summer rainfall because in their desert environment, rainfall is a more reliable indicator of future food for offspring than are changes in photoperiod. Rainfall also appears to be the cue that stonechats in Kenya use to time their breeding in the unpredictable savanna environment in which they live (Dittami and Gwinner 1985). Rainfall-induced gonadal growth and reproduction have also been observed in small ground finches (*Geospiza fuliginosa*) from the Galápagos Islands (Hau et al. 2004b), one of the places on Earth with the most erratic and unpredictable rainfall. Small ground finches also have lower levels of LH outside the breeding season, but as in rufous-winged sparrows, levels rapidly increase when the rains begin. Thus, in arid and semiarid areas of the world where food supplies depend on the rains, it is the start of the rainy season that cues animals' breeding behavior.

For two seed-eating finch species from the temperate zone of North America, the white-winged crossbill (*Loxia leucoptera*) and red crossbill (*L. curvirostra*), it is food intake itself that acts as the primary cue for breeding (Benkman 1990). Even though we typically think of temperate environments as having predictable cycles from winter to summer, the seeds that crossbills are specialized to feed on are not actually a predictable resource. Craig Benkman has found that in years when conifer seed production is so high that the birds can find ample food for themselves and their brood in almost any month, crossbills take advantage of the good times to breed. This pattern can be linked to a distinctive feature of crossbill hormonal physiology, which is that birds that have been on long photoperiod days (20L:4D) do

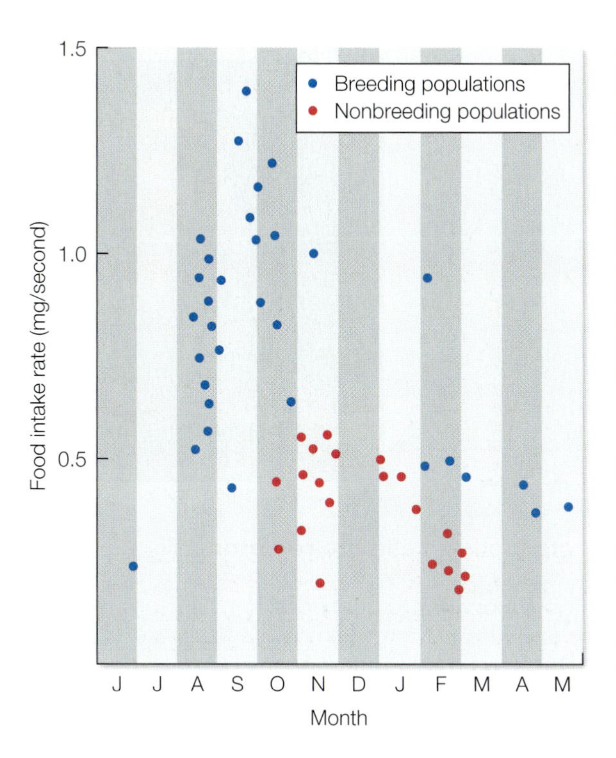

White-winged crossbill

FIGURE 5.18 Food intake and reproductive timing in the white-winged crossbill. White-winged crossbills are more likely to breed when food is plentiful and food intake rates are higher. Each point on the graph represents a breeding or nonbreeding population of white-winged crossbills. (After Benkman 1990; photograph © Timothey Kosachev/ShutterStock.)

not completely shut down their gonadotropin-releasing neurons, which is also the case in related birds such as common redpolls (*Carduelis flammea*) and pine siskins (*Spinus pinus*) (Pereyra et al. 2005). Instead, crossbills appear to retain the capacity to stimulate the release of reproduction-regulating hormones should conditions for reproduction become especially favorable.

This flexibility, however, does not mean that crossbills ignore all environmental cues except seed abundance (Hahn et al. 2004). In studies of crossbills in the wild, Thomas Hahn noticed a break in breeding in December and January (**FIGURE 5.18**), even in environments where food was plentiful. Therefore, although these birds are more flexible and opportunistic than the average songbird, Hahn wondered if crossbills still have an underlying reproductive cycle that is dependent on photoperiod. When he held red crossbills at a constant temperature with unlimited access to food while letting the birds experience natural photoperiod changes, he found that male testis length fluctuated in a cyclical fashion (**FIGURE 5.19**), becoming smaller during October through December of each year, even when the birds had all the seeds that they could hope to eat. In addition, free-living red crossbills also always undergo declines in gonad size and in the concentrations of sex hormones in their blood at this time of year (Hahn 1998). Therefore, the reproductive opportunism of these birds is not absolute, but rather is superimposed on the photoperiod-driven timing mechanism inherited from a common ancestor of temperate-zone songbirds, a point that applies to some other opportunistically breeding bird species as well (Hahn et al. 2008).

The partial persistence of the standard timing system in the flexibly breeding crossbills could be explained at the ultimate level of analysis in at least two ways: (1) the photoperiod-driven mechanism might be a nonadaptive holdover from a distant ancestor with the standard pattern; or (2) crossbills might derive reproductive benefits from the retention of a physiological system that reduces the likelihood that they will attempt to reproduce at times when other costly, competing activities take priority (such as the need to molt and replace their feathers in the fall) (Deviche and Sharp 2001). However, by maintaining their gonadal equipment in at least partial

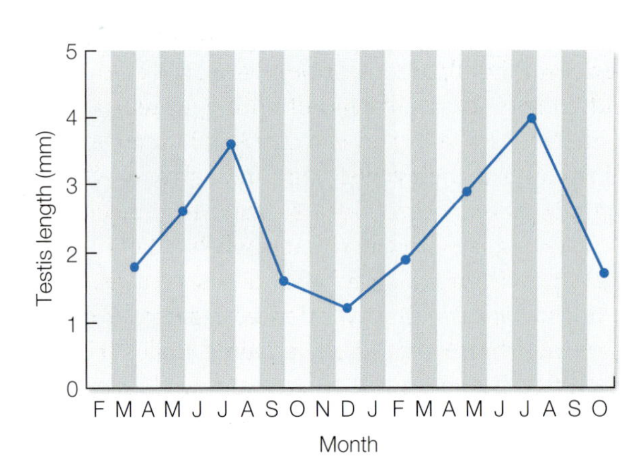

FIGURE 5.19 Photoperiod affects testis size in the red crossbill. Six captive birds were held under natural photoperiods, which changed over the seasons, but temperature and food supply were held constant. Data are the average testis length among these birds at different times of the year. (After Hahn 1995; photograph © Robert L. Kothenbeutel/Shutterstock.com)

readiness for much of the year, crossbills and other birds that encounter unpredictable variation in their environments are able to react more quickly to a chance to reproduce (Tökölyi et al. 2012).

reproductive readiness hypothesis Animals maintain their gonads in at least partial readiness for much of the year so that if they encounter favorable conditions in their environments, they are able to react more quickly to a chance to reproduce.

Nicole Perfito and colleagues tested this **reproductive readiness hypothesis** by comparing Australian populations of the archetypal opportunistically breeding bird, the zebra finch (*Taeniopygia guttata*) (Perfito et al. 2007). The researchers compared reproductive activation and suppression in zebra finches from two different habitats: (1) an unpredictable environment in arid central Australia where breeding is closely tied to aperiodic rainfall, which can occur during any month of the year; and (2) a more predictably seasonal habitat in southern Australia where breeding occurs during the same months each year. By measuring gonad size and LH concentrations, the researchers found consistent differences between the populations in their degree of reproductive readiness. They found that males in the predictable environment had regressed gonads during the nonbreeding season and lower levels of LH compared with during the breeding season. Likewise, females had smaller ovarian follicles and lower LH concentrations during the nonbreeding compared with the breeding season. However, males living in the unpredictable environment had gonads of intermediate size and LH levels that were similar to those in the breeding season (and higher than those of nonbreeding males in the predictable environment). Females in the unpredictable environment had small follicles, but like males, LH levels that were similar to those in females during breeding (and higher than those of nonbreeding females in the predictable environment). Thus, for adult male and female zebra finches living in the predictable environment, reproductive readiness declined from the breeding to the nonbreeding state. In contrast, for birds living in the unpredictable environment, more nonbreeding birds maintained a higher level of readiness, even though they were in poorer body condition. Together, these results are consistent with the reproductive readiness hypotheses and the idea that opportunistically breeding species of birds living in unpredictable environments maintain a state of reproductive readiness that enables them to rapidly take advantage of appropriate changes in rainfall and food availability.

Social Conditions and Changing Priorities

As we have just seen, different features of the physical or ecological environment (such as moonlight, day length, the tides, rainfall, or food supply) are used by different species to set their priorities and thereby forage instead of hiding, or reproduce instead of waiting some more. In addition, animals can use changes in their social environments to make adaptive adjustments in their physiology and behavior. Thus, for example, when Hahn and several coworkers performed an experiment on crossbills in which some captive males were caged with their mates while others were forced into bachelorhood but were kept within sight and sound of the paired red crossbills in a neighboring aviary, the bachelor males experienced a slower return to reproductive condition after the winter break than did the paired males (Hahn 1995). Even more dramatically, when free-living female song sparrows (*Melospiza melodia*) were given implants containing the sex hormone estrogen, which greatly prolonged their sexual receptivity, their hormonally untreated mates also remained in reproductive mode for months longer than males paired with control females at this location. The sexual and territorial responses of male song sparrows in this experiment demonstrate the sensitivity of animals to their social environment (Runfeldt and Wingfield 1985).

Much the same thing occurs in house mice (*Mus domesticus*). When female mice are given a chance to show a preference for a potential mate, they spend time sniffing the male of their choice when placed in the center cubicle with males in two compartments at either end of the cage. Females that have had prior experience only with a subordinate male do not exhibit a preference when given a choice between sniffing a dominant and a subordinate male. But those that have had prior experience with the odor of a dominant male spend much more time near the dominant individual (**FIGURE 5.20**). Exposure to the scent of the dominant male promotes the addition of neurons in two regions of the female brain, effectively rewiring her neural machinery such that she can identify a dominant male mouse should her social environment provide her with such a partner (Mak et al. 2007).

Dominant male house mice are also programmed to permit social experience to alter their behavioral decisions. For example, after a male house mouse mounts a female and ejaculates, he immediately becomes highly aggressive toward mouse

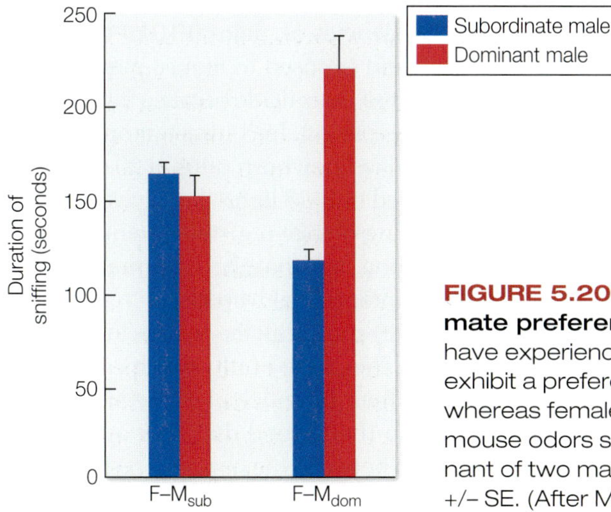

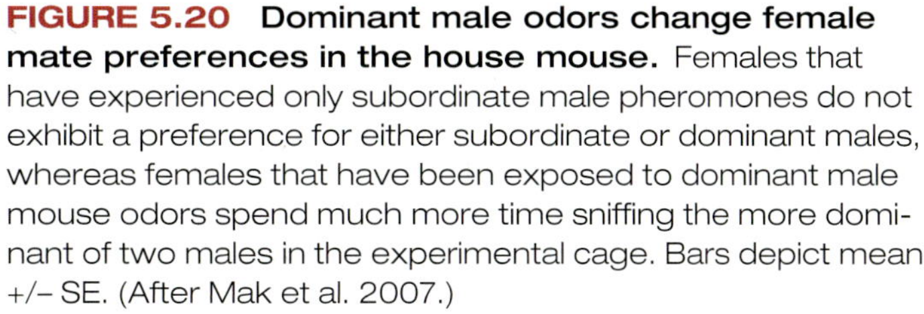
FIGURE 5.20 Dominant male odors change female mate preferences in the house mouse. Females that have experienced only subordinate male pheromones do not exhibit a preference for either subordinate or dominant males, whereas females that have been exposed to dominant male mouse odors spend much more time sniffing the more dominant of two males in the experimental cage. Bars depict mean +/– SE. (After Mak et al. 2007.)

pups, killing any he finds. For almost 3 weeks after mating, he is likely to commit infanticide, but after that time, he becomes more and more likely to protect any young pups he encounters. When about 7 weeks have passed since ejaculation, he becomes infanticidal once again (Perrigo et al. 1990). This remarkable cycle has clear adaptive value. After a male transfers sperm to a partner, 3 weeks pass before she gives birth. Attacks on pups during these 3 weeks will invariably be directed against a rival male's offspring, with all the benefits attendant on their elimination. After 3 weeks, a male that switches to paternal behavior will almost always care for his own neonatal offspring. After 7 weeks, his weaned pups will have dispersed, so once again he can practice infanticide advantageously.

Hormonal Mechanisms Underlying Behavioral Change

As you have just seen, hormones play a key role in linking environmental information to rapid shifts in behavior. Additionally, hormones can work during development to alter the nervous system and permanently change physiology and behavior. The ability to measure hormone levels in blood or other tissues has greatly increased the power of behavioral biologists to study the mechanistic bases of a variety of behaviors in the field or lab (**BOX 5.2**). Below, we explore how hormones can act during development and throughout life to alter behavior in response to environmental cues.

Organizational versus Activational Effects of Hormones on Behavior and Development

At the proximate level, what kind of mechanism could enable a male mouse to switch from being infanticidal to paternal just 3 weeks after a mating? One possible explanation involves an internal timing device that records the number of days since the male last copulated. If such a sexually activated timing mechanism exists, then an experimental manipulation that either increases or decreases the length of a "day," as perceived by the mouse, should have an effect on the absolute amount of time that passes before the male makes the transition from killer to caregiver.

Glenn Perrigo and his coworkers manipulated day length by placing groups of mice under two different laboratory conditions, one with "fast days," in which 11 hours of light were followed by 11 hours of darkness (11L:11D) to make a 22-hour "day," and another with "slow days" (13.5L:13.5D) that lasted for 27 hours (Perrigo et al. 1990). As predicted, the total number of light–dark cycles, not the number of 24-hour periods, controlled the infanticidal tendencies of males (**FIGURE 5.21**). Thus, when male mice in the fast-day group were exposed to mouse pups 20 real days after mating, only a small minority committed infanticide because these males had experienced 22 light–dark cycles during this period. In contrast, more than 50 percent of the males in the slow-day group attacked newborn pups at 20 real days after mating, because these males had experienced only 18 light–dark cycles during this time. These results demonstrate that a timing device registers the number of light–dark cycles that have occurred since mating and that this information provides the proximate basis for the control of the infanticidal response.

Some sort of chemical or physiological mediator must link the change in daylight to the change in male aggression. The most obvious type of physiological mediator is a hormone; if this is the case, then during their infanticidal phase, male mice might have high concentrations of hormones in their blood that alter aggressive behavior. Given the well-established relationship between the hormone testosterone and male aggression, this steroid hormone seems a likely candidate. An alternative

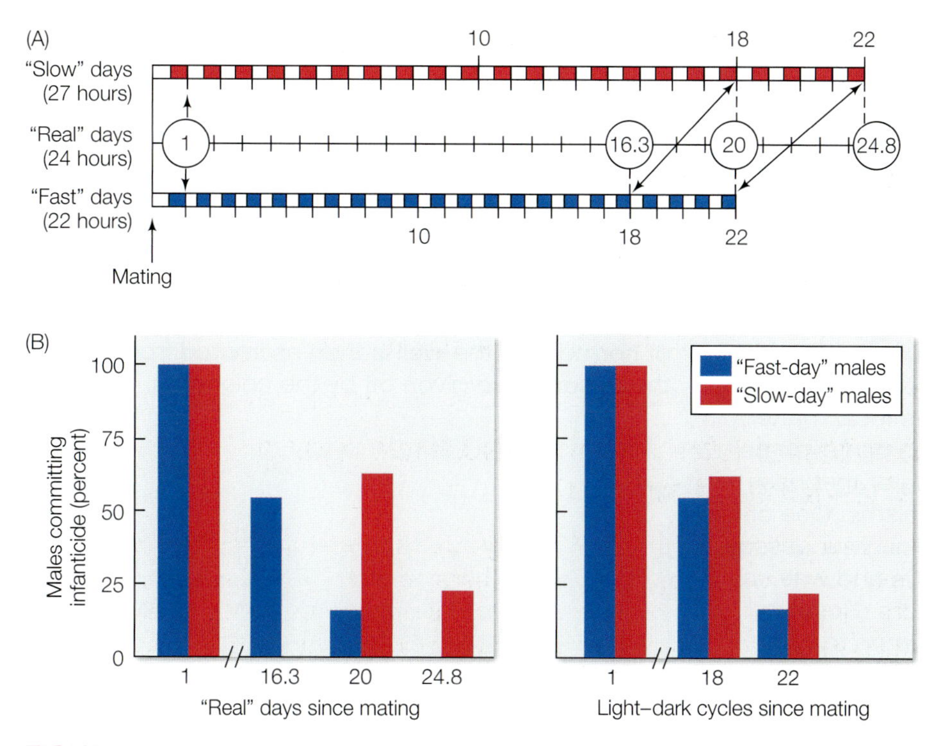

FIGURE 5.21 Regulation of infanticide by male house mice. (A) Male mice were held under artificial "slow-day" and "fast-day" experimental conditions. The timeline illustrates the relationship between absolute time and the increasing desynchronization of light/dark cycles experienced by both groups during the experiment. The alternating red/blue and white bars on the slow and fast time scales represent the dark phase of the repeating light/dark cycles. By 20 absolute days after mating, the fast day males had experienced four more light/dark cycles than the slow days males. (B) Most of the males held under fast-day conditions had stopped being infanticidal by 20 real days (22 fast days) after mating. In contrast, males experiencing slow days did not show the same decline in infanticidal behavior until nearly 25 real days had passed. (After Perrigo et al. 1990.)

hypothesis, however, is that extreme male aggression toward mouse pups occurs when the males are under the influence of another steroid hormone, progesterone, which is known to suppress parental behavior in female rodents. Steroid hormones such as testosterone and progesterone are known to alter behavior rapidly in a matter of seconds to minutes. This immediate but temporary regulatory effect of a hormone is called an **activational effect**. But not all hormonal effects are temporary. Some hormones—including these and other steroids—can also permanently affect the nervous system during development and cause permanent changes to physiology and behavior, a phenomenon called an **organizational effect**.

Although this book focuses mainly on activational effects of hormones, it is important to realize how hormones affect behavioral phenotypes during development. One of the best examples of how hormones organize behaviors is seen in their effect on sexual partner preference. Experiments in mammals as diverse as ferrets, rats, mice, hamsters, dogs, and pigs, as well as in birds, have demonstrated that exposure to hormones during development can actually lead to a reversal in sexual partner preference (Adkins-Regan 2005). For example, female zebra finches treated with the steroid hormone estradiol as nestlings behave like males as adults. Indeed, a team led by Elizabeth Adkins-Regan demonstrated that estradiol-treated nestlings grew up to have a masculinized song system and prefer females rather than males when pairing (Mansukhani et al. 1996).

Working with an unusually aggressive strain of laboratory mice, Jon Levine and his coworkers used genetic knockout techniques to test the alternative proximate

BOX 5.2 INTEGRATIVE APPROACHES

Measuring hormones in animals

The ability to measure levels of hormones from animals has given behavioral biologists a physiological window into how animals perceive the world. Scientists are able to measure not only hormone levels in blood, but also levels in other fluids, including urine and saliva, as well as in other tissues or forms of excrement that can be reconstituted as fluid (such as fecal matter, hair, nail, and feather). Hormones can even be measured from fluids such as water that has held fish and other aquatic organisms. One of the first hormone assays, called radioimmunoassay (RIA), was developed in the 1950s and was so revolutionary that its three creators (Rosalyn Yalow [the second American woman to win a Nobel Prize], Roger Guillemin, and Andrew Schally) shared the Nobel Prize in Physiology or Medicine in 1977. In an RIA, an antibody specific to the hormone of interest is mixed with both a sample of fluid from the organism with an unknown amount of antigen (or hormone) and with a known amount of purified standard hormone that has been labeled with a radioactive isotope. The non-radioactive hormone occurring naturally in the sample fluid competes with the radioactive-labeled hormone to bind to the antibody, thus leaving some radioactive-labeled hormone free. The amount of free radioactive hormone is then measured to determine how much of the original hormone was in the fluid sample.

Like RIA, the enzyme-linked immunosorbent assay (ELISA), or simply enzyme immunoassay (EIA), allows one to measure hormone levels in a fluid. Instead of resulting in radioactivity, however, ELISA results in a colored end product where the intensity of the color change is correlated with the amount of hormone present in the original sample (**FIGURE A**). Whether for RIA or ELISA, high radioactivity or lots of color equates to low hormone levels in the unknown sample. There are many types of ELISA, including competitive binding assays that work in much the same way as RIA. In the typical ELISA, an antibody designed to bind to the antigen of interest (the hormone) is fixed to the bottom of each well of a plastic plate. The sample being measured is then added to a well. Once the hormone binds to the antibody, an enzyme-conjugated detection antibody is added to the solution and binds to the antibody–hormone complex. After all of the unbound antibody and enzyme have been washed away, a chromographic substrate that changes color when it comes in contact with the enzyme is added to the solution. The amount of hormone in the well is then estimated from the absorbance given off by the coloration.

(A)

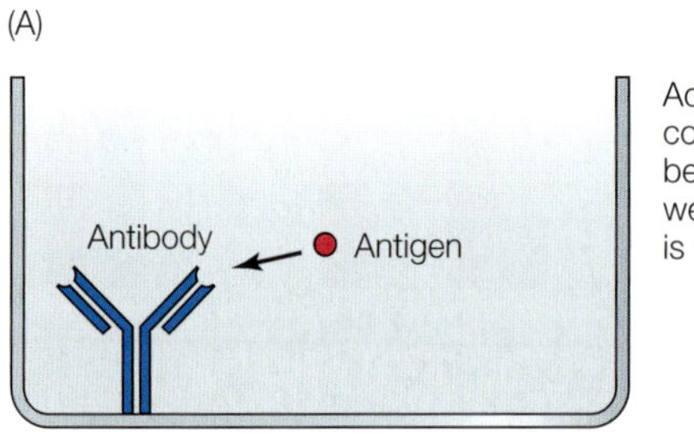

Add sample solution containing antigen to be measured to the well where antibody is solidified.

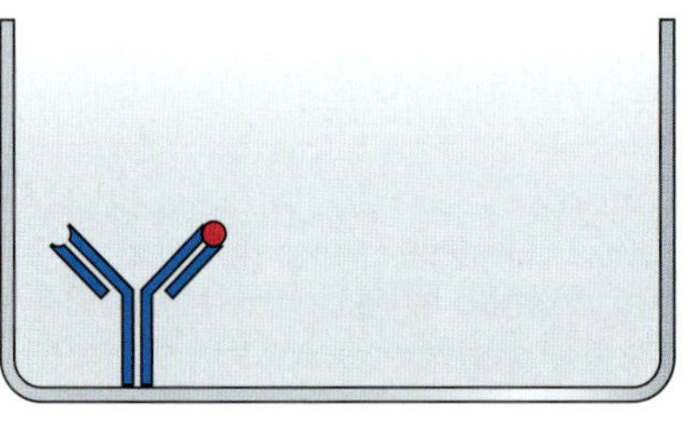

Antigen is captured by antibody. Wash out excessive substances.

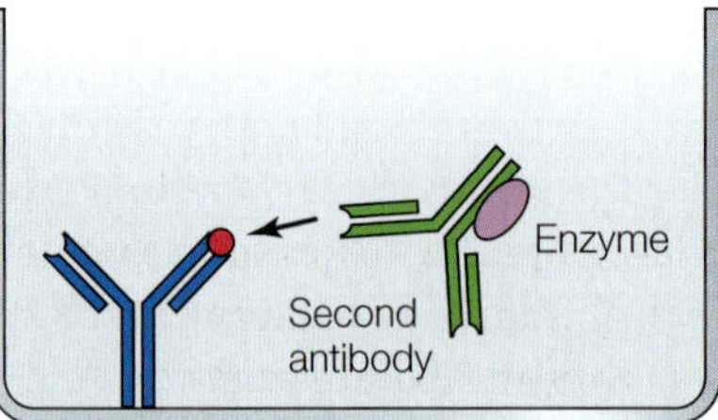

Add enzyme-labeled second antibody to the well.

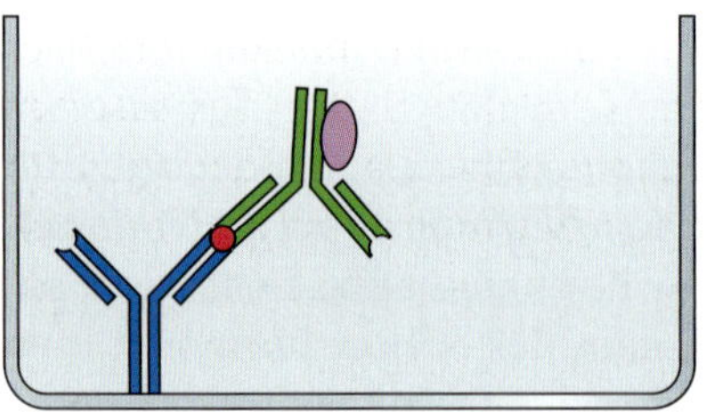

The enzyme-labeled second antibody binds to the captured antigen. Wash out excessive substances.

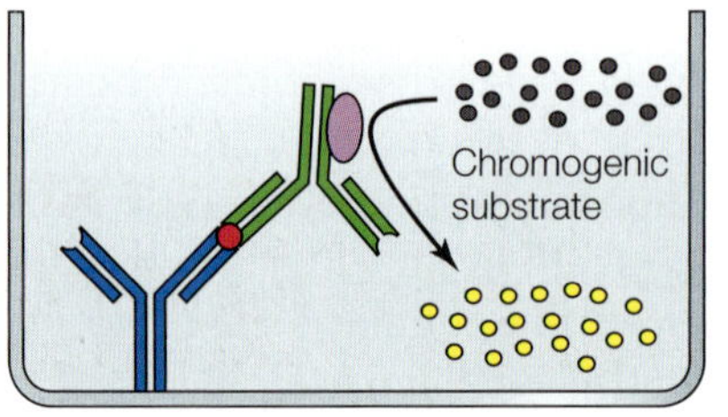

Add a chromogenic substrate of the enzyme that shows coloration by enzyme. The amount of the antigen is estimated from the absorbance.

Diagram of how basic ELISA works.

BOX 5.2 INTEGRATIVE APPROACHES (continued)

Thinking Outside the Box

Both RIA and ELISA can be used to measure hormones levels in samples collected from animals. How are they similar? How are they different? ELISA has become the preferred method of assay for many hormones, largely because it does not require radioactivity and because it is relatively cost-effective, requires lower sample volumes and has lower detection limits than RIA (something that is often critical to field biologists working with small animals and rare samples), and is both fast and efficient. However, some hormones—including many peptide hormones—still require RIA.

For both methods, how do you think a researcher actually determines the amount of hormone that is in a sample based on the amount of radioactivity or light given off? In both types of assay, standards of known hormone concentration are used. In fact, scientists create a standard curve, where the known hormone concentration for a handful of standards is plotted against the radioactivity or absorbance of those standards (**FIGURE B**). Look at the standard curve plotted here. If you have two unknown samples that have an absorbance of 0.35 and 0.55, what would the concentration of hormone be in those samples? What happens if the absorbance of the samples occurs above or below the absorbance of all of the standards used to make your standard curve?

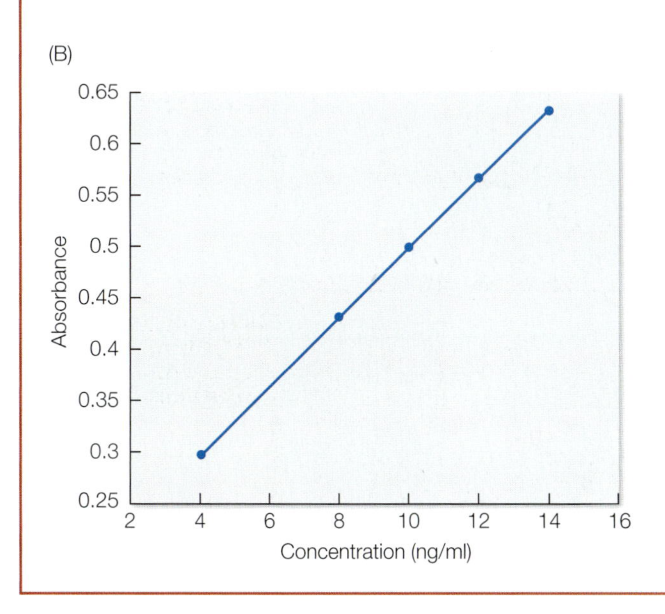

A typical standard curve used to calculate the concentrations of unknown hormone values in an ELISA. The hormone concentration is plotted against absorbance for five standards.

hypotheses that either testosterone or progesterone influences male aggressive behavior. The researchers created a population of laboratory mice in which males lacked progesterone receptors and thus could not detect the progesterone in their bodies. When the knockout males were exposed to a pup, they never attacked it, whereas over half of the males from a population that had not been genetically modified assaulted a test infant (**FIGURE 5.22**). The researchers also looked at circulating hormone levels, finding no significant differences in testosterone or progesterone levels between the two groups of mice—only a difference in the ability of their brain

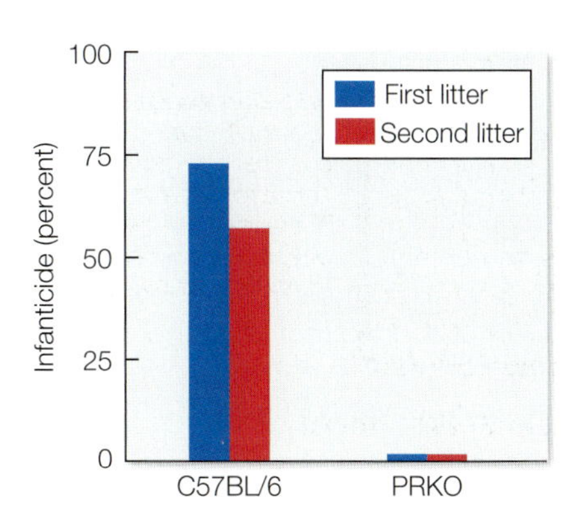

FIGURE 5.22 A hormonal effect on infanticidal behavior in laboratory mice. Males of the strain C57BL/6 are highly likely to attack their own first or second litters, rather than caring for their young. If, however, the progesterone receptor gene is removed from their genome, the resulting progesterone knockout (PRKO) mice cannot detect the progesterone in their bodies, and they do not exhibit infanticidal behavior toward their offspring. (After Schneider et al. 2003, © 2003 National Academy of Sciences, USA.)

cells to detect progesterone. This work suggests that when progesterone is present in certain concentrations in an intact male, the mouse is primed to be an infanticidal killer (Schneider et al. 2003). As progesterone levels slowly fall (or sensitivity to the hormone declines in brain cells) after mating, the paternal capacity of the male may slowly increase in time, which has the adaptive effect of keeping him from harming his own brood. Similarly, steroid hormones influence parental care behavior in other mice species (**BOX 5.3**), as do some neuropeptides, as you saw in Chapter 3.

Hormones have many regulatory (or activational) effects that enable individuals to modify their response to key social stimuli in adaptive ways. A dramatic example is provided by male Japanese quail (*Coturnix japonica*), which behave differently after mating with a female. Prior to copulating, a male housed next to a female in a two-compartment cage spends relatively little time peering through a window between the compartments to look at her. But after mating, a male appears positively fascinated by his partner, to the extent that he will stare at her for hours on end (**FIGURE 5.23A**) (Balthazart et al. 2003). This behavior, which presumably encourages unconfined males to get together with a receptive female for yet

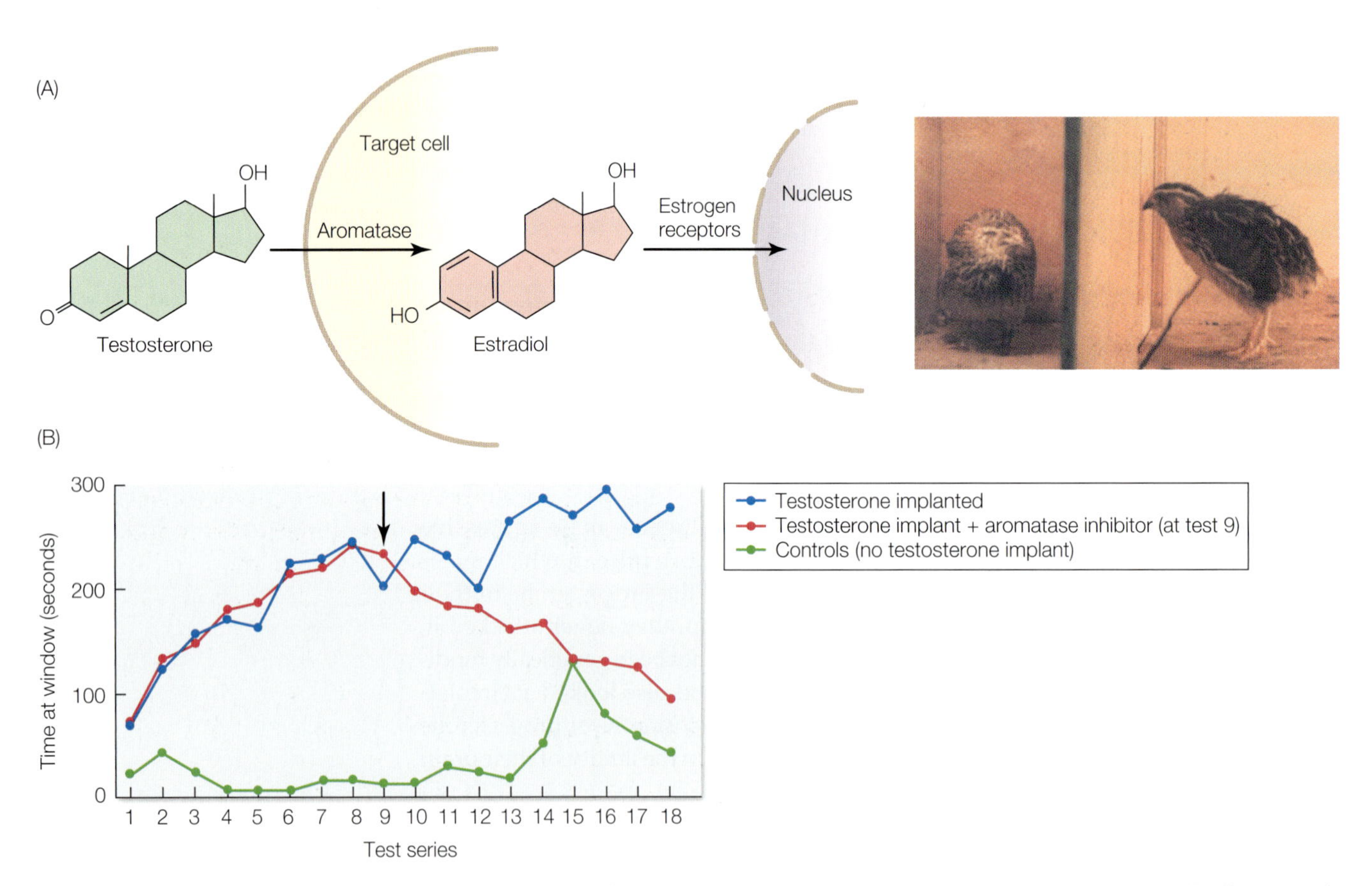

FIGURE 5.23 Testosterone and the control of sexual motivation. (A) Hormonal changes in male Japanese quail affect how sexually experienced males respond to the opportunity to view a mature female quail. Testosterone is released from the testes and is captured by target cells in the brain, where the enzyme aromatase converts the hormone to estradiol. That substance then binds with an estrogen receptor to form an estrogen–receptor complex, which is transferred to the target cell nucleus, where it promotes chemical changes that ultimately cause a male quail to stare at a female through a window in their two-compartment cage. (B) Castrated males that have received testosterone implants exhibit the staring-at-female response, unlike castrated controls that have not been given testosterone implants. But when implanted males are injected with an aromatase inhibitor (at test 9, as indicated by the arrow), they lose the response, presumably because testosterone can no longer be converted to estradiol, the essential hormone for modulating male sexual motivation. (From Balthazart et al. 2003.)

BOX 5.3 EXPLORING BEHAVIOR BY INTERPRETING DATA

Do steroid hormones modulate male parental behavior in California mice?

In the California mouse (*Peromyscus californicus*), males are highly paternal. Explanations for this behavior include the proximate hypothesis that progesterone is responsible, as we explored earlier for laboratory mice. Alternatively, it is possible that decreased levels of testosterone are responsible for male parental behavior. In laboratory experiments, Brian Trainor and colleagues found that differences in testosterone levels between inexperienced males (with neither mates nor offspring) and fathers (with a mate and offspring) are not statistically significant (**FIGURE A**; Trainor et al. 2003). In contrast, progesterone levels are far higher in the blood of inexperienced males than in males with offspring (**FIGURE B**).

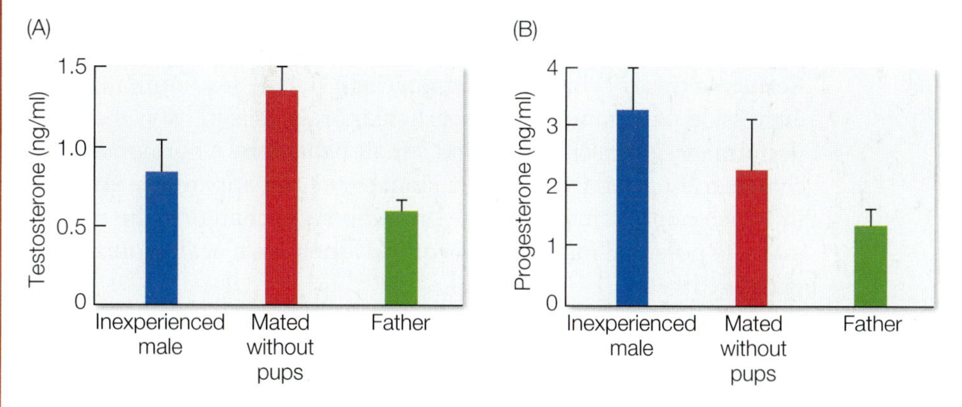

California mouse

Comparison of steroid hormone levels. (A) Testosterone and (B) progesterone levels in the blood of male California mice. Bars depict mean +/– SE. (A, B after Trainor et al. 2003; photograph © Miles Barton/Minden Pictures.)

Thinking Outside the Box

Consider the alternative proximate hypotheses that two different steroid hormones are responsible for the behavioral differences in males. Which of these hypotheses do the data support? Given these results, how would you design a different type of experiment to verify them? Hint: consider manipulating hormones rather than simply measuring them.

another copulation, is heavily influenced by the presence of testosterone in particular regions of the male's brain. This conclusion is based in part on the finding that removal of the male's testes eliminates all sexual behavior—unless the male receives an implant containing testosterone, in which case he regains his motivation to remain close to a sexual partner.

Interestingly, testosterone itself is not the signal that turns a male Japanese quail into an apparently lovelorn individual when separated from his mate. Instead, testosterone is converted into the steroid hormone estradiol, an estrogen, in target cells in the preoptic area of the brain. The conversion requires an enzyme called aromatase, which is coded for by a gene that becomes much more active after a male quail has mated. The estrogen produced in the presence of aromatase then binds with an estrogen receptor protein. The resulting estrogen–receptor complex relays a signal to the nucleus of the cell, leading to further biochemical events, which ultimately translate into the neural signals that cause a male to stare intently at a currently unreachable sexual partner (**FIGURE 5.23B**).

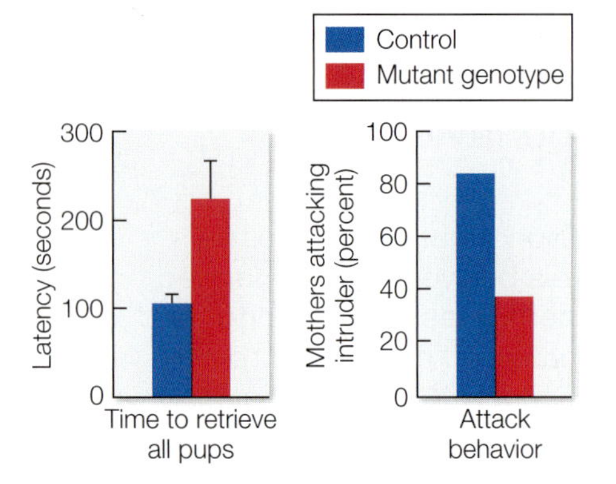

FIGURE 5.24 One gene's effect in activating two maternal behaviors in mice. Females with a mutation in the *Irs4* gene take longer to retrieve a group of pups moved away from them. They also are less likely to attack an intruder that approaches their pups. Bars depict mean +/– SE. (After Xu et al. 2012.)

The way in which hormones activate specific behaviors in the lab mouse has been worked out by a team of researchers studying the interrelationship between hormones and genes in certain regions of the mouse brain (Xu et al. 2012). The team began by using microarray analysis to identify a set of genes that differed between the sexes in their expression in the hypothalamus, a portion of the brain that plays a major role in controlling sexual behavior in many vertebrates. Having identified several genes that had the potential to influence the reproductive behavior of mice, the researchers then compared the behavior of individuals with and without the typical (wild-type) alleles of these genes. Males missing wild-type alleles of the gene *Sytl4* were much slower to copulate with a receptive female than were nonmutant males, although the sexual behavior of the mutants was otherwise largely normal. In contrast, females with the mutant *Sytl4* genotype behaved normally, but those without the wild-type allele of a different gene (*Irs4*) were much slower in retrieving displaced pups and much less likely to attack intruders than were wild-type mothers (**FIGURE 5.24**). Results of this sort indicate that male and female sex hormones affect the expression of certain genes in particular brain cells in ways that affect the performance of specific male and female behaviors. A hormonally induced change in the activity of even a single gene can apparently alter the way in which neural circuits in the brain operate, changing the response of males to potential mates or how quickly mothers react to young that have wandered away.

Hormones and Reproduction

Unlike Japanese quail, many other animals do not form long-lasting sexual attachments. Indeed, soon after a fruit fly female mates, she becomes completely unreceptive, rejecting not only her previous partner but all other males, should they attempt to court her. The female fly then spends her time laying eggs, an adaptive shift in behavior given that she can fertilize all her mature eggs with stored sperm received from a single mating. As it turns out, the female's dramatic switch from sexual receptivity to sexual refusal is regulated by a hormone—not one that she manufactures herself, but one that she receives from the seminal fluid the male transfers to her during copulation.

The male-donated hormone is called sex peptide (SP). Using RNA interference, it is possible to block the specific gene in male fruit flies that codes for SP. The blocked males are normal, except for the fact that they cannot make SP, and therefore, when they copulate with females, they are unable to donate SP to their mates. If SP is indeed the critical signal that stops females from responding to courting males, then females mated with SP-deficient males should remain receptive following copulation, despite having received sperm and seminal fluid. This prediction is correct (**FIGURE 5.25**) (Chapman et al. 2003). As noted earlier, a chemical messenger acts

(B)

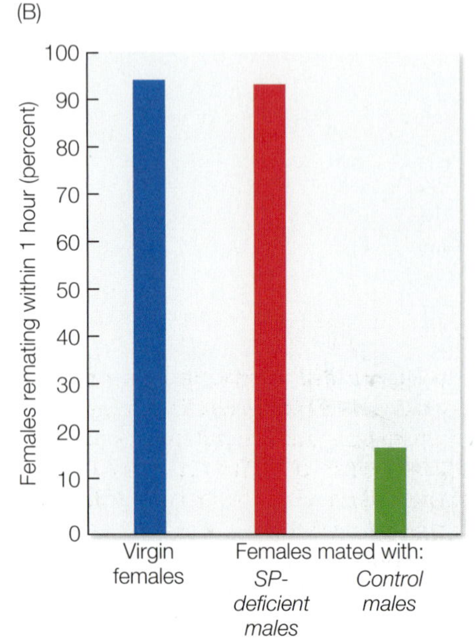

FIGURE 5.25 What controls female receptivity in a fruit fly? (A) A sex peptide (SP) transferred to females in the male seminal donation appears to be a critical factor. (B) Female fruit flies mated to males unable to supply SP are as likely to copulate with a new male 48 hours after mating with an SP-deficient male as are virgin females to mate on their first sexual encounter. (A, photograph by T. Chapman/CC BY 2.0; B after Chapman et al. 2003, © 2003 National Academy of Sciences, USA.)

by binding with receptor molecules on the surface of a target cell. The fruit fly hormone SP should therefore bind with a specific receptor protein, and accordingly, the appropriate molecule, unimaginatively called sex peptide receptor (SPR), has been found. Here too it is possible to block the single gene that codes for SPR, creating "mutant" females whose cells are incapable of binding with the male-donated hormone. When this is done experimentally, SPR-deficient females that have mated once copulate again when given a chance, just as if they were still virgins (Yapici et al. 2008).

Hormonal signals that regulate animal sexual behavior often lead to increases in both gamete production and sexual activity. This **associated reproductive pattern** (**FIGURE 5.26**) is, for example, evident in the seasonal reproductive behavior of British red deer (*Cervus elaphus*). Stags that have been living peacefully with one another all summer become aggressive as September approaches prior to the mating season. At this time, their testes generate sperm and testosterone. Behavioral endocrinologists have long used a classic experimental paradigm in vertebrates to test the hypotheses like the one that steroid hormones generated in the testes (of males) or ovaries (of female) activate sexual and reproductive behavior. Specifically, the reproductive organs are removed from an individual and then hormones are replaced, typically via an implant. For example, adult male red deer that have been castrated prior to the fall rutting season show little aggression and do not try to mate with sexually receptive females. If the behavioral differences between castrated and intact males stem from an absence of circulating testosterone in the castrated stags, then testosterone implants should restore their aggressive and sexual behavior during the rut, and they do (Lincoln et al. 1972).

Likewise, in the green anole (*Anolis carolinensis*), when males first become active after a winter dormant phase, their circulating concentrations of the hormone testosterone are very low. As this hormone begins to be produced in greater amounts, however, the males' testes grow in size, and mature sperm are made. At this time some males begin to defend territories and court females (Jenssen et al. 2001). These individuals tend to be larger and more powerful biters, have much more testosterone in their blood than smaller male anoles, and have larger dewlaps with which to court females and threaten rivals (Husak et al. 2007). As expected, castration of male anoles prevents courtship, but castration followed by testosterone implants returns the normal male courtship behavior (Crews 1974).

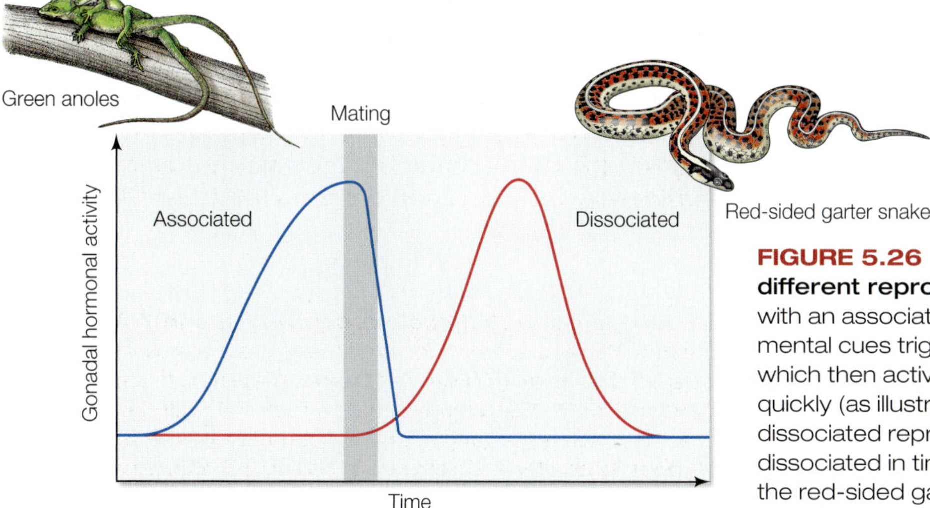

FIGURE 5.26 Gonadal hormones and two different reproductive patterns. In species with an associated reproductive pattern, environmental cues trigger internal hormonal changes, which then activate behavioral responses relatively quickly (as illustrated by the green anole). In a dissociated reproductive pattern, mating may be dissociated in time from hormonal activity (as in the red-sided garter snake). (After Crews 1984.)

These activational effects of testosterone are in part dependent on the season of the year, as Jennifer Neal and Juli Wade showed by dividing captive male green anoles with implanted testosterone capsules into two groups: one that was exposed to conditions that mimicked the normal summer breeding season environment (warm temperatures and long day lengths), the other kept under cooler conditions and shorter day lengths (Neal and Wade 2007). The males in the "breeding season" group were much more likely to display to females and to copulate with them than were males in the "nonbreeding season" group. The brains of "breeding season" males were also different from those of "nonbreeding season" males in that neurons in their amygdalae were larger, suggesting that here, too, specific regions of the brain are under hormonal regulation. Thus, the associated reproductive pattern of the green anole is dependent on the season of the year.

Testosterone and Reproductive Behavior

Many animal species, whether closely related or not, have much the same set of hormones, a result of the conservative nature of evolution in which ancestral molecules are often modified only slightly during evolutionary time as they take on very different roles in different species. In Chapter 3, we discussed how the neuropeptide vasopressin is one such ancient and conserved hormone. Testosterone is another prime example. Although white-crowned sparrows, green anoles, and red deer all produce similar forms of the hormone, the sparrows are *not* testosterone-dependent when it comes to mating. Even without his testes, a male white-crown will mount females that solicit copulations, provided that he has been exposed to long photoperiods (Moore and Kranz 1983). In addition, when male white-crowned sparrows mate with females to produce a second or third clutch of fertilized eggs in the summer, they do so without much testosterone in their blood (**FIGURE 5.27**), further evidence that high levels of circulating testosterone are not essential for male sexual behavior in this species.

The male white-crowned sparrow is not the only bird in which courtship and mating do not require high levels of circulating testosterone (Hau et al. 2010). This

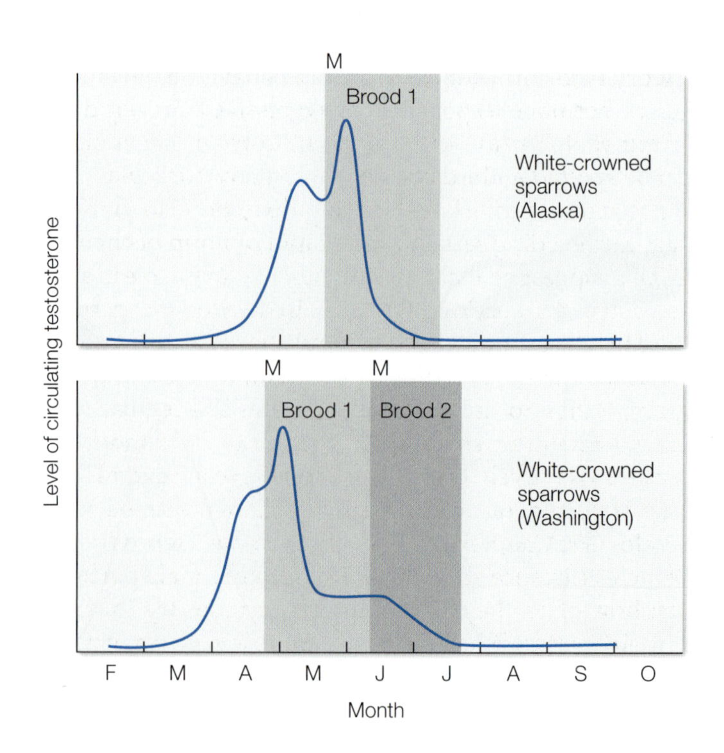

FIGURE 5.27 Hormonal and behavioral cycles in single-brood and multiple-brood populations of white-crowned sparrows. Testosterone concentrations in the blood of male white-crowned sparrows peak shortly before the time when the males mate with females (M) in their first breeding cycle of the season. In populations that breed twice in one season, however, copulation also occurs during the second breeding cycle, at a time when testosterone concentrations are declining. (After Wingfield and Moore 1987.)

is not to say, however, that testosterone is unimportant for reproduction in male white-crowned sparrows. Instead, testosterone appears to be more important for territorial behavior than for sexual behavior in males (Nelson et al. 1990), a phenomenon that appears widespread in birds. Perhaps in species of this sort, testosterone facilitates aggression rather than sexual behavior? If this hypothesis is correct, then we can predict that in seasonally territorial birds, testosterone concentrations should be especially high early in the breeding season, when males are aggressively defending a territory against rivals—as is the case for white-crowned sparrows (see Figure 5.27) and some other birds (Wingfield and Moore 1987). John Wingfield and colleagues proposed a model called the **challenge hypothesis** for explaining the relationship between testosterone and aggression (Wingfield et al. 1990). According to this hypothesis, testosterone promotes aggression only when it should be most beneficial for reproduction, such as for mate guarding or repelling rival males.

challenge hypothesis Testosterone promotes aggression only when it should be most beneficial for reproduction, such as for mate guarding or repelling rival males.

Why don't birds maintain high testosterone levels at other times of the year? Using your Darwinian reasoning, you might hypothesize that birds may pay a cost for high levels of testosterone in their blood at times of year when they do not need to reproduce or be aggressive. Recall from our discussion above that spotted antbirds in Panama live in the kind of uniform environment associated with the equator. Unlike white-crowned sparrows and other temperate birds that migrate *toward* the equator in winter, antbirds do not need to go anywhere during the nonbreeding season. Instead, they guard their food-rich territories year-round. Although testosterone has many beneficial effects on reproductive behavior (such as increased displays and aggression) and reproduction itself (such as spermatogenesis), elevated levels of the hormone can also compromise the immune system and decrease survival. Males of many species may be able to deal with these potential costs for short periods of time during the breeding season, but for species that guard territories year-round and have extended breeding seasons, the costs of maintaining elevated levels of testosterone may be too great.

In spotted antbirds, males have undetectable levels of testosterone during the nonbreeding season but possess elevated levels of a sex steroid precursor called dehydroepiandrosterone (DHEA). In nonbreeding males, Michaela Hau and colleagues found that levels of DHEA were related positively with aggressive vocalizations in response to simulated territorial intrusions conducted in both the field and lab using decoy birds and audio playbacks of song. The researchers hypothesized that DHEA is the hormone responsible for aggressive behavior during the nonbreeding season (Hau et al. 2004a). But what about testosterone in males of this species? Although male spotted antbirds do not exhibit any particular pattern of seasonal increase in testosterone during breeding, they do respond to simulated territorial intrusions during the breeding season with a rapid buildup of circulating testosterone (Hau et al. 2000), suggesting that testosterone still plays a role in aggression during the breeding season in this species (when it is also needed for spermatogenesis).

The challenge hypothesis and the idea that testosterone is linked to aggression during the breeding season lead to another prediction, which is that when competition for mates among females is a regular feature of a bird's life history, then aggressive females, which occur in the dunnock (*Prunella modularis*), should possess relatively high testosterone levels. Several female dunnocks often live in the same general area, where they share the sexual favors of one or more resident males (see Chapter 10). Because a female's reproductive success depends on how much assistance she can secure from her male partner(s), which in turn depends on how often they mate with her, females try to keep other females away from "their" mates. When the testosterone levels of aggressive females in multi-female groups were compared with those of unchallenged females that were each paired with a single male, the results indicated that testosterone does make females more

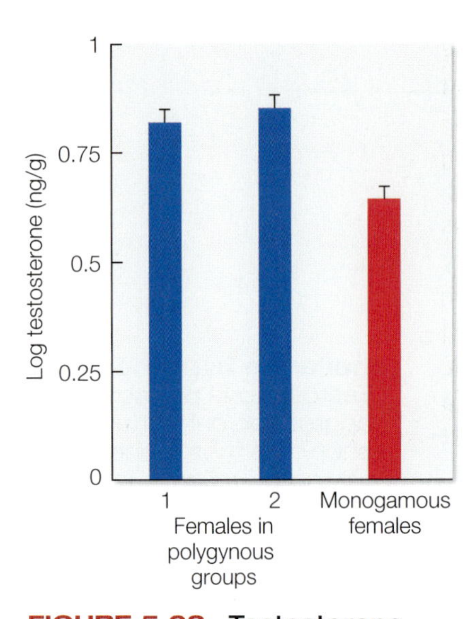

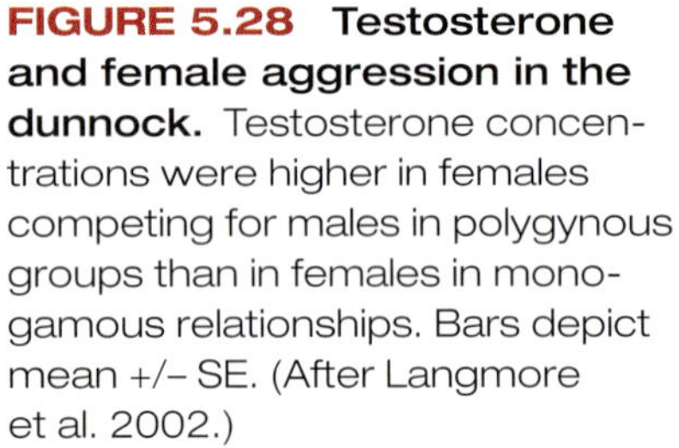

FIGURE 5.28 Testosterone and female aggression in the dunnock. Testosterone concentrations were higher in females competing for males in polygynous groups than in females in monogamous relationships. Bars depict mean +/– SE. (After Langmore et al. 2002.)

aggressive (**FIGURE 5.28**) (Langmore et al. 2002), though it is possible that fighting causes testosterone levels to rise, rather than the other way around. This study illustrates another important point about hormones, namely that there are not "male-specific" or "female-specific" hormones. Although it is true that certain steroid hormones play a role in sperm production or pregnancy, as we have just seen, both sexes can have estradiol and testosterone, which are sometimes incorrectly referred to female and male sex hormones, respectively.

The Costs of Hormonal Regulation

As you know by now, adaptationists think about both the benefits *and* the costs of the traits that interest them, and testosterone, for example, has many disadvantages to consider. As we mentioned above, testosterone can interfere with the immune response (Zuk et al. 1995, Greives et al. 2006), which may explain why males of so many vertebrates are more likely than females to be infected by viruses, bacteria, and parasites (Wingfield and Ramenofsky 1997, Klein 2000). Moreover, in a population of chimpanzees (*Pan troglodytes*), the higher-ranking males with higher testosterone levels supported a greater diversity of helminth parasites, a group that includes the familiar tapeworms (Muehlenbein and Watts 2010).

In addition, the direct behavioral effects of testosterone can be costly, leading individuals under the influence of the hormone to expend energy at a much higher rate than they otherwise would (Marler et al. 1995). For example, when male barn swallows (*Hirundo rustica*) had their chest feathers painted a darker red, they became a target of other aggressive males. As a result, their testosterone concentrations went up—and their body mass went down over time, either because the hormone induced the birds to become more physically active or the painted males increased their responses as a result of the increased attacks by other males (Safran et al. 2008). Testosterone-saturated males can also become so focused on mating with females or fighting with rivals that they become sitting ducks for predators, which may be why in some species like the common sideblotched lizard (*Uta stansburiana*), males with higher testosterone concentrations are less likely to survive than those with more modest amounts of the hormone in their blood (**FIGURE 5.29**) (Sinervo et al. 2000). Even if a testosterone-soaked male overcomes

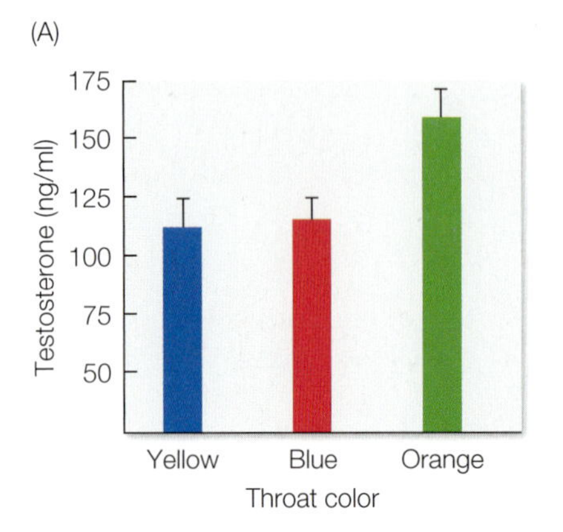

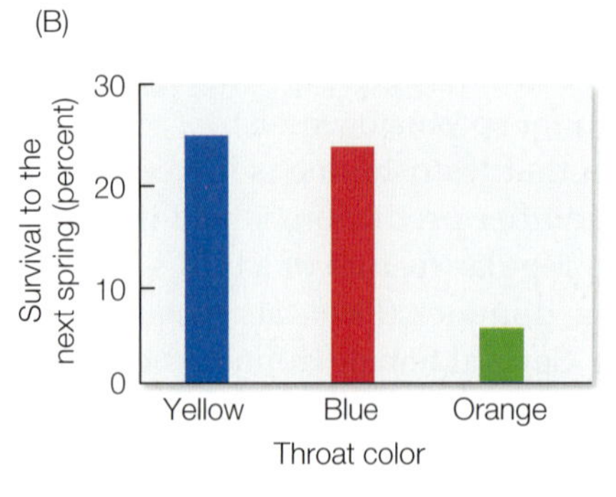

FIGURE 5.29 A survival cost of testosterone. (A) The three different forms of the common side-blotched lizard have different throat colors and, on average, different amounts of testosterone circulating through their blood. Bars depict mean +/– SE. (B) The form with the highest testosterone concentrations also has a very high annual mortality rate. (After Sinervo et al. 2000; photograph by John Alcock.)

these handicaps, he may spend more time fighting with other males than caring for his offspring. For example, although male dark-eyed juncos (*Junco hyemalis*) that received testosterone implants did not experience higher mortality than control birds, the implanted males did feed their young less often (Ketterson and Nolan 1999), to the detriment of these dependent offspring (Reed et al. 2006). Perhaps the trade-off between aggression and parental care is one reason why in humans, testosterone levels decline significantly after a man becomes a father (Gettler et al. 2012).

The paternal costs of testosterone can occur even when the increases in the hormone are only brief and transitory. A team led by Ellen Ketterson injected captured male juncos with gonadotropin-releasing hormone (GnRH), a peptide hormone produced in the hypothalamus that triggers a hormonal cascade resulting in the release of luteinizing hormone from the pituitary (another brain region) and eventually of testosterone from the gonads. This cascade of reproductive hormones is referred to as the **hypothalamic-pituitary-gonadal (HPG) axis (FIGURE 5.30)**. This treatment in male juncos induced a brief but variable increase in testosterone concentrations in the birds, which were then released back to their nesting territories. Those males whose testosterone concentrations were relatively high after the GnRH challenge behaved aggressively toward a simulated intruder (a caged male junco placed in the center of the resident's territory). However, the males that increased their testosterone concentrations the most in response to the GnRH challenge fed their nestlings at the lowest rate, a result that matches the finding that implants of testosterone reduce male paternal behavior in this species (McGlothlin et al. 2007). Thus, even a short period of elevated testosterone concentrations apparently has the potential to make a male a less helpful parent. The various disadvantages of testosterone may also help explain why after becoming aggressive (and often sexually motivated) early in the breeding season, males of many bird species reduce their circulating levels of testosterone dramatically over time (Wingfield et al. 2001). Thus males may reduce any of several disadvantages of prolonged exposure to the hormone, such as a reduction in their parental behavior.

Testosterone is not just costly in males of species that perform parental care. Although male red-sided garter snakes (*Thamnophis sirtalis*) do not care for their young in any way, they too provide evidence that testosterone is an expensive tool to use in regulating

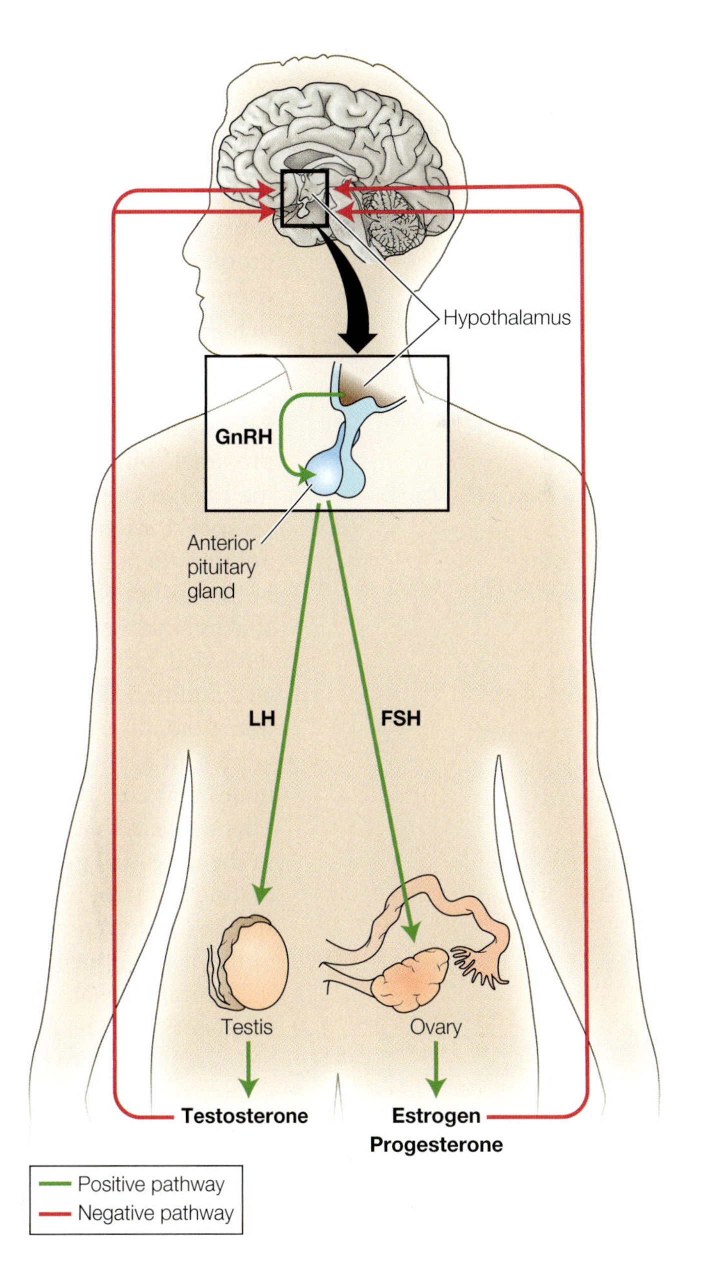

FIGURE 5.30 The hypothalamic-pituitary-gonadal (HPG) axis. Gonadotropin-releasing hormone (GnRH) is secreted from the hypothalamus and stimulates the release of luteinizing hormone (LH) and follicle-stimulating hormone (FSH) from the anterior pituitary gland, which in turn leads to the production of estrogen and progesterone in the ovaries of females and testosterone in the testes of males. This hormonal cascade forms a positive pathway from the brain to the gonads. Elevated levels of steroid hormones (testosterone, estrogen, progesterone) in the blood regulate GnRH in the brain via a negative feedback loop.

FIGURE 5.31 Spring mating aggregation of red-sided garter snakes. The males search for and avidly court females emerging from hibernation. (Photograph © Cindy Creighton/Shutterstock.com.)

reproduction and aggression (Soma et al. 2000). This snake lives as far north as southern Canada, where it spends much of the year dormant, sheltered underground. On warm days in the late spring, the snakes begin to stir, and they soon emerge, sometimes by the thousands, from their hibernacula (**FIGURE 5.31**). Before a group of snakes leaves their retreat, they engage in an orgy of sexual activity, with males slithering after females and attempting to copulate with them to the exclusion of all else. Later in the season, when the odds of finding receptive females are very low, male snakes are more likely to forage for food, a nice example of how proximate physiological mechanisms enable animals to resolve competition among their behavioral options (O'Donnell et al. 2004). During the mating frenzy, males compete for females by trying to contact receptive partners before other males do, but they do not fight with one another for the privilege of copulation. Examination of the sex hormone concentrations in their blood reveals that these nonaggressive snakes have almost no circulating testosterone. Yet they have no trouble mating, so red-sided garter snakes are animals with a **dissociated reproductive pattern** (see Figure 5.26) (Crews and Greenberg 1981). However, there is some evidence that aromatase, the enzyme responsible for converting testosterone into estrogen, is present at relatively high levels in certain parts of the snake's brain in those males that are actively seeking females in the spring (Krohmer et al. 2002, 2010). This suggests that there are elements of an associated reproductive pattern at work in the red-sided garter snake. If an estrogen, rather than an androgen, influences mating behavior in male garter snakes, what then is the role of testosterone in this species? Testosterone production, which normally begins in the male snake's second or third year, appears to be necessary for the full development and maintenance of the neural mechanisms that control the snake's sexual behavior (**FIGURE 5.32**) (Crews 1991). The fact that the enzyme that converts testosterone into estrogen also becomes more common in particular regions of the snake's brain in the fall is consistent with the hypothesis that seasonal changes in estrogen (derived from testosterone) prepare the key neural circuits to promote sexual activity in the spring. So once again we have an example of a proximate mechanism, in this case an endocrine–neuronal system, that has surely been retained in part from a distant ancestor but is used in a distinctive manner to achieve adaptive goals for a modern species.

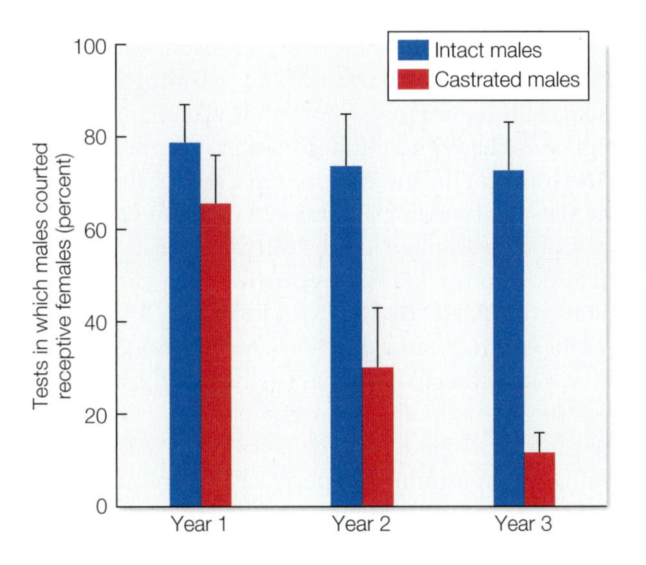

FIGURE 5.32 Testosterone and the long-term maintenance of mating behavior. Male red-sided garter snakes whose testes were removed shortly before the breeding season in year 1 remained sexually active during that breeding season, despite the absence of testosterone. But in years 2 and 3, these males became less and less likely to court receptive females, compared with males that still possessed their testes. Bars depict mean +/– SE. (After Crews 1991.)

Glucocorticoids and Responding to Environmental Change

Up until this point of the chapter, our discussion of hormones has primarily focused on the social environment and how steroid hormones such as testosterone, estradiol, and progesterone (often referred to as "reproductive hormones") influence reproduction and aggression. Hormones such as these are among the primary ways in which information from the environment leads to rapid behavioral change. In other words, hormones are modulators that both cause changes in behavior and themselves change in response to different behaviors. We saw an example of this bidirectional nature of the hormone–behavior relationship in the context of reproductive behavior, in which as predicted by the challenge hypothesis, testosterone increases during the time of year when male–male aggressive interactions can be anticipated but also when interactions between males can lead to a rapid rise in testosterone in the blood. In other words, testosterone can cause behavior appropriate for a social environment, and in turn, the social environment can generate increased hormone production, providing strong evidence for the bidirectional relationship between steroid hormones and the social environment.

Hormones are not only influenced by social factors—they can also be affected by ecological factors such as food availability, severe weather, and even the presence of predators. Glucocorticoids are another class of steroid hormone that can be influenced by both social and ecological factors. Although glucocorticoids are often referred to as "stress hormones," because they can be released in response to a variety of stressors, they also play an essential role in metabolism, immune function, and various other physiological processes. If anything, they should really be referred to as "anti-stress" hormones because they enable animals to respond rapidly to stressors by adjusting their behavior and physiology.

To understand the role that glucocorticoids play in shaping behavior, let's think about what happens when an animal is exposed to a stressor. Upon sighting a hawk, for example, an American robin (*Turdus migratorius*) may immediately begin to fly up from the ground to find safety high in a tree. This behavioral response requires the diversion of energy from other tasks to flight muscles, which in itself requires an increase in blood flow and the enhancement of the cardiovascular system. The robin might also experience cognitive changes as it focuses on the predator and tunes out everything else around it. Although there are several chemicals involved in this flight response (hormones such as adrenaline and neurotransmitters such as norepinephrine), glucocorticoids play a central and rapid role, acting on the scale of seconds to minutes. Now imagine another ecological stressor: a storm. If a spring snowstorm disrupts that robin's first breeding attempt of the season, what will happen? The bird may experience a stimulation of its immune system, an inhibition of its reproductive physiology and behavior, or a decrease in its appetite as it conserves energy reserves until the snow melts. Instead of preparing for reproduction, the robin may instead return to its nonbreeding activities. As with the predator example, these changes in behavior and physiological function are at least partially under the control of glucocorticoid hormones and the **hypothalamic-pituitary-adrenal** (**HPA**) **axis** more generally (**FIGURE 5.33**) (Sapolsky et al. 2000).

The predatory hawk and the unexpected snowstorm are examples of **acute stressors** than can directly and rapidly affect an animal's behavior through the release of glucocorticoids. The resulting behavioral responses are clearly adaptive if they prevent the robin from being eaten or expending all of its energy on a breeding attempt that may be doomed to fail. But some animals experience **chronic stressors** that result in continued and often long-term activation of the HPA axis. If predation risk or severe weather—or food limitation, or social conflict, and so on—become prolonged, chronically elevated glucocorticoids can have negative physiological effects. In much the same way that elevated testosterone can be costly for long periods of time, high levels of glucocorticoids for extended periods of time can impair immune function, cause sickness, and even lead to death (Sapolsky et al. 2000). Therefore, animals will often adjust their behavior, thereby leading to a reduction in glucocorticoids—another illustration of the bidirectional relationship between steroid hormones and behavior. Thus, glucocorticoids can be adaptive in the short-term by allowing animals to respond to environmental stressors, but maladaptive in the long-term if individuals are continually exposed to chronic stress.

Although ecological factors can greatly influence circulating levels of glucocorticoids, it is also true that social factors can affect them. For example, high population density or frequent territorial intrusions are associated with elevated glucocorticoids in several vertebrate species, ranging from mammals to birds to fish and reptiles (Creel et al. 2012). Chronically elevated glucocorticoid levels in high-density populations may negatively affect immune function and survival, particularly in the very young or very old individuals, or those in poor body condition. For

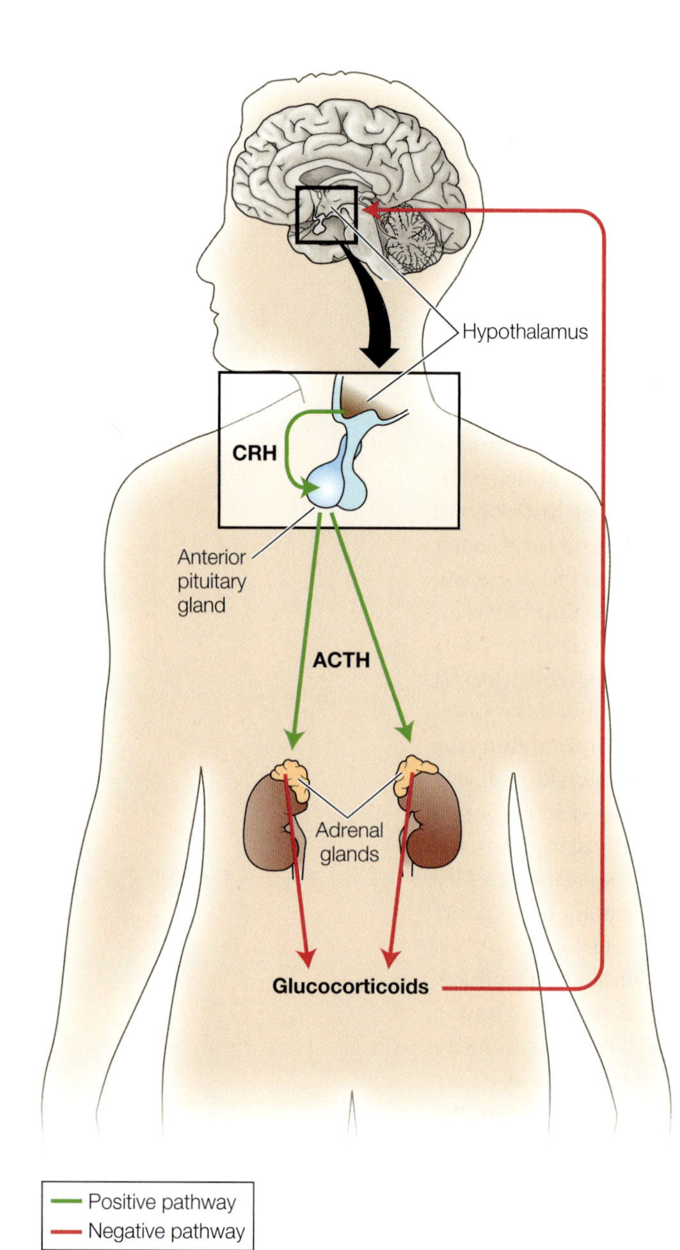

FIGURE 5.33 The hypothalamic-pituitary-adrenal (HPA) axis. Corticotropin-releasing hormone (CRH) is secreted from the hypothalamus and stimulates the release of adrenocorticotropic hormone (ACTH) from the anterior pituitary gland, which in turn leads to the production of glucocorticoids (including cortisol and corticosterone) in the adrenal cortex. This hormonal cascade forms a positive pathway from the brain to the gonads. Elevated levels of glucocorticoids in the blood regulate CRH in the brain via a negative feedback loop.

species that are highly social and live not just in dense populations but in groups, there is often a relationship between dominance rank and glucocorticoid levels. However, the direction of this relationship depends on the stability of the social hierarchy and the way in which high rank is obtained and maintained. That is, in species where dominant individuals can easily maintain their status without having to fight with subordinates, then dominants typically have lower levels of glucocorticoids than subordinates. However, when dominants must continually fight with subordinates to maintain their status, then the dominants often have higher levels of stress hormones (Goymann and Wingfield 2004). In research led by Tim Clutton-Brock on cooperatively breeding meerkats (*Suricata suricatta*), which live in groups with a single dominant breeding female and several non-breeding females that aid in rearing the dominant's offspring, dominant females have been shown to reproductively suppress subordinates (Young et al. 2006) (see Chapter 13). This reproductive suppression of subordinate females by the dominant breeder results in elevated glucocorticoids in the subordinates, which in turn inhibits reproductive hormones and results in reduced rates of conception and increased rates of miscarriage.

Individuals living in social groups not only vary in dominance rank, but also in the number and strength of ties to other individuals. In short, some individuals in group-living species have more friends than others. In humans and other animals, strong affiliative social relationships can have positive effects on health and fitness, whereas weak relationships can result in the opposite. Christopher Young and colleagues showed in Barbary macaques (*Macaca sylvanus*) that males with more affiliative (rather than aggressive) relationships with other males had lower long-term levels of glucocorticoid metabolites in their feces (Young et al. 2014). The authors argued that social bonds in this species buffer males against both social (aggressive interactions) and environmental stressors (changes in food availability), as reflected by their hormone levels.

For all animals, life is a combination of predictable and unpredictable events. Natural selection has shaped the behavior of migratory birds just as it has their flight responses to potential predators. Migratory behavior has evolved to deal with the changing seasons, a relatively predictable phenomenon. Similarly, daily light–dark and tidal cycles are highly predictable events that animals use as cues to adjust their behavior. In response to these predictable cycles, the nervous systems of animals as diverse as mammals and insects have been shaped to produce clock mechanisms regulated by similar sets of genes and neurons. These endogenous rhythms are regulated by a variety of chemical messengers, including hormones. Hormones represent one way in which information from the environment is translated into behavioral change. In additional to circadian behavior, hormones also play a key role in regulating other behaviors, including reproductive behavior, aggression, and social behavior. In vertebrates, steroid hormones—both reproductive hormones such as testosterone and estradiol and so-called stress hormones such as glucocorticoids—modulate behaviors in response to a variety of social and ecological factors. Levels of steroid hormones can be regulated by both predictable events, such as the changing seasons that influence the switch from a nonbreeding to a breeding life history stage, as well as by unpredictable events such as social challenges, territorial intrusions, or severe weather events. In other words, these physiological mediators are often the key mechanistic link between an animal's behavioral response and its environment. And like any phenotypic trait, hormones can be targets of natural selection. Thus, studies designed to determine how an animal's behavior is shaped by its local environment—things such as food availability or density of conspecifics—can be enriched by considering the physiological basis of that behavior.

SUMMARY

1. Because an animal's environment provides various stimuli that can trigger contradictory responses, and because its physical and social environment often change over time, animals have evolved proximate mechanisms that help individuals set behavioral priorities, which they can adjust to changing conditions.
2. Biological clock mechanisms, acting in conjunction with neural and hormonal systems, enable individuals to shift their behavior in accordance with predictable changes in the environment that occur over periods ranging from a day to a year.
3. Circadian and circannual clocks have environment-independent components, but they can also adjust their performance by acquiring information from the environment about local conditions, such as the time of sunrise or sunset or of high or low tide. This information can entrain (reset) the clock so that changing behavioral priorities do not get out of synchrony with the physical environment.
4. The endocrine system also contributes to the operation of neurophysiological mechanisms that respond to predictable changes in the physical environment (such as seasonal changes in photoperiod) as well as to unpredictable changes in the social environment (such as the presence of potential mates) or ecological environment (severe weather events). Hormones produced by endocrine organs often set in motion a cascading series of physiological changes that make reproductive activity the top priority at times when the production of surviving offspring is most likely to occur.
5. The fact that so many species employ essentially the same hormonal tool kit is a reflection of the conservative consequences of evolution by descent with modification (from ancestral species). Like any phenotypic trait, hormones themselves can be targets of natural selection.
6. Hormonal control, like all proximate mechanisms, comes with costs as well as potential benefits. The many damaging effects of testosterone or glucocorticoids—things such as a compromised immune system that can cause sickness and lead to death—provide a case in point. The cost-benefit approach helps explain why different animal species use the same hormone, such as testosterone, for very different functions.

COMPANION WEBSITE

Go to **ab11e.com** for discussion questions and all of the audio and video clips.

CHAPTER

6

Avoiding Predators and Finding Food

Recall that natural selection is the outcome of the differential survival and reproduction of individuals due to differences in their hereditary phenotype, including their behavior. We will spend a great deal of time discussing reproductive behavior in subsequent chapters, but here our focus is on survival, which for many animals comes down to two factors: avoiding predators and finding food. After all, unless an individual is able to find food and avoid predators, it will not be able to survive long enough to reproduce. To aid in our understanding of these survival strategies, we introduce two key theoretical tools that behavioral biologists have at their disposal: evolutionary game theory and optimality theory. These mathematical approaches will feature prominently throughout the remaining chapters of the book. Game theory enables behavioral biologists to model the interactions

Chapter 6 opening photograph courtesy of Giangiorgio Crisponi.

between different individuals or species. One of the crucial aspects of game theory modeling is that selection among alternative behaviors depends on what others are doing. Therefore, since game theory emphasizes the dynamics of behavioral interactions, it can be used to predict how an individual should behave in a given social context (**BOX 6.1**). In contrast, optimality theory attempts to predict the best way (in terms of greater fitness benefits than costs) for an individual to behave, often irrespective of what others are doing. Both game theory and optimality theory enable researchers to generate testable hypotheses about the behavior of their subjects.

By the end of this chapter, you will have a better understanding of the behaviors that many animals use to survive long enough to reproduce and pass their genes on to the next generation. First, we discuss predator avoidance, introducing both the social and solitary tactics that animals use to escape detection by, or to deter, predators. We discuss active group defense, and then present a series of hypotheses for passive social defense in group-living species. We then consider how solitary individuals use camouflage and warning coloration to avoid or signal to predators, respectively. Next, we introduce the topic of foraging behavior and highlight optimal foraging theory. Clearly, avoiding predators and finding food go hand in hand, particularly for species that are themselves the potential prey of other species. Many species of animals must make trade-offs about where to go to find food in order to balance the risk of becoming a meal for someone else. We integrate these topics by talking about the landscape of fear, before ending with a discussion of the alternative approaches that individuals of the same species use to find food or escape predators.

Avoiding Predators

Animals generally have two approaches when it comes to escaping predation: they can do it alone or in a group. Animals in a group may work together passively or actively to avoid or deter predators. Of course, forming a group may also make animals more conspicuous to predators, so one must always consider both the costs and benefits of social defenses. The potential survival costs of joining a group are one reason why some organisms remain solitary and rely on other means to avoid predators. Solitary individuals may try to blend in with their environment and hide, or stand out and warn potential predators not to come near them. Below we discuss the different ways that animals try to avoid predators, beginning with those that form groups.

Social Defenses

As we just noted, social, or group, defenses against predators can be passive or active. Communal mobbing behavior is an example of an active social defense against predators. Recall that in Chapter 1 we discussed gulls mobbing potential predators near their breeding colonies. As Hans Kruuk demonstrated in black-headed gulls (*Chroicocephalus ridibundus*), this behavior can be adaptive if chasing away predators from nests leads to increased offspring survival (Kruuk 1964). Many species of birds mob predators, and this form of active social defense is common in other group-living species as well. For example, humpback whales (*Megaptera novaeangliae*) are known to mob attacking killer whales (*Orcinus orca*), which increases the likelihood that the humpbacks will escape (Pitman et al. 2017), and vespid wasps (**VIDEO 6.1**) and Africanized honey bees (*Apis mellifera scutellata*) will en masse sting intruders to the hive (Vetter et al. 1999).

VIDEO 6.1
Group nest defense by vespid wasps
ab11e.com/v6.1

BOX 6.1 INTEGRATIVE APPROACHES

Evolutionary game theory

Game theory is typically used in animal behavior to determine which among a series of behavioral strategies is likely to evolve in a given context or population. When, for example, should individuals cooperate? When should they not? When a single strategy for a given population cannot be invaded by an alternative strategy that is initially rare, we call it an **evolutionarily stable strategy** (**ESS**). The ESS concept is a refinement of the Nash equilibrium, named after the Nobel Prize–winning mathematician John Nash, which is a solution (or optimal strategy) for how competing individuals, corporations, or governments should act to maximize their gains. John Maynard Smith and George Price not only introduced the idea of the ESS, they also described one of the first evolutionary game theory models, the Hawk–Dove game, which is a contest between two individuals over food resources (Maynard Smith and Price 1973). You could imagine a scenario in which individuals of the same species are foraging and come into conflict over the same resource, and each needs to decide how to behave in order to maximize its own fitness. Should these individuals share the resources or fight for control?

The Hawk–Dove game provides an answer to this question. Individual contestants in the game can adopt one of two behavioral strategies: (1) a hawk strategy, in which the individual first displays aggression and then escalates into a fight until it either wins or is injured (and loses the fight); or (2) a dove strategy, in which the individual first displays aggression and then either retreats to safety if faced with an escalation or attempts to share the resources if not faced with an escalation. The costs and benefits to each behavioral strategy can be shown as payoffs in a matrix where V is the value of the resource (the benefit) and C is the cost of losing a fight (**FIGURE**). It is generally assumed that the value of the resource is less than the cost of a fight ($C > V > 0$). These payoffs mean that if a hawk meets a dove, the hawk gets all of the resources (V). However, if a hawk meets a hawk, half the time it wins and half the time it loses; the average payoff is therefore $(V - C)/2$. If a dove meets a hawk, the dove gets nothing (0), but if a dove meets a dove, they share the resources equally ($V/2$). The actual payoff that each individual receives is a function of how often it is likely to meet a hawk or a dove. In other words, it depends on the frequency of each strategy in the population (a topic we will return to at the end of this chapter).

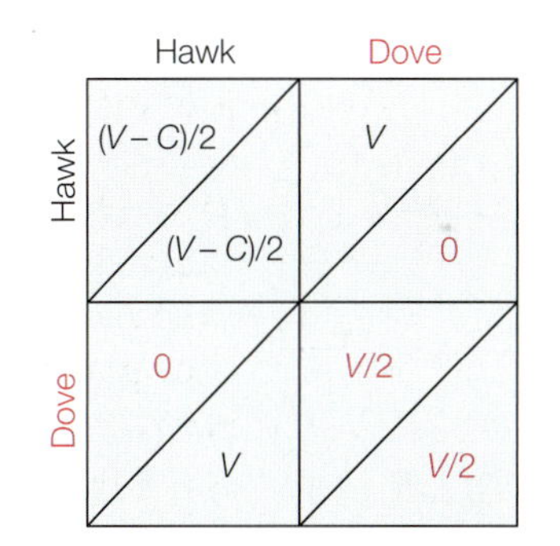

Payoff matrix for the Hawk–Dove game

Thinking Outside the Box

To solve the Hawk–Dove game, we first need to determine if either the hawk or the dove strategy is an ESS. In other words, are either of these strategies impervious to invasion by another? Clearly the dove will not fare well against a hawk, and a population of all doves could be invaded by hawk strategy. But what about the other way around? In a population of all hawks, could a dove strategy invade? Although doves do not gain resources when they meet a hawk, they also do not pay a cost (their payoff is 0 in our matrix above). Therefore, they can persist in the population, and if they happen to meet a dove instead of a hawk, they can gain resources. Thus, neither hawks nor doves are sole ESSs, or what we call pure ESSs. Instead, both of these strategies are part of what is called a mixed ESS. To further understand the Hawk–Dove game, try solving it by using different values for the payoffs. Start with $V = 6$ and $C = 18$. What happens if you violate our initial assumption that the value of the resource is less than the cost of a fight ($C > V > 0$)? Try reversing the values and this time use $V = 18$ and $C = 6$. What happens to the mixed ESS?

(A)

(B)

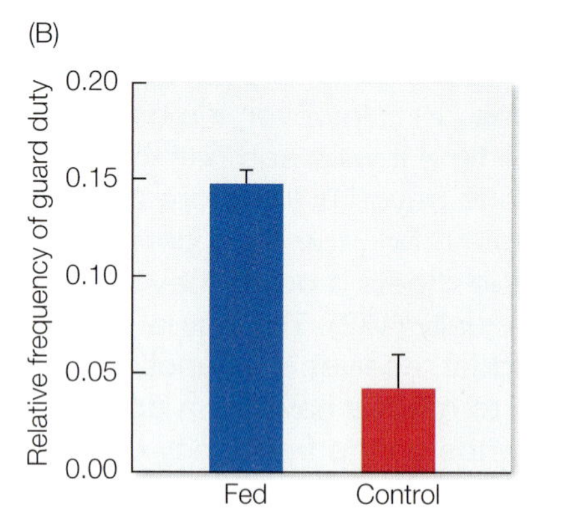

FIGURE 6.1 Meerkats perform guard duties after eating. (A) Meerkats perform guard duty from high but relatively safe locations. (B) Who are these sentinels? A series of experiments demonstrated that sentinels are not individuals with specific set tasks. Instead, all members in the group spend part of their time guarding. Experimentally fed individuals were more likely to guard than control animals. Bars depict mean +/– SE. (After Clutton-Brock et al. 1999; photograph © EcoPrint/Shutterstock.com.)

Despite the obvious benefits to gulls, whales, bees, and other species of deterring predators, mobbing can also be costly. Not only can mobbing lead to injury or death, it may also attract more predators than it deters. For example, many species of birds that mob predators produce mobbing calls to recruit other mobbers, but this may inadvertently attract additional predators as well. To test the hypothesis that predators eavesdrop on these mobbing calls, recorded mobbing calls of the pied flycatcher (*Ficedula hypoleuca*) were played at nest boxes (Krama and Krama 2005). The researchers found that nests with mobbing calls were predated significantly more by pine martens (*Martes martes*) than were control nests with blank playbacks.

Another form of active social defense is the shared or improved vigilance associated with additional sets of eyes to search for predators. At the most basic level, more eyes mean a greater likelihood that a predator will be spotted. However, some cooperatively breeding species that form kin-based social groups have taken shared vigilance to the extreme with the adoption of specialized sentinels, whose job is to spot potential predators and produce alarm calls to warn other group members of danger. One might assume that these sentinels are more likely to be targeted by predators because they are often visible and the alarm call may attract attention, but there is little evidence to support the idea that these individuals are at greater risk than any other individual in the group. Although in many cooperatively breeding species it is subordinate individuals that are relegated to the sentinel role, in meerkats (*Suricata suricatta*) it is the individuals that have just finished eating that take on the role of guard duty (Clutton-Brock et al. 1999). By experimentally feeding some group members hard-boiled eggs, a team led by Tim Clutton-Brock showed that fed animals went on guard duty three times as often as control animals did (**FIGURE 6.1**). An interesting twist on alarm calling and predation risk is seen in another South African native, the fork-tailed drongo (*Dicrurus adsimilis*), a midsize perching bird. This species exploits fear in other bird species by mimicking those other species' alarm calls to scare them away from food so that the drongos can steal the bounty (Flower et al. 2014). Thus, active social defenses such as mobbing, alarm calling, and improved vigilance can have both fitness benefits and costs, though the costs are not always as obvious as we might assume.

Not all social defenses actively involve individuals searching for, responding to, or attacking predators. Some types of social defense are passive (**HYPOTHESES 6.1**). Consider the harvester ants (*Pogonomyrmex* spp.) that come together to mate only a few days each year when huge numbers of individuals briefly form a dense aggregation (**FIGURE 6.2**). Although predatory dragonflies and birds can locate these swarms and capture and eat some ants, the odds that any one ant will be targeted (the fitness cost of gathering together in conspicuous masses) are small given the ratio of ants to predators (and of

HYPOTHESES 6.1

Non-mutually exclusive hypotheses to explain passive social defense in groups

dilution effect hypothesis
Associating in groups makes it less likely that any one individual will be depredated.

confusion effect hypothesis
Moving as a group may reduce the likelihood of predators capturing prey because of their inability to single out and attack individual prey.

selfish herd hypothesis
Individuals in a group (or herd) attempt to reduce their predation risk by putting other individuals between themselves and the predator.

FIGURE 6.2 Why do harvester ants mate in groups? (A) Harvester ants form huge but brief mating aggregations on hilltops. So many individuals come to these rendezvous sites that the local predators, such as (B) this dragonfly (which has captured one unlucky ant), cannot possibly consume them all. (Photographs by John Alcock.)

course there are also fitness benefits to being part of a mating aggregation). According to this **dilution effect hypothesis**, associating in groups makes it less likely that any one individual will be depredated. This hypothesis may also explain why many butterflies create tightly packed groups when drinking from mud puddles, where they may be securing fluids high in nitrogen or sodium (Molleman 2011). One would think that these groups would be likely to be spotted by butterfly-eating birds (but see Ioannou et al. 2011). However, any costs imposed on puddling butterflies arising from increased conspicuousness of the groups may be offset by a dilution in the chance that any one individual will be killed by an attacker (Turner and Pitcher 1986). Imagine that only a few predators (say, five birds) occupy the space that contains a puddling aggregation of butterflies. Further imagine that each of the five predators kills two prey per day there. Under these conditions, the risk of death for a member of a group of 1000 butterflies is 1 percent per day, whereas it is ten times higher for members of a group of 100 butterflies. The **dilution effect** has been confirmed by Joanna Burger and Michael Gochfeld, who showed that any butterfly puddling by itself or with only a few companions would be safer if it moved to an even slightly larger group (**FIGURE 6.3**) (Burger and Gochfeld 2001).

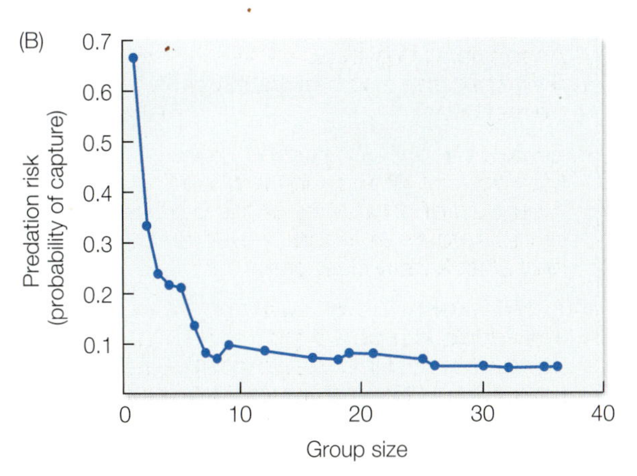

FIGURE 6.3 The dilution effect in butterfly groups. (A) A group of butterflies drinking fluid from a Brazilian mud puddle. (B) Individual butterflies that "puddle" in large groups experience a lower risk of daily predation than those that suck up fluids from the ground by themselves or in small groups. (A, photograph by John Alcock; B after Burger and Gochfeld 2001.)

The dilution effect has also been tested in the mayfly *Dolania americana*, which synchronizes the change from aquatic nymph to flying adult such that most individuals emerge from the water during just a few hours on a few days each year (Sweeney and Vannote 1982). This temporal bottleneck should limit the access of predators to the mayflies. If so, the higher the density of emerging individuals from a stream, the lower the risk to any one individual of being inhaled by a trout as the mayflies make the risky transition to adulthood. To test this prediction of the dilution effect hypothesis, Bernard Sweeney and Robin Vannote placed nets in streams to catch the cast-off skins of mayflies, which molt on the water's surface as they change into adults, leaving their discarded cuticles to drift downstream on the current (Sweeney and Vannote 1982). Counts of the molted cuticles revealed how many adults emerged on a given evening from a particular segment of stream. The nets also caught the bodies of females that had laid their eggs and then died a natural death; a female's life ends immediately after she drops her clutch of eggs into the water, provided a predator does not consume her first. Sweeney and Vannote measured the difference between the number of molted cuticles of emerging females and the number of intact corpses of spent adult females that washed into their nets on different days. The greater the number of females emerging together on a given day, the better the chance each mayfly had to live long enough to lay her eggs before expiring, as predicted from the dilution effect hypothesis (**FIGURE 6.4**).

(A)

(B)

(C)

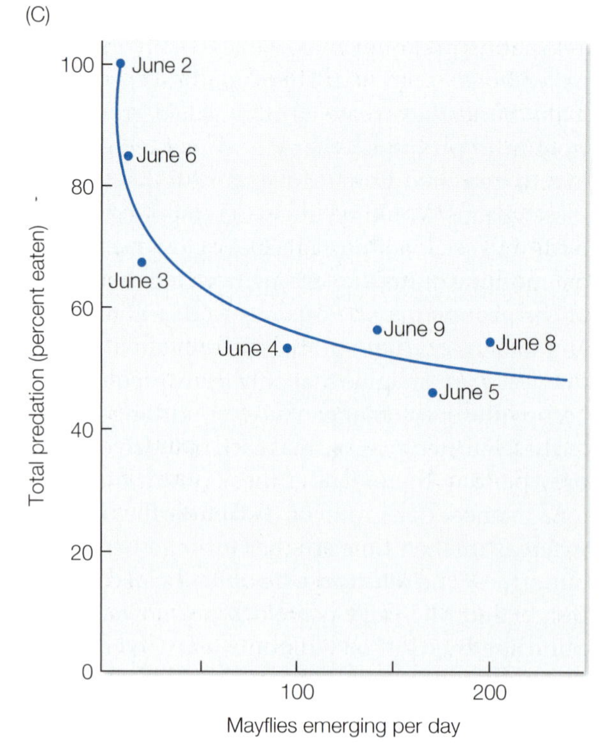

FIGURE 6.4 The dilution effect in mayflies. (A) Mayflies emerge from the water (B) in vast numbers in the spring. (C) The more female mayflies that emerge together, the less likely any individual mayfly is to be eaten by a predator. (After Sweeney and Vannote 1982; A, photograph © iStock.com/shurub; B, photograph © iliuta goean/Shutterstock.com.)

In many parts of the world, European starlings (*Sturnus vulgaris*) form enormous flocks during the nonbreeding season. If you are lucky, you might observe their famous aggregations at dusk as a flock tries to find a roosting site for the evening (see Video 13.1). Although this collective behavior has long fascinated behavioral biologists and generated numerous hypotheses to explain its existence, it was only recently that researchers were able to definitively conclude that predation risk seems to be the reason starlings perform their dances at dusks. Using **citizen science data**—observations made by the public and collated by scientists to test hypotheses—Anne Goodenough and colleagues collected data on size, duration, habitat, temperature, and predator presence from more than 3000 starling flocks (Goodenough et al. 2017). The researchers found that the presence of predatory hawks was positively correlated with group size and how long the flock stayed together, especially when hawks were near to or chasing starlings. Although these data are consistent with the dilution effect hypothesis, another hypothesis that may explain the relationship between this behavior and predation risk is that the aggregations may serve to confuse predators that have difficulty tracking a single object. The **confusion effect hypothesis** proposes that moving as a group may reduce the attack-to-kill ratio experienced by a predator because of an inability to single out and attack individual prey within a group (Krause and Ruxton 2002). Using a three-dimensional computer simulation of European starlings to test the effects of flock size and density on the ability of a (human) predator to track and capture a target starling within a flock, Benedict Hogan and colleagues demonstrated that starlings are safer in larger and denser flocks because flocking confuses the surrogate predators (Hogan et al. 2017). The researchers showed that as flock size increased, targeting errors also increased. In this way, the researchers were able to tease apart the dilution and confusion effect hypotheses and show that large groups of starlings not only the reduce the risk that any one starling will be attacked (dilution effect), but that the groups also serve to confuse predators and reduce their efficiency (confusion effect).

confusion effect hypothesis Moving as a group may reduce the likelihood of predators capturing prey because of their inability to single out and attack individual prey.

Another form of passive social defense is the simple ability to use others to hide behind (Hamilton 1971). According to the **selfish herd hypothesis**, individuals in a group (or herd) attempt to reduce their predation risk by putting other individuals between themselves and the predator. Imagine, for example, a population of antelope grazing on an African plain in which all individuals stay well apart, thereby reducing their conspicuousness to their predators. Now imagine that a mutant individual arises in this species, one that approaches another animal and positions itself so as to use its companion as a living shield for protection against attacking predators. The social mutant that employs this tactic would incur some costs. For example, two animals may be more conspicuous to predators than one and so attract more attacks than scattered individuals, as has been demonstrated in some cases (Watt and Chapman 1998). But if these costs were consistently outweighed by the survival benefit gained by social individuals, the social mutation could spread through the population. If it did, then eventually all members of the species would be aggregated, with individuals jockeying for the safest position within their group, actively attempting to improve their odds at another individual's expense. The result would be a **selfish herd**, whose members would actually be safer if they could all agree to spread out and not try to take advantage of one another. But since populations of solitary individuals would be vulnerable to invasion by an exploitative social mutant that takes fitness from its companions, social behavior could spread through the species—a clear illustration of why we define adaptations in terms of their contribution to the fitness of individuals relative to that of other individuals with alternative traits.

selfish herd hypothesis Individuals in a group (or herd) attempt to reduce their predation risk by putting other individuals between themselves and the predator.

Game Theory and Social Defenses

The example above of the hypothetical African antelope introduced the idea of two inherited behavioral strategies that individuals in a population might adopt to help reduce their risk of predation. This is really a verbal application of **evolutionary game theory**, a mathematical approach that can be used to study behavioral adaptations such as these. As we discussed in Box 6.1, users of this theory treat individuals as participants in a contest in which the success of one competitor is dependent on what its rivals are doing. Under this approach, decision making is treated as a game, just as it is by the economists who invented game theory to understand the choices made by people as they compete with one another for consumer goods or wealth. Game theoretical economists have shown that individuals using one strategy, which works well in one situation, may come up short when matched against other individuals using another strategy. In evolutionary biology, "strategy" designates an inherited behavioral trait, such as a decision-making ability of some sort, not the consciously adopted game plans of the sort humans often employ (Dawkins 1989). The fact that all organisms, not just humans interested in how to spend their money or get ahead in business, are really competing with one another in a reproductive sweepstakes means that the game theory approach is a natural match for what goes on in the real world. The fundamental competition of life revolves around getting more of one's genes into the next generation than someone else, something that requires surviving long enough to reproduce. Winning this game almost always

BOX 6.2 **EXPLORING BEHAVIOR BY INTERPRETING DATA**

Game theory and the selfish herd

The concept of a selfish herd can be illustrated with a game theory matrix that shows the fitness of individuals that adopt different strategies. In a population entirely composed of solitary prey, the fitness payoff to individuals is *P*. But then mutant individuals arise that use others as living shields. When a solitary individual is found and used by a social type for protection, the solitary animal loses some fitness (*B*) to the social type. The costs (*C*) to social individuals include the time spent searching for another individual to hide behind. When two social types interact, we will say that they each have one chance in two of being the one that happens to hide behind the other when a predator attacks.

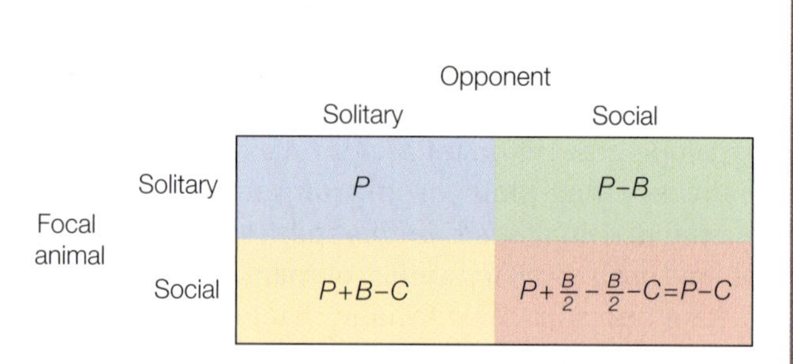

A game theoretical model of the selfish herd. In this model, the fitness gained by a solitary or social animal is dependent on the behavior of its opponent, which may be either a solitary or a social individual.

Thinking Outside the Box

If *B* is greater than *C*, what behavioral type will come to predominate in the population over time? Now compare the average payoff for individuals in populations composed entirely of solitary versus social types. If the average fitness of individuals in a population of social types is less than that of individuals in a population composed of solitary types, can hiding behind others be an adaptation?

depends on what those other individuals in the population are doing, which is why game theory is so popular with behavioral biologists as they develop hypotheses to explain the behaviors they observe.

The selfish herd hypothesis generates testable predictions that can be applied to any prey species that forms groups (**BOX 6.2**). Thus, we expect that there will be competition among group members for the safest positions in the selfish herd, which generally means that individuals will jockey for a place in the center of the group (Morrell et al. 2011). In bluegill (*Lepomis macrochirus*) nesting colonies that form on the bottoms of lakes and ponds, the larger, more dominant fish do indeed come to occupy the central positions, forcing smaller, subordinate males to build their nests on the periphery, where they are subject to more frequent attacks by predatory fishes (Gross and MacMillan 1981). Likewise, if the feeding flocks of redshanks (*Tringa totanus*), which are Eurasian sandpipers, constitute selfish herds, we would expect that those individuals targeted by predatory sparrowhawks (*Accipiter nisus*) should be relatively far from the shelter provided by their companions. John Quinn and Will Cresswell collected the data needed to evaluate this prediction by recording 17 attacks on birds whose distance to a nearest neighbor was known. Typically, a redshank selected by a sparrowhawk was about five body lengths farther from its nearest companion than that companion was from its nearest neighbor (**FIGURE 6.5**) (Quinn and Cresswell 2006). Thus, birds that moved a short way from their fellow redshanks, presumably to forage with less competition for food, put themselves at greater risk of being singled out for attack.

Blending In

Although up until this point we have emphasized social defenses against predators, there are vast numbers of solitary creatures, and they too have much to gain by staying alive longer than the average for their species if their greater life span translates into greater reproductive success. As a result, solitary prey often blend in with their environment by evolving camouflaged color patterns along with behavioral tactics that enhance their concealment. Indeed, background matching is the most common camouflage strategy. One potential tactic of this sort is for the

FIGURE 6.5 Redshanks form selfish herds. Redshanks that are targeted by hawks are usually standing farther from their nearest neighbors in the flock than those birds are standing from their nearest neighbors. Bars depict mean +/– SE. (After Quinn and Cresswell 2006; top photograph © David Dohnal/Shutterstock; bottom photograph © iStock.com/Mike Lane45.)

cryptically colored animal to select the kind of resting place where it is especially hard to see (**FIGURE 6.6**). A classic example involves the peppered moth (*Biston betularia*). The melanic (black) form of this moth, once extremely rare in Great Britain, almost completely replaced the once abundant whitish salt-and-pepper form in the period from about 1850 to 1950 (Grant et al. 1996). Most biology undergraduates have heard the standard explanation for the initial spread of the melanic form (and the genetic basis for its mutant color pattern): as industrial soot darkened the color of forest tree trunks in urban regions, the whitish moths living in these places became more conspicuous to insectivorous birds, which ate the whitish form and thereby literally consumed the allele for this color pattern.

Despite some claims to the contrary (Coyne 1998), this story remains valid (Grant 1999, Rudge 2006), particularly with respect to the conclusions of H. B. D. Kettlewell, who pinned fresh specimens of the black and the whitish forms of the moth on dark and pale tree trunks (Kettlewell 1955). When he returned later, Kettlewell found that whichever form was more conspicuous to humans had been taken by birds much more often than the other form. In other words, paler individuals were at special risk of attack when they were placed on dark backgrounds, and vice versa. Michael Majerus repeated Kettlewell's classic experiment but with living specimens and vastly more of them over a 7-year period. His data confirm that the form more conspicuous to human observers is also at significantly higher risk of attack and

(A)

(B)

FIGURE 6.6 Cryptic coloration depends on where the animal chooses to rest. (A) The Australian thorny devil (*Moloch horridus*) is camouflaged, but only when the lizard is motionless in areas littered with bark and other varicolored debris, not on roads. (B) The moth *Biston betularia* comes in two forms, the "typical" salt-and-pepper morph and the melanic type, each of which is well hidden on a different kind of background. (A, photographs by John Alcock; B, photograph © IanRedding/Shutterstock.com.)

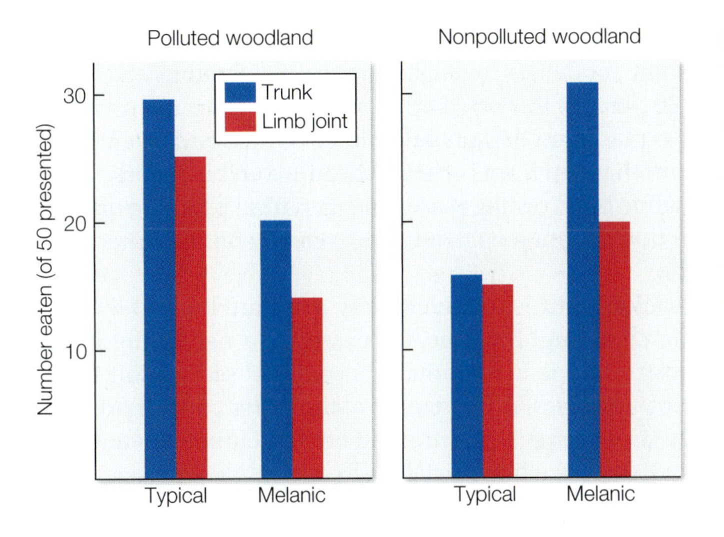

FIGURE 6.7 Predation risk and background selection by *Biston betularia*. Specimens of typical and melanic forms of the moth were attached to tree trunks or to the undersides of limb joints. Birds were more likely to find and remove moths on tree trunks than on limb joints, and this was true for both types of moths. But overall, melanic forms were less often discovered by birds in polluted (darkened) woods, while typical forms "survived" better in unpolluted woods. (After Howlett and Majerus 1987.)

capture by birds, nine species of which were seen taking the moths as they perched on trees during the daytime (Cook et al. 2012).

In nature, *B. betularia* tends to settle within the shaded patches just below where branches meet the trunk. If the moth's actual perch site selection is also adaptive, then we can predict that individuals resting underneath tree limbs should be better protected against predators than if they were to choose an alternative location. Majerus and a colleague attached samples of frozen moths to open trunk areas and to the undersides of limb joints. They found that birds were particularly likely to overlook moths on shaded limb joints (**FIGURE 6.7**) (Howlett and Majerus 1987), another demonstration that a putative behavioral adaptation almost certainly does provide a survival benefit, which in this case enables the moths to ultimately reproduce more.

Like *B. betularia*, many species of moths make decisions about where to perch during the day that are driven by predation risk. For example, the whitish moth *Catocala relicta* usually perches head-up, with its whitish forewings over its body, on white birch and other light-barked trees (**FIGURE 6.8**). When given a choice of resting sites, the moth selects birch bark over darker backgrounds (Sargent 1976). If this behavior is truly adaptive, then birds should overlook moths more often when the insects perch on their favorite substrate. To evaluate this prediction, Alexandra Pietrewicz and Alan Kamil used captive blue jays (*Cyanocitta cristata*), photographs of moths on different backgrounds, and operant conditioning techniques (Pietrewicz and Kamil 1977). That is, they trained the blue jays to respond to slides in which cryptically colored moths were shown positioned on an appropriate background. When a slide flashed on a screen, the jay had only a short time to react. If the jay detected a moth in the picture, it pecked at a key, received a food reward, and was quickly shown a new slide. But if the bird pecked incorrectly when a slide appeared with a scene without a moth, it not

FIGURE 6.8 Cryptic coloration and body orientation. The orientation of a resting *Catocala* moth determines whether the dark lines in its wing pattern match up with the dark lines in birch bark. (Photograph © Sylvie Bouchard/Shutterstock.com.)

only failed to secure a food reward but had to wait a minute for the next chance to evaluate a slide and get some food. The caged jays' responses demonstrated that they saw the moth 10 to 20 percent less often when *C. relicta* was pinned to pale birch bark than when it was placed on darker bark. Moreover, the birds were especially likely to overlook a moth when it was oriented head-up on birch bark. Thus, the moth's preference for white birch resting places and its typical perching orientation appear to be antidetection adaptations that thwart visually hunting predators such as blue jays.

Blending into a solid background is relatively easy compared with blending into one that is constantly changing. Imagine, for example, a fish living in the open ocean where the background is in continual flux as light scatters off water molecules and microscopic particles. The movement of the water and the progression of the sun, as well as its occasional positioning behind clouds, means that the background color for a fish is rarely predictable. To determine whether open-ocean fishes have evolved a cryptic reflectance strategy for their heterogeneous light environment, a team led by Molly Cummings compared open-ocean and coastal fishes against different backgrounds (Brady et al. 2015). The researchers found that open-ocean fishes exhibited camouflage that was superior to that of both nearshore fishes and mirrorlike surfaces, with significantly higher crypsis at angles associated with predator detection and pursuit. In other words, open-ocean fishes seem to have adapted to the constantly changing light angles in their environment by camouflaging themselves to predators in ways that nearshore fishes are unable to do. But how were these open-ocean fishes able to do this? By examining fish skin using scanning electron microscopy to probe deep within the scales, the researchers demonstrated that specific arrangements of reflective pigments called guanine platelets in the fish skin allowed the fish to blend into the consistent vertical light intensity of the open ocean. Thus, the researchers not only demonstrated that fish are able to camouflage themselves in one of the most heterogeneous and constantly changing environments on Earth, they also identified a mechanism that has been selected to shape reflectance properties in this complex environment.

Cryptic behavior comes with costs, notably the time and energy that individuals expend in finding the right background on which to rest, and for many species, the time spent immobile during the day when visual predators might spot them if they were to move. Given the apparent advantages of remaining hidden, it is surprising that some camouflaged insects do things that make them conspicuous. For example, the walnut sphinx caterpillar (*Amorpha juglandis*), a big green species that rests on green leaves, can produce a whistling squeak up to 4 seconds long by forcing air out of one pair of its respiratory spiracles (Bura et al. 2011). Likewise, the peacock butterfly (*Aglais io*), which hibernates during the winter in sheltered sites, flicks open its wings repeatedly when approached by a potential predator, such as a mouse, and in so doing, generates hissing sounds with ultrasonic clicks (Olofsson et al. 2011). Although these actions announce the presence of otherwise cryptic insects, and so would seem costly, the two prey species only do so when under attack or about to be attacked. The sounds they produce frighten at least some of their enemies. For example, yellow warblers (*Setophaga petechia*) that pecked the walnut sphinx caterpillar departed in a hurry when the caterpillar whistle-squeaked in response. Similarly, when infrared-sensitive cameras were used to film mice approaching peacock butterflies at night, the predators were often seen to back off promptly as soon as the butterflies began their wing flicking. The unexpected sounds made by these edible prey species appear to startle their predators into retreat.

Standing Out

Many animals are not camouflaged at all. Instead of blending in to their surroundings, they seem to go out of their way to make themselves obvious to their predators (**FIGURE 6.9**). Conspicuous **aposematic**, or warning, coloration is often used by noxious organisms to signal their unpalatability to potential predators. Take, for example, the monarch butterfly (*Danaus plexippus*), whose orange-and-black wing pattern makes it very easy to spot. Since bright coloration of this sort is correlated with greater risk of attack in some cases (Stuart-Fox et al. 2003), how can it be adaptive for monarchs to flaunt their wings in front of butterfly-eating birds? To overcome the costs of conspicuous coloration, the trait must have substantial benefits, and in the case of the monarch, it does. These benefits appear to be linked to the ability of monarch larvae to feed on poisonous milkweeds, from which they sequester an extremely potent plant poison in their tissues (Brower and Calvert 1984). After vomiting up a noxious monarch just once, a surviving blue jay will steadfastly avoid this species thereafter (Brower 1958, Brower and Calvert 1984). Moreover, monarchs recycle poisons from their food plants to make themselves taste so bad that most birds will release any monarch immediately after grabbing it by the wing (Brower and Calvert 1984). In fact, when caged birds are offered a frozen but thawed monarch, many pick it up by the wing and then drop it at once, evidence that in nature living monarchs could benefit personally from being highly unpalatable, a fact advertised by their color pattern. Thus, the conspicuously colored wings of the monarch are a type of warning that signals to a potential predator that the animal is unpalatable.

Although aposematic coloration is rarer in mammals than in insects and other vertebrates, mammals have some of the most dramatic aposematic coloration in

(A)

(B)

FIGURE 6.9 Warning coloration and toxins. Animals that are chemically defended typically are conspicuous in appearance and behavior. (A) Monarch butterflies that feed on toxic milkweeds as caterpillars store the cardiac glycosides acquired from their food in their bodies and wings when they become adults. (B) Blister beetles, which have blood laced with cantharidin, a highly noxious chemical, often mate conspicuously for hours on flowering plants. (A, photograph by David McIntyre; B, photograph by John Alcock.)

the animal world. For example, the bold black and white patterns of skunks and crested porcupines (*Hystrix cristata*) warn potential predators of their noxious anal gland secretions and sharp spines, respectively. In a comparative analysis of more than 200 mammalian carnivores, Theodore Stankowich and colleagues found that boldly colored and dichromatic species such as these tend to use anal gland secretions for defense, and are also stockier and live in more exposed habitats where other forms of antipredator defense are limited (Stankowich et al. 2011). Mammals that live in these open habitats also tend to use other morphological adaptations beyond noxious sprays—such as spines, quills, and dermal plates—to deter predators, whereas those that live in more closed habitats tend to be smaller and rely on crypsis to escape predators (Stankowich and Campbell 2016).

As you saw with social defenses, investing in elaborate antipredator defenses may entail costs. In an analysis of nearly 650 mammal species, Stankowich's team found that species with active defenses had smaller brains (Stankowich and Romero 2016). The researchers speculated that this pattern could be due at the proximate level to the expense of producing brain tissue, or at the ultimate level to the need for advanced cognitive abilities of larger brained species for constant assessment of environmental predation risk, especially in open environments. Thus, warning coloration in mammals seems to be closely linked to other types of morphological and behavioral adaptations used to deter predators, and investment in these defenses can result in costly trade-offs, including in brain size and cognitive ability.

Not all animals with aposematic colors are noxious. In some cases, a non-noxious species mimics a noxious one to deceive predators. When an edible species resembles a distasteful or dangerous one, it is called **Batesian mimicry**. Batesian mimicry gets its name from the English naturalist Henry Bates, who collected hundreds of species of butterflies in the Brazilian rainforest and hypothesized that the close resemblance between groups of unrelated species had an antipredator adaptation. Another group of organisms common to the Americas that has long been thought to use Batesian mimicry is the species of snakes that resemble in color and patterning the poisonous coral snakes. Using museum records from nearly 300,000 specimens from over 1000 New World snake species to integrate geographic, phenotypic, and phylogenetic data, Alison Davis Rabosky and colleagues confirmed that Batesian mimicry has occurred in this group (**FIGURE 6.10**) (Davis Rabosky et al. 2016). The researchers found that shifts to mimetic coloration in nonvenomous snakes are highly correlated with the presence of coral snakes in both space and time. In other words, poisonous coral snakes and their nonpoisonous mimics often co-occur, and mimetic coloration evolved independently many times in New World snakes, together suggesting that the patterning really is an adaptation in nonpoisonous species to resemble poisonous species in warning coloration. In an experimental test of Batesian mimicry in these snakes, David Pfennig and colleagues showed that replicas of a mimetic kingsnake were attacked more by predators in geographic areas where their model, a coral snake, was absent (Pfennig et al. 2001).

Not all examples of aposematic mimicry involve a noxious template species and a non-noxious mimic. Occasionally, two or more distasteful or dangerous species resemble each other, a pattern referred to as **Müllerian mimicry**. Müllerian mimicry gets its name from the German naturalist Johan Müller who, like Bates, studied butterflies in Brazil. He was particularly interested in the species groups that contained multiple apparently distasteful species. Müller's insight was that both species can benefit from converging on the same type of warning coloration, a mutualistic rather than an exploitative arrangement. A potential example of Müllerian mimicry comes from the poison frogs in the family Dendrobatidae, which sequester toxins in their skin and are thought to warn potential predators with their bright, aposematic coloration (see Figure 11.15). Despite great variation in warning

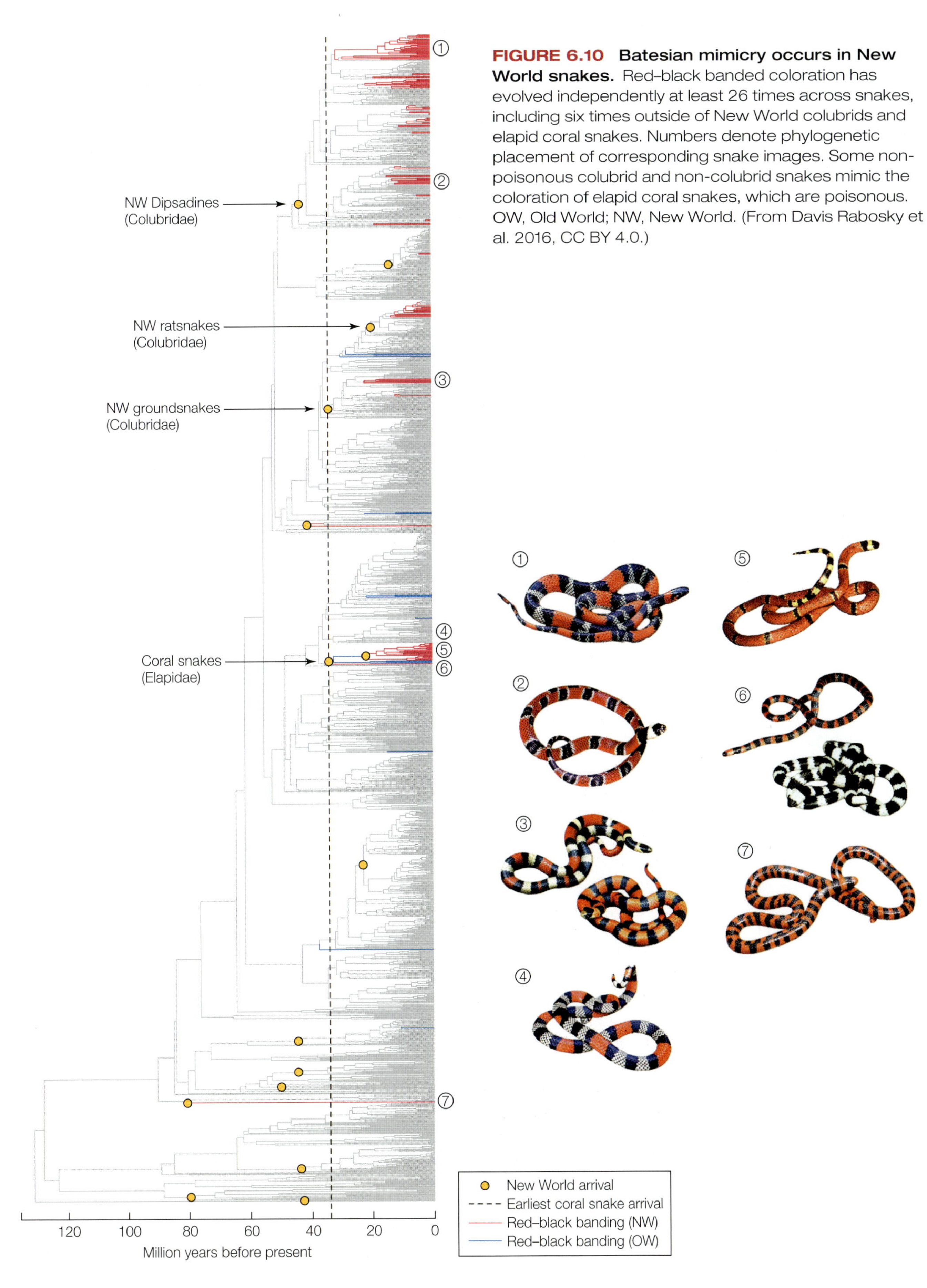

FIGURE 6.10 Batesian mimicry occurs in New World snakes. Red–black banded coloration has evolved independently at least 26 times across snakes, including six times outside of New World colubrids and elapid coral snakes. Numbers denote phylogenetic placement of corresponding snake images. Some non-poisonous colubrid and non-colubrid snakes mimic the coloration of elapid coral snakes, which are poisonous. OW, Old World; NW, New World. (From Davis Rabosky et al. 2016, CC BY 4.0.)

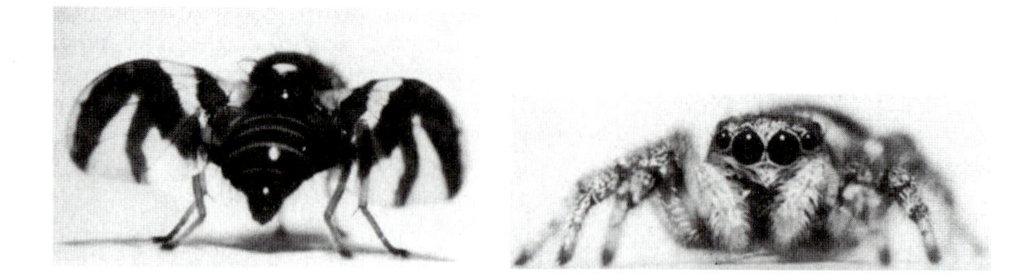

FIGURE 6.11 Why behave conspicuously? This tephritid fly (left) habitually waves its banded wings, which gives it the appearance of a leg-waving jumping spider (right). When the spiders wave their legs, they do so to threaten one another. The fly mimics this signal to deter attack by the spider. (From Mather and Roitberg 1987.)

coloration among species in this group, many species of co-occurring poison frogs have converged on similar color patterns. A team led by Kyle Summers studied learned avoidance of frogs by naïve predators in a test of the Müllerian mimicry hypothesis (Stucker et al. 2014). By examining two species of co-occurring Peruvian frogs from a complex of four in the genus *Ranitomeya*, Summers's team showed that young, naïve chickens exposed to either stimulus species demonstrated reciprocal learned avoidance. Thus, chickens that learned to avoid one poison frog species would also avoid the other similarly colored one, and vice versa.

Bright coloration is not the only means by which a prey species can make itself conspicuous to predators. Take the tephritid fly *Zonosemata vittigera* that habitually waves its banded wings as if trying to catch the attention of its predators. This puzzling behavior attracted two teams of researchers who noticed that the wing markings of the fly resemble the legs of jumping spiders, important consumers of flies. The biologists proposed that when the fly waves its wings, it creates a visual effect similar to that of the aggressive leg-waving displays that the spiders themselves use (see Chapter 8) (**FIGURE 6.11**) (Greene et al. 1987, Mather and Roitberg 1987). Thus, the fly's appearance and behavior might activate escape behavior on the part of the predator, an idea referred to as the **signal deception hypothesis**.

signal deception hypothesis When a prey's appearance or behavior activates escape behavior on the part of the predator.

To test the signal deception hypothesis, one group of researchers exchanged wings between clear-winged houseflies (*Musca domestica*) and pattern-winged tephritid flies. After the operation, the tephritid flies behaved normally, waving their now plain wings and even flying about their enclosures. But these modified tephritids with their housefly wings were soon eaten by the jumping spiders in their cages. In contrast, control tephritids whose own wings had been removed and then glued back on repelled their enemies in 16 of 20 cases. Houseflies with tephritid wings gained no protection from the spiders, showing that it is the combination of leglike color pattern and wing movement that enables the tephritid fly to deceive its predators into treating it as a dangerous opponent rather than a meal (Greene et al. 1987).

FIGURE 6.12 An advertisement of unprofitability to deter pursuit? Thomson's gazelles, a type of small antelope, leaps into the air when threatened by a predator, just as springboks do. (Photograph © Denis-Huot/Minden Pictures.)

Like the tephritid flies, some vertebrates also behave in ways that paradoxically make them easier to spot. Take the Thomson's gazelle (*Eudorcas thomsonii*), for example. When pursued by a cheetah (*Acinonyx jubatus*), this antelope may leap several feet into the air while flaring its white rump patch (**FIGURE 6.12**). Any number of possible explanations exist for this behavior, which is called stotting. Perhaps a stotting gazelle sacrifices speed in escaping from one detected predator to scan ahead for other as-yet-unseen enemies, such as lions (*Panthera leo*), lying in ambush (Pitcher 1979). This **antiambush hypothesis** predicts that stotting will not occur on short-grass savannas but will instead be reserved for tall-grass or mixed grass-and-shrub habitats, where predator detection could be improved by jumping into the air. But gazelles feeding in short-grass habitats do stot regularly, so we can reject the antiambush hypothesis (Caro 1986a, b). Tim Caro generated 10 additional hypotheses to explain how stotting could be used by gazelles to signal to potential predators or to conspecifics (Caro 1986a, 1986b). We address four of these alternative hypothesis (**HYPOTHESES 6.2**), whose predictions are listed in **TABLE 6.1**. In testing these predictions, Tim Caro learned that a single gazelle will sometimes stot when a cheetah approaches, an observation that helps eliminate the **alarm signal hypothesis**, since if the idea is to communicate with other gazelles, then lone gazelles should not stot. Caro was also able to rule out the **confusion effect hypothesis** because the confusion effect can occur only when a group of animals flees together. We cannot rule out the **social cohesion hypothesis** on the grounds that solitary gazelles stot, because there is the possibility that solitary individuals stot to attract distant gazelles to join them. But if the goal of stotting is to communicate with fellow gazelles, then stotting individuals, solitary or grouped, should direct their conspicuous white rump patch toward other gazelles. Stotting gazelles, however, orient their rumps away from other gazelles and toward the predator. Only the **attack deterrence hypothesis** remains: Gazelles stot to signal to a predator that they have seen it and that they have plenty of energy and so will be hard to capture. Cheetahs get the message, since they are more likely to abandon hunts when the gazelle stots than when the potential victim does not perform the display (Caro 1986a).

For stotting to be an honest form of communication, the gazelle and the cheetah must both benefit from their interaction (see Chapter 8). If this is such a system, the signal should indicate that stotters will be hard to capture, which therefore makes it advantageous for cheetahs to accept the attack deterrence signal and to call off the hunt to avoid wasting valuable time and energy. Although this prediction has not been tested in gazelles, it has been tested in the lizard *Anolis cristatellus*, which

HYPOTHESES 6.2

Alternative hypotheses to explain stotting in the Thomson's gazelle

alarm signal hypothesis
Gazelles stot to communicate predation risk to conspecifics.

confusion effect hypothesis
If several gazelles stot simultaneously in a group while fleeing, then the predator may become confused.

social cohesion hypothesis
Gazelles stot to attract conspecifics to join them.

attack deterrence hypothesis
Gazelles stot to signal their vigor to potential predators.

antiambush hypothesis
Gazelles stot to look for other predators that may be lying in wait.

TABLE 6.1 Predictions derived from four alternative hypotheses on the adaptive value of stotting by Thomson's gazelle

Prediction	Alternative hypotheses			
	Signal directed to conspecifics		Signal directed to predators	
	Alarm signal	Social cohesion	Confusion effect	Attack deterrence
Solitary gazelle stots	No	Yes	No	Yes
Grouped gazelles stot	Yes	No	Yes	No
Stotters show rump to predators	No	No	Yes	Yes
Stotters show rump to gazelles	Yes	Yes	No	No

FIGURE 6.13 Are push-up displays an honest signal of a lizard's physiological condition? (A) The lizard *Anolis cristatellus* performs a push-up display when it spots an approaching snake. (B) The time an individual lizard spent running until exhaustion was positively correlated with the number of push-ups it performed under perceived threat from a model snake. (A, photograph by Manuel Leal; B after Leal 1999.)

performs a series of push-up displays when it spots a predatory snake coming its way. To test the attack deterrence hypothesis, Manuel Leal counted the number of push-ups that each individual lizard performed in the laboratory when exposed to a model snake. He then took each lizard to a circular runway, where he induced it to keep running by lightly tapping it on its tail. The total running time sustained by tapping was proportional to the number of push-ups the lizard performed in response to the model of its natural predator (**FIGURE 6.13**) (Leal 1999). Thus, as the attack deterrence hypothesis requires, predators could derive accurate information about the physiological state of a potential prey by observing its push-up performance. Since this lizard sometimes does escape when attacked, it could pay predatory snakes to make hunting decisions based on the signaling behavior of potential victims.

Optimality Theory and Antipredator Behavior

Behavioral biologists have a variety of ways of testing their various adaptationist explanations for behaviors of interest. Much work of this sort in the past has been dedicated to demonstrating that the behavior under examination provides fitness benefits for individuals. Other cost–benefit approaches, however, can potentially yield precise quantitative predictions, rather than the more general qualitative ones that we have been discussing up to this point. This information may permit a researcher to determine whether or not a current trait is likely to have been "better" than alternatives over evolutionary time. One technique of this sort relies on **optimality theory**, the notion that adaptations have greater benefit-to-cost ratios than the putative alternatives that have been replaced by natural selection. We will illustrate the theory by examining the costs (*C*) and benefits (*B*) of four alternative hereditary behavioral traits in a hypothetical species (**FIGURE 6.14**). Of these four traits (W, X, Y, and Z), only one (Z) generates a net loss in fitness ($C > B$) and is therefore obviously inferior to the others. The other three traits all generate net fitness gains, but just one, trait X, is an adaptation, because it produces the greatest net benefit of the four alternatives in this population. Of course, in a different population, another of the traits may be adaptive and generate the greatest net benefit.

Given the occurrence of trait X in a population, the alleles associated with this trait will spread at the expense of the alternative alleles underlying the development of the alternative traits in this population. Behavior X can be considered optimal

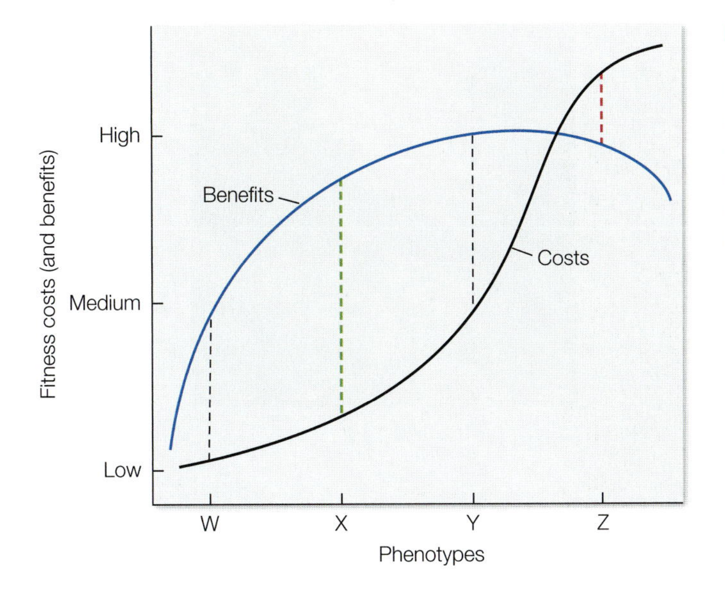

FIGURE 6.14 An optimality model. If one can measure the fitness costs and benefits associated with four alternative behavioral phenotypes in a population, then one can determine which trait confers the greatest net benefit on individuals in that population. In this case, phenotype X is the adaptation—an optimal trait that would replace competing alternatives, given sufficient evolutionary time.

because the difference between benefits and costs is greatest for this trait, and because this trait will spread while all others are declining in frequency (as long as the relationship between their costs and benefits remains the same). If it were possible to measure *B* and *C* for a set of reasonable alternatives, one could predict that the characteristic with the greatest net benefit would be the one observed in nature. Unfortunately, it is often difficult to secure precise measures of *B* and *C* in the same fitness units, although as you will see, some studies of foraging behavior have been able to measure the calories gained from collected foods (the benefits) and calories expended in collecting those foods (the costs).

One application of optimality theory in animal behavior has been to antipredator behaviors. For example, in the midwestern United States, northern bobwhite quail (*Colinus virginianus*) spend the winter months in small groups, called coveys, which range in size from 2 to 22 individuals. Most coveys are of intermediate size and contain about 11 individuals. These groups almost certainly form to gain protection against predators because members of larger coveys are safer from attack, judging from the fact that overall group vigilance (the percentage of time that at least one member of the covey has its head up and is scanning for danger) increases with increasing group size, but then levels off around a group size of 10. Moreover, in aviary experiments, members of larger groups reacted more quickly than members of smaller ones when exposed to a silhouette of a predatory hawk (Williams et al. 2003).

The antipredator benefits to bobwhite quail of being in a large group are almost certainly offset to some degree by the increased competition for food that occurs in larger groups. This assumption is supported by evidence that relatively large groups move more each day than coveys of 11. (Small groups also move more than groups of 11, probably because these birds are searching for other groups with which to amalgamate.) The mix of benefits and costs associated with coveys containing different numbers of individuals suggests that birds in intermediate-size coveys have the best of all possible worlds, and in fact, daily survival rates are highest during the winter in coveys of this size (**FIGURE 6.15**) (Williams et al. 2003). This work demonstrates not just that the quail form groups to deal with their predators effectively, deriving a benefit from their social behavior, but that they attempt to form groups of the optimal size. If they succeed in joining such a group, they will derive a greater net benefit in terms of their survival than by joining a smaller or larger group.

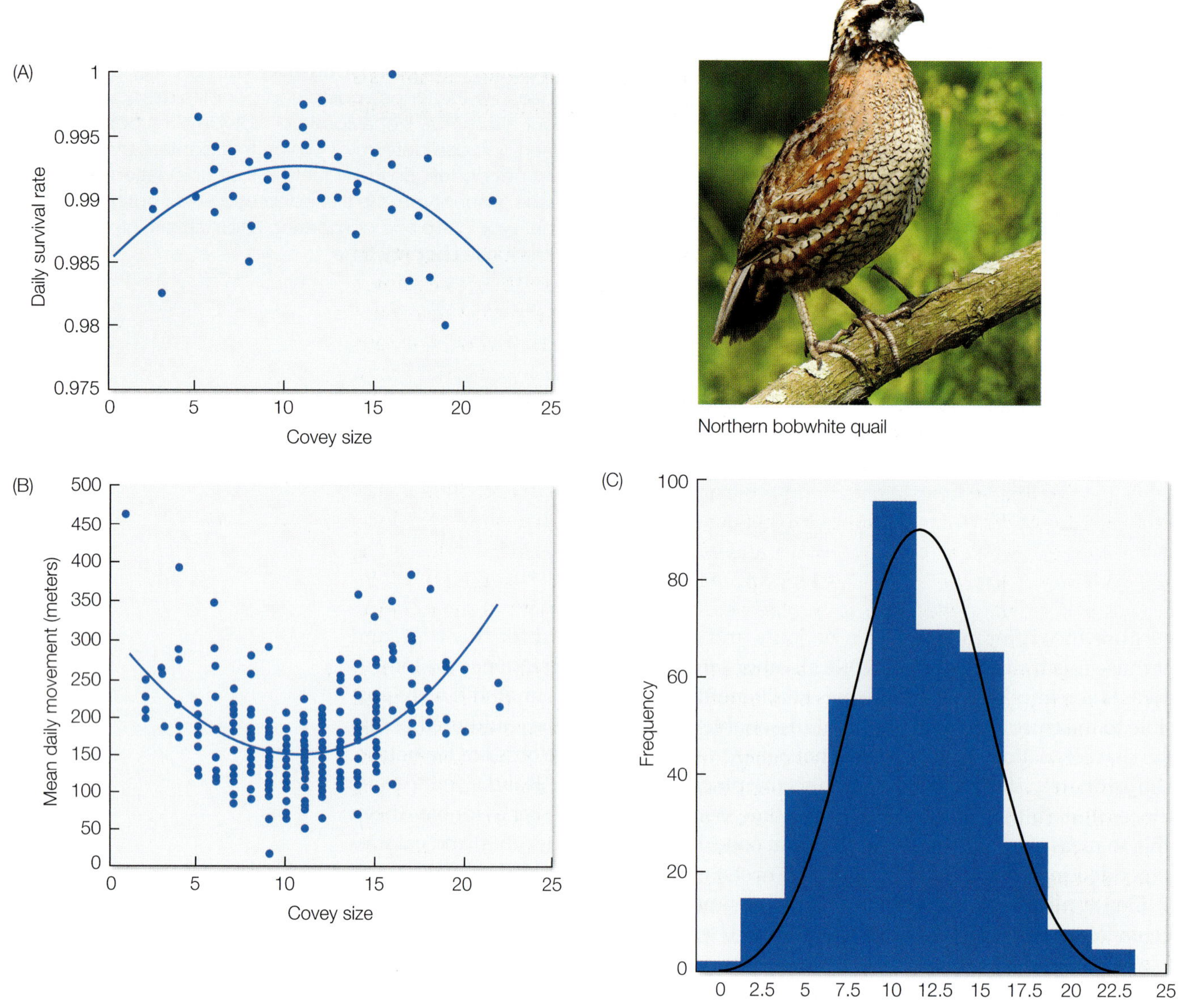

FIGURE 6.15 Optimal covey size for northern bobwhite quail is a function of the costs and benefits of belonging to groups of different sizes. (A) The probability that an individual will survive on any given day (its survival rate) is highest for birds in coveys of about 11 birds. (B) Distance traveled, a cost, is lowest for individuals in coveys of about 11 birds. (C) The most common covey size was about 11 in this bobwhite population. (After Williams et al. 2003; photograph © iStock.com/staden.)

Finding Food

Escaping predation only helps an organism survive if it can also find food. After all, sitting in hiding most of one's life would not aid in the quest to find the resources necessary for reproduction, let alone for homeostasis. How do individuals find the food they need, and do so while avoiding predators? As you will see, these strategies often go together. The foraging cycle typically involves searching for, pursuing, and handling food resources. Animals must make trade-offs in the time they spend in each of these phases. For example, when should an individual stop pursuing or handling one prey item and set its sights on another? Optimality theory provides a framework within which to examine this type of question. We therefore begin with a discussion of optimality theory applied to foraging behavior. We then link foraging behavior back to predator avoidance with a discussion of landscapes of fear. By the end of this chapter, you should have an understanding of the decisions that animals make when it comes to both finding a meal and making sure they don't become someone else's meal.

Optimality Theory and Foraging Decisions

Perhaps the most common application of optimality theory to studies of animal behavior has been to foraging decisions, where the theory has come to be known as **optimal foraging theory**. Optimal foraging theory makes predictions about how animals maximize fitness while foraging. Animals use all manner of ingenious methods to secure the food they need to survive and reproduce, ranging from a blood-consuming leech that specializes on hippopotamuses (in whose rectums the leeches assemble to find mates) (Oosthuizen and Davies 1994) to several Australian hawks that consume the tongues of road-killed cane toads (*Rhinella marina*) (the safest part of this generally toxic non-native animal) (Beckmann and Shine 2011). Whatever the tactics that animals use to get enough food to survive and reproduce, the various behaviors all come with costs and benefits. It is not surprising, therefore, that both optimality and game theoretic approaches have helped behavioral biologists produce and test hypotheses on the possible adaptive value of the foraging decisions of assorted animals.

Optimal foraging models address several related topics, including the optimization of food type, patch choice, time spent in different patches, or patterns and speed of movement (Pyke et al. 1977). The basis of all optimal foraging models is that food has obvious benefits in terms of energy intake, but finding food comes with costs in terms of the time and energy spent locating it (search time) and accessing it (handling time). Therefore, energy is the currency used in most optimal foraging models, and energy gain per cost (time or energy spent searching or handling) is what is being optimized (**BOX 6.3**). For example, optimal diet models are used to predict whether an animal that encounters different prey items should decide to eat what it finds or continue searching for more profitable prey items. Optimal patch choice models work in similar ways to optimal diet models, with animals choosing among places to forage rather that choosing among specific food items. Animals must also decide when to leave one patch and search for a more profitable one. One type of optimality model, which has been termed the **marginal value theorem**, predicts that an animal should leave a patch when its rate of food intake in that patch drops below the average rate for the habitat, and that this marginal capture rate should be equalized over all patches within a habitat (Krebs et al. 1974). Some animals leave a home base in search of food items or patches, but once they locate resources, they must bring it back to their home. These "central place foragers" include species such as nesting birds or ants that live in large colonies. Central place foraging models therefore examine how organisms can optimize foraging rates while traveling through patches to locate food, before returning home with those food items. Finally, some animals must optimize their ability to encounter food, often by changing speed or direction.

Consider a northwestern crow (*Corvus caurinus*) out hunting for food on a tidal flat in British Columbia. When the bird spots a clam, snail, mussel, or whelk, it sometimes picks it up, flies into the air, and then drops its victim onto a hard surface. If the mollusk's shell shatters on the rocks below, the bird flies down to the prey and plucks out the exposed flesh. The adaptive significance of the bird's behavior seems straightforward. Since it cannot use its beak to open the extremely hard shells of certain mollusks, it instead opens the shells by dropping them on rocks. At first glance, this behavior appears to be adaptive. However, Reto Zach was much more ambitious in his analysis of the crow's foraging decisions (Zach 1979). When

BOX 6.3

EXPLORING BEHAVIOR BY INTERPRETING DATA

Territoriality and feeding behavior in golden-winged sunbirds

The golden-winged sunbird (*Drepanorhynchus reichenowi*) feeds exclusively on nectar from certain flowers during the winter in South Africa. Sometimes the birds are territorial at patches of flowers, but at times they are not. Frank Gill and Larry Wolf devised a way to measure the rate of nectar production per bloom at a given site (Gill and Wolf 1975a,b). With this information and previously published data on the caloric costs of resting, flying from flower to flower, and chasing intruders, they were able to calculate when it would be advantageous for the birds to be territorial (**TABLE**).

Benefits of territoriality to golden-winged sunbirds under different conditions

Nectar production[a]			
In undefended site (foraging time in hours)	**In territory (foraging time in hours)**	**Rest gained by territorial birds (in hours)[b]**	**Energy saved (in calories)[c]**
1 (8)	2 (4)	8 – 4 = 4	2400
2 (4)	3 (2.7)	4 – 2.7 = 1.3	780
4 (2)	4 (2)	2 – 2 = 0	0

Source: Gill and Wolf 1975a.

[a]Measured as microliters of nectar produced per blossom per day.

[b]Calculated from the number of hours in foraging time needed to meet daily caloric needs, which depends on nectar production:

Nectar production (microliters/blossom)	1	2	3	4
Foraging time (hours)	8	4	2.7	2

[c]For each hour spent resting instead of foraging, a sunbird expends 400 calories instead of 1000.

Thinking Outside the Box

Gill and Wolf assumed that the nonbreeding birds' goal was to collect sufficient nectar to meet their daily survival needs. Was this assumption reasonable? Based on the table above, how many minutes of territorial defense would be worthwhile if a territory owner had access to a 2-microliters-per-blossom-per-day site while other flower patches were producing nectar at half that rate?

a hungry crow is searching for food, it has to decide which mollusk to select, how high to fly before dropping the prey, and how many times to keep trying if the mollusk does not break on the first try. Zach made several observations while watching foraging crows. First, he noted that the crows selected and flew up with only large whelks about 3.5 to 4.4 centimeters long. Second, he observed that the crows flew up about 5 meters to drop their chosen whelks. And third, he noted that the crows kept trying with each chosen whelk until it broke, even if many flights were required.

Zach sought to explain the crows' behavior by determining whether the birds' decisions were optimal in terms of maximizing whelk flesh available for consumption per unit of time spent foraging (Zach 1979). To do this, he first had to determine whether the birds' behavior was efficient in terms of generating broken whelks. If so, then the following predictions should apply. First, large whelks should be more likely than small ones to shatter after a drop of 5 meters. Second, drops of less than 5 meters should yield a reduced breakage rate, whereas drops of much more than 5 meters should not greatly improve the chances of opening a whelk. And third, the probability that a whelk will break should be independent of the number of times it has already been dropped.

Zach tested each of these predictions in the following manner. He erected a 15-meter pole on a rocky beach and outfitted it with a platform whose height could be adjusted and from which whelks of various sizes could be pushed onto the rocks below. He then collected samples of small, medium, and large whelks and dropped them from different heights (**FIGURE 6.16**). He found that large whelks required significantly fewer 5-meter drops before they broke than medium-size or small whelks did. He also discovered that the probability that a large whelk would break improved sharply as the height of the drop increased—up to about 5 meters—but that going higher did little to improve the breakage rate. Finally, he learned that the chance that a large whelk would break was not affected by the number of previous drops and was instead always about one chance in four on any given drop. Therefore, a crow that abandoned an unbroken whelk after a series of unsuccessful drops would not have a better chance of breaking a replacement whelk of the same size on its next attempt. Moreover, finding a new prey would take time and energy.

Zach then calculated the average number of calories required to open a large whelk (0.5 kilocalories), a figure he subtracted from the food energy present in a large whelk (2.0 kilocalories), for a net gain of 1.5 kilocalories. In contrast, medium-size whelks, which require many more drops, would yield a net loss of 0.3 kilocalories. And trying to open small whelks would have been even more disastrous in terms of energy loss (Zach 1979). Thus, the crows' rejection of all but large whelks was adaptive, assuming that fitness is a function of energy gained per unit of time. However, another group of researchers argued later that the crows would have been better off calorically if they had dropped their prey from greater heights (Plowright et al. 1989). Thus, it remains a puzzle why Zach's crows chose to fly up only 5 meters rather than to maximize caloric intake by flying higher to drop differently sized whelks.

The approach behind Zach's research included the simple assumption that the caloric cost of gathering food had to be less than the caloric benefits gained by eating

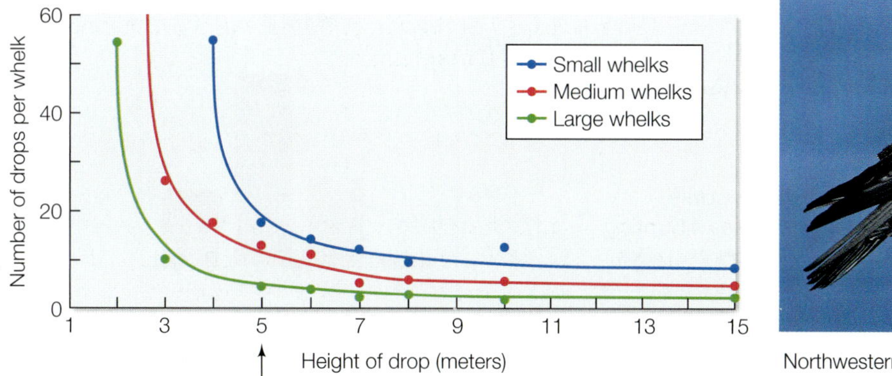

Northwestern crow

FIGURE 6.16 Optimal foraging decisions by northwestern crows when feeding on whelks. The curves show the number of drops at different heights needed to break whelks of different sizes. Northwestern crows pick up only large whelks, increasing the calories available to them, and they drop their whelks from a height of about 5 meters (indicated by the arrow), thereby minimizing the energy they expend in opening their prey. (After Zach 1979; photograph © 7877074640/ShutterStock.)

the food. The same premise has been examined in many other species of animals (**BOX 6.4**), including in blue whales (*Balaenoptera musculus*), which feed by diving underwater for up to 15 minutes in their search for krill, shrimplike animals that form huge swarms in Antarctic waters. A feeding blue whale surges into a shoal of krill, engulfing huge amounts of water before pushing the water out through the baleen plates in its mouth, an action that traps the edible krill contained in the water. Whale biologists had long assumed that open-mouthed blue whales paid a heavy price to force their way through the water at a high enough speed to catch enough krill to pay for their foraging movements. When a team of blue whale experts outfitted a whale with hydrophones, pressure sensors, and accelerometers, which provided the data needed to estimate the cost of a lunge, they estimated that

BOX 6.4 **EXPLORING BEHAVIOR** BY **INTERPRETING DATA**

Optimal foraging by pike cichlid fish

The pike cichlid (*Crenicichla saxatilis*) is a predatory fish that tends to attack and consume relatively large individual Trinidadian guppies (*Poecilia reticulata*), its primary prey in the rivers in Trinidad where both species live. A research team measured the time it took for pike cichlids to detect, approach, stalk, and attack guppies of four size classes (10, 20, 30, and 40 millimeters long) (Johansson et al. 2004). They also recorded the capture success rate for each size class, as well as noting the time it took to handle and consume those prey that were actually captured. With these data, the researchers constructed two models of prey value as measured by mass of prey consumed per unit of time. Model A considered only the time to attack and the capture rate, whereas model B incorporated these two factors plus the post-capture handling time in calculating the weight of food consumed per unit of time (**FIGURE**).

Pike cichlid

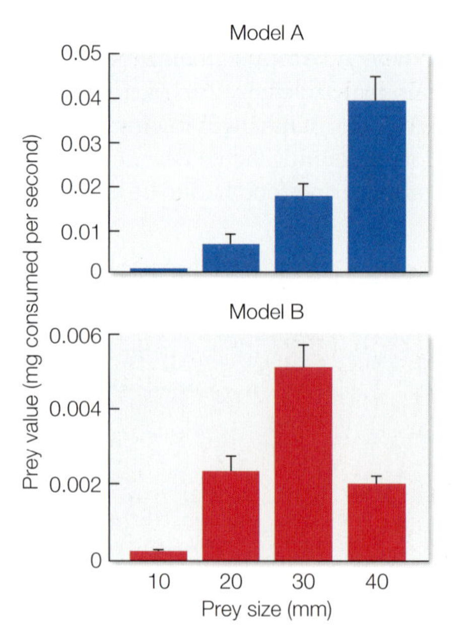

Two optimal foraging models. These models make different assumptions about the value of hunting guppies of different sizes by a predatory fish, the pike cichlid. Bars depict mean +/– SE. (After Johansson et al. 2004; photograph © Jane Burton/Naturepl.com.)

Thinking Outside the Box

Which of the two models strikes you as the most realistic optimal foraging model, and why? In light of the two models, what evolutionary issues are raised by the actual preference of pike cichlids for 40-millimeter-long guppies? Can you think of any other factors that might improve the model further?

the energy expended on the typical dive in which three to four underwater lunges occurred was about 60,000 kilojoules. So what does a blue whale secure in return for its expenditure of metabolic and mechanical energy? Of course, the answer depends on how many krill a whale can capture, which is a function of how much water it engulfs and the density of krill in a given volume of water. Based on the size of blue whale jaws and published data on krill densities, the researchers estimated that a single underwater lunge could yield anywhere from 30,000 to almost 2 million kilojoules. Given that during the typical dive, a blue whale makes several lunges, which in total yield about 5 million kilojoules on average, krill-hunting whales come out way ahead, with the food ingested paying for a standard dive nearly 80 times over (Goldbogen et al. 2011).

Both northwestern crows and blue whales gain more energy from their foraging activities than they expend in the attempt to secure food. But do these and other animals achieve maximum reproductive success by maximizing the number of calories ingested per unit of time, a key assumption of the standard optimality model? The relationship between whelk-opening efficiency and fitness has not been established for crows, nor do we know if blue whale foraging efficiency translates into maximum reproductive gain. However, in an experiment with captive zebra finches (*Taeniopygia guttata*) given the same kind of food under different feeding regimes so that some birds had higher foraging costs than others, the individuals with the highest daily net caloric gains survived best and reproduced sooner and produced more offspring than their compatriots (Lemon and Barth 1992, Wiersma and Verhulst 2005). In another experiment, male red crossbills (*Loxia curvirostra*) were placed on both sides of a divided cage that contained either a branch with pinecones from which most of the edible seeds had been removed or a branch whose pinecones were unaltered. Female crossbills that observed the two categories of foraging males associated preferentially with males they had seen securing many seeds (**FIGURE 6.17**). Female crossbills can evidently assess the feeding rate of male foragers, and they use this information to choose potential mates accordingly. In other words, male foraging efficiency leads to increased reproductive opportunities in this species (Snowberg and Benkman 2009).

FIGURE 6.17 Red crossbills feeding on seeds hidden in a pinecone. Thanks to their crossed bills, these birds are capable of twisting open the closed elements of pinecones to remove seeds, one at a time, from within the cones. Here a male (the redder individual) pauses while a female (the greenish bird) probes the pinecone with her specialized beak. (Photograph by Craig Benkman.)

As you have seen, the assumptions that go into building optimal foraging models are critical for fitting empirical data to the models' predictions. Another good example of how empirical data can help refine model assumptions to improve the fit with the data comes from the Eurasian oystercatcher (*Haematopus ostralegus*). Two researchers, P. M. Meire and A. Ervynck, developed a calorie maximization hypothesis to apply to oystercatchers feeding on mussels (Meire and Ervynck 1986). As Zach did with his crows, Meire and Ervynck calculated the profitability of different-size prey, based on the calories contained in the mussels (a fitness benefit) and the time and energy required to open them (a fitness cost). They discovered that even though mussels over 50 millimeters long require more time to hammer or stab open, they provide more calories per minute of opening time than smaller mussels. Therefore, the model predicts that oystercatchers should focus primarily on the largest mussels. But in real life, the birds do not prefer the really large ones (**FIGURE 6.18**). Why not?

The researchers proposed two hypotheses to explain why their observational data were not consistent with the model (Meire and Ervynck 1986). First, they hypothesized that the profitability of very large mussels is less than the maximum calculated because some cannot be opened at all, reducing the average return from handling these prey. In their initial calculations of prey profitability, the researchers had considered only those prey that the oystercatchers actually opened (**FIGURE 6.19**, Model A). That is, they made a faulty assumption. As it turns out, oystercatchers occasionally select some large mussels that they find impossible to open, despite their best efforts. The handling time wasted on these large, impregnable mussels reduces the average payoff for dealing with this size class of prey. When this factor is taken into account, a new optimality model yields the prediction that the oystercatchers should concentrate on mussels 50 millimeters in length, rather than on the very largest size classes (see Figure 6.19, Model B). The oystercatchers, however, actually prefer mussels in the 30- to 45-millimeter range. Therefore, time wasted in handling large, invulnerable mussels fails to explain the oystercatchers' food selection behavior.

Eurasian oystercatcher

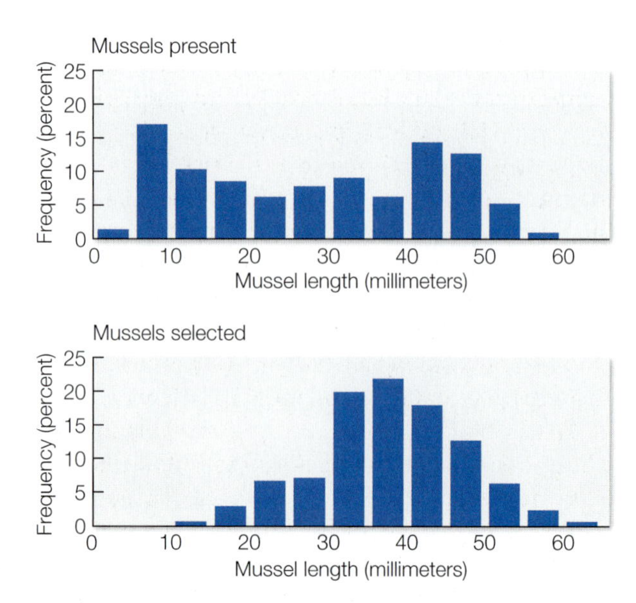

FIGURE 6.18 Available prey versus prey selected. Although foraging Eurasian oystercatchers tend to choose fairly large mussels, they do not prefer the very largest mussels. (After Meire and Ervynck 1986; photograph © Rui Saraiva/ShutterStock.)

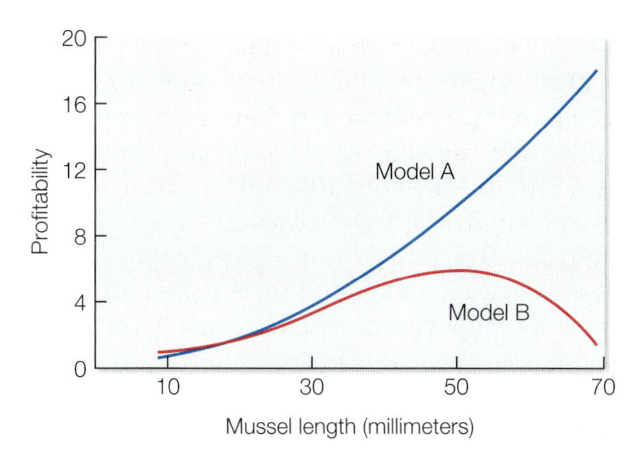

FIGURE 6.19 Two optimal foraging models yield different predictions because they calculate prey profitability differently. Model A calculates the profitability of a mussel based solely on the energy available in opened mussels of different sizes divided by the time required to open these prey. Model B calculates profitability with one added consideration, namely that some very large mussels must be abandoned after being attacked because they are impossible to open. (After Meire and Ervynck 1986.)

Next the researchers hypothesized that many large mussels are not even worth attacking because they are covered with barnacles, which makes them impossible to open (Meire and Ervynck 1986). This additional explanation for the apparent reluctance of oystercatchers to feast on large, calorie-rich mussels is supported by the observation that oystercatchers never touch barnacle-encrusted mussels. The larger the mussel, the more likely it is to have acquired an impenetrable coat of barnacles, which eliminates these prey from consideration. According to a mathematical model that factors in prey-opening time, time wasted in trying but failing to open a mussel, and the actual size range of realistically available prey, the birds should focus on 30- to 45-millimeter mussels—and they do. Thus, by using the scientific method to refine their models with additional assumptions and data, the researchers were able to show that Eurasian oystercatchers do indeed appear to forage optimally.

Criticisms of Optimal Foraging Theory

By developing and testing optimality models, researchers have concluded that northwestern crows and Eurasian oystercatchers, among other species, choose prey that tend to provide the maximum caloric benefit relative to time spent foraging. But some scientists have criticized the use of optimality theory on the grounds that animals apparently do not always select or hunt for food as efficiently as possible. Consider, for example, the tropical jumping spider *Bagheera kiplingi* that feeds largely on the specialized leaf tips of certain acacias—a food high in fiber but low in nitrogen, a valuable nutrient found in much larger quantities in the insects preyed on by thousands of other jumping spiders (Meehan et al. 2009). How can leaf-tip feeding be the most efficient way for this one spider to secure large amounts of nitrogen? Why doesn't it also feed on insects that would provide it with the nitrogen it needs? These questions raise interesting problems, but remember that optimality models are constructed not to make statements about perfection in evolution but rather to make it possible to test whether one has correctly identified the variables that have shaped the evolution of an animal's behavior.

As you have also seen, the assumptions that go into an optimality model have a large effect on the predictions that follow. If an oystercatcher is assumed to treat every mussel in a tidal flat as a potential prey item, then it will be expected to make different foraging decisions than if the modeler assumes that oystercatchers simply ignore all barnacle-covered mussels. If the predictions of an optimality model are based on faulty assumptions and fail to match reality, researchers will reject that model. This does not mean the approach is wrong, however. It may simply mean that a better model is needed. This is exactly what happened in the study of oystercatchers, as the researchers modified their model with new information, made better assumptions about their study system, and subjected the model to a new test. As a result of their application of the scientific method to the problem, they gained an improved understanding of oystercatcher feeding behavior and were able to demonstrate that the birds forage optimally.

If ecological factors other than caloric intake affect oystercatcher foraging behavior, for example, then a caloric maximization model will fail its test, as it should. And for most foragers, foraging behavior does indeed have consequences above and beyond the acquisition of calories and certain nutrients. In addition to caloric needs, predators can also shape the evolution of an animal's foraging behavior. In species where predation risk influences foraging behavior, the kind of optimality model one might choose to construct and test would not focus solely on calories gained versus calories expended; it must also take into account predation risk.

If foraging exposes an animal to the risk of sudden death, then when that risk is high, we would expect foragers to sacrifice short-term caloric gain for long-term survival (Walther and Gosler 2001, Johnsson and Sundström 2007; but see Urban 2007). A sacrifice of this sort has been demonstrated for dugongs (*Dugong dugon*), which are large, relatively slow-moving marine mammals that are preyed on by tiger sharks (*Galeocerdo cuvier*) in Shark Bay, Western Australia. There the dugongs feed on sea grasses in two different ways: cropping, in which the herbivores quickly strip leaves from standing sea grasses, and excavation, in which the foragers stick their snouts into the sea bottom to pull up the sea grasses by their "roots," which are called rhizomes. Dugongs that eat sea grass with rhizomes attached secure more energy per time spent feeding. But when the animals have their heads partly buried in the sandy ocean floor, they cannot see well and are at greater risk of attack from sharks. In contrast, when they are cropping sea grass, they can more easily scan for approaching enemies, reducing the risk of shark attack. Researchers have shown that dugongs in Shark Bay stop excavating sea grass rhizomes when tiger sharks

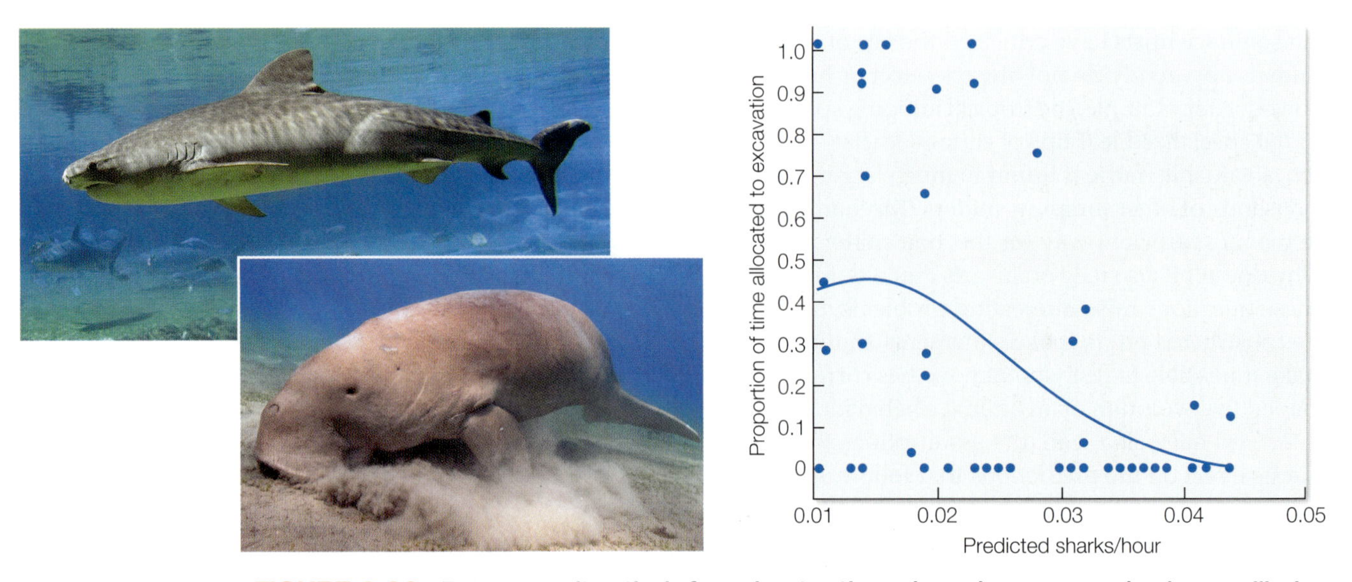

FIGURE 6.20 Dugongs alter their foraging tactics when dangerous sharks are likely to be present. The time these large, slow-moving sea mammals spend excavating sea grass from the ocean bottom declines in relation to the probability that tiger sharks will be in the area. (After Wirsing et al. 2007; top photograph © Ian Scott/ShutterStock.com; bottom photograph © Kristina Vackova/Shutterstock.com.)

are relatively common in the area, as measured by the shark catches made by local fishermen (**FIGURE 6.20**) (Wirsing et al. 2007). Thus, an optimality model that failed to consider the trade-offs between foraging success and predation risk would fail to predict the dugongs' behavior accurately.

Landscapes of Fear

Dugongs are not the only animals that alter their behavior in the face of predation risk in ways that entail both costs and benefits for individuals. The spatially explicit elicitation of fear in prey by the perceived risk of predation has been termed the **landscape of fear** (Gallagher et al. 2017). For example, elk (*Cervus canadensis*) living in Yellowstone National Park have changed their foraging behavior considerably following the reintroduction of wolves (*Canis lupus*) into the area. Over a 5-year period, female elk increased their vigilance behavior by 50 percent (Laundre et al. 2001). Moreover, instead of feeding comfortably in open meadows where their preferred foods are present in abundance, elk now leave the grasslands when wolves arrive and move into wooded areas where they are less easily spotted and chased by their enemies (Creel et al. 2005). Although they are safer in forests, elk pay a price for altering their foraging behavior. In winter, analysis of fecal samples suggests that elk in areas with wolves are eating 25 percent less than elk in wolf-free sites (Christianson and Creel 2010). In summer, the probability that a given cow elk will give birth to a calf has fallen sharply, and the likelihood that her calf will still be with her by the time winter rolls around has also declined (**FIGURE 6.21**). To reduce the risk of being killed, the animals lower their energy intake, which increases the chances that they will be malnourished and reduces the odds of producing and tending a calf. If we were to fail to consider the consequences of ignoring the presence of wolves, we might conclude that elk in some places were foraging in a suboptimal manner. Instead, they are making trade-offs between reproduction and their own survival.

The African savanna, where the densities of predators and prey are among the highest anywhere on Earth, offers a prime example of a landscape of fear. In a study of more than 600 plains zebras (*Equus quagga*) in a fenced conservancy in Kenya, a

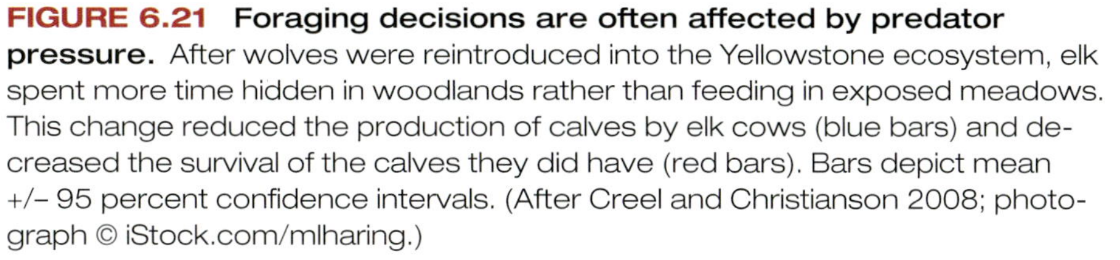

FIGURE 6.21 Foraging decisions are often affected by predator pressure. After wolves were reintroduced into the Yellowstone ecosystem, elk spent more time hidden in woodlands rather than feeding in exposed meadows. This change reduced the production of calves by elk cows (blue bars) and decreased the survival of the calves they did have (red bars). Bars depict mean +/– 95 percent confidence intervals. (After Creel and Christianson 2008; photograph © iStock.com/mlharing.)

team led by Daniel Rubenstein found that zebra abundance on or near a grassland patch (where zebras find the best grazing) is lower if lions have also been observed there the same day (Fischhoff et al. 2007). Because predation pressure is highest in grasslands during the night when lions are more active, zebras also reduce their use of grassland habitat at night, instead using more woodland habitat where there is less food. Using global positioning system (GPS) collars, the researchers found that zebras move faster and make sharper turns in grasslands at night, presumably to avoid detection and capture by lions. Lions have similar effects on many other species, including African wild dogs (*Lycaon pictus*) (Gallagher et al. 2017). Although wild dogs are a predatory species, conflict with lions is a common cause of death for wild dogs. In Tanzania, wild dogs avoid areas heavily used by lions, which causes them to move from preferred hunting areas to more deciduous woodland, where rates of encounter with prey are nearly two-thirds lower.

Although most studies of predator–prey interactions, like those of the zebras and wild dogs with lions, consider single sets of species, prey often must deal with many potential predators at the same time. Different predators operate at different times of day and at different scales, meaning that even for the same species of prey, predation risk can vary in both time and space. Moreover, it is not clear whether prey respond to short-term (the predator is close by) or long-term (there are often predators in this area) risks of predation. Egil Droge and colleagues took a community-based approach to study the landscape of fear in Zambia at long and short temporal scales and over a broad geographic range (Droge et al. 2017). The team focused on three prey species—plains zebra, wildebeest (*Connochaetes gnou*), and oribi (*Ourebia oribi*)—and all of the predators in the area: African wild dogs, spotted hyenas (*Crocuta crocuta*), lions, and cheetahs. To the researchers' surprise, all three prey species were more vigilant in areas that were heavily used by predators over the long term, but they were not more vigilant if they were closer to the nearest predator at the time of observation. Thus, predators really do create a landscape of fear that has long-standing implications for the behavior in prey animals, including foraging, movement, and grouping behavior.

Game Theory and Feeding Behavior

Earlier in this chapter, we used game theory to analyze the possible antipredator function of certain social behaviors. Here we return to game theory in order to identify the adaptive basis of certain foraging behaviors, using the tiny wormlike larvae of fruit flies as an example. *Drosophila* larvae come in two forms, active rovers and sedentary sitters, which differ in how far they maneuver through fruit fly medium in the laboratory in search of food. The odds that rovers and sitters will secure exactly the same net caloric benefits per unit of time spent foraging seem vanishingly small. If one type of larva did even slightly better on average than the other, the genes specifically associated with that trait should spread and replace any alternative alleles linked to the less effective food-acquiring behavior. So why are both types still reasonably common in some places?

The two kinds of fruit fly larvae differ genetically (de Belle and Sokolowski 1987), and so game theorists would say that they employ two different behavioral strategies. This means that if one strategy were to confer higher fitness than the other in a population, eventually only the superior strategy will persist. But under some special circumstances, two strategies can coexist indefinitely, thanks to the effects of **frequency-dependent selection**. This kind of selection occurs when the fitness of one strategy is a function of its frequency relative to the other inherited behavioral trait. The fitness of a genotype can either increase (positive frequency-dependent selection) or decrease (negative frequency-dependent selection) as the genotype frequency in the population increases. For example, when the fitness of one type increases as that type becomes rarer, then that type will become more frequent in the population—but only until such time as it has the same fitness as individuals playing the other strategy. Frequency-dependent selection will act against either type if it becomes even a little more common, pushing the proportion of that form back toward the equilibrium point at which both types have equal fitness. At equilibrium, the two types of individuals will coexist indefinitely and have equal fitnesses.

In the case of rover and sitter fruit flies, experiments have shown that when food resources are scarce, the odds that an individual of the rarer phenotype will survive to pupation (a correlate of fitness) are greater than the corresponding odds for the more common type in the population (**FIGURE 6.22**) (Fitzpatrick et al. 2007). The effect of this kind of negative frequency-dependent selection is to lead to an increase in the frequency of the rarer behavioral type, which keeps it in the population.

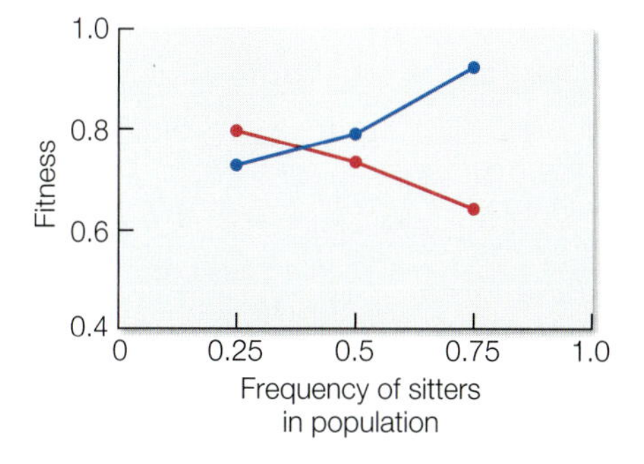

FIGURE 6.22 Negative frequency-dependent selection. When resources are scarce for fruit fly larvae, the fitness of a sedentary sitter (red line) versus an active rover (blue line) depends on which of the two types is rarer. (After Fitzpatrick et al. 2007.)

Another example of negative frequency-dependent selection in action involves an African cichlid fish, *Perissodus microlepis*. This cichlid comes in two forms, one with the jaw twisted to the right and the other with the jaw twisted to the left. The fish makes a living by darting in to snatch scales from the bodies of other fishes in Lake Tanganyika. Adults with the jaw twisted to the right always attack the prey's left flank, while those with the jaw twisted to the left hit the right side (**FIGURE 6.23**) (Hori 1993). Right-jawed parents usually produce offspring that inherit their jaw shape and behavior, and vice versa for left-jawed parents. This suggests that the difference between the two forms is hereditary (although some recent studies have raised questions about this conclusion, including the possibility that the degree of asymmetry in jaw shape is enhanced by the attack behavior of individuals; Palmer 2010). Michio Hori proposed that the reason why both kinds of predators occur in this species is because the prey this fish attacks could

FIGURE 6.23 Two hereditary forms of an African cichlid fish. Right-jawed and left-jawed *Perissodus microlepis* fish have asymmetrical mouths that they employ to snatch scales from the left and right sides, respectively, of the fish they prey on.

learn to expect a raid on their scales from the left if most attacks were directed at the prey's left flank. Thus, in a population of predators in which the right-jawed form predominated, a lefty would have an advantage because its victims would be less vigilant with respect to scale snatchers darting in on the right flank. This advantage would translate into higher reproductive success for the rarer phenotype and an increase in its frequency until left-jawed fish made up half the population (Hori 1993). With a 50:50 split, the equilibrium point would have been reached. If Hori's hypothesis is correct, the frequency of any one phenotype should oscillate around the equilibrium point. Hori confirmed that this prediction was true by measuring the relative frequencies of the two types over a decade (**FIGURE 6.24**).

Positive frequency-dependent selection can also occur in the context of predation. For example, when rarer forms are more visible to prey, they are more likely to be eaten than the more common forms. This might occur in a school of fish if individuals of a rare color are more visible to a predator, effectively canceling out any benefit of the dilution effect. Similarly, in systems with aposematism, individuals without warning coloration are more likely to be killed and eaten than are those with bright warning coloration (Sherratt 2001). In both of these cases, the survival of individuals of the rare form should be lower than that of individuals of the more common form.

Not every example of the coexistence of multiple foraging types has been attributed to frequency-dependent selection. There are numerous examples of behavioral types that do not have equal fitnesses. Consider, for example, the ruddy turnstone (*Arenaria interpres*), a small sandpiper with many different ways of finding prey items on beaches, ranging from pushing strands of seaweed to one side, to turning stones over, to probing in mud and sand for little mollusks. Although some individuals specialize in one foraging method while others prefer a different technique, individual turnstones are rarely committed to just one way of finding food (Whitfield 1990). This observation suggests that the differences between the sandpipers are not caused by genetic differences but instead are due to an environmental difference. Philip Whitfield wondered if that environmental factor might be the dominance status of the foragers. Turnstones often hunt for food in small flocks, and the individuals in these groups form a pecking order. The birds at the top can displace subordinates merely by approaching them, thereby keeping them from exploiting the richer portions of their foraging areas. Dominant individuals use their status to monopolize patches of seaweed, which they push about and flip over; subordinate individuals keep their distance and are often forced to probe into the sand or mud instead of feasting on the invertebrates contained in seaweed litter (Whitfield 1990).

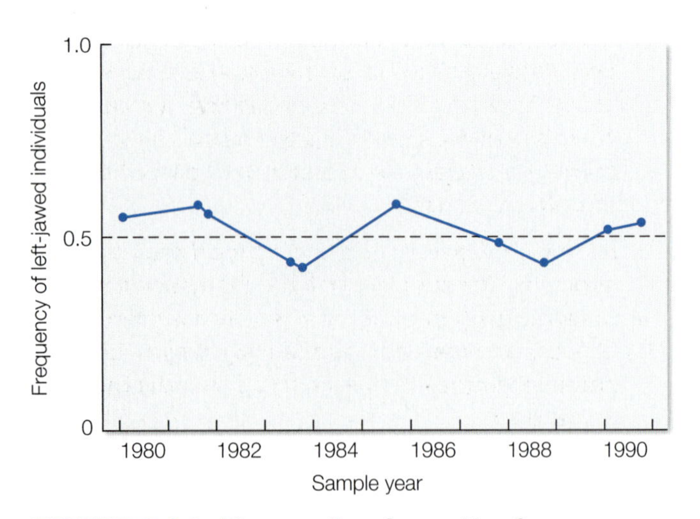

FIGURE 6.24 The results of negative frequency-dependent selection in *Perissodus microlepis*. The proportion of the left-jawed form in the population oscillates from slightly above to slightly below 0.5 because whenever it is more common than the alternative phenotype, it is selected against (and becomes less numerous); whenever it is rarer than the alternative phenotype, it is selected for (and becomes more numerous). (After Hori 1993.)

The turnstones exhibit flexibility in their foraging behavior, as individuals are apparently capable of adopting feeding methods in keeping with their ability to control different sources of food on the beach. The capacity to be flexible is provided by what game theoreticians have labeled a conditional strategy with alternative mating tactics (see Chapter 9). In other words, these alternative mating tactics (or phenotypes) are the results of a single genetically based program or strategy (Dawkins 1980, Eberhard 1982, Gross 1996) that is an inherited mechanism that gives the individual the ability to alter its behavior adaptively in light of the conditions it confronts (such as having to deal with socially dominant competitors on a beach). We will explore these types of tactics in greater detail in Chapter 9 in the context of reproductive behavior. Unlike the left- and right-jawed cichlids, which are locked into one particular behavioral phenotype, turnstones can almost certainly switch from one feeding tactic to another. But subordinate turnstones tend to stick with the probing technique because if they were to challenge stronger rivals for seaweed patches, the low-ranking birds would probably lose, in which case they would have wasted their time and energy for nothing while also running the risk of being injured by an irate dominant. Instead, by knowing their place in the pecking order, these subordinates make the best of a bad situation and presumably secure more food than they would if they tried without success to explore seaweed patches being searched by their superiors. Thus, it can be adaptive to concede the better foraging spots to others if you are a turnstone with more powerful rivals.

SUMMARY

1. Behavioral biologists have several mathematical tools at their disposal to generate testable hypotheses about the behavior of their subjects. These include evolutionary game theory and optimality theory. Game theory views behavioral decisions as a game played by competitors. Here the better strategy (an inherited behavior) for making decisions about how to achieve reproductive success is one that takes into account the competing strategies of other individuals. In contrast, optimality theory attempts to predict the best way (in terms of greater fitness benefits than costs) for an individual to behave, often irrespective of what others are doing. Optimality theory is based on the premise that optimal traits are characteristics with a better benefit-to-cost ratio than any alternative traits that have arisen during the evolution of a species.
2. To be able to survive long enough to reproduce, animals need to avoid predators and find food. Animals can escape predators alone or in groups, and either actively or passively. Active social defense includes mobbing behaviors and group vigilance, whereas passive social defense includes diluting or reducing risk, or confusing predators. Solitary defense mechanisms include camouflage to blend in, or aposematic warning coloration to stand out and announce your distastefulness to most or all predators.
3. Conspicuous coloration may signal to predators that individuals are toxic. However, some species use warning coloration to mimic other species. Batesian mimicry occurs when a nontoxic species mimics a toxic one, whereas Müllerian mimicry occurs when two toxic species mimic each other.
4. Optimal foraging theory makes predictions about how animals maximize fitness while foraging, including the optimization of food type, patch choice, time spent in different patches, or patterns and speed of movement. Energy is the currency used in most optimal foraging models.
5. The assumptions that go into optimal foraging models are critical for assessing how well the data fit the models' predictions. Adjusting the assumptions can often help improve a model's fit. This process of refining the model based on the data available is a standard part of the scientific process.
6. Avoiding predators and finding food are linked, especially for species that are themselves the potential prey of other species. The landscape of fear describes how the fear by prey of being eaten influences their foraging and movement behavior.

COMPANION WEBSITE

Go to **ab11e.com** for discussion questions and all of the audio and video clips.

CHAPTER 7

Territoriality and Migration

Each spring, in temperate-zone forests around the world, something amazing happens: millions of songbirds appear as if out of nowhere. One day the forests are quiet, and seemingly the next, bird song fills the canopy. Most birds arrive after the spring snowmelt to take advantage of the bountiful resources and to breed. In many species, males establish territories and sing to attract mates. Females often build their nests on male territories, and after breeding, all of the birds migrate south, returning to their wintering grounds as the days become shorter and the weather colder. Where do these birds go for the winter? How do they navigate back to the same forests each year, sometimes even to the same territories? How do they choose a territory? These are some of the questions that we address in this chapter.

Chapter 7 opening photograph

In Chapter 6 we discussed the decisions that animals make to increase their chances of surviving by avoiding predators and finding food. Here we introduce a series of other important decisions that animals make, all of which influence their survival and reproduction: where to live and whether to defend a patch of real estate or not, and whether to stay put in a particular location or to leave it, either temporarily or permanently. In particular, we explore the causes and consequences of territoriality, dispersal, and migration. Despite the potential benefits of guarding a territory or dispersing to a new patch of habitat, the decisions about the use of space all come with obvious and substantial costs. Once again, we employ the cost–benefit approach and use evolutionary game theory to examine why individuals make these settlement and movement decisions.

Where to Live

Chapter 6 introduced the concept of the landscape of fear and discussed how perceived predation risk influences where animals go and live. In addition to predation risk, two other critical factors that influence where animals live have to do with finding food (or other resources) and mates. When it comes to finding food, many species will choose habitat based not only where the food is, but on where the competitors for that food are not. Oftentimes, finding food and finding mates are one and the same, as males defend territories with resources to attract mates, and females choose mates that give them access to food or other resources.

Habitat Selection

The rule that certain species live in particular places applies to all kinds of animals, perhaps because in so many cases, the opportunities for successful reproduction by members of any given species are much better in one type of habitat than another. Given the importance of the "right" habitat for successful reproduction, we might expect an animal's preferred habitat to be the one where its breeding success is the greatest. For example, in the blackcap warbler (*Sylvia atricapilla*), a European songbird, males settle first along stream edges in deciduous forests, demonstrating a preference for that particular streamside habitat. Males that settle later occupy mixed coniferous woodlots away from water. But the average number of offspring produced by blackcaps in those two habitats is exactly the same (Weidinger 2000). So why don't warblers in the preferred environment always do better than those that breed in the second-choice habitat?

IDEAL FREE DISTRIBUTIONS One answer to the question above has been provided by Steve Fretwell and his colleagues, who used game theory to develop what they called **ideal free distribution theory** (Fretwell and Lucas 1969). You may recall from Chapter 6 that game theory is a kind of optimality approach in which behavioral decisions are treated as if the individuals involved in the "game" are attempting to maximize their reproductive success (or proxies for that success, such as the number of mates secured or the amount of food ingested). Ideal free distribution theory enables behavioral biologists to predict what animals should do when choosing between alternative habitats of different quality in the face of competition for space, food, or other critical resources. Fretwell and Lucas demonstrated mathematically that if individuals were free to distribute themselves spatially in relation to resource quality and the intensity of competition from conspecifics, then as the density of resource consumers in the superior habitat increased, there would come a point at which an individual could gain higher fitness by settling for a lower-ranked habitat that had fewer conspecific occupants (Fretwell and Lucas 1969). In line with this approach, Karel Weidinger found that the density of nesting blackcaps was four times higher in the preferred stream-edge habitat than in the second-ranked habitat away from streams (Weidinger 2000). Thus, these birds must make habitat selection decisions based not just on the nature of the vegetation and other indices of insect productivity, but also on the intensity of intraspecific competition as reflected by the density of nesting birds in a location (a process referred to as **density-dependent habitat selection**).

Ideal free distribution theory requires that individuals move about to evaluate the quality of different habitats. Implicit in ideal free distribution theory are the assumptions that individuals are free to move among patches that vary in quality and that they can assess the value of patches. Moreover, as the number of individuals in a patch increases, the quality of the patch is assumed to decline. The theory's key prediction is that individuals will settle on the sites where their reproductive success is maximized. Gordon Orians used this approach in a study of red-winged blackbirds (*Agelaius phoeniceus*), a polygynous species in which a male can attract up to a dozen females to settle on his **territory**, an area where the owner has sole or prioritized access to resources such as food or shelter (Orians 1969). Since female red-winged blackbirds are attracted to territories that have safer nesting sites, Orians predicted that female red-winged blackbirds would choose territories of lower quality when better territories with safer nesting sites were densely occupied by other females. In so doing, a new resident in a low-quality territory would have a better nest site or access to more food or more male parental care than she would have secured if she had chosen to live elsewhere with a larger number of competitors. This is exactly what Orians found.

The same theory has been used to determine whether wintering red knots (*Calidris canutus*), a medium-sized shorebird, move about among seven large tidal areas in the United Kingdom, the Netherlands, and France in ways that equalize the amount of food eaten per individual (Quaintenne et al. 2011). Red knots feed on a small mollusk that occurs in varying densities in the mud (**VIDEO 7.1**). By combining data on the numbers of knots present in these seven wintering sites with information on the density of mollusks over 5 years, a research team was able to show that the birds were not distributed uniformly over the seven areas. Instead, as predicted by ideal free distribution theory, individual birds achieved equal food intakes by shifting from site to site depending on the availability of mollusks and the density of their fellow red knots in the area (**FIGURE 7.1**). Thus, over both large and small spatial scales, animals appear to minimize resource competition in ways that maximize their probable fitness.

VIDEO 7.1

Foraging red knots
ab11e.com/v7.1

RESOURCE SELECTION FUNCTIONS Ideal free distribution theory emphasizes the habitat that animals should choose to optimize their reproductive success. This theoretical approach can predict how animals should distribute themselves among patches. However, to link patch quality and the resources within a patch using empirical data, other approaches are needed. One such approach is the use of **resource selection** functions, which are any functions that are proportional to the probability of use of a resource unit (Manley et al. 2004). If one knows the distribution of resources on which an animal relies, then the distribution of that animal can be characterized by resource selection functions. For example, survey or remote sensing data can be compiled into a geographic information system (GIS) to empirically estimate the probabilities of occurrence for different individuals or populations (Boyce and McDonald 1999). This approach has been used to make predictions about how population density is influenced by ecology, making it an important tool for conservation managers as well as behavioral biologists (McLoughlin et al. 2010).

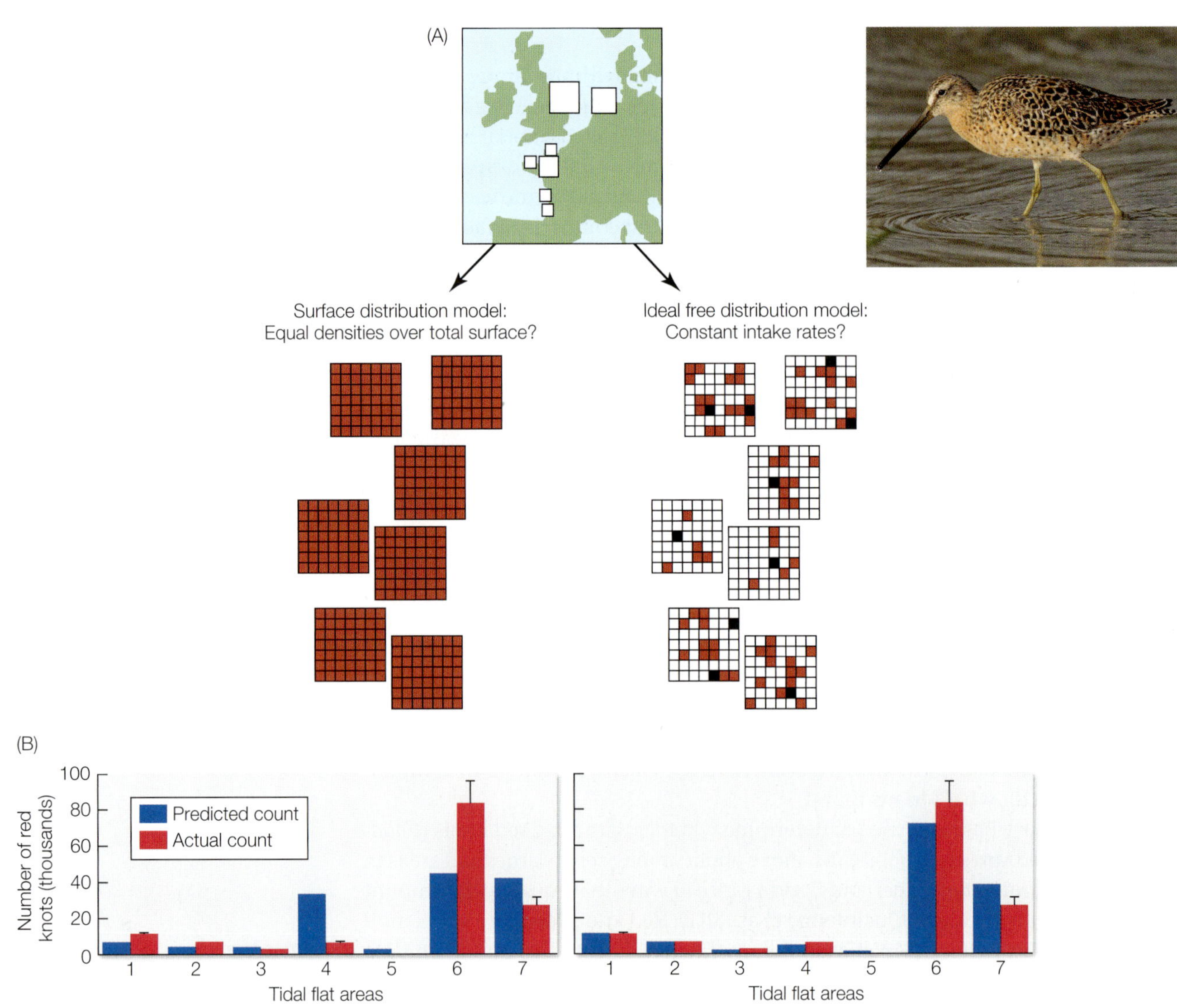

(B)

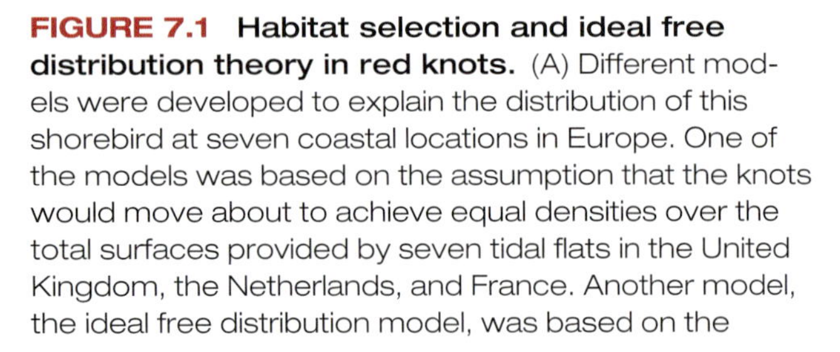

FIGURE 7.1 Habitat selection and ideal free distribution theory in red knots. (A) Different models were developed to explain the distribution of this shorebird at seven coastal locations in Europe. One of the models was based on the assumption that the knots would move about to achieve equal densities over the total surfaces provided by seven tidal flats in the United Kingdom, the Netherlands, and France. Another model, the ideal free distribution model, was based on the assumption that individual knots were free to move about so as to achieve a constant intake of prey from the seven tidal flats where mollusks ranged in local abundance from low (white squares) to moderate (red squares) to high (black squares). (B) The ideal free distribution model made predictions that more accurately matched the actual numbers of knots counted at the seven locations. Bars depict mean +/– SE. (After Quaintenne et al. 2011; photograph © visceralimage/Shutterstock.)

Territoriality and Resource-Holding Potential

In some cases, such as the red knot example, individuals can distribute themselves freely across their environment. In red-winged blackbirds, females distribute themselves freely across male territories to maximize fitness, but what about the males? Free movement seems unlikely to maximize fitness for territorial males that attempt to monopolize high-quality sites, forcing losers to look elsewhere for breeding sites. Thus, while female red-winged blackbirds may distribute themselves freely to avoid competition, males should compete for the best territories in order to attract the most females (see Chapters 9 and 10).

Although high-quality territories are often limited, there are many potential benefits to be gained if a male succeeds in securing one. However, defending territories also has costs, including energy expenditure and the risk of injury. The term **economic**

defensibility describes the trade-off in costs versus benefits for maintaining a territory, and individuals are only predicted to defend territories when the benefits outweigh the costs (Brown 1964). One of the primary benefits of territorial behavior is the ability to use the resources on a territory without interference from others. For example, young male collared lizards (*Crotaphytus collaris*) that acquire a territory on the death of an older territory holder have significantly more opportunities to court females than do those males unable to secure a territory (Baird and Curtis 2010).

The potential benefits of possessing a territory can also be seen in our nearest living relative, the chimpanzee (*Pan troglodytes*). Males patrol the boundaries of their territorial community living space in African forests. If a group on patrol encounters a smaller number of chimps from another territory during one of these forays, they will attack and even kill their neighbors. Male chimpanzees from one such band killed 21 fellow chimpanzees over a 10-year period, depleting the size of one neighboring group so much that the aggressors were eventually able to expand their territory by more than 20 percent, giving them access to food resources once controlled by their neighbors (**FIGURE 7.2**) (Mitani et al. 2010). This example illustrates the many types of competition that lead to territory conflict, which may not always be about taking over an entire territory from the previous owners, but may also occur as

(A)

(B)

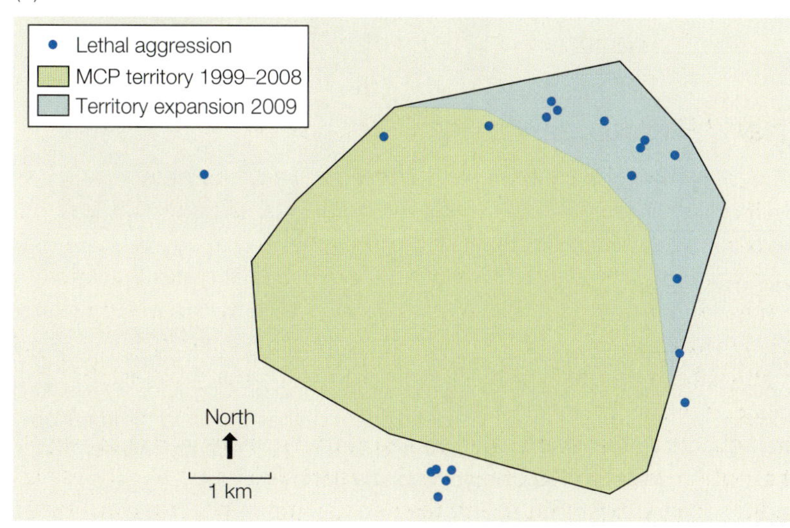

FIGURE 7.2 Male chimpanzees patrolling their territory. (A) This group of males will march along the boundaries of their band's territory (in Uganda, Africa), where they will attack and even kill the members of other groups. (B) The effect of this behavior, if successful, is the expansion of the attacking band's territory, labeled MCP by the research team. (A, photograph by John C. Mitani; B after Mitani et al. 2010.)

individuals attempt to change the shape of the territory or to intrude on it (Hinsch and Komdeur 2017).

To be adaptive, the fitness benefits of having a territory (such as having more offspring) have to be weighed against the very real costs of the aggression and patrolling required to establish and maintain the territory. These costs include the time and energy expended by males that also risked injury and even death when violently attacking their neighbors. In several species, males reduce or even completely forego foraging during the breeding season when they are guarding territories throughout the day. Thus, the costs of trying to control a territory are substantial, a point that Catherine Marler and Michael Moore made experimentally in their studies of Yarrow's spiny lizard (*Sceloporus jarrovii*) (Marler and Moore 1989, 1991). Marler and Moore made some nonterritorial male lizards territorial by inserting small capsules containing testosterone beneath the skin of males they captured in June and July, a time of year when the reptiles are typically only weakly territorial. These experimental animals were then released back into a rock pile high on Mount Graham in southern Arizona. The testosterone-implanted males patrolled more, performed more push-up threat displays, and expended almost one-third more energy than control lizards that had been captured at the same time

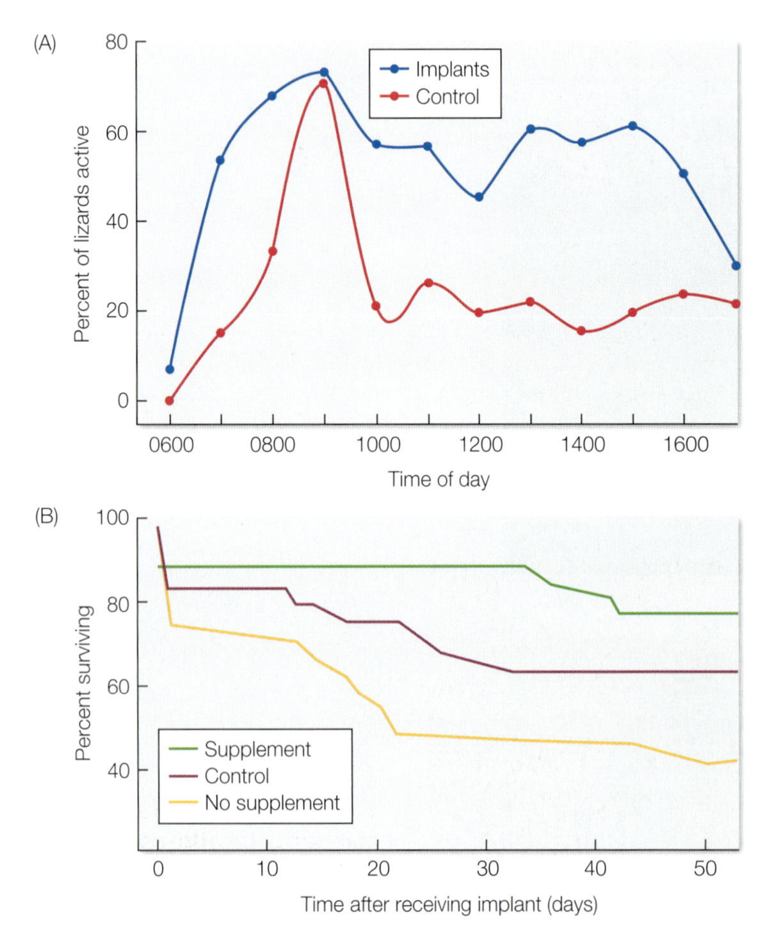

Yarrow's spiny lizard

FIGURE 7.3 Energetic costs of territoriality. (A) Male Yarrow's spiny lizards that received an experimental implant of testosterone spent much more time moving about than did control males. (B) Testosterone-implanted males that did not receive a food supplement disappeared at a faster rate than control males. Since testosterone-implanted males that received a food supplement (mealworms) survived as well as or better than controls, the high mortality experienced by the unfed group probably resulted from the high energetic costs of their induced territorial behavior. (A after Marler and Moore 1989; B after Marler and Moore 1991; photograph by John Alcock.)

and given a chemically inert implant. As a result, the hyperterritorial males used up their energy reserves and died sooner than males with normal concentrations of testosterone (**FIGURE 7.3**).

Because territoriality is costly, we can predict that peaceful coexistence on an undefended living space or home range should evolve when the benefits of owning a resource do not outweigh the costs of monopolizing it. In Yarrow's spiny lizards, for example, males do live together in relative harmony in June and July, when females are not sexually receptive, but when fall comes and females are ready to mate, the males become highly aggressive. This connection between territoriality and breeding season is seen in many other male animals that compete for mates. For example, the life cycle of the tiny tropical pseudoscorpion *Cordylochernes scorpioides* involves periods of colonization of recently dead or dying trees, followed in a few generations by dispersal to fresh sites from the now older trees. Dispersing pseudoscorpions make the trip under the wing covers of a huge harlequin beetle (*Acrocinus longimanus*) (**FIGURE 7.4**), disembarking when the beetle touches down

(A)

(B)

FIGURE 7.4 Dispersal by territorial beetle-riding pseudoscorpions. (A) Two male harlequin beetles battle for position on a tree trunk. (B) Individuals of the pseudoscorpion *Cordylochernes scorpioides* compete for space on a beetle's back before it flies from one tree to another. (A, photograph by Jeanne A. Zeh; B, photograph © Piotr Naskrecki/ Minden Pictures.)

on an appropriate tree. The pseudoscorpions mate both on trees and on the backs of beetles, but much-larger-than-average male pseudoscorpions attempt to control real estate only when they occupy a beetle's inch-long abdomen, which can be defended in its entirety at less expense than a larger swath of tree trunk (Zeh 1997). A male riding on a harlequin beetle may be joined by a harem of receptive females traveling with him to their new destination, much to the male's reproductive benefit (Zeh and Zeh 1992).

The cost–benefit approach to territoriality also predicts that if individuals vary in their territory-holding abilities or their **resource-holding potential** (the ability of a territorial animal to win a fight), then those with superior competitive ability should be found in the highest-quality habitat. A case in point is the American redstart (*Setophaga ruticilla*), a warbler that breeds in North America and that competes for territories during the nonbreeding season on its wintering grounds in the Caribbean and Central America. In Jamaica, males tend to occupy black mangrove forests along the coast, while females are more often relegated to second-growth scrub inland, in a form of sexual segregation. Field observations reveal that older, heavier males in mangrove habitat attack intruding females and younger males, apparently forcing these birds into second-rate habitats (Marra and Holmes 2001). If older, dominant males derive a net benefit from their investment in territorial aggression, then there should be survival (and eventually reproductive) advantages for individuals that succeed in occupying the favored habitat. In fact, redstarts living in mangroves retained their weight over the winter, whereas those in the apparently inferior scrub habitat generally lost weight (Marra and Holmes 2001).

These differences in the quality of the wintering habitat not only influence American redstart body mass and condition, they also influence the structure of the genome. Telomeres, which are repetitive DNA sequences at the end of each chromosome that protect the chromosome during cell replication and play a role in cellular aging, differ in length between redstarts wintering in the two types of habitats. From one year to the next, telomeres shorten more in birds wintering in the low-quality habitat than they do in birds wintering in the high-quality habitat. Moreover, birds with longer telomeres tend to have greater return rates each winter, suggesting a relationship between survival, telomere length, and winter habitat quality (Angelier et al. 2013). Since telomere length is related to aging in many species, including humans, this means that the proximate cause of aging may be related to habitat or resource quality in other animals as well.

This competition for high-quality wintering territories and its effect on American redstart survival, body condition, and telomere length has carryover effects from the nonbreeding season to the breeding season and influences reproductive success. Probably because they have more energy reserves, male territory holders in mangroves leave their wintering grounds for their distant breeding sites far to the north sooner than birds living in second-growth scrub (**FIGURE 7.5**). But do the birds that leave the wintering grounds earlier also arrive earlier on the breeding grounds? In many migrant songbird species, including the American redstart, early male arrivals secure a reproductive advantage by claiming the best territories and gaining quick access to females when they arrive (Hasselquist 1998). Female redstarts also have something to gain by reaching their breeding grounds early (and in good condition), because they will have only a few months to raise their offspring before turning around and migrating south. Indeed, early birds produce the most fledglings in this species (**FIGURE 7.6**) (Norris et al. 2004).

Peter Marra and colleagues were able to show that American redstarts that depart the wintering grounds earlier do indeed arrive first on the breeding grounds (Marra et al. 1998). They made their discovery not by following individual birds from their Caribbean wintering grounds to their North American breeding grounds in the

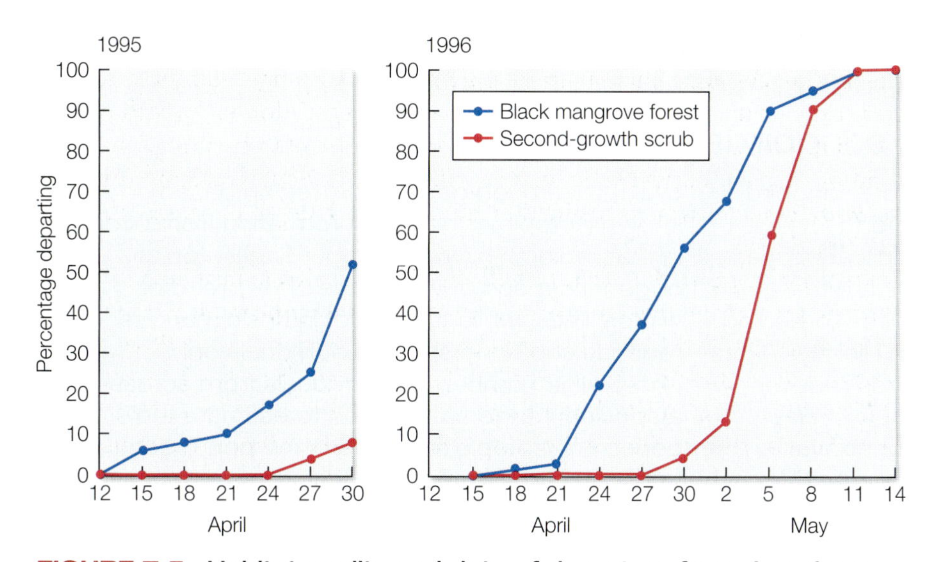

Female American redstart

Male American redstart

FIGURE 7.5 Habitat quality and date of departure from Jamaican wintering grounds by American redstarts. Birds occupying the preferred black mangrove habitat leave sooner for the northern breeding grounds than do birds forced into second-rate, second-growth scrub. (After Marra et al. 1998; top photograph © Paul Reeves Photography/Shutterstock.com; bottom photograph © Rick & Nora Bowers/Alamy.)

forests of New Hampshire, but by examining stable carbon isotopes in blood and muscle of departing and arriving birds (**BOX 7.1**). Since carbon isotopic signatures in the blood of birds captured on arrival in New Hampshire reflect the diet and habitat occupied on the wintering grounds over the previous 6 to 8 weeks, the researchers were able to link the first arriving birds in New Hampshire to those that wintered in the high-quality mangrove habitats, simply through the habitat-specific isotopic signatures in their blood.

During the breeding season, the benefits of territoriality can include not only access to resources but also access to mates. In the arctic ground squirrel (*Urocitellus parryii*), for example, males compete with one another for control of patches of meadows in the Canadian Arctic. Female ground squirrels also live in these meadows, and at the start of the breeding season they mate with one or several males. Given that a territorial male does not necessarily monopolize sexual access to the females living within his plot of grassy meadow, it seems odd that he invests considerable time and energy keeping other males away from his home ground. But Eileen Lacey and John Wieczorek knew that even though a female arctic ground squirrel regularly mates with several males, the male that is first to copulate with her is almost always the only one to fertilize her eggs (Lacey and Wieczorek 2001). With this knowledge, they predicted that territorial males would be more likely to mate first with females on their territories, and as predicted, in 20 of 28 cases, a sexually

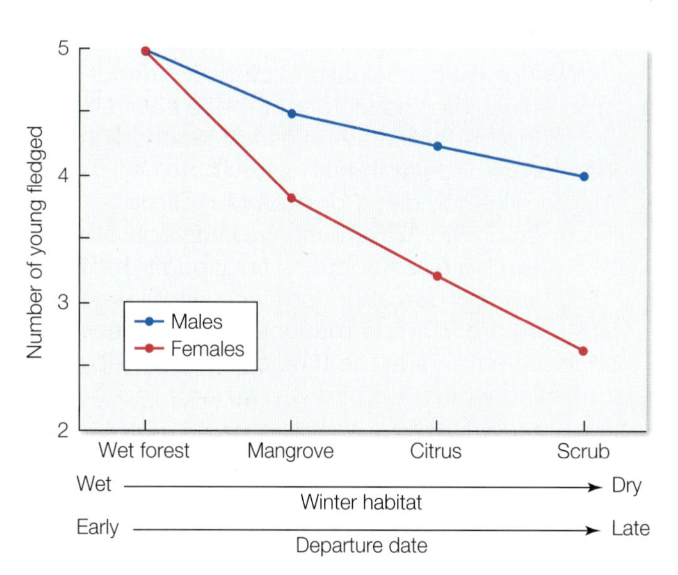

FIGURE 7.6 American redstart winter territoriality affects reproductive success. The estimated number of fledglings produced by American redstarts in relation to where they spent the winter. Birds that wintered in the most productive habitat (wet forest) had the highest reproductive success. (After Norris et al. 2004.)

BOX 7.1 INTEGRATIVE APPROACHES

How to track migratory songbirds

Over the past few decades, migratory songbird populations have declined across North America and other parts of the temperate zone. Determining whether these declines are the results of events occurring on the birds' breeding or wintering grounds is essential for their conservation. Tracking migratory birds requires following them throughout their annual cycle, during which, for example, they might travel south thousands of miles from their breeding grounds in the Northern Hemisphere. **Migratory connectivity** describes the movement of individuals between summer and winter populations, including the stopover sites between the breeding and wintering grounds (Webster et al. 2002). A variety of new technologies has made studying migratory connectivity easier. These technologies can be divided into two types, those that use markers that are intrinsic to the birds to indirectly track movements, and those that require attaching extrinsic markers to the birds to more directly track movements.

Intrinsic markers that indirectly track movement include stable isotopes and genetic markers. Stable isotopes are naturally occurring forms of elements that animals incorporate into their tissues through the food they eat or the water they drink. Since stable isotopes vary naturally and predictably in the environment due to a variety of biological and biogeochemical processes, information about these naturally occurring patterns can be used to identify the location of birds at the time they obtained these markers in their diet. For example, carbon isotopes provide information about vegetation, hydrogen isotopes about water, and strontium and lead isotopes about soil type. Stable isotopes are fixed permanently into inert tissues such as feathers at the time of molt, thereby indicating where the bird was when it molted. Additionally, stable isotopes are incorporated into other tissues such as blood, muscle, and liver, where they turn over (or are replaced) at varying timescales and provide information about migration over shorter time frames. In contrast, genetic markers rely largely on genetic variation among avian populations. Both stable isotopes and genetic markers involve assigning birds of unknown origin (for example, captured during migration or at particular stages of the annual cycle) to a rough geographic region based on known isotopic or genetic information from birds in that region. This requires building isotopic or genetic maps of where the birds breed or winter. Thus, while the use of intrinsic markers requires capturing specific individuals only once, large numbers of individuals from across a species' range are needed to estimate the genetic or isotopic structure in the population.

In contrast to intrinsic markers, many extrinsic markers require recapturing the same individual at least twice. The first type of extrinsic marker to directly track movement was a uniquely numbered metal ring that could be attached primarily to the leg of a bird. Although ringing large numbers of birds allowed researchers to recapture individuals that returned to the same breeding or wintering sites annually, establishing migratory connectivity by recapturing a bird banded during the breeding season on its wintering grounds was like finding a needle in a haystack. For example, of more than 1 million pied flycatchers (*Ficedula hypoleuca*) ringed in Europe, only six were recaptured on their African wintering grounds (Webster et al. 2002). Instead, extrinsic markers that record information throughout the migration are necessary. Light-level geolocators, for example, are used to estimate location based on seasonal and locational differences in the timing of day and night. Geolocators record the amount of ambient light using a light sensor at a regular interval throughout the life of the unit at specific time points. Once the geolocator is retrieved from an individual bird, the data can be analyzed to recreate when and where the bird traveled.

Unlike geolocators, which require recapturing birds, satellite tracking of birds requires attaching global positioning system (GPS) tags and then simply retrieving information remotely (**FIGURE**). Historically, battery constraints have made GPS tags too large and heavy for most migratory songbirds, though that is changing as new technologies are making batteries smaller and lighter (Wikelski et al. 2007, Kays et al. 2015). Ultimately, these and other new technologies will allow scientists to better define the migratory patterns of songbirds, which continue to decline across the world due to habitat loss on both their breeding and wintering grounds.

INTEGRATIVE APPROACHES *(continued)*

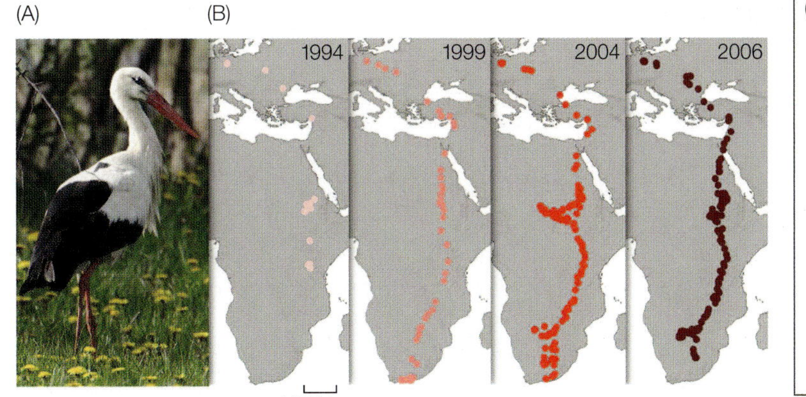

GPS tracking of migratory storks. GPS tags have long provided movement data on larger species of migratory birds. (A) For example, “Princess,” a white stork (*Ciconia ciconia*), was tagged with a GPS tracking device as a 3-year-old in Germany in 1994 and was tracked until her death in 2006. (B) Four generations of satellite tags were used to track her, and she had to be recaptured and retagged multiple times. (C) More recent tracking efforts have documented the migratory behavior of 11 different populations of white storks across their ranges, discovering unexpected stationary populations in Uzbekistan and new migratory behaviors in Tunisian storks, which cross the Sahara multiple times per year. (From Kays et al. 2015.)

Thinking Outside the Box

Consider the costs and benefits associated with the different approaches to tracking migratory songbirds, which are quite small and can travel vast distances. When would it be more appropriate to use one approach or type of technology over another? Do you think you could pair multiple approaches to get more information and do so in a more cost-effective manner?

receptive female was mated first by the owner of the patch of ground where the female resided. Thus, the male ground squirrels gained a fertilization payoff for their investment in territorial behavior.

Why Give Up Quickly When Fighting for a Territory?

Studies of territoriality in numerous species have found that winners in the competition for territories appear to gain reproductive success as a result. Therefore, it seems paradoxical that in many species, when an intruder challenges a territory holder, the intruder usually gives up quickly—often within seconds—rather than fighting intensely to take over the habitat patch. Why do challengers concede defeat so quickly if there are large fitness benefits to be gained by being territorial? Below we address a series of hypotheses to explain this observation (**HYPOTHESES 7.1**).

RESIDENTS ALWAYS WIN To answer the question of why challengers concede defeat so quickly, let’s first take a game theoretical approach based on a logical demonstration that a rule for resolving conflicts between residents and intruders could be an evolutionarily

HYPOTHESES 7.1

Non-mutually exclusive hypotheses to explain why challengers concede defeat so quickly if there are large fitness benefits to be gained by being territorial

arbitrary contest resolution hypothesis The resident always wins territorial battles.

resource-holding potential hypothesis Residents have an edge in physical combat.

payoff asymmetry hypothesis Residents place a higher value on the territory than do rivals.

arbitrary contest resolution hypothesis The resident always wins territorial battles.

VIDEO 7.2 Male speckled wood butterflies fighting **ab11e.com/v7.2**

stable strategy (ESS), which as you will recall from Box 6.1, is one that, when adopted by a certain proportion of the population, cannot be replaced by an alternative strategy over evolutionary time. One simple and arbitrary rule is that the resident always wins. According to the **arbitrary contest resolution hypothesis**, if all competitors for territories were to adopt the resident-always-wins rule so that intruders always gave up and residents never did, a mutant with a different behavioral strategy would not spread through the population by natural selection. For example, a mutant that always challenged a resident would sometimes pick on a better fighter and be injured during the contest. Intruders that immediately gave up would never be damaged by a tougher resident. Therefore, the resident-always-wins strategy could persist indefinitely (Maynard Smith and Parker 1976).

When the resident-always-wins version of the arbitrary conflict resolution rule was first introduced, it was thought that males of the speckled wood butterfly (*Pararge aegeria*) provided a supportive example. Territorial male butterflies defend small patches of sunlight on the forest floor, where they encounter receptive females (**VIDEO 7.2**). Males that successfully occupy sun-spot territories mate more frequently than those that do not, judging from an experiment in which males and females were released into a large enclosure. Under these conditions, sun-spot males secured nearly twice as many matings as their nonterritorial rivals (Bergman et al. 2007). Based on these results, a logical prediction might be that there would be stiff competition for sun spots. However, Nick Davies had previously found that territorial males always quickly defeated intruders, which invariably departed rather than engaging in lengthy territorial conflicts (Davies 1978).

The butterflies were capable of prolonged fights, as Davies showed by capturing and holding territorial males in insect nets until new males had arrived and claimed the seemingly empty sun-spot territories. When a prior resident was released, he returned to "his" territory, only to find it taken by a new male. This new male, having claimed the site, reacted as if he were the resident, with the result that the two males had what passes for a fight in the butterfly world. The combatants circled one another, flying upward, occasionally clashing wings, before diving back to the territory, sometimes repeating this maneuver. Eventually the previous resident gave up and flew away, leaving the site under the control of the replacement resident, evidence that the "resident always wins" (**FIGURE 7.7**) (Davies 1978).

Although these experimental results were consistent with the hypothesis that male butterflies use an arbitrary rule to decide winners of territorial contests, Darrell Kemp and Christer Wiklund were unconvinced. They repeated the experiment but without subjecting the captured initial resident to the potentially traumatic effects of being held in an insect net before release. Instead, they put their captives in a cooler, retrieving them after replacements had taken up residence in their sun-spot territories for about 15 minutes. After placing the original resident on the ground near his old territory, the researchers steered the new male toward his rival by tossing a wood chip above him in order to provide a visual stimulus that captured his attention. When the perched replacement male saw the incoming initial resident, he reacted in the manner of a territorial speckled wood butterfly by engaging his opponent in a spiral flight. But these flights lasted much longer than the ones that Davies had observed, and their outcomes were strikingly different, with the original sun-spot holders winning 50 of 52 contests with new residents (Kemp and Wiklund 2003). This time, the resident-always-wins rule did not hold.

Why did the results of the two experiments differ so greatly and lead to completely different conclusions? Kemp and Wiklund hypothesized that in Davies's experiment, the original resident males that were held in nets until their release just wanted to escape rather than fight for their old territories, hence the relatively short interactions that led to victory by the new residents. Their study shows just how

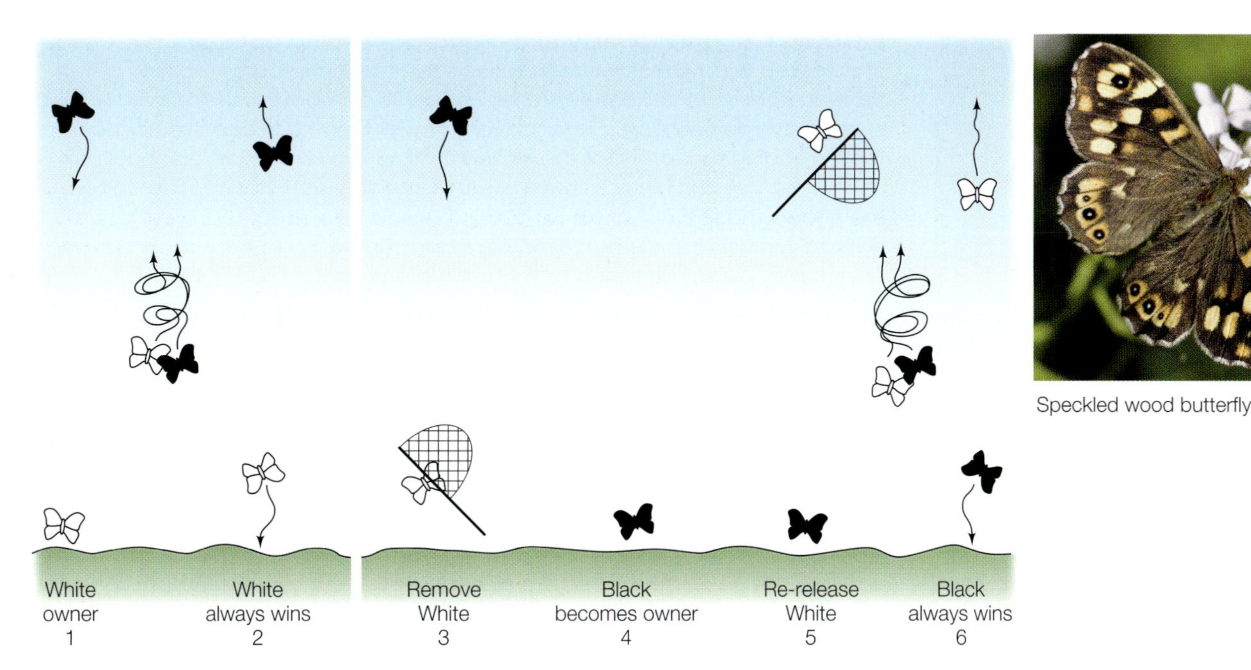

Speckled wood butterfly

FIGURE 7.7 Does the resident territory owner always win? An experimental test (now known to be flawed) of the hypothesis that territorial residents always win conflicts with intruders was performed with males of the speckled wood butterfly. When one male ("White") is the resident, he always defeats intruders (1, 2). But when White is temporarily removed (3), permitting a new male ("Black") to settle on his sun-spot territory (4), then Black always defeats White upon his return after release from captivity (5, 6). (After Davies 1978, drawing by Tim Halliday; photograph © Christian Musat/ShutterStock.)

important experimental design can be. Thanks to Kemp and Wiklund's replication of the original experiment with a slight design tweak, we now know that the current resident speckled wood butterfly does not always win, a result that not only eliminates the arbitrary contest resolution hypothesis for this species, but demonstrates the scientific value of revisiting the research of others.

DIFFERENCES IN RESOURCE-HOLDING POTENTIAL Although the arbitrary contest resolution hypothesis was mathematically viable, work on the speckled wood butterfly and other species has convinced behavioral researchers that territorial disputes in the animal kingdom are rarely, if ever, decided on arbitrary grounds. Instead, a more popular hypothesis is one in which winners are said to be decided on the basis of having an edge in physical combat, an idea referred to as the **resource-holding potential hypothesis**. John Maynard Smith and Geoffrey Parker modeled this type of interaction in the classic Hawk–Dove game (see Box 6.1) (Maynard Smith and Parker 1976). As you learned in Chapter 6, there are two strategies in this game: that of the hawk, which is aggressive and fights until it wins or loses, and that of the dove, which is less aggressive and gives up in the face of a superior aggressor (a hawk) or tries to share the resources when it meets another dove. As in any game theory model, the fitness payoff for each of these strategies depends on the frequency of hawks and doves in the population. Maynard Smith and Parker demonstrated that when the cost of losing is greater than the benefit of winning, both strategies can coexist in the population (a mixed ESS), though the more aggressive hawks will occur in greater frequency. In species ranging from rhinoceroses (Rachlow et al. 1998) to fiddler crabs (Bergman et al. 2007) to wasps (O'Neill 1983), territory holders are relatively large and aggressive

resource-holding potential hypothesis Residents have an edge in physical combat.

Male red-shouldered widowbird

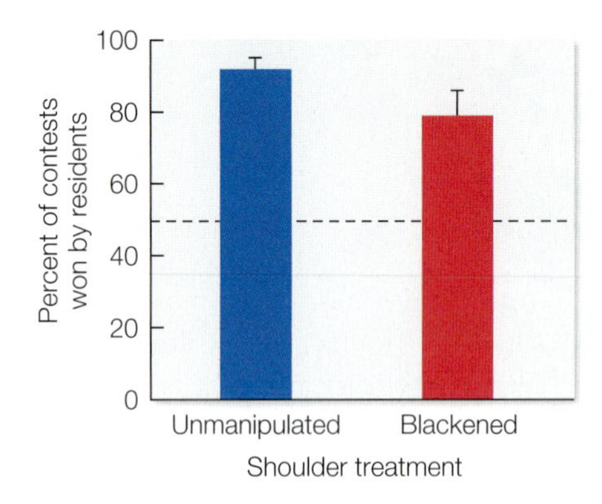

FIGURE 7.8 Territoriality and resource-holding potential. Male red-shouldered widowbirds with territories have higher resource-holding potential than nonterritorial floaters. When residents compete with floaters for food in captivity, residents usually win, even though they have been removed from their breeding territories for the experiment, and even if their red shoulders have been painted black to eliminate this signal of dominance. The dotted line shows the result that would occur if residents and floaters were equally likely to win dominance contests. Bars depict mean +/– SE. (After Pryke and Andersson 2003; photograph by Francesco Veronesi/CC BY-SA 2.0.)

individuals, apparently stronger than their smaller rivals, and therefore capable of securing and defending the territorial resource against weaker opponents (see Chapters 9 and 10).

Being larger is not the key to territorial success in every species, however. For example, in the red-shouldered widowbird (*Euplectes axillaris*), an African bird that looks remarkably like the North American red-winged blackbird, males with bigger and redder shoulder patches are more likely to hold territories than are rivals with smaller, duller shoulders. The less gaudy males become nonterritorial "floaters," hanging around the territories of other males, conceding defeat whenever challenged but ready to assume control of vacant territories should a resident disappear. Territorial males and floaters do not differ in size, but even so, when males in the two categories were captured and paired off in a cage unfamiliar to either individual, the ex–territory holder almost always dominated the floater, even when the resident's red shoulders had been painted black (**FIGURE 7.8**) (Pryke and Andersson 2003). These results suggest that some intrinsic feature of male quality other than body weight is advertised by the size and color of the male shoulder patches and is expressed even in the absence of this signal (see Chapter 8), which enables these males to win fights even without their red shoulders.

Just as is true for red-shouldered widowbirds, males of some damselfly species can be divided into territory holders and floaters. Although territorial male damselflies (*Calopteryx* spp.) are not necessarily larger or more muscular than nonterritorial males, those that win the lengthy back-and-forth aerial contests that occur over streamside mating territories almost always have a higher fat content than the males they defeat (**FIGURE 7.9**) (Marden and Waage 1990, Plaistow and Siva-Jothy 1996). In these insects, contests do not involve physical contact but instead consist of what has been called a war of attrition, with the winner outlasting the other, and the ability to continue flying is related to the individual's energy reserves (Maynard Smith 1974).

Differences in resource-holding potential, whether based on size or on energy reserves, do not account for every case in which territory owners have a huge territorial advantage. For example, older female Mediterranean fruit flies (*Ceratitis capitata*), which are almost certainly not as physically fit as younger flies, can still keep younger females away from the fruits in which the insects lay their eggs (Papaj and Messing 1998). By the same token, older male song sparrows (*Melospiza melodia*) respond more intensely to simulated intrusion of their territories via song playback than do younger males (Hyman et al. 2004), and some older male butterflies are more willing to persist in their reproductive attempts than their younger rivals are (**FIGURE 7.10**) (Kemp 2002, Fischer et al. 2008). Findings of this sort do not support the hypothesis that territorial winners, at least in these species, are in better condition or are more physically imposing than losers. So what is going on in these species?

FIGURE 7.9 Fat reserves and resource-holding potential. Fat reserves determine the winner of territorial conflicts in black-winged damselflies (*Calopteryx maculata*). In this species, males sometimes engage in prolonged aerial chases that determine ownership of mating territories on streams. (A) Larger males, as measured by dry thorax weight, do not exhibit a consistent advantage in territorial takeovers. (B) However, males with greater fat content almost always win. (A and B after Marden and Waage 1990; photograph by John Alcock.)

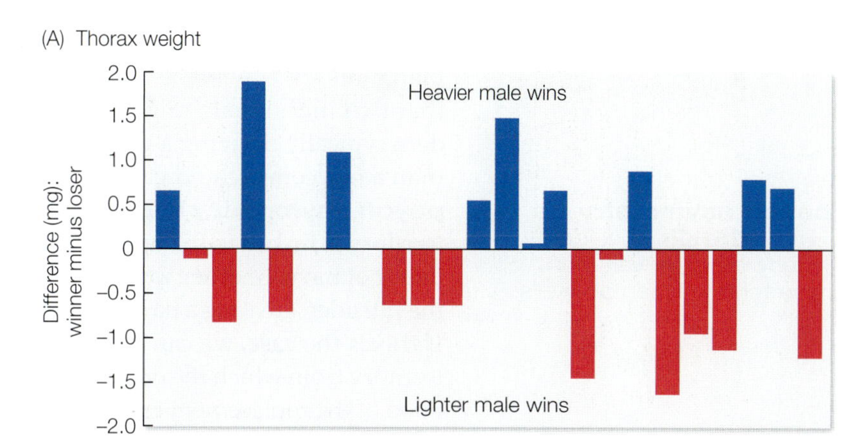

Male black-winged damselfly

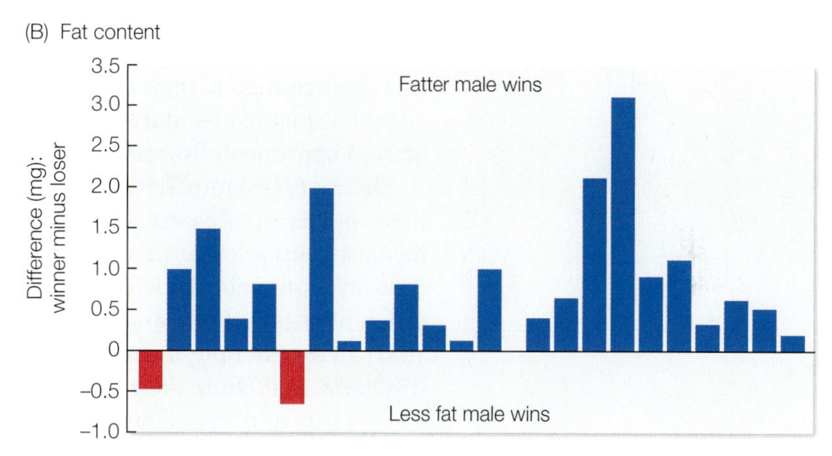

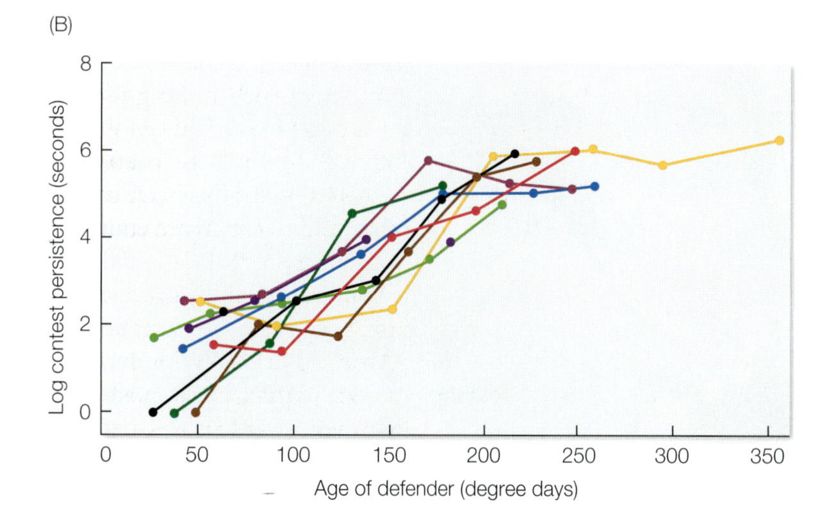

FIGURE 7.10 Motivation affects territorial success. Older males fight harder to control territories in the eggfly butterfly (*Hypolimnas bolina*). (A) An old male missing large chunks of his hindwings but still defending his perching territory in Queensland, Australia. (B) Results of a series of encounters in the laboratory between a hand-reared male of known age and a fresh young rival placed in his cage. In the ensuing aerial clashes, the resident persisted longer as he grew older. Each color represents a different individual resident. (A, photograph courtesy of Darrell J. Kemp; B after Kemp 2002.)

RESOURCE VALUE AND PAYOFF ASYMMETRIES A different kind of explanation for why some males win and others lose derives from a realization that two individuals might value the same resource differently, with the resident typically deriving a greater payoff from maintaining control of his territory than a newcomer could gain by taking the property from its current owner. This **payoff asymmetry hypothesis** suggests that the value a resident places on his territory is linked to the male's familiarity with a location, something an intruder does not have. In other words, the territory holder values the territory more than the intruder, creating a payoff asymmetry if the two were to fight for that territory. If this is the case, we can predict that when a newcomer is permitted to claim a territory from which the original resident has been temporarily removed, the likelihood of the replacement resident winning a fight against the original resident will be a function of how long the replacement has occupied the site. This experiment has been done with birds, such as the red-winged blackbird, as well as with some insects and fishes. In red-winged blackbirds, when captive ex–territory holders are released from an aviary and allowed to return to their home marsh to compete again for their old territories, they are more likely to fail if the new males have been on the experimentally vacated territories for some time (Beletsky and Orians 1989).

payoff asymmetry hypothesis Residents place a higher value on the territory than do rivals.

The payoff asymmetry hypothesis also predicts that contests between an ex-resident and his replacement will become more intense as the tenure of the replacement increases, since longer tenure boosts the value of a site to the current holder and thus his motivation to defend it. This prediction has been supported in animals as different as tarantula hawk wasps (Alcock and Bailey 1997) and songbirds (Krebs 1982). If, for example, one removes a male *Hemipepsis ustulata* tarantula hawk wasp (**FIGURE 7.11**) from the peak-top shrub or small tree that he is defending and pops him into a cooler, his vacant territory will often be claimed within a few minutes. If the ex–territory holder is quickly released, he usually returns promptly to his old site and chases the newcomer away in less than 3 minutes. But if the ex–territory holder is left in the cooler for an hour, then when he is warmed up and released to hurry back to his territory, a battle ensues. The newcomer resists eviction, and the two males engage in a long series of ascending flights in which they climb rapidly up into the sky side by side for many meters before diving back down to the territory, only to repeat the activity again and again until finally one male—usually the replacement wasp—gives up and flies away. The mean duration of these contests is about 25 minutes, and some go on for nearly an hour (Alcock and Bailey 1997).

FIGURE 7.11 Residency provides an advantage to a territorial wasp. A male *Hemipepsis ustulata* tarantula hawk wasp perched on his territory, a large creosote bush, where he waits for arriving females. Intruder males almost never oust the resident. (Photograph by John Alcock.)

Although examples of this sort support the payoff asymmetry hypothesis and explain why residents usually win a territorial battle, it is possible that lengthier contests occur after replacements have been on territories for some time because the removed residents have lost some resource-holding potential while in captivity. This possibility was tested in a study of European robins (*Erithacus rubecula*) by taking the first replacement away and permitting a second replacement to become established for a short time before releasing the resident. In this way, Joe Tobias was able to match 1-day replacements against residents that had been held captive for 10 days. Under these circumstances, the ex–territory holders always won, despite their prolonged captivity (Tobias 1997). In contrast, when ex-residents that had been caged for 10 days went up against replacements that had been on territories for 10 days, the original territory holders always lost. Therefore, contests between replacements and ex-residents were decided by how long the replacement had been on the territory, not by how long the ex-resident had been in captivity, a result that is consistent with the payoff asymmetry hypothesis.

The Dear Enemy Effect

One of the reasons why established territory holders may have more to gain by hanging on to their real estate than intruders is that boundary disputes with neighbors usually get settled with the passage of time, with the result that neighbors treat familiar rivals as "dear enemies" (Fisher 1954). For example, territorial male Broadley's flat lizards (*Platysaurus broadleyi*) charge after an intruder when he is far away, whereas they permit familiar neighbors to come much closer. Should the resident chase a neighbor, the pursuit covers only a few centimeters, whereas a territory holder dashes after an unfamiliar intruder for a meter and a half on average (Whiting 1999). Thus, once a territory owner and his neighbors have learned who is who, they no longer need to expend time and energy in lengthy chases—just one of the several advantages associated with becoming familiar with one's living space (Piper 2011). If an established resident is ousted, the new territory owner will have to fight intensely for a time with his unfamiliar neighbors in order to settle who owns what. The original resident therefore has more to lose should he be ousted than the new intruder can secure by acquiring his territory, given the expenditures associated with being a new kid on the block.

In fishes, mounting a challenge to a known versus unknown challenger not only results in reduced aggressive behavior, but also in a reduced androgen response (Aires et al. 2015). Therefore, treating known rivals differently can result in both lower energy expenditure and a reduction in the production of costly steroid hormones.

Although there can be a clear reduction in the energetic and physiological costs associated with having familiar enemies, determining if there are also reproductive benefits was particularly important to a team of conservation biologists interested in moving groups of highly endangered Stephens' kangaroo rats (*Dipodomys stephensi*) to new nature reserves in California. Initial attempts to establish new breeding populations failed. Thinking that perhaps the failures stemmed from the absence of dear enemies in the reconstituted groups, Debra Shier and Ron Swaisgood moved 99 kangaroo rats, permitting about half to retain their familiar neighbors and mixing unfamiliar pairs in the remainder. The group composed of dear enemies fought less and produced more offspring than those kangaroo rats that were saddled with unfamiliar opponents (Shier and Swaisgood 2011). Here is a wonderful example of how basic research in behavioral biology can help conservation biologists do their work more effectively (see also Caro 1998).

To Stay or Go

The cost–benefit approach has helped us make sense of the behavioral diversity exhibited by animals as they select places in which to settle and as they defend these areas against intruders—or leave potential rivals alone. We employ the same approach in the analysis of why migratory animals often leave areas where they have been living only to come back again, sometimes to the same place after an interval of many months. Although some animals, such as clown fishes (Buston 2004), Australian sleepy lizards (*Tiliqua rugosa*) (Bull and Freake 1999), and terrestrial salamanders (Marvin 2001) spend their entire lives in one spot, it is common for a young animal to leave its birthplace and disperse elsewhere. Dispersal and migration not only occur at different stages in an animal's life, but they also represent two very different scales of movement, both of which can be examined with the cost–benefit approach.

Ruffed grouse

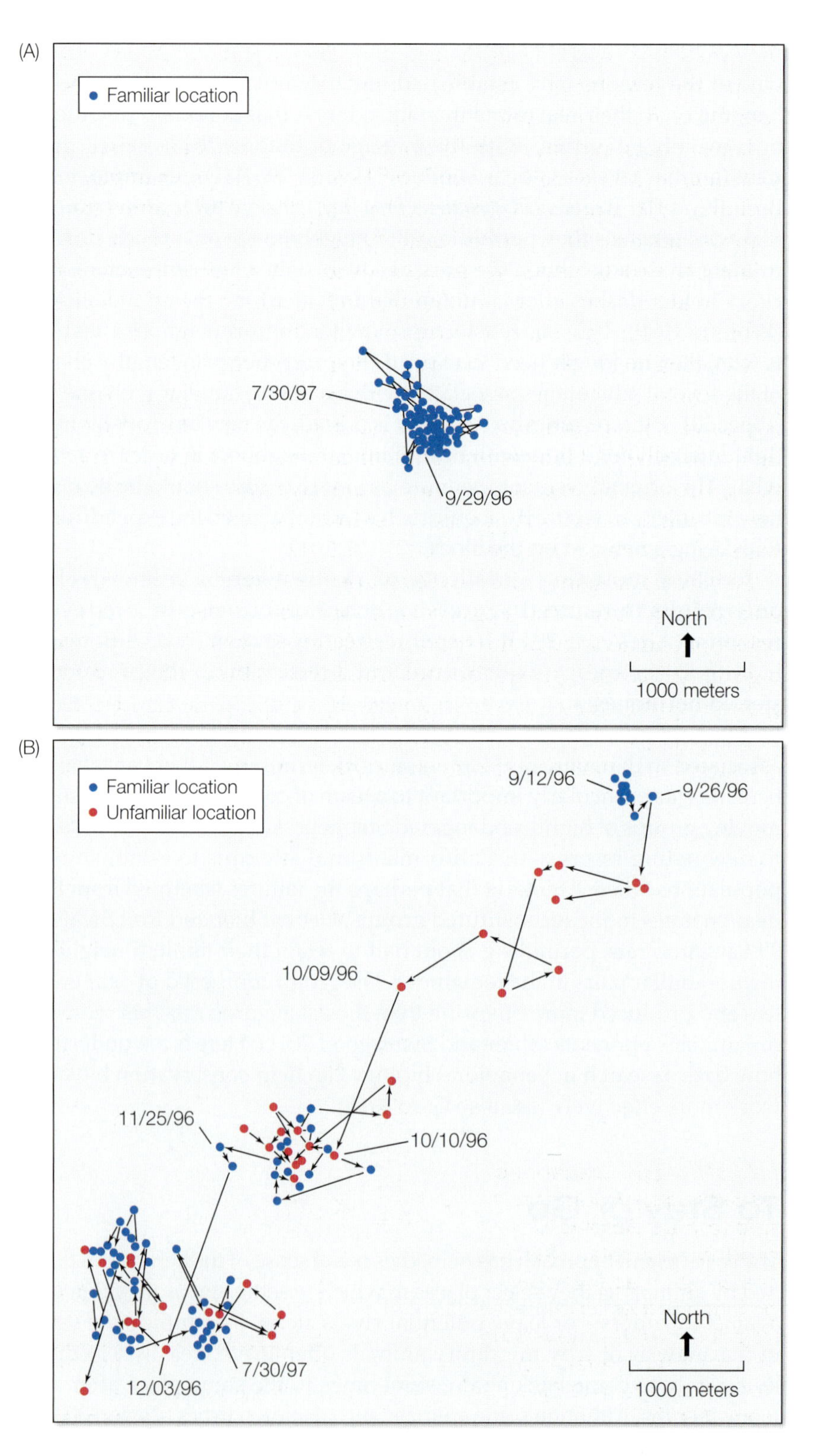

FIGURE 7.12 Two patterns of movement by radio-tracked ruffed grouse. (A) This bird stayed largely within the same fairly small home range for many months. (B) Another individual alternated bouts of staying put with substantial dispersal movements through unfamiliar terrain, a risky business for a ruffed grouse. (After Yoder et al. 2004; photograph © John Kirinic/ShutterStock.)

Dispersal

Dispersal, the permanent movement from the birthplace to somewhere else, can be costly because individuals not only have to secure extra energy for their travels (energetic costs), but they may also run the risk of falling prey to predators in the unfamiliar area through which they are moving (risk costs). There are also costs in terms of time that cannot be invested in other activities (time costs) and the potential for reduced fitness when not selecting the best habitat (opportunity costs) (Bonte et al. 2012). After all, dispersers in many species do not know where they are going, and do not have a map. They not only have to avoid predators while searching for new habitat, but they also have to deal with aggressive conspecifics and still find food.

James Yoder and his coworkers examined movement in ruffed grouse (*Bonasa umbellus*) by following both adult and young birds that had been captured and outfitted with radio transmitters (Yoder et al. 2004). This enabled the researchers to map the movements of individuals precisely and to find any grouse whose transmitter signaled that the bird had not moved for 8 daytime hours, a very good indication that the bird was dead. The researchers discovered that some birds stayed for months near the place where they had been captured, while others moved at considerable intervals from one location to another (**FIGURE 7.12**). Being in a new area increased the risk of being killed by a hawk or other predator at least threefold compared with birds that stayed in locations with which they had become familiar, demonstrating that movement into unfamiliar areas is risky. Numerous studies of birds, insects, amphibians, fishes, and more have shown that survival is particularly low for young animals dispersing away from their natal habitat in search of a place to settle and attempt to breed (Bonte et al. 2012).

So why are so many animals willing to leave home if doing so might increase their risk of dying? Users of the cost–benefit approach begin by thinking how the social behavior and ecology of the species might affect the balance between the advantages and disadvantages of dispersal. Consider that the typical young male Belding's ground squirrel (*Urocitellus beldingi*) travels about 250 meters from the safety of his mother's burrow before he settles in a new burrow, whereas a young female usually goes only 50 meters or so from where she was born (**FIGURE 7.13**) (Holekamp

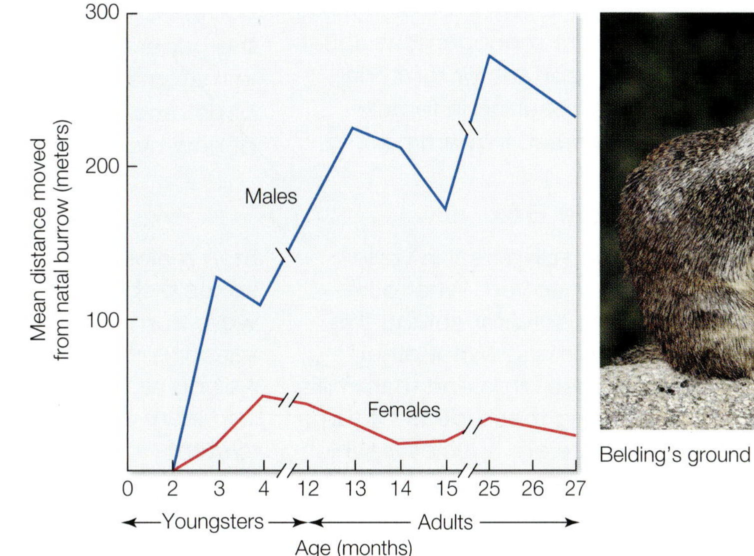

Belding's ground squirrel

FIGURE 7.13 Distances dispersed by young male and female Belding's ground squirrels. Males searching for a new home go much farther on average from their natal burrows than females. (After Holekamp 1984; photograph © iStock.com/4FR.)

1984). In other words, Belding's ground squirrels, like most other animals, exhibit **sex-biased dispersal**, in which one sex disperses farther than the other.

Why should young male Belding's ground squirrels disperse farther than their sisters, a pattern observed in other mammals (**BOX 7.2**)? According to the

BOX 7.2 **DARWINIAN PUZZLE**

Opposite patterns of sex-biased dispersal in mammals and birds

One of the more puzzling phenomena related to sex-biased dispersal is that mammals and birds exhibit very different patterns (Greenwood 1980). In birds, females tend to disperse farther from the natal territory than males, whereas in mammals the reverse is true. What causes this difference in sex-biased dispersal between mammals and birds?

One obvious difference between mammals and birds that could be related to sex-biased dispersal is how sex is determined at the chromosomal level. In birds, females are the heterogametic sex (they have a Z and a W sex chromosome), whereas males are homogametic (they have two Z chromosomes). In contrast, male mammals are heterogametic (they have an X and a Y chromosome) and females are homogametic (they have two X chromosomes). Although the homogametic sex is the one that tends to disperse farther in both mammals and birds, Greenwood dismissed the role of sex determination in underlying patterns of sex-biased dispersal, and instead focused on the role of local competition for resources and mates (see Chapter 6). He argued that in birds, in which most species are monogamous, local resources should be important for territorial males to attract mates, resulting in female-biased dispersal. In contrast, in mammals, in which most species are polygynous, male-biased dispersal is likely to be more common because local resources are important for females to rear young and because male–male competition for mates is intense.

Nearly 40 years after Greenwood's observation and proposed explanation for the contrasting patterns in mammals and birds, Trochet and colleagues used the comparative approach and a database of more than 250 species to test this idea (Trochet et al. 2016). Partially consistent with Greenwood's hypothesis, the researchers found associations among the sex doing the dispersing, mating systems, and territoriality. When male mammals are polygynous, dispersal tends to be male-biased, presumably to reduce competition among males. But there is no relationship between mating system and sex-biased dispersal in birds. The researchers did, however, find relationships among sex-biased dispersal, parental care, and sexual asymmetry in morphology in birds, suggesting that the explanation for sex-biased dispersal in animals may be more complex than originally hypothesized. Thus, the general pattern of male-biased dispersal in mammals and female-biased dispersal in birds remains a Darwinian puzzle in search of new hypotheses and additional tests.

Thinking Outside the Box

The puzzle of sex-biased dispersal in vertebrates remains largely unsolved. What other types of data would be useful for solving this puzzle in birds and mammals? What other taxonomic groups besides birds and mammals might be useful for testing these ideas? It turns out that in primates, there are species in which males tend to disperse farther than females, species where females tend to disperse farther than males, and still other species where both sexes disperse from the groups in which they were born. What factors might explain this variation in dispersal among primate species? As you read additional chapters of this book and learn about other genetic sex-determining systems in social species, remember this Darwinian puzzle and think about the patterns of dispersal in those species.

inbreeding avoidance hypothesis, sex-biased dispersal by juvenile animals of many species reduces the chance of inbreeding, which often affects fitness negatively (Pusey and Wolf 1996). When two closely related individuals mate, the offspring they produce are more likely to carry two copies of damaging recessive alleles than are offspring produced by unrelated individuals. The risk of associated genetic problems often reduces the average fitness of inbred offspring (**FIGURE 7.14**) (Ralls et al. 1979, Jiménez et al. 1994, Margulis and Altmann 1997). The prairie vole (*Microtus ochrogaster*) is an example of a species in which females appear to have evolved a preference for unfamiliar partners with which to produce a litter, a preference that should prevent them from reproducing with litter mates, thereby increasing the odds of adaptive outbreeding (Lucia and Keane 2012). Likewise, in the spotted hyena (*Crocuta crocuta*), most young males leave their natal clans because females prefer sexual partners that have either recently joined the group from another clan or have been with the group for a very long time, and are therefore not likely to be their close relatives. As a result, females generally avoid males that were born in their clan, making it adaptive for these males to leave to search for unrelated females elsewhere (Höner et al. 2007).

inbreeding avoidance hypothesis In the context of dispersal, sex-biased dispersal reduces the chance of inbreeding.

But if avoidance of inbreeding is the primary benefit of dispersing, then one might expect as many female as male ground squirrels or hyenas to travel away from where they were born. But they do not, perhaps because female ground squirrels or hyenas that stay put receive assistance from their mothers and so may eventually achieve high status within their groups with all the attendant benefits (Greenwood 1980). If so, the advantages of remaining on familiar ground are greater for females than for males, and this difference can contribute to the evolution of sex differences in dispersal in ground squirrels, hyenas, and other mammals (Holekamp 1984).

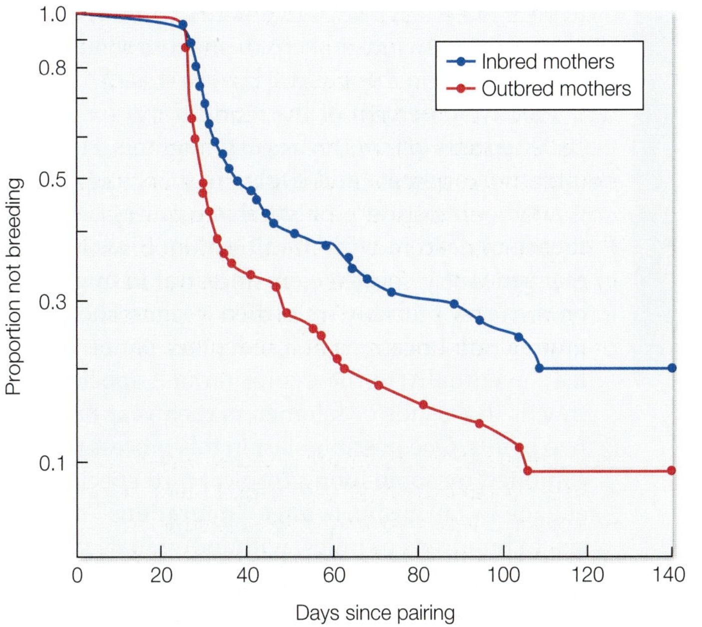

Female oldfield mouse

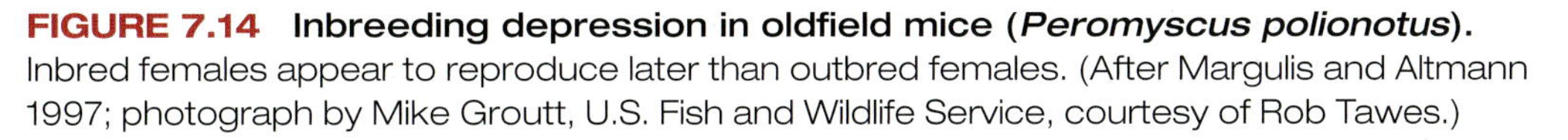

FIGURE 7.14 Inbreeding depression in oldfield mice (*Peromyscus polionotus*). Inbred females appear to reproduce later than outbred females. (After Margulis and Altmann 1997; photograph by Mike Groutt, U.S. Fish and Wildlife Service, courtesy of Rob Tawes.)

Migration

Unlike dispersal, which is a one-way trip, **migration** involves movement away from and subsequent return to the same location on an annual basis. As you read at the beginning of this chapter, migration is a phenomenon strongly associated with temperate-zone songbirds (Cox 1985). However, migration apparently also occurred in some extinct dinosaurs (Fricke et al. 2011) and is seen today in many mammals, fishes, sea turtles, and even some insects (Coombs 1990, Carpenter et al. 2003). As we discussed in Chapter 4, the monarch butterfly (*Danaus plexippus*) is famous for its ability to fly several thousand miles from Canada to Mexico in the fall (Calvert and Brower 1986). Some dragonflies also make an impressively long trip from southern India to Africa across the Indian Ocean in the fall, only to return in the spring (Anderson 2009).

Birds make some of the most spectacular migrations of any animals on Earth. Among the standout migrant birds is the tiny ruby-throated hummingbird (*Archilochus colubris*), which despite weighing only about as much as a penny flies nonstop 850 kilometers across the Gulf of Mexico twice a year, on its way from as far north as Canada to as far south as Panama (Welty 1982). When the much larger bar-tailed godwit (*Limosa lapponica*) flies in autumn nonstop all the way from Alaska to New Zealand, it covers a distance of 11,000 kilometers in just 8 sleepless days (Hedenström 2010). The total distance champion for a migratory bird may be the sooty shearwater (*Ardenna grisea*), which travels well over 60,000 kilometers a year in a figure-eight journey over the whole Pacific Ocean (**FIGURE 7.15**). These shearwaters breed in New Zealand and then move north to feeding areas off Japan, Alaska, or California, only to loop back down to their breeding grounds in time for another breeding season (Shaffer et al. 2006). Then there is the bar-headed goose (*Anser indicus*), which makes one of the highest and most iconic transmountain migrations in the world. En route from sea level in India to their breeding grounds in central Asia, these geese fly over the Himalayas in a single day, climbing up to 6000 meters in only 4 to 6 hours without the aid of upslope tailwinds to help them cross some of the highest peaks in the world (Hawkes et al. 2011).

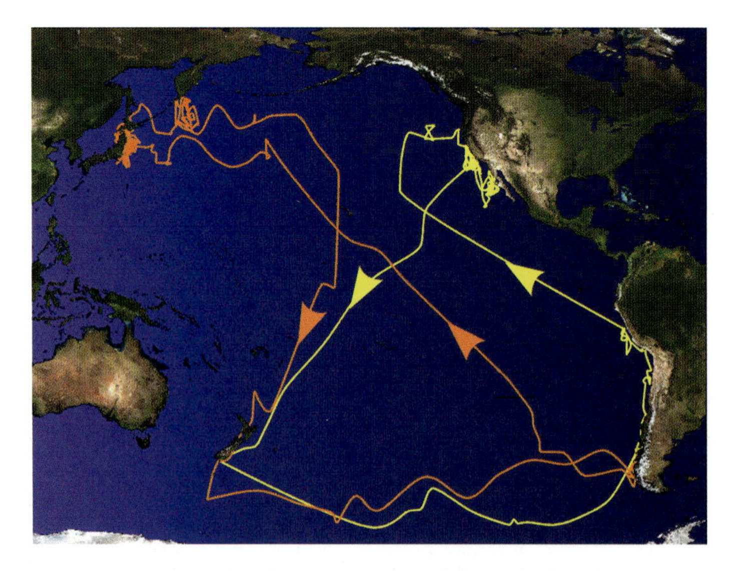

FIGURE 7.15 The long migratory route of the sooty shearwater. The sooty shearwater travels around the whole of the Pacific Ocean each year. Shown here are the routes taken by two shearwaters outfitted with GPS tracking tags as they left their breeding place in New Zealand and traveled across the Pacific Ocean to South America. One bird then went north to California, where it spent considerable time foraging before flying back to New Zealand. The other bird went northwest to Japan to its summer foraging ground before returning to nest in New Zealand. (From Shaffer et al. 2006, © 2006 National Academy of Sciences, USA; courtesy of Scott Shaffer.)

Nearly 40 percent of the world's approximately 10,000 species of birds are regular migrants. How did migration originate and evolve in this group? Here is another historical problem that requires Darwin's theory of descent with modification. If we assume that sedentary species were ancestral to migratory ones, as they probably were, then we must show how gradual modifications of a sedentary pattern could lead eventually to the evolution of a species that travels thousands of kilometers each year between two points. One possible start in this process may be exhibited by some living tropical bird species that engage in fairly short-range "migrations" of dozens to hundreds of miles, with individuals moving up and down mountainsides or from one region to another immediately adjacent one. The three-wattled bellbird (*Procnias tricarunculatus*), for example, has an annual migratory cycle that takes it from its breeding area in the mid-elevation forests on the mountains of north-central Costa Rica to lowland forests on the

(A)

(B)

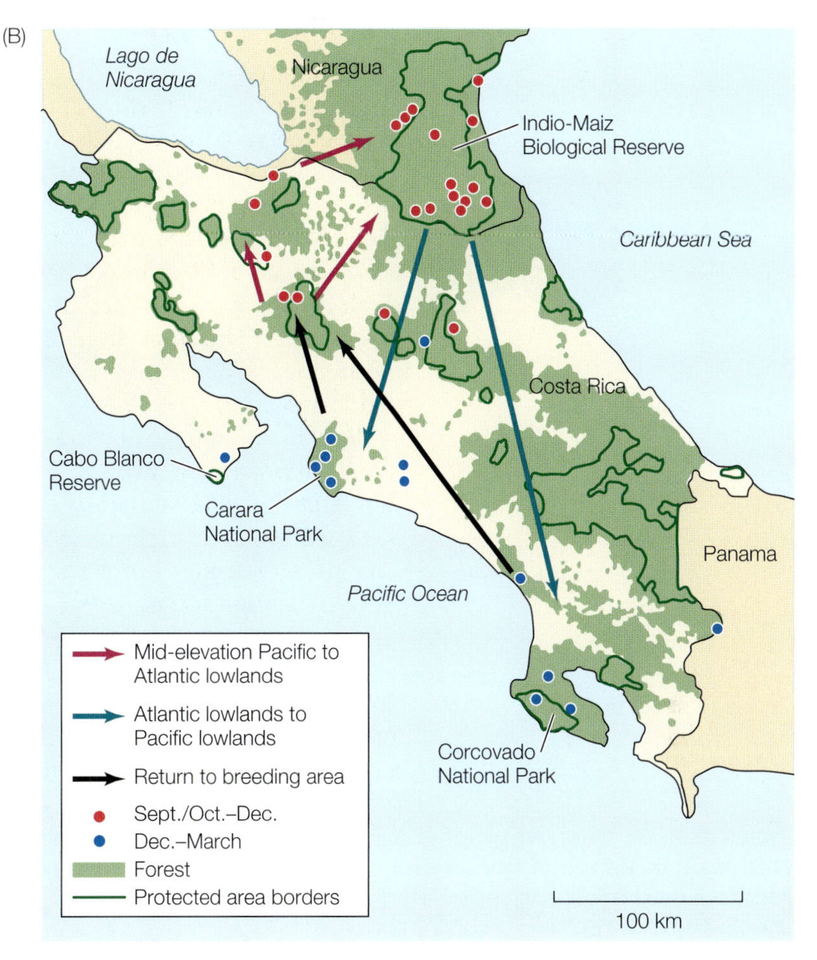

FIGURE 7.16 Short-range migration in the three-wattled bellbird. (A) A male three-wattled bellbird calling from a perch in a Costa Rican forest. (B) After breeding in the mountains of north-central Costa Rica, bellbirds first head to the north and east, then go south and west to reach forests on the Pacific coast before returning north to the mountains. (A, photograph by Ryan Kozie, CC BY 2.0; B after Powell and Bjork 2004, © 2004 Blackwell Publishing and the Society for Conservation Biology.)

Atlantic side of Nicaragua, then to coastal forests on the Pacific side of southwestern Costa Rica, from which the bird returns to its mountain breeding area (**FIGURE 7.16**) (Powell and Bjork 2004). The distances between any two locations visited by migrating bellbirds are substantial (up to 200 kilometers), but not breathtaking like those of the bar-tailed godwit or sooty shearwater.

For more than a century, there have been two alternative scenarios for how avian migration evolved (**HYPOTHESES 7.2**). On the one hand, seasonal migration could have evolved through a geographic shift from the tropics to the temperate breeding grounds to take advantage of the rich bounty of food resources for reproduction (**tropical origins hypothesis**). On the other hand, migration could have evolved through a shift toward the tropical wintering grounds from the temperate breeding grounds to take advantage of the milder climate during the nonbreeding season (**temperate origins hypothesis**). Results from comparative tests of these hypotheses have been mixed, though as more and better data have accumulated over time, the story is becoming clearer.

The earliest analyses of the evolutionary origins of avian migration were consistent with the tropical origins hypothesis and the idea that migration evolved in tropical birds as they moved north to find resources needed for reproduction. For example, Douglas Levey and Gary Stiles point out that short-range migrants occur in nine families

HYPOTHESES 7.2

Alternative hypotheses to explain the evolution of migration in birds

tropical origins hypothesis
Migration evolved through a geographic shift from the tropics to the temperate breeding grounds to take advantage of the rich bounty of food resources for reproduction.

temperate origins hypothesis
Migration evolved through a shift toward the tropical wintering grounds from the temperate breeding grounds to take advantage of the milder climate during the nonbreeding season.

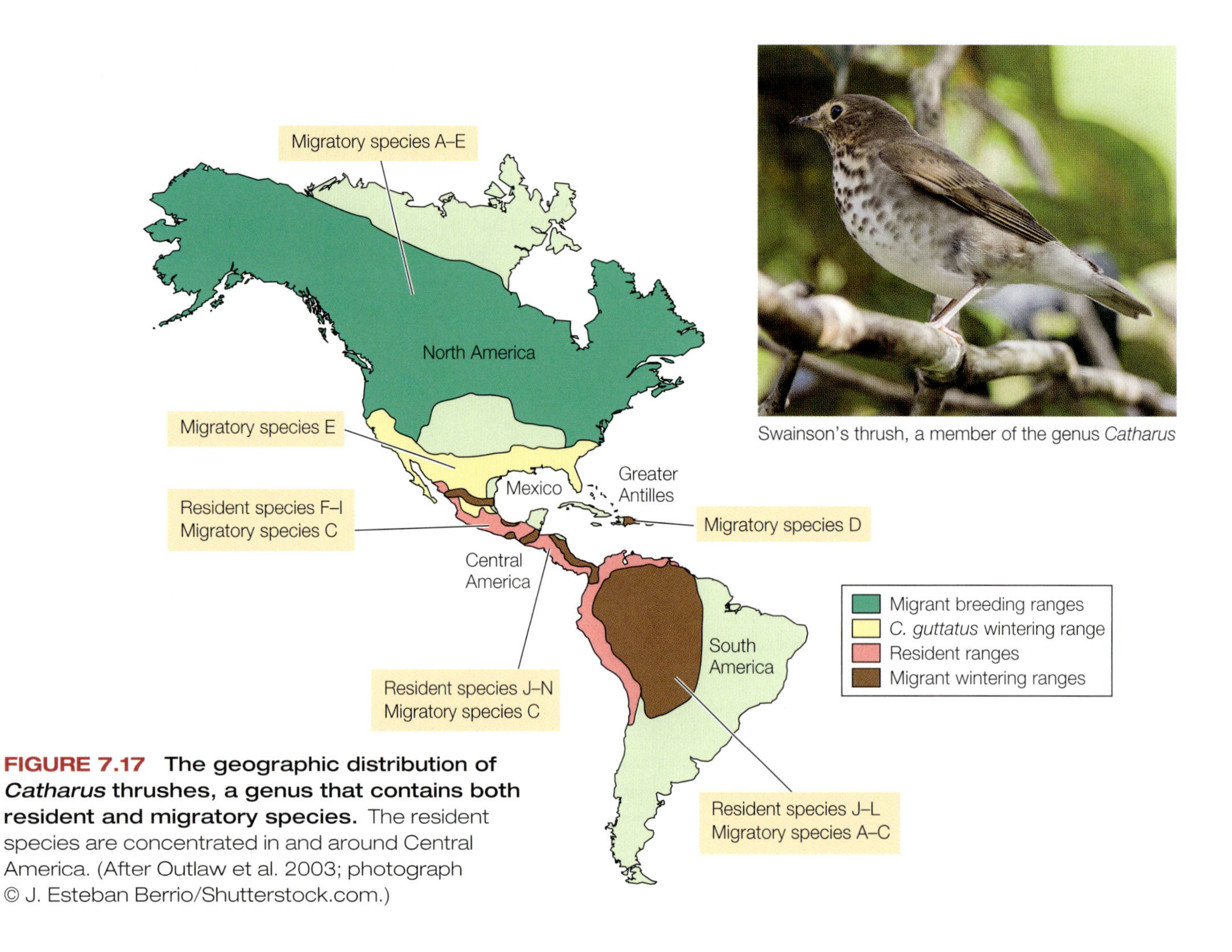

FIGURE 7.17 The geographic distribution of *Catharus* thrushes, a genus that contains both resident and migratory species. The resident species are concentrated in and around Central America. (After Outlaw et al. 2003; photograph © J. Esteban Berrio/Shutterstock.com.)

of songbirds believed to have originated in the tropics (Levey and Stiles 1992). Of these nine families, seven also include long-distance migrants that move thousands of kilometers from tropical to temperate regions. The co-occurrence of short-range and long-distance migrants in these seven families suggests that short-range migration preceded long-distance migration, setting the stage for the further refinements needed for the impressive migratory trips of some species. Additional data in support of this hypothesis come from one genus of birds, the *Catharus* thrushes. The genus *Catharus* contains 12 species, 7 of which are year-round residents in areas from Mexico to South America; the other 5 are migratory species that travel between breeding areas in northern North America and wintering zones far to the south, especially in South America (**FIGURE 7.17**). These observations suggest that the ancestors of the migratory species lived in Mexico or Central America. Moreover, the most parsimonious interpretation of a phylogeny of this genus is that migratory behavior has evolved three times (Winker and Pruett 2006), with subtropical or tropical resident species giving rise to migratory lineages each time (**FIGURE 7.18**) (Outlaw et al. 2003).

A series of more recent comparative analyses, however, is consistent with the temperate origins hypothesis and the idea that migration evolved in temperate birds as they moved south for the winter. For example, a phylogeny of the New World warblers, which belong to the family Parulidae, indicates that migratory behavior has been lost repeatedly during their evolution, with sedentary species having been derived from migratory ones (Winger et al. 2011). Moreover, in an analysis of more

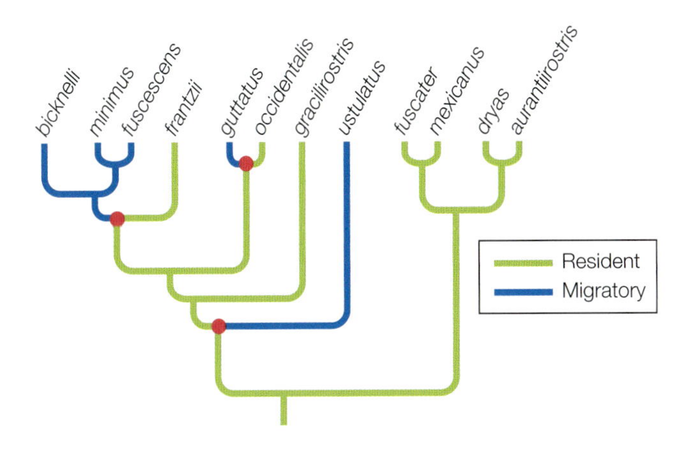

FIGURE 7.18 The long-distance migratory trait overlain on the phylogeny of *Catharus* thrushes. The phylogenetic tree was constructed on the basis of similarities among the species with respect to mitochondrial DNA. There appear to have been three independent origins of migratory behavior in this cluster of species. (After Outlaw et al. 2003.)

than 800 species in the emberizoid passerines, the largest radiation of New World birds, Benjamin Winger and colleagues found that long-distance migration in this group evolved primarily through evolutionary shifts of geographic range south for the winter out of North America (Winger et al. 2014). Thus, this example illustrates how with more data and the inclusion of additional species, comparative analyses can generate improved results when testing long-standing hypotheses, in this case some that have been debated for more than 100 years. Currently, the data appear to support the temperate origins hypothesis and the idea that migratory behavior evolved through a shift toward the wintering grounds from the temperate breeding grounds to take advantage of the milder climate during the nonbreeding season.

The Costs and Benefits of Migration

Whatever the evolutionary history of migratory behavior for a given species, the trait will only be maintained if its fitness benefits exceed its fitness costs, which are not trivial. The primary benefit for most migratory species is increased access to resources such as food and water, as moving from one location to another enables individuals to escape deteriorating conditions (typically driven by seasonal cycles) and find resources elsewhere. However, there are also many costs to moving vast distances in search of resources, including the greater metabolic demands and likelihood of mortality due to increased risk of pathogen transmission, higher predation risk, or flying over large bodies of water (Alerstam et al. 2003).

COSTS OF MIGRATION Although some species of birds may be able to migrate without expending much energy (Mandel et al. 2008), migratory behavior is energetically demanding for most animals (**BOX 7.3**). For birds, the costs of migration include the extra weight that migrants have to gain in order to have the energy reserves they will need to fly long distances. The storage of fuel is only one of a battery of costly physiological changes that make migration possible for birds. Among these changes are the temporary atrophy of reproductive organs, increases in muscle contraction efficiency, and an altered metabolism that enables the bird to process stored fats more quickly (Weber 2009).

The red knot illustrates the importance of securing the extra energy reserves (**FIGURE 7.19**). This bird, which migrates in both the New and Old Worlds, can add more than 70 grams to its initial base weight of about 110 grams after arriving, for example, at Delaware Bay from South America, provided that there are plenty of horseshoe crab (*Limulus polyphemus*) eggs on the beaches of that staging area. The knots need these fat-rich eggs to fuel the long trip to their Arctic breeding grounds

BOX 7.3 **EXPLORING BEHAVIOR** BY **INTERPRETING DATA**

Behaviors to reduce the costs of flying during migration

Although flying long distances can be energetically costly, some species of birds have developed ways to reduce these costs. The **FIGURE** at right shows the heart rates and wingbeat frequencies of great white pelicans (*Pelecanus onocrotalus*) flying in V-formation as opposed to flying alone. Birds traveling behind others can take advantage of updrafts created by the wingbeats of their companions, which enables them to cut their energetic costs by about 10 percent. Perhaps this is also why so many large birds, such as Canada geese (*Branta canadensis*), typically fly in V-formation when migrating.

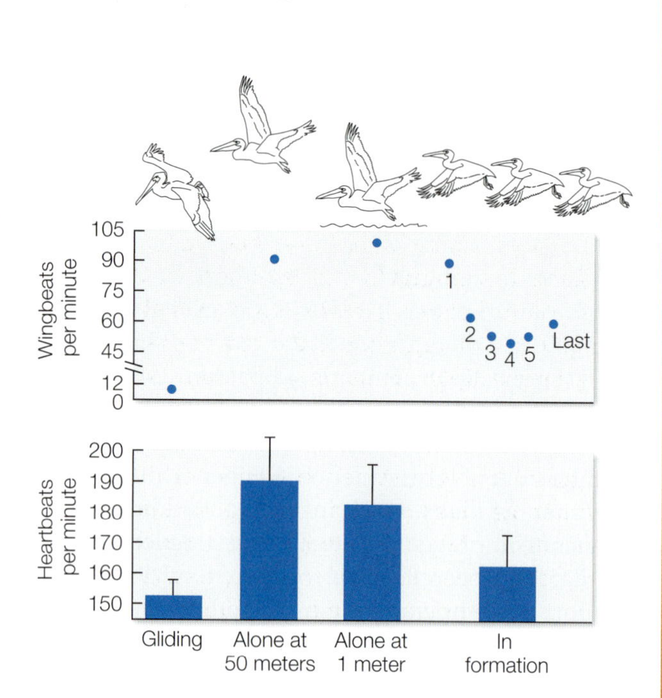

Flying in V-formation can be an energy saver. Data on wingbeat frequency and heart rate are presented for four flight options available to the great white pelican. Bars depict mean +/– SE. (After Weimerskirch et al. 2001.)

Thinking Outside the Box

What can you conclude from looking at the bottom graph? Now look at the right side of the top graph, where the order of the birds in flight formation is indicated. What do these data show? How do you think flight formation is determined in this species? Use game theory to consider two different behavioral strategies about being a "leader" that is in position 1, or a "follower" that is in positions 2 through 6 ("Last"). Develop a behavioral strategy that would favor the evolution of the V-formation, rather than just solo flight.

FIGURE 7.19 Red knots on migration at a critically important staging area. Delaware Bay once had huge numbers of horseshoe crabs whose eggs fed red knots at a key point along their migratory route. The current scarcity of horseshoe crabs threatens the well-being of migrant red knots. (Photograph © Prisma Bildagentur AG/Alamy.)

and to provide for the physiological changes that make egg laying possible after they arrive. If horseshoe crab eggs are the essential fuel for migrant knots, we can predict that a decline in horseshoe crab eggs should have negative effects on the population of red knots that use Delaware Bay during their migration. In fact, as horseshoe crab egg levels fell by 90 percent due to their rampant overharvesting as bait by fishermen, the red knot population at this site decreased by 75 percent (Niles et al. 2009). As the population was falling, knots that left Delaware Bay for the far north at relatively low body weights were less likely to be recaptured in the following year, which suggests that they were more likely to die than well-fed birds (Baker et al. 2004). Thanks to their knowledge of the basic biology of knot migration through Delaware Bay, behavioral biologists have helped develop management plans designed to rebuild the horseshoe crab population and sustain the population of red knots that migrates from southern South America north along the eastern seaboard of the United States (McGowan et al. 2011).

Another cost of migration is the risk of dying during the trip. Mortality rates in raptors, for example, are six times higher during migration seasons than during stationary periods, accounting for more than half of these species' annual mortality (Klaassen et al. 2014). However, if birds did not migrate, they might have even higher mortality if they stayed to experience the deteriorating conditions associated with winter. This is true of the common blackbird (*Turdus merula*), a partially migratory species in which some individuals remain in Germany throughout the year, whereas others migrate 800 kilometers south (Schwabl 1983). Using a combination of capture-mark-recapture and radio telemetry, a group of researchers found that migrant blackbirds had a 16 percent higher probability of surviving the winter than residents did (Zuniga et al. 2017). Thus, despite the potential survival costs associating with migration, the survival costs associated with not migrating might actually be higher in many species.

The risk of mortality on migration may be especially high for those birds that attempt long crossings over water. Indeed, the survival advantages of reducing the overwater component of a migratory trek explain why so many small songbirds travel east to west across all of Europe before crossing the Mediterranean at the narrow point between southern Spain and northern Africa (Berthold et al. 1992). Although this route greatly lengthens the total journey for birds headed to central Africa, it may prevent some individuals from drowning at sea. If this hypothesis is true, then we would expect other songbirds to make migratory decisions that decrease the risks of mortality during the trip. Red-eyed vireos (*Vireo olivaceus*) migrating in the fall from the eastern United States to the Amazon basin of South America must either cross a large body of water, the Gulf of Mexico, or stay close to land, moving in a southwesterly direction along the coast of Texas to Mexico and then south. The trans-Gulf flight is shorter, but vireos that cannot make it all the way to Venezuela are dead ducks, so to speak. In light of this danger, Ronald Sandberg and Frank Moore predicted that red-eyed vireos that happened to have low fat reserves would be less likely than those with considerable body fat to risk the long journey due south across the Gulf of Mexico. They captured migrating vireos in the fall on the coast of Alabama, classified each individual as lean or fat, and placed the birds in orientation cages (**FIGURE 7.20**). Consistent with their hypothesis, vireos with less

FIGURE 7.20 A funnel cage used in migration experiments. As the captive bird tries to begin its nighttime flight during the migratory season, it jumps up from an inkpad floor and leaves ink marks on the lining of the funnel cage that reveal the direction in which it intended to travel. (Photograph by Jonathan Blair/National Geographic.)

than about 5 grams of body fat showed a mean orientation at sunset toward the west-northwest, whereas vireos that had been classified as having more fat tended to head due south (**FIGURE 7.21**) (Sandberg and Moore 1996).

Given that many songbirds appear to avoid long overwater journeys, it is very surprising that some blackpoll warblers (*Setophaga striata*) voluntarily make a nonstop flight of 3000 kilometers from Canada to South America over the Atlantic Ocean (**FIGURE 7.22**) (Williams and Williams 1978). Surely these small warblers should take the safer passage along the East Coast of the United States and down through Mexico and Central America. However, the blackpoll warblers that manage this overwater trip have substantially reduced some of their costs of getting to South America. First, the sea route from Nova Scotia to Venezuela is about half as long as a land-based trek, although admittedly it requires an estimated 50 to 90 hours of continuous flight. Second, there are very few predators lying in wait in mid-ocean or on the islands of the Greater Antilles that the transoceanic blackpolls reach. Third, the birds leave the Canadian coast only when a west-to-east cold front can push them out over the Atlantic Ocean for the first leg of their journey, after which the birds use the westerly breezes typical of the southern Atlantic to help them reach an island (Latta and Brown 1999, McNair et al. 2002).

Many species are vulnerable to predators when they migrate. For example, predation risk by falcons (*Falco* spp.) is thought to have influenced both the timing and route of the southern migration in several shorebird species that breed in the Arctic (Lank et al. 2003). Predation risk—along with cooler temperatures and less turbulent skies—is also one of the reasons that many birds may migrate at night instead of during the day. Similarly, migrating juvenile sockeye salmon (*Oncorhynchus nerka*) in British Columbia are vulnerable to a variety of predators, including mergansers, gulls, trout, and otters, and in one population, 32 percent died while swimming a 14-kilometer stretch of river when migrating to the ocean (Furey et al. 2016). This high risk of predation has led to two adaptations in sockeye salmon; some populations have begun migrating at night when it is safer (Clark et al. 2016),

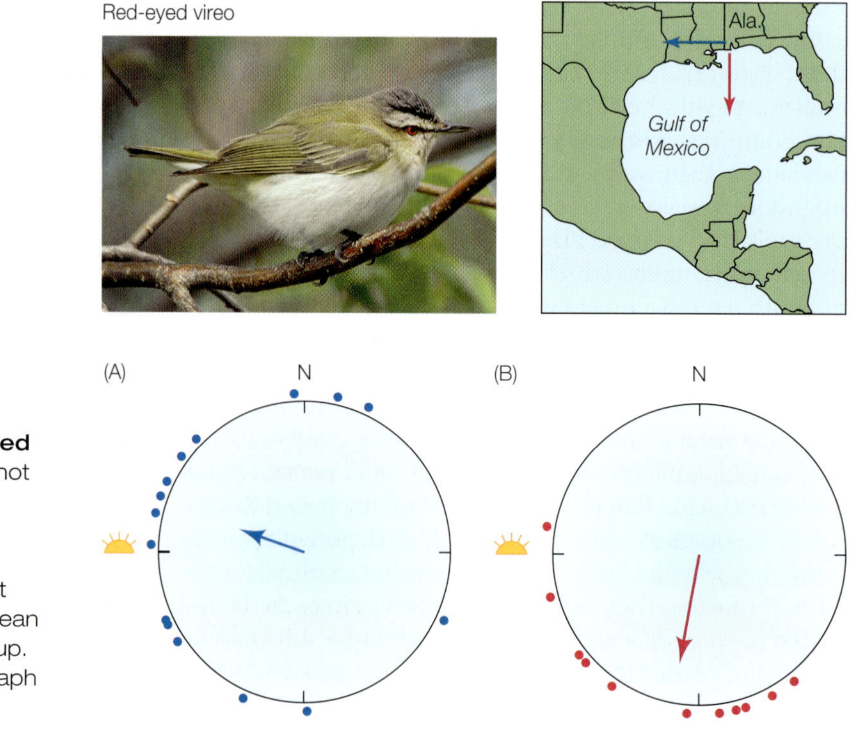

FIGURE 7.21 Body condition affects the migratory route chosen by red-eyed vireos. (A) Birds with low fat reserves do not head south toward the Gulf of Mexico but instead head west (sunset symbol) as if to begin an overland flight toward Mexico. (B) Birds with ample energy reserves orient due south. The central arrow shows the mean orientation for the birds tested in each group. (After Sandberg and Moore 1996; photograph © iStock.com/creighton359.)

Blackpoll warbler

FIGURE 7.22 Transatlantic migratory path of blackpoll warblers. Blackpoll warblers travel across the Atlantic Ocean as they fly from southeastern Canada and New England to their South American wintering grounds. (Map courtesy of Janet Williams; photograph © iStock.com/impr2003.)

as well as waiting in relatively predator-free areas for other individuals to come before migrating en masse (Furey et al. 2016). The movement of migratory individuals together in high densities is called **predator swamping**, which leads to the dilution effect that we discussed in Chapter 6 as an antipredator adaptation. Predator swamping occurs not only in fishes, but also in birds such as Arctic-breeding shorebirds and many others.

BENEFITS OF MIGRATION For all of their navigational and meteorological skills, migrating blackpolls and other birds cannot eliminate the costs of their travels altogether. What ecological conditions might elevate the benefits of migration enough to outweigh these costs, leading to the spread and maintenance of migratory abilities by natural selection? One answer for many songbirds in the Americas may lie in the immense populations of protein-rich insects that appear in the northern United States and Canada in the summer, when long days fuel the growth of plants on which herbivorous insects feed (Brown 1975). Moreover, the many hours of summer daylight mean that breeding migrant songbirds can search for food for longer each day than tropical bird species, which have only about 12 hours each day to harvest prey for their offspring. But a summertime food bonanza cannot be the only factor favoring migration, given that many migrants abandon areas where food is still plentiful to winter elsewhere (Bell 2000).

Resources other than food can also vary in availability seasonally, making migration adaptive. In Serengeti National Park in Tanzania, more than a million wildebeests (*Connochaetes gnou*), plains zebras (*Equus quagga*), and antelopes move from south to north and back again each year. The move north appears to be triggered by the dry season, while the onset of the rains sends the herds south again. It might be that the herds are tracking grass production, which is dependent on rainfall. Eric Wolanski and his colleagues, however, have tested another hypothesis, namely that a decline in water and an increase in the saltiness of the water in drying rivers and shrinking water holes is the critical factor underlying migration. If one knows the salinity of the water available to the great herds, one can predict when they will leave on their march north (Wolanski et al. 1999), although the precise route they follow north is influenced by the availability of vegetation, which in turn is influenced by rainfall patterns in the Serengeti (Musiega et al. 2006).

When food and other resources influence migration (or any other behavior), we call these ecological drivers **bottom-up forces**. However, as you saw earlier migration can also be driven from the top-down by predation risk (**top-down forces**). To determine the role of these bottom-up and top-down forces in influencing migration in Serengeti wildebeests and zebras, Grant Hopcraft and colleagues used GPS collars to follow the migratory herd as it moved between Tanzania and Kenya over an 8-year period from 2000 to 2008 (Hopcraft et al. 2014). The researchers found that both species of grazers move more each day when resources are abundant. However, wildebeests tend to move in response to food quality and pay little attention to predators, whereas zebra movements are influenced by both predation risk and access to high-quality food. Thus, bottom-up and top-down forces can interact to shape migratory behavior.

The monarch butterfly is a species whose migratory behavior has nothing to do with finding food. When monarchs fly in fall from the eastern half of North America (see Chapter 4), they head for central Mexico, where they spend the winter roosting (not feeding) in oyamel fir forests high in the mountains northwest of Mexico City (Urquhart 1960, Brower 1996). Whereas red knots use expensive flapping flight to get where they are going, monarchs use favorable winds to help them glide and soar relatively cheaply toward their destination. True, as they make their long journey south, monarchs must find enough flower nectar to fuel their flight, but unlike red knots that create and expend large fat reserves on their migration, migrating monarchs carry only small quantities of lipids for much of their trip. Only when monarchs get fairly close to the oyamel forests do they collect large amounts of nectar for conversion to the lipid energy reserves they will need for the long months of cold storage in their winter roosts (Brower et al. 2006).

But why go to the trouble of flying up to 3600 kilometers to reach a tree in the cold, high mountains of Mexico? Even if the butterflies keep the costs of the journey relatively low by migrating primarily on days when the winds facilitate inexpensive soaring flight, still one would think that they could spend the winter roosting in places much closer to the milkweed-producing areas that female monarchs visit to produce their offspring the next spring and summer. But perhaps not, since killing freezes occur regularly at night throughout eastern North America during winter. In contrast, freezes are very rare in the Mexican mountain refugia used by the monarchs. In these forests, at about 3000 meters elevation, temperatures rarely drop below 4°C, even during the coldest winter months. Occasionally, however, snowstorms do strike the mountains, and when this happens, as many as 2 million monarchs can die in a single night of subfreezing temperatures. The risk of freezing to death could be completely avoided in many lower-elevation locations in Mexico. But William Calvert and Lincoln Brower note that monarchs would quickly use up their water and energy reserves in warmer and drier areas. By remaining moist and cool—without freezing to death—the butterflies conserve vital resources, which will come in handy when they start back north after their 3 months in the mountains (Calvert and Brower 1986).

The hypothesis that the stands of oyamel fir used by the monarchs provide a uniquely favorable microclimate that promotes winter survival is being unwittingly tested in an unfortunate manner. Even in supposedly protected reserves, an alarming amount of woodcutting and logging has occurred (Ramirez et al. 2003). Brower and his associates believe that timber removal causes butterfly mortality, even when some roosting trees are left in place. Opening up the forest canopy increases the chances that the butterflies will become wet, which increases the risk that they will freeze (**FIGURE 7.23**). Thus, even partial forest cutting may destroy the conditions needed for the survival of monarch aggregations. If the loss of a relatively small number of oyamel firs causes the local extinction of overwintering monarch populations, it will be a powerful but sad demonstration of the value of a specific

(A)

(B)

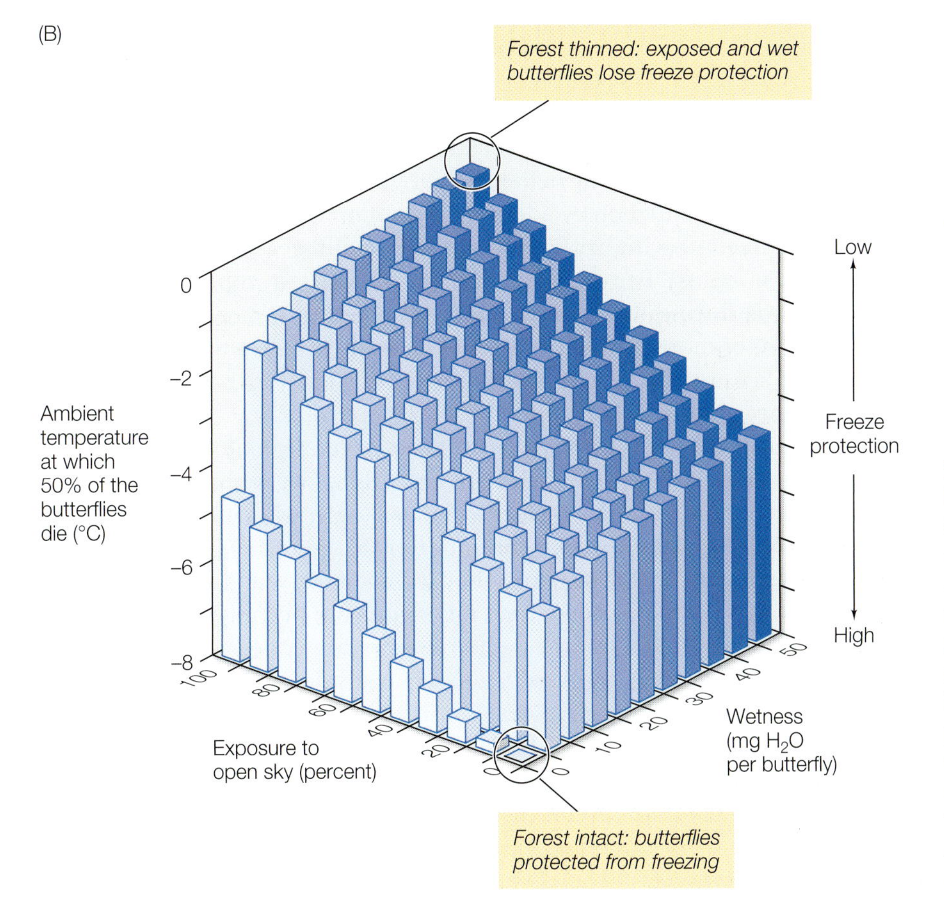

FIGURE 7.23 Monarch butterfly habitat selection. (A) Vast numbers of monarch butterflies spend the winter resting in huge clusters on fir trees in a few Mexican mountain sites. (B) Habitat quality correlates with survival of overwintering monarchs. Protection from freezing in the high Mexican mountains depends on a dense tree canopy that reduces wetting of the butterflies by rain or snow and their exposure to open sky. (A, photograph © Ingo Arndt/Minden Pictures; B after Anderson and Brower 1996.)

habitat for this migratory species (Anderson and Brower 1996). It will also illustrate once again how the world is afflicted by the destructive activities of humans, which include the conversion of milkweed-containing fields in the United States, depriving monarchs of nectar and egg-laying sites.

Variation in Migratory Behavior

Although we have accumulated good evidence that migratory behavior in birds has a strong hereditary component, and we are beginning to identify genes that contribute to this behavior (see Chapter 3), researchers have also observed rapid shifts in migratory behavior in several species, as well as individual variation in migratory behavior in birds from the same population. One species that appears to be able to rapidly change its migratory behavior is the barn swallow (*Hirundo rustica*), which breeds across the Northern Hemisphere. North American subspecies of barn swallows winter throughout South America, reaching the tip of Argentina. David Winkler and colleagues recently discovered a population of barn swallows that stopped migrating to North America and instead began to breed in Argentina, all within the past 35 years (Winkler et al. 2017). Using geolocators, the researchers discovered that the Argentinian swallows migrate only within South America. Thus, in a very short period of time, evolutionarily speaking, these birds have completely shifted their migratory behavior.

Both migratory and nonmigratory individuals occur in some species, as we mentioned previously for the common blackbird (Schwabl 1983). The fact that an individual can switch from the migrant strategy to the resident strategy from one year to another indicates that the two behavior patterns are not hereditary. Indeed, in a genomic comparison of migratory and nonmigratory populations of common blackbirds, Franchini and colleagues found very few genetic differences and observed that only 4 of more than 10,000 genes were differentially expressed (Franchini et al. 2017). Perhaps then, all the blackbirds in a population share the same conditional strategy that provides them with the flexibility to choose whether to migrate or stay put, depending conceivably on their social status. Socially dominant individuals should be in a position to select the better of the two options, forcing subordinates to make the best of a bad situation by adopting the option with a lower reproductive payoff (but one that confers higher fitness than they could get from futile attempts to behave like dominants). For example, perhaps an area can support only a few resident blackbirds during the winter. Under such circumstances, subordinates faced with a cadre of more powerful residents might do better to migrate away from the competition, returning in the spring to occupy territories made vacant by the deaths of some rivals over the winter.

Given the logic of this hypothesis, we can make several predictions: (1) blackbirds should have the ability to switch between migratory tactics (or available behavioral options), rather than being locked into a single behavioral response; (2) socially dominant birds should adopt the superior tactic; and (3) when choosing freely between tactics, individuals should choose the option with the higher reproductive payoff. In light of these predictions, it is significant that migratory common blackbirds head off in the fall at times when dominance contests are increasing in frequency (Lundberg 1985). Moreover, when blackbirds do drop the migratory option in favor of staying put, they typically do so when they are older and presumably more dominant (Lundberg 1988). Based on this limited evidence, the variation in common blackbird migratory behavior appears to be due to conditional tactics (**FIGURE 7.24**). Thus, common blackbirds are an informative species to study the behavioral plasticity of migratory behavior, and ultimately to link the fitness consequences of migration to the underlying genome (Franchini et al. 2017).

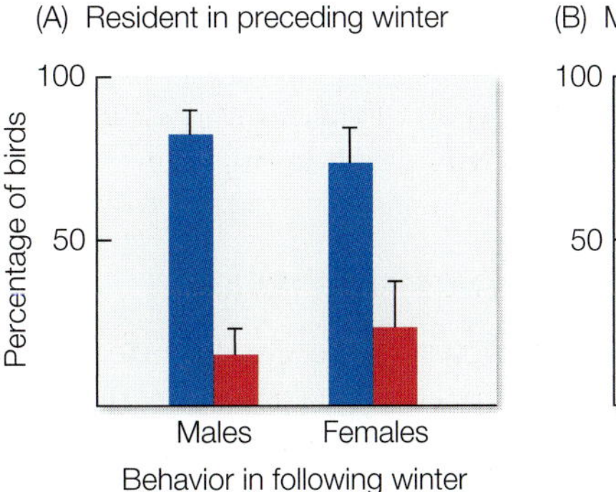

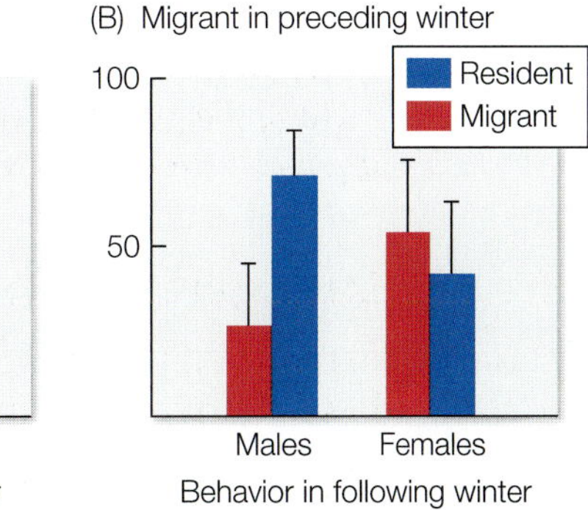

FIGURE 7.24 A conditional strategy controls the migratory behavior of common blackbirds. (A) Birds that were residents in the preceding winter tended to be nonmigratory the next winter as well. (B) In contrast, birds that were migrants in the preceding winter often switched to the resident option the following winter. Bars depict mean +/– SE. (After Schwabl 1983; photograph © Karel Brož/Shutterstock.)

In addition to exhibiting individual variation in migratory behavior, some species exhibit consistent population differences in migratory behavior (**BOX 7.4**). For example, black-throated blue warblers (*Setophaga caerulescens*) migrate to islands in both the western and eastern Caribbean. Using stable carbon and hydrogen isotopes in feathers, which as we noted earlier reflect diet and location at the time of molt and then become chemically inert and locked into keratin tissue (see Box 7.1), researchers showed that keratin-containing feathers of black-throated blue warblers in the western Caribbean are isotopically distinct from the feathers of conspecifics that winter in the eastern Caribbean. The isotopic signatures in the two populations have been traced back to where the feathers were produced in the United States, where birds molt at the end of the breeding season before starting their treks south. As a result, we know that the western Caribbean black-throated blue warblers breed in the northeastern United States, while those in the eastern Caribbean breed in the mountains of the southern United States (Rubenstein et al. 2002). In this case, different populations—rather than different individuals in the same population—exhibit variation in migratory behavior and migratory connectivity in particular.

Population differences in migratory behavior, as well as the degree of migratory connectivity between the breeding and wintering grounds, can have important consequences for conserving declining populations of songbirds. Gunnar Kramer and colleagues studied two species of warblers—blue-winged warblers (*Vermivora cyanoptera*) and golden-winged warblers (*V. chrysoptera*)—that are so closely related that they hybridize where their ranges meet in North America (Kramer et al. 2018). One of these species, the golden-winged warbler, has undergone greater breeding population declines than the other, but only in parts of its range. Using geolocators, the researchers showed that much like black-throated blue warblers, golden-winged warblers exhibited strong migratory connectivity, with different breeding populations wintering in distinct areas of Central and South America (**FIGURE 7.25**). In contrast, blue-winged warblers from across the breeding range mixed in the same Central American wintering areas. By integrating the long-term ecological data collected on population abundance on the breeding grounds, the researchers were able to show that the more southern Appalachian Mountain golden-winged warbler breeding population has declined precipitously, corresponding with severe conversion of native forest to other land uses in this population's northern South American

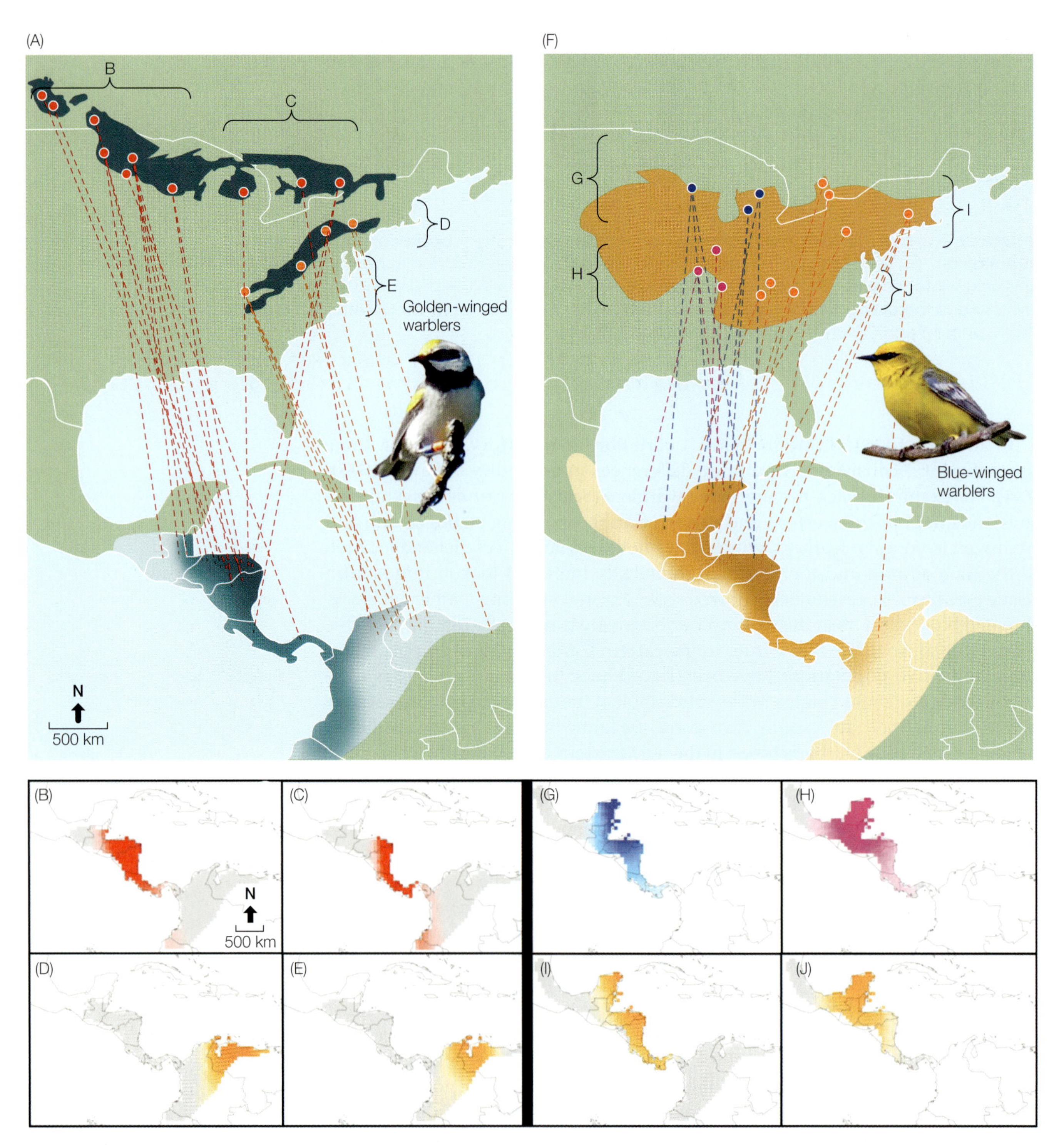

FIGURE 7.25 Migratory connectivity in warblers. Migratory patterns based on geolocator data for (A) golden-winged warblers and (F) blue-winged warblers. Golden-winged warblers exhibit strong migratory preferences, with different breeding populations wintering in distinct areas of Central and South America, whereas blue-winged warblers from across the breeding range mix in Central America. Brackets in (A) and (F) define sites used to create population-level averages of wintering (B–E) golden-winged warblers and (G–J) blue-winged warblers. Darker colors in the wintering grounds correspond to higher probability of use; the bottom 50 percent of probabilities are shown in gray to aid in visualization of core use areas indicated by the colors. (After Kramer et al. 2018.)

wintering areas (Sherry 2018). In contrast, the golden-winged (and blue-winged) warblers wintering in Central America, where forests have been better preserved, were not experiencing population declines. Ultimately, information about migratory connectivity in species such as these that are experiencing population declines in portions of their breeding ranges might help pinpoint locations in the tropics that are in need of the greatest conservation efforts.

BOX 7.4 **EXPLORING BEHAVIOR BY INTERPRETING DATA**

Migratory pathways of Swainson's thrush

Swainson's thrush (*Catharus ustulatus*) breeds in a large region across North America. Those birds that breed in northwestern North America do not all follow the same migratory route. Some birds go down the Pacific coast and winter in Central America. But others travel all the way to eastern North America before flying south to winter in South America (**FIGURE**).

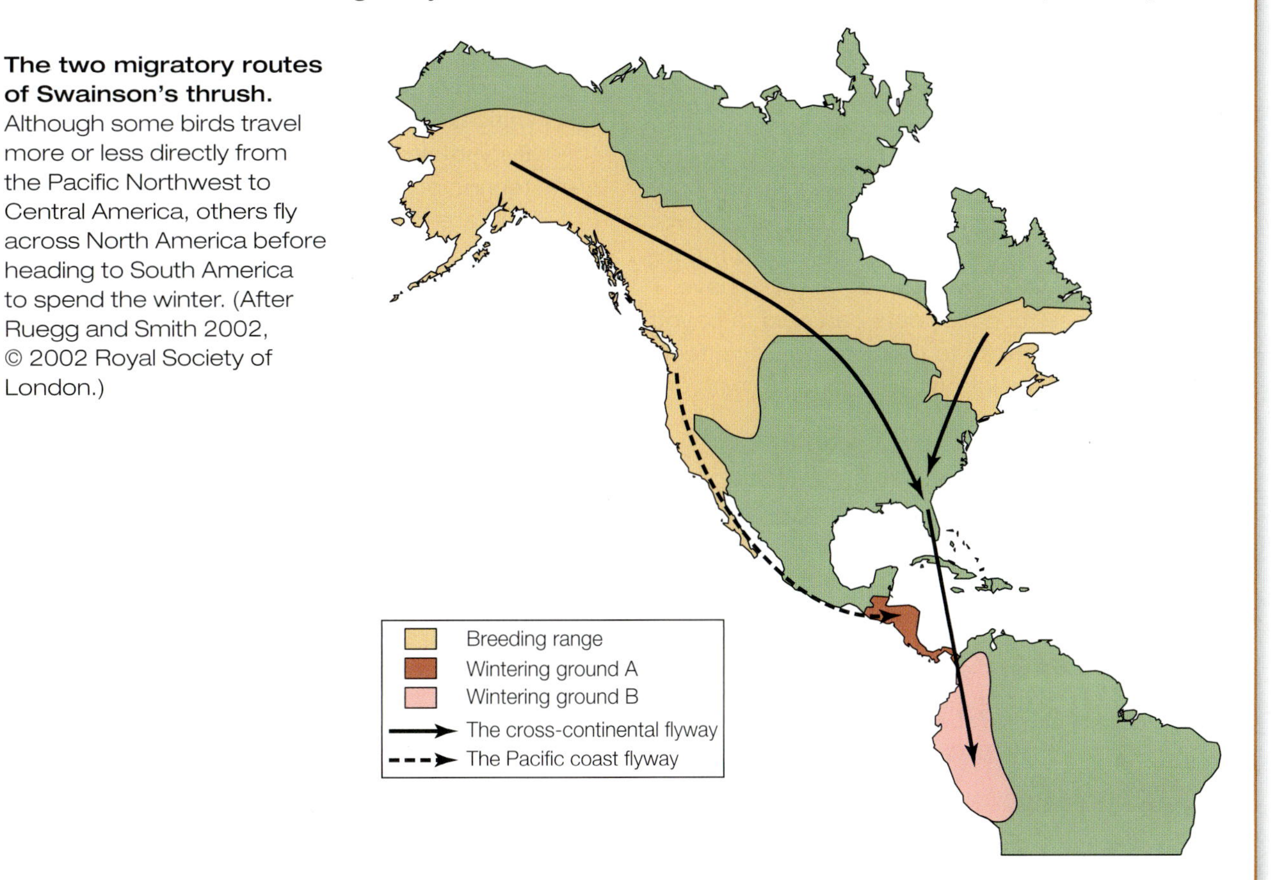

The two migratory routes of Swainson's thrush. Although some birds travel more or less directly from the Pacific Northwest to Central America, others fly across North America before heading to South America to spend the winter. (After Ruegg and Smith 2002, © 2002 Royal Society of London.)

Thinking Outside the Box

What methods would have allowed researchers to reach these conclusions? One hypothesis to account for the behavior of the thrushes taking the long way south is that these birds are descendants of those that expanded the species' range from the East Coast far out to the west and north after the retreat of the glaciers about 10,000 years ago. What kind of evolutionary hypothesis is this? Is the behavior maladaptive? How can you account for the persistence of the trait?

SUMMARY

1. In choosing where to live, many animals actively select certain places over others. Ideal free distribution theory deals with the surprising observation that some animals may occupy less favored areas rather than competing for prime habitat. Hypotheses based on ideal free distribution theory predict that animals that choose areas where there is less competition may have the same fitness as those that have joined many others in using a favored habitat.
2. Territorial behavior results in clear costs to individuals that attempt to monopolize a patch of real estate against rivals. Defense of living space evolves only when individuals can gain substantial benefits from holding a territory—such as greater access to food or mates. As predicted, individuals abandon their territories when the costs exceed the benefits.
3. Territorial contests are usually won quickly by owners. The competitive edge held by territorial residents over intruders may stem from superior physical strength or energy reserves (which give residents higher resource-holding potential), or it may exist because residents have more to lose than intruders can gain (a payoff asymmetry) thanks to the dear enemy effect, in which familiar neighbors stop fighting with one another over territorial boundaries.
4. In many species, young individuals disperse from territories defended by their parents. Dispersal, like territoriality, comes with fitness costs such as increased mortality. Major benefits can include the avoidance of inbreeding with close relatives, escape from aggressive competitors, and the ability to find and exploit resources in short supply in the natal territory or home range.
5. Migration is the temporary movement of individuals between two well-separated areas. The ability to migrate very long distances may have originated in populations that had acquired the capacity for short-range migrations. The behavior may have adaptive value because migrants keep the costs of travel low while gaining major benefits from their long-distance journeys, such as access to greater resources that are available only during certain seasons.
6. Studying migratory connectivity (linking populations between the breeding and wintering grounds) has the potential to help identify regions of conservation importance that are critical to a species' survival. Migratory connectivity in birds can be examined using a variety of technologies, including those that use markers that are intrinsic to the birds to indirectly track movements, as well as those that require the attachment of extrinsic markers on the bird to more directly track movement.
7. Some species of birds show variation in migratory behavior, both among populations as well as within populations. Understanding what drives this behavioral variation may enable scientists to help conserve species that are showing population declines only in portions of their breeding ranges.

COMPANION WEBSITE

Go to **ab11e.com** for discussion questions and all of the audio and video clips.

CHAPTER 8

Principles of Communication

If you are ever fortunate enough to go on an African safari, you might see a spotted hyena (*Crocuta crocuta*) endowed with what looks like an erect penis approach another hyena, which would then inspect the organ of its companion with apparent interest (**FIGURE 8.1**) (**VIDEO 8.1**). Who are these individuals? What are they doing? Why are they doing it? You might hypothesize that the individual with the enlarged genitalia is the dominant male of the clan. But you would likely be wrong. What if we told you that many of the penis presenters are actually female spotted hyenas, not males! But female spotted hyenas do not have a penis. Instead, they have what is called a pseudopenis. The female hyena's clitoris is shaped like a penis, and her labia are fused to form a pseudoscrotum. What is even more puzzling is that she possesses no external vaginal

Chapter 8 opening photograph © Tui De Roy/Minden Pictures.

(A)

(B)

FIGURE 8.1 The pseudopenis of the female spotted hyena. (A) The pseudopenis is not a penis at all. Instead, it is an enlarged clitoris that can be erected. (B) The erect pseudopenis is often presented to a companion—often a female but sometimes a male—in the greeting ceremony of this species. (A, photograph © Winfried Wisniewski/MInden Pictures; B, photograph by © Martin Mecnarowski/Shutterstock.com.)

opening. Instead, females give birth—and urinate and copulate—through the pseudopenis. As one might imagine, copulation can be difficult, as the male must insert his erect penis into the female's flaccid pseudopenis, and birth can be both painful and dangerous (Holekamp and Dloniak 2010).

Knowing that these two hyenas are females still doesn't answer the questions about what they are doing and why they are doing it. In this chapter we introduce a framework that will enable you to dissect this strange display and determine not only what these females are doing with their pseudopenis display, but also how this type of communication system could evolve among females. This framework for **communication**—the transfer of information from one individual to another that affects current or future behavior and the fitness of one or both individuals—emphasizes how the information exchanged between two individuals influences their genetic success. After defining communication, we then explore the evolutionary history of communication systems and how signals may originate and evolve. We discuss cases in which an individual produces a signal from a preexisting trait, as well as those in which preexisting sensory mechanisms already present in an individual lead to the evolution of animal signals. Throughout this chapter, we discuss how information is encoded into messages and transferred between individuals in both reproductive and non-reproductive contexts. Finally, we explore some of the adaptive reasons why signaling systems evolve to honestly convey information. We also highlight cases of deception, as well as how eavesdropping on others evolves. Although we largely focus on a single signal in each species, we introduce the idea that individuals might use multiple traits to signal to the same or different individuals, a phenomenon referred to as multimodal signaling. Spotted hyenas feature prominently throughout this chapter, but we also discuss communication in insects, birds, fishes, and reptiles.

Communication and Animal Signals

If the hyena's genital display is really a form of communication, then the sender (the hyena with an erection) is using its genitalia as a **signal** (a specially evolved message that contains information) to modify the behavior of the receiver (the inspector hyena). Central to this definition of communication is the "specially evolved signal," which eliminates cases where the behavior of one animal influences that of another without an evolved message being involved. For example, a mouse making rustling sounds while foraging at night is not communicating with an owl that is hunting it by listening for the mouse's sounds. To understand which types of interactions constitute communication and which don't, and to determine how communication systems evolve, we must recognize that the transfer of information from a sender to a receiver—whether intentional or not—can positively or negatively affect the fitness of both individuals.

VIDEO 8.1

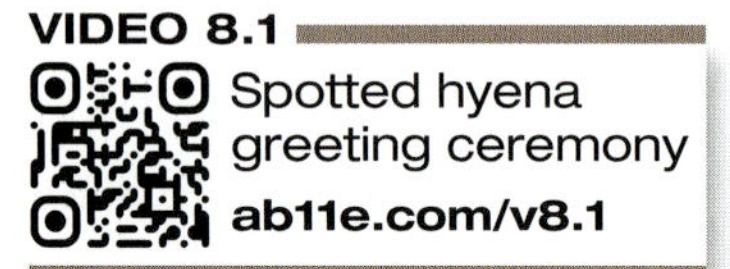

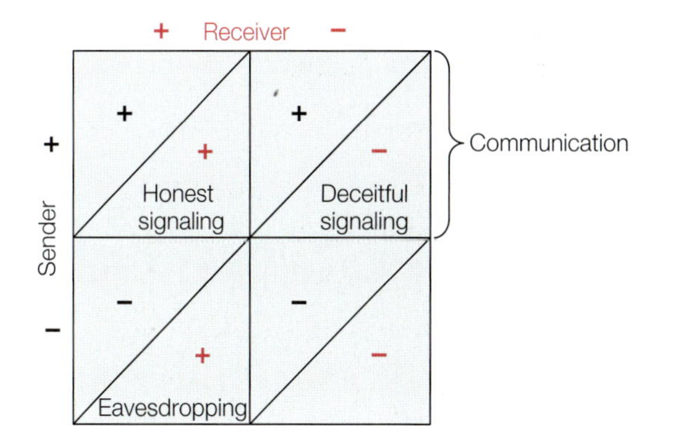

FIGURE 8.2 A cost-benefit framework for communication. Honest interactions are those in which both the sender and the receiver obtain a fitness benefit (+/+). Deceitful interactions are those in which a sender uses a specially evolved signal to manipulate the behavior of a receiver such that the sender receives a fitness benefit but the receiver pays a fitness cost (+/–). Both honesty and deceit are forms of communication. In contrast, eavesdropping occurs when the sender pays a fitness cost but the receiver receives a fitness benefit (–/+). Communication systems in which both individuals pay a fitness cost (–/–) are unlikely to evolve.

FIGURE 8.2 outlines a framework for understanding communication from this cost–benefit perspective. When both the sender and the receiver obtain a fitness benefit (+/+), we refer to this interaction as **honest signaling** (which is sometimes referred to as true communication; Bradbury and Vehrencamp 2011). Honest signaling evolves as a cooperative interaction between two individuals. When a sender uses a specially evolved signal to manipulate the behavior of a receiver such that the sender receives a fitness benefit but the receiver pays a fitness cost (+/–), we call this **deceitful signaling** (or manipulation). Both honest and deceitful signaling are forms of communication because in both cases the sender uses a specially evolved signal to positively influence its own fitness. When the sender pays a fitness cost but the receiver receives a fitness benefit (–/+), we refer to this as **eavesdropping**. In the example of the mouse and the owl, the owl was eavesdropping on the mouse, whose rustling sounds were an incidental transfer of information to the owl called a **cue**, an unintentional transfer of information (which is different from the intentional transfer of information in a signal). Unlike honest and deceitful signaling, eavesdropping involves information transfer via cues rather than signals. Communication systems in which both individuals pay a fitness cost (–/–) are unlikely to evolve. And you should know why.

Information Use and Animal Signals

Information is critical to survival and reproduction because it helps individuals adjust to changing social and ecological circumstances (Dall et al. 2005). For example, individuals need to decide where to settle, where to forage, and with whom to breed, and having more information allows them to make more knowledgeable decisions. Animals communicate information about the world around them (such as where to find food), as well as information about themselves that can be stable over time or that varies over longer or shorter time periods (Bradbury and Vehrencamp 2011). Stable information includes things such as species identity, sex, and toxicity, whereas states such as physiological condition (e.g., hunger, health status) and dominance rank can change from season to season or even minute to minute. As you learned above, information transfer is critical to communication because it can directly influence the fitness of both the sender and the receiver. Senders may be attempting to transfer either motivational information about themselves (for example, when courting a potential mate) or referential information about a specific object (for example, when warning others about a predator; Donaldson et al. 2007). Despite some criticism that it is too restrictive and should instead acknowledge the different roles and often divergent interests of signalers and senders that can yield fundamental asymmetries in signaling interactions (Rendall et al. 2009), an information-based framework of communication integrates signal reliability with receiver decoding, decision making, and fitness consequences into one measure that is subject to selection (Bradbury and Vehrencamp 2011).

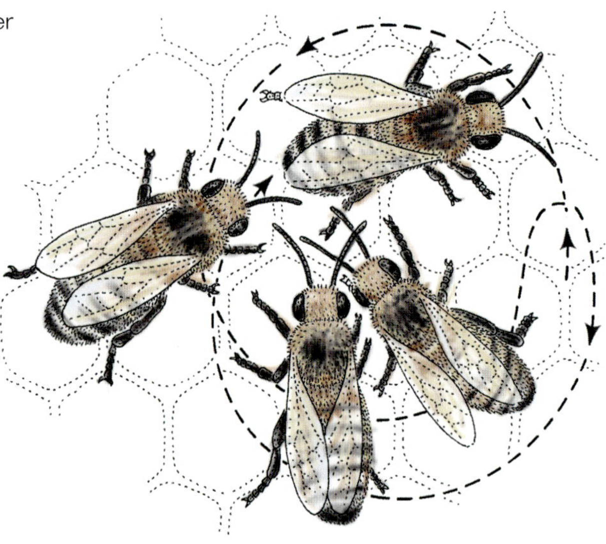

FIGURE 8.3 The round dance of honey bees. The dancer (the uppermost bee) is followed by three other workers, which may acquire information that a profitable food source is located within 50 meters of the hive. (From von Frisch 1965.)

The communication framework detailed above gives behavioral biologists the ability to identify the information being transmitted and how it affects both the sender's and the receiver's fitness. Take for example, the elaborate dances of the honey bee *Apis mellifera.* Dissecting this form of referential signal allows us to explore not only who is communicating, but what they are communicating about (the message) and how this influences the bees' fitness. These dances are performed by foragers (senders) when they return to their hive after finding good sources of pollen or nectar and are directed toward other workers (receivers) (von Frisch 1967). As the dancers move about on the vertical surface of the honeycomb in the complete darkness of the hive, they attract other bees, which follow them around as they move through their particular routines. Researchers watching dancing bees in special observation hives have learned that their dances contain a surprising amount of information about the location of a new food source, such as a patch of flowers. If the forager executes a round dance (**FIGURE 8.3**), she has found food fairly close to the hive—say, within 50 meters of it. If, however, the forager extends the round dance into a waggle dance (**FIGURE 8.4**), she has found a nectar or pollen source more than 50 meters away. The longer the waggle-run portion lasts, the more distant the food. Thus, the target distance—up to 10,000 meters—is encoded in the duration of the waggle run.

Knowing how far the food is from the hive only matters if the bees know which direction to fly. Luckily, this too is encoded in the waggle dance. By measuring the angle of the waggle run with respect to the vertical, an observer bee (or human) can tell the direction to the food source. A foraging bee on its way home from a distant but rewarding flower patch notes the angle between the flowers, hive, and sun. The bee transposes this angle onto the vertical surface of the comb when she performs the waggle-run portion of the waggle dance. If the bee walks directly up the comb while waggling, the flowers will be found by flying directly toward the sun. If the bee waggles straight down the comb, the flower patch is located directly away from the sun. A patch of flowers positioned 20 degrees to the right of a line between the hive and the sun is advertised with waggle runs pointing 20 degrees to the right of the vertical on the comb. In other words, when outside the hive, the bees' directional reference is the sun, whereas inside the hive, their reference is gravity.

The conclusion that the dances of honey bees contain information about the distance and direction to good foraging sites was reached by Karl von Frisch after

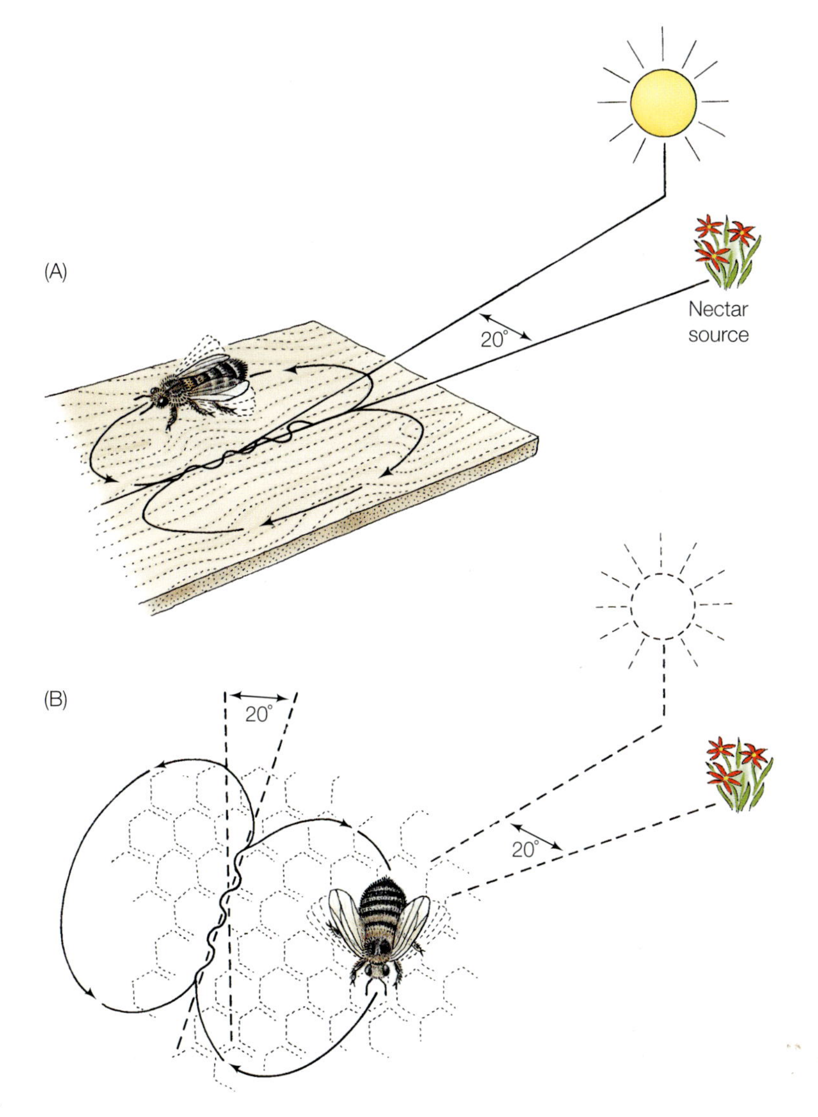

FIGURE 8.4 The waggle dance of honey bees. As a forager performs the waggle-run portion of the dance, she shakes her abdomen from side to side. The duration and orientation of the waggle runs contain information about the distance and direction to a food source. In this illustration, workers attending to the dancer learn that food may be found by flying 20 degrees to the right of the sun when they leave the hive. (A) The directional component of the dance is most obvious when the dance is performed outside the hive on a horizontal surface in the sunlight, in which case the bee uses the sun's position in the sky to orient her waggle runs directly toward the food source. (B) On the comb, inside the dark hive, dances occur on vertical surfaces, so they are oriented with respect to gravity; the deviation of the waggle run from the upward vertical equals the deviation of the direction to the food source from a line between the hive and the sun.

years of experimental work (von Frisch 1967). His basic research protocol involved training bees (which he marked with dots of paint for individual identification) to visit feeding stations that he stocked with concentrated sugar solutions. By watching the dances of these trained bees, von Frisch saw that their behavior changed in highly predictable ways depending on the distance and direction to a feeder. More important, his dancing bees were able to direct other bees to a feeder they had found (**FIGURE 8.5**), leading von Frisch to believe that bees use the information in the dances of their hive mates to find good foraging sites. Many years later, Jacobus Biesmeijer and Thomas Seeley were able to show that more than half the young worker bees that were just beginning their careers as pollen or nectar gatherers spent some time following dancing bees before launching their collecting flights (Biesmeijer and Seeley 2005). This finding suggests that dance information really is useful to bees about to start foraging for food.

In addition to obtaining distance and directional information from the waggle dance, bees may obtain other information from the dance. There is some evidence that recruits learn the odors of collected food from the dancers they follow. Receivers make contact with the dancer's body parts where food odors are most intense, such as the mouthparts and hindlegs. Some bees appear to rely more on the signals in the dance, while others rely more on the food odors. The waggle dance also helps

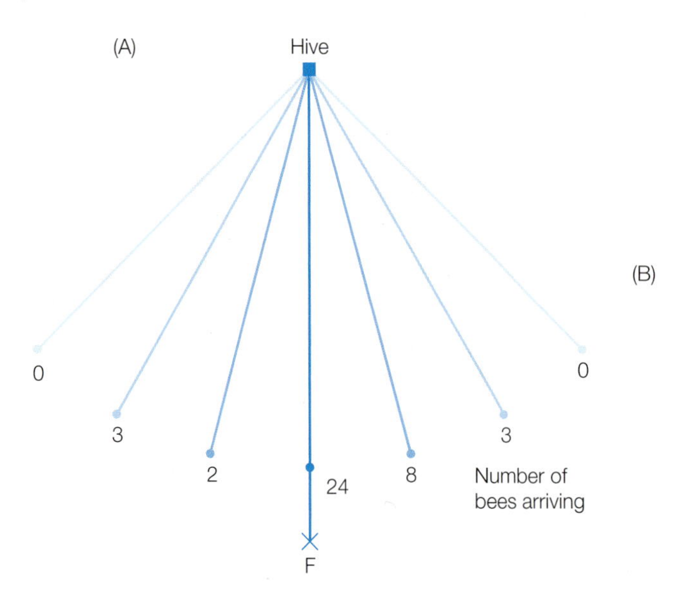

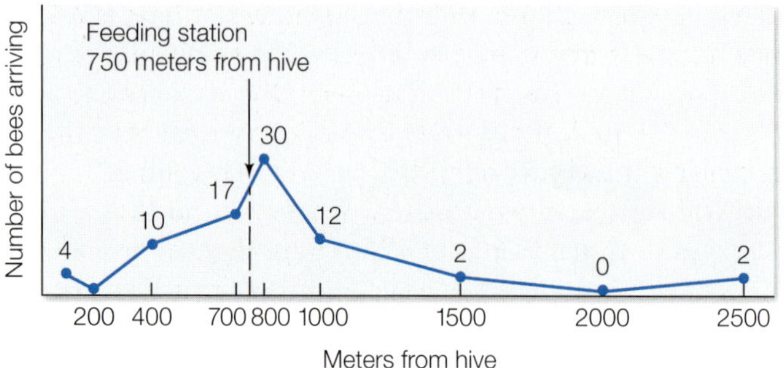

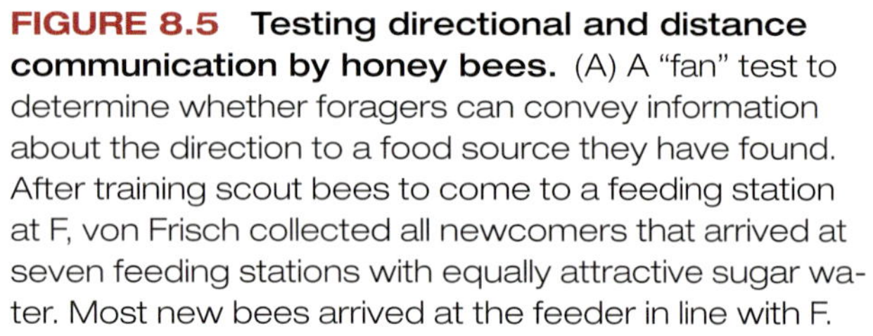

FIGURE 8.5 Testing directional and distance communication by honey bees. (A) A "fan" test to determine whether foragers can convey information about the direction to a food source they have found. After training scout bees to come to a feeding station at F, von Frisch collected all newcomers that arrived at seven feeding stations with equally attractive sugar water. Most new bees arrived at the feeder in line with F. (B) A test for distance communication. After training scouts to come to a feeding station 750 meters from the hive, von Frisch collected all newcomers arriving at feeders placed at various distances from the hive. In this experiment, 17 and 30 newcomers were captured at the two feeders closest to 750 meters, far more than were caught at any other feeder. (After von Frisch 1965.)

modulate the readiness of bees to respond to the spatial information encoded in the dance (Gruter and Farina 2009).

The dances of the honey bee illustrate how animals communicate referential information about their environment. This includes not only searching for food in the case of the bees, but things such as producing alarm calls to warn others about approaching predators. One of the best studied of these referential alarm calls is that of the vervet monkey (*Chlorocebus pygerythrus*). Robert Seyfarth, Dorothy Cheney, and Peter Marler demonstrated that vervet monkeys living in the Kenyan savanna have distinctive alarm calls that signal different types of predators and classes of external danger (Seyfarth et al. 1980). By recording alarm calls made in the presence of leopards, eagles, and snakes, and then playing those calls back in the absence of predators, the researchers demonstrated that monkeys responded very differently to the different types of alarms: when played a leopard alarm call, monkeys ran into trees; when played an eagle alarm call, monkeys looked up or ran for cover; and when played a snake alarm call, monkeys looked down or approached the source of the signal, presumably to mob a snake if they encountered one. Thus, each call conveyed different information about the type of predator, and receivers interpreted these varied signals and responded differently to each one.

The Evolution of Animal Signals

Now that you understand a bit about how communication systems are structured and the vital role that information transfer plays in animal signaling, we can ask how such systems evolve. Doing so requires considering the coevolutionary interactions between senders and receivers and thinking about how new signals might originate in the first place. There are two general hypotheses for how a signal might evolve:

from **preexisting traits** in senders or from **preexisting biases** in receivers (**HYPOTHESES 8.1**). Preexisting traits are behavioral, physiological, or morphological characteristics that already provide informative cues to receivers, and if the sender benefits from the receiver's response as a result of these cues, then the cue can be modified into a signal via a process called ritualization (Bradbury and Vehrencamp 2011). Piloerection, the fluffing of feathers or hair, is a thermoregulatory response that has been ritualized in some animals. For example, a dog may raise the hair on its back if it is excited or aggressive, or a Gambel's quail (*Callipepla gambelii*) may raise its head crest during aggressive interactions (**FIGURE 8.6**) (Hagelin 2002). Additionally, pupil dilation in birds can signal aggression, and yawning in primates can signal a threat display. Each of these signals provides information to receivers, even though the behaviors themselves evolved for other (physiological) reasons.

In order for these initial cues to become signals, they must be detected by the receiver. Since signal transmission varies with the environment (for example, sound travels differently in air than underwater), many signals are fine-tuned to the environment in a process called **sensory drive** (Endler 1992, 1993). The term *sensory drive* comes from the fact that animals use their sensory system to stay alive—to find food and to avoid becoming prey themselves. Sensory systems are therefore under natural selection to optimize these behaviors within the constraints of each species' habitat. One byproduct of this sensory tuning is that sensory systems have biases such that they detect some features of an organism's world better than others. These preexisting biases can influence the design properties of signals. In other words, receivers tend to have preexisting detection biases to certain stimuli in the environment that arose before the evolution of the signal.

HYPOTHESES 8.1

Alternative hypotheses for how signals evolve

preexisting trait hypothesis
Signals evolve through senders' preexisting traits, which are behavioral, physiological, or morphological characteristics that already provide informative cues to receivers and can be modified into a signal in a process called ritualization.

preexisting bias hypothesis
Signals evolve through receivers' preexisting biases, which are biases in their sensory systems that detect some features of the world better than others and that can be exploited by sender signals in a process called sensory exploitation.

(A)

(B)

FIGURE 8.6 Male Gambel's quail use their feathered head crests to signal during aggressive interactions. (A) Dominant individuals have erect crests that hang forward, whereas subordinate males (B) have flattened crests that are lowered and point backwards. (A, photograph © vagabond54/Shutterstock.com; B, photograph © Deep Desert Photography/Shutterstock.com.)

When communication signals originate in actions that activate sensory abilities and biases of receivers that are already in place, the result is often due to what is called **sensory exploitation** (Basolo 1990, Ryan 1998, Grether 2010). As you will see below, there are many examples of signals that are thought to have arisen via sensory exploitation. Although signals derived from sensory biases are usually used in mate attraction (Endler and Basolo 1998, Ryan and Cummings 2013), when a signal that mimics the features of environmental stimuli exploits the associated receiver response, it is referred to as a *sensory trap* (Bradbury and Vehrencamp 2011). An example of a sensory trap is the light lures used by several deep-sea anglerfishes that mimic the bioluminescent prey of other fishes in order to manipulate those fishes to come close enough to be eaten. Thus, animal signals evolve via preexisting traits or preexisting biases. Below we discuss several examples of each mechanism, beginning with a thought experiment in hyenas and followed by a summary of empirical studies from other animal species.

Preexisting Traits and the Development of a Strange Display

To understand how signals evolve via preexisting mechanisms, let's begin with the female spotted hyena's pseudopensis and do a thought experiment. Imagine that the costs to a pseudopenis presenter were greater than the benefits of the behavior. Over time, if there were some individuals with a hereditary aversion to pseudopenis presentation, the distinctive alleles of these hyenas would be favored by natural selection, with the result that eventually pseudopenile signaling would disappear from the population. Likewise, if receivers were in some way harmed by responding to the signal provided by the pseudopenis, then female hyenas with a hereditary predisposition to ignore pseudopenis presentations would have higher fitness, and eventually responders would no longer exist. Thus, both senders and receivers are required if a communication system is to evolve. So let's assume that because female spotted hyenas regularly engage in pseudopenis presenting and pseudopenis sniffing, we are dealing with a cooperative interaction and a form of honest signaling, an evolved communication system with benefits to both parties.

How would such a system evolve in the first place? To answer this question, we turn to Darwin's theory of descent with modification—the idea that traits providing a reproductive advantage become more common in a population over time. In 1939, L. Harrison Matthews proposed that the pseudopenis might have originated as the developmental result of high levels of male sex hormones in female hyenas while they were still in their mother's uterus (Matthews 1939). Thus, the pseudopenis was the result of developmental and physiological processes, and only later was this preexisting trait co-opted for signaling. Stephen Jay Gould termed this proximate hypothesis the **extra androgen hypothesis** (Gould 1981). As Matthews and Gould knew, the male penis and the female clitoris of mammals develop from the same embryonic tissues. The general mammalian rule is that if these tissues are exposed very early on to androgens (such as testosterone), as they almost always are in male embryos, a penis is the end result. If the same target cells do not interact with androgens, as is the case for the typical embryonic female mammal, then a clitoris develops. But when a female embryo of most mammals comes in contact with testosterone, either in an experiment or because of an accident of some sort, her clitoris becomes enlarged and rather penislike (Ewer 1973). This effect has been observed in our own species in the daughters of pregnant women who received medical treatment that inadvertently exposed their offspring to extra testosterone (Money and Ehrhardt 1972), as well as in females whose adrenal glands produce more testosterone than normal (Place and Glickman 2004).

extra androgen hypothesis The female pseudopenis develops as a result of the exposure in utero to elevated androgens.

Given the general pattern of development of mammalian external genitalia, Gould thought it likely that female embryos in the spotted hyena must be exposed to unusual concentrations of androgens. In support of this descent with modification hypothesis, he pointed to a 1979 paper in which P. A. Racey and J. D. Skinner reported that captive female spotted hyenas had circulating levels of testosterone equal to those of males, unlike other hyenas and unlike mammals generally (Racey and Skinner 1979). This discovery appeared to support the extra androgen hypothesis and the idea that an unusual hormonal environment during development led to the origin of the female spotted hyena's pseudopenis, setting the stage for the origin of the penis-sniffing greeting ceremony.

Work on the extra androgen hypothesis continued decades later, with some researchers measuring androgen levels in wild female spotted hyenas. In contrast to captive hyenas, the researchers found that testosterone levels in free-living females are lower than in adult males (Goymann et al. 2001). Given the role that testosterone plays in reproduction and spermatogenesis in males (see Chapter 5), this is perhaps not terribly surprising in hindsight. However, *pregnant* female spotted hyenas have higher testosterone levels than lactating females (**FIGURE 8.7**) (Dloniak et al. 2004). This finding leaves the door open for the extra androgen hypothesis (Drea et al. 2002), as does that from another mammal with a penile clitoris, the ring-tailed lemur (*Lemur catta*), in which females boost the concentration of androgens in their blood at the onset of pregnancy (Drea 2011). Because the placenta of a pregnant spotted hyena converts certain androgens to testosterone, her female embryos could conceivably be exposed to masculinizing levels of this hormone (Glickman et al. 1993). However, during the time that embryonic clitoral development is likely to be most sensitive to testosterone, the mother hyena's placental cells are producing substantial amounts of the enzyme aromatase which inactivates testosterone and turns it into estrogen. In humans and most other mammals, the same enzyme prevents masculinization of the genitalia of embryonic females (Conley et al. 2006).

The extra androgen hypothesis also leads to the prediction that experimental administration of anti-androgenic chemicals to pregnant adult spotted hyenas should abolish the pseudopenis in their subsequent daughters while also feminizing the external genitalia of their newborn sons. In reality, however, when anti-androgens are administered to pregnant hyenas, the daughters of the treated females retain the elaborate pseudopenis, albeit in somewhat altered form (Glickman et al. 2006). Results such as this are at odds with the view that early exposure to male sex hormones is sufficient in and of itself to cause female embryos of the spotted hyena to develop their unusual genitalia (Drea et al. 1998, Place et al. 2011).

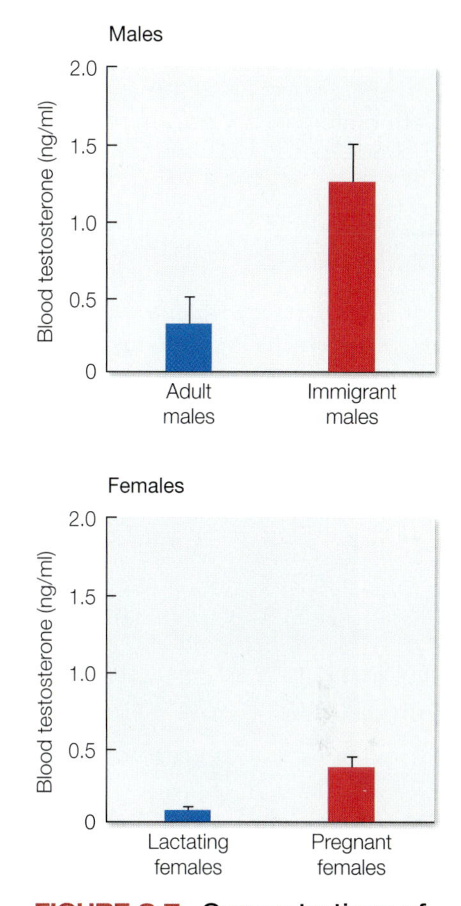

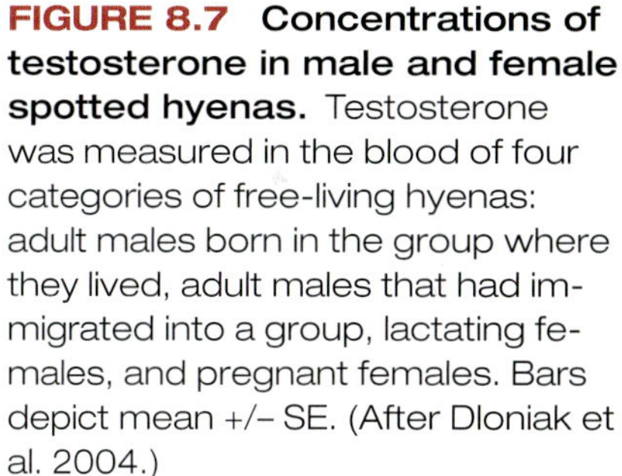

FIGURE 8.7 Concentrations of testosterone in male and female spotted hyenas. Testosterone was measured in the blood of four categories of free-living hyenas: adult males born in the group where they lived, adult males that had immigrated into a group, lactating females, and pregnant females. Bars depict mean +/– SE. (After Dloniak et al. 2004.)

The Panda Principle and Preexisting Traits

To further understand the role that preexisting traits play in the evolution of signals, it might help to step back and think a bit more broadly about how evolution by natural selection operates. If natural selection had been given the job of coming up with a jet plane, for example, it would have had to start with what was already available, presumably a propeller-driven plane, changing this "ancestral" structure piece by piece, with each modified airplane flying better than the preceding one, until a jet was built (Dawkins 1982). In nature, evolutionary transitions must occur this way because selection does not start from scratch but instead acts on what already exists. In other words, evolution is a tinkerer, not an engineer (Jacob 1977).

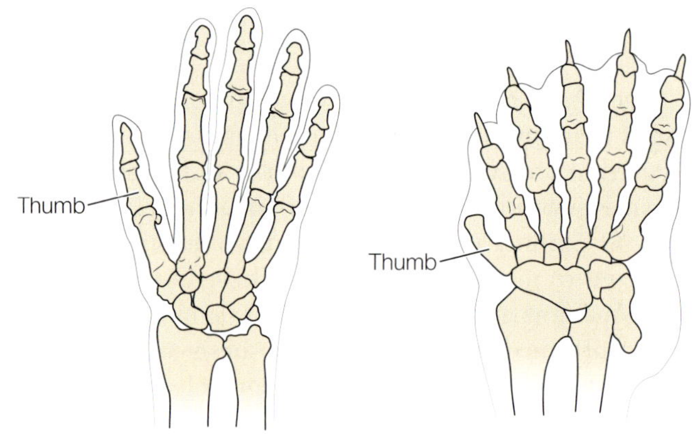

FIGURE 8.8 The panda principle. The human hand (left) has five digits, four fingers and a thumb. In contrast, the panda's forepaw has the same five digits as most vertebrate hands, plus a small thumblike projection from the radial sesamoid bone.

FIGURE 8.9 Descent with modification is responsible for the sexual behavior of a parthenogenetic whiptail lizard. In the left column, a male of a sexual species engages in courtship and copulatory behavior with a female. In the right column, two females of a closely related parthenogenetic species engage in very similar behavior. (From Crews and Moore 1986, © 1987 by Prentice-Hall Inc., Englewood Cliffs, NJ.)

A good example of this type of evolutionary tinkering is the thumb of the giant panda (*Ailuropoda melanoleuca*). The panda's thumb is not a "real" finger at all, but instead a highly modified wrist bone. The panda's forepaw has the same five digits that most vertebrate hands have plus a small thumblike projection from the radial sesamoid bone, which is present, although as a much smaller bone, in the forepaws of bears, dogs, raccoons, and the like (**FIGURE 8.8**) (Gould 1986). Why do pandas have their own special thumb? According to Stephen Jay Gould, pandas evolved from carnivorous ancestors whose first digit had become an integral part of a paw used in running. As a result, when pandas evolved into herbivorous bamboo eaters, the first digit was not available to be employed as a thumb in stripping leaves from bamboo shoots. Instead, selection acted on variation in the panda's radial sesamoid bone, which is now used in a thumblike manner by the bamboo-eating panda.

What Gould called "the panda principle," Darwin labeled "the principle of imperfection." No matter the title, the phenomenon is widespread. Consider, for example, the persistence of sexual behavior in species of whiptail lizards that are composed entirely of parthenogenetic (asexual) females. In these species, a female may be "courted" and mounted by another female. Why would females do this? It turns out that females subjected to pseudomale sexual behavior are much more likely to produce a clutch of eggs than if they do not receive this sexual stimulation from a partner (**FIGURE 8.9**) (Crews and Moore 1986). The effect of courtship on fecundity in unisexual lizards obviously exists because these reptiles are derived from sexual ancestors. The parthenogenetic females retain characteristics, such as an acceptance of courtship, that their non-parthenogenetic ancestors possessed.

As we detailed earlier, the waggle dance of the honey bee *Apis mellifera* enables the transfer among workers of information critical to the survival of the colony. Perhaps such a complicated method of communication evolved according to the panda principle and the idea of modification of preexisting traits (behaviors) in ancestral bee species? Martin Lindauer was the first to propose a hypothetical historical scenario for the extraordinary dances of honey bees that involved the retention of ancestral traits (Lindauer 1961). Lindauer began his work by comparing three other members of the genus *Apis*, which he found to perform dance displays identical to those of *A. mellifera*, except that in one species, *A. florea*, the bees dance on the horizontal surface of a comb built in the open over a tree branch (**FIGURE 8.10**). To indicate the direction to a food source, a forager of this species simply orients her waggle run directly toward the food's location. Because this is a less sophisticated maneuver than the transposed pointing done in the dark on vertical surfaces by *A. mellifera*, it may resemble a form of dance communication that preceded the dances of *A. mellifera*.

Lindauer then looked to stingless tropical bees in other genera for recruitment behaviors that might

FIGURE 8.10 The nest of an Asian honey bee, *Apis florea*, is built out in the open around a branch. Dancing foragers on the flat upper surface of the nest can run directly toward the food source when performing the waggle dance. (Photograph © Martin Mecnarowski/Shutterstock.com.)

FIGURE 8.11 Communication by scent marking in a stingless bee. In this species (*Scaptotrigona* [formerly *Trigona*] *postica*), foragers that found food on the side of a pond opposite from their hive could not recruit new foragers to the site until Martin Lindauer strung a rope across the pond. Then the scouts placed scent marks on the vegetation hanging from the rope and quickly led others to their find. (Photograph by Martin Lindauer.)

provide hints about the steps preceding waggle dancing. These comparisons led him to suggest the following historical sequence:

A possible first stage: Workers of some species of stingless bees in the genus *Trigona* run around excitedly, producing a high-pitched buzzing sound (a form of preexisting trait) with their wings when they return to the nest from flowers rich in nectar or pollen. This behavior arouses their hive mates, which detect the odor of the flowers on the dancers' bodies. With this information, the recruits leave the nest and search for similar odors. The same kind of behavior also occurs in some species of bumblebees, which form small colonies with "dancing" scouts that do not provide signals containing directional or distance information but that do alert their fellow bees to the existence of resource-rich flowers in the neighborhood of the colony (Dornhaus and Chittka 2001).

A possible intermediate stage: Workers of other species of *Trigona* do convey specific information about the location of a food source. In these species, a worker that makes a substantial find marks the area with a pheromone produced by her mandibular glands. As the bee returns to the hive, she deposits pheromone on tufts of grass and rocks every few meters. Inside the hive entrance, other bees wait to be recruited. The successful forager crawls into the hive and produces buzzing sounds that stimulate her companions to leave the hive and follow the scent trail she has made (**FIGURE 8.11**).

A still more complex pattern: Several stingless bees in the genus *Melipona* (a group related to *Trigona* bees) convey distance and directional information, but they do so separately. A dancing forager communicates information about the distance to a food source by producing pulses of sound; the longer the pulses, the farther away the food is. To transmit directional information, she leaves the nest with several followers and performs a short zigzag flight that is oriented toward the food source. The scout returns and repeats the flight numerous times before flying straight off to the food source, with the recruited bees in close pursuit.

Lindauer was not suggesting that some stingless bees behave in a more adaptive manner than others. He simply used the diversity of existing traits in this group to provide possible clues about the behaviors of now extinct bees whose communication systems were modified in species derived from these ancestral bees. On the basis of what extant species do, Lindauer hypothesized that communication about

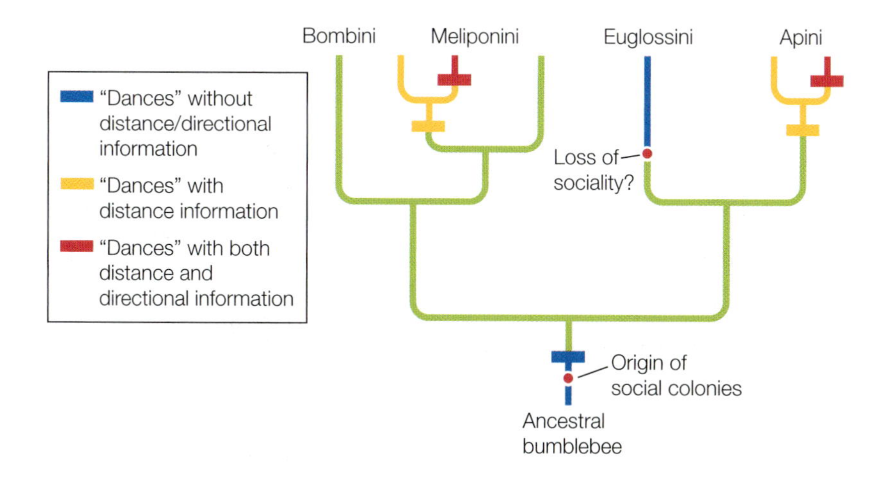

FIGURE 8.12 Evolutionary history of the honey bee dance communication system. A phylogeny of four closely related bee groups, one of which (the Apini) includes *Apis mellifera.* According to this tree, the closest relative of the Apini is the Euglossini (the orchid bees, which are not social), not the Meliponini (the social tropical stingless bees). This phylogeny indicates that complex dance communication has evolved independently in the Meliponini and Apini. (After Cardinal and Danforth 2011.)

the *distance* to a food source was initially encoded in the degree of activity by a returning food-laden worker (Nieh 2004). Subsequently, selection acting in some species may have favored standardization of the sounds and movements made by successful foragers, which set the stage for the round and waggle dances of *Apis* bees (Lindauer 1961, Wilson 1971). Moreover, Lindauer's comparisons indicated that communication about the *direction* to a food source might have originated with a worker guiding a group of recruits directly to a nectar-rich area. This behavior has evolved gradually to contain a less and less complete performance of the guiding movements, beginning with partial leading (as in some *Melipona*) and later consisting of pointing in the proper direction with a waggle run on a horizontal surface (as in *A. florea*). From antecedents such as these came the transposed pointing of *A. mellifera,* in which the direction of flight relative to the sun is converted into a signal (the waggle run) oriented relative to gravity.

The historical sequence of events outlined by Martin Lindauer can be tested if we have an accurate phylogeny of four closely related groups of bees: the bumble bees, the orchid bees, the tropical stingless bees, and the *Apis* bees. Sophie Cardinal and Bryan Danforth have developed just such a phylogeny that makes use of a large data set of molecular similarities and differences among the four groups of bees as well as sophisticated statistical procedures (**FIGURE 8.12**) (Cardinal and Danforth 2011). The researchers' phylogeny, however, does not support the hypothesis that the dance behavior of *Apis mellifera* was derived with descent by modification from bees that gave rise to today's tropical stingless bees. Indeed, evidence from the fossil record suggests that the dance language could be older than 40 million years and may predate the genus *Apis,* or it could be only 20 million years old, a time when honey bees diverged from other bees. Perhaps, then, the ancestral honey bee dance may have had nothing to do with finding food. Instead, some researchers have argued that the honey bee dance originally evolved to allow individuals to select a nest site, and only later was co-opted for foraging (Beekman et al. 2008). In honey bees, dancing continues to play a role during swarming and nest-site selection (Seeley 2010). Though we may not yet know how the honey bee's waggle dance evolved, researchers have done a remarkable job of deciphering its complex message.

Preexisting Biases and the Evolution of Animal Signals

Sensory drive is likely to be important for signal evolution when the constraints on the signaling environment are extreme (**BOX 8.1**). Therefore, it is not surprising that there is an abundance of evidence for sensory mechanisms in aquatic habitats where the light environment dramatically changes due to interactions with the

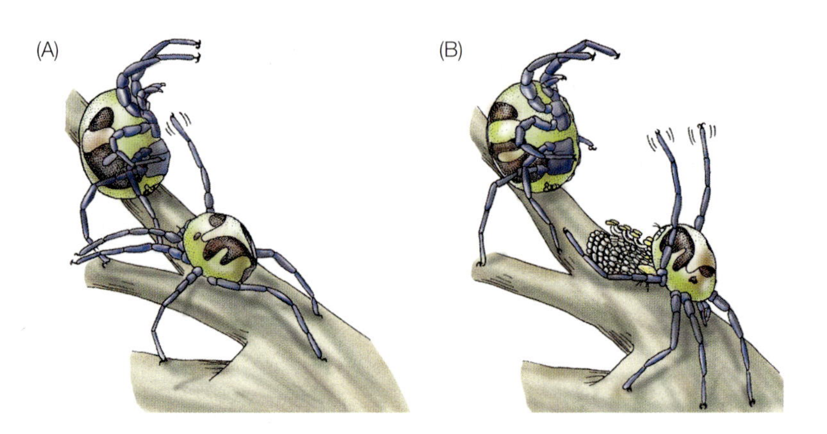

FIGURE 8.13 Sensory exploitation and the evolution of a courtship signal in the water mite *Neumania papillator*. (A) The female (on the left) is in her prey-catching position (the net stance). The male approaches and waves a trembling foreleg in front of her, setting up water vibrations similar to those a copepod might make. The female may respond by grabbing him, but she will release him unharmed. (B) The male then deposits spermatophores on the aquatic vegetation in front of the female before waving his legs over them. (After Proctor 1991.)

water medium. In the predatory water mite *Neumania papillator*, for example, courting males appear to have tapped into the preexisting sensory abilities of females (Proctor 1992). While a female is waiting for an edible copepod to make the mistake of bumping into her, she adopts a particular pose, called the net stance. A male that encounters a female in this position may vibrate a foreleg in front of her, a behavior that generates vibrational stimuli in the water similar to those provided by an approaching copepod. The female water mite, in turn, may grab the male mite, using the same response that she uses to capture her prey, although she will release the captured male unharmed. He may then turn around and place **spermatophores** (packets of nutrients and sperm) near the female, which she will pick up in her genital opening if she is receptive (**FIGURE 8.13**) (Proctor 1991, 1992). If males trigger the prey-detection response of females, then unfed, hungry female water mites held in captivity should be more responsive to male signals than well-fed females. In a series of experiments, Heather Proctor showed that hungry females are in fact more responsive to males than are well-fed females, providing support for the contention that once the first ancestral male happened to use a trembling signal, the male's behavior spread because it effectively activated a preexisting prey-detection mechanism in females (Proctor 1991, 1992).

There are numerous accounts that fish visual sensitivity varies according to optical properties of the water, and this in turn results in predictable male signaling traits and female mate preferences. For example, females in some populations of the Trinidadian guppy (*Poecilia reticulata*) prefer to mate with males that have bright orange markings (**FIGURE 8.14**) (Grether 2000). Male Trinidadian guppies cannot synthesize the orange pigments that go into their body coloration and instead have to acquire these carotenoids from the plants they eat, primarily a type of unicellular alga. Those males that get enough carotenoids from their food are more attractive to females, but why? One idea

(A)

(B)

FIGURE 8.14 Food, carotenoids, and female mate preferences in the Trinidadian guppy. (A) Males have to acquire orange pigments from the foods they eat, such as this *Clusia* fruit that has fallen into a stream where the guppies live. Males that secure sufficient amounts of carotenoids incorporate the chemicals into ornamental color patches on their bodies. (B) Females (such as the larger fish on the right) find males with large orange patches more sexually appealing than males without them. (Photographs by Greg Grether.)

BOX 8.1 **EXPLORING BEHAVIOR BY INTERPRETING DATA**

Spiders hunting prey at night

Many spiders hunt at night. Some species, such as this giant golden orb weaver spider (*Nephila pilipes*), have conspicuous body coloration on one side of their body but not the other (**FIGURE A**). What do you think this bright coloration is used for, and why is it on only one side of the body?

FIGURE A The dorsal (left) and ventral (right) sides of the giant golden orb weaver spider. (Left photograph by imagebroker/Alamy Stock Photo; right photograph by National Geographic Creative/Alamy Stock Photo.)

Thinking Outside the Box

Researchers studying another species of orb weaver spider, *Neoscona punctigera*, hypothesized that the difference in coloration was related to the spider's diurnal and nocturnal behavior (Chuang et al. 2008). Can you generate a hypothesis to explain the difference in this nocturnally hunting spider's dorsal (brown) and ventral (brown with bright spots) coloration? The researchers found that the spiders tended to be camouflaged to diurnal hymenopteran predators. They quantified prey interception rates when the spider was present and not present on the web at night (**FIGURE B**), as well as when the bright belly spots were covered with brown paint (**FIGURE C**). Look at the graphs depicting the results from these experiments and interpret them in relation to your hypothesis. If necessary, revise your hypothesis to make it consistent with these experimental findings. What do you think spiders that hunt during the day would look like?

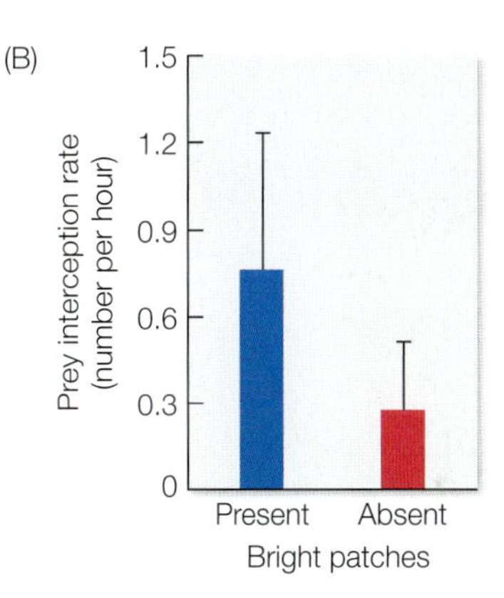

FIGURE B Prey interception rates of webs with spider present versus not present. In webs built by *Neoscona punctigera*, the prey interception rate was 2.5 times higher when the spider was present versus when it was not present. Bars depict mean +/– SE. (After Chuang et al. 2008.)

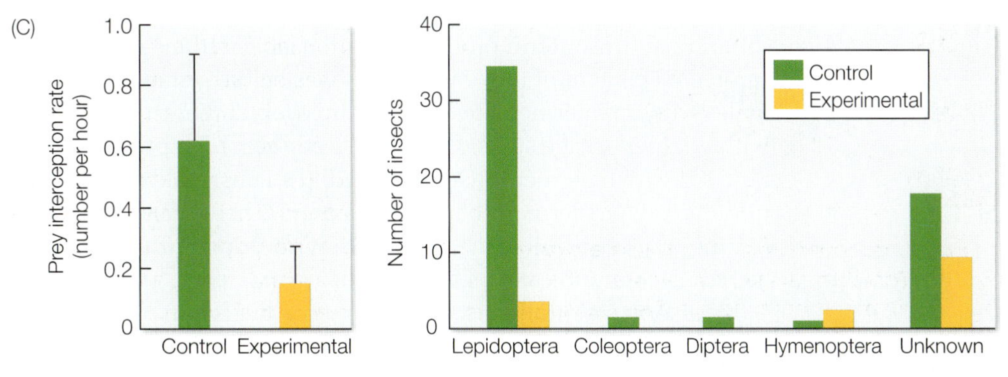

FIGURE C Prey interception rates of webs with control versus experimental spiders. *Neoscona punctigera* in the experimental group had their bright belly spots painted brown, whereas those in the control group had the brown part of their belly painted brown. The prey interception rate was 3 times higher in webs with control spiders than in webs with experimental spiders. Bars depict mean +/– SE. (After Chuang et al. 2008.)

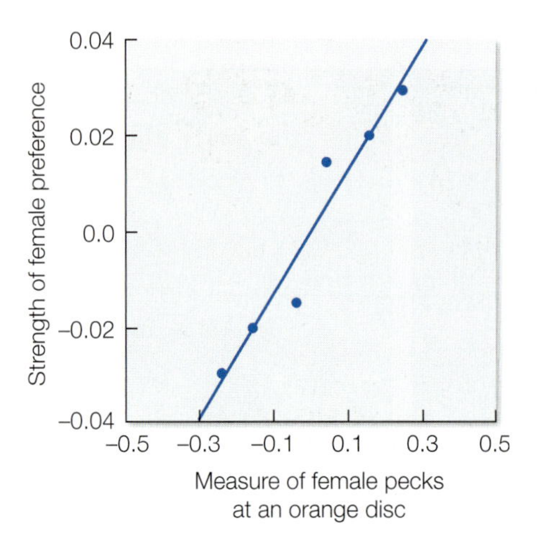

FIGURE 8.15 Sexual preferences for orange spots match foraging preferences for orange foods by female Trinidadian guppies. The strength of the female response to orange spots on males (vertical axis) varies from population to population and is proportional to the strength of the foraging response to an orange disc. Each point represents a different guppy population living in a different stream. (After Rodd et al. 2002.)

about the origin of the female preference for orange-spotted males suggests that when this mating preference first appeared, it was a by-product of a sensory preference that had evolved in another context. This sensory exploitation idea has received support from observations of female guppies feeding avidly on rare but nutritionally valuable orange fruits that occasionally fall into the Trinidadian streams where these guppies live (Rodd et al. 2002). Thus, it is possible that females evolved visual sensitivity to orange stimuli because of the foraging benefits associated with this ability, not because of any fitness benefits from selective mate choice. If this is the case, then female guppies can be predicted to respond more strongly to orange stimuli than to other colors when feeding. In fact, female guppies do approach and try to bite an inedible orange disc far more often than discs of any other color. Moreover, the strength of the mate choice preference of females from different populations was matched by the relative rates at which females snapped at orange discs presented to them (**FIGURE 8.15**).

Of course, there is more to the Trinidadian guppy story than just sensory exploitation. Although the guppies' orange spots are largely the result of carotenoid pigments obtained from their diet, the spots also contain a bit of a red pigment called drosopterin, something that guppies *can* synthesize (Grether et al. 2001). One might predict that when carotenoids are scarce, males instead synthesize more drosopterin to make the orange spots. But in fact, males seem to produce more—not less—drosopterin in streams with higher carotenoid availability. Why might they do this? By bringing guppies from six different Trinidadian streams into the lab and rearing them together through the second generation of offspring, Grether and colleagues showed that geographic variation in drosopterin production has a large hereditary basis (Grether et al. 2005). In nature this results in a geographical pattern whereby the carotenoid content of the orange spots of males is counterbalanced by genetic variation in drosopterin production, resulting in a uniform pigment ratio across all populations. What would cause such a pattern and developmental system to evolve? Grether hypothesized that female preference for color shading (termed hue) could produce this pattern, since hue is directly affected by the ratio of the two pigments. By crossing two populations differing in drosopterin production in the lab, Grether's team found that when the carotenoid content of the orange spots was held constant, female guppies preferred males with an intermediate drosopterin level. Rather than simply preferring males with brighter orange spots, females instead seemed to prefer males with a particular ratio of the two pigments (Deere et al. 2012). Thus, male guppy signals used to communicate to females have both a dietary and a hereditary component, another good reminder of how both genes and the environment influence all phenotypic traits, including communication signals.

If sensory exploitation is a major factor in the origin of effective signals, then one key prediction is that it should be possible to create novel experimental signals that trigger responses from animals that have never encountered those stimuli before (Rodríguez and Snedden 2004). Scientists should then be able to create new types of mating signals to determine if they tap into a female's preexisting sensory bias. To test this prediction, researchers have played sounds to frogs that contain acoustic

(A)

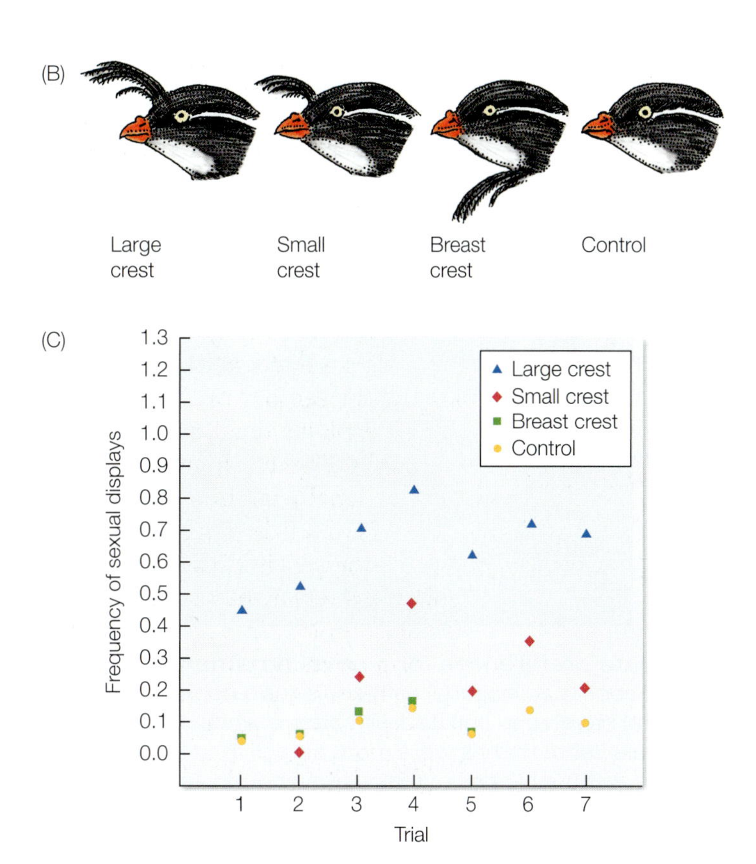

FIGURE 8.16 The response of least auklets to three novel artificial signals. (A) A stuffed male least auklet (*Aethia pusilla*) with an artificial crest of the sort used in an experiment on female sexual preferences in this species. (B) A diagram of some of the heads of models used in the experiment (from right to left): the control (which lacks a crest, as do males of the least auklet), a breast crest model, a small crest model, and a large crest model. (C) The models with large crests elicited the highest frequency of sexual displays by the females during a standard presentation period. (A, photograph by I. L. Jones; B, C after Jones and Hunter 1998.)

elements not present in the species' natural calls (Ryan et al. 1990, 2010), attached strips of yellow plastic to the shorter tails of male platyfish (Basolo 1990), supplied male sticklebacks with bright red spangles to add to their nests (Östlund-Nilsson and Holmlund 2003), and added feather crests to the crestless heads of auklets (**FIGURE 8.16**) (Jones and Hunter 1998) as well as to the heads of some crestless Australian finches (**FIGURE 8.17**) (Burley and Symanski 1998). In all of these cases, researchers found that the artificial attributes elicited stronger reactions from females than the natural traits did. Thus, in a diversity of taxonomic groups, it is possible to elicit a preexisting bias in female mate choice, demonstrating that these receiver biases are likely to be mechanisms that generate new signals, particularly in a reproductive context. In some cases, however, males manipulate female biases for their own good (**BOX 8.2**).

Preexisting Traits versus Preexisting Biases

It is often unclear whether an animal signal is the result of a preexisting trait in the sender or a preexisting bias in the receiver. After all, the experiments above suggest that novel signals could originate when senders happened to possess mutant attributes that turned on preexisting sensory preferences that had evolved for other purposes. An alternative to the sensory exploitation explanation is that some artificial signals may elicit responses because the ancestors of the tested species used

FIGURE 8.17 Mate preferences for a novel ornament. (A) A male long-tailed finch (*Poephila acuticauda*) (left) and a male zebra finch (*Taeniopygia guttata*) (right) have been outfitted with bizarre white plumes. (B) The "white-plumed" male zebra finches were more attractive to females than were control males without plumes or males given headdresses of red or green feathers. Bars depict mean +/– SE. (A, photograph by Kerry Clayman, courtesy of Nancy Burley; B after Burley and Symanski 1998.)

similar signals during courtships in the past and today's descendants have retained the sensory preferences of their ancestors (Ryan and Wagner 1987). This conjecture is plausible, given the numerous cases in which elaborate male traits used in courtship and aggression have been lost after having once evolved. In these cases, the close relatives of the ornament-free species possess the ornaments in question, which suggests that their mutual common ancestor did as well (Wiens 2001). Yet in all of these scenarios, we are still left with the question: which came first, the trait or the preference?

An example suggesting that shared ancestral response may explain responses to novel stimuli is provided by the lizard *Sceloporus virgatus*, which lacks the large blue abdominal patches of many other members of its genus. Males in *Sceloporus* species with blue abdominal patches use them in a threat posture, which involves the elevation of the lizard's body and its lateral compression, making the abdomen visible. Although this signal has apparently been lost in *S. virgatus*, the behavioral response to the signal has not, as was shown when some male *S. virgatus* lizards had blue patches painted on them (Quinn and Hews 2000). Males that observed a rival giving a threat display experimentally enhanced by the blue paint were far more likely to back off than were males that saw a displaying opponent that did not have the added blue patches (**FIGURE 8.18**).

Another example where the case for sensory exploitation versus shared ancestry is unclear occurs in fish. What are we to make of the fact that the fish *Xiphophorus maculatus* does not have an elaborate elongated swordtail like its relative, the swordtail fish *X. hellerii*, and yet when Alexandra Basolo endowed males of *X. maculatus* with an artificial yellow sword, females found males with this novel trait highly attractive (Basolo 1995)? Furthermore, the longer the sword, the greater the

BOX 8.2

DARWINIAN PUZZLE

Why do female moths mate with males that produce ultrasonic mimetic signals similar to those produced by predatory bats?

A puzzling example of apparent sensory exploitation comes from the "ears" of male and female Australian whistling moths (*Hecatesia exultans*), which are capable of detecting ultrasonic signals generated by male moths (**FIGURE A**). Researchers interested in the evolutionary history of these unusual structures have argued that they represent modifications of ears that were used by the ancestors of this species to detect bats. As we discussed in Chapter 4, predatory bats locate and track moth prey by listening to the ultrasonic echoes from the high-frequency calls bats produce when out hunting (Fullard 1998). If the ancestor of today's whistling moths could hear bat cries and take defensive action, it could also have heard the ultrasound produced when the first perched male whistling moth beat its wings together in a way that generated an ultrasonic signal. If this were the case, a bat-detecting system in moths could have been co-opted as a communication system with mate-searching females (receivers) locating acoustic males (senders) by hearing moth-produced ultrasound (Alcock and Bailey 1995).

This hypothesis has been tested with females of another species of noctuid moth, the tobacco cutworm (*Spodoptera litura*), in which males also use ultrasonic courtship signals to induce sexual receptivity in potential partners. Recordings of these signals as well as those of simulated bat calls did equally well in triggering female receptivity, as one would predict from the hypothesis that a sensitivity of females to bat ultrasound preceded and facilitated the evolution of ultrasonic courtship by male moths (Nakano et al. 2010). In other words, females do not distinguish between male songs and bat calls, consistent with the idea that the acoustic signaling by males evolved from a preexisting sensory mechanism.

(A)

FIGURE A Ultrasonic signaling in moths. Male Australian whistling moths produce ultrasounds by striking together the two "castanets" on their forewings. (Photograph by John Alcock.)

Thinking Outside the Box

This behavior seems like a dangerous proposition for female moths. It is one that appears maladaptive if females are really unable to distinguish a mate from a predator. In a related moth species, the Asian corn borer (*Ostrinia furnacalis*), females respond to bat ultrasounds by exhibiting evasive behaviors such as diving, looping flight, or steep turns when the potential predators are close by and produce ultrasonic calls of high intensity, but by freezing when the potential predators are far away and produce ultrasonic calls of low intensity (Nakano et al. 2003, 2008, 2015). Males of this species produce a quiet courtship ultrasound song of lower intensity by rubbing specialized scales on the wing against scales

(Continued)

BOX 8.2 **DARWINIAN PUZZLE** *(continued)*

Thinking Outside the Box

on the thorax (**FIGURE B**) (Nakano et al. 2008). Mimicking bats that are far away, the male's quiet ultrasound makes females freeze. The male's song therefore suppresses the escape behavior of the female, thereby increasing the likelihood that the male moth can copulate with the female. Indeed, males that sing tend to copulate more often than those that do not (Nakano et al. 2008). Where does this example fit into the communication framework we introduced earlier (see Figure 8.2)? In the long run, one might expect selection to promote bat-conspecific discrimination. Why do think this has not yet happened in this system? In another species of moth, the Japanese lichen moth (*Eilema japonica*), females *can* distinguish between ultrasound produced by males of their own species and by bats (Nakano et al. 2003). Moreover, females respond differently to the two types of ultrasound by emitting defensive ultrasonic clicks against bats to jam their calls, but not against the calls from male moths. Contrast this type of communication system with that of the Asian corn borer and explain why female discrimination has evolved in one species but not the other.

(B)

FIGURE B Ultrasonic signaling in moths. Male Asian corn borers produce ultrasounds by rubbing specialized wing scales against scales on the thorax. (Photograph by Kembangraps.)

female's desire to stay close to the altered male. Basolo concluded that a sensory preference for long tails must have preceded the eventual evolution of swordtails in some relatives of *X. maculatus* belonging to the genus *Xiphophorus*, in which the male's swordtail currently plays an important role in courtship (**FIGURE 8.19**). Just what adaptive aspect of the female's sensory system was responsible for its original swordtail preference is not known for sure, but Basolo suggests that, for example, the female's sensory bias might have evolved in the context of recognizing conspecific males (which have a large, vaguely swordlike gonopodium for the transfer of sperm to their mates).

Despite a plausible story to explain the response of female *X. maculatus* to males with artificial swords, Axel Meyer and his colleagues were not convinced. They argued on the basis of their own detailed molecular phylogeny of *Xiphophorus* that swords have evolved and been lost several times in this genus (Meyer et al. 1994). Therefore, even though *X. maculatus* does not currently have a swordtail, it is conceivable that some of its ancestors did and that the female preference for this trait is an evolutionary holdover from the past, a reflection of the tendency for sexual signals to be gained and lost rapidly over time (Borgia 2006). In fact, Meyer hypothesized that in *Xiphophorus* and other animals with elaborate traits used in mate choice, male traits may be more evolutionarily labile than the female preferences for those traits (Meyer 1997). Thus, it is possible that the female preferences for abnormal male stimuli in fishes, birds, lizards, and other animals may simply be due to the fact that males are evolving more quickly than females.

Sceloporus virgatus

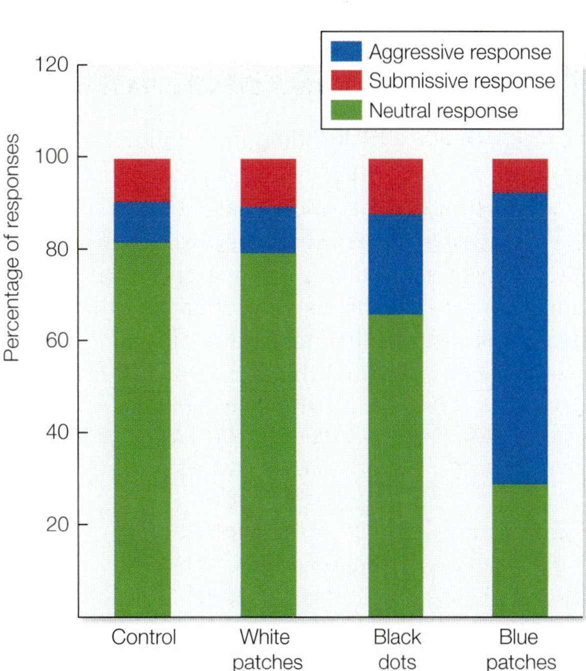

FIGURE 8.18 Receivers can respond to an ancestral signal not present in their species. Lizards of a species whose relatives have blue patches on the abdomen are more likely to abandon a conflict when confronted by a conspecific rival with blue patches painted on its abdomen than when seeing the threat display of an unmanipulated control individual or one that has had white patches or black dots painted on the abdomen. (After Quinn and Hews 2000; photograph © Matt Jeppson/Shutterstock.com.)

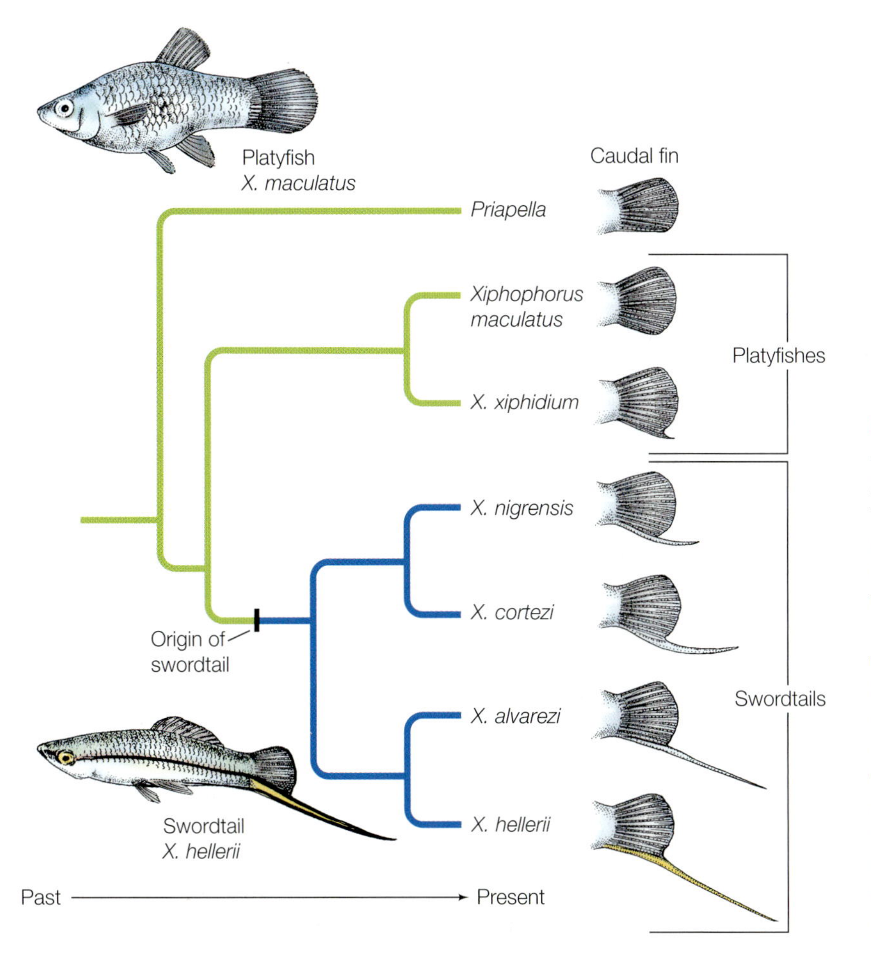

FIGURE 8.19 Sensory exploitation and swordtail phylogeny. The genus *Xiphophorus* includes the swordtails, which have elongated caudal fins, and the platyfishes, a group without tail ornaments. Because the closest relatives of the platyfishes and swordtails belong to a genus (*Priapella*) in which males lack long tails, the ancestor of *Xiphophorus* probably also lacked this trait. The long tail apparently originated in the evolutionary lineage that diverged from the platyfish line. Even so, females of the platyfish *X. maculatus* find males of their species with experimentally lengthened tails more attractive, suggesting that the females possess a sensory bias in favor of elongated tails. (After Basolo 1990.)

Both of these fish studies were trying to determine which came first: the sword or the preference for the sword. In a series of experiments, Gil Rosenthal and Christopher Evans tried to answer this question by showing female swordtails a series of computer-altered video sequences depicting a courting male (Rosenthal and Evans 1998). The researchers found that females preferred sequences of intact males to otherwise identical sequences in which portions of the sword had been deleted selectively. Yet there was no difference between responses to an isolated sword and to a swordless male of comparable length. Furthermore, female preference for a sworded male was abolished by enlarging the image of a swordless male to compensate for the reduction in length caused by removing the ornament. Ultimately, the researchers concluded that the female preference had nothing to do with the sword at all. Instead, they argued that female mate choice in swordtails was mediated by a general preference for large males rather than by specific characters of the sword. They concluded that the sword simply helped amplify a male's size. Whether this female preference was originally due to some sort of ancient sensory bias still remains unclear.

The Function of Animal Signals

This chapter's focus now shifts from describing the evolution of communication signals (*how* new signals originated) to detailing their function (*why* natural selection resulted in the spread of some signal changes over time). For example, even though the evolutionary origin of the female spotted hyena's pseudopenis remains unknown, we can use natural selection theory to explore the adaptive value of the trait. This task is challenging, however, given that the early forms of the pseudopenis probably had substantial fitness costs. Females in our own species that are experimentally exposed to much higher than normal amounts of androgens as embryos not only develop an enlarged clitoris, as we mentioned above, but also may become sterile as adults. Indeed, pregnant women who have naturally high levels of circulating testosterone produce babies of low birth weight, a lifetime handicap for the infants so affected (Carlsen et al. 2006). If developmental costs of this sort were experienced by hormonally distinctive ancestors of the spotted hyena, how could the genes for a pseudopenis spread through the species by natural selection (Frank et al. 1995b, Muller and Wrangham 2002)? We begin with a discussion of the potential adaptive function of the spotted hyena's greeting display, before using the communication framework we introduced earlier (see Figure 8.2) to explore honest signaling, deceptive signaling, and eavesdropping in greater detail.

HYPOTHESES 8.2

Non-mutually exclusive hypotheses to explain the evolution of the pseudopenis in female spotted hyenas

by-product hypothesis
The pseudopenis is not adaptive per se, but instead developed as a by-product of some other adaptive change that had positive fitness effects and that may have also resulted in high testosterone levels in females.

submission hypothesis
The pseudopenis signals subordination or willingness to bond with other individuals.

social-bonding hypothesis
The pseudopenis promotes the formation of cooperative coalitions.

The Adaptive Function of a Strange Display

Before we examine the many potential adaptive explanations for the pseudopenis display (**HYPOTHESES 8.2**), we must consider the possibility that the pseudopenis is not adaptive per se but instead developed as a by-product of a change that had other kinds of positive fitness effects, and that may have also resulted in high testosterone levels in females (**by-product hypothesis**). Perhaps the pseudopenis is simply a developmental side effect of the hormones present in newborn spotted hyenas that are favored by natural selection because they promote extreme aggression between siblings, enabling them to compete for control of their mother's parental care? Baby spotted hyenas are born with eyes that are open and erupted teeth, which they often use on one another. Infant males have a large penis and infant females have

FIGURE 8.20 Aggression in spotted hyenas. Competition for food is fierce among spotted hyenas, which may favor highly aggressive individuals. A hyena clan can consume an entire giraffe in minutes. (Photograph © Andrew Parkinson/Naturepl.com.)

a large pseudopenis, which could be the by-product of whatever hormonal features produce rapid development, the foundation for neonatal aggression, which leads in some cases to the death of one of the two siblings (East et al. 1993, Golla et al. 1999).

Alternatively, a hereditable change of some sort that resulted in an enlarged clitoris might have spread because it helped make adult females larger and more aggressive, but not because it helped them acquire a pseudopenis (Glickman et al. 2006). By all accounts, female spotted hyenas are indeed unusually aggressive, at least by comparison with immigrant males, which are usually subordinate and deferential to the opposite sex. Females not only keep these males in their place, sometimes with threats or attacks, but they also compete with other females in their clan for high rank, with winners gaining priority of access to the food resources that their clan kills or steals from lions (**FIGURE 8.20**) (Frank et al. 1995a, Watts and Holekamp 2007). Moreover, with high rank comes greater reproductive success, as the sons and daughters of high-ranking females grow faster, are more likely to survive, and are more likely to become high-ranking individuals themselves than are the offspring of subordinate hyenas (**FIGURE 8.21**). Many of the advantages enjoyed by the offspring of dominant females come from the support they receive from their mother and those hyenas in her coalition of helpers (East et al. 2009, Hofer and East 2003). Even after leaving their mothers behind, the dispersing sons of dominant females are more likely to be accepted in clans where the males' fitness prospects are highest, and they begin reproducing sooner than the sons of lower-ranking females (Höner et al. 2010). All of these benefits could possibly support some costly developmental side effects for females, such as having a pseudopenis.

There are clearly many problems with the by-product hypothesis. For one thing, if the pseudopenis is strictly a liability, why hasn't selection favored mutant alleles that happened to reduce the negative clitoral effects that androgens (or other developmentally potent biochemicals) have on female embryos? So let's consider the possibility that the pseudopenis might have adaptive value in and of itself, despite the apparent costs associated with its development or with having to give birth through it. At some point, a female ancestor of today's spotted hyenas must have been the first female ever to greet a companion by offering her a chance to inspect her genitals, perhaps during a bout in which the receiver was analyzing chemical signals from the anogenital region of the sender. Indeed, all four modern species

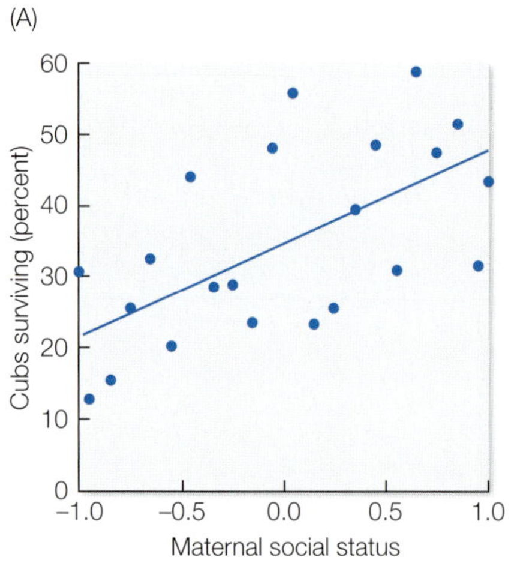

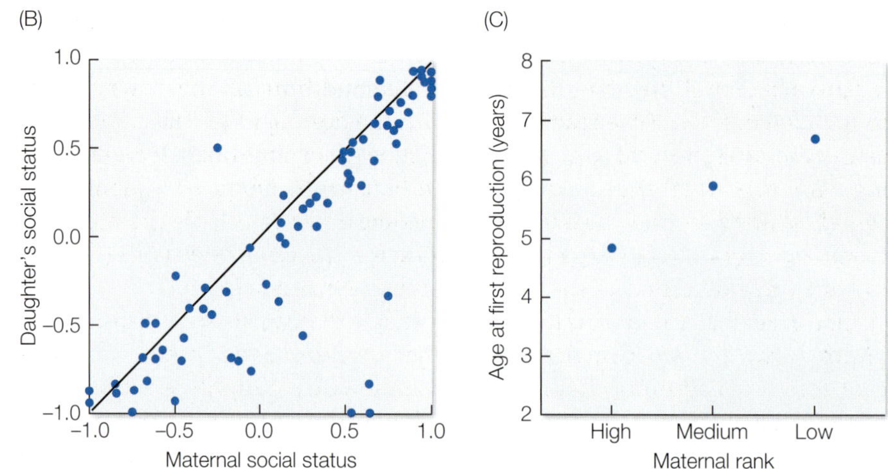

FIGURE 8.21 Dominance greatly advances female reproductive success in the spotted hyena. A mother's social status is directly linked to (A) cub survival to age 2 years and (B) the dominance rank of her daughters, as determined by observation of interactions between pairs of hyenas. (C) Maternal social status (rank) also affects when a female's sons begin to reproduce. (A, B after Hoefer and East 2003; C after Höner et al. 2010.)

of hyenas engage in chemical communication via anal scent glands (Kruuk 1972, Watts and Holekamp 2007). Female spotted hyenas appear to use the anal gland secretions of other individuals to evaluate their social rank (Burgener et al. 2009). In a predecessor of spotted hyenas, females added inspection of the erect penises of their fellow hyenas, male and female alike, when inspecting a companion. Could the female's enlarged clitoris have been so useful in this encounter as to outweigh any reproductive handicaps imposed by the structure?

One answer to this question comes from considering the possibility that sensory exploitation played a role in the origins of the display. Imagine that the female-dominant, male-subordinate relationship was well established before the first pseudopenis appeared. Furthermore, imagine that one of the cues that sexually motivated, subordinate male hyenas supplied to dominant females during courtship was an erect penis. Presentation of this organ to the female would clearly signal the suitor's male identity and intentions and thus his subordinate, nonthreatening status. Such a signal might have encouraged females on occasion to accept the male's presence rather than rejecting or attacking him. Once this kind of male–female communication system was in place, mutant females with a pseudopenis would have been able to tap into an already established system to signal subordination or willingness to bond with the other individuals, or both (Watts and Holekamp 2007). Evidence in support of the **submission hypothesis** comes from the observation that subordinate females and youngsters are far more likely than dominant animals to initiate interactions involving pseudopenile display, indicating that they gain by transferring information of some sort (probably a willingness to accept subordinate status) to more dominant individuals (East et al. 1993). Perhaps this kind of signal was and is still adaptive even for hyena cubs as they struggle for dominance with their siblings. Young hyenas begin to use their penises and pseudopenises very early in life. As for adult hyenas, dominant individuals sometimes force subordinates to participate in a greeting display (East et al. 1993), which suggests that subordinates sometimes have to demonstrate their acceptance of lower status to avoid punishment

submission hypothesis The pseudopenis signals subordination or willingness to bond with other individuals.

by their superiors. The greeting display may enable lower-ranking individuals to remain in the clan, where their chances of survival and eventual reproduction are far greater than if they had to leave the clan and hunt on their own. Permission to remain could be more readily granted if the dominant female could accurately assess the physiological (e.g., hormonal) state of a subordinate female by inspecting her erect, blood-engorged clitoris. Perhaps truly submissive subordinates are signaling that they lack the hormones (or some other detectable chemicals) needed to initiate a serious challenge to the dominant individual, which can then afford to tolerate these others because they do not pose a threat to her.

One thing that is clear about spotted hyenas is that female social rank is extremely important. Dominant females have higher reproductive success, and since rank is at least partially inherited, a dominant female's daughters are also likely to have higher fitness. The highly competitive feeding environment typical of spotted hyenas selects for large, aggressive females that are better able to secure food resources for themselves and their dependent young (Frank 1986). A team led by Kay Holekamp expanded this idea to focus specifically on a morphological trait that distinguishes hyenas from other carnivores: their highly specialized skull that allows them to crack open bones to secure the fatty marrow inside them (Watts et al. 2009). The researchers argued that the need by mothers in this species to continue to help feed young after weaning led to the evolution of female dominance. The slow development of cubs' skulls to the point where they can crack bone on their own necessitates extended maternal care, which is facilitated by maternal aggression at kills. Thus, with the evolution of a female dominance hierarchy may have come the evolution of a signal of social dominance, the pseudopenis.

The **social-bonding hypothesis** is similar to the submission hypothesis in that this odd mode of greeting could promote the formation of cooperative coalitions within the clan, an effect that requires the willingness of dominant individuals to accept somewhat lower-ranking individuals as associates. Coalition members cooperate with one another in ways that appear to raise the direct and indirect fitness of these individuals (Smith et al. 2010). Moreover, coalitions within a clan of spotted hyenas are constantly shifting, which might help explain why the animals so often engage in their bizarre greeting ceremonies (Smith et al. 2011). In addition, this social-bonding hypothesis is congruent with the observation that after a conflict, some individuals initiate a greeting ceremony to reconcile their differences with previous opponents (Hofer and East 2000).

social-bonding hypothesis The pseudopenis promotes the formation of cooperative coalitions.

There are still other hypotheses on the possible adaptive value of the pseudopenis than the ones we detailed here. For example, because the vagina is reached through the clitoris in spotted hyenas, copulation is a tricky and delicate matter, requiring complete cooperation by the female (Holekamp and Dloniak 2010). The pseudopenis gives the female spotted hyena a great degree of control in the selection of a sexual partner, which could contribute to the maintenance of this feature (East et al. 2003). Perhaps the most important point to be derived from our discussion of the hyena pseudopenis is that much more remains to be learned about spotted hyenas, even though this animal has been carefully studied for decades. Alas, after nearly 100 years, the story of the female spotted hyena's bizarre pseudopenis remains a Darwinian puzzle.

Honest Communication and Threat Displays

Recall that honest signaling benefits both the sender and the receiver. One way that we can explore this form of communication is to ask why animals often resolve key conflicts with harmless threat displays rather than by fighting. For example, when one hyena threatens another by approaching with tail raised while walking in a stiff-legged manner, the interaction often ends without any physical contact between the two animals, let alone an all-out fight. Instead, one hyena usually runs off or crouches down submissively, in effect conceding whatever resource stimulated one or both individuals to initiate the threat display. Moreover, the "winning" individual often accepts signals of submission by a rival and does not follow up with a physical attack on the other animal.

The use of noncontact threat displays to resolve conflicts is common in the animal kingdom, although violent interactions also occur. Rather than assaulting a rival, males of many bird species sometimes settle their disputes over a territory or a mate with much singing and plumage fluffing, without ever so much as touching a feather of the opponent. Even when genuine fighting does occur between animals, the "fighters" often appear to be auditioning for a comic opera. For example, after a body slam or two, a subordinate male northern elephant seal (*Mirounga angustirostris*) generally lumbers off as fast as it can, inchworming its blubbery body across the beach to the water, while the victor bellows in noisy, but generally harmless, pursuit (**VIDEO 8.2**).

VIDEO 8.2

Male northern elephant seals fighting
ab11e.com/v8.2

Let's do a thought experiment and consider why animals might use noncontact threat displays to resolve conflicts rather than escalate to an all-out fight (Estes 2012). Using a cost–benefit approach, we might recognize that while winning could be beneficial in terms of gaining access to mates, losing could be quite costly. Take, for example, a male impala (*Aepyceros melampus*). During the breeding season, some males guard territories and a group (or harem) of females with which they mate. Other males live in small groups as bachelors, waiting for their chance to take over a territory and gain access to a harem of females. Most territorial males are able to maintain their territory only for a few months because doing so is costly. Territorial males spend a lot of time herding their females or repelling rivals, leaving little time to forage. Over the breeding season, territorial males lose body mass, while bachelor males spend their days fattening up and practicing fighting. Because male impalas fight by ramming, locking horns, and twisting, it is not uncommon to find a male missing one horn. Imagine the likelihood of a one-horned male taking over or defending a harem of females. It's probably not high. A territorial male will often approach a bachelor male, snorting and shaking his head in a threat display. By assessing the size of a rival male's horns—a good proxy for body size—relative to his own during this threat display, a bachelor may be able to determine if it is worth escalating to a fight. From his perspective, the potential costs of getting wounded in a fight only make sense if he has an opportunity to win, something likely only when the rival is of similar or smaller size. From the territorial male's perspective, there is little to be gained from a fight because he already possesses the bounty: the territory and the harem of females. In this way, the information communicated through threat displays can prevent an escalation to a fight that might be costly to both individuals.

Much like impalas, males of certain bizarre antlered fly species (*Phytalmia* spp.) convey information about body size to their opponents by standing directly in front of each other in a position that enables the two flies to compare the relative size of their antlers (**FIGURE 8.22A**). The size of these structures is correlated with the fly's body size, and smaller (and presumably weaker) males quickly abandon the field of battle to the larger, more powerful males (Dodson 1997, Wilkinson and Dodson 1997). Antler size is also linked to body size in roe deer (*Capreolus capreolus*), one of many vertebrates in which opponents can determine their size relative to a rival by examining his weapons (**FIGURE 8.22B**) (Vanpé et al. 2007). For male collared lizards (*Crotaphytus collaris*), a signal of fighting ability is provided by the

(A)

(B)

(C)

FIGURE 8.22 Honest visual signals of size and strength. (A) Males of this Australian antlered fly in the genus *Phytalmia* confront each other head to head, permitting each fly to assess his own size relative to the other's size. (B) Antler size in male fallow deer (*Dama dama*) provides information about male fighting ability. (C) The larger the patch of ultraviolet-reflecting skin by the open mouth of a gape-displaying collared lizard the harder the reptile can bite an opponent. The UV-reflecting skin is outlined in the image on the right. (A, photograph © Mark Moffett/Minden Pictures; B, photograph © iStock.com/Schaef1; C from Lappin et al. 2006; photograph by A. K. Lappin.)

ultraviolet-reflecting patches near the opponent's mouth (Lappin et al. 2006); the larger the patch, the harder the bite that the lizard can deliver to a rival (**FIGURE 8.22C**). Male barking geckos (*Ptenopus garrulus garrulus*) use an acoustic display that signals body size (**FIGURE 8.23**) (Hibbitts et al. 2007).

If threat displays and the typical response to them are the result of natural selection, it should be possible to demonstrate that the losers—those signal receivers that run off or give up without a fight—benefit from their reaction to the signals of others. This task can be difficult. Consider the European or common toad (*Bufo bufo*), in which males compete for receptive females. When a male finds another male mounted on a female, he may try to pull his rival from her back. The mounted male croaks as soon as he is touched, and often the other male immediately concedes defeat and goes away, leaving his noisy rival to fertilize the female's eggs. How can it be adaptive for the signal receiver in this case to give up a chance to leave descendants simply on the basis of hearing a croak?

Because male European toads come in different sizes, and because body size influences the pitch of the croak produced by a male, Nick Davies and Tim Halliday proposed that males can judge the size of a rival by his acoustic signals (Davies and

FIGURE 8.23 Honest acoustic signals of size in a barking gecko. The larger the male, the lower the frequency of his calls. (After Hibbitts et al. 2007; photograph Bernard Dupont/CC BY-SA 2.0.)

Halliday 1978). If a small male can tell, just by listening, that he is up against a larger opponent, then the small male should give up without getting involved in a fight he cannot win. Therefore, deep-pitched croaks (made by larger males) should deter attackers more effectively than higher-pitched croaks (made by smaller males). To test this prediction, the two researchers placed mating pairs of toads in tanks with single males for 30 minutes. Each mounted male, which could be large or small, had been silenced by a rubber band looped under his arms and through his mouth. Whenever a single male touched a pair, a speaker supplied a 5-second call of either low or high pitch. Small paired males were much less frequently attacked if the interfering male heard a deep-pitched croak (**FIGURE 8.24**). Thus, deep croaks do deter rivals to some extent, although tactile cues also play a role in determining the frequency and persistence of attacks.

In addition to using vocalizations as a threat display, some species of frogs use coloration as a warning to other males. Poison frogs in the family Dendrobatidae use bright coloration to warn potential predators of their toxicity (see Chapter 6). To determine whether these bright warning signals also serve to mediate disputes among males, Laura Crothers and Molly Cummings observed male strawberry poison-dart frogs (*Oophaga pumilio*) in their native territories and in experimental contests to assess the influences of warning signal brightness and body size on the outcomes of territorial interactions (Crothers and Cummings 2015). Although neither body length nor body mass predicted male aggressiveness, color brightness did predict a male's willingness to initiate aggressive interactions. This is likely to be adaptive because brighter males (although surprisingly, not bigger or heavier males) were quicker to orient toward a rival and begin making agonistic calls, perhaps suggesting their willingness to fight.

Honest Signaling

European toads and strawberry poison-dart frogs signal to their rivals something about their fighting ability using vocalizations and color displays, respectively. So why don't small male European toads, for example, pretend to be large by giving low-pitched calls? Perhaps they would if they could, but they can't. A small male toad simply cannot produce a deep croak, given that body mass and the unbendable rules of physics determine the dominant frequency (or pitch) of the signal that he can generate. Thus, male European toads have evolved a warning signal that accurately announces their body size, and this signal cannot be faked or cheated. By attending to this **honest signal**, a fellow male can determine something about the size of his rival and thus his probability of winning an all-out fight with him. Honest signals are

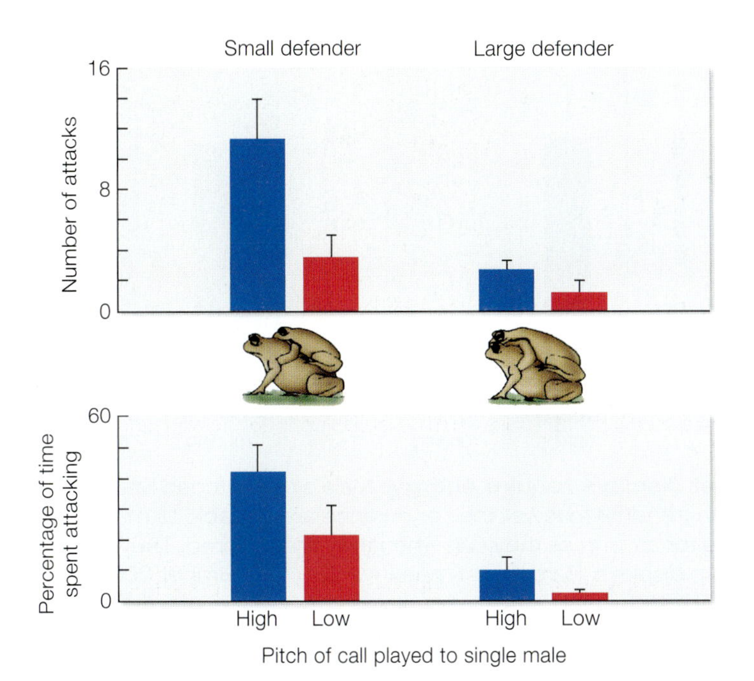

FIGURE 8.24 Deep croaks deter rival toads. When a recording of a low-pitched croak was played to them, male European toads made fewer contacts with, and interacted less with, silenced mating rivals than when a higher-frequency call was played. However, tactile cues also play a role, as one can see from the higher overall attack rate on smaller toads. Bars depict mean +/– SE. (After Davies and Halliday 1978.)

inherently honest because they indicate the quality of the sender, facilitate decision making by a receiver, and cannot be easily produced by the sender.

To understand the value of a quick concession, imagine two kinds of aggressive individuals in a population: one that fought with each opponent until he was either victorious or physically defeated, and another that checked out the rival's aggressive potential and then withdrew as quickly as possible from fighters that were probably superior. If this argument is correct, then in a species in which there are nondisplaying individuals, when two such individuals fight each other, the level of aggression should be higher than when two signal-producing males engage in an aggressive encounter. Researchers found support for this prediction in a population of crickets *Teleogryllus oceanicus*, in which most males on the Hawaiian island of Kauai no longer sing, whereas males from the Big Island sing normally (Logue et al. 2010). Thus, in species in which combatants provide honest signals of competitive ability, any "fight no matter what" types would eventually encounter a powerful opponent that would administer a serious thrashing. The "fight only when the odds are good" types would be far less likely to suffer an injurious defeat at the hands of an overwhelmingly superior rival (Maynard Smith 1974, West-Eberhard 1979).

As you learned in Chapter 1, individuals within a population vary. It is this morphological or behavioral variation that may lead to the process of evolution by natural selection. But in the context of honest signaling, why can't animals cheat the system and make a better signal? Imagine two kinds of superior fighters in a population: one that generated signals that other, lesser males could not produce, and another whose threat displays could be mimicked by smaller males. As mimics became more common in the population, natural selection would favor receivers that ignored the easily faked signals, reducing the value of producing them. This, in turn, would lead to the spread of the hereditary basis for some other type of honest signal that could not be devalued by deceitful senders.

If honest signal theory is correct, we can predict that male threat displays should be relatively easy for large males to perform but more difficult for small, weak, or unhealthy males to imitate, or cheat. As predicted, in the side-blotched lizard *Uta stansburiana*, both the duration of a male's threat display and the number of

Male side-blotched lizard

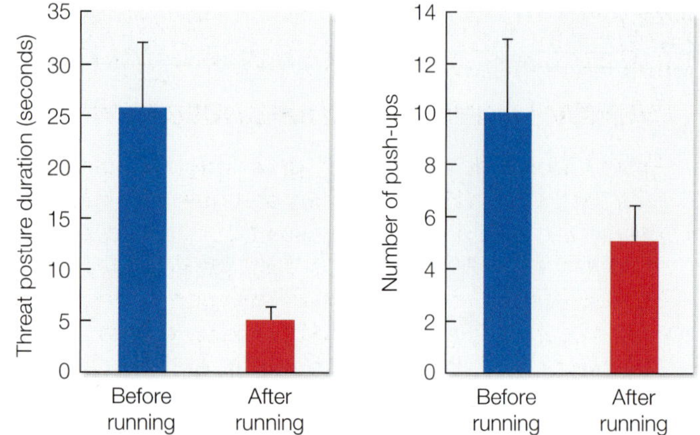

FIGURE 8.25 Threat displays require energy. Male side-blotched lizards that have been run on a treadmill to lower their endurance are not able to maintain their threat posture for as long as they can when they are not tired. They also generate fewer push-up displays. Bars depict mean +/– SE. (After Brandt 2003; photograph © iStock.com/KateLeigh.)

push-ups he performs falls markedly after he has been forced to run on a treadmill for a time (**FIGURE 8.25**). Conversely, males that first perform a series of displays before being placed on the treadmill run for a shorter time than when they have not been displaying. In other words, a male's current condition is accurately reflected by his push-up performance (Brandt 2003), which enables him to advertise his true fighting capacity to rival males.

The inability of some males to produce the push-up display is an example of a **production cost**. Production costs are one way that honest signals can be enforced, or not cheated. Recall our earlier discussion about orange coloration in guppies and carotenoid pigments. Fish cannot synthesize these pigments and instead have to acquire them in their diet, which can be costly if the pigments are rare and hard to find. Fish are not the only species that use carotenoid pigments to communicate. Indeed, carotenoids confer color to most invertebrate phyla and all vertebrate classes, and as you will see in Chapter 9, they are particularly common in the feathers of many bird species (**BOX 8.3**). Since carotenoids can act as antioxidants by scavenging unpaired electrons from damaging products produced during metabolism, they are thought to be an honest signal in several avian species of not only foraging ability but also of health (McGraw 2006b).

Female *Polistes dominula* paper wasps, which signal their fighting ability not by an acoustic signal or display but via the markings on their faces, illustrate another way in which honest signals can be enforced. Bigger wasps tend to have more black spots on their faces, and more dominant wasps tend to have more black coloring on their faces (Tibbetts and Dale 2004). These types of signals are called **badges of status**, and they typically reveal information about an individual's size or dominance status. Although badge of status signals are typically not costly to produce, they may be costly to maintain due to social enforcement, a phenomenon referred to as a **maintenance cost**. To test the hypothesis that wasp face markings impose maintenance costs, Elizabeth Tibbetts and colleagues performed a series of experiments in captive paper wasps. In staged contests between pairs of unfamiliar wasps, subordinate wasps with experimentally increased black markings on their faces received more aggression from dominants than did controls (Tibbetts and Dale 2004). This experiment suggested that dishonest signaling imposes maintenance costs. But do wasps enforce these signals in nature? In a

BOX 8.3 INTEGRATIVE APPROACHES

Mechanisms and measurement of animal coloration

From iridescent butterfly wings to brightly colored bird feathers to jet-black mammal hair, animals are capable of producing tissues of nearly every shade of the rainbow. The colors they produce in their tissues result from two categories of mechanisms: pigments and structural colors. **Pigments** are molecules that differentially absorb and emit wavelengths of visible light. Some pigments (such as melanins) can be produced by most animals, whereas others (such as carotenoids) cannot be produced by vertebrates and must instead be absorbed from their diet. Because **carotenoids** play a role in scavenging dangerous free radicals such as antioxidants that have been linked to disease and aging, these pigments have direct links to health and immune function (McGraw 2006a). For this reason, these dietary pigments are thought to play a key role in honest signaling in many animals, transmitting information related to immune status, foraging ability, and so forth. In contrast to carotenoids, **melanins** are among the most abundant pigments in animals, found in everything from feathers to skin to hair (McGraw 2006b). The two main forms of melanin are eumelanin, which produces brown and black coloration, and pheomelanin, which produces red and yellow coloration. Human hair, for example, varies in coloration at least partly because of different proportions of eumelanin and pheomelanin. Melanin pigments play a key role in many vertebrate and invertebrate badges of status, which are signals of dominance or fighting ability.

In contrast to pigments that absorb and emit light, **structural colors** are produced by light interacting physically with the nanometer-scale arrangement of tissues and air. The different colors that we observe in the iridescent scales of some butterflies or in the plumage of some birds are the result of how these structures are layered and how those layers interact with light. This is why, for example, if you look at a hummingbird from different angles, the colors you see can be very different. The production costs of structural colors are thought to be lower than those of pigment-based colors, though there have been surprisingly few studies on this topic (Prum 2006).

Over the past few decades, spectrometric color quantification has become a standard tool in behavioral biology, both in the field and the lab (Andersson and Prager 2006). This has enabled researchers to examine color in a diversity of free-living species, lab organisms, and even museum specimens. Most behavioral biologists use a reflectance spectrometer to measure the reflectance of colorful tissues. In a matter of seconds, precise measurements can be made of the complete visible light spectrum from wavelengths of 300 to 700 nanometers, which include the ultraviolent wavelengths that many animals—though not humans—can see. Reflectance spectra can then be decomposed using statistical analyses to determine which individuals, tissue types, or body regions are more colorful than others. In this way, coloration can be quantified consistently, inexpensively, and perhaps most important, repeatably.

Thinking Outside the Box

Given what you just read about the different mechanisms of animal coloration, which types of colors are more likely to be used in signaling to mates, rivals, or predators? Think about the production costs of making the different types of colors and how they may influence the evolution of color signals. Do you think some types of colors are more likely to be used in honest signals than others? To answer this question, think about how individuals might be able cheat in the production of signals using the different kinds of mechanisms.

subsequent experiment, Tibbetts and colleagues showed that females on a paper wasp nest test each other from time to time, and in so doing, they can detect cheaters whose facial signals are not in synchrony with their real fighting ability. Individuals whose faces have been painted to signal dominance but whose capacity

FIGURE 8.26 Honest visual signals of dominance in the paper wasp *Polistes dominula*. In this species, dominance is linked to dark patches on the face of the wasp. Notice the variation among wasps in the size of black patch from left (where it is completely absent) to right (where there is nearly as much black as yellow). (Photographs courtesy of Elizabeth Tibbetts.)

for combat is actually low are attacked more often by their colony mates than are dominant females whose faces have been painted but not altered in appearance (**FIGURE 8.26**) (Tibbetts and Izzo 2010). Thus, dishonest senders in this wasp system are punished by their companions and pay a special price for the mismatch between facial markings and behavior, an outcome that in turn creates selection for the evolution of accurate signaling.

Tibbetts and colleagues explored the relationship between fitness and facial markings in wild wasps (Tibbetts et al. 2015). Reproductive success was quantified as the number of nest cells built at the beginning of the breeding season. Overwinter survival was assessed by measuring the face and weight of spring foundresses and compared with temperature during overwintering and the summer development period. This allowed the researchers to determine if wasps with certain types of faces were more or less likely to survive cold winters. They found that wasps with more black on their faces had larger nests and were more likely to survive harsh conditions. In other words, socially enforced signals directly influence both reproduction and survival.

When Multiple Honest Signals Are Better Than One

In addition to production and maintenance costs, honesty can be enforced through signal trade-offs. Up until this point, we have focused primarily on a single type of signal in each species. Yet many species use multiple signals simultaneously, often so senders can communicate with several different types of receivers—a process called **multimodal signaling**. One of the first studies of honest signaling was Malte Andersson's study of tail length in long-tailed widowbirds (*Euplectes progne*) (Andersson 1982). As their name would suggest, male long-tailed widowbirds have a long tail that can extend many times their body length. By experimentally shortening or lengthening the tails of some males in the wild, Andersson demonstrated that tails are used in signaling to females and not to other males, and males with longer tails had higher mating success than those with shorter tails (**FIGURE 8.27**). However, in addition to having a long tail, male widowbirds also have a collar of red, carotenoid-based feathers. In a closely related species, the red-collared widowbird (*Euplectes ardens*), a team led by Andersson showed that these red feathers are used in male–male encounters to establish territories (see Chapter 7) (Andersson et al. 2002). The researchers also showed that like their cousins, female red-collared widowbirds use tail length to

FIGURE 8.27 Mating success in male long-tailed widowbirds whose tail length was manipulated. Before the experiment, there were only minor differences in mating success between males that were to be treated and not treated (not shown). However, after the treatment, males with elongated tails ended up having more new active nests on their territories than did control males or males with shortened tails. Control I males had their feathers cut off but re-attached to be the same length, whereas Control II males were untouched. (After Andersson 1982; photograph © jirisykoracz/Shutterstock.com.)

make mating decisions (Pryke et al. 2001). Given this information, you might think that the best males would have the reddest feathers, which they use to establish territories, *and* the longest tails, which they use to attract females. However, this is not the case. Male red-collared widowbirds exhibit a trade-off—a negative relationship—between the size of the red collar and the length of the tail, such that males with a long tail tend to have a duller red collar, and vice versa (**FIGURE 8.28**). Why, then, can't some males produce a red collar and a long tail? If we assume there is a limited pool of resources that males can invest in feather growth, individuals apparently cannot maximize both types of traits. Instead, they must invest in signaling to one type of receiver (a rival male or a potential mate) more than another. Having to make this trade-off in how males invest in signaling traits helps enforce honesty

FIGURE 8.28 Signal trade-off between tail length and collar area in red-collared widowbirds. There was a negative relationship between the size of the male's red collar and the length of his tail. The trends were similar for territory-holding "resident" males (blue circles) and non-territory-holding "floater" males (red circles). (After Andersson et al. 2002; photograph © iStock.com/AOosthuizen.)

because males appear to be incapable of cheating the system and investing a lot in both types of signals.

The examples from widowbirds illustrate how two different feather ornamental signals can be used to advertise to two different types of receivers. But many species use completely different modalities to signal to the same receiver. In fact, multimodal signaling turns out to be much more common in nature than once realized (Partan and Marler 1999). Multimodal signals are diverse and use a variety of signaling modalities, including tactile, visual, acoustic, chemical, and more. Eileen Hebets and colleagues developed a framework to categorize the many types of multimodal signals, which they describe as being either redundant, degenerate, or pluripotent (Hebets et al. 2016). A redundant, or backup, signal increases its robustness, often through repetition. A degenerate signal increases the robustness and functionality of the signal, often with two components serving similar functions. A pluripotent signal is one in which one structure serves multiple functions, something that can increase signaling efficiency. A single species can produce multimodal signals that serve multiple functions. Take, for example, the barn swallow (*Hirundo rustica*), a common bird in North America, Europe, and Asia. Males communicate to both males and females using song and red color patches on their chests. These two distinct signaling modalities are an example of degeneracy, and the fact that both signals are directed at both sexes is an example of pluripotency. And since the male's song has many repeated notes, it can also be considered an example of redundancy (Wilkins et al. 2015).

Spiders have mastered the art of multimodal signaling, and among arthropods, the peacock spiders have evolved some of the most elaborate displays involving both visual and vibrational signals (Girard et al. 2011). Your textbook cover illustrates one modality that that these spiders use to signal. Male peacock spiders unfurl from underneath their abdomen what looks like the tail of a male peacock. They display this ornamented fan toward interested females while waving their legs in an intricate and coordinated dance (**VIDEO 8.3**). At the same time, they produce vibrations by oscillating their abdomen. What are these different modalities signaling? Each modality could be informative to females in a different way (**multiple message hypothesis**), or the different modalities may independently convey the same information, providing insurance for signaling errors or environmental noise (**redundant signal hypothesis**) (**HYPOTHESES 8.3**).

VIDEO 8.3
Dance of the peacock spider
ab11e.com/v8.3

In a series of lab experiments, a group led by Damian Elias demonstrated that in the Australian peacock spider *Maratus volans*, visual and vibrational signals were both important to female choice and male mating success, though females were more likely to mate with males that spent more time displaying visually (Girard et al. 2015). Why, then, has multimodal signaling evolved in peacock spiders? The authors argue that multimodal signaling facilitates quicker and more reliable decision making, something that may be important for peacock spiders because many species often co-occur in the same places. Further experiments are clearly needed to distinguish between the multiple message and redundant signal hypotheses for this and other species, as well as to determine why these complex displays have evolved.

HYPOTHESES 8.3

Alternative hypotheses to explain multimodal signaling

multiple message hypothesis
Different signaling modalities convey different information.

redundant signal hypothesis
Different modalities independently convey the same information, providing insurance for signaling errors or environmental noise.

Deceitful Signaling

Recall that honest signaling involves an interaction in which the sender and the receiver both benefit. Yet clearly not all interactions involving information transfer between two individuals are mutually

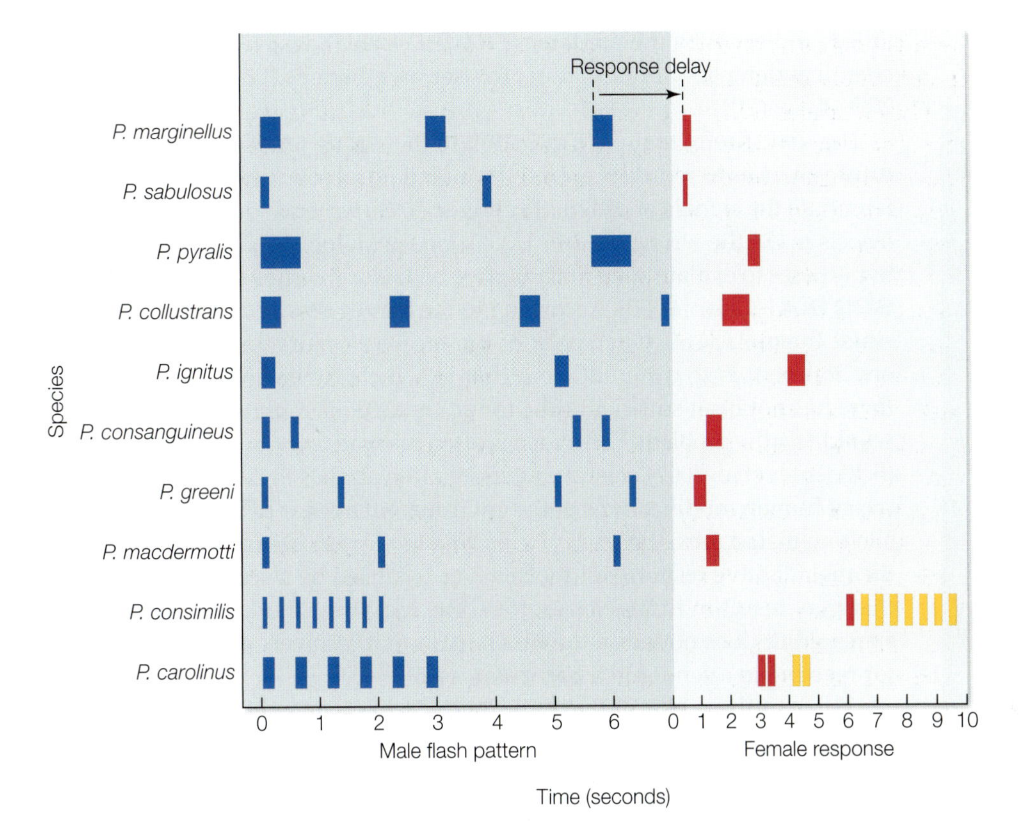

FIGURE 8.29 The diversity of communication flashes by different species of fireflies. The timing (in seconds) of the male flash pattern is shown for ten North American species of *Photinus* fireflies in the blue portion of the diagram. The timing of the female response flash pattern is shown in the lighter portion of the diagram. A female provides her flash response to the last pulse of light from a male of her species. The differences among species lead to reproductive isolation of the various species. (Based in part on Lloyd 1966; diagram after Lewis 2016.)

beneficial. Interactions in which the sender manipulates the receiver such that the sender receives a fitness benefit but the receiver incurs a cost are referred to as being deceitful (or manipulative). There are many cases of deceit among animals, including predatory margay cats (*Leopardus wiedii*) imitating the calls of their prey, juvenile pied tamarin monkeys (*Saguinus bicolor*) (Calleia et al. 2009); an octopus that can change its appearance to look like a toxic marine flatfish (Huffard et al. 2010); and a butterfly whose larvae are raised in ant colonies where the caterpillars mimic the sounds that ant queens make to receive preferential care from ant workers (Barbero et al. 2009).

Firefly "femme fatales" provide another example of deceitful communication (Lloyd 1965). Females of these nocturnal predatory insects in the genus *Photuris* respond to firefly flashes given by males in the genus *Photinus* as these males try to locate and attract conspecific mates (**FIGURE 8.29**) (Lewis and Cratsley 2008). The predatory female answers her prey's signal with a flash of light, perfectly timed to resemble the response given by the right *Photinus* female. Some *Photuris* females can supply the differently timed answering flashes of three different *Photinus* species (Lloyd 1975). Typically, a male *Photinus* that sees a receiver's signal of the appropriate type flies toward the light flash, thinking he has successfully found a willing mate. But if he encounters a deceptive *Photuris* female, he may well be grabbed,

FIGURE 8.30 A firefly femme fatale. This female *Photuris* firefly has killed a male *Photinus* firefly that she lured to her by imitating the timed flash response given by females of his species. (Photograph by James E. Lloyd.)

killed, and eaten by the predator (**FIGURE 8.30**), which recycles the toxic chemical defenses of her victims for her own benefit (Lewis and Cratsley 2008, Lewis 2016).

The male firefly that pays attention to the wrong female signal loses any future chances to reproduce. Even though natural selection has acted to differentiate the signals of different firefly species, why do some males of these insects make the wrong choice? Evolutionary biologists have proposed two hypotheses to explain such maladaptive behavioral responses (**HYPOTHESES 8.4**) (Crespi 2000). According to the **novel environment hypothesis**, the maladaptive behavior of the receiver occurs because the current environment is so different from that in which the behavior evolved that there has not been sufficient time for advantageous mutations to occur that would "fix the problem." This idea is often referred to as an evolutionary trap (Schlaepfer et al. 2002), a term that more often applies to cases in which very recent human modifications of the environment appear to be responsible for eliciting maladaptive behavior. According to the **net benefit hypothesis**, the maladaptive response of the receiver is caused by a sensory mechanism that may result in fitness losses for some individuals under some circumstances, but does not erase the fitness gain that receivers derive on average for reacting to a sender in a particular way.

Which of these alternative hypotheses is likely to explain the maladaptive response to femme fatales by male *Photinus* fireflies? Although the novel environment hypothesis seems unlikely to account for the nature of interactions between *Photinus* males and predatory *Photuris* females, artificial light interference does lead to a reduction in flashing activity of up to 70 percent in some species of *Photinus* fireflies, as well as reduced courtship behavior and mating success (Firebaugh and Haynes 2016). Yet instances in which prey respond maladaptively to their predator's signals have been explained by arguing that, on average, the response of a male *Photinus* to certain kinds of light flashes increases his fitness, even though one of the costs of responding is the (small) chance that he will be devoured by a *Photuris* femme fatale. Males that avoid the deceptive *Photuris* signals might well live longer, but they would probably ignore females of their own species as well, in which case they would leave few or no descendants to carry on their cautious behavior.

HYPOTHESES 8.4

Alternative hypotheses to explain maladaptive responses to signals

novel environment hypothesis The current environment is sufficiently different from that in which the behavior evolved that there has not been enough time for a species to adapt.

net benefit hypothesis A sensory mechanism that may result in fitness losses for some individuals under some circumstances does not erase the fitness gain that receivers derive on average for reacting to a sender in a particular way.

The net benefit hypothesis for deceptive communication highlights the definition of adaptation employed by most behavioral biologists, which is that an adaptation need not be perfect, but it must contribute more to fitness *on average* than alternative traits do. The male firefly that responds to the signals of a predatory female of another species possesses a mechanism of mate location that clearly is not perfect, but it is better than those alternatives that improve a male's survival chances at the cost of making him a sexual dud. In effect, the evolved communication system involving male *Photinus* senders and female *Photinus* receivers has been taken advantage of by predatory fireflies in another genus. These illegitimate senders presumably reduce but do not eliminate the net fitness benefits of participating in a generally advantageous communication system (Otte 1974). Cases of this sort are usually ones in which a deceptive sender of one species exploits a receiver of another species, but there are possible examples of intraspecific deception as well, including in the slender crayfish (*Cherax dispar*), in which males display their

FIGURE 8.31 A dishonest signal of strength? Male slender crayfish display their front claws to opponents, and males with larger claws dominate those with smaller claws. However, claw size is not correlated with the strength of the claw's grip. Could claw size instead be related to another trait linked to fighting? (Photograph courtesy of Robbie Wilson, from Wilson et al. 2007.)

claws to opponents but claw size is not correlated with the strength of the claw grip (**FIGURE 8.31**) (Wilson et al. 2007).

If this adaptationist hypothesis is correct, then deception by an illegitimate sender should exploit a response that has clear adaptive value under most circumstances (Christy 1995). Consider the interaction between male thynnine wasps (*Thynnoturneria* spp.) and the elbow orchid (*Spiculaea ciliata*) of Western Australia. A wasp in search of a mate will sometimes fly to the bizarre lip petal of an elbow orchid flower, which it will then grasp and attempt to fly away with. The lip petal is jointed, and so any wasp that flies while holding onto the object at the end of the petal moves upward in an arc. This trajectory quickly brings the male into contact with the column, the part of the flower where packets of pollen are located. By this time, the male will have released the lip petal, but the orchid does not let him get away immediately. Instead, it traps him because the wasp's wings slip into the two hooks on the underside of the column. There the wasp is held temporarily while his thorax touches the sticky pollen packets of the orchid. When the wasp manages to struggle free from his floral captor, he is likely to carry with him the orchid's pollen (**FIGURE 8.32**). Should he be deceived again, the wasp will transfer the pollen packets on his back to the pollen-receiving surface of the column of another plant. Thus, the wasp becomes a pollinator for the elbow orchid thanks to the illegitimate signals offered by the elbow orchid flower (Alcock 2000). The benefits of this interaction accrue entirely to the orchid, though at least the orchid does not kill the wasp, unlike a firefly femme fatale. However, the pollinating wasp does waste his time and energy when interacting with the elbow orchid, as is true for many other species of wasps that are fooled by other species of orchids (Schiestl 2010). In at least one wasp–orchid system, the male thynnine finds the orchid's female decoy so stimulating that he also wastes his sperm by ejaculating after grasping the lip petal (Gaskett and Herberstein 2008).

Why are male thynnine wasps prone to the fitness-reducing deception practiced by elbow orchids and other Australian orchids? To answer this question, we need to observe the interactions between female and male wasps. The wingless females of this species come to the surface of the ground after searching underground for

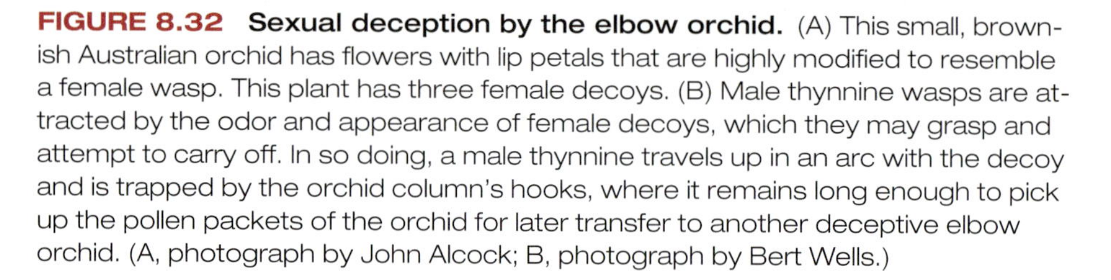

FIGURE 8.32 Sexual deception by the elbow orchid. (A) This small, brownish Australian orchid has flowers with lip petals that are highly modified to resemble a female wasp. This plant has three female decoys. (B) Male thynnine wasps are attracted by the odor and appearance of female decoys, which they may grasp and attempt to carry off. In so doing, a male thynnine travels up in an arc with the decoy and is trapped by the orchid column's hooks, where it remains long enough to pick up the pollen packets of the orchid for later transfer to another deceptive elbow orchid. (A, photograph by John Alcock; B, photograph by Bert Wells.)

beetle larvae to paralyze with a sting before laying eggs on their victims. Once above ground, females crawl upward a short distance on a plant or twig before perching and releasing a sex pheromone, a chemical signal that females use to attract males (**FIGURE 8.33**). Males patrol areas where females may appear, and they rush to calling females, which they grab and carry away for a mating elsewhere (Ridsdill-Smith 1970). The elbow orchid unconsciously exploits the males' ability to respond to the olfactory and visual cues associated with the right kind of female wasp. The lip petal offers odors similar enough to those released by calling females to attract sexually motivated males to the plant. Once nearby, the flying wasp can see the lip petal, which looks vaguely like a wingless female wasp of his own species. The male wasp that pounces is deceived into becoming a potential pollinator (Alcock 2000).

Let's imagine what would happen if there were mutant males that tried to discriminate between deceptive orchids and female wasps. These careful males almost certainly would be slower to approach elbow orchids—and females of their own species. They might not be deceived by elbow orchids very often, but they would also be somewhat less likely to reach a calling female wasp before other males. Competition for females is stiff among thynnine wasps, and males scramble to find mates as quickly as they can (see Chapter 10). The first male to grab a legitimate sender is very likely to win the chance to inseminate his mate,

FIGURE 8.33 Why the elbow orchid's sexual deception works. (A) A female thynnine wasp releasing a sex pheromone from her perch. (B) A male of the same species grasps the calling female. (C) A copulating pair of this thynnine species on a eucalyptus flower. Note the orchid pollinia attached to the thorax of the male, which has been deceived at least once. (Photographs by John Alcock.)

whereas males arriving even a few seconds later wind up with nothing. So as is true for *Photuris* fireflies, the elbow orchid shows that illegitimate senders can evolve if there is a legitimate communication system to exploit. In a way, the plant is exploiting a preexisting sensory bias in male wasps, which first evolved a sensitivity to the appearance and odor of females of their species, which the orchid later mimicked via the evolution of female decoys that look and smell like female wasps. Even plants that feed rather than deceive their pollinators appear to have evolved their come-and-get-it signals *after* the sensory capacities of their nectar- and pollen-consuming pollinators had evolved. These perceptual skills originated earlier for some other purpose, further evidence in support of the importance of preexisting biases in the evolution of communication signals, including deceptive ones (Schiestl and Dötteri 2012).

Eavesdropping on Others

Just as light-signaling male *Photinus* fireflies may come to an unhappy end when they are "overheard" and deceptively lured to their deaths by predators, so too do calling túngara frogs (*Engystomops pustulosus*) sometimes have the great misfortune to attract a fringe-lipped bat (*Trachops cirrhosus*) rather than a female of their own species (Ryan et al. 1981). Predators that take advantage of acoustic, visual, and olfactory (pheromonal) signals provided by their prey are illegitimate receivers, listening in on individuals that lose fitness by having their message reach the wrong target (Leech and Leonard 1997, Wyatt 2009). When predatory illegitimate receivers succeed in using their prey's signals to secure a meal, the consumed signal givers lose all future opportunities to reproduce, which one would think would select strongly against signaling. This exploitation of a signal by an illegitimate receiver that benefits the receiver but is costly to the sender is called eavesdropping. It is important to point out a key difference between the *Photinus* firefly and the túngara frog examples. In the fireflies, the femme fatale predator eavesdrops on the male *Photinus* but then deceives him with a flash of her own to lure him to his death (becoming an illegitimate sender), whereas the fringe-lipped bat simply eavesdrops on the male túngara frog and then eats him (becoming an illegitimate receiver).

Earlier we introduced a population of crickets on the Hawaiian island of Kauai where most males no longer sing to attract mates (Zuk et al. 2006). By remaining silent, these males are less likely to be visited by a fatal parasitoid fly, *Ormia ochracea*, which uses the chirps of crickets as a guide to their location (Gray and Cade 1999). Upon finding a cricket, the female fly then stealthily deposits larvae on the unfortunate cricket, which the immature flies eventually consume from the inside out. In most areas, field crickets are noisy, but on Kauai, where the lethal fly has only recently arrived, silent males now make up a large proportion of the cricket population, a change that took only 20 generations to evolve (Zuk et al. 2006). Still, in most places with *O. ochracea*, calling male crickets are not uncommon, just as the presence of the fringe-lipped bat has not resulted in the evolution of silent male túngara frogs. Yet in Kauai, where silent males are common, how do males attract females? When males originating from Kauai were reared in silence in the laboratory, Nathan Bailey and Marlene Zuk found that they moved sooner and spent more time walking during the silent periods than did males reared with calling song (Bailey and Zuk 2008). However, the same experiments conducted on crickets from other islands, where male song was common, showed no effect of the acoustic environment on movement behavior. Thus, although males have been selected to be silent, they have simultaneously become more active. This may help them not only escape parasitoid flies, but also secure mates.

The same argument presented when discussing illegitimate senders can help explain why some animals provide illegitimate receivers with information that can be used against them. When an evolved communication system contributes to the fitness of legitimate senders and legitimate receivers, the system can be exploited by outsiders. When predators or parasites are eavesdropping, they impose costs on the nearby legitimate senders. If these costs are very high, signaling can be all but eliminated, as shown by the case of field crickets on Kauai. But if the reproductive costs do not exceed the benefits of signaling on average, legitimate senders will persist in the population. Consider, for example, the túngara frog system. Males that fail to call might well live somewhat longer on average than callers, but any quiet males incur a reproductive disadvantage. Indeed, one of the things that Mike Ryan and his associates have learned about túngara frogs is that many males that produce longer, louder calls composed of an introductory whine followed by a series of chucks are more likely to attract females (**AUDIO 8.1**) (Baugh and Ryan 2011)—just as males of some fireflies that produce their flashes at a relatively high rate are more likely to be answered by females of their species (Lewis and Cratsley 2008). However, both male frogs and fireflies that generate the most attractive calls or flashes are also likely to attract predators, namely eavesdropping bats (**FIGURE 8.34**) (Ryan 1985) and *Photuris* fireflies (Lewis and Cratsley 2008), respectively. The trade-off between attracting females and attracting predators has had an effect on the male túngara frogs' calling strategy (and probably on vulnerable male fireflies too). Because frogs calling in small assemblages are at greater danger of becoming a bat's meal, males in small groups are more likely to drop the chucks from their calls than are males in large groups (Ryan et al. 1981, Bernal et al. 2007).

AUDIO 8.1

(A)

(B)

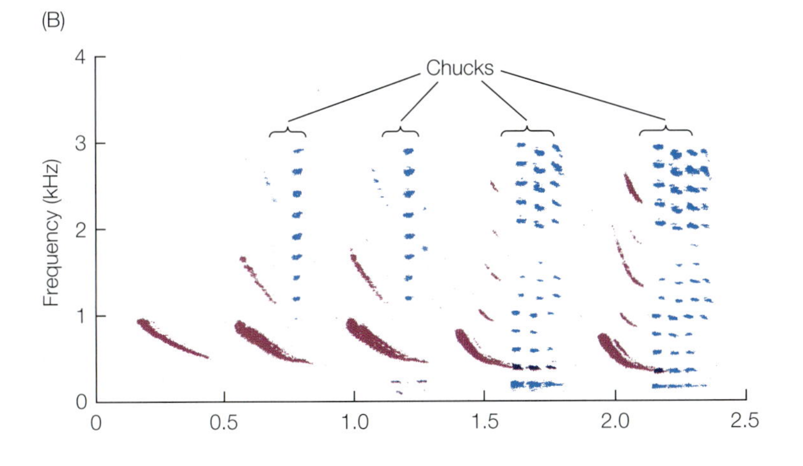

FIGURE 8.34 Illegitimate receivers can detect the signals of their prey. (A) A calling male túngara frog may inadvertently attract an illegitimate receiver—the fringe-lipped bat, a deadly predator. (B) The risk of attack is greater if the male's call includes one or more chucks (blue in the sonograms) as well as the introductory whine (red in the sonograms). (A, photograph by Merlin D. Tuttle, Bat Conservation International/Science Source; B after Ryan 1983.)

The behavior of senders in many other species has also been shaped by the risk of attracting eavesdropping predators. For example, nestling songbirds produce loud cheeps and peeps that could give a listening predator information on the location of a nest full of edible nestlings. When recordings of begging tree swallows (*Tachycineta bicolor*) were played at an artificial swallow nest containing a quail's egg, the egg in that "noisy" nest was taken or destroyed by predators before the egg in a nearby quiet control nest in 29 of 37 trials (Leech and Leonard 1997). The predator that eavesdrops on the communication taking place between a brood of baby tree swallows and their parents uses information produced by the nestlings to the detriment of these individuals and their parents.

The begging calls of birds vary in the degree to which they can be exploited by listening predators, suggesting that natural selection has shaped these signals. For example, ground-nesting warblers would seem to be more exposed to raccoons, skunks, and the like than those species that nest in relative safety high in trees (Haskell 1999). The young of ground-nesting warblers produce begging cheeps of higher frequencies than their tree-nesting relatives (**FIGURE 8.35A**). These higher-frequency sounds do not travel as far and so should better conceal the individuals that produce them. David Haskell tested this prediction by creating artificial

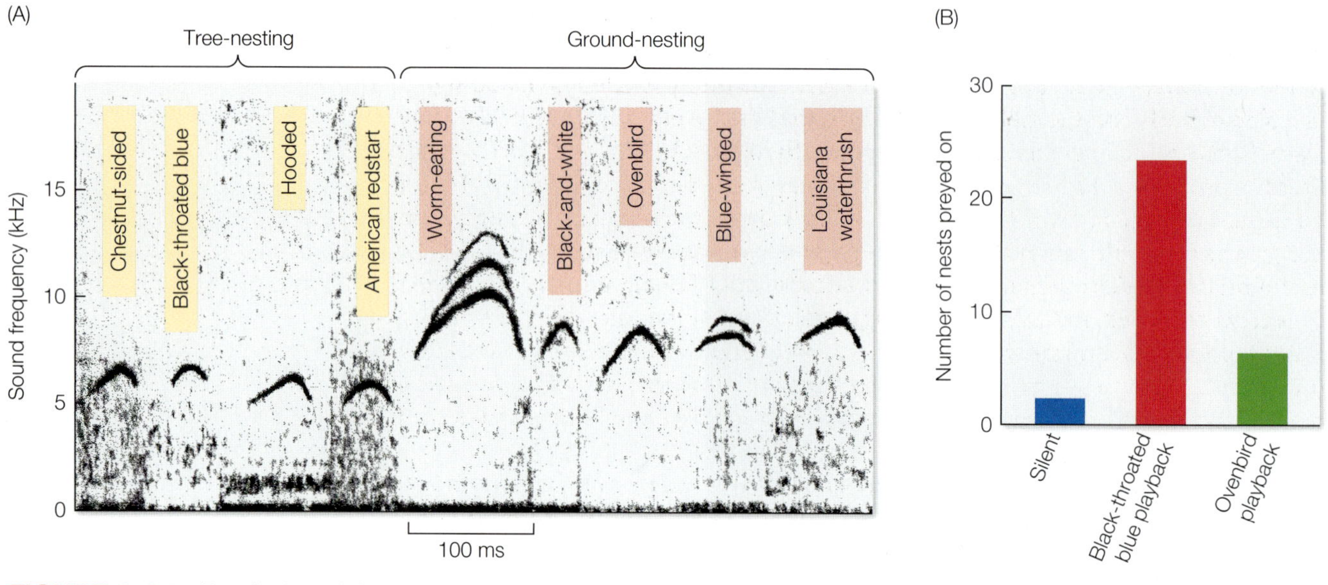

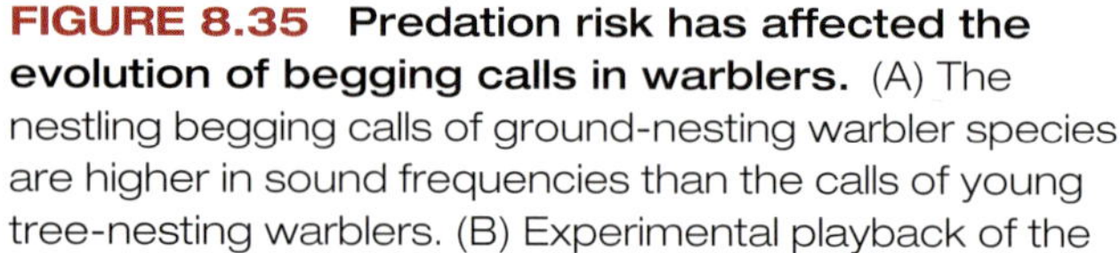

FIGURE 8.35 Predation risk has affected the evolution of begging calls in warblers. (A) The nestling begging calls of ground-nesting warbler species are higher in sound frequencies than the calls of young tree-nesting warblers. (B) Experimental playback of the begging calls of the black-throated blue warbler (a tree-nesting species) at artificial nests placed on the ground resulted in higher rates of discovery by predators than did playback of begging calls of the ground-nesting ovenbird. (After Haskell 1999.)

nests with clay eggs, which he placed on the ground near a speaker that played the begging calls of either tree-nesting or ground-nesting warblers. The eggs advertised by the begging calls of the tree-nesting black-throated blue warbler (*Setophaga caerulescens*) were found and bitten significantly more often by eastern chipmunks (*Tamias striatus*) than were the eggs associated with the calls of the ground-nesting ovenbird (*Seiurus aurocapilla*) (**FIGURE 8.35B**) (Haskell 1999). These results suggest that nest predation may be responsible for maintaining some of the interspecific differences in the acoustic structure of avian begging calls.

The risk of exploitation by an illegitimate receiver may also be responsible for the differences between the rasping mobbing call and the "seet" alarm call of the

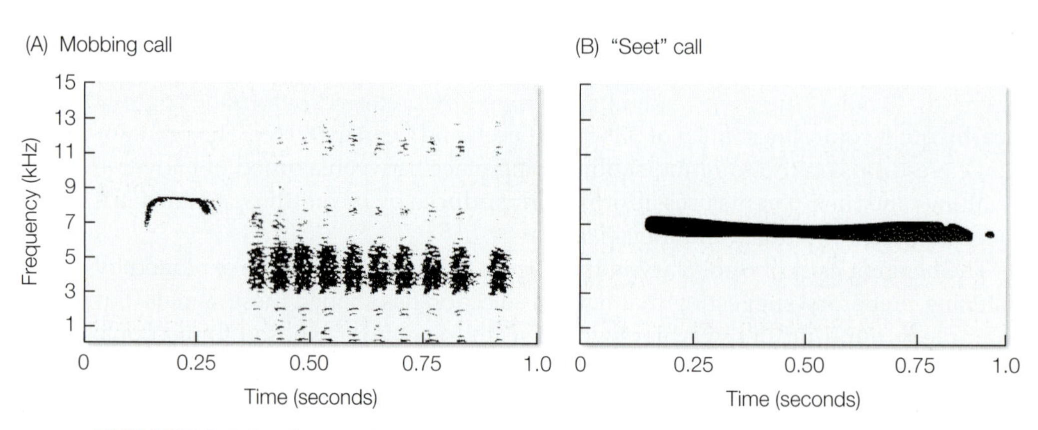

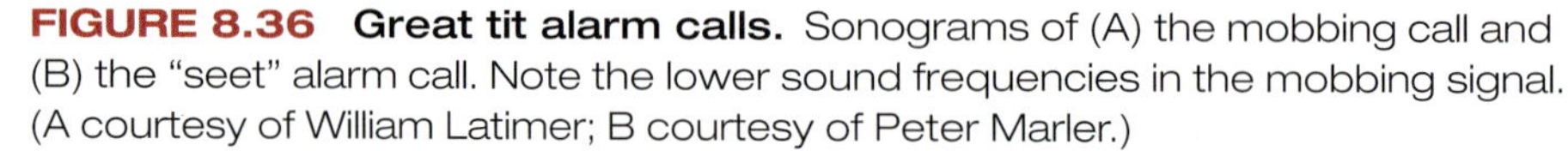

FIGURE 8.36 Great tit alarm calls. Sonograms of (A) the mobbing call and (B) the "seet" alarm call. Note the lower sound frequencies in the mobbing signal. (A courtesy of William Latimer; B courtesy of Peter Marler.)

great tit (*Parus major*) (**FIGURE 8.36**) (Marler 1955). These small European songbirds sometimes approach a perched hawk or owl and give a loud mobbing call whose dominant frequency is about 4.5 kHz. This easily located acoustic signal helps other birds find the mobbers and join in the harassment of their mutual enemy (see Chapter 6). If, however, a great tit spots a flying hawk, it gives a much quieter "seet" alarm call, which appears to warn mates and offspring of possible danger while making it harder for the predator to track down the sender. Interestingly, in the Japanese population of the great tit, parents use a harsh signal rather like the mobbing call to warn their chicks of an approaching Japanese rat snake (*Elaphe climacophora*), which induces the nestlings to leap out of the nest; the adults give a much softer call in a narrow range of sound frequencies when a large-billed crow (*Corvus macrorhynchos*) is nearby, which induces the nestlings to crouch down in the nest cavity, thereby making the baby birds harder for the crow to find and extract (Suzuki 2011).

The dominant frequency of the European great tit's "seet" calls lies within 7 to 8 kHz, so the sound attenuates (weakens) after traveling a much shorter distance than the mobbing signal. The rapid attenuation of the "seet" call compromises its effectiveness in reaching distant legitimate receivers, but it also lowers the chance that a dangerous predator on the hunt will be able to tell where the caller is. Moreover, the frequencies of the "seet" call lie outside the range that hawks can hear best, while falling within the range of peak sensitivity of the great tit (**FIGURE 8.37**). As a result, a great tit can "seet" to a family member 40 meters away but will not be heard by a sparrowhawk (*Accipiter nisus*) unless the predator is less than 10 meters distant (Klump et al. 1986). If the "seet" call of the great tit has evolved properties that reduce the risk of detection by its enemies, then unrelated species should also have evolved alarm signals with similar properties. The remarkable convergence

(A)

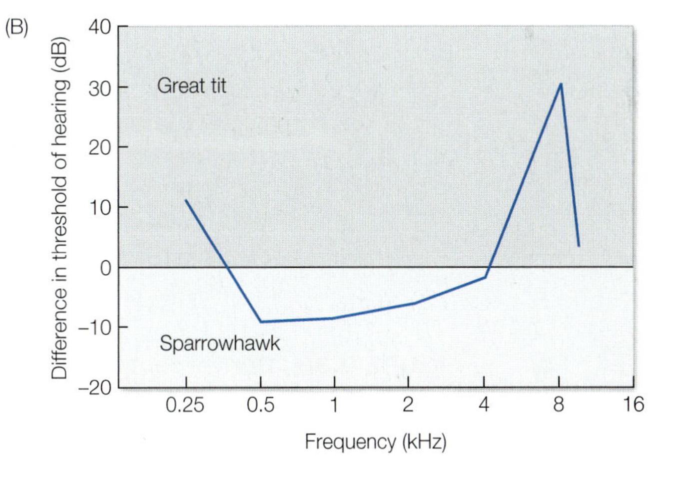

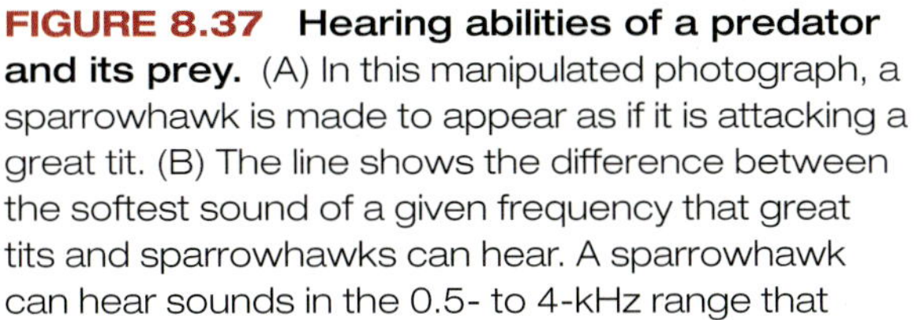
FIGURE 8.37 Hearing abilities of a predator and its prey. (A) In this manipulated photograph, a sparrowhawk is made to appear as if it is attacking a great tit. (B) The line shows the difference between the softest sound of a given frequency that great tits and sparrowhawks can hear. A sparrowhawk can hear sounds in the 0.5- to 4-kHz range that are fainter (5 to 10 dB lower in intensity) than those that great tits can hear, but great tits can detect an 8-kHz sound (in the range of the "seet" call) that is fully 30 dB fainter than any 8-kHz sound that a sparrowhawk can detect. (A, photograph © Andrew Darrington/Alamy; B after Klump et al. 1986.)

in the "seet" calls of many unrelated European songbirds suggests that predation pressure by bird-eating hawks has favored the evolution of alarm calls that are hard for hawks to hear (**FIGURE 8.38**) (Marler 1955). Here we have one more example of how productive the theory of natural selection can be for researchers interested in the adaptive nature of communication.

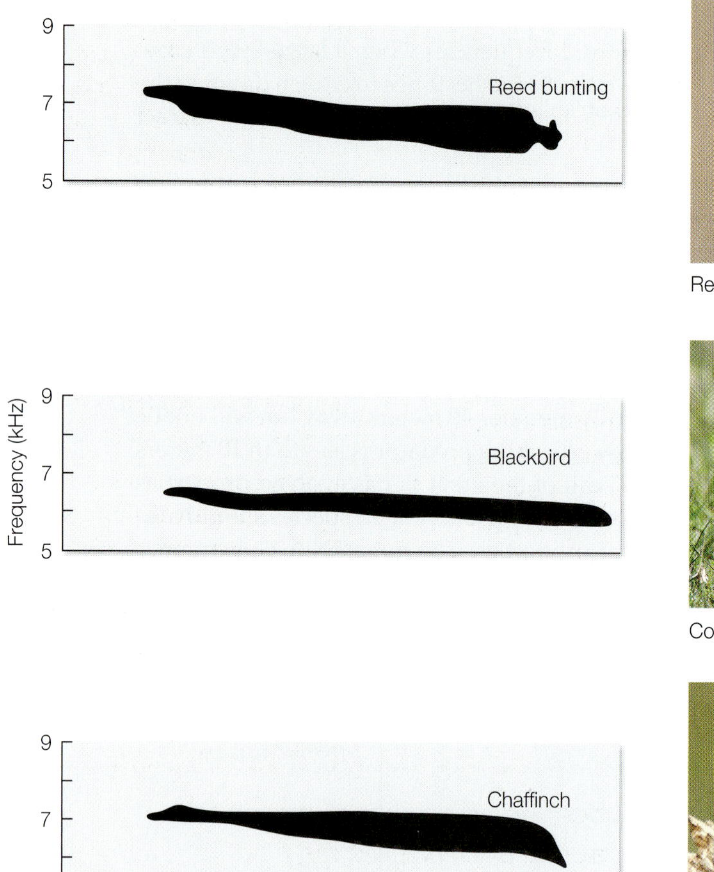

Reed bunting

Common blackbird

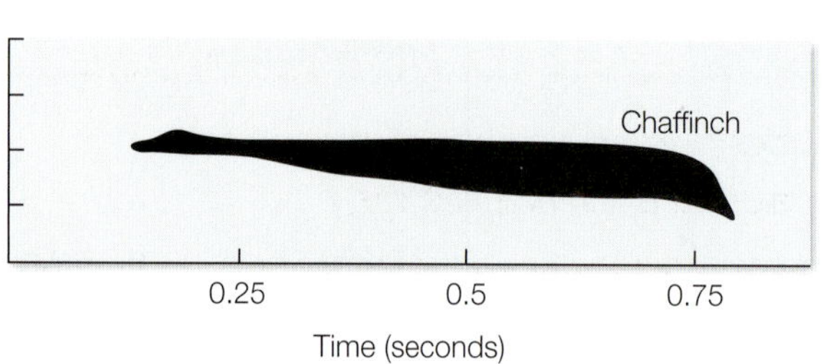

Chaffinch

FIGURE 8.38 Convergent evolution in a signal. The great tit's high-pitched "seet" alarm call (see Figure 8.36) is very similar to the alarm calls given by other, unrelated songbirds when they spot an approaching hawk. This includes the reed bunting (*Emberiza schoeniclus*), the common blackbird (*Turdus merula*), and the chaffinch (*Fringilla coelebs*) (After Marler 1955. Photographs from top to bottom: © Gertjan Hooijer/ShutterStock; © Karel Brož/ShutterStock; © Andrew Howe/istockphoto.com.)

SUMMARY

1. Animal interactions involving information transfer occur in a variety of contexts. Whether these interactions constitute communication or something else depends on the costs and benefits that senders and receivers incur. Honest signaling occurs when both the sender and the receiver obtain fitness benefits. Deceitful signaling results in costs to the receiver. Eavesdropping results in benefits for the receiver but costs for the sender. Communication includes cases when the sender receives fitness benefits (honest and deceitful signaling).
2. Animal signals are critical to evolved communication systems because they contain information that affects the fitness of senders and receivers. Animal signals evolve via preexisting traits or preexisting biases. Preexisting traits are behavioral, physiological, or morphological characteristics that already provide informative cues to receivers, which can be modified into a signal via a process called ritualization. Preexisting biases are biases in the sensory systems that detect some features of an organism's world better than others and that can be exploited by sender signals in a process called sensory exploitation.
3. Darwin's theory of descent with modification tells us that the adaptive traits of an animal living today (including its communication system) are the products of past changes layered on still older changes that occurred during the history of the species.
4. Changes will spread through populations by selection when individuals with the modified signal (or response) gain fitness from their altered behavior. Awareness of this rule helps researchers identify attributes that seem to reduce rather than increase the individual's fitness, such as a female spotted hyena's use of a pseudopenis (an enlarged clitoris) in greeting displays with other members of her clan.
5. Honest signals are those that indicate the quality of the sender, facilitate decision making by a receiver, and cannot be easily produced by the sender. Honesty is enforced either because signals are costly to produce or to socially maintain or because of trade-offs between the signal and other communication traits.
6. Deception and eavesdropping reduce the fitness of receivers and senders, respectively. Despite deceptive senders and exploitative eavesdroppers, some evolved communication systems can persist because of the net benefits they provide to legitimate senders and legitimate receivers.

COMPANION WEBSITE

Go to **ab11e.com** for discussion questions and all of the audio and video clips.

CHAPTER

Reproductive Behavior

Visitors to an eastern Australian forest might come upon a small stick structure decorated with flowers, pebbles, and even brightly colored pieces of trash. This fairylike structure is the handiwork of a male satin bowerbird (*Ptilonorhynchus violaceus*), and it looks more like something a precocious child might have built rather than a construction by a bird not much larger than a robin (**FIGURE 9.1**). The male decorates his bower with all of the colorful items he can find. In satin bowerbirds, males tend to prefer blue items, though other bowerbird species, such as the greater bowerbird shown in the opening photograph, prefer other colors. But why do males decorate their bowers with colorful items? They do so to attract females. If and when a female arrives at the bower, the male begins with a preamble of chortles and squeaks, followed by an elaborate courtship that has him dancing across the entrance to his

Chapter 9 opening photograph
© Konrad Wothe/Minden Pictures.

FIGURE 9.1 Bowerbird courtship revolves around the bower. A male satin bowerbird, with a yellow flower in his beak, courts a female that has entered the bower that he has painstakingly constructed and ornamented with blue objects. (Photograph by Bert and Babs Wells.)

VIDEO 9.1

Bowerbird reproductive behavior
ab11e.com/v9.1

bower while opening and closing his wings in synchrony with a buzzing trill. This dance may be followed by another in which the male bobs up and down while imitating the songs of several other species of birds (**VIDEO 9.1**). Yet despite this elaborate male display—a stereotyped action used to communicate with others—most courtships end with the abrupt departure of a seemingly indifferent female (Borgia 1986). In fact, female satin bowerbirds initially visit several bowers scattered throughout their home range, but just to look, not yet to mate with any bower builder (Uy et al. 2001). After the first round of visitations, the female takes a break of several days to construct a nest before returning to various bowers, during which time she usually observes the full courtship routines of many males. Finally, after several weeks, the female chooses one male and enters his bower, where she is courted again before she crouches down to invite the male to copulate. Afterward she flies off and will have no further contact with her partner, incubating her eggs and rearing her nestlings all by herself. Her mate stays at or near his bower for most of the 2-month breeding season, courting other females that come to inspect his creation and copulating with any that are willing.

Not only do the two sexes of satin bowerbirds look different (see Figure 9.1), their reproductive approaches are also very dissimilar. Satin bowerbirds are not unusual in this regard; in most animals, males do the courting and females do the choosing, whether we are talking about bowerbirds or belugas, aardvarks or zebras. This pattern is so widespread that biologists ever since Darwin have tried to provide an evolutionary explanation for sex differences in reproductive behavior. In this chapter, we explore reproductive behavior by focusing on the reasons why males and females differ in their behaviors and traits. We introduce the concept of sexual selection, using satin bowerbirds to illustrate it, much as we used spotted hyenas to illustrate the concept of communication in Chapter 8. The chapter begins with a discussion of what sexual selection is and how it results in sex differences in behavior and morphology. We then review two mechanisms of sexual selection (intra- and intersexual selection) and some hypotheses to explain how elaborate traits evolve, typically in only one sex or the other. Finally, we end with an exploration of sexual conflict—conflict between males and females over mate choice. As we have done in previous chapters, we integrate function and mechanism in our discussions of trait evolution—an approach that ultimately offers a better understanding of the tactics, strategies, and traits that males and females use to obtain access to mates and to pass their genes on to the next generation.

Sexual Selection and the Evolution of Sex Differences

What does a male satin bowerbird gain by spending so much time constructing his bower, gathering decorations (often taken from other males' bowers), and defending his display site from rival males while also destroying the bowers of his neighbors? Here we turn to Darwin's theory of evolution by sexual selection to get at the possible adaptive value of both the male bowerbird's courtship behavior, as well as the female's mate choice behavior. **Sexual selection** theory was Darwin's solution to a major evolutionary puzzle, some might say the first Darwinian puzzle: why extraordinarily extravagant courtship behaviors and morphological traits had evolved (Darwin 1871). The evolution of these elaborate morphological traits—**ornaments** used in mate attraction and **armaments** (weapons) used in intrasexual competition—did not seem to fit nicely into the framework of natural selection. Darwin realized that bower building and ornaments such as the huge tails of male peacocks, widowbirds, and so many other birds must surely make the males with these costly traits more likely to die young (**FIGURE 9.2**). Why then would natural selection favor traits that might

(B)

(C)

FIGURE 9.2 Elaborate costly traits exhibited by male birds. These characteristics evolved despite their probable negative effect on the survival of individuals. (A) This displaying male Goldie's bird of paradise (*Paradisaea decora*) is showing off his extraordinary plumage to a visiting female of his species. (B) The male greater sage grouse (*Centrocercus urophasianus*) can hold his own with other examples of birds that possess conspicuous feather ornaments. (C) The elongated tail of the male resplendent quetzal (*Pharomachrus mocinno*) complements the bird's exquisite body plumage. (Photographs: A, © Phil Savoie/Naturepl.com; B, © Tom Reichner/Shutterstock.com; C, © iStock.com/Thorsten Spoerlein.)

ultimately decrease rather than increase survival? After all, males of many species are more likely to be captured and eaten by predators while searching for females or trying court them with conspicuous displays or stunning bowers than when not trying to obtain mates. Darwin proposed the theory of sexual selection to help explain why individuals, usually males, evolved elaborate traits that lowered their survival chances. Darwinian sexual selection could favor such traits, provided that these attributes helped one sex (usually males) gain an advantage over others of the same sex in acquiring mates, an end that could be achieved either because these males were especially attractive to females, or because they intimidated or outcompeted their rivals in order to monopolize access to females. If males lived shorter lives but reproduced more, then the ornaments, armaments, displays, and behaviors that reduced longevity could nevertheless spread over time. Sexual selection, therefore, is a form of natural selection that favors the evolution of elaborate traits—and preferences for those traits in the opposite sex—if they increase the mating success of those that bear them or those that choose mates that bear them.

Since sexual selection most often favors the evolution of elaborate traits in one sex and not the other, the end result in many species is not only a sex difference in the trait itself, but also in many of the associated reproductive behaviors. Bower building in satin bowerbirds is a clear example of a sex difference in reproductive behavior, but why does such a sex difference tend to evolve in species that exhibit extreme traits or behaviors used in mating? Because most female satin bowerbirds mate with just one male—often the same individual that is popular with other females—male reproductive success in any given breeding season is variable, with some males getting many mates and others very few, if any (**FIGURE 9.3**) (Borgia 1985, Uy et al. 2001). This unequal partitioning of reproduction within a population or social group is called **reproductive skew** (see Chapters 12 and 13). In the case of satin bowerbirds, there is high skew among males, but much lower skew among females because most females will reproduce, whereas most males will not. The fact that reproduction tends to be shared less equitably in one sex than the other has great significance for a key puzzle that we have yet to address: why do male,

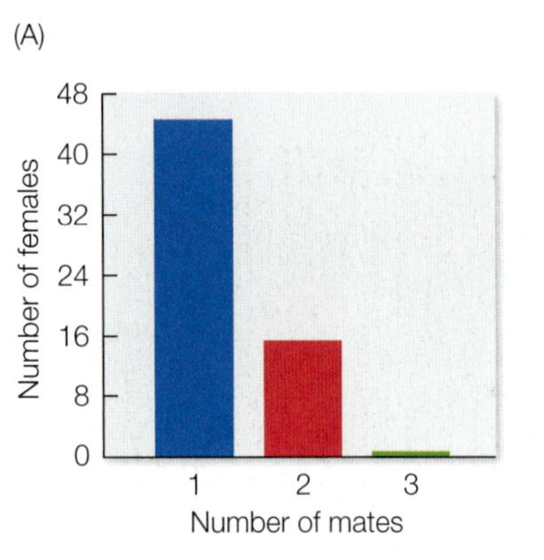

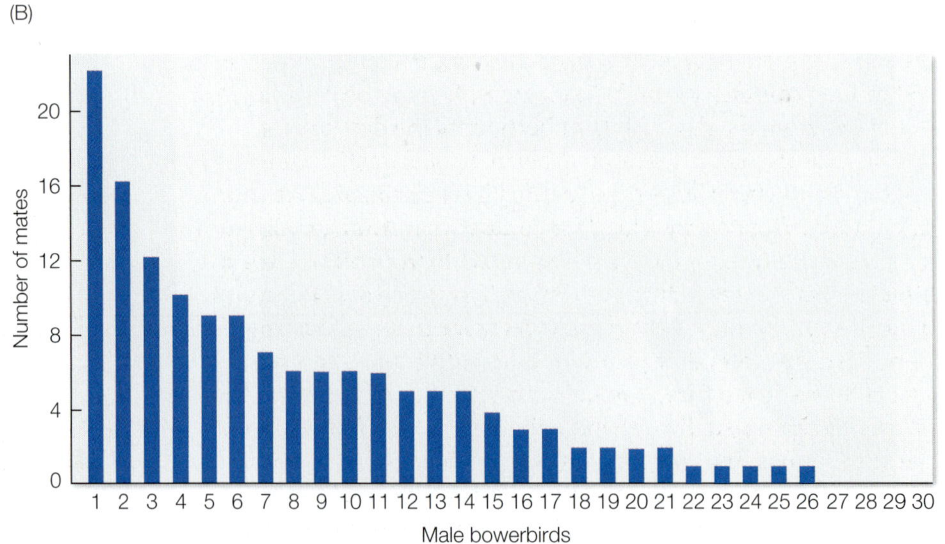

FIGURE 9.3 Reproductive variance is greater for males than for females in the satin bowerbird. (A) Very few female bowerbirds have more than two mates per breeding season, and few, if any, use the sperm of more than one male to fertilize their eggs. (B) Some male bowerbirds, however, mate with more than 20 females in a single season, while others do not mate at all. (After Uy et al. 2001.)

but not female, satin bowerbirds build courtship display structures, and why do females evaluate male performance instead of the other way round? Indeed, it is common throughout the animal kingdom for males to try to court and mate with many females, while the objects of their desire are typically content with fewer matings, albeit with a male or males they have carefully chosen.

Sex Differences in Reproductive Behavior

Why is it more common for males to do the courting, and females to do the choosing? To understand the widespread differences between the way males and females look, behave, and share reproduction, we need to start at the beginning—the developmental beginning for all sexually reproducing organisms. The widespread pattern of sex differences is almost certainly related to a fundamental difference between the sexes, which is that males produce small sperm (and many of them) and females produce large eggs (and relatively few of them) (Schärer et al. 2012). The **fundamental asymmetry of sex** argues that eggs are expensive but sperm are cheap (**BOX 9.1**). In sexual species, eggs are by definition larger than sperm, which are usually just big enough to contain the male's DNA and enough energy to fuel the journey to an egg. The fusion of these two gametes that differ so greatly in size is called **anisogamy**, a process that generates many of the sex differences observed in satin bowerbirds and other species. Even in species in which males produce oversize sperm—such as the fruit fly *Drosophila bifurca*, in which males make sperm that are (when uncoiled) nearly 6 centimeters long, or 20 times the length of the fly's body (Bjork et al. 2007)—the mass of an egg is still vastly greater than that of a sperm (**FIGURE 9.4**). Moreover, a single bird egg may constitute up to about 20 percent of the female's body mass (Lack 1968). By contrast, a male splendid fairy-wren (*Malurus splendens*), which weighs less than 10 grams, may have as many as 8 billion sperm in his testes at any given moment (Tuttle et al. 1996). The same pattern applies to coho salmon (*Oncorhynchus kisutch*), in which males shower about 100 billion sperm on a typical batch of 3500 eggs (Casselman and Montgomerie 2004). Likewise, a woman has only a few hundred cells that can ever develop into mature eggs during her lifetime (Daly and Wilson 1983), whereas a single man could theoretically fertilize all the eggs of all the women in the world, given that just one ejaculate contains on the order of 350 million minute sperm. What's more, he may be capable of doing this from the day he hits puberty until the day he dies.

The critical point is that small sperm usually vastly outnumber the many fewer large eggs available for fertilization in any population. This sets the stage for competition among males to fertilize those rarer eggs (Kokko et al. 2006). As you will see a bit later, it is this reproductive competition in one sex (typically males)—but not the other—that sets the stage for sexual selection to act on traits that improve that individual's reproductive chances. Therefore, a male's contribution of genes to the next generation usually depends directly on how many sexual partners he has: the more mates, the more eggs fertilized, the more descendants produced, and the greater the male's fitness relative to less sexually successful individuals—and the greater the effect of sexual selection on the evolution of male attributes. This point was first made in 1948 by the English geneticist Angus Bateman in classic experiments with *Drosophila* showing that a male's reproductive success increases with his number of mates (Bateman 1948).

FIGURE 9.4 Male and female gametes differ greatly in size. A hamster sperm fertilizing a hamster egg (magnified 4000x) illustrates the proportionally small contribution of materials to the zygote by the male. (Photograph by David M. Phillips/The Population Council.)

Although male *Drosophila* can increase their reproductive success by mating with additional females, Bateman showed that the same was not true for females (Bateman 1948). Whereas male animals usually try to have many sexual partners, females generally do not, since their reproductive success is typically limited by the number of eggs they can manufacture, not by any shortage of willing mates.

BOX 9.1

DARWINIAN PUZZLE

Are sperm always cheap?

Since Bateman's classic experiments in *Drosophila*, biologists have also learned that while sperm are relatively cheaper to produce than eggs, spermatogenesis is far from limitless and ejaculation can be costly (Wedell et al. 2002). In many species, males need to recover after ejaculating, and even when they do recover, they often have fewer sperm in subsequent ejaculations. In Soay sheep (*Ovis aries*), following males through multiple stages of the rut showed that large males that copulated frequently transferred fewer sperm per ejaculate (**FIGURE A**) (Preston et al. 2001).

Males also may not be able to produce limitless amounts of sperm throughout their lives, at least not high-quality sperm. Researchers studying captive houbara bustards (*Chlamydotis undulata*) found that paternal aging reduces both the likelihood that eggs will hatch and the rate at which chicks grow, with older males producing the lightest offspring as measured one month after hatching (Preston et al. 2015). Interestingly, the effect of paternal aging on hatching success was nearly identical to that of maternal aging (**FIGURE B**).

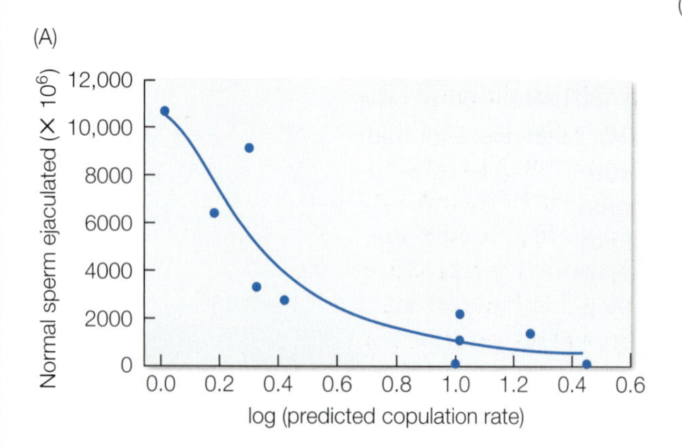

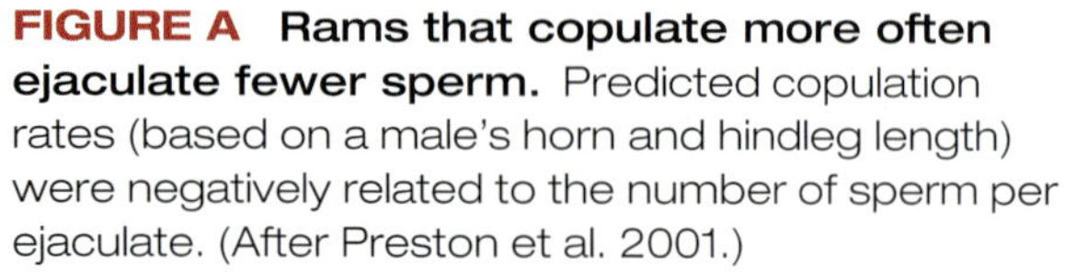
FIGURE A Rams that copulate more often ejaculate fewer sperm. Predicted copulation rates (based on a male's horn and hindleg length) were negatively related to the number of sperm per ejaculate. (After Preston et al. 2001.)

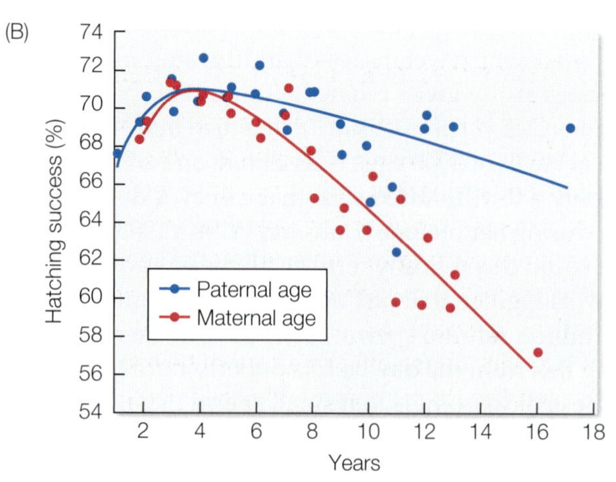

FIGURE B The effect of parental aging on the hatching success of bustard eggs. Hatching success varies according to both the age of the male that inseminates the female as well as the age of the female that lays the egg. (After Preston et al. 2015, CC BY 4.0.)

Thinking Outside the Box

In the soil nematode *Caenorhabditis elegans*, sex significantly reduces life span (Van Voorhies 1992). Can you come up with a series of hypotheses to explain why the act of having sex might cause individuals to live shorter lives? It turns out that in this species spermatogenesis, and not the act of mating, is the reason that life span is decreased in males. Can you design an experiment that would allow you to reach this conclusion?

Eggs are costly to produce because they are large, which means that females have to secure the resources to make them. And this increased cost to females relative to males does not stop with the production of eggs. After one batch of eggs has been fertilized, a female may spend still more time and energy than a male caring for the resultant offspring (see below). Thus, a female's reproductive success is far more dependent on the quality of her partner (and her own ability to produce eggs and care for offspring) than on the number of males she mates with. As a result of these sex differences in mating behavior, males tend to have higher reproductive variance than females, a phenomenon referred to as **Bateman's principle**.

More recently, however, this classic principle first identified by Bateman has been challenged. As we have mentioned previously in this book, one of the basic tenets of science is that experiments can and should be replicated. More than 60 years after Bateman's classic work with *Drosophila*, a team led by Patricia Gowaty tried to replicate his work using the exact same methods; they could not (Gowaty et al. 2012). When Gowaty and colleagues tried to replicate the experiments by culturing heterozygous adult *Drosophila* (those with one mutant and one wild-type allele) with phenotypes that were easy to score visually (traits such as eye color), they found that offspring with two copies of the mutant phenotype allele were inviable and died. They concluded that Bateman's method overestimated subjects with no mates, underestimated subjects with one or more mates, and produced systematically biased estimates of offspring number by each sex. Although this team was unable to replicate Bateman's results, their findings do not eliminate the important principle that Bateman outlined, as his ideas have been shown to apply to many other species. Instead, these newer results demonstrate the need to rigorously replicate scientific experiments and continually challenge ideas, even long-standing ones.

Sex Differences and Parental Investment

Parental care is surprisingly rare in many taxonomic groups, but in those species where parents do care for their young, why are females more likely than males to care for the offspring that come from their fertilized eggs? To understand investment in both gametes and offspring care, one must apply the cost–benefit approach. On the one hand, putting resources into large gametes and helping offspring become adults increases the probability that the offspring will live long enough to reproduce and pass on the parental genes to the next generation. On the other hand, what a parent supplies to one offspring cannot be used to make additional offspring down the road, and so the parent must make a potential trade-off in current versus future reproduction. This parental game of weighing the costs and benefits of caring for young is dominated by expenditures of time and energy and risks taken by a parent to help existing offspring at the expense of reducing future reproductive opportunities (Trivers 1972), a phenomenon that Robert Trivers labeled **parental investment** (**FIGURE 9.5**).

In most species, females are more likely than males to derive a net benefit from taking care of offspring. One reason for this sex difference in the fitness benefits associated with parental care is that the offspring a female cares for are extremely likely to carry her genes. In contrast, a male's paternity is often less certain, given that females of many species are inseminated by more than one male (see Chapter 10), and therefore males have less incentive to be parental if paternal males lose fertilization opportunities. Moreover, as Bateman showed, if a male can mate with several females, it pays him to do so, particularly if he has attributes that give him an edge in the race to fertilize eggs (Queller 1997). When some males do better than others and monopolize reproduction (in other words, when there is high reproductive skew among males), there will be high reproductive variance among

FIGURE 9.5 Parental investment takes many forms. (A) A male frog carries his tadpole offspring on his back. (B) A male katydid gives his mate an edible spermatophore containing orange carotenoid pigments that will be incorporated into her eggs. (C) An eared grebe (Podiceps nigricollis) protects its young by letting them ride on its back. (D) The nutritious coiled spermatophore of a male Photinus firefly (shown here greatly enlarged) is donated to the female along with the male's sperm during copulation. (Photographs: A, © iStock.com/Sand1983; B, Klaus Gerhard-Keller; C, © Piotr Kamionka/Shutterstock.com; D, Sara Lewis.)

males. In contrast, since one female's success will not influence the likelihood of another mating, reproductive variance among females will generally be low. Thus, although males and females have the same average reproductive success (because all offspring of diploid species have a mother and a father), they are likely to differ in their reproductive variance (Brennan 2010). Moreover, because females that have already mated typically have less to gain than males by copulating again (or because it is more energetically expensive to replenish eggs than sperm), there are generally many fewer sexually active females than males at any given time, creating a male-biased **operational sex ratio** (the ratio of sexually active males to sexually receptive females) (Emlen and Oring 1977). Thus, key behavioral differences between the sexes have apparently evolved in response to the difference in what they produce (in size and number of gametes), a response that is often amplified by variation in the degree to which a female and male provide parental care for their putative offspring.

A Reversal in Sex Differences

We have reviewed a theory of sex differences that focuses on the role of gametic differences and inequalities between the sexes in parental investment (**FIGURE 9.6**). The theory postulates that as a result of sex differences in mating behavior, as well as in reproductive variance and a skewed operational sex ratio, sexual selection operates more strongly in the sex where reproductive competition for mates is greatest. As we have mentioned repeatedly, in most species this is the male sex. One way we can test this theory is by finding unusual cases in which males make the larger parental investment or engage in other activities that cause the operational sex ratio to become reversed, so that there are more sexually active females than males. For example, in some species, males make contributions other than sperm toward the welfare of their offspring (or their mates) because if they do not, they may not get a chance to fertilize any eggs at all, a precondition for male reproductive success. For species of this sort, we can predict female competition for mates and careful mate choice by males—in other words, a **sex role reversal** with respect to which sex competes for mates and which does the choosing. Such a reversal occurs in the mating swarms of certain empid flies, in which the operational sex ratio is heavily female-biased because most males are off hunting for insect prey to bring back to the swarm as a mating inducement called a **nuptial gift** (Svensson 1997). When a male enters the swarm bearing his nuptial gift, he may be able to choose among females advertising themselves with (depending on the species) unusually large and patterned wings or decorated legs (Gwynne et al. 2007) or bizarre inflatable sacs on their abdomens (**FIGURE 9.7**) (Funk and Tallamy 2000).

Just as in some insects, males of some fish species offer their mates something of real value, namely a brood pouch in which the female can place her eggs. For example, in the pipefish *Syngnathus typhle*, "pregnant" males provide nutrients and oxygen to a clutch of fertilized eggs for several weeks, during which time the average female produces enough eggs to fill the pouches of two males. Females evidently compete for the opportunity to donate eggs to parental males, which pay a price when they are pregnant because they feed and grow less while brooding eggs. As a result, at the start of the breeding season, males in an aquarium experiment that were given a choice between allocating time to feeding or responding to potential mates showed more interest in feeding than in mating (Berglund et al. 2006). In the lab and in the ocean, large males with free pouch space actively choose among mates, discriminating against small, plain females in favor of large females, which can provide selective males with larger clutches of eggs to fertilize (Berglund et al. 1986, Rosenqvist 1990). In a controlled lab experiment with the related gulf pipefish (*Syngnathus scovelli*), researchers found that offspring survival was at least partly influenced by female body size (Paczolt and Jones 2010), reinforcing the idea that large-bodied females are preferred by males in this and other pipefish species. Furthermore, male gulf pipefish seem to prefer not just larger females, but those with a temporary striping on their

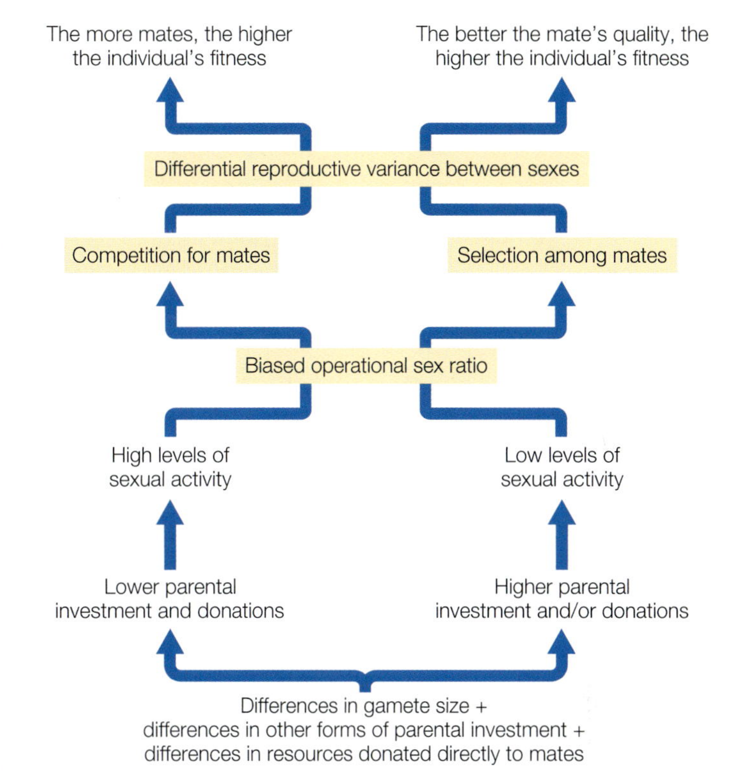

FIGURE 9.6 A theory of sex differences. The sexual behavior of males and females may differ because of differences in parental investment that affect the rate at which individuals can produce offspring. The sex that can potentially leave more descendants gains from high levels of sexual activity, whereas the other sex does not. An inequality in the number of receptive individuals of the two sexes leads to competition for mates within one sex, while the opposite sex can afford to be choosy.

FIGURE 9.7 A sex role reversal. In the long-tailed dance fly (*Rhamphomyia longicauda*), females fly to swarms where they advertise themselves while waiting for the arrival of a gift-bearing male. Perhaps the female's inflated abdomen and dark, hairy legs make her appear as large as possible to discriminating males. (From Funk and Tallamy 2000; photograph by David Funk.)

sides, a trait that is thought to emphasize body size during female–female aggressive interactions (Berglund and Rosenqvist 2001, Partridge et al. 2013). By statistically testing for the strength of sexual selection on these traits in the gulf pipefish, a team led by Adam Jones showed that female–female completion exerts strong selection on the ornament (Flanagan et al. 2014), suggesting that it may be an honest signal of female body size.

Another example of a species in which females sometimes compete for males is the flightless Mormon cricket (*Anabrus simplex*), which despite its common name has no religious affiliation (and is commonly referred to as a katydid). Darryl Gwynne demonstrated that when Mormon cricket population densities are high, fights between females increase and females exhibit mate choice among males (Gwynne 1984). However, as Mormon cricket densities increase, food availability decreases, which leads to a reduction in the number of sexually active males (a reduced operational sex ratio) (Gwynne 1993). When male Mormon crickets mate, they transfer to their partners another kind of nuptial gift, an enormous edible spermatophore that, in addition to sperm, gives a female nutrients for the production of more eggs (**FIGURE 9.8**) (Gwynne 1981). Given that the spermatophore constitutes 25 percent of the male's body mass, most male Mormon crickets probably cannot mate more than once. In contrast, some females are able to produce several clutches of eggs, but they have to persuade several males to mate with them if all their eggs are to be fertilized. This leads to competition among females for access to mates, and therefore increased sexual selection for female body size (Gwynne 1993).

Female competition for mates in Mormon crickets is evident during the times when huge numbers of these katydids leave home and march across the countryside, devouring farmers' crops (and one another) (Simpson et al. 2006). In a band of Mormon crickets, a male may announce his readiness to mate, which he does by producing an acoustic signal that causes females to come running, jostling with one another for the opportunity to climb onto the stridulating male, a prelude to insertion of the male's genitalia and transfer of a spermatophore. Males, however, may refuse to transfer spermatophores to lightweight females. A choosy male that rejects a female weighing 3.2 grams in favor of one weighing 3.5 grams fertilizes about 50 percent more eggs (Gwynne 1981). So here, too, when there are more receptive

FIGURE 9.8 Mormon cricket males give their mates an edible nuptial gift. Here a mated female carries a large white spermatophore just received from her partner. (Photograph by John Alcock.)

FIGURE 9.9 Black swans exhibit mutual ornamentation. Both males and females have curled wing feathers that serve a sexual function. (Photograph © iStock.com/steveo73.)

females than males (a reversal in the typical operational sex ratio), competition for mates takes place among females rather than among males.

The critical point to remember from these examples is that not all species exhibit the typical pattern of greater offspring investment by females than by males, more sexually active males than females within the population, and higher reproductive variance in males than in females. When these trends are reversed, sexual selection is predicted to act more strongly on female than male traits. In these cases of sex role reversal, females rather than males are more likely to be ornamented, to be larger, to guard territories, and to provide little parental care. And of course, there are even some species where sexual selection acts strongly on both sexes. This is particularly common in cooperatively breeding species (see Chapter 13) in which both males and females exhibit high reproductive skew and competition for access to mates (Rubenstein and Lovette 2009). When sexual selection operates strongly in both sexes, the end result may be mutual ornamentation and mate choice. An example of this mutual ornamentation occurs in the black swan (*Cygnus atratus*), in which both sexes grow curled feathers on their wings that are displayed prominently in a range of social interactions (**FIGURE 9.9**) (Kraaijeveld et al. 2004).

Intrasexual Selection and Competition for Mates

As we have just discussed, males in most species, with their minute sperm, can potentially have a great many offspring. Yet if they are to achieve even a small fraction of their potential, they must deal both with other males, which are attempting to mate with the same limited pool of receptive females, and with the females themselves, which often have a great deal to say about which male fertilizes their eggs. This scenario describes the way in which Darwin viewed sexual selection. He theorized that there were two independent, though related, mechanisms of sexual selection: (1) **intrasexual selection**, typically male–male combat and threat (using armaments); and (2) **intersexual selection**, generally male courtship of females (using ornaments) that can influence a female's choice of mates. Although these two mechanisms describe the majority of species, it is important reemphasize that intrasexual selection can involve female–female competition in some species (including those with sex role reversal), just as male mate choice, or even mutual mate choice by both males and females, is important in several species (Rubenstein 2012). Below we discuss these two mechanisms of sexual selection in greater detail, focusing largely on intrasexual selection in males and intersexual selection by females.

Competition and Access to Mates

Let's first consider those male traits that appear to be adaptations that help males compete effectively with other males in order to gain access to mates. We mentioned that male satin bowerbirds dismantle each other's bowers when they have a chance (Pruett-Jones and Pruett-Jones 1994). Because males with destroyed bowers lose opportunities to copulate, males have been sexually selected to keep a close eye on their display territories and to be willing to fight with rival intruders. Outright fighting among males is one of the most common features of life on earth because winners of male–male competition generally mate more often, something that is true for everything from giraffes (**FIGURE 9.10**) (Pratt and Anderson 1985, Simmons and Scheepers 1996) to digger bees (Alcock et al. 1977).

FIGURE 9.10 Males of many species fight for females. Here two male giraffes slam each other with their heavy necks and clubbed heads. Note that the long necks of giraffes may be as much a product of sexual selection as of natural selection for access to treetop food. (Photograph by Gregory Dimijian/Photo Researchers, Inc.)

Sexual selection for fighting ability often leads to the evolution of large body size because larger males tend to be able to beat up smaller ones. Furthermore, when males regularly fight for access to mates, they tend to be larger than the females of their species (Blanckenhorn 2005). In other words, when intrasexual selection on male body size is strong, **sexual dimorphism** in body sizes evolves. Thus, in species in which sexual selection acts strongly on males, males tend to be larger than females (and vice versa in many role-reversed species where sexual selection acts more strongly on females). In addition, males are the sex that tends to develop armaments. Animals ranging from spiders to dinosaurs, and from beetles to rhinoceroses, have evolved weapons in the form of horns, tusks, antlers, clubbed tails, and enlarged spiny legs, which they use when fighting with other males over females (**FIGURE 9.11**) (Emlen 2008). Other types of competition traits include the badges of status that we introduced in Chapter 7, traits such as breast coloration that are used to signal dominance rank but are not actually used in fighting.

Although males of many species battle for females, conflicts among males of some species are not about an immediate mating opportunity but instead have to do with the male **dominance hierarchy**. Once individuals have sorted themselves out from top dog to bottom mutt, the alpha male need only move toward a lower-ranking male to have that individual hurry out of the way or otherwise signal submissiveness. If the costly effort to achieve high status (and the priority of access to useful resources that is associated with dominance) is adaptive, then high-ranking individuals should be

FIGURE 9.11 Convergent evolution in male weaponry. Rhinoceros-like horns have evolved repeatedly in unrelated species because of competition among males for access to females. The species illustrated here are 1, narwhal (*Monodon monoceros*); 2, chameleon (*Chamaeleo* [*Trioceros*] *montium*); 3, trilobite (*Morocconites malladoides**); 4, unicornfish (*Naso annulatus*); 5, ceratopsid dinosaur (*Styracosaurus albertensis**); 6, horned pig (*Kubanochoerus gigas**); 7, protoceratid ungulate (*Synthetoceras* sp.*); 8, dung beetle (*Onthophagus raffrayi*); 9, brontothere (*Brontops robustus**); 10, rhinoceros beetle (*Allomyrina* [*Trypoxylus*] *dichotomus*); 11, isopod (*Ceratocephalus grayanus*); 12, horned rodent (*Epigaulus* sp.*); 13, giant rhinoceros (*Elasmotherium sibiricum**); (*extinct species). (Courtesy of Doug Emlen.)

rewarded reproductively. In keeping with this prediction, among mammalian species, dominant males generally mate more often than do subordinates (Cowlishaw and Dunbar 1991).

The relationship between dominance and sexual access to mates has been studied in groups of yellow baboons (*Papio cynocephalus*), in which males compete intensely for high social status. Opponents are willing to fight to move up the dominance hierarchy (**FIGURE 9.12A**), and as a result, individual males get bitten about once every 6 weeks, an injury rate nearly four times greater than that for females (Drews 1996). So does it pay males to risk damaging bites and serious infections to secure a higher ranking? When Glen Hausfater first attempted to test this proposition, he counted matings in a troop of baboons and found that, contrary to his expectation, males of low and high status were equally likely to copulate (Hausfater 1975). Hausfater subsequently realized, however, that he had made a questionable assumption, which was that any time a male mated, he had an equal chance of fathering an offspring. But copulation only results in fertilization if a female is ovulating. When Hausfater reexamined the timing of copulations, he found that dominant males had indeed monopolized females during the few days when they were fertile. The low-ranking males had their chances to mate, but typically only when females were in the infertile phase of their estrous cycles.

Since Hausfater's pioneering research, other researchers have followed up with similar studies in other populations of baboons. A summary of data gathered on

(A)

(B)

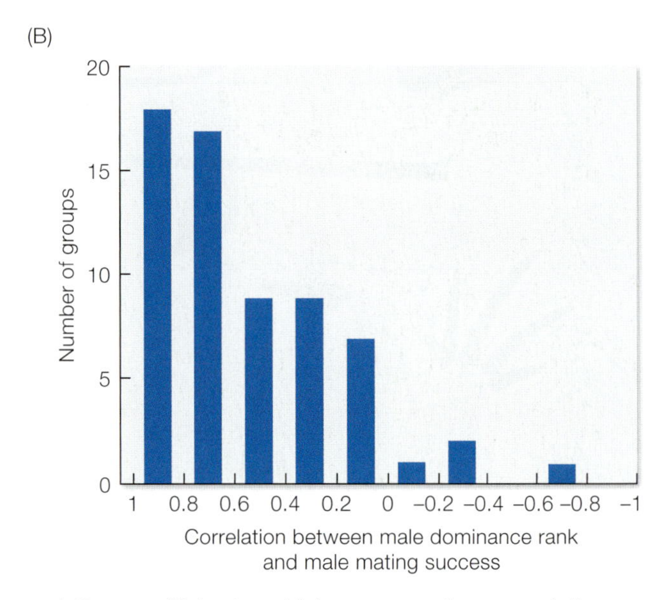

FIGURE 9.12 Dominance and mating success in yellow baboons. (A) Male baboons fight for social status. (B) In a Kenyan reserve in which many different troops were followed over several breeding seasons, the relationship between male dominance rank and male mating success generally yielded strongly positive correlation coefficients, which measure the association between the two variables. The closer the correlation coefficient is to +1, the closer the match between male rank and mating success. (A, photograph by Joan Silk; B after Alberts et al. 2003.)

troops of yellow baboons observed in Kenya over many years reveals that male dominance is almost always positively correlated with male copulatory success with fertile females (**FIGURE 9.12B**) (Alberts et al. 2006). A male that copulates with an ovulating female while keeping all other males away from her would seem all but certain to be the father of any offspring that his partner subsequently produces. But we no longer need to guess about paternity in these cases, thanks to developments in molecular-based parentage analysis that make it possible to determine with nearly complete certainty whether a given male has fathered a given baby. And yes, a male's dominance rank predicts not only his mating success but also his genetic success (Alberts et al. 2006). Dominant male baboons sired more offspring than subordinate ones by virtue of their ability to identify receptive females that were highly likely to conceive (Gesquiere et al. 2007), and they also kept other males away from these fertile females (Alberts et al. 2006), as also happens in other primate species (Widdig et al. 2004).

Coexistence of Conditional Mating Tactics

Although research on baboons and many other animals has confirmed that high dominance status yields paternity benefits for males able to secure the alpha position, the benefits are often not as large as one would expect (**FIGURE 9.13**) (Alberts et al. 2003). As it turns out, socially subordinate baboons can compensate to some extent for their inability to physically dominate others in their group. For one thing, lower-ranking males can and do develop friendships with particular females, relationships that depend less on physical dominance than on the willingness of a male to protect a given female's offspring (Palombit et al. 1997). Once a male, even a moderately subordinate one, has demonstrated that he will protect a female and her infant, that female may seek him out when she enters estrus again (Strum 1987). Male baboons also form friendships with other males. Through these alliances, they can

(A)

(B)

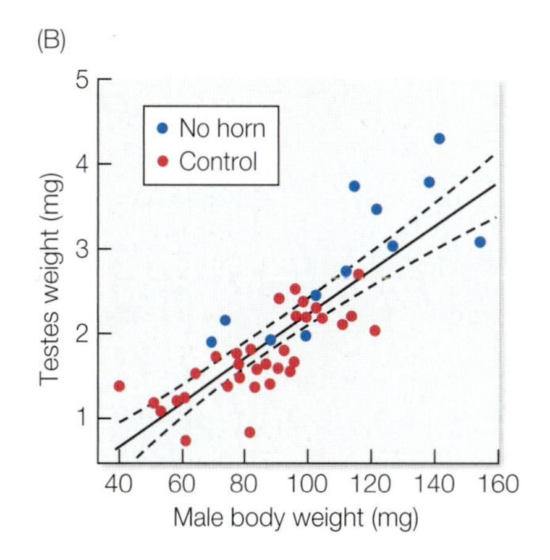

FIGURE 9.15 Large horns mean smaller testes. (A) In the horned scarab beetle *Onthophagus nigriventris*, "major" males are large and horned, whereas "minor" males are smaller and have a small to nonexistent horn. (B) When males are experimentally induced to develop as hornless males, they tend to grow larger than horned controls and invest in relatively large testes for their body size. The regression line for both data sets combined is shown, along with the dotted 95 percent confidence limit lines. (A, photograph by Doug Emlen; B after Simmons and Emlen 2006, © 2006 National Academy of Sciences, USA.)

bigger horns. Therefore, when a male beetle larva is poorly nourished, it shifts its investment of resources away from growing horns and into what will become his sperm-producing testes. Such a "minor" male has small to nonexistent horns as an adult but larger testes than his bigger opponents with long horns. Thus, these beetles exhibit a conditional mating tactic because whether a larva grows into a larger, horned major male or a small, well-endowed minor male depends on the diet (the size of the dung ball) during development of the male.

Given these developmentally driven morphological differences between major and minor beetle males, what are the behavioral tactics that these two types of males use to mate? In keeping with his body plan, the minor male sneaks into a burrow where a female is being guarded by a big-horned major male and attempts to inseminate the female on the sly, passing large amounts of sperm to her should he be successful in evading detection by her consort (Hunt and Simmons 2002). Large males cannot build both large weapons and large testes (**FIGURE 9.15**) (Simmons and Emlen 2006), just as it has been suggested that male bats cannot afford both large testes and large brains (Pitnick et al. 2006). A large-horned male *O. nigriventris* with his relatively small testes can lose egg fertilizations to a smaller rival, as the smaller male's large ejaculate can swamp the sperm the female has received from her more heavily armed consort (see below).

Let's apply this approach to examine the three alternative mating tactics exhibited by some *Panorpa* scorpion flies (**FIGURE 9.16**) (Thornhill 1981). Many *Panorpa* males aggressively defend dead insects, a food resource highly attractive to receptive females. However, other males secrete saliva on leaves and wait for occasional females to come and consume this nutritional gift, and still other males offer females nothing at all but instead grab them and force them to copulate. In experiments with caged groups of ten male and ten female scorpion flies, Randy Thornhill showed that the largest males monopolized the two dead crickets placed

FIGURE 9.16 A male *Panorpa* scorpion fly. This insect has a strange scorpion-like abdomen tip that it can use to grasp females in a prelude to forced copulation, one of three mating tactics available to males of this species. (Photograph © Mario Saccomano/Shutterstock.com.)

in the cage, which gave these males easy access to females and about six copulations on average per trial. Medium-size males could not outmuscle the largest scorpion flies in the competition for the crickets, so they usually produced salivary gifts to attract females but gained only about two copulations each. Small males were unable to claim crickets, nor could they make salivary presents, so these scorpion flies instead forced some females to mate but averaged only about one copulation per trial.

Thornhill proposed that in this case, all the males, large and small alike, possessed a single conditional strategy that enabled each individual to select one of three mating tactics based on his relative social standing. This best-of-a-bad-situation argument predicts that the differences among the behavioral phenotypes are environmentally caused, not based on genetic differences among individuals, and that males should switch to a tactic yielding higher reproductive success if the social conditions they experienced make the switch possible. To test these predictions, Thornhill removed the large males that had been defending the dead crickets. When this change occurred, some males promptly abandoned their salivary mounds and claimed the more valuable crickets. Other males that had been relying on forced copulations hurried over to stand by the abandoned secretions of the males that had left them to defend dead crickets. Thus, a male *Panorpa* goes with whichever tactic gives him the highest possible chance of mating, given his current competitive status. These results clinch the case in favor of conditional mating tactics as the explanation for the coexistence of three mating forms in *Panorpa* scorpion flies (Thornhill 1981).

Coexistence of Alternative Mating Strategies

As we have just discussed, the ability of an individual to shift their approach throughout their lives is referred to as alternative mating tactics. In many cases, the adoption of one or another mating tactic is conditional as a male attempts to make the best of a bad situation (Dawkins 1980, Eberhard 1982, Gross 1996). Yet not all animals can adopt alternative approaches to mating that can change flexibly with social rank or body condition. Instead, some individuals exhibit an **alternative mating strategy**, or a type of behavioral polymorphism that has a strong genetic component and is therefore inflexible and fixed throughout an individual's life. As an example of a species exhibiting a behavioral polymorphism with (three) alternative mating strategies, recall our discussions in Chapter 3 of the ruff (*Philomachus pugnax*), in which males can be categorized as territorial "independents," "satellites" that join an independent on his territory, or "faeders," which are female mimics (**FIGURE 9.17**). The three ruff types are hereditary, controlled by a set of non-recombining supergenes that evolved independently (see Figure 3.33) (Küpper et al. 2015, Lamichhaney et al. 2015). These supergenes contain a variety of genes related to the behavioral differences among male types,

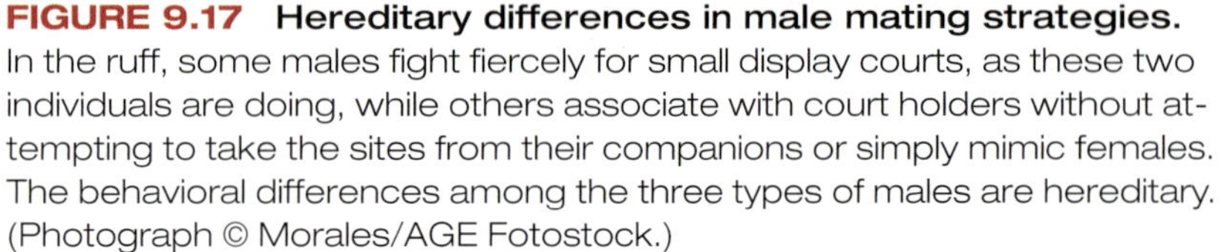

FIGURE 9.17 Hereditary differences in male mating strategies. In the ruff, some males fight fiercely for small display courts, as these two individuals are doing, while others associate with court holders without attempting to take the sites from their companions or simply mimic females. The behavioral differences among the three types of males are hereditary. (Photograph © Morales/AGE Fotostock.)

including those for reproductive hormones and feather color. Thus, the three ruffs types are hereditary strategies and each persists in the population (at very different frequencies) presumably because they may have equal average lifetime reproductive success (Widemo 1998).

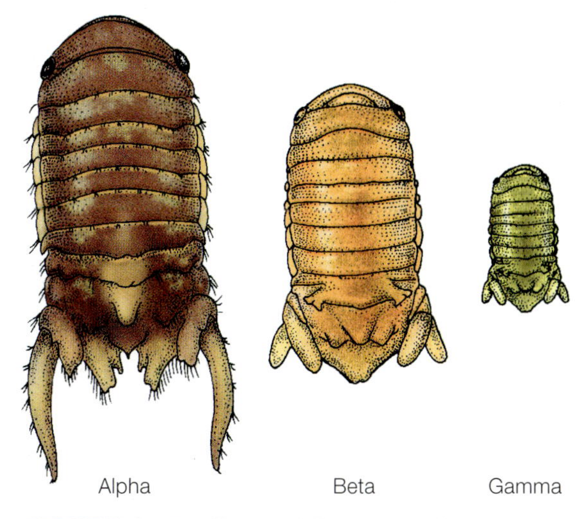

FIGURE 9.18 Three different male types of the sponge isopod: the large alpha male, the female-size beta male, and the tiny gamma male. Each type not only has a different size and shape but also uses a different hereditary strategy to acquire mates. (After Shuster 1992.)

The marine isopod *Paracerceis sculpta* is another species with males that exhibit three types of reproductive roles. Although this creature vaguely resembles the more familiar terrestrial sow bugs and pill bugs that live in moist debris in suburban backyards, it resides in sponges found in the intertidal zone of the Gulf of California. If you were to open up a sufficient number of sponges, you would find females, which all look more or less alike, and an assortment of males that come in three dramatically different sizes: large (alpha), medium (beta), and small (gamma) (**FIGURE 9.18**), each with its own behavioral phenotype. The big alpha males attempt to exclude other males from the interior cavities of sponges where one or several females live. If a resident alpha encounters another alpha male in a sponge, a battle ensues that may last for hours before one male gives way. Should an alpha male find a tiny gamma male, however, the larger isopod will simply grasp the smaller one and throw him out of the sponge. Not surprisingly, gammas avoid alpha males as much as possible while trying to sneak matings with the females living in their sponges (Shuster 1989). When an alpha and a medium-size beta male meet inside a sponge cavity, the beta behaves like a female, and the male courts his rival ineffectually. Through female mimicry, the female-size beta males coexist with their much larger and stronger rivals and thereby gain access to the real females that the alpha males would otherwise monopolize.

In the marine isopod, therefore, we have three different types of males, and one type has the potential to dominate others in male–male competition. If the three types represent three distinct *strategies*, then the differences among them should be traceable to genetic differences, but the mean reproductive success of the three types should be equal, a requirement of frequency-dependent selection (see Chapter 6). If, however, alpha, beta, and gamma males use three different *tactics* resulting from the same conditional strategy, then the behavioral differences among them should be the result of different environmental conditions, not different genes, and the mean reproductive success of males using the alternative tactics need not be equal.

Steve Shuster and his coworkers collected the information needed to determine whether this species exhibits alternative reproductive tactics or strategies (Shuster and Wade 1991, Shuster 1992). First, they showed that the size and behavioral differences among the three types of male isopods are the hereditary result of differences in a single gene represented by three alleles. Second, they measured the reproductive success of the three types in the laboratory by placing various combinations of males and females in artificial sponges. The males used in this experiment had distinct genetic markers—distinctive characteristics that could be passed on to their offspring—that enabled the researchers to identify which male had fathered each baby isopod that each female eventually produced. Shuster and Wade found that the reproductive success of a male depended on how many females and rival males lived with him in a sponge. For example, when an alpha and a beta male lived together with one female, the alpha isopod fathered most of the offspring. But when this male combo occupied a sponge with several females, the alpha male could not control them all, and the beta male outdid his rival, siring 60 percent of the progeny. In still other combinations, gamma males out-reproduced the others. For each combination, it was possible to calculate an average value for male reproductive success for alpha, beta, and gamma males. Shuster and Michael Wade then returned to the Gulf of California to collect a large random sample of sponges, each one of which they opened to count the isopods within (Shuster and Wade 1991). Knowing how often alphas, betas, and gammas lived in various combinations with competitors and females enabled Shuster and Wade to estimate the average reproductive success of the three types of males, given the laboratory results gathered earlier. They estimated that alpha males in nature had mated with 1.51 females on average, while betas checked in at 1.35 and gammas at 1.37 mates. Since these means were not significantly different statistically, Shuster and Wade concluded that the three genetically different types of males had essentially equal fitness in nature, and thus the requirements for an alternative reproductive strategy explanation had been met.

FIGURE 9.19 Three different egg fertilization behaviors coexist in the bluegill sunfish. (A) A territorial male guards a nest that may attract gravid females. (B) A small sneaker male waits for an opportunity to slip between a spawning pair, releasing his sperm when the territory holder does. (C) A slightly larger satellite male with the body coloration of a female hovers above a nest before slipping between the territorial male and his mate when the female spawns.

Sperm Competition

Males that adopt different mating tactics or strategies differ in how they go about securing copulatory partners. But the reproductive competition need not stop with the behavioral differences that influence how males access mates. When females mate with more than one male in a short period of time, the males rarely divvy up a female's eggs evenly. In the case of bluegill sunfish (*Lepomis macrochirus*), for example, older males nesting in the interior of the colony produce ejaculates with more sperm, and their sperm swim faster than those of younger bluegills, suggesting that older males should generally enjoy a fertilization advantage (Casselman and Montgomerie 2004). However, when both a sneaker male bluegill and a guarding territorial male release their sperm over a mass of eggs (**FIGURE 9.19**), the sneaker male fertilizes a higher proportion of the eggs than the nest-guarding male, in part because the sneaker gets closer to the egg mass before spawning (Fu et al. 2001). What we have here is evidence of competition among males with respect to the fertilization success of their sperm. **Sperm competition** is a common phenomenon in the animal kingdom, no matter whether fertilization is external (as in bluegills and many other fishes) or internal (as in some fishes, insects, and all mammals, birds, and reptiles). If the sperm of some males have a consistent advantage in the race to fertilize eggs, then counting up a male's spawnings or number of copulatory partners will not measure his fitness accurately (Birkhead and Møller 1992). Thus, males must not only compete with other males for access to mates; in many cases, their sperm must also compete with those of other males once they have mated. Sexual selection can act on male traits that improve the success of males in either or both contexts.

Sperm competition occurs in most animal groups, including in many insects (Parker 1970, Simmons 2001). In the black-winged damselfly (*Calopteryx maculata*) of eastern North America, males try to win the sperm wars by physically removing rival gametes from their mate's body before transferring their own (Waage 1979), a common mechanism of sperm competition in damselflies and dragonflies (Córdoba-Aguilar et al. 2003). Male black-winged damselflies defend territories containing floating aquatic vegetation in which females lay their eggs. When a female flies to a stream to lay her eggs, she may visit several males' territories and copulate with each site's owner, laying some eggs at each location. The female's behavior creates competition among her partners to fertilize her eggs (Parker 1970), and the resulting sexual selection has endowed males with an extraordinary penis.

To understand how the damselfly penis works, we need to describe the odd manner in which damselflies (and dragonflies) copulate. First, the male

(A)

(B)

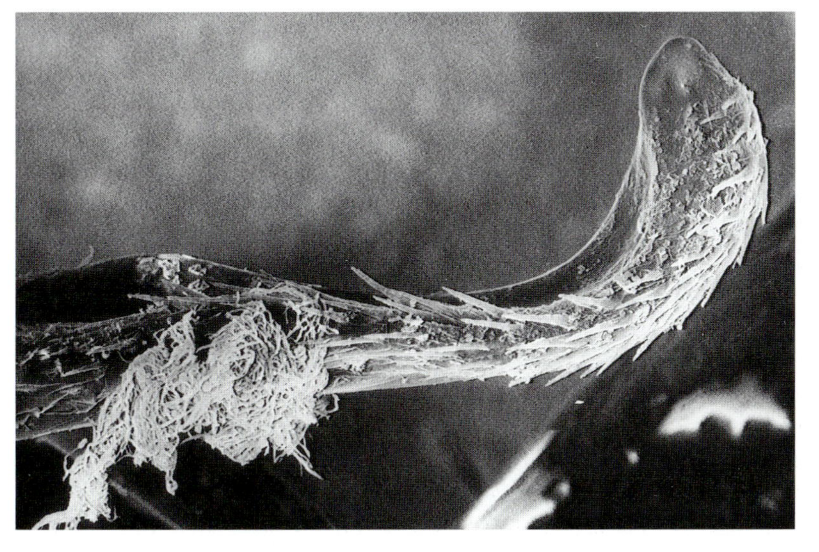

FIGURE 9.20 Copulation and sperm competition in the black-winged damselfly. (A) The male (on the right) has grasped the female with the tip of his abdomen; the female bends her abdomen forward to make contact with her partner's sperm-removing and sperm-transferring penis. (B) The male's penis has lateral horns and spines that enable him to scrub out a female's sperm storage organ before passing his own sperm to her. (A, photograph by John Alcock; B, photomicrograph by Jon Waage, from Waage 1979, © 1979 by the American Association for the Advancement of Science.)

catches the female and grasps the front of her thorax with specialized claspers at the tip of his abdomen. A receptive female then swings her abdomen under the male's body and places her genitalia over the male's sperm transfer device, which lies on the underside of his abdomen near the thorax (**FIGURE 9.20A**). The male damselfly then rhythmically pumps his abdomen up and down, during which time his spiky penis acts as a scrub brush, catching and drawing out any sperm already stored in the female's sperm storage organ (**FIGURE 9.20B**). Jon Waage found that a copulating male *C. maculata* removes from 90 to 100 percent of any competing sperm before he releases his own gametes, which he earlier transferred from his testes on the tip of his abdomen to a temporary storage chamber near his penis (Waage 1979). After emptying the female's sperm storage organ, the male lets his own sperm out of storage and into the female's reproductive tract, where they remain for use when she fertilizes her eggs—unless she mates with yet another male before ovipositing, in which case the first male's sperm will be extracted in turn.

Another way that males of some species attempt to gain an advantage in sperm competition is to simply produce more gametes. In many species this can be done by producing larger testes, as larger testes tend not only to produce more sperm, but produce it more quickly. A comparative study in fishes examined the relationship between the intensity of sperm competition and gonad size in 89 species (Stockley et al. 1997). The researchers collected data on fish mating systems and type of fertilization (internal versus external) to estimate the intensity of sperm competition for each species. They assumed that sperm competition intensity increases with the degree of polygamy or communal spawning and that at comparable levels of polygamy and communal spawning, internal fertilization carries a lower risk of sperm competition than does external fertilization. As predicted, the researchers found a positive association between the intensity of sperm competition and testis size (as well as sperm number).

Males can also gain an advantage in sperm competition by producing more competitive sperm. In primates and rodents, for example, males from species in which females mate promiscuously have longer sperm—which swim faster—than do species in which females mate with one male (Gomendio and Roldan 1991). In some species of mice, males have evolved another way to produce competitive sperm: their sperm cooperate. For example, the sperm of the wood mouse (*Apodemus sylvaticus*) form aggregations, or "trains," of hundreds or thousands of cells that significantly increase sperm motility and allow them to reach the egg more quickly (Moore et al. 2002). Heidi Fisher and Hopi Hoekstra showed in two species of *Peromyscus* mice that sperm aggregate more often with conspecific than with heterospecific sperm, and with sperm from closely related individuals versus more distantly related ones (**FIGURE 9.21**) (Fisher and Hoekstra 2010). These results suggest that sperm can discriminate between genetically related and unrelated sperm, though the mechanism for how they do this remains unknown. In fact, the researchers showed that sperm can not only recognize other sperm from the same male, but also from his relatives (since relatives share genes and so cooperation among related sperm should potentially result in higher genetic success; see Chapter 12).

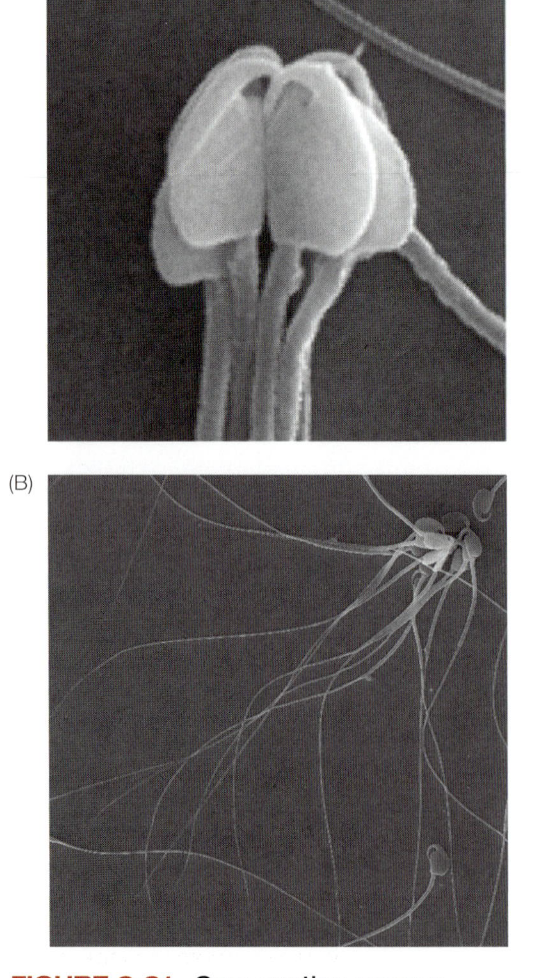

FIGURE 9.21 Cooperation among deer mice sperm. (A) Scanning electron micrographs illustrate the head orientation of sperm in a typical aggregate with hooks facing inward. (B) Thirteen sperm forming a cooperative aggregation. (From Fisher et al. 2014, CC BY 4.0.)

To test the hypothesis that sperm competition is what drives sperm to cooperate with relatives, Fisher and Hoekstra compared sperm behavior in the oldfield mouse (*Peromyscus polionotus*), in which females mate with only a single male (no sperm competition), with that in the highly promiscuous deer mouse *Peromyscus maniculatus*, in which females mate with multiple males (resulting in sperm competition). As the researchers predicted, sperm were more likely to aggregate in the species with sperm competition. In a series of follow-up experiments, Fisher's team modeled the optimal sperm train shape to improve its swimming velocity and then showed that the species with high sperm competition does in fact produce sperm train shapes that result in optimal swimming velocity (Fisher et al. 2014). The researchers then examined differences in sperm morphology between the two mice species, finding that the sperm of the deer mouse (the species with sperm competition) has a longer midpiece section (the middle of the flagellum, or tail). Midpiece length is controlled by a gene called *Prkar1a*, which encodes the regulatory region of a protein called Protein Kinase A (PKA), which among other functions plays a role in metabolism. Apparently, genetic variation at this gene predicts male reproductive success in *Peromyscus* mice (Fisher et al. 2014). In other words, small changes to genes that regulate metabolism also influence the morphology of sperm, which facilitates cooperation among sperm of the same or related individuals. Ultimately, this cooperative behavior improves a male's chances of fertilizing the egg in the face of competition with sperm from other—particularly unrelated—males.

Mate Guarding and Paternity Assurance

Many species of animals form monogamous pairs (see Chapter 10) for mating and rearing offspring. However, one or both of the pair often engage in **extra-pair copulations** with individuals outside of their pair-bond. Extra-pair copulations are especially common in birds, where extra-pair offspring were found in nearly 90 percent of the 150 species examined (Griffith et al. 2002). Sperm are stored in female birds in tiny tubules where in some species the sperm can survive for more than a month (**FIGURE 9.22**). Since the last sperm in is typically the first one out when sperm are moved out of storage to be used in fertilization, this allows females great control over fertilization. For example, females of the collared flycatcher (*Ficedula albicollis*) are able to use their storage system to bias male fertilization chances by copulating in ways that give one male's sperm a numerical advantage over another's. By controlling when and with which male she mates, a female flycatcher can manipulate the number of sperm from different males within her reproductive tract, a phenomenon referred to as **cryptic female choice** (a topic we explore further later in this chapter). A female that stopped mating with her social partner for a few days and then mated with an attractive neighbor would have five times as many of her extra-pair mate's sperm available for egg fertilizations as she retained from earlier matings with her social partner. This imbalance would give the extra-pair

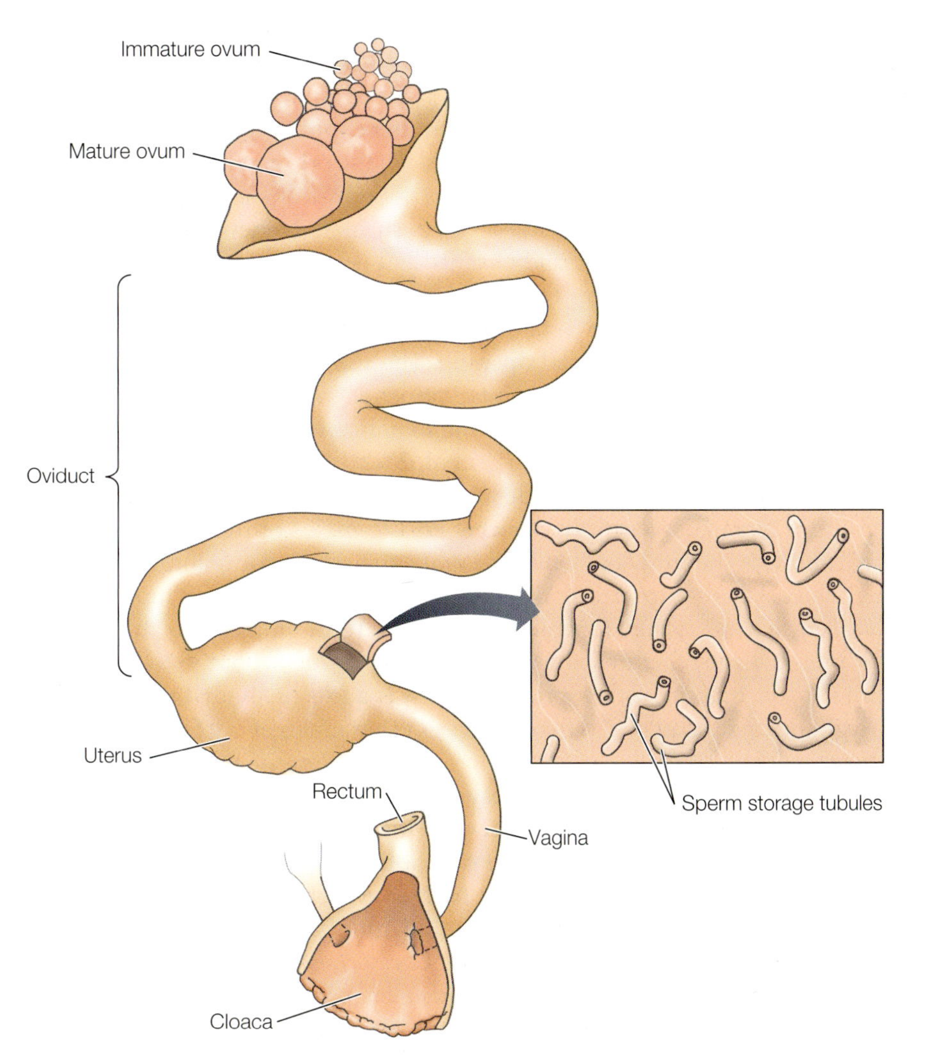

FIGURE 9.22 The reproductive anatomy of fertilization in birds. When a mature ovum produced in an ovary is released, it becomes fertilized before traveling down the oviduct. Viable sperm received from males can be stored for lengthy periods in small sacs or tubules in the interior wall of the uterus where it meets the vagina. The sperm gradually move out of the tubules over time and migrate up the oviduct to meet freshly released ova. The hard shell of an egg is added in the uterus before the egg is laid via the cloaca. (After Birkhead and Møller 1992, Bakst 1998.)

male a big fertilization advantage. In fact, a female collared flycatcher paired with a male with a small white forehead patch is likely to secure sperm around the time of egg laying by slipping away for a tryst with a nearby male sporting a larger white patch. Therefore, females of this species seem to play a major role in determining whose sperm will fertilize their eggs (Michl et al. 2002), a conclusion that may apply to many animal species (Eberhard 1996). Here, as is so often the case, not everyone agrees that females manage extra-pair matings to their advantage; the alternative view is that in some species, females are essentially forced to mate more often than they would otherwise by sexually motivated males that are the fitness beneficiaries of these matings (Arnqvist and Kirkpatrick 2005).

One way for a male to assure that his sperm fertilize a female's eggs is to make sure she mates with no one else, a behavior called mate guarding. For example, males of the bee *Centris pallida* quickly inseminate their partners but remain with them for several minutes afterward, during which time the pair flies to a nearby tree or bush where the male strokes his partner vigorously with his legs and antennae. Mate guarding, however, can be costly to the male because while he is engaged in these postcopulatory activities, he might miss finding another virgin female whose eggs he could fertilize. Yet despite this kind of potential cost, males of many species in which females do all of the offspring care stay with partners after copulation, sometimes for hours or days, rather than resuming the search for additional mates. In some of these cases, mated males deceptively lure new suitors away from their sexual partners (Field and Keller 1993), while in other cases, males seal their mates' genitalia with various secretions (Dickinson and Rutowski 1989, Polak et al. 2001) or they induce their mates to fill their genital openings themselves (Aisenberg and Eberhard 2009, Aisenberg and Barrantes 2011). The most extreme commitments of this sort occur when males sacrifice themselves after copulating (Miller 2007, Uhl et al. 2011), as happens, for example, when the male of the orb-weaving spider *Argiope aurantia* expires after inserting both pedipalps (sperm-transferring appendages) into a female's paired genital openings. The dead male is then carried about by the female spider as a kind of morbid chastity belt (Foellmer and Fairbairn 2003). Somewhat less extreme examples of this phenomenon in spiders include cases in which the male amputates his own sperm-transferring pedipalps during copulation so that it blocks the female's genital opening. The emasculated male nonetheless can gain fitness when the sperm-containing appendage not only acts as a mating plug but continues to pump sperm into the female, even if the rest of the male has been eaten by his predatory mate (Li et al. 2012). As predicted, such extreme mate-guarding behaviors tend to occur when the probability that a male will find another virgin female is very low. Another cost to males of their postinsemination associations is the time lost while the male guards his companion and fights off other males. Because guarding males of the western Mexican whiptail lizard (*Cnemidophorus neomexicanus*) forage much less and fight much more, they pay an energetic price to keep other males at bay (Ancona et al. 2010). For the New Zealand stitchbird (*Notiomystis cincta*), the price to a male is a loss of about 4 percent of his body weight during the period when he chases other males away from a previous sexual partner (Low 2006).

If remaining with an inseminated partner is an adaptation, the behavior must provide fitness benefits to males that outweigh the several costs they pay. The key benefit appears to be a reduction in the probability that a female will copulate with another male, thereby diluting or removing the sperm the guarding male has already donated to his partner (**FIGURE 9.23**) (Andersson 1994). Janis Dickinson tested this idea in the blue milkweed beetle (*Chrysochus cobaltinus*), in which the male normally remains mounted on the female's back for some time after copulation. When Dickinson pulled pairs apart, about 25 percent of the separated males

(A)

(B)

(C)

FIGURE 9.23 Mate guarding occurs in many animals. (A) A small male red damselfly (*Ceriagrion tenellum*) grasps his mate in the tandem position so she cannot mate with another male. (B) The male blue-banded goby (*Lythrypnus dalli*) closely accompanies his mate wherever she goes. (C) A male tropical harvestman (right) stands guard by a female ovipositing in his territory. (Photographs: A, by John Alcock; B, © Juniors Bildarchiv/AGE Fotostock; C, courtesy of Glauco Machado.)

found new mates within 30 minutes. Thus, remaining mounted on a female after inseminating her carries a considerable cost for the guarding male, since he has a good chance of finding a new mate elsewhere if he just leaves his old one after copulating. However, nearly 50 percent of the females whose guarding partners were plucked from their backs also acquired a new mate within 30 minutes. Since mounted males cannot easily be displaced from a female's back by rival males, a mounted male reduces the probability that an inseminated partner will mate again, giving his sperm a better chance to fertilize her eggs. Dickinson calculated that if the last male to copulate with a female fertilized even 40 percent of her eggs, he would gain fitness by giving up the search for new partners to guard his current one (Dickinson 1995).

In general, the benefits of mate guarding increase with the probability that unguarded females will mate again and use the sperm of later partners to fertilize their eggs. But how do you figure out what unguarded females would do in a species in which all the available females are guarded? Dickinson's technique involved the simple removal of a male from a female. Other studies have tested the prediction that experimental removal of the male would lead to higher rates of copulation by his mate with other males (Wolff et al. 2002, Johnsen et al. 2008). Jan Komdeur and his colleagues tricked male Seychelles warblers (*Acrocephalus sechellensis*) into

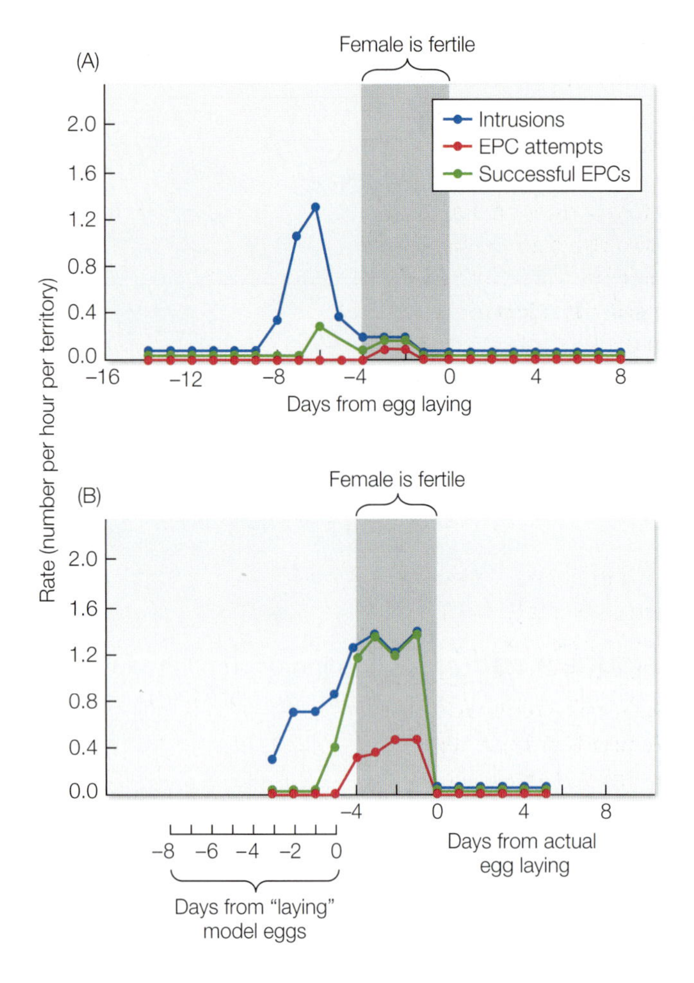

Seychelles warbler

FIGURE 9.24 Adaptive mate guarding by the Seychelles warbler. The graphs show the rates of intrusions, attempted extra-pair copulations (EPCs), and successful EPCs by males other than the female's social mate in relation to the female's fertile period (shaded area). (A) Control pairs, in which the female's mate was present throughout her fertile period. (B) Pairs in which the female's mate was experimentally induced to leave her unguarded by the placement of a false egg in the nest. (A, B after Komdeur et al. 1999; photograph © M. D. England/ardea.com/Mary Evans Picture Library Ltd./aeg fotostock.)

ending their guarding prematurely by placing a false warbler "egg" in a nest a few days before the male's partner was due to lay her one and only egg. The male warblers used the false egg cue to stop mate guarding at a time when their partners were still fertile. In short order, many of these unguarded females copulated with neighboring males (**FIGURE 9.24**) (Komdeur et al. 1999) and used the extra-pair sperm to fertilize their eggs. Indeed, the probability that a nestling would be sired by a male other than the female's social partner increased in relation to the number of days that the female's "mate" neglected to guard her during her fertile period (Komdeur et al. 2007). Mate guarding provides clear fitness benefits for male Seychelles warblers. In general, male mate-guarding behavior is flexible, and the intensity of mate guarding is influenced not only by female behavior, but also by the number of other reproductively active males and females in the population.

Intersexual Selection and Mate Choice

Up until this point, we have focused primarily on intrasexual interactions among males and male alternative mating approaches. Yet discriminating mate choice (usually by females) can also enable members of the opposite sex with certain traits to reproduce more successfully than others (Waage 1997, Zeh and Zeh 2003). As a result, the favored traits of males spread through the species in much the same way that they do when used for intrasexual competition. As you learned in Chapter 8, traits that honestly signal male quality to prospective mates can help facilitate female choice. In fact, these traits are honest precisely because they cannot be faked or cheated, and therefore enable females to make accurate mating decisions. However, in some cases females choose males that are simply the winners of intrasexual contests, and thus separating intra- and intersexual selection components can be difficult, as they often act in concert.

To understand why females are choosy when it comes to deciding on a mate, we must consider the potential fitness benefits they can receive. There are two types of benefits that females can receive by being choosy about their mate: direct and indirect benefits (Birkhead and Møller 1992, Griffith et al. 2002). **Direct benefits** increase the fitness of the choosy female through material advantage in the form of parental care, access to resources, and occasionally other things such as safety from predators or reduced harassment by other males. In contrast, **indirect (genetic) benefits** do not benefit the female directly, but instead increase the fitness of her offspring.

Female Mate Choice for Direct Benefits

PARENTAL CARE Since males of many animal species may help care for offspring, we would expect females in these species to prefer to mate with males that provide high-quality care. In keeping with this expectation, females of the fifteen-spined stickleback (*Spinachia spinachia*), a small fish with nest-protecting males, associate more with males that shake their body relatively frequently when courting. Males that behave in this way also perform more nest fanning after courtship is over and eggs have been laid in their nests. Nest fanning sends oxygenated water flowing past the eggs, which increases gas exchange and, ultimately, egg hatching success (Östlund and Ahnesjö 1998). The female fifteen-spined stickleback that evaluates the courtship display of a male sees him behaving in ways that are linked to his parental capacities.

Males of a related species, the three-spined stickleback (*Gasterosteus aculeatus*), also appear to offer females cues of paternal helpfulness—in the form of a colorful belly. As a rule, males with redder bellies are more attractive to potential mates (Rowland 1994) and have fewer parasites (Milinski and Bakker 1990). The reddish pigment that colors a male's skin comes from the carotenoids he has consumed, which as we discussed in Chapter 8 can reduce oxidative stress and improve health and survival. Males with red bellies are for some reason—perhaps because they are healthier or in better condition from having had a better diet—able to fan their eggs for longer periods under low-oxygen conditions than are paler males whose diets are short in carotenoids (Pike et al. 2007). Thus, a male stickleback's appearance advertises his ability to supply oxygen for the eggs he will brood.

Could the carotenoid pigments in a male bird's plumage or bill also be an indicator of his capacity for paternal behavior? Certainly females of many animal species are especially attentive to the reds and yellows in male color patterns (Grether 2000, McGraw and Hill 2000), which might reveal something about the health of a male if, as has been argued, the quality of an individual's immune system is enhanced by a diet rich in carotenoids or if only healthy males can use their dietary carotenoids to make bright feathers (Hill and Montgomerie 1994). A healthy male might well be a better provider for his offspring. When male zebra finches (*Taeniopygia guttata*) were provided with extra carotenoids in their food, they had more of these pigments in their blood, brighter red beaks, and stronger immune responses (McGraw and Ardia 2003). Moreover, female zebra finches found experimentally carotenoid-enhanced males more attractive than males that received normal diets (Blount et al. 2003). Female zebra finches are so drawn to the color red that they even prefer males that have been given red leg bands over those with green bands; female zebra finches who mate with red-banded males lay eggs with greater mass than those they produce after mating with less attractive males (Gilbert et al. 2012). In nature, female zebra finches that acquire more brightly colored mates might benefit by investing more in their offspring because having a healthy mate might mean that he would provide superior care for these nestlings.

Much the same thing may be happening in the blue tit (*Cyanistes caeruleus*), another small songbird with a carotenoid-based plumage ornament, a bright yellow breast. Male blue tits collect and deliver food for their nestlings, usually in the form of carotenoid-containing caterpillars. If the amount or quality of food supplied by a male is related to how bright his yellow feathers are, then the offspring of brilliantly yellow males should be larger and healthier when they fledge than the youngsters of less brightly colored males. Indeed, the offspring of brighter fathers are in better condition and have stronger immune systems than those of less yellow fathers (Hidalgo-García 2006).

But wait a minute. This same result might occur if bright yellow males were themselves large and healthy at fledging, thanks to their genetic makeup, in which case their offspring would simply inherit these traits from their father rather than deriving these benefits from the male's high-quality care. A team of behavioral biologists recognized this problem, and wisely controlled for heredity through a cross-fostering experiment. They took complete clutches of eggs and transferred them between nests, moving the offspring of one set of parents to another pair's nest. The foster parents were willing to rear these genetic strangers, and when their adopted chicks had reached the age of independence, the size of the fledglings was a function of the brightness of their foster father's yellow plumage, not of their genetic father's yellow plumage. Bright yellow foster males produced larger fledglings. If the foraging capabilities of bright yellow males really are greater than those of duller individuals, as they appear to be (García-Navas et al. 2012), females could benefit by choosing their mates on the basis of the males' plumage (Senar et al. 2002).

RESOURCES In many species, the direct benefit that females receive from males is not offspring care but access to resources or a nutritious meal. For example, in chimpanzees (*Pan troglodytes*), males hunt and kill other, smaller primates; females that receive meat from successful hunters are more likely to copulate with these individuals over the long term (Gomes and Boesch 2009). In some dung beetles, the male constructs a ball of dung from a cowpat or other source and then rolls it away to a burrow to provide nutrients for the young. A female may accompany him to the burrow (**FIGURE 9.25**), where mating will probably occur, with the female rewarding her partner for his contribution of food to her or her offspring (Sato 1998). In the red-winged blackbird (*Agelaius phoeniceus*), pair-bonded females that copulate with an extra-pair male on a neighboring territory are allowed to feed on his territory (Gray 1997).

An analogous case comes from insects, where males commonly present females with a nutritious meal in the form of a protein rich, sperm-containing spermatophore, as mentioned previously in the case of the Mormon cricket. The black-tipped hangingfly (*Hylobittacus apicalis*) is one such insect in which female acceptance of a male depends on the nature of his nuptial gift (**FIGURE 9.26A**) (Thornhill 1976). In this species, a male that tries to persuade a female that an unpalatable ladybird beetle is a good mating present is out of luck. Even males that transfer edible prey items to their mates will be permitted to copulate only for as long as the meals last. If the nuptial gift is polished off in less than 5 minutes, the female will separate from her partner without having accepted a single sperm from him. When, however, the nuptial gift is large enough to keep the copulating female feeding for 20 minutes, she will depart with a full complement of the gift giver's sperm (**FIGURE 9.26B**). Males of many other animals provide food presents before or during copulation

FIGURE 9.25 Female mate choice and male contributions. In some dung beetles, the female mates with the male after he has made a dung ball and has rolled it to a distant burrow where he turns the present over to the female. She may accompany him on his journey, as in this species. (Photograph by John Alcock.)

(Vahed 1998), including male fireflies that add packets of protein to the ejaculates that they donate to their mates. In at least one firefly species, females evaluate mates on the basis of the duration of their light flashes, which correlates with the size of the spermatophore the male will give to his partner (Lewis and Cratsley 2008).

The spermatophore has been well studied in a small butterfly called the cabbage white (*Pieris rapae*). Male cabbage white butterflies are more colorful on their underside than are females, and during an elaborate courtship flight, males display to females their ventral colors, which are the result of nitrogen-rich pigments called

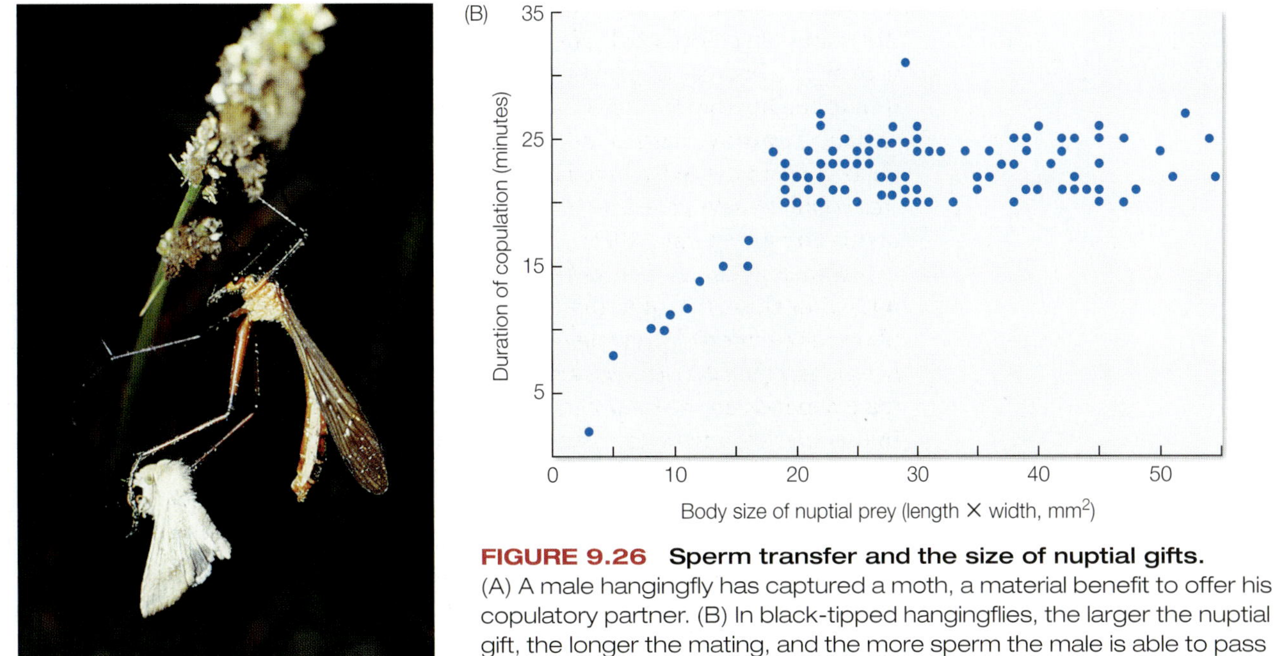

FIGURE 9.26 Sperm transfer and the size of nuptial gifts. (A) A male hangingfly has captured a moth, a material benefit to offer his copulatory partner. (B) In black-tipped hangingflies, the larger the nuptial gift, the longer the mating, and the more sperm the male is able to pass to the female. (A, photograph by John Alcock; B after Thornhill 1976.)

pterins (Morehouse and Rutowksi 2010b). Since this species is strongly nitrogen-limited during development, male coloration may not only be an honest signal of foraging ability, but the pigments themselves could be beneficial to females if they are part of the spermatophore the males donate during mating (Morehouse and Rutowksi 2010a). Nathaniel Morehouse and colleagues studied the nutrients in the male cabbage white's spermatophore, finding that these structures make up nearly 13 percent of the male's body mass (Morehouse and Rutowksi 2010a). This means that males cannot mate indiscriminately and instead must make each copulation count. In fact, males with limited resources tend to make smaller spermatophores and often break down their own muscles and organs to ensure that they can produce the necessary nuptial gift to entice females. Thus, males are likely to be able to mate only a few times, since producing spermatophores is costly. Although one would think that females would be more free to mate multiply than males, Morehouse and Rutowkski showed that this was not really the case because the spermatophore, though a source of rich nutrients for females, is also problematic for females because it cannot be fully digested and instead clogs their reproductive tract.

Why would such an awkward system of nutrient-rich spermatophores clogging female reproductive tracts evolve? As we will explore later in this chapter, this takes us into the territory of sexual conflict, or conflict between males and females over mating, and in this case, associated nuptial gifts. Since the spermatophore often plugs the female's genital tract and prevents other males from mating with her, perhaps females have evolved ways to deal with these difficult spermatophores. In butterflies, for example, the female reproductive tract, or bursa copulatrix, both receives and helps break down the male spermatophore. After mating, the sperm leave the spermatophore and travel to a specialized sperm storage organ called the spermatheca. The quicker that females are able to remove the spermatophore once the sperm have migrated out, the more quickly they can remate with other males.

In the cabbage white, females produce at least nine different protease enzymes to help break down the spermatophore (Plakke et al. 2015). These enzymes are similar to what one might find in the stomach, but here they are produced in the

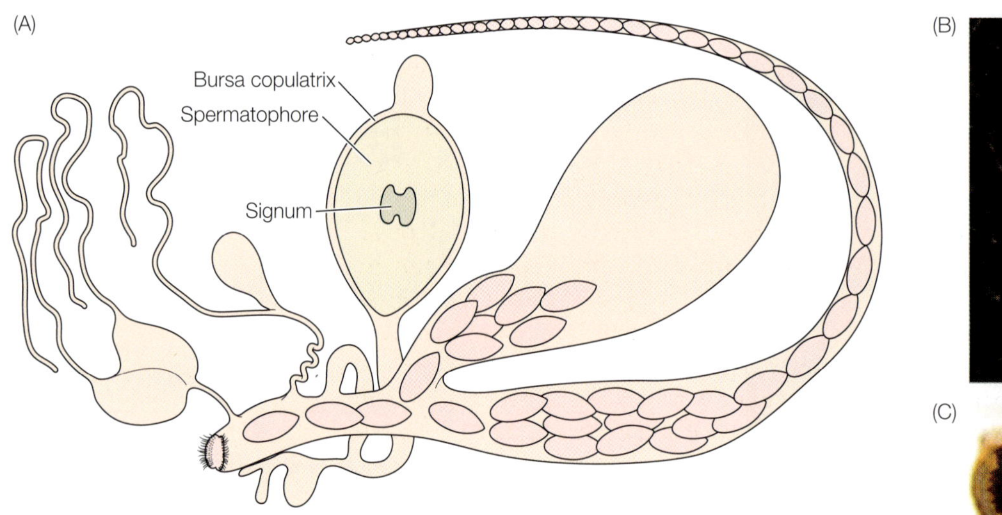

FIGURE 9.27 Reproductive tract of the female cabbage white butterfly. (A) Complete reproductive tract with bursa copulatrix, spermatophore, and signum. (B) Close-up of the male spermatophore. (C) Close-up of the female signum, which exhibits a "chewing" behavior. Scale bars: 1 mm (A and B), 200 mm (C). (After Meslin et al. 2015.)

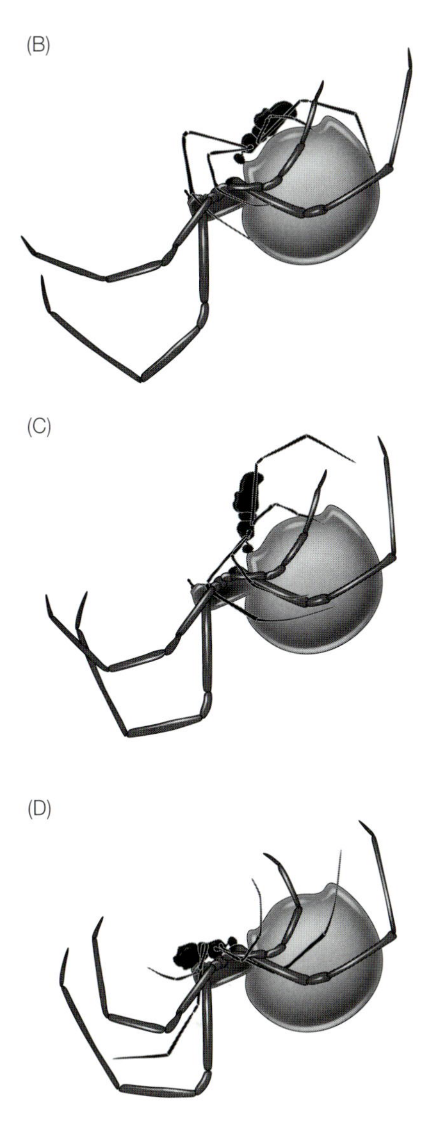

FIGURE 9.28 Sexual suicide in the redback spider. (A) A female redback spider. (B) The smaller male first aligns himself facing forward on the underside of the female's abdomen while he inserts his sperm-transferring organs into her reproductive tract. (C) He then elevates his body and (D) flips over into the jaws of his partner. She may oblige by consuming him while sperm transfer takes place. (A, photograph by © Peter Waters/Shutterstock.com; B–D after Forster 1992.)

genital tract. In fact, the genes underlying these enzymes were co-opted from other insect tissues such as the digestive track and salivary glands, where the proteins they produce normally help break down food, not leftover nuptial gifts (Meslin et al. 2015). What's more, females also have a specialized morphological structure in their bursa called a signum, which aids in "chewing" apart the tough exterior of a spermatophore (**FIGURE 9.27**) (Meslin et al. 2015). This combination of morphological and physiological adaptations allows a female to remove the spermatophore from her reproductive tract as quickly as possible so that she can attempt to remate.

Sometimes female choice for a good meal means not what forage or spermatophore the male can provide for the female, but whether he is willing to provide himself to a cannibalistic mate. Some researchers have suggested that this is a special class of nuptial gift giver—those males, mainly mantids and spiders, that wind up being eaten by their mates—and should be recognized as such (see Figure 4.13). Indeed, it could be adaptive, under some special circumstances, for a male to wrap up a copulation by turning himself into a meal for his sexual partner (Buskirk et al. 1984, Prete 1995). It is true that males of several species of mantids avoid or very cautiously approach poorly fed females, suggesting that males of these species typically do not benefit from sacrificing themselves to a mate (Lelito and Brown 2008, Maxwell et al. 2010). By contrast, male redback spiders (*Latrodectus hasselti*) often do make it easy for their mates to eat them (Andrade 1996, Stoltz et al. 2008). After the male has transferred sperm into both of the female spider's sperm receptacles, which he has blocked with his two sperm-transferring organs, the male redback performs a somersault, literally throwing his body into his partner's jaws (**FIGURE 9.28**). About two-thirds of the time, the female accepts the invitation and devours her sexual companion (Andrade 1996).

Why does a male redback sacrifice himself to his mate? Since a male redback weighs no more than 2 percent of what a female weighs, he does not make much of a meal for a cannibalistic partner. But once eaten, the deceased male redback does in fact derive substantial benefits from what might seem like a genuinely fitness-reducing experience. Maydianne Andrade showed that cannibalized males fertilized more of their partners' eggs than uneaten males did, partly because a fed female redback is less likely to mate again promptly (Andrade 1996). Moreover, the cost of a postcopulatory death is very low for male redbacks. Young adult males in search of mates are usually captured by predatory ants, or other spider predators, long before they find webs with adult females. The intensity of predation pressure on wandering male redbacks is such that fewer than 20 percent manage to locate even one mate, suggesting that the odds that a male could find a second partner are exceedingly low even if he should survive his initial mating (Andrade 2003). Under these circumstances, males need very little benefit from sexual suicide to make the trait adaptive.

sexual suicide hypothesis Male self-sacrifice can enhance fertilization success if it means females are less likely to mate again.

This **sexual suicide hypothesis** can be tested by predicting that males of other, unrelated spiders will also make the ultimate sacrifice in those species in which the male pedipalps are used to plug the female's genital openings during the male's first (and usually only) copulation. Jeremy Miller has found that males are complicit in their cannibalism or die of "natural causes" soon after mating in five or six lineages of spiders. In all but one of these, the male's pedipalps become broken or nonfunctional in the course of an initial mating (Miller 2007), providing support for the hypothesis that sexual suicide occurs when males have almost no chance of mating again in the future.

The redback spider example illustrates how male self-sacrifice can enhance fertilization success because females are less likely to mate again. However, this is not the only way that self-sacrifice could benefit a male. This suicidal behavior could also enhance fitness through an increase in offspring number or quality via the nutrients the male's body provides. A team led by Eileen Hebets studying the dark fishing spider (*Dolomedes tenebrosus*)—a species in which males spontaneously die after mating and are typically eaten by the female—tested this idea and demonstrated that when females were allowed to consume their partners, there were large increases in the number of offspring produced (nearly twice as many), their size (14 to 20 percent larger), and their survivorship (44 to 63 percent higher) (Schwartz et al. 2016). To ensure that these benefits were not simply due to an increase in nutrients from the male meal, the authors fed all of the females before and after the experiment. And to really determine if these fecundity benefits were the result of the nutrients in the meal itself or something about eating their particular mate, the researchers fed another group of females a cricket meal instead of their mate. In this experiment, they found none of the same fecundity benefits, suggesting that allowing oneself to be eaten really does increase a male's fitness by increasing the number and quality of offspring he produces, though the mechanism remains unknown. But could being eaten also enhance a male's fertilization success by making females less likely to remate, as is the case in redback spiders? In a previous study, the authors showed that eating a male does not reduce a female's probability of remating, disproving this alternative hypothesis for male self-sacrifice in dark fishing spiders (Schwartz et al. 2014).

Female Mate Choice for Indirect Benefits

Some of the examples we have just presented are consistent with the idea that females choose males for direct benefits in the form of parental care, which explains aspects of male color, ornamentation, and courtship behavior as sexually selected indicators of a male's capacity to provide parental care, a factor of obvious significance for female reproductive success. In other cases, females receive direct benefits in the form of a nutritious meal or other resources. However, in the satin bowerbird, as well as many other species (Kodric-Brown 1993), the males do not provide food or any other material benefit to their mates or their offspring. Even so, female bowerbirds prefer males with more ornaments (in their bowers) and the ability to court more intensely (Borgia 1985, Patricelli et al. 2003). Let's explore some of the types of indirect benefits that females get from being choosy (**TABLE 9.1**).

Earlier in this chapter we saw that signals used in mating by male bowerbirds (bower building and complex displays) might be indicators of male quality unrelated to parental investment. If this hypothesis is true, then we expect male mating success to be correlated with some features of the bower that vary from male to male, such as the skill with which the bower has been constructed and decorated, or perhaps the number of blue feathers at its entrance (Wojcieszek et al. 2007). In fact, even humans can detect differences among the bowers built by different males. Some contain neat rows of twigs lined up to create a tidy, symmetrical bower, while others are messy and disorganized. Bowers also differ markedly in how well they are decorated with feathers and the like. Female bowerbirds evidently notice these differences too, because they tend to remain calmer at well-constructed bowers, a response that often leads the female to mate with the bower builder (Patricelli et al. 2003, 2004). Female behavior helps explain why bower quality correlates with male mating success in this (Borgia 1985) and other bowerbird species (Madden 2003a, 2003b). In the satin bowerbird, male mating success translates directly into male genetic success, because females typically mate with only one male, whose sperm fertilize all of her eggs (Reynolds et al. 2007).

To understand why female satin bowerbirds choose males that build the best bowers, we must determine which males build the best bowers. Perhaps attractive, well-decorated bowers are built by males that are superior in some way to birds that cannot construct a top-flight bower. This idea has been formalized in the **good genes model of sexual selection**, which proposes that preferences for certain male ornaments and courtship displays enable females to choose partners whose genes will help their offspring have higher fitness (Andersson 1994). In other words, if a female chooses a high-quality male, she is more likely to produce high-quality offspring. In the case of satin bowerbirds, the bower's

good genes model of sexual selection Preferences for certain male ornaments and courtship displays enable females to choose partners whose genes will help their offspring have higher fitness.

TABLE 9.1 Three models of how indirect benefits could lead to the evolution of extreme male ornamentation and striking courtship displays, even if males provide no direct benefits to their mates

Model	Females prefer trait that is:	Primary adaptive value to choosy females
Good genes	Indicative of male quality or viability	Offspring may inherit advantages of their father
Runaway selection	Sexually attractive	Sons inherit trait that makes them sexually attractive; daughters inherit the majority mate preference
Chase-away selection	Exploitative of preexisting sensory biases	Females derive no benefits by being choosy

quality could be an indicator (an honest signal) of a male's cognitive ability. To test the idea that females should prefer males with greater cognitive skills, researchers tried to measure relative cognitive abilities in individual male satin bowerbirds (Keagy et al. 2011). They evaluated the ability of male satin bowerbirds in nature to solve several problems, such as whether they would be able to cover up a red item experimentally inserted and fixed within the bower, given that red objects are not tolerated at the bowers of this species. This team found that the average cognitive ability of the bowerbirds tested on six such problems was highly correlated with the males' mating success (**FIGURE 9.29**). Thus, there is some evidence that brainier bowerbirds experience greater reproductive success than their less with-it rivals and that the bower is an honest signal of a male's cognitive ability (but see Healy and Rowe 2007). Another possible example of the good genes model of sexual selection involves one of the iconic species of sexual selection theory: the Indian peacock (*Pavo cristatus*) (**BOX 9.2**).

A male's appearance could also be correlated with his hereditary resistance to parasites, a valuable attribute to pass on to offspring. Stéphanie Doucet and Bob Montgomerie found that male satin bowerbirds that build better bowers have fewer ectoparasitic feather mites than males that make less appealing display structures (Doucet and Montgomerie 2003; see also Borgia et al. 2004). In other species, females might be able to evaluate (unconsciously) the strength of a male's immune system by his courtship displays. One possible example involves the cricket *Teleogryllus oceanicus*, in which females prefer to approach artificial male songs that have been manipulated to sound like those sung by males with strong immune systems, as opposed to songs that sound like those sung by males with weak immune systems (Tregenza et al. 2006). In all of these cases, the "good genes" are those that underlie strong immune function. William D. Hamilton and Marlene Zuk explored this idea and hypothesized that selection for honest signals of noninfection would lead bird species with numerous potential parasites to evolve strikingly colored plumage (Hamilton and Zuk 1982). In what has become known as the **Hamilton and Zuk hypothesis**, the researchers argued that brightly colored feathers are difficult to produce and maintain when a bird is parasitized, because parasitic infection causes physiological stress. Hamilton and Zuk found the predicted correlation between plumage brightness and the incidence of blood parasites in a large sample of bird species, supporting the view that males at special risk of parasitic infection engage in a competition that signals their condition to choosy females (Hamilton and Zuk 1982).

Hamilton and Zuk hypothesis The expression of particular male traits is associated with resistance to parasites or other pathogens.

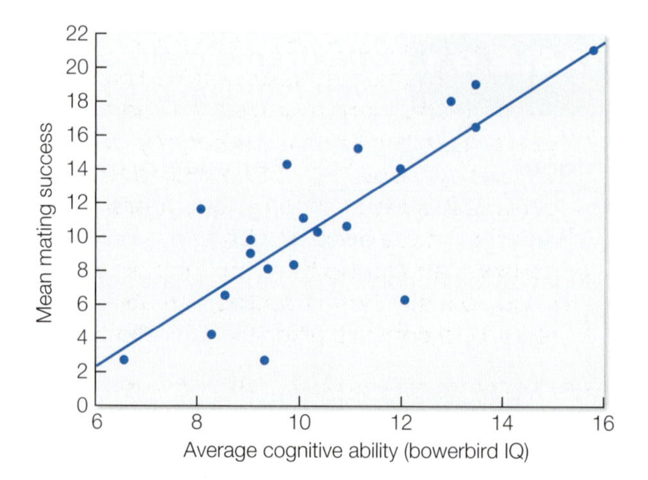

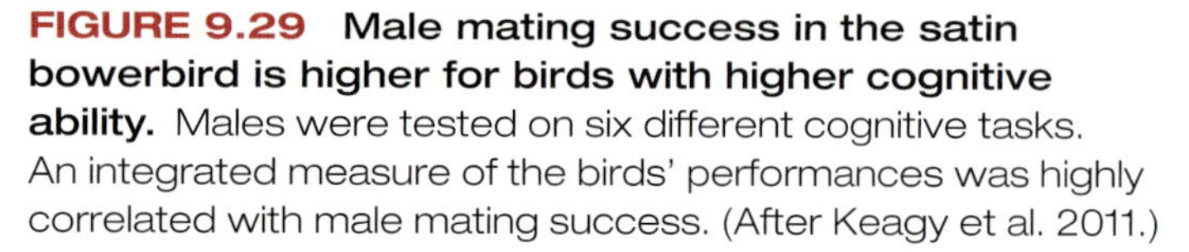
FIGURE 9.29 Male mating success in the satin bowerbird is higher for birds with higher cognitive ability. Males were tested on six different cognitive tasks. An integrated measure of the birds' performances was highly correlated with male mating success. (After Keagy et al. 2011.)

BOX 9.2

EXPLORING BEHAVIOR BY INTERPRETING DATA

Sexual selection in the peacock

Marion Petrie applied the good genes model to peacocks and derived the following predictions: (1) males should differ genetically in ways related to their survival chances; (2) male behavior and ornamentation should provide accurate information on the survival value of the males' genes; (3) females should use this information to select mates; and (4) the offspring of the chosen males should benefit from their mother's mate choice (Petrie 1992, Petrie 1994). In other words, males should signal their genetic quality in an accurate manner, and females should pay attention to those signals because their offspring would derive hereditary benefits as a result (Zahavi 1975, Kodric-Brown and Brown 1984).

Petrie studied a captive but free-ranging population of peacocks in a large forested English park, where she found that males killed by foxes had significantly shorter tails than their surviving companions. Moreover, she observed that most of the males taken by predators had not mated in previous mating seasons, suggesting that females could discriminate between males with high and low survival potential (Petrie 1992). The peahens' preferences translated into offspring with superior survival chances, as Petrie showed in a controlled breeding experiment. She took a series of males with different degrees of ornamentation from the park and paired each of them in a large cage with four females chosen at random from the population. The young of all the males were reared under identical conditions, weighed at intervals, and then eventually released back into the park. The sons and daughters of males with larger eyespots on their ornamented tails weighed more at day 84 of life and were more likely to be alive after 2 years in the park than the progeny of males with smaller eyespots (**FIGURE**) (Petrie 1994). To confirm this result, in another experiment Petrie removed male eyespots between breeding seasons and found that manipulated males showed a significant decline in mating success relative to controls (Petrie and Halliday 1994).

Male peacock

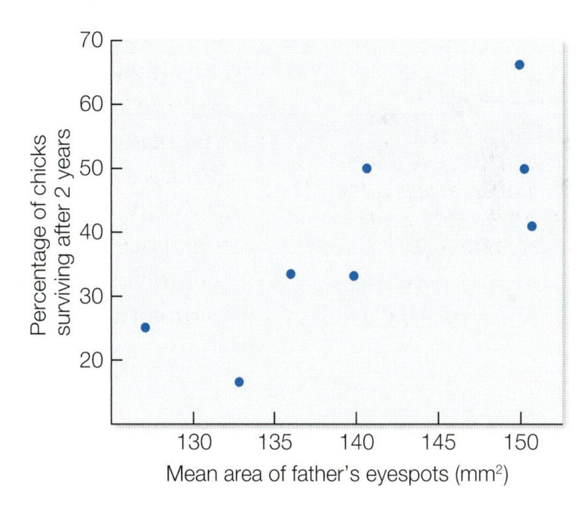

Do male ornaments signal good genes? Peacocks with larger eyespots on their tails produced offspring that survived better when released from captivity into an English woodland park. (After Petrie 1994; photograph © John Kirinic/Shutterstock.com.)

(Continued)

BOX 9.2 EXPLORING BEHAVIOR BY INTERPRETING DATA *(continued)*

Since the peacock is viewed as an iconic example of sexual selection, Petrie's studies have been replicated many times in different populations, often with conflicting results. In a study of free-ranging peafowl in France, Adeline Loyau and colleagues showed that female choice is based not just on the number of eyespots, but also on male display activity (Loyau et al. 2005b). Loyau and colleagues further demonstrated that males with both high display rates and a large number of eyespots had better health status (Loyau et al. 2005a). However, a 7-year study of peafowl in Japan failed to find a relationship between mating success and the number of eyespots (Takahashi et al. 2008). Similarly, a study of feral peafowl in North America found little variation in eyespot number among males, and what little variation there was, was the result of feather loss (natural wear and tear) (Dakin and Montgomerie 2011). Thus, the authors argued that variation in mating success of feral peacocks cannot be explained by natural variation in the number of eyespots. Indeed, Petrie failed to find significant heritability in eyespot number, suggesting that the environment may be more important than genes in affecting this trait (Petrie et al. 2009). However, Petrie's team did find heritability in train length (Petrie et al. 2009), something that has been shown to influence the ability of males to obtain central mating territories (Loyau et al. 2005b).

Thinking Outside the Box

Clearly, further work is needed to understand signaling in peacocks, but to the extent that Petrie's earlier studies hold up, what would you conclude? If peahens appear to be choosing males on the basis of criteria that are tied to offspring survival, female mate choice in this species could be of the good genes variety. But perhaps healthy mate benefits are involved as well if, for example, peahen mate choice helps females avoid parasitized males and so reduce their risk of acquiring parasites they would pass on to their offspring. How would you design a series of follow-up experiments to test alternative ideas about sexual selection in the peacock?

immunocompetence handicap hypothesis Because the hormone testosterone stimulates the development of traits used in sexual selection (such as the production of colors, displays, and vocalizations) as well as reduces immunocompetence, testosterone-dependent signals may honestly advertise health to females.

What explains the relationship between bright plumage and parasite number? Ivar Folstad and Andrew Karter extended the Hamilton and Zuk hypothesis to include a potential mechanism (Folstad and Karter 1992). Their **immunocompetence handicap hypothesis** proposed that because the hormone testosterone stimulates the development of traits used in sexual selection (such as the production of colors, displays, and vocalizations) as well as reduces immunocompetence, testosterone-dependent signals may honestly advertise health to females. In other words, only healthy individuals are capable of producing elaborate, testosterone-dependent traits because they are also capable of withstanding the concomitant immune suppression. Evidence in support of the immunocompetence handicap hypothesis comes from numerous species of birds, reptiles, and mammals, including our own species. In humans, for example, there are positive relationships among testosterone, facial attractiveness, and immune function (as measured by an antibody response to a hepatitis B vaccine) in males (Rantala et al. 2012).

In addition, "good genes" derived from males could be involved in the development of other fitness-advancing traits besides resistance to parasites and diseases. For example, if females had a way to avoid genetically similar males or to identify males with high levels of heterozygosity, these choosy females might well help their offspring avoid the developmental problems that can occur when individuals have two copies of damaging recessive alleles, a common result of inbreeding. One set of genes that appears to function in this way in vertebrates is the set for the **major histocompatibility complex** (**MHC**), a group of cell surface proteins critical for the immune system to recognize foreign molecules. Each individual has many different MHC alleles, and those with more diverse MHC alleles may have a more robust immune system (Jordan and Bruford 1998, Ziegler et al. 2005). But how do females judge a male's MHC diversity? Work with mice suggests that variation in the MHC genes contributes to unique individual odors in mouse urine (Singer et al. 1997). In addition to female preferences for males with MHC diversity, female preferences for males with relatively heterozygous genomes have also been documented for several bird species, including house sparrows (*Passer domesticus*) (Griggio et al. 2011), sedge warblers (*Acrocephalus schoenobaenus*) (Marshall et al. 2003), and superb starlings (*Lamprotornis superbus*) (Rubenstein 2007) (see Chapter 10).

Another way that females may assess the quality of a potential mate and his genes is through the quality of his display. For example, in the canary *Serinus canaria*, a female's choice of a mate appears to be heavily influenced by his ability to sing a certain portion of the male song, the "A phrase," composed of many two-note syllables (**FIGURE 9.30**). When females primed with estradiol (the hormone that induces female sexual receptivity) hear an A-phrase trill that packs many syllables into a second of song, they readily adopt the precopulatory position (Vallet et al. 1998). Passing a female canary's song test requires that males not only generate a rapid trill but also make the individual syllables in the trill cover a

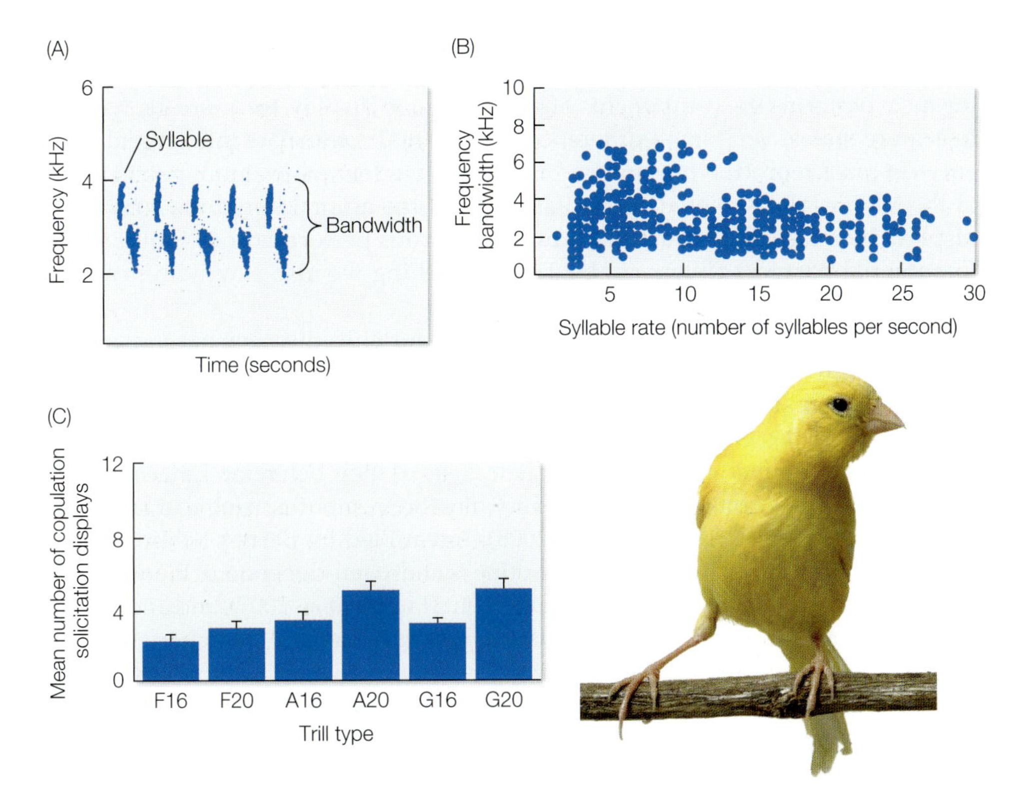

FIGURE 9.30 Mate choice based on a physiologically challenging task. (A) Male canaries sing songs that contain a special trill comprising a series of syllables, each composed of two notes that require the coordinated action of the syrinx and the respiratory system to produce. Trill syllables have frequency bandwidths that range from small (e.g., 2–4 kHz, shown here) to large (e.g., 2–6 kHz). (B) The rate at which the syllables are produced also can vary. There is an upper limit on the rate at which males can sing syllables of a given frequency bandwidth. (C) Females prefer trills composed of broad-bandwidth syllables sung at a very fast rate. Recordings of three trill types were played to females: F had the narrowest bandwidth and G the greatest. Each trill type was played at two rates: 16 syllables per second and 20 syllables per second. Females responded with significantly more copulation solicitation displays to A20 and G20 trills, those with 20 syllables per second. Bars indicate mean +/– SE. (After Drăgăniou et al. 2002, © 2002 by the Royal Society of London; photograph © ene/istockphoto.com.)

relatively wide range of sound frequencies (the bandwidth of the trill). We know this because of the responses of females to tapes of artificial trills, including some that were impossibly exaggerated versions of the A phrase. The most extreme versions of the trill elicited the most copulation solicitation displays from listening female canaries. Because there is an upper limit for male canaries with respect to how rapidly they can sing syllables of a given bandwidth, the female preference in effect favors males able to sing at the upper limit of their ability. Much the same is true for Alston's singing mouse (*Scotinomys teguina*), in which mechanical constraints mean that the faster a male produces his trills, the lower the range of frequencies he can generate in each note. When researchers produced artificial songs with exaggerated trill rates and broad bandwidths and female mice were given a choice between approaching a speaker playing the "enhanced" song and another playing a "normal" song, the females preferred the enhanced song that males could never actually sing (Pasch et al. 2011).

Experiments of this sort suggest that the greater the sensory stimulation provided by a courting male, the more attractive he becomes to females. This point has been established for the golden-collared manakin (*Manacus vitellinus*), a small tropical bird with an extraordinarily complex courtship display with maneuvers that are carried out at high speed. The male's display is multifaceted and involves dancing and producing sound simultaneously. In this species, however, the sound comes not from vocalizing but from the snapping of the wings above the male's back to produce a loud firecracker-like sound (Bodony et al. 2016), something he does while darting between saplings on the forest floor (**VIDEO 9.2**). The faster the male performs the components of his jump-snap display, the more likely he is to acquire mates. Very small differences among males translate into large differences in male reproductive success because only a few males monopolize most of the females (Barske et al. 2011). Throughout the animal kingdom, courtship displays by males are characterized by the vigorous performance of motor skills to potential partners (Byers et al. 2010), something we also saw with peacock spiders in Chapter 8.

VIDEO 9.2

Courtship by male golden-collared manakins

ab11e.com/v9.2

The widespread occurrence of demanding courtship displays has suggested that these behaviors and allied structural features evolved to enhance the demonstrations of male quality to observant females. In the golden-collared manakin, for example, males engaged in the jump-snap display have relatively high metabolic rates, indicative of the large energetic costs of their behavior. Indeed, males may perform these complex displays every day for six months, a job that is clearly energetically taxing (Schlinger et al. 2008). A team led by Barney Schlinger has demonstrated that the male's display routine is androgen-dependent. Blocking the androgen receptor decreases display movement (Fusani et al. 2007), and androgens themselves affect metabolism by facilitating many behaviors (Fuxjager and Schlinger 2015). To determine if these androgen-influenced effects on motor skills and metabolism are are uniquely related to this species' energetically taxing display, Schlinger's team compared male golden-collared manakins with other species of manakins and other less related species of birds. The researchers found high levels of androgen receptor expression in the wing muscles of the golden-collared manakin but not of the other species (Fuxjager et al. 2015). Thus, by basing their sexual preferences on the males' displays, females secure males that are able to afford the energetic costs associated with their unusual motor skills. Of course, it is also possible that male displays have evolved to play on some aspect of female sensory capabilities, a topic we will discuss shortly.

If male ornaments have evolved via female choice, then experimentally augmenting a male's courtship traits should enhance his copulatory success. We discussed an experiment that did exactly this in Chapter 8 when we talked about signaling in long-tailed widowbirds (*Euplectes progne*). Recall that Malte Andersson captured male widowbirds, then shortened the tails of some of them by removing a segment of tail feathers, only to glue them onto the tails of other males, thereby lengthening those ornaments (Andersson 1982). The tail-lengthened males were far more attractive to females than were males that had much-reduced ornaments. Moreover, the tail-lengthened males also had higher reproductive success than controls, some of whose tails were untouched, while others had been cut and then put back together without changing their length.

All of these studies suggest that some feature of the male's ornamentation or display signals something about his genetic quality or viability. No matter what the basis for female preferences for one male's courtship or copulatory skill over another, one needs to demonstrate that female choosiness translates into fitness gains for females if one is to show that the behavior is adaptive. In other words, we must understand what indirect benefits a female receives from a male that does not provide any direct benefits. Of course, it could be that females generally do not benefit by being choosy but instead are being manipulated by showy males, which gain if they are persuasive, while the females that select them actually lose fitness as a result. This idea is rendered less likely if we apply the thinking that has convinced most behavioral biologists that honest communication must benefit signal receivers as well as signal senders if it is to persist (see Chapter 8). If courtship and copulatory behavior are part of an evolved communication system, then choosy females should leave more surviving descendants as a result of their responses to the signals provided by courting males. In many species, this seems to be the case.

Runaway versus Chase-away Sexual Selection

Often male traits become more exaggerated than we might expect for a straightforward quality signal. How does such trait exaggeration come about? One possible answer is the **runaway selection model of sexual selection**, which is a mechanistic argument based on the fact that female choice creates a genetic link between mate choice by females and the male trait and, because of this correlation, leads to the evolution of preferences for ever more extreme traits over time (Fisher 1930, Arnold 1983, Dawkins 1986, Prum 2010). As a result, a sexual preference for the elaborate song of male canaries could be adaptive for females if their sons inherit the capacity to sing attractive songs, even if they are costly to produce, because these males may be especially appealing to females in the next generation.

runaway selection model of sexual selection When female mating preferences for certain male attributes create a positive feedback loop favoring both males with these attributes and females that prefer them.

Because the runaway selection alternative is the least intuitively obvious explanation for extreme male courtship displays, let's sketch the argument underlying the mathematical models of Russell Lande (Lande 1981) and Mark Kirkpatrick (Kirkpatrick 1982). Imagine that a slight majority of the females in an ancestral population had a preference for a certain male characteristic, perhaps initially because the preferred trait was indicative of some survival advantage enjoyed by the male. Females that mated with preferred males would have produced offspring that inherited the genes for the mate preference from their mothers, as well as the genes for the attractive male character from their fathers. Sons that expressed the preferred trait would have enjoyed higher fitness, in part simply because they possessed the key cues that females found attractive. In addition, daughters that responded positively to those male cues would have gained by producing sexy sons with the trait that many females liked. Thus, female mate choice genes as well as genes for the preferred male attribute could be inherited together. This pattern could generate a runaway process in which ever more extreme female preferences and male attributes spread together as new mutations. The runaway process would end only when natural selection against costly or risky male traits balanced sexual selection in favor of traits that appealed to females. For example, think about the long tails of male long-tailed widowbirds. As the tails become longer and longer, males become poorer fliers and are more likely to be caught and eaten by predators. Thus, at some point, predation pressure (natural selection) will limit tail length (and runaway sexual selection) even in the face of strong female choice. This tension between sexual selection acting via female choice and natural selection acting via predation pressure or resource availability plays a role in the evolution of many sexually-selected traits, from body size to ornament size to coloration.

But how do female preferences for these types of traits originate in the first place? If, for example, female satin bowerbirds originally preferred males with well-decorated bowers because such males could forage efficiently, they might later favor males with extraordinary bowers because this mating preference had taken on a life of its own, resulting in the production of sons that were exceptionally attractive to females and the production of daughters that would choose these extraordinary males for their own mates. In fact, it is likely that the original connection between female mate choice and male foraging skill or any other useful trait would have disappeared. Indeed, the Lande–Kirkpatrick models demonstrate that, right from the start of the process, female preferences need not be directed at male traits that are utilitarian in the sense of improving survival, feeding ability, and the like. Any preexisting preference of females for certain kinds of sensory stimulation (see Chapter 8) could conceivably get the process under way. As a result, traits opposed by natural selection because they reduced viability could still spread through the population by runaway selection (Lande 1981, Kirkpatrick 1982). Instead of mate choice based on genes that promote the development of useful characteristics in offspring, runaway selection could yield mate choice for arbitrary characters that are a burden to individuals in terms of survival, a disadvantage in every sense except that females mate preferentially with males that have them!

According to Brett Holland and Bill Rice, extreme ornaments and elaborate courtship displays could also be the result of chase-away selection, a process that begins when a male happens to have a mutation for a novel display trait that manages to tap into a preexisting sensory bias that affects female mate preferences in his species (see Chapter 8) (Holland and Rice 1998). According to this **chase-away selection model of sexual selection**, such a male might induce individuals to mate with him even though he might not provide the material or genetic benefits offered by other males of his species (Ryan et al. 1990). In other words, the chase-away selection model suggests that some male traits that attract females may actually be disadvantageous to the female, and that females gain neither material nor genetic benefits by mating with males with exceptionally stimulating courtship moves or stunning ornaments. The resulting spread over time of the traits of such truly exploitative males would create selection on females favoring those that were psychologically resistant to the purely attractive display trait. As females with a higher threshold for sexual responsiveness to the exploitative trait spread, selection would then favor males able to overcome female resistance, which might be achieved by mutations that further exaggerated the original male signal. A cycle of increasing female resistance to, and increasing male exaggeration of, key characteristics could ensue, leading gradually to the evolution of costly ornaments of no real value to the female and useful to the male only because without them, he would have no chance of stimulating females to mate with him (**FIGURE 9.31**).

chase-away selection model of sexual selection An antagonistic coevolutionary relationship between males and females in which some male traits that attract females may actually be disadvantageous to the female.

In the light of chase-away model of sexual selection, it is revealing that female fruit flies actually reduce their fitness by preferring to mate with larger males, which they choose either because they find large body size an attractive feature in and of itself, or because large body size is correlated with some other attractive feature, such as more persistent courtship. Whatever the reason for their preference, mate choice by females based on this characteristic lowers their longevity and also reduces the survival of their offspring (Friberg and Arnqvist 2003). These negative effects on female fitness may arise from the physiological costs of dealing with increased rates of courtship by "preferred" mates and possibly the increased quantities of toxins received from the seminal fluid of "attractive" males (Pitnick and Garcia-Gonzalez 2002). These findings can be taken as evidence that, for the moment, large males are ahead in a chase-away arms race between the sexes, a topic we will explore further below.

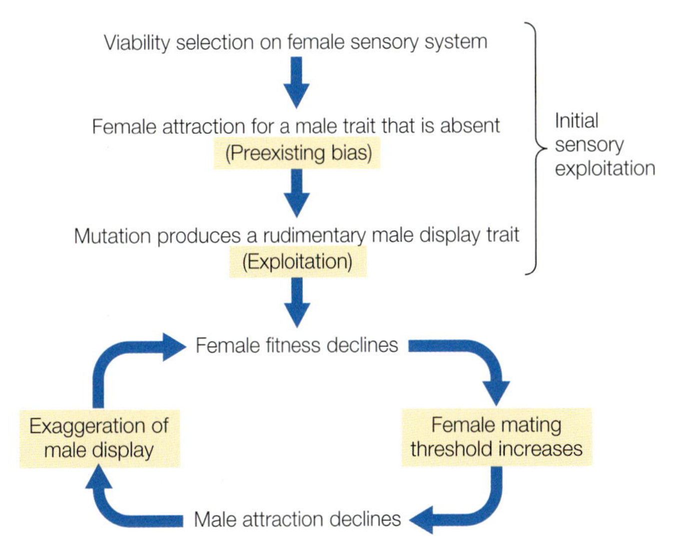

FIGURE 9.31 The chase-away selection model of sexual selection. The evolution of extreme male ornaments and displays may originate with exploitation of females' preexisting sensory biases. If sensory exploitation by males reduces female fitness, the stage is set for a cycle in which increased female resistance to male displays leads to ever-greater exaggeration of those displays. (After Holland and Rice 1998.)

Cryptic Female Choice

Just as sperm competition represents a less overt form of intrasexual selection, females exhibit a form of less overt mating preference referred to as cryptic female choice. In species with internal fertilization, cryptic female choice represents a female-driven bias in fertilization that occurs after mating or the release of gametes (Birkhead 1998). Cryptic choice can begin immediately after mating through the manipulation of sperm, or continue well into the period of development through the manipulation of fetal growth or survival. For example, female *Drosophila melanogaster* can discriminate among sperm from different males and preferentially eject sperm from specific partners (Clark et al. 1999). Different species of *Drosophila* are able to exhibit different levels and timing of control of such so-called sperm precedence, indicating that sperm precedence is not associated with a last male fertilization advantage in all species (Lupold et al. 2013, Manier et al. 2013). Sperm precedence via sperm ejection also occurs in many bird species (**FIGURE 9.32**) (Davies 1983, Pizzari and Birkhead 2000). Female chickens (*Gallus gallus domesticus*), for example, eject a larger proportion of the semen received from low-ranking males, indicating that they are able to bias the fertilization of their eggs by giving an advantage to the dominant males (Dean et al. 2011). In another example of sperm manipulation, female

FIGURE 9.32 Sperm competition may require female cooperation. A male dunnock (*Prunella modularis*) pecks at the cloaca of his partner after finding another male near her; in response, she ejects a droplet of sperm-containing ejaculate just received from the other male. (After Davies 1983.)

ocellated wrasse (*Symphodus ocellatus*) release ovarian fluid with eggs that changes the outcome of sperm competition by decreasing the importance of sperm number, thereby increasing the relative importance of sperm velocity (Alonzo et al. 2016). Males in this species adopt one of three conditional mating tactics (Taborsky et al. 1987), and female ovarian fluid likely increases the paternity of the preferred parental male phenotype (nesting males) because these males release fewer but faster sperm (Alonzo et al. 2016).

Cryptic female choice also occurs after the egg and sperm have fused. For example, in zebra finches, females invest more resources in offspring that are sired by preferred males than in those fathered by less attractive males (Gilbert et al. 2012). In some avian species, the female adjusts the amount of testosterone she contributes to a fertilized egg in relation to the attractiveness of her mate (Schwabl et al. 1997, Gil et al. 2004, Loyau et al. 2007). Likewise, female mallards (*Anas platyrhynchos*) make larger eggs after copulating with unusually attractive males (Cunningham and Russell 2000), while female black grouse (*Lyrurus tetrix*) produce and lay more eggs subsequent to mating with top-ranked males (Rintamäki et al. 1998). Whether it occurs before or after the sperm and egg fuse, cryptic choice illustrates not only another level of female choice, but also a mechanism by which females can maintain some level of control in the face of sexual conflict, particularly when copulations are forced on them, a topic we explore further below.

Sexual Conflict

The model of chase-away selection illustrates how far we have come from the once popular view of sexual reproduction as a gloriously cooperative enterprise designed to perpetuate the species or to maximize the fitness of both participants. When Darwin proposed the concept of sexual selection in Victorian England, the prevailing view of the day was that reproduction was a harmonious process between males and females. Although cooperation between the sexes is often the case in sexually reproducing species, particularly those in which both males and females rear the young together, **sexual conflict**, conflict between the sexes, is more common than one might think (Parker 2006). Indeed, many behavioral biologists now see reproduction as an activity in which the two sexes battle for maximum genetic advantage, even if one member of a pair loses fitness as a result (**TABLE 9.2**). Consider that females of many species often turn down sexually motivated males (**FIGURE 9.33**), as seen in the typical response of female satin bowerbirds to displaying males (Bruning et al. 2010). In fact, one can argue that conflict is always expected because the evolutionary interests of males and females are rarely if ever the same. Although the reproductive interests of a female and a potential mate are often in conflict—for example when a male is not chosen as a mate, or a female biases paternity against him—when females are free to exercise their choice of mate, both the chosen males and the choosy females are expected to derive fitness gains from their interactions. However, often it is the case that males can manipulate or force females to act against their interests, resulting in a loss in female fitness that often leads to bizarre sexual interactions.

The Manipulation of Female Choice

Sexual conflict can become even more unpleasant when, for example, males kill the infants of females in order to cause them to become sexually receptive again (see Chapter 1) or to force apparently unwilling females to mate. In these cases, it is

TABLE 9.2 Ways in which females and males attempt to control reproductive decisions in typical mating situations where males compete for access to choosy females

Ways in which females influence male reproduction
Egg investment: May influence what materials, and how much of them, to place in an egg
Mate choice: May decide which male or males will be granted the right to be sperm donors
Egg fertilization: May decide which sperm will actually fertilize the female's eggs
Offspring investment: May influence how much maintenance and care goes to each embryo and offspring
Ways in which males influence female reproduction
Resources transferred to female: May influence egg investment, mate choice, or egg fertilization decisions by female
Elaborate courtship: May influence mate choice or egg fertilization decisions by female
Sexual coercion: May overcome female preferences for other males
Infanticide: May overcome female decisions about offspring investment

Source: After Waage 1979.

FIGURE 9.33 Rejection behavior by females. A female cane toad (*Rhinella marina*) inflates her body to make it difficult for a male to grasp her properly so that he can fertilize her eggs when she releases them. (Photograph by Crystal Kelehear, courtesy of Rick Shine.)

FIGURE 9.34 Forced copulation in a bird. Stichbird males sometimes assault females on neighboring territories and force them to the ground, where the male transfers sperm to his "partner" while she is on her back. Note the tail of the pinned female in the lower left. (Photograph by Matt Low.)

difficult—but not impossible—to believe that the male's behavior is somehow reproductively advantageous to the female. When a male black-tipped hangingfly grabs a female by a wing and mates with her without providing a food gift, she loses a meal (Thornhill 1981). When a male stitchbird chases a female paired to another male and eventually drives her into the ground for a face-to-face copulation (**FIGURE 9.34**), a most unusual pattern in birds, the female has at the very least expended time and energy for what certainly looks like an unnecessary copulation (Low 2005). Likewise, when female orangutans (*Pongo pygmaeus abelii*) finally copulate with the sexually eager young males that have been harassing them for days on end, it appears that those females are mating against their will. Given a choice, female orangutans seek out and copulate with huge, older adult males that protect them from those annoying smaller males (Fox 2002). Similar preferences are exhibited by female chimpanzees, which do not synchronize their estrous cycles with other group members, perhaps to ensure that at least one dominant male will be available to guard them against sexual harassers of lower rank (Matsumoto-Oda et al. 2007). Nonetheless, dominant and subordinate male chimpanzees alike often sexually assault fertile females (Muller et al. 2007). Some violent copulations are fatal for females, as is true for our own species as well (Brownmiller 1975) (see Chapter 14).

Sexual conflict can often be observed in the ways that males try to manipulate female mate choice and the mating process more generally. For example, males of many insects transfer chemicals along with their sperm that increase the delay in remating by their partners (Himuro and Fujisaki 2008). Although receiving these chemicals may be advantageous to females in some instances, in other cases females are damaged by male seminal fluid. When male *Drosophila melanogaster* copulate, they transfer to their mates a protein (Acp62F) (Lung et al. 2002) that boosts male fertilization success (perhaps by damaging rival sperm) at the expense of females, whose lives are shortened, whose sleep patterns are disrupted (Kim et al. 2009), and whose fecundity is lowered (Pitnick and García-González 2002). Despite the negative long-term effects that toxic protein donors have on their mates, males still gain because they are unlikely to mate with the same female twice. Under these circumstances, a male that fertilizes a larger proportion of one female's current clutch of eggs can derive fitness benefits even though his chemical donations reduce the lifetime reproductive success of his partner.

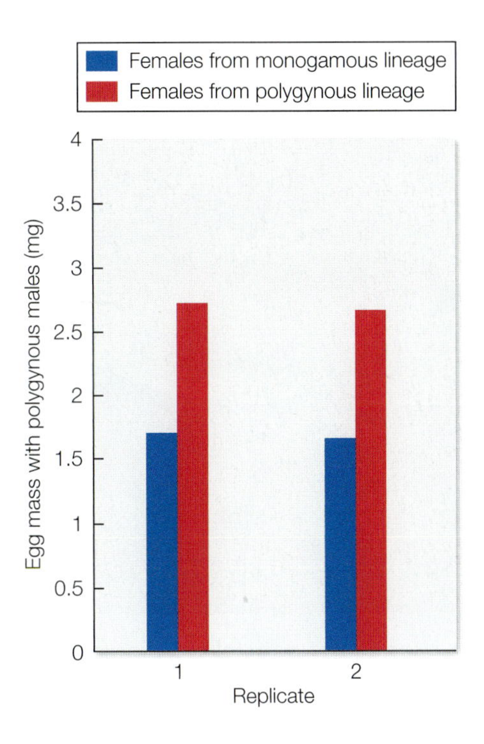

FIGURE 9.35 Sexual selection and the evolution of male traits harmful to females. Females from an experimental monogamous lineage of fruit flies have lost much of their biochemical resistance to the damaging chemicals present in the seminal fluids of polygynous male fruit flies. Therefore, monogamous females lay fewer eggs when mated with males from a polygynous lineage than do control females that evolved with polygynous males. The results of an experiment repeated twice are shown here as replicates 1 and 2. (After Holland and Rice 1999, © 1999 National Academy of Sciences, USA.)

If male fruit flies really do harm their mates as a consequence of their success in sperm competition, then placing generations of males in laboratory environments in which no male gets to mate with more than one female should result in selection against donors of damaging seminal fluid. Under these conditions, any male that happened not to poison his mate would reap a fitness benefit by maximizing his single partner's reproductive life span and output. In turn, a reduction in the toxicity of male ejaculates should result in selection for females that lack the chemical counteradaptations to combat the negative effects of the seminal fluid proteins. Indeed, after more than 30 generations of selection in a one male–one female (a monogamous lineage) environment, females from the population that were mated once with seminal fluid–donating "control" males taken from a typical multiple-mating population (a polygynous lineage) laid fewer eggs and died earlier than females that had evolved with many seminal fluid-donating males (**FIGURE 9.35**) (Holland and Rice 1999).

Sexual Arms Races

At this point, you may be wondering why such harmful male traits spread despite their negative impact on female fitness. From the male's perspective, since evolution by natural selection proceeds through individual advantage, the genes for violent or harmful behavior will spread in the population as long as these traits increase the reproductive success of those males that bare them. Females, however, are not powerless in an evolutionary sense to these harmful male traits. Indeed, conflict between males and females over mating often sets up an evolutionary arms race, as males evolve traits used in manipulating or forcing mates, and females evolve traits to circumvent this often violent process. A good example of a **sexual arms race** may be taking place in bedbugs (Reinhardt et al. 2003) and some other creatures (Rezac 2009, Hoving et al. 2010) that employ traumatic insemination or its equivalent. In the bedbug, this method of sperm transfer takes place when the male uses his spearlike intromittent organ to stab a female in her abdomen before injecting

(A)

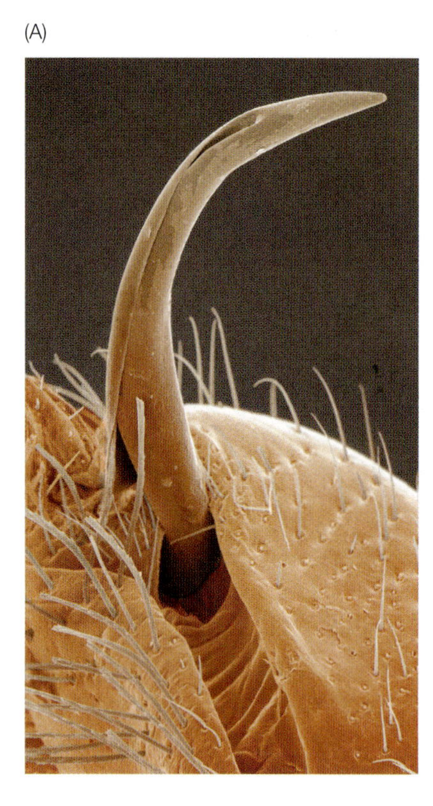

(B)

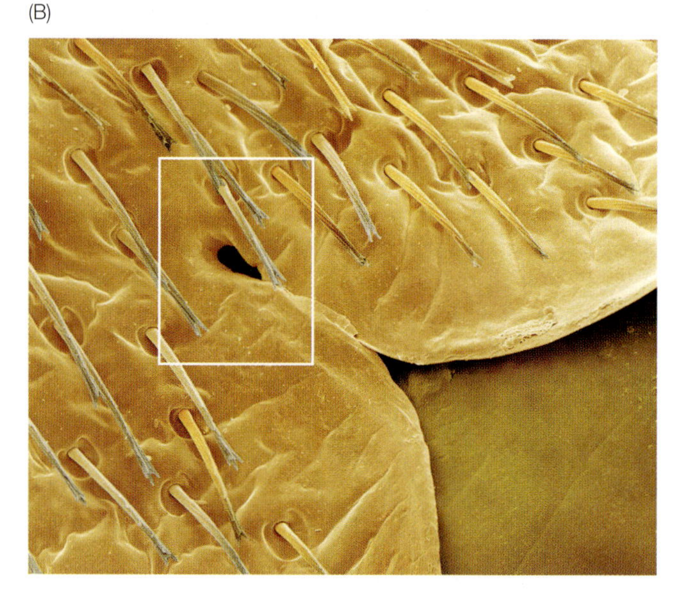

FIGURE 9.36 A product of conflict between the sexes? (A) Male bedbugs have evolved a saberlike penis that they insert directly into the abdomen of a female prior to injecting her with sperm. The trait may have originated when males benefited by employing traumatic insemination to overcome resistance to mating by unwilling females. (B) The white box shows the groove over the pleural membrane, the location on the female's abdomen where the male penetrates her with his penis. (Photographs © Andrew Syred/Science Source.)

his sperm directly into her circulatory system (**FIGURE 9.36**) (Stutt and Siva-Jothy 2001). This odd method of insemination presumably originated as a result of males injecting their sperm into sexually resistant females that had already acquired sperm from earlier partners via the traditional, less damaging mode of insemination. If so, then traumatic insemination should be costly to females, and it is, given that females mating at high frequencies live fewer days and lay fewer eggs (Stutt and Siva-Jothy 2001). However, females that mated for a minute with three males per week laid eggs at a higher rate for a longer period than females that were permitted just one 60-second copulation per week (Reinhardt et al. 2009). Something in the male ejaculate appears to boost female fitness in bedbugs, at least for those that experience traumatic insemination in moderation.

For another example of sexual conflict and genital evolution, consider waterfowl. Actually, most male birds do not have external genitalia at all; only about 3 percent of the more than 10,000 avian species have a phallus, or male intromittent organ. In waterfowl, most males have a phallus that varies across species in length and degree of elaboration, ranging from smooth to covered with spines and grooves (Briskie and Montgomerie 1997). Penis size and structure reflect the degree of sexual conflict among waterfowl species. That is, in some waterfowl species males and females are faithful, while in others, males forcefully copulate (occasionally to frequently) with females that are already paired to other males, removing female choice from the mating equation. In species in which forced extra-pair copulations are most frequent, males tend to have large phalluses covered with many knobs and ridges (Coker et al. 2002). These male adaptations are thought to improve the likelihood of fertilization and help males overcome any female choice. So how do females counteract these phallus adaptations and exert some control during mating? Patricia Brennan and colleagues discovered that females have evolved elaborate vaginas whose complexity rivals that of male phalluses (Brennan et al. 2007). In what appears to be unique among birds, female waterfowl vaginas have dead-end sacs and clockwise coils. The coils are in the opposite direction to the coiling of the male phallus, and together with the dead-end sacs, the female morphological adaptations presumably make it challenging for males to copulate successfully. Using high-speed video and glass models, Brennan's team demonstrated that the female morphology does indeed make male intromission less successful (**VIDEO 9.3**) (Brennan et al. 2010). Thus, male and female genitalia seem to be coevolving in a sexual arms race, in which females have regained some control of their reproductive choices.

VIDEO 9.3

Male duck genitalia in action

ab11e.com/v9.3

The bedbug and waterfowl cases illustrate not only the complexity of interactions between males and females, but their consequences for trait evolution. Let's accept that in at least some instances, there are mating costs to females, such as a reduction in longevity or the developmental damage done to a female's daughters as a result of having genes that are beneficial only when expressed in males (Pischedda and Chippindale 2006). But could these and other costs be outweighed by the exceptional reproductive success of a female's adult sons (Parker 2006)? If so, the females that permitted the fathers of these sons to mate with them might be more than compensated by the increased reproduction of their male offspring. Females might even accept a reduction in their lifetime reproductive success if they produced sons that inherited effective manipulative, or even coercive, tactics from their fathers—tactics that might make their sons unusually successful reproducers, thereby endowing their mothers with extra grand-offspring (Cordero and Eberhard 2003). However, this argument appears to fall apart when you consider the effect on daughters, not just sons. Although females may benefit from successful manipulation of sons, they will also lose through their daughters which are likely to suffer from the manipulation of those males in the population, particularly when there are direct costs involved, as is often the case with forced copulations (Brennan and Prum 2012).

In some species, apparent physical conflict among males and females may even be the way that females judge the capacity of males to supply them with sons that will also be able to overcome the resistance of females in the next generation (Eberhard 2005). Alternatively, aggressive courtship by males may be the way in which females choose males whose sons will do well in aggressive competition with other males (Borgia 2006). These kinds of hypotheses suggest that females could be "winning by losing" when they mate with males that appear to be forcing them to copulate or to be blocking their apparent preference for other males, at least in species where females do not otherwise exert positive female choice for a different kind of male. In any event, it is clear that in most species the interests of males and females are rarely if ever the same, which results in sexual conflict. These sex differences begin even before conception, with the size asymmetry of eggs and sperm. These differences carry through to the periods of mate choice and of parental care. As we have seen, the end result of all of these and other sex differences is the evolution of elaborate traits that Darwin explained in his theory of sexual selection. And as you will see in the next chapter, these sex differences influence the type of mating system that a species adopts.

SUMMARY

1. Darwin's sexual selection theory explains why certain traits can spread through populations even though these attributes appear to lower individual survival, a result that would seem disfavored by natural selection. But sexual selection can overcome natural selection for improved survival if the sexually selected characteristics promote success in securing mates.
2. Sexual selection results in sex differences in reproductive behavior and morphology. These standard sex roles arise because males usually make huge numbers of very small gametes and often try to fertilize as many eggs as possible, while providing little or no care for their offspring. In contrast, females make fewer, larger gametes and often provide parental care as well. As a result, receptive females are scarce, becoming the focus of male competition while ensuring that females have many potential partners to choose among.
3. Typically, males compete with each other for mates while trying to convince females to mate with them. The intrasexual or competition-for-mates component of sexual selection has led to the evolution of many characteristic male reproductive behaviors, including a readiness to fight over females or for social dominance. In addition, this form of sexual selection has resulted in the evolution of mating tactics that enable some males to make the best of bad situations created by their rivals as well as the ability of males to give their sperm an edge in the competition for egg fertilizations, especially by guarding their mates for some time after copulation.
4. In a typical species, females exercise mate choice (intersexual selection) because they control egg production, egg fertilization, and the care of offspring. Females often seek direct benefits in the form of parental care or access to resources. Males of some species seek to win favor with females by offering them material benefits, including nuptial gifts or parental care. In these cases, males with better gifts or superior paternal capabilities often produce more descendants.
5. Mate choice by females occurs even in some species in which males provide no material benefits of any sort. In these cases, females seek indirect genetic benefits that enhance the viability of their offspring (good genes model). However, extravagant male features could spread through a population in which even arbitrary elements of male appearance or behavior became the basis for female preferences. Exaggerated variants of these elements could be selected strictly because females preferred to mate with individuals that had them (runaway selection model). Yet another possibility is that the extreme ornaments of males evolve as a result of a cycle of conflict between the sexes, with males selected for ever-improved ability to exploit female perceptual systems and females selected to resist those males ever more resolutely (chase-away selection model).
6. Interactions between the sexes can be viewed as a mix of cooperation and conflict as males seek to win fertilizations in a game whose rules are set by the reproductive mechanisms of females. Conflict between the sexes (sexual conflict) is widespread and includes sexual harassment, manipulation, and forced copulations, which often result in sexual arms races and demonstrate that what is adaptive for one sex may be harmful to the other.

COMPANION WEBSITE

Go to **ab11e.com** for discussion questions and all of the audio and video clips.

CHAPTER 10

Mating Systems

Male satin bowerbirds (*Ptilonorhynchus violaceus*), as you saw in Chapter 9, are capable of copulating with many females in a single breeding season, though they rarely are able to do so (Borgia 1985). Males that mate with multiple females are called **polygynous**. In contrast, female satin bowerbirds are almost always **monogamous**, typically mating with just one male per nesting attempt. Yet the combination of potentially polygynous males interacting with monogamous females is only one of a variety of sexual arrangements found in the animal kingdom. In fact, different mating systems can be found even among the bowerbirds. For example, both males and females of the green catbird (*Ailuroedus crassirostris*), a close relative of the satin bowerbird, pair off monogamously for a given breeding season and rear offspring together (Frith and Frith 2001). In some other birds, such as the spotted sandpiper (*Actitis macularius*), a relative of the ruff (*Philomachus pugnax*) that we discussed in previous chapters, females are **polyandrous** and copulate with two or three males in a breeding season (Oring and Knudson 1973), often using the sperm of more than one male to fertilize their eggs. Males of these sex-role reversed species are generally monogamous. A still more extreme version of polyandry is exhibited by the honey bee *Apis mellifera*, in which young queens fly out from their hives into aerial swarms of drones that pursue, capture, and mate with them in midair. The average queen is highly polyandrous, coupling with many males and storing the sperm of perhaps a dozen or so to use during her lifetime of egg laying (Tarpy and Nielsen 2002). In contrast, drones never mate with more than one queen because a drone violently propels his genitalia into his first and only mate, a suicidal act that ensures that he is both monogamous and, shortly thereafter, dead (Woyciechowski et al. 1994). Finally, both males and females of some species mate with multiple partners. When these matings involve pair-bonds between the males and females, often necessary to rear the young, it is referred to as **polygynandry**. In contrast, when the matings are seemingly random and there is no association between mates beyond sperm transfer, the mating system is referred to as **promiscuity**.

Among this diversity of animal mating systems, the key feature that distinguishes one type from another is the number of mates that males and females have (**FIGURE 10.1**). As we discussed in Chapter 9, both sexes must weigh the costs and benefits of their mating decisions, and therefore males and females often have different motivations for attempting to mate multiply. This means that most mating systems are plastic and can vary among populations of the same species. Stephen Emlen and Lewis Oring developed a socio-ecological framework to explain the emergence of different animal mating systems (Emlen and Oring 1977). Their framework builds from the principles of sexual selection theory, namely that a population's mating system is the result of the ability of one sex to gain—and even monopolize—access to the other sex, either by associating with that sex directly or by defending territories and other resources that it requires for breeding. Because of sex differences in parental investment and in the operational sex ratio, the sex that invests more in reproduction (typically females) should distribute themselves in ways that maximize their ability to obtain essential resources. Since this sex is therefore likely to be limiting, it is the one that typically is courted and fought over by the other (limited) sex (typically males). Members of the limited sex should therefore attempt to control access to members of the limiting sex, often by regulating the resources they need to survive and reproduce. In many species, the spatial dispersion of food or other key resources, as well as the temporal availability of receptive mates, influences one sex's ability to do this. In general, the greater the potential to monopolize mates or resources, the greater the likelihood that mating systems in which one sex mates multiply will evolve. Thus, polygyny should evolve when males can monopolize access to multiple females, and polyandry should evolve when females can monopolize access to multiple males. In contrast, monogamy should evolve when neither sex can monopolize

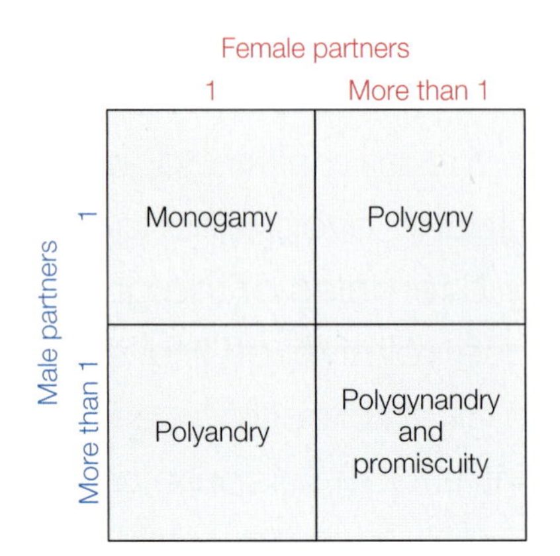

FIGURE 10.1 Animal mating systems are defined by the number of mates each sex has. Monogamy occurs when males and females both only have a single mate. When males have multiple partners but females have just one, the mating system is termed polygyny. In contrast, when females have multiple partners but males have just one, the mating system is termed polyandry. When both sexes have multiple partners, the mating system is termed polygynandry if there is a pair-bond between males and females, and promiscuity if there is no association between mates beyond sperm transfer.

access to mates or resources. When resources or access to mates drive both sexes to live together because neither sex is able to gain an advantage over the other, polygynandry can evolve, but when resources are unnecessary for breeding (or parenting), promiscuity can evolve. Thus, the form of mating system that evolves will depend on which sex is limiting and the degree to which the other sex controls resource access or monopolizes mates (Emlen and Oring 1977).

In this chapter, we build from Emlen and Oring's classic framework to examine each form of mating system in detail. In discussing the different types of mating systems, we use examples from both invertebrates and vertebrates to illustrate not only the similarities in mating systems across disparate animal taxa, but to demonstrate how resource distribution, costs of parental care, and sexual conflict between males and females over mating decisions interact to influence the type of mating system that a population or species exhibits. As you will see, there are numerous hypotheses to explain why males and females mate in one way or another. We will explore many of these hypotheses and develop a rich picture of the costs and benefits that both sexes experience by mating once or more than once in a breeding bout.

Monogamy: A Lack of Multiple Mating

Based on everything you have read thus far about sexual selection and mating systems, you might think that monogamy is common. In fact, monogamy is extremely rare in nature. For example, less than 5 percent of mammals are monogamous, and even though more than 90 percent of birds appear to be monogamous, they are not. We now know, thanks to advances in molecular genetics, that one or both mates in most birds that form monogamous pair-bonds are not as faithful as they seem. **Social monogamy** (the pairing of male and female) does not equate to **genetic monogamy** (when pairs produce and rear only their own genetic offspring). In fact, in nearly 90 percent of all bird species, one or both members of a pair-bond engage in extra-pair fertilizations (Westneat et al. 1990, Griffith et al. 2002). In the most extreme case, 95 percent of superb fairy-wren (*Malurus cyaneus*) nests examined contained extra-pair young (Mulder et al. 1994). This is not to say, however, that all birds seek extra-pair fertilizations. One of the most faithful of all avian species is the Florida scrub-jay (*Aphelocoma coerulescens*), in which nearly 100 percent of nestlings were shown to be produced by the social parents (Quinn et al. 1999, Townsend et al. 2011). Determining why, for example, monogamy has evolved in Florida scrub-jays but not in superb fairy-wrens requires considering sex differences in the costs and benefits that animals incur by mating multiply.

Why Be Monogamous?

One answer to the question "Why be monogamous?" is that since the costs to mating multiply may be prohibitively high for both males and females, being monogamous may be safer and more rewarding. Costs of mating multiply and include the time and energy spent searching for and mating with several partners, the risk of being killed by a predator during these forays and when mating, and the chance of acquiring a sexually transmitted disease from some mates. If the risk of sexually transmitted diseases is greater for polyandrous or polygynous individuals than for monogamous individuals, then we would expect the immune systems of species that mate multiply to be stronger than those of related species that have a greater tendency toward monogamy. One research team led by Charles Nunn tested this idea using comparative data from primates. The team took advantage of the fact that there are extremely polyandrous primates, such as the Barbary macaque (*Macaca*

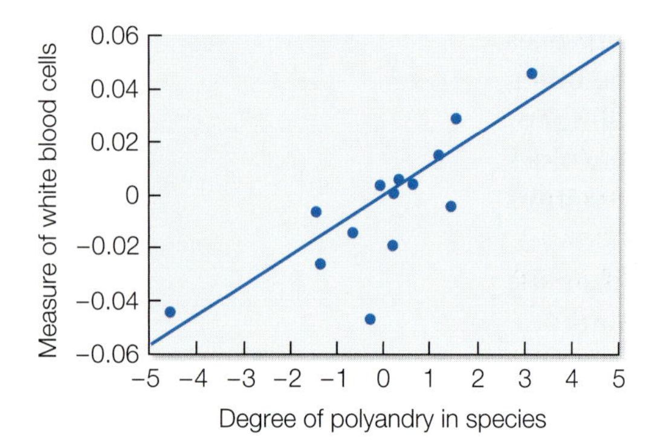

FIGURE 10.2 Polyandry has fitness costs. In primates, the number of mates accepted by females is correlated with the investment made in white blood cells, a component of an animal's immune system. (The higher the measure of white blood cells, the stronger the immune system.) This result suggests that the more polyandrous the species, the greater the challenge to the female's immune system. Each data point represents a different primate species. Only species with well-defined periods of female mating activity were included in the analysis. Negative values are the result of controlling for other variables and for evolutionary history in the analysis. (After Nunn et al. 2000.)

sylvanus), in which females mate with as many as ten males in one day. At the other end of the spectrum, there are monogamous species, such as the gibbons. Nunn and his colleagues measured immune system strength across the spectrum of mating systems by looking at data on white blood cell counts from large numbers of adult females of a variety of primate species, most of which were held in zoos where their health is regularly monitored. Consistent with the idea that sexually transmitted disease risk is higher in species where males or females mate multiply than in monogamous species, white blood cell counts of healthy individuals were indeed higher (but within the normal range) in females of the more polyandrous species (**FIGURE 10.2**) (Nunn et al. 2000). Similar results have been observed in birds. For example, Aldo Poiani and Colin Wilks tested for the presence of sexually transmitted bacteria and yeast in four species of birds that differ in their mating systems, two monogamous and two polyandrous species (Poiani and Wilks 2000). The researchers found that the polyandrous species had higher levels of both bacteria and yeast in cloacal swabs than the monogamous species did.

As you saw above, monogamy is predicted to evolve when neither sex can monopolize access to mates or resources. This can occur when potential mates do not form groups and roam widely, making them costly to locate (**mate limitation hypothesis**), when individuals have the ability to restrict mating behavior in their partner (**mate guarding hypothesis**), when resources are so critical to successful reproduction that both parents are necessary to rear young (**mate assistance hypothesis**), or when the risk of infanticide is high and having a partner can help provide protection against infanticidal males (**infanticide hypothesis**) (**HYPOTHESES 10.1**).

HYPOTHESES 10.1

Non-mutually exclusive hypotheses to explain why monogamous mating systems evolve

mate limitation hypothesis
Monogamy is likely to evolve when potential mates do not form groups and roam widely, making them costly to locate.

mate guarding hypothesis
Monogamy is likely to evolve when individuals have the ability to restrict mating behavior in their partner.

mate assistance hypothesis
Monogamy is likely to evolve when resources are so critical to successful reproduction that both parents are necessary to rear young.

infanticide hypothesis
Monogamy is likely to evolve when the risk of infanticide is high and a partner can provide protection against infanticidal males.

MATE GUARDING AND ASSISTANCE As you saw in Chapter 9, many males guard their mates when they are receptive to prevent them from seeking extra-pair fertilizations, a behavior consistent with the mate guarding hypothesis. Recall also that mate guarding by males results in monogamy for certain species of spiders, in which the male gives his all, quite literally, to his first mate. Those males that commit sexual suicide by breaking off their genital appendages in a partner's reproductive tract or by offering themselves as food to a partner (Andrade 1996, Miller 2007) may increase their fertilization success with that one female (of course, the trade-off is that they will never mate with another female). This gain from what might be called the male's posthumous mate guarding may exceed the cost of his actions if both his mate has the potential to remain receptive after one mating and the male's probability of finding a second female is extremely low. These conditions appear to be met in some spiders with male monogamy (Andrade 2003, Miller 2007), as well as in some other species (Emlen and Oring 1977).

Parental care plays a central role in the mate assistance hypothesis, which proposes that a male remains with a single female to help her in various ways but generally because paternal care and protection of offspring are especially advantageous (Emlen and Oring 1977). Of course, guarding a mate does not necessarily preclude helping her, and in fact, a study of crickets in a Spanish field using video cameras to monitor burrows showed that guarded females were much less likely than unguarded ones to fall victim to predators, primarily birds. Why was this the case? The researchers found that when a predator approached, the male permitted his mate to dash into the burrow first, a self-sacrificing act that increased his odds of being killed (**FIGURE 10.3**) (Rodríguez-Muñoz et al. 2011). Yet because the male had copulated frequently with the female, most of the offspring that she produced carried his genes. Thus, by mating monogamously, a male not only assured his paternity but also increased the survival odds of the female carrying his offspring. Indeed, paternity assurance is one of the primary potential benefits to males in many monogamous species.

Males are not the only sex that mate guards. In some species, a female may block her partner's polygynous intentions to monopolize his parental assistance, a form of mate guarding often referred to as female-enforced monogamy. For example, paired females of the burying beetle *Nicrophorus defodiens* are aggressive toward intruders of their own sex. After a mated male and female work together to bury a dead mouse or shrew, which will feed their offspring once they hatch from the eggs the female lays on the carcass, the male may climb onto an elevated perch and release a sex pheromone to call a second female to the site. If another female added her clutch of eggs to the carcass, her larvae would compete for food with the first female's offspring, reducing their survival or growth rate. Thus, when the paired

(A)

(B)

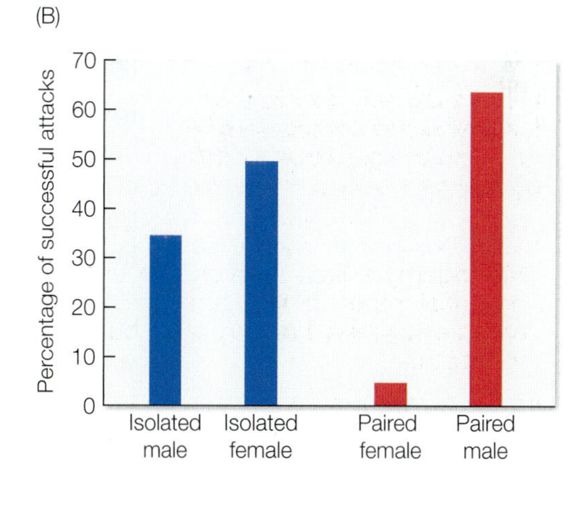

FIGURE 10.3 Mate guarding and mate assistance in the cricket *Gryllus campestris*. (A) A male that has acquired a mate remains with her in his territory. If the pair is attacked, the male permits the female to rush into the burrow first, at great cost to his personal safety. (B) Solitary territorial males are much more likely to survive an attack. A female that lives alone is less likely to survive than one paired with a self-sacrificing partner. (A, photograph courtesy of WildCrickets.org; B after Rodríguez-Muñoz et al. 2011.)

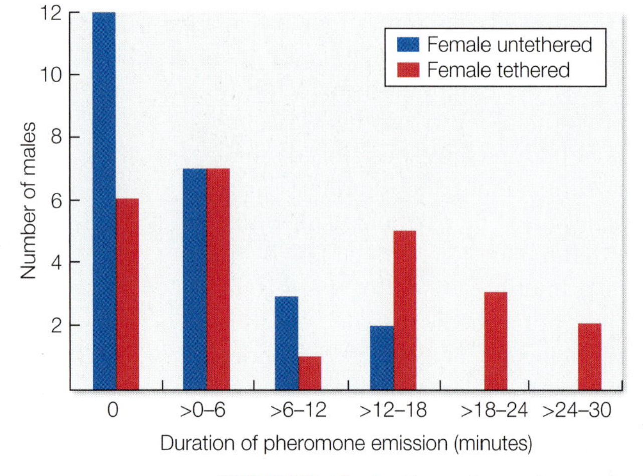

FIGURE 10.4 Female-enforced monogamy in burying beetles. When a paired female beetle is experimentally tethered so that she cannot interact with her partner, the amount of time he spends releasing sex pheromones rises dramatically. (After Eggert and Sakaluk 1995; photograph © Henrik Larsson/Shutterstock.com.)

female smells her mate's pheromone, she hurries over to push him from his perch. These attacks reduce his ability to signal, as Anne-Katrin Eggert and Scott Sakaluk showed by tethering paired females so that they could not suppress their partners' sexual signaling (Eggert and Sakaluk 1995). Freed from controlling partners, the experimental males released scent for much longer periods than did control males that had to cope with untethered mates (**FIGURE 10.4**). Thus, male burying beetles may often be monogamous not because it is in their genetic interest but because females force them to be—another example of the kinds of sexual conflicts that occur between males and females (see Chapter 9).

Monogamy in Species with Paternal Care

In species of birds in which males and females form partnerships for one or more breeding seasons (Lack 1968), males tend to contribute in a major way to the welfare of the offspring produced by their mates (Orians 1969). In the yellow-eyed junco (*Junco phaeonotus*), for example, the male takes care of his mate's first brood of fledglings while the female incubates a second clutch of eggs. The paternal help provided is essential for the survival of these young and highly inept foragers (Weathers and Sullivan 1989). The value of male assistance has also been documented for European starlings (*Sturnus vulgaris*). In a population in which some males helped their mates incubate their eggs and others did not, the clutches with biparental attention stayed warmer (**FIGURE 10.5**) and so could develop more rapidly. Indeed, 97 percent of the eggs that had been incubated by both parents hatched, compared with just 75 percent of eggs that had been cared for by mothers alone (Reid et al. 2002). Thus, in many bird species the high costs of parental care that require both parents to share the workload (see Chapter 11) may promote (social) monogamy.

Demonstrations of the importance of biparental care in birds include some studies in which females have been experimentally "widowed" and left to rear broods on their own. For example, widowed snow buntings (*Plectrophenax nivalis*) usually produce three or fewer young, whereas control pairs often fledge four or more (Lyon et al. 1987). Reproductive success is therefore tightly tied to paternal care behavior in many monogamous species. Consider another example: when male spotless starlings (*Sturnus unicolor*) are given extra testosterone, they become less

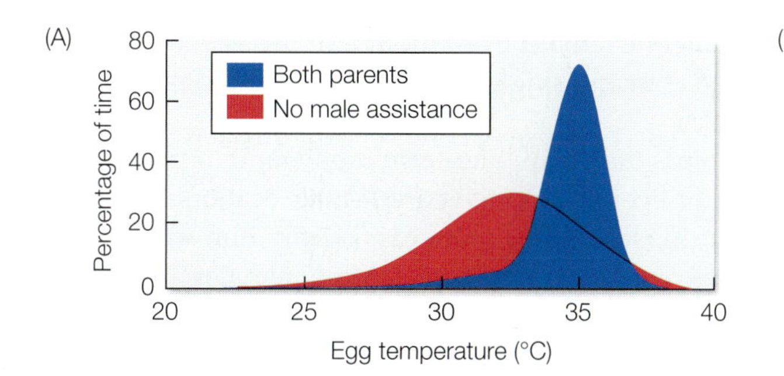

FIGURE 10.5 Paternal male European starlings help their mates. Males keep their clutches warmer by sharing in incubation duties. (A) Eggs that were incubated by both parents were kept at about 35°C most of the time, whereas eggs incubated by the female alone were often several degrees cooler. (B) Both male and female starlings incubate the eggs. In this nest, all the eggs hatched successfully. (A after Reid et al. 2002; B, photograph © Bengt Lundberg/Naturepl.com.)

willing to feed nestlings, whereas males that receive an anti-androgenic chemical that blocks the effects of naturally circulating testosterone feed their offspring at increased rates. The mean number of fledged young per brood was lowest for spotless starlings with extra testosterone and highest for those with the testosterone blocker (**FIGURE 10.6**) (Moreno et al. 1999).

Fathers provide care for young in a variety of other taxonomic groups besides birds, including in the extreme in the seahorse *Hippocampus whitei*, in which males take on the responsibility of "pregnancy" by carrying a clutch of eggs in a sealed brood pouch for about 3 weeks (**FIGURE 10.7**). Each male has a durable relationship with one female, which provides him with a series of clutches. Pairs even greet one another each morning before moving apart to forage separately; they will ignore any members of the opposite sex that they happen to meet during the day (Vincent and Sadler 1995). Since a male's brood pouch can accommodate only one clutch of eggs at a time, he gains nothing by courting more than one female at a time. In fact, in another species of *Hippocampus*, genetic data indicate that males do

Male spotless starling

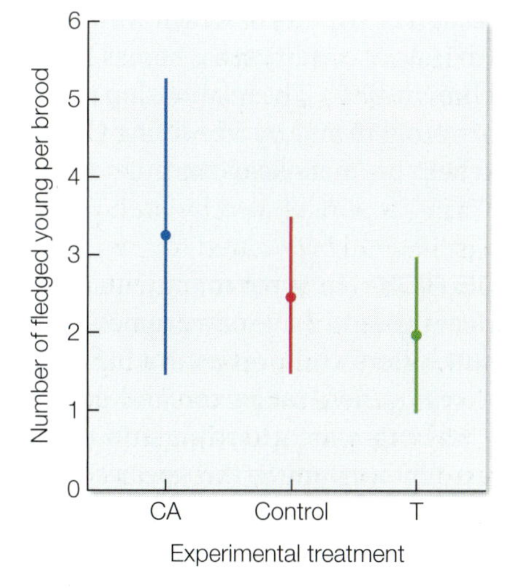

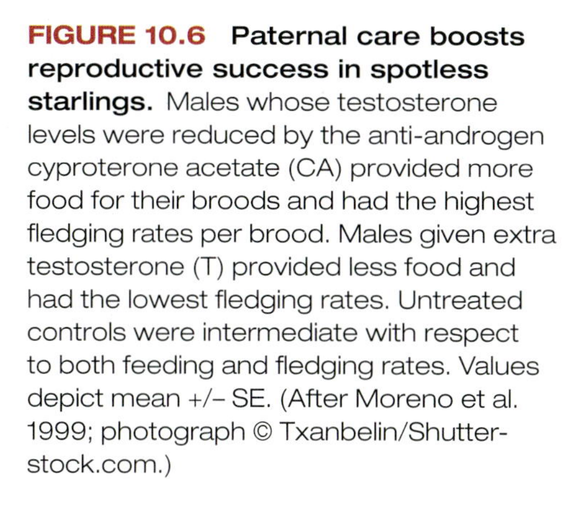

FIGURE 10.6 Paternal care boosts reproductive success in spotless starlings. Males whose testosterone levels were reduced by the anti-androgen cyproterone acetate (CA) provided more food for their broods and had the highest fledging rates per brood. Males given extra testosterone (T) provided less food and had the lowest fledging rates. Untreated controls were intermediate with respect to both feeding and fledging rates. Values depict mean +/– SE. (After Moreno et al. 1999; photograph © Txanbelin/Shutterstock.com.)

FIGURE 10.7 Monogamy and paternal care in a seahorse. A pregnant male is giving birth to his single partner's offspring. Two young can be seen emerging from their father's brood pouch. (Photograph © Dr. Paul A. Zahl/Science Source.)

not accept eggs from more than one female, even though in this species, groups of females are sometimes seen courting single males (Wilson and Martin-Smith 2007).

Just as in many species of birds that form long-term monogamous pair-bonds for multiple breeding events or even for many years, a male seahorse may not benefit by switching mates if his long-term mate can supply him with a new complement of eggs as soon as one pregnancy is over. Many females apparently do keep their partners pregnant throughout the lengthy breeding season but cannot produce batches of eggs so quickly that they have some to give to other males (Vincent et al. 2004). By pairing off with one female, a male might be able to match his reproductive cycle with that of his partner. When the two individuals were in sync, the male would complete one round of brood care just as his partner had prepared a new clutch. Therefore, he would not need to spend time waiting or searching (often a costly endeavor that can increase predation risk and lead to reduced survival) for alternative mates, but could immediately secure a new clutch from his familiar mate. If this is true, the experimental removal of a female partner should lengthen the interval between spawning for the male, which it does in at least some species related to seahorses. For example, males of a paternal pipefish that had to change mates took 8 more days to acquire replacement partners with clutches of mature eggs than did males permitted to retain their mates from one clutch to the next (Sogabe et al. 2007).

Monogamy When Paternal Care Is Rare

Although monogamy is uncommon in most animals, it is exceptionally rare in mammals, a group notable because internal gestation and lactation are largely limited to females (with the exception of one species of bat in which males lactate; Francis et al. 1994). Since most male mammals can therefore provide little parental care, sexual selection theory suggests that they should instead try to be polygynous—and they do. However, exceptions to the rule are useful for testing alternative hypotheses for male monogamy in mammals. For example, if the mate assistance hypothesis for monogamy applies to mammals, then males of those rare mammalian species that exhibit paternal behavior should tend to be monogamous (Woodroffe and Vincent 1994). One monogamous mammal with paternal males is the Djungarian hamster (*Phodopus sungorus*), in which males actually help deliver their partner's pups (**FIGURE 10.8**) (Jones and Wynne-Edwards 2000). Male parental care contributes to offspring survival in this species and also in the monogamous California mouse (*Peromyscus californicus*), pairs of which were consistently able to rear a litter of four pups under laboratory conditions, whereas single females did not do nearly as well (Cantoni and Brown 1997). The relationship also held under natural conditions, with the number of young reared by free-living California mice falling when a male was not present to help his mate keep the pups warm (**FIGURE 10.9**) (Gubernick and Teferi 2000).

INFANTICIDE RISK In some mammals, one helpful behavior a paternal male might do for females is to drive other males away from his brood, thereby blocking potential infant killers and preventing infanticide. In the mate-guarding prairie vole (*Microtus ochrogaster*), males can and do defend their offspring against infanticidal rivals, so both mate guarding and mate assistance may contribute to the tendency toward monogamy in this species (Getz and Carter 1996). Infanticide risk may also favor monogamy in many primate species, including in the white-handed gibbon (*Hylobates lar*), in which there was a much lower risk of infant disappearance in groups composed of a monogamous pair living with their offspring than

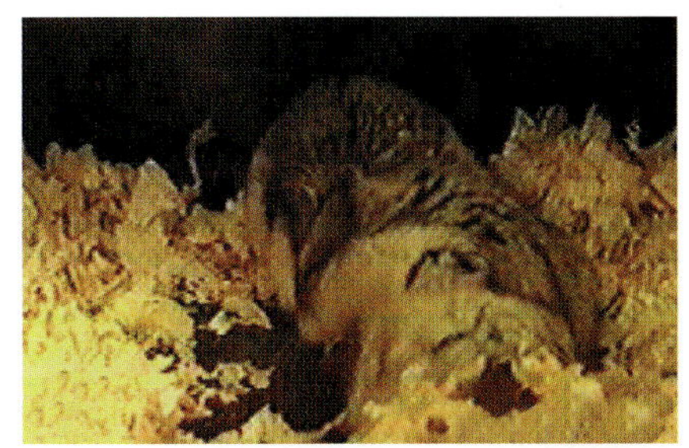

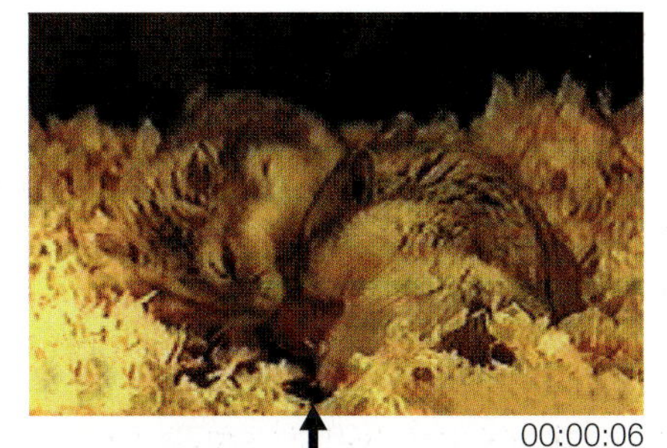

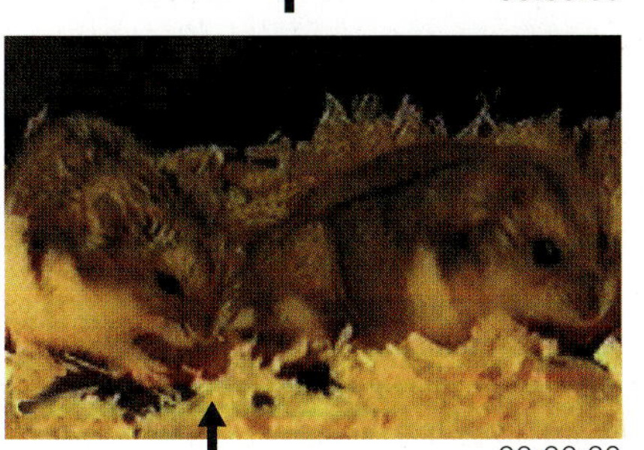

FIGURE 10.8 An exceptionally paternal mammal, the Djungarian hamster. A male Djungarian hamster may pull newborns from his mate's birth canal and then clear the infants' airways by cleaning their nostrils, as shown in these photographs (the male is the hamster on the left; arrows point to the pink newborn). (Photographs © Katherine E. Wynne-Edwards, Faculty of Veterinary Medicine, University of Calgary.)

in polyandrous groups consisting of an adult female and several adult males (Borries et al. 2011). Christopher Opie and colleagues examined the role of infanticide in driving the evolution of monogamy in 230 primate species (Opie et al. 2013). By reconstructing the behaviors of species long since extinct, the authors demonstrated that the presence of infanticide in a species precedes the initial shift from polygyny to monogamy. In other words, monogamy has tended to evolve only in species that faced strong infanticidal pressures by males, making the defense by a long-term partner beneficial to males and females alike.

Despite strong comparative evidence for the relationship between the risk of male infanticide and monogamy in primates, evidence supporting this idea in mammals more broadly is lacking. Dieter Lukas and Tim Clutton-Brock tested the infanticide hypothesis in more than 2500 species of mammals and, to their surprise, found no relationship between monogamy and risk of male infanticide (Lukas and Clutton-Brock 2013). In fact, Lukas and Clutton-Brock argue that paternal care is a consequence rather than a cause of monogamy in mammals. Their work shows that monogamy has evolved in mammal species where breeding females are intolerant of each other and where female density is low. Such results are consistent with the mate guarding hypothesis we introduced above, and correspond to earlier comparative work demonstrating that males tend to live with females in two-adult units more often when females live well apart from one another in small territories (Komers and Brotherton 1997). Under these circumstances, it becomes very hard for a male to gain and defend access to multiple females, so the optimal strategy may be to simply find one female and spend all of his time with her.

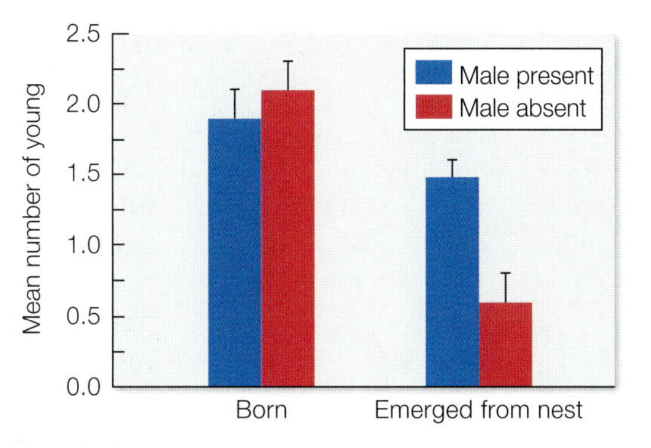

FIGURE 10.9 Male care of offspring affects fitness in the California mouse. The number of offspring reared by female mice falls sharply in the absence of a helpful male partner. Bars depict mean +/– SE. (After Gubernick and Teferi 2000.)

MATE LIMITATION The monogamous rock-haunting ringtail possum (*Petropseudes dahli*) of northern Australia is a case in point of how low female density and broad dispersion influence monogamy in mammals, the central premise of the mate limitation hypothesis. A female, her mate, and their young live along the edges

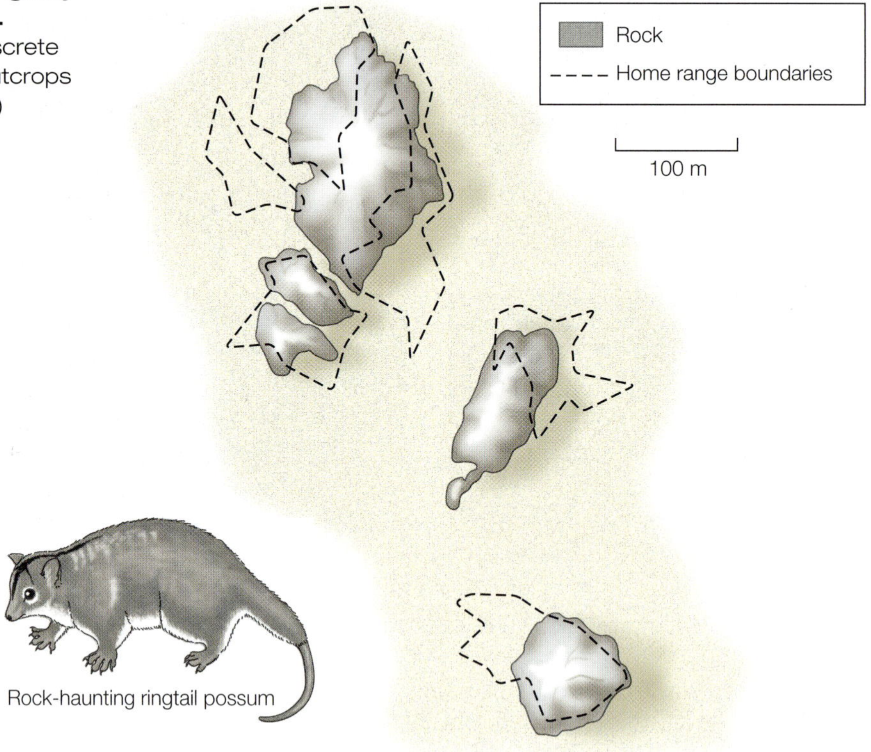

FIGURE 10.10 Mate-guarding monogamy in the rock-haunting ringtail possum. Females of this species occupy small, discrete home ranges along the edges of rock outcrops in northern Australia. (After Runcie 2000.)

of rock outcrops in territories of about 100 meters by 100 meters (**FIGURE 10.10**), an area about one-sixth the size of that occupied by other herbivorous mammals of equivalent weight (Runcie 2000). In these small territories, a male possum can effectively monitor the activities of one female without expending a great deal of energy. Any male that attempted to move back and forth between the home ranges of several females would run the risk that intruder males would visit and mate with those females that he had temporarily left behind. Thus, ecological factors that enable females to live in small, defensible territories tilt the cost–benefit equation toward mate guarding, which then leads to male monogamy. How female choice drives male decisions and the evolution of mating systems is a topic that we will return to a bit later in this chapter when we discuss different forms of polygyny.

Polyandry: Multiple Mating by Females

In Chapter 9 we introduced several potential benefits to females of being choosy. These include both indirect benefits in the form of genetic advantages to a female's offspring, as well as direct benefits (such as parental care, access to resources, safety from predators, or reduced harassment by other males) that can increase a female's fitness. We now explore these benefits not just in the context of a female choosing a mate, but in her deciding whether or not to choose multiple mates.

Monogamous Males and Polyandrous Females

That social monogamy by males often coexists with genetic polyandry by their mates greatly surprised ornithologists who had always assumed that males helped their mates care for their mutual offspring, not for the progeny of other males. We now know that polyandry is much more common than monogamy, not just in birds but in most taxonomic groups. A recent review led by Nina Wedell found evidence that mating with more than one male per female occurs in all of the 14 major taxonomic groups explored—from sea spiders to mammals—and in 89 percent of populations examined (Taylor et al. 2015). We will deal with the puzzles associated with socially monogamous females that accept and use sperm from more than one male in the next section, but for the moment let's consider why a male attaches himself to a single female (**BOX 10.1**) only to have her fertilize some or all of her eggs with sperm from one or more extra-pair partners. One explanation for this is that the male may be able to take advantage of his own opportunity for extra-pair copulations with females other than his social partner. A polygynous male of this sort may be able to fertilize at least some of the eggs of his social partner while also inseminating other females whose offspring will receive care, but not from him. These males, which often exhibit ornaments and displays thought to be indicative of superior physical condition (Griffith et al. 2002), avoid some or most of the costs of monogamy. Therefore, it falls to the genetically monogamous males, which are more likely to be cuckolded by their mates, to suffer the disadvantages of monogamy. The question remains, why do these genetically monogamous males fail to engage in fitness-boosting extra-pair fertilizations? As you learned in Chapter 9, one answer is that females have something to say about who gets to mate with them. If female mate choice revolves around parental contributions from a social partner and honest indicators of quality from extra-pair partners, then males that are unwilling to be paternal and are also unable to offer cues of good condition will likely find themselves between a rock and a hard place. They could drop out of the reproductive competition altogether but would obviously fail to leave descendants for as long as they remained on the sidelines. The only real option for these monogamous males is to help social partners that may fertilize at least some eggs with their sperm. Females may then have the motivation and the opportunity to mate multiply. Thus, sexual conflict is potentially present even in monogamous mating systems.

The existence of constraints imposed by choosy females and competitors may also help explain the evolution of polyandrous mating systems in which most breeding females form social bonds with several males, rather than having just one social partner and a variable number of extra-pair mates. The mating system of the Galápagos hawk (*Buteo galapagoensis*), for example, ranges along a continuum from monogamy to extreme polyandry (Faaborg et al. 1995). Polyandry appears to be associated with a scarcity of suitable territories, which leads to a highly male-biased operational sex ratio since males outnumber the limited number of territorial, breeding females. The intense competition for these females and the territories on which they live has favored males capable of forming a cooperative defense team to hold an appropriate site. A breeding female may acquire as many as eight mates prepared to pair-bond with her for years, forming a male harem that will help her rear a single offspring per breeding event. In this hawk and another bird, the pukeko (*Porphyrio melanotus*), all the males that associate with one female appear to have the same chance of fertilizing her eggs (Jamieson 1997). If females do give all of their mates an equal chance at fertilizations, the mean fitness of each mate is certainly higher than that of males that fail to be part of a team.

In some other polyandrous species, several mates share the reproductive output of the female with the other members of her harem. For example, in the wattled jacana (*Jacana jacana*), a sex-role reversed species in which females use spikes at the tip of their wings to aggressively fight for and defend territories that can accommodate several males, a territorial female mates with all of the members of her harem, providing each male with his own clutch of eggs, which he cares for exclusively by himself. However, males that pair monogamously with a polyandrous female are

BOX 10.1 DARWINIAN PUZZLE

Sexual parasitism, dwarf males, and the evolution of gigolos

In some species, males don't just attach to females in a pair-bond, they literally attach themselves to females. This type of mating system has been referred to as sexual parasitism, and the obvious puzzle here is why such a bizarre form of reproductive behavior has evolved. The poster child for sexual parasitism is the deep-sea anglerfishes. Some ceratioid anglerfishes, often called sea devils, exhibit extreme size dimorphism, with gigantic females and so-called dwarf males. These males appear to be specially adapted for one purpose: finding females. Relative to their bodies, males have large and long tails for improved swimming, forward-facing eyes, and large and well-developed nasal organs that are sensitive to pheromones, all for locating females in the vast, deep ocean. Males also have pincerlike teeth for grasping females if and when they find a potential mate. Once a male attaches himself to a female, he fuses his epidermal tissue to her, eventually uniting their circulatory systems to receive her blood-transported nutrients. With the parasitic male fused to the female permanently, she essentially becomes a self-fertilizing hermaphrodite (**FIGURE**). At this point, the male's only purpose is to produce sperm, something he cannot do until he has fused with a female. In some anglerfishes, males are only 6 to 7 millimeters long, making them among the smallest known sexually mature vertebrates. In anglerfish species with parasitic males, females are also not thought to be able to reach reproductive maturity and produce eggs until a male has attached himself to her body. Although some sexually parasitic anglerfish species appear to be monogamous, females of other species have multiple parasitic males attached, suggesting that these species are polyandrous (Pietsch 2005).

(A)

(B)

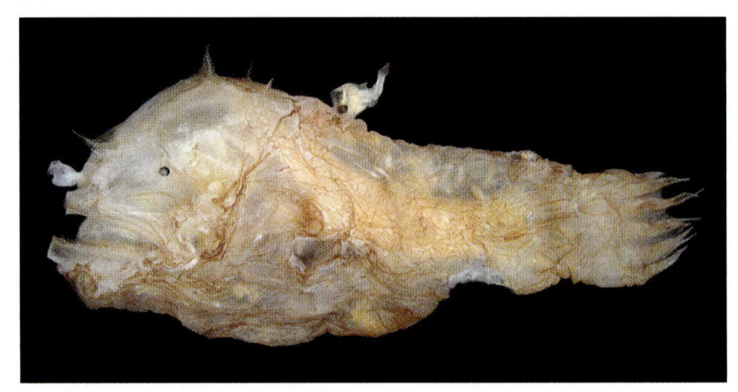

Female anglerfishes with parasitic males attached. (A) A 75-mm female *Melanocetus johnsonii* with a 23.5-mm parasitic male. (B) A 46-mm female *Photocorynus spiniceps* with an 6.2-mm parasitic male. (A, photograph from Pietsch 2005, courtesy of Dr. Edith Widder, ORCA; B, courtesy of Theodore W. Pietsch, University of Washington.)

also likely to care for offspring sired by other males, since 75 percent of the clutches laid by polyandrous female jacanas are of mixed paternity (Emlen et al. 1998). The same sort of thing happens in the red phalarope (*Phalaropus fulicarius*), a shorebird in which females are both larger and more brightly colored than males (**FIGURE 10.11**).

BOX 10.1 DARWINIAN PUZZLE (*continued*)

Anglerfishes live up to 1500 meters or more below the ocean surface. At these depths, the sea is vast, dark, and nutrient-poor. Anglerfishes don't move fast or far, and they occur at extremely low densities (Pietsch 1975). Females of some species attract mates by releasing a pheromone into the water, and when a male finds a potential mate, it clearly benefits him to latch on and never let go. Although the ecological conditions that some anglerfishes experience seem to promote sexually parasitic dwarf males, dwarf males are not restricted to anglerfishes. Males that are on average 50 percent or less of female size (the definition of dwarf males) occur in a variety of other animal species, including spiders in which dwarf males closely cohabitate with a female and derive sustenance from her excretions. Fritz Vollrath has argued that these males should not be called parasites at all (Vollrath 1998). Instead, he suggests that they be referred to as "gigolos" because these tiny males provide something critical for the females with which they associate: sperm. Indeed, gigolism might be favored in spiders if females require reproductive diversity and sperm for future reproductive bouts. Dwarfism might be favored because it makes males less likely to be eaten after sex (at least in some species) and so that they are not a drain on resources. And in the spider *Nephila clavipes*, the evolution of this mating system might also be related to the fact that females are sedentary and difficult to locate, just as they are in anglerfishes. In other words, selection will favor a harem of small males that do not eat much, do not look particularly appetizing after sex, but are capable of providing plenty of sperm for females that are difficult to locate. Whether such a system is referred to as sexual parasitism or gigolism, there are many parallels between the bizarre mating systems of the spiders and the anglerfishes, as well as some other organisms such as barnacles.

Thinking Outside the Box

Sexual parasitism appears to have evolved multiple times in the anglerfishes. Yet not all species of anglerfishes have males that fuse permanently to females, or females that produce eggs only after a male has fused to her body. Free-living males and females of some species have well-developed gonads, suggesting that parasitism is not necessary for reproduction in these species. In still other species, parasitism may be facultative and temporary. Even in sexually parasitic species that require male fusion for reproduction, the location of where the male fuses on the female body varies greatly among species. Although males in most species fuse upside down on the underside of the belly, males in some species may be found anywhere on the head or body, oriented in any direction (Pietsch 2005). Given this variation in anglerfish mating behavior, develop a socio-ecological model to explain why some species are obligately parasitic, others are facultatively parasitic, and still others are not parasitic at all.

Female phalaropes fight for males, and a male provides all the parental care for the clutch deposited in his nest, whether or not the eggs have been fertilized by other males (Dale et al. 1999). These cases illustrate a cost to males that are monogamously bonded with a polyandrous female, but male jacanas and phalaropes may have to accept the eggs their partners lay if they are to leave any descendants at all. And in some cases, these males are able to gain additional reproductive success by having other males care for their eggs in other nests.

In many polyandrous species, males maintain at least some control over their mating decisions and can reduce the probability that they will care for eggs fertilized by other males. For example, males of the red-necked phalarope (*Phalaropus*

FIGURE 10.11 Female red phalaropes are polyandrous. After securing one male partner, a female phalarope may attract another, donating a clutch of eggs to each male in turn. The more brightly colored bird (on the left) is the female. (Photograph © Hanne and Jens Eriksen/Minden Pictures.)

lobatus), like those of its relative the red phalarope, care for broods on their own. Females of this species produce two clutches of eggs in sequence. Polyandrous females almost always draw their second mates from those males that have lost a first clutch to predators. Such a male, however, favors his original sexual partner over a novel female more than nine times out of ten (Schamel et al. 2004). By copulating with this female again and then accepting her eggs, the male reduces the risk that the eggs he receives will have been fertilized by another male's sperm.

The same applies to the spotted sandpiper, another sex-role reversed species (Oring and Knudson 1973). In addition to taking the lead in courtship, females are the larger and more combative sex. Moreover, although in most migrating birds males arrive before females and establish territories, in the spotted sandpiper, females arrive on the breeding grounds before males. Once on the breeding grounds, female spotted sandpipers fight with one another for access to territories (**FIGURE 10.12**). A female's territorial holdings may attract first one and then later a second male. The first male mates with the female and gets a clutch of eggs to incubate and rear on his own in her territory, which she continues to defend while producing a new clutch for a second mate. However, this second mate may not fertilize all of the eggs he broods if his mate uses sperm stored from the first male to fertilize some of her second clutch. In this species, a few females achieve higher reproductive success than the most successful males (Oring et al. 1991), an atypical result for animals generally. Recall from Chapter 9 our discussion of Bateman's principle and the idea that males tend to have higher reproductive variance than females. In sex-role reversed species such as the spotted sandpiper, the typical pattern of reproductive variance is also reversed.

An understanding of the mating system of spotted sandpipers is advanced by recognizing that in all sandpiper species, females never lay more than four eggs at a time, presumably because five-egg clutches cannot be incubated properly (Oring 1985). Egg addition experiments have shown that sandpipers given extra eggs to incubate sometimes damage their clutches inadvertently and also lose eggs more often to predators (Arnold 1999). If female spotted sandpipers are "locked into" a four-egg maximum, then they can capitalize on rich food resources only by laying more than one clutch, not by increasing the number of eggs laid in any one clutch. To do so, however, they must acquire more than one mate to care for their sequential

FIGURE 10.12 Polyandrous female spotted sandpipers fight for males. Two female spotted sandpipers (A) about to fight and (B) fighting for a territory that may attract several monogamous, paternal males to the winner. (Photographs by Stephen Maxson.)

clutches, making this a rare case in which female fitness is limited more by access to mates than by production of gametes.

Male spotted sandpipers may be forced into monogamy by the confluence of several unusual ecological features (Lank et al. 1985). First, the adult sex ratio is slightly biased toward males, meaning that the opportunities for polygny are limited for males and that females may need to compete for access to mates. Second, spotted sandpipers nest in areas with immense mayfly hatches that provide superabundant food for young and may allow for multiple clutches to be laid. Third, a single parent can care for a clutch about as well as two parents, in part because the young are precocial, meaning they can move about, feed themselves, and thermoregulate shortly after hatching. This combination of excess males, abundant food, and precocial young means that female spotted sandpipers that desert their initial partners can find new ones without harming the survival chances of their first broods. Once a male has been deserted, however, he is stuck. Were he also to leave the nest, the eggs would fail to develop, and he would have to start all over again. If all females are deserters, then a male single parent presumably experiences greater reproductive success than he would otherwise, even if his partner acquires another mate to assist her with a second clutch. Furthermore, the first male to mate with a female spotted sandpiper may provide her with sperm that she stores and uses much later to fertilize some or all of her second clutch of eggs, as noted above (Oring et al. 1992). Once again, the males that get the short end of the stick are the less competitive individuals, the ones slower to arrive on the breeding grounds. Thus, male spotted sandpipers must deal with polyandrous females if they are to have any chance of reproducing, even if the female's ability to control the reproductive process puts the male at a major disadvantage.

Polyandry and Indirect Genetic Benefits

As you have just seen, males may pay a reproductive cost if their partner mates multiply. But for females, there are a variety of potential benefits from mating multiply, particularly for their offspring. Indeed, females can potentially obtain several indirect genetic benefits by mating polyandrously and obtaining sperm (and

HYPOTHESES 10.2

Non-mutually exclusive hypotheses to explain the potential genetic benefits that females receive from being polyandrous

good genes hypothesis Females mate polyandrously to produce offspring of higher genetic quality or viability.

genetic compatibility hypothesis Females mate polyandrously to increase the odds of receiving genetically complementary sperm.

genetic diversity hypothesis Females mate polyandrously to increase the heterozygosity (genetic diversity) of either individual offspring or of the group of offspring produced in a single bout.

inbreeding avoidance hypothesis In the context of mating, females mate polyandrously to avoid inbreeding with their social partner.

genes) from multiple males (**HYPOTHESES 10.2**) (Forstmeier et al. 2014). Below, we discuss each of these potential indirect benefits in greater detail.

GOOD GENES One well-studied genetic benefit linked to a polyandrous mating system has to do with improved offspring quality. In the grey foam nest treefrog (*Chiromantis xerampelina*), for example, tadpoles of polyandrous females whose clutches had been fertilized by at least ten males, a common occurrence in nature for more than 90 percent of females, were significantly more likely to survive to metamorphosis than were tadpoles of females that had had just a single mate (**FIGURE 10.13A**) (Byrne and Whiting 2011). According to this **good genes hypothesis**, females can receive a genetic benefit in terms of increased offspring quality or viability by mating multiply. The same factor may be at work in a wild guinea pig, the yellow-toothed cavy (*Galea musteloides*), in which females will copulate with more than one male when given the opportunity. The payoff for polyandry in this species appears to be a reduction in stillbirths and losses of babies before weaning (**FIGURE 10.13B**) (Hohoff et al. 2003).

But polyandry may provide for more than just improved early survival for the offspring of females that mate with several males. In the dark-eyed junco (*Junco hyemalis*), a common North American songbird, not only are the genetic offspring of females and their extra-pair mates more likely to survive to reproduce, but when they do reproduce, they do better than the offspring of females and their social partners (Gerlach et al. 2012). The sons of extra-pair liaisons appear to boost their lifetime reproductive success by mating with females outside of the pair-bond, and the daughters of these arrangements are more fecund.

Because extra-pair male juncos provide only genes to their sexual partners, the fitness benefits that females gain from polyandry appear to be due to those genes. However, there is an alternative explanation for effects of this sort, namely that when a female songbird mates with an extra-pair male, she subsequently invests more in the offspring of such a relationship. For example, females can, and sometimes do, make larger than usual eggs to be fertilized by the sperm of their extra-pair mates. The greater supply of nutrients can give extra-pair offspring an early developmental boost, enabling these young to become high quality sons and fecund daughters (Tschirren et al. 2012). In other words, in some species with polyandrous females, the production of fitter offspring is due to the contributions of the mother to her progeny (referred to as maternal effects; see Chapter 3), not because females have acquired good genes from one or more males.

The superb fairy-wren is an animal for which the good genes hypothesis is a plausible explanation for polyandry. We introduced this bird earlier in the chapter when we told you that it has the highest rate of extra-pair paternity of any avian species studied. In this colorful Australian cooperative breeder (see Chapter 13), a socially bonded pair lives on a territory with several subordinate helpers, usually males. When the breeding female is fertile, she regularly leaves her family before sunrise and travels to another territory, where she often mates with a dominant male before returning home (**VIDEO 10.1**) (Double and Cockburn 2003). Tiny radio transmitters attached to the females revealed what they were up to, and genetic analyses of offspring demonstrated that the females' social partners typically lost paternity to distant rivals that congregated to mate with females (Cockburn et al. 2009). When social mates produced offspring, it was generally because they sired extra-pair young with

VIDEO 10.1

Male superb fairy-wren in the morning

ab11e.com/v10.1

(A)

(B)

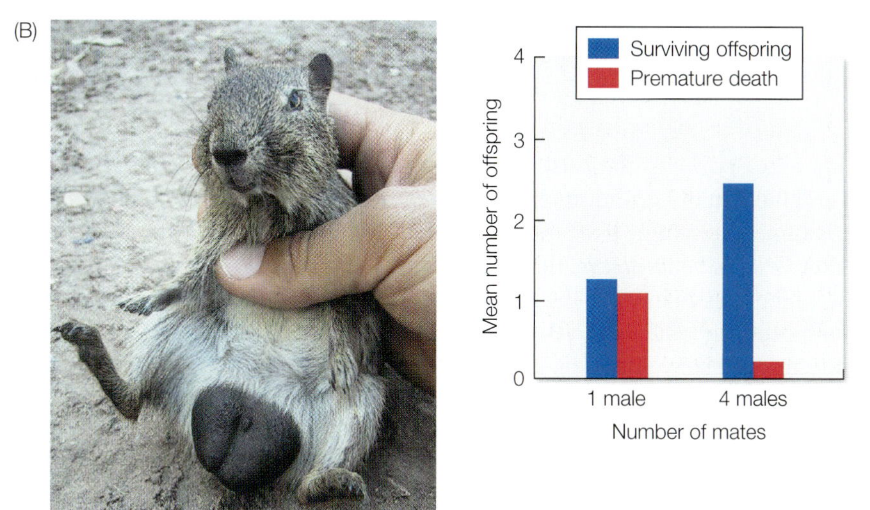

FIGURE 10.13 Polyandry has fitness benefits. (A) Females of the grey foam nest treefrog lay clutches of eggs that may be fertilized by anywhere from 1 to 12 or more males. The offspring of polyandrous females are more likely to survive than are those of females mated to only one male. (B) A male yellow-toothed cavy, whose very large testes speak to the intense sperm competition in this species, which occurs because females usually mate with several males. When females were experimentally restricted to a single mate, they had fewer surviving offspring than females allowed to mate freely with four males. (A, photograph © Michael and Patricia Fogden/Minden Pictures; B after Keil and Sachser 1998; photograph by Matthias Asher, courtesy of Norbert Sachser.)

females other than their social partner (**FIGURE 10.14**) (Webster et al. 2007). For example, 10 of the 14 fledglings fathered by one dominant male were the result of extra-pair copulations, which were all cared for by other males.

Thus, for both dark-eyed juncos and superb fairy-wrens, a major component of the fitness benefits gained by females through extra-pair copulations could be the production of sons that are so attractive that they repeat the extra-pair success of their fathers (Weatherhead and Robertson 1979). If sons inherit the very traits that made their fathers sexually appealing (and if their daughters inherit an affinity

FIGURE 10.14 Extra-pair paternity and male reproductive success. In the superb fairy-wren, males vary greatly in their reproductive success in part because of differences in the number of offspring they have with females other than their social partners. Shown are the number of young produced per year by dominant males with social partners and by auxiliary helper males that live in groups with a "breeding" pair. (After Webster et al. 2007; photograph © sompreaw/Shutterstock.com.)

for seeking extra-pair copulations), these "sexy sons" increase the odds that their mothers will have many grandoffspring. As required by this argument, male attractiveness is hereditary in at least some species (Taylor et al. 2007). Indeed, in the field cricket *Gryllus bimaculatus*, the sons of sexually successful males were more appealing to females than were the sons of males that failed to acquire a mate in an experimental setting (**FIGURE 10.15**) (Wedell and Tregenza 1999).

To really test the sexy son idea, one needs to rule out potential maternal effects driving the pattern. Helga Gwinner and Hubert Schwabl did exactly that in European starlings by studying sons of monogamous and polygynous males, investigating both their nest-site-acquisition behavior and the hormones in their eggs (Gwinner and Schwabl 2005). Consistent with the idea that polygynous males were of higher quality, the researchers found that sons of polygynous fathers defended more nest boxes and produced more courtship song than did sons of monogamous fathers. However, the researchers found that maternally derived levels of steroid hormones (androstenedione, 5α-dihydrotestosterone, testosterone, and estradiol) did not differ in eggs fathered by polygynous versus monogamous males. Together

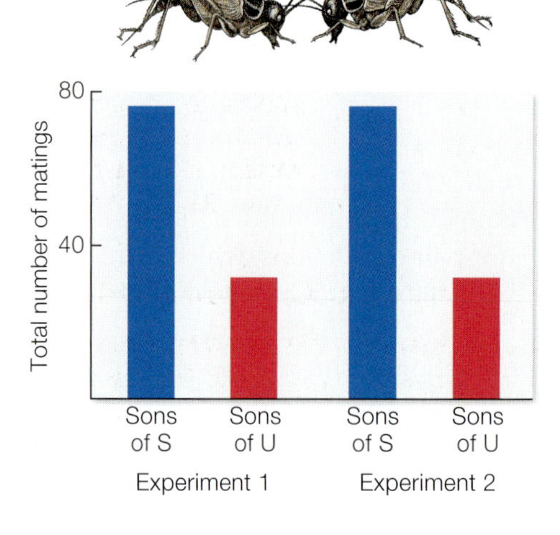

FIGURE 10.15 A father's mating success can be transmitted to his sons. In experiment 1, two male field crickets were given an opportunity to compete for a female; one male (S) mated successfully, while the other male (U) was unsuccessful. When the sons of male S were placed in competition for a female with the sons of male U (which had been given a female to mate with after failing to win the initial competition), the sons of S were about twice as likely to mate with the female as were the sons of U. In experiment 2, a male that had won a mating competition was later allocated a female at random for breeding, as was a male that had lost the competition. The sons of the two males were then placed in an arena with a female, and as before, the sons of S were much more likely to mate with the female than were the sons of U. (After Wedell and Tregenza 1999.)

new males rather than with previous partners (provided that the sperm of these new mates are more likely to fertilize her eggs). Indeed, when a female was given an opportunity to mate with the same male 90 minutes after an initial copulation, she refused to accept his spermatophore in 85 percent of the trials. Instead, if the partner was new to her, she would usually accept his gametes (Zeh et al. 1998).

DIVERSE GENES One reason that females might choose more genetically compatible males is to produce offspring that are more heterozygous (Westneat et al. 1990). According to this **genetic diversity hypothesis**, females benefit by increasing the heterozygosity (genetic diversity) of either individual offspring or of the group of offspring produced in a single bout. At the individual level, those offspring with two different forms of a given gene often enjoy an advantage over homozygotes, which are more likely to carry two recessive alleles that might cause defects (see Fossøy et al. 2008). At the group level, a greater diversity of individuals might allow at least some members of a colony to withstand a disease outbreak or some offspring produced in a single breeding bout survive in a range of environmental conditions. If genetically similar pairs are in danger of producing inbred offspring with genetic defects, then we can predict that females that are socially bonded with genetically similar individuals will be prime candidates to mate with other males that are either genetically dissimilar or highly heterozygous themselves (which means these males are likely to carry rare alleles that, when donated to offspring, will increase the odds that the offspring will also be heterozygous) (Tregenza and Wedell 2000).

genetic diversity hypothesis Females mate polyandrously to increase the heterozygosity (genetic diversity) of either individual offspring or of the group of offspring produced in a single bout.

The genetic diversity hypothesis has been tested in several bird species. For example, Dustin Rubenstein found that when female superb starlings (*Lamprotornis superbus*) went looking for partners outside their social groups, they tended to have social mates that were less heterozygous than they were (Rubenstein 2007). The offspring fathered by the extra-pair males were more heterozygous on average than were those produced by the females' social partners (**FIGURE 10.17**). In the polyandrous songbird the bluethroat (*Luscinia svecica*), extra-pair offspring were also more heterozygous than those sired by the females' social partners (Fossøy et al. 2008). In this study, the authors went further and examined whether offspring heterozygosity enhanced a fitness-related trait: immune function. Indeed, extra-pair offspring enjoyed stronger immune systems, a point established by injecting a foreign substance that triggers an

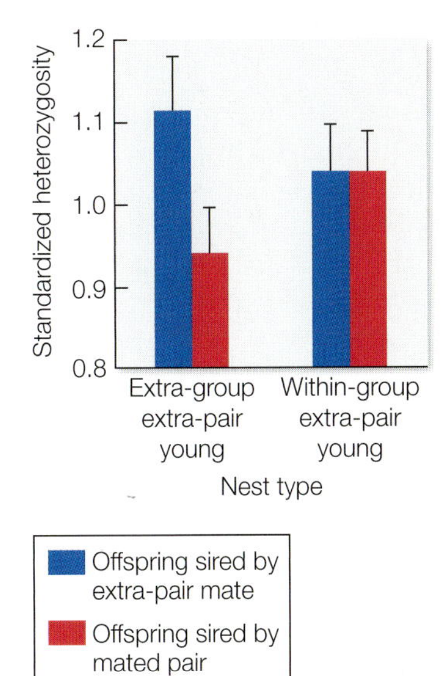

Superb starling

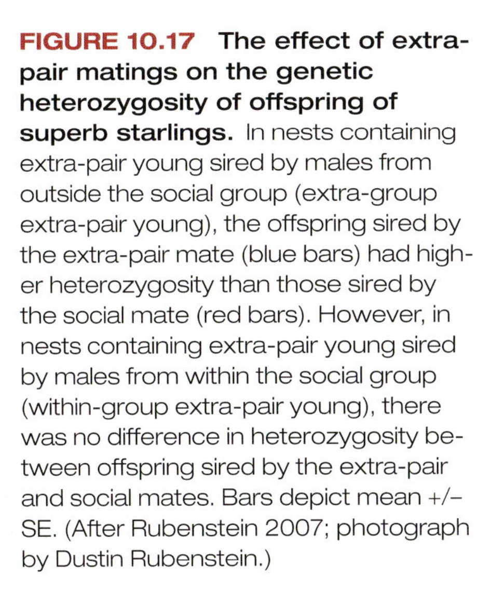

FIGURE 10.17 The effect of extra-pair matings on the genetic heterozygosity of offspring of superb starlings. In nests containing extra-pair young sired by males from outside the social group (extra-group extra-pair young), the offspring sired by the extra-pair mate (blue bars) had higher heterozygosity than those sired by the social mate (red bars). However, in nests containing extra-pair young sired by males from within the social group (within-group extra-pair young), there was no difference in heterozygosity between offspring sired by the extra-pair and social mates. Bars depict mean +/– SE. (After Rubenstein 2007; photograph by Dustin Rubenstein.)

Male bluethroat

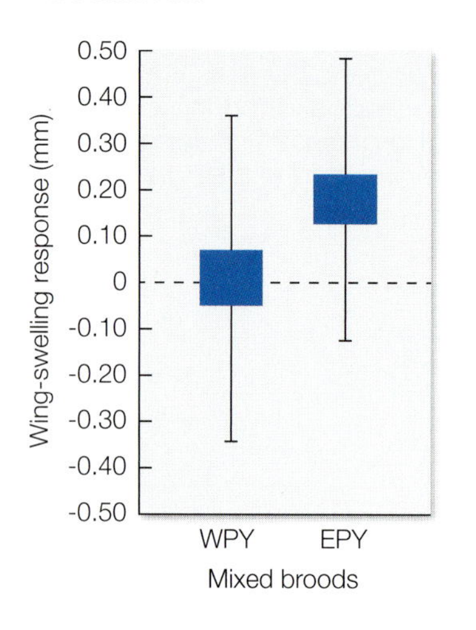

FIGURE 10.18 Extra-pair matings can boost the immune responses of offspring in the bluethroat. The mean wing-swelling response (an indicator of immune system strength) of young sired by their mother's social partner (within-pair young, or WPY) was less than that of young birds sired by their mother's extra-pair partner (extra-pair young, or EPY) in broods of mixed paternity. Bars depict mean +/– SE. (After Johnsen et al. 2000; photograph © iStock.com/PrinPrince.)

immune response into the wing webs of nestling bluethroats and then measuring the swelling at the injection site (**FIGURE 10.18**).

We have now seen evidence from both insects and birds that females of some species choose males that may be genetically compatible or that help them produce more heterozygous offspring (or both). Does this also occur in mammals? Male mandrills (*Mandrillus sphinx*) are stunning monkeys with bright red and blue facial skin. Female mandrills tend to prefer mates with brightly colored faces, suggesting that this may be a sexually selected honest signal (Setchell 2005). Moreover, females tend to reproduce with genetically compatible males, namely those whose MHC (major histocompatibility complex) genes are dissimilar to their own, the better to produce heterozygous offspring with stronger immune systems (Setchell et al. 2010). As we mentioned in Chapter 9, MHC genes code for proteins that contribute to immune system functioning, so mate choice by mandrills appears to have converged on adaptive criteria similar to those used by bluethroats.

The role of MHC variation in mate choice decisions has been examined in the Seychelles warbler (*Acrocephalus sechellensis*) by David Richardson and colleagues, who found that females were more likely to obtain extra-pair fertilizations when their social mate had low MHC diversity (Richardson et al. 2005). Female Seychelles warblers chose extra-pair males that had higher MHC diversity than the cuckolded males, producing extra-pair offspring with higher MHC diversity than within-pair offspring. But do these patterns translate into fitness benefits? In a 10-year study, Richardson's team found a positive association between MHC diversity and juvenile survival, suggesting that extra-pair fertilizations do produce genetic benefits in the form of a more diverse immune system for offspring that positively influence survival (Brouwer et al. 2010).

The genetic diversity hypothesis has been extended in eusocial insects, in which polyandry is common (see Chapter 12). For example, some members of the genus *Apis*, a group to which the familiar honey bee *A. mellifera* belongs, are highly polyandrous, with females of *A. florea* accepting and using sperm from about 10 males on average and *A. dorsata* females acquiring sperm from an average of 63 males (Seeley and Tarpy 2007). As we mentioned earlier, virgin queens in this genus fly out from the hive on one or more nuptial flights, during which time they are "captured" and mated by several males in quick succession. What makes this business particularly puzzling is that in most bees other than those in the genus *Apis*, including some from large eusocial colonies, the queens mate just once (Strassmann 2001). In the typical bee, a single mating suffices to supply the female with all the sperm she will need for a lifetime of egg fertilizations (queens of stingless bees can live for several years), thanks to her ability to store and maintain the abundant sperm she receives from her single partner. The widespread nature of bee monogamy indicates that this trait was the ancestral mating pattern, with polyandry in the genus *Apis* a more recently derived characteristic.

What selective pressures might have led to the replacement of monogamy by extreme polyandry in some bee species? Benjamin Oldroyd and Jennifer Fewell reviewed a dozen potential answers to this question, ranging from hypotheses about individual-level benefits (as we discussed above in all of our examples from other species) such as that long-lived queens do not run out of sperm, to those of colony-level benefits whereby the genetic diversity among individuals in a colony is more important than the diversity within any single individual (see Chapter 12) (Oldroyd and Fewell 2007). Potential colony-level benefits of increased genetic diversity include enhanced disease or parasite resistance and increased productivity because of a varied workforce with a range of skills. Some of the best support for these group-level benefits of genetic diversity includes the results of an experiment in which some honey bee colonies were set up with queens that had been artificially inseminated just once and some set up with queens that had been supplied with the sperm of ten drones. Both types of colonies were then infected with a bacterium that causes the disease American foulbrood, a killer of honey bee larvae. The colonies with polyandrous queens not only had milder forms of foulbrood but also had larger populations of workers on average than the colonies headed by once-mated queens (Seeley and Tarpy 2007).

But what is the mechanism underlying this colony-level immunity? A group led by David Tarpy explored the idea that colony-level disease resistance in honey bees is the result of *individual* differences in innate immunity, or the nonspecific defense mechanisms of the immune system (Simone-Finstrom et al. 2016). By extracting mRNA from pools of larvae and measuring transcript levels of the genes encoding the antimicrobial peptides hymenoptaecin and abaecin, the researchers found that within-colony variance among samples in the upregulation of hymenoptaecin, but not of abaecin, in response to immune challenge significantly decreased with increasing genetic diversity. In other words, more genetically diverse colonies exhibited less variation in individual immune responses. Thus, individual larval immunocompetence alone may be the mechanism explaining colony-level immunity in honey bees and other social insects.

The experimental results described here are supportive of an anti-disease function associated with queen polyandry in the honey bee. However, the advantages shown by the colonies with polyandrous queens might have arisen in part or in whole from the ability of a more genetically diverse worker force to carry out the full range of worker tasks more efficiently than a less diverse worker force. To test this idea, Heather Mattila and Thomas Seeley again used artificial insemination techniques to create two categories of colonies: those headed by single-insemination queens and those whose queens had received sperm from many males (Mattila and Seeley 2007). The colonies were kept free from disease through antibacterial and antiparasitic treatments to eliminate this variable from the experiment. The genetically uniform and genetically diverse colonies were also monitored to measure such things as the total weight of the bees, the foraging rates of workers, and the mean area of comb in the hive (where offspring are reared and food is stored) (**FIGURE 10.19**). None of the genetically uniform colonies made it through the winter, whereas about one-fourth

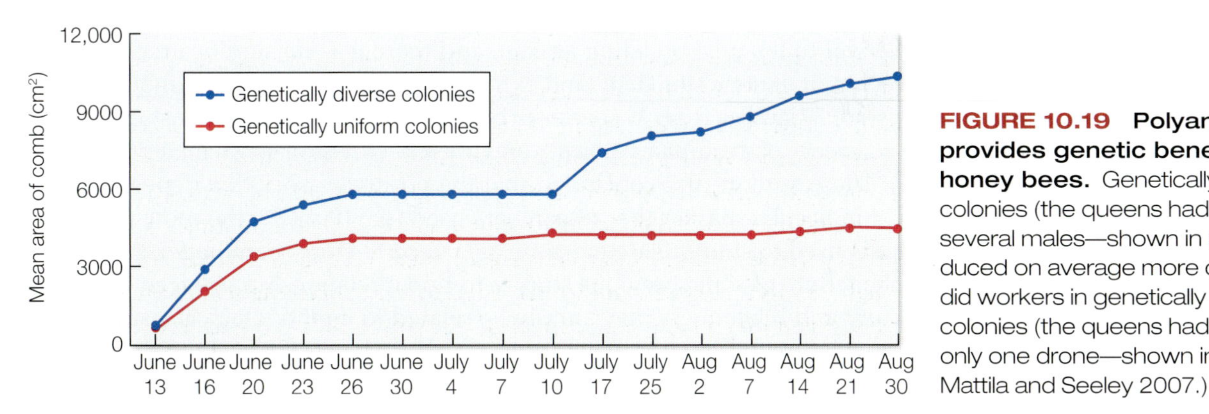

FIGURE 10.19 Polyandry provides genetic benefits in honey bees. Genetically diverse colonies (the queens had mated with several males—shown in blue) produced on average more comb than did workers in genetically uniform colonies (the queens had mated with only one drone—shown in red). (After Mattila and Seeley 2007.)

of the colonies led by polyandrous queens survived. These and other similar findings demonstrate that the colony's genetic diversity, which is boosted when the queen mates with several males, elevates the fitness of the queen. Thus, the puzzle of polyandry for the honey bee and its relatives is at least partly solved by the demonstration that a genetically diverse workforce is better than one of limited genetic variety.

inbreeding avoidance hypothesis In the context of mating, females mate polyandrously to avoid inbreeding with their social partner.

INBREEDING AVOIDANCE Both the genetic compatibility hypothesis and the genetic diversity hypothesis may ultimately be related to the idea that females choose extra-pair mates to avoid inbreeding, which would likely result in the production of less heterozygous offspring. Studies from two species of fairy-wrens are consistent with this **inbreeding avoidance hypothesis**. Work in red-backed fairy-wrens (*Malurus melanocephalus*) showed that females paired to genetically similar males were more likely to produce young sired by extra-pair males, and that those offspring were less inbred (more heterozygous) than within-pair offspring (Varian-Ramos and Webster 2012). A study in purple-crowned fairy-wrens (*Malurus coronatus*) enabled researchers to go a step further and more directly test the effects of inbreeding on the likelihood of extra-pair matings because many pairs in this species are incestuous. Because of limited opportunities for dispersal, parents in this species are often forced to pair-bond with their own offspring if their mate dies, though these incestuous pairings typically result in divorce. Before these pairings end in divorce, however, Sjouke Kingma and colleagues found that these incestuous pairs also motivate females to engage in frequent extra-pair fertilizations (Kingma et al. 2014). Indeed, 46 percent of incestuous pairs resulted in extra-pair offspring versus only 3 percent in non-incestuous pairs, suggesting that avoiding inbreeding plays a role in female reproductive decisions in this species.

In summary, we have now seen several examples where multiple mating by females may result in indirect genetic benefits to a female's offspring. This includes studies comparing fitness-related traits in extra-pair versus social mates, extra-pair paternity rates in pairs that vary in their degree of genetic similarity or inbreeding, and genetic diversity or fitness-related traits in extra-pair versus within-pair offspring (Forstmeier et al. 2014). Yet there are also many studies that have failed to find such relationships, and meta-analyses have for this reason also generated mixed results. For example, an analysis of 16 avian species found that polyandry was not significantly beneficial for any single offspring performance trait (growth rate, survival, adult size) (Slatyer et al. 2011). Similarly, an analysis of 55 avian species by Erol Akçay and Joan Roughgarden found little support for the good genes or genetic compatibility hypotheses in more than half of the species studied (Akçay and Roughgarden 2007). Although these researchers showed that extra-pair males are on average larger and older than within-pair males, males were not different in terms of secondary sexual traits, body condition, or genetic relatedness to the female. Moreover, no difference was found between extra-pair and within-pair young in survival to the next breeding season, and there was no significant correlation between pair genetic similarity and extra-pair paternity. However, another meta-analysis of 39 studies from 33 species of birds found a positive relationship between the occurrence of extra-pair paternity and the relatedness of social mates (Arct et al. 2015). These authors also concluded that the positive association depends on the type of molecular marker that researchers used (see Box 12.2; the same molecular methods used to study relatedness are also used to study parentage), suggesting that the failure of some previous studies to find a relationship between extra-pair paternity and relatedness may simply be related to methodological issues. Thus, there is at least some good evidence from a variety of species to suggest that indirect benefits promote the evolution of polyandrous mating systems.

Polyandry and Direct Benefits

Our focus thus far on the potential indirect genetic benefits of polyandry should not obscure the possibility that females sometimes mate with several males to secure direct benefits rather than genes alone. These direct benefits can take many forms, including access to additional resources, access to additional parental care, better protection from sexually harassing males (see below), and reduced risk of infanticide (**HYPOTHESES 10.3**). Below, we discuss each of these potential direct benefits in greater detail.

HYPOTHESES 10.3

Non-mutually exclusive hypotheses to explain the potential direct benefits that females receive from being polyandrous

additional resources hypothesis Females mate polyandrously to gain access to additional resources from their partner.

additional care hypothesis Females mate polyandrously to gain more caregivers to help rear young.

infanticide reduction hypothesis Females mate polyandrously to create greater uncertainty about the paternity of offspring in order to reduce the risk of infanticide.

ADDITIONAL RESOURCES Females often mate multiply to gain access to resources that are necessary for successful reproduction (**additional resources hypothesis**). As we discussed in Chapter 9, female red-winged blackbirds (*Agelaius phoeniceus*) may be allowed to forage for food on the territories of males with which they have engaged in extra-pair copulations, whereas genetically monogamous females are chased away (Gray 1997). Similarly, females of some bees must copulate with territorial males each time they enter a territory if they are to collect pollen and nectar there (**FIGURE 10.20**) (Alcock et al. 1977). It is even possible that males of some insects pass sufficient fluid in their ejaculates to make it worthwhile for water-deprived females to copulate in order to combat dehydration (Edvardsson 2007).

Females in many insect species have an incentive to mate several times in order to receive direct benefits in the form of food gifts such as nutritious spermatophores from their partners. The spermatophores of highly polyandrous butterfly species contain more protein than the spermatophores of monogamous species (Bissoondath and Wiklund 1995). Males of polyandrous butterfly species could bribe females into mating with them by providing them with nutrients that can be used to make more eggs. In some of these species, the male's spermatophore can represent up to 15 percent of his body weight, making him a most generous mating partner. If this argument is valid, then the more polyandrous a female is, the greater her reproductive output should be. Christer Wiklund and his coworkers used the comparative

FIGURE 10.20 Polyandry can yield direct benefits in bees. By mating with many males, females of this megachilid bee gain access to pollen and nectar in those males' territories. (Photograph by John Alcock.)

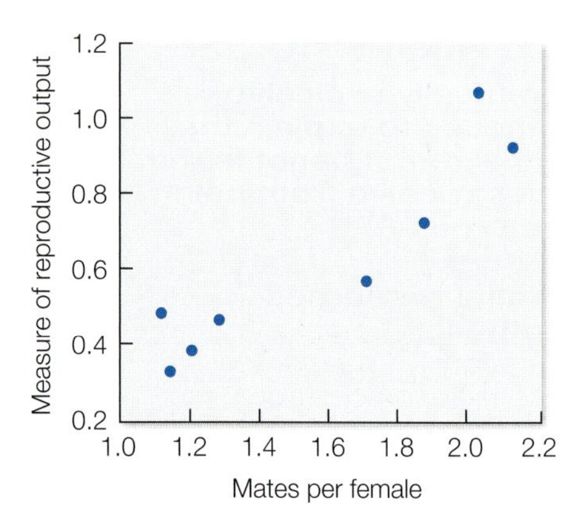

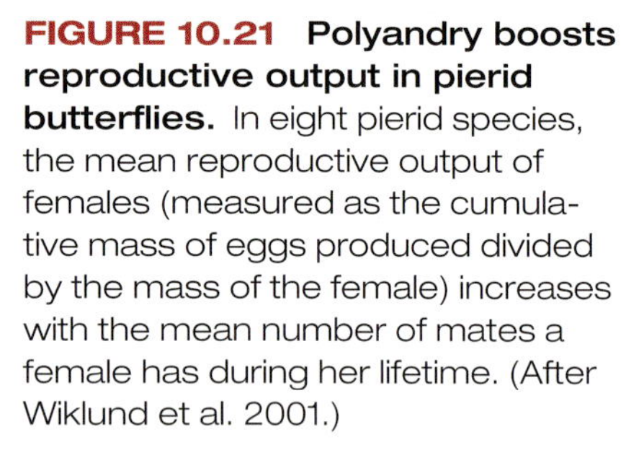

FIGURE 10.21 Polyandry boosts reproductive output in pierid butterflies. In eight pierid species, the mean reproductive output of females (measured as the cumulative mass of eggs produced divided by the mass of the female) increases with the mean number of mates a female has during her lifetime. (After Wiklund et al. 2001.)

approach to test this prediction by taking advantage of the fact that several closely related butterfly species in the same family (Pieridae) differ substantially in the number of times they copulate over their lifetimes. The researchers used the mean number of copulations for females of each species as a measure of its degree of polyandry. To quantify reproductive output, they measured the mass of eggs produced by a female in the laboratory and divided that figure by the body mass of the female in order to control for differences among species in weight. As predicted, across these eight butterfly species, the more males a female mated with on average, the more spermatophores she received, and the greater her production of eggs (**FIGURE 10.21**) (Wiklund et al. 2001).

A female's desire for nutrient-rich spermatophores seems to have been taken to the extreme in a genus of sex-role reversed cave insects. In *Neotrogla* barklice that live inside Brazilian caves, females have an elaborate, penislike structure called a gynosome, whereas males lack an intromittent organ altogether (**FIGURE 10.22**) (Yoshizawa et al. 2014). Some *Neotrogla* species have a gynosome that is covered in bristled spines, likely the coevolutionary result of intense sexual conflict between males and females (see Figure 10.22C). Females insert this weaponlike gynosome deep into the male to receive a spermatophore and anchor their bodies together for up to 3 days of copulation. Males appear to be incapable of breaking the bond once copulation begins, as researchers' attempts to disconnect the pair resulted in severing the male before severing the gynosome connection. Since all known *Neotrogla* species live in caves and feed on scarce bat carcasses and guano, nutritious nuptial gifts may cause strong selection for multiple mating by females. Over the course of her lifetime, a polyandrous female may receive up to 11 spermatophores. Thus, in a system where spertmatophores could mean life or death for hungry females, manipulating multiple males to mate may be one way to obtain this type of potentially lifesaving direct benefit.

additional care hypothesis Females mate polyandrously to gain more caregivers to help rear young.

ADDITIONAL CARE In some animals, polyandry enables females to indirectly access resources through greater parental assistance from their several mates (**additional care hypothesis**). For example, in the dunnock (*Prunella modularis*), a small European songbird, a female that often lives in a territory controlled by one (alpha) male may actively encourage another, subordinate male to stay around by seeking him out and copulating with him when the alpha male is elsewhere. Female dunnocks are prepared to mate as many as 12 times per hour, and hundreds of times in total, before laying a complete clutch of eggs. The benefit to a polyandrous female of distributing her copulations between two males is that both sexual partners will help her rear her brood—provided that they have both copulated often enough with her. Female dunnocks ensure that this paternal threshold is reached by actively soliciting matings from whichever male, alpha or beta—usually the latter—has had less time in their company (**FIGURE 10.23**) (Davies 1985). Moreover, in a population of introduced dunnocks in New Zealand, polyandrous groups had higher hatching and fledging success than monogamous pairs (Santos and Nakagawa 2013).

Likewise, a female superb starling that is pair-bonded with one male sometimes mates with another unpaired male in her social group; the second male then sometimes helps rear the brood of his mate (Rubenstein 2007). By copulating with an additional male, the polyandrous female has in effect made it advantageous for the "other male" to invest parentally in her offspring. Recall from earlier that female superb starlings also gain indirect genetic benefits in the form of increased offspring heterozygosity when they mate with males from outside their social group—males

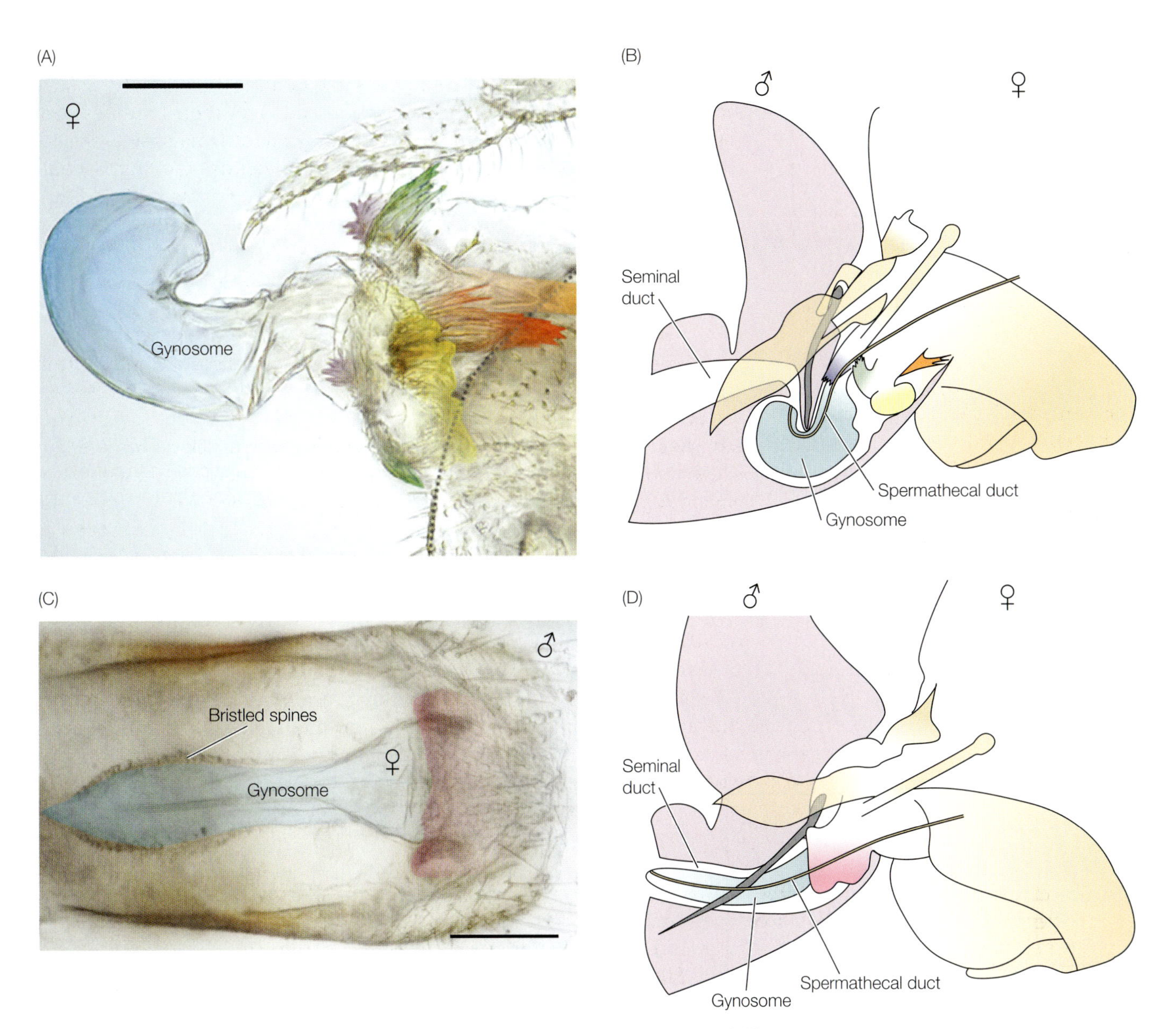

FIGURE 10.22 Female *Neotrogla* barklice have an elaborate penislike structure called a gynosome, whereas males lack a penis altogether. (A) A gynosome of a female *Neotrogla curvata* (B) inserted into a male. (C) A gynosome of a female *N. truncata* (D) inserted into a male. Notice the bristled spines on the tip of the gynosome in (C). Scale bars represent 0.1 mm. (From Yoshizawa et al. 2014.)

that would be unable to contribute any parental care (Rubenstein 2007). Dunnocks too receive some indirect genetic benefits in the form of an amelioration of the negative effects of inbreeding (Santos et al. 2015), in addition to the direct benefits. Together, these studies remind us that females can and do receive both direct and indirect benefits from mating with several males.

REDUCED INFANTICIDE It is also possible that by mating with several males a female may encourage all of her sexual partners to leave her next newborn alone (**infanticide reduction hypothesis**). Potentially infanticidal males of the Hanuman langur (*Semnopithecus entellus*) generally do ignore a female's baby if they have mated with the mother prior to the birth of the infant (Borries et al.

infanticide reduction hypothesis Females mate polyandrously to create greater uncertainty about the paternity of offspring in order to reduce the risk of infanticide.

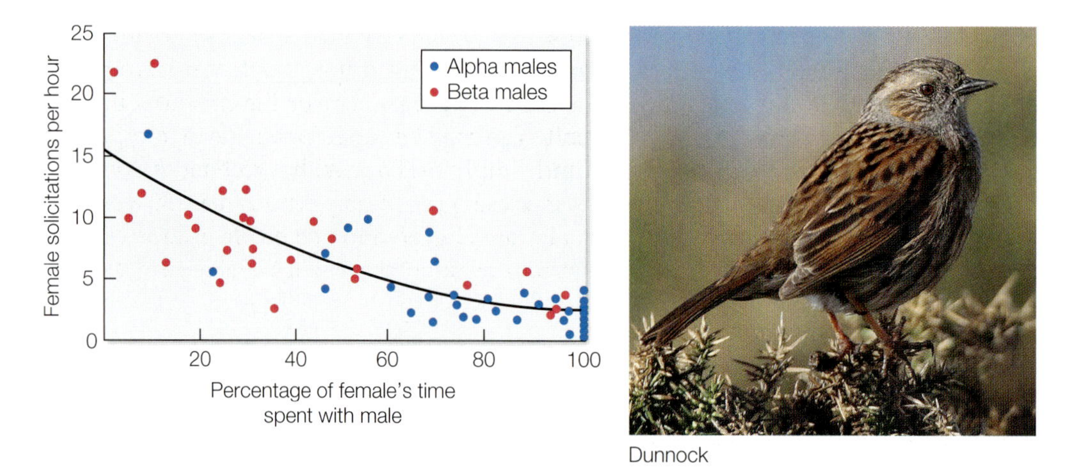

FIGURE 10.23 Adjustment of copulation frequency by polyandrous female dunnocks. A female living in a territory with two males solicits copulations more often from the male that has spent less time with her, whether that is the alpha or the beta male. (After Davies et al. 1996; photograph © Joe Gough/ShutterStock.com.)

1999). The fact that females will copulate with more than one male even when they are not ovulating—in fact, even when they are pregnant—suggests that polyandry promotes the female langur's interests by lowering the risk of infanticide, another type of direct benefit (Heistermann et al. 2001). It is important to recall from earlier in this chapter that infanticide risk is also a potential reason why females mate monogamously.

HYPOTHESES 10.4

Alternative hypotheses to explain why males are polygynous

female defense polygyny hypothesis When resources are evenly distributed in space but females form groups to better access those resources or to help dilute the risk of predation, males will follow and guard a group of females.

resource defense polygyny hypothesis When resources are clumped, attract multiple females, and are easily defensible, males will guard the resources—and by extension the females—by setting up a territory.

lek polygyny hypothesis When resources are distributed heterogeneously and females are widespread and do not form groups, males will wait for females to come to them.

scramble competition polygyny hypothesis When resources are distributed heterogeneously and females are widespread and do not form groups, males will seek out females.

Polygyny: Multiple Mating by Males

We have discussed why females may mate with more than one male, but let's now consider multiple mating in males (polygyny), which is by far the most common form of mating system in nature. One of the main themes of mating system theory in general is that in the vast majority of species, female resource needs and reproductive behaviors create the circumstances that determine which competitive and display maneuvers will provide payoffs for males. As a result, male mating tactics are an evolved response to female reproductive needs and to the ecological factors that determine the spatial distribution of receptive females. **HYPOTHESES 10.4** lists four hypotheses to explain why males are polygynous. When resources are evenly distributed in space but females form groups to better access those resources or to help dilute the risk of predation, males can simply follow and guard a group of females, more easily gaining access to reproductively ready females at the appropriate time (**female defense polygyny hypothesis**). In contrast, when resources are clumped, attract multiple females, and are easily defended by males, males will simply guard the resources—and by extension the females—by setting up a territory (**resource defense polygyny hypothesis**). However, when resources are distributed heterogeneously and females are widespread and do not form groups, the only chance a male may have to find more than one female is to wait for females to come in search of him and his competitors (**lek polygyny hypothesis**) or simply to seek them out (**scramble competition polygyny hypothesis**). And of course, when other factors such as the costs of parental care or the difficulty in finding even a single mate become prohibitive, it may be better for a male to forego trying to reproduce with many females and simply find one with which he can be monogamous. Ultimately, the differences in these types of polygynous mating systems stem from the degree to which females are clustered in defensible groups as a result of predation pressure or food distribution, a point we will develop further below (Emlen and Oring 1977).

Female Defense Polygyny

When receptive females occur in defensible clusters, males will compete directly for those clusters and female defense polygyny will result. For example, nests of *Cardiocondyla* ants produce large numbers of virgin queens, leading to lethal combat among the males of the colony, the survivors of which get to copulate with dozens of freshly emerged females (Heinze et al. 1998). But one does not have to be hyperaggressive to practice female defense polygyny. For example, males of tiny siphonoecetine amphipods construct elaborate cases composed of pebbles and fragments of mollusk shells found in the shallow ocean bays where they live. They move about in these houses and capture females by gluing the females' houses to their own, eventually creating an apartment complex containing up to three potential mates (**FIGURE 10.24**) (Just 1988).

Female defense polygyny is also quite common in vertebrates, particularly in species in which females form groups, which makes defending multiple females at the same time much easier. Indeed, social monogamy in mammals never occurs when females live in groups (Komers and Brotherton 1997). In mammals as different as bighorn sheep (*Ovis canadensis*) and western gorillas (*Gorilla gorilla*), groups of females that form in part for protection against predators attract males that compete to control sexual access to the entire group (Harcourt et al. 1981). Likewise, male lions (*Panthera leo*) fight over prides of females, which have gathered together for defense of permanent hunting territories and protection against infanticidal males (Packer et al. 1990). Thus, given the existence of female groups, female defense

FIGURE 10.24 Female defense polygyny in a marine amphipod. (A) A male without his house. (B) A male that has glued the houses of two females to his house. (Drawings courtesy of Jean Just.)

FIGURE 10.25 Female defense polygyny in the greater spear-nosed bat (*Phyllostomus hastatus*). A large male (bottom right) guards a roosting cluster of smaller females. A successful male may sire as many as 50 offspring with his harem females in a year. (Photograph © Merlin D. Tuttle/Science Source.)

polygyny is the standard male tactic in mammals (**FIGURE 10.25**) (McCracken and Bradbury 1981).

The same concept also applies to birds that form groups. For example, males of the Montezuma oropendola (*Psarocolius montezuma*), a tropical blackbird, try to control clusters of nesting females, which group their long, dangling nests in specific trees. The dominant male at a nest tree may secure up to 80 percent of all matings by driving subordinates away (Webster 1994). You might think that males guard critical nesting trees that females need to breed. However, these dominant males shift from nest tree to nest tree, following females rather than defending a nesting resource per se, demonstrating that they are employing the female defense tactic (rather than a resource defense tactic) when the level of competition makes this possible (**FIGURE 10.26**) (Webster and Robinson 1999).

In most polygynous species, it is critical to realize that it is the behavior of females that dictates the behavior of males, and ultimately the type of mating system that forms. Take, for example, the plains zebra (*Equus quagga*). Male plains zebras attempt to defend a harem of females that forms to reduce the risk of predation, as well as to avoid harassment by bachelor males that do not have their own harems. This type of female defense polygyny is sometimes referred to as harem defense polygyny because females are socially bonded together in harems, making it easier for males to monopolize a group. Like most other ungulates, female zebras need to spend most of their lives eating grass, which often requires covering great distances each day in search of enough forage. The average female zebra spends 65 percent of every hour grazing, totaling nearly 18 hours a day of feeding (Rubenstein 1994). In order to spend this much time foraging, females do not want to be disturbed, either through harassment by males or by having to keep an eye out for

FIGURE 10.26 Female defense polygyny in the Montezuma oropendola. (A) Male Montezuma oropendolas attempt to monopolize females in (B) small colonies of nesting females. (C) As colony size increases, mating attempts are often disrupted by rivals. (D) As a result of these disruptions, frequency of copulations per hour at the colony site decreases. (A, photograph © Eduardo Rivero/Shutterstock.com; B, photograph © Ammit Jack/Shutterstock.com; C, D after Webster and Robinson 1999.)

predators. It turns out that female zebras that associate with dominant males are able to feed for 10 percent longer per day than those that associate with subordinate males, since dominant males successfully keep other males away. In other words, females directly benefit from joining a group controlled by a dominant male, and he in turn benefits by mating with a large number of females in his harem. Thus, it is the resource needs of females—in the case of plains zebras, their need to feed—that dictate the formation of male-defended groups. Since males are unable to defend territories (and resources directly) because females need to cover great distances in order to find enough grass each day, they instead follow the females and chase away subordinate and bachelor males. Since females that group together have more eyes to spot predators, and those that associate with dominant (harem) males can forage longer without being harassed by other males, they are likely to join the harems of strong males (Rubenstein 1994).

Resource Defense Polygyny

In many animal species, females do not live together permanently, but a male may still become polygynous if he can control access to resources that females visit on occasion. This type of mating system occurs in Grevy's zebra (*Equus grevyi*), a close relative of the plains zebra. Female associations in Grevy's zebra are much shorter lived than those in the plains zebra, and groups of females are not overseen by a male. What explains this difference in group formation and mating system in the two species? To answer this question, Daniel Rubenstein considered the behavioral ecology of these grass-feeding species (Rubenstein 1994). Plains zebras occur in slightly wetter regions of Africa where grass and water sources are more readily accessible and evenly spread on the landscape, whereas Grevy's zebras live in more arid regions where most females wander far and wide in search of water and grass that is both scarce and more heterogeneously distributed. Although Grevy's females are arid-adapted animals, females that have recently given birth and have young foals need to drink daily, and therefore must remain closer to water than females without young. It is these postpartum females that are also sexually receptive and therefore most attractive to males. Plains zebra males can monopolize both receptive and (temporarily) nonreceptive females by guarding groups for extended periods of time. But since receptive Grevy's females rarely associate with nonreceptive females, it pays males to set up territories near water where the receptive females spend most of their time. Dominant Grevy's males are able to defend the best territories where more females are likely to be, and therefore are more likely to mate with the most females (Rubenstein 1994).

Once again, we can see how the distribution of females, which is affected by their resource needs and grouping decisions, dictates male reproductive decisions and the adoption of different forms of polygynous mating systems. Resource defense polygyny occurs not only in zebras and other mammals, but in many other vertebrates, including the African cichlid fish *Lamprologus callipterus*. Females of this species deposit a clutch of eggs in an empty snail shell and then swim inside to remain with the eggs and hatchlings until they are ready to leave the nest. Territorial males of this species, which are much larger than their tiny mates, not only defend suitable nest sites but also collect shells from the lake bottom and steal them from the nests of rival males to create middens of up to 86 shells (**FIGURE 10.27**). Because

(A)

(B)

FIGURE 10.27 Resource defense polygyny in the African cichlid fish *Lamprologus callipterus*. (A) A territorial male bringing a shell to his midden. The more shells there are in a male's midden, the more nest sites are available for females to use. (B) The tail of one of the territorial male's very small mates can be seen in a close-up of the shell (lower left) that serves as her nest. (Photographs by Tetsu Sato, Research Institute for Humanity and Nature, Japan.)

as many as 14 females may nest simultaneously in different shells in one male's midden, the owner of a shell-rich territory can have extraordinary reproductive success (Sato 1994).

Resource defense polygyny is also common in insects, in which a safe location for eggs often constitutes a defensible resource. Some male black-winged damselflies (*Calopteryx maculata*), for example, defend floating vegetation that attracts a series of sexually receptive females, each of which mates with the male and then lays her eggs in the vegetation he controls (Waage 1973). Likewise, male antlered flies (*Phytalmia* spp.) fight for small rot spots on fallen logs or branches of certain tropical trees because these locations attract receptive, gravid females that mate with successful territory defenders before laying their eggs (**FIGURE 10.28**) (Dodson 1997). Ultimately, the more of this resource a territorial male holds or the longer he holds it, the more likely he is to acquire several mates. This rule also applies to some Neotropical harvestman (*Pseudopucrolia* sp.), which are relatives of the more familiar daddy longlegs. One Brazilian species occupies and defends nest burrows in muddy banks that attract gravid, sexually receptive females, particularly if there are already eggs present being protected by a guarding male. In effect, the male's parental care, in addition to the safe burrow that he defends, constitutes a resource for receptive females, enabling some nest- and egg-guarding males to be polygynists (Nazareth and Machado 2010).

If the distribution of females is controlled by the distribution of key resources, and if male mating tactics are in part dictated by this fact of life, then it should to be possible to alter the mating system of a species by moving resources around, thereby altering where females are located. This prediction was tested by Nick Davies and Arne Lundberg, who first measured male territories and female foraging ranges in the dunnock. In this drab little songbird, females hunt for widely dispersed food items over such large areas that the ranges of two males may overlap with those of several females, creating a polygynous–polyandrous mating system (also known as polygynandry; see below). However, when Davies and Lundberg gave some females supplemental oats and mealworms for months at a time, female home ranges contracted substantially (Davies and Lundberg 1984). Those females with the most reduced ranges had, as predicted, fewer social mates than other females whose ranges had not diminished in size (**FIGURE 10.29**). In other words, as a female's home range contracted, the capacity of one male to monitor her activities increased, with the result that female polyandry tended to be replaced by female monogamy. These results support the general theory that males attempt to monopolize females within the constraints imposed on them by the spatial distribution of potential partners, which in turn may be affected by the distribution of food or other resources.

What should a female do when the selection of an already mated male means that she must share the resources under

FIGURE 10.28 Resource defense polygyny in an Australian antlered fly. Males compete for possession of egg-laying sites that can be found only on certain species of recently fallen trees. (Photograph by Gary Dodson.)

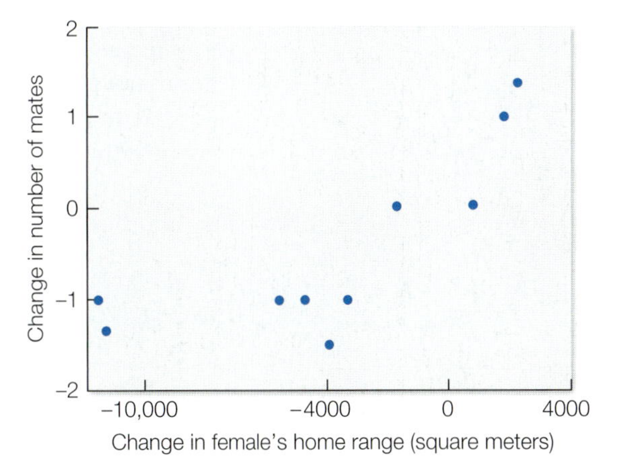

FIGURE 10.29 A test of the female distribution theory of mating systems. When food supplementation reduces the size of a female dunnock's home range, a male can monopolize access to that female and reduce the number of males with which she interacts. (After Davies and Lundberg 1984.)

his control with other females? As we noted in Chapters 7 and 9, there is a point at which a female can gain more by mating with a polygynist on a good territory than by pairing off with a male occupying a resource-poor or predator-vulnerable territory (Orians 1969). Stanislav Pribil and William Searcy tested the idea that there is a "polygyny threshold" (after Emlen and Oring's "polygamist threshold"; Emlen and Oring 1977) at which a female will gain higher fitness by mating with an already mated male than with a bachelor. The researchers took advantage of their knowledge that female red-winged blackbirds in an Ontario population almost always choose unmated males over mated ones. The females' usual refusal to mate with a paired male was adaptive, judging from the fact that on those rare occasions when females did select mated males, they produced fewer offspring than females that had their monogamous partners' territories all to themselves. Pribil and Searcy predicted that if they could experimentally boost the quality of territories held by already mated males while lowering the value of territories controlled by unmated males, then mate-searching females should reverse their usual preference. They tested this prediction by manipulating pairs of red-winged blackbird territories in such a way that one of the two sites contained one nesting female and some extra nesting habitat (added cattail reeds rising from underwater platforms), while the other territory had no nesting female but had supplemental cattail reed platforms placed on dry land. Female red-winged blackbirds prefer to nest over water, which offers greater protection against predators. For 14 pairs of territories, the first territory to be settled by an incoming female was the one where she could nest over water, even though this meant becoming part of a polygynous male's harem. But was there a fitness benefit to this choice of mate? Females that made this choice reared almost twice as many young on average as latecomers that had to make do with an onshore nesting platform in what became a monogamous male's territory (Pribil and Searcy 2001).

This work with red-winged blackbirds demonstrates that when there is a free choice between a superior territory and an inferior one, it can pay a female to pick the better site even if she has to share it with another female. But why, then, do some female red-winged blackbirds, as well as females of many other avian species, ever accept a second-rate territory, as this usually means that the second female rears fewer offspring than the first female? Svein Dale and colleagues found that unmated female pied flycatchers (*Ficedula hypoleuca*) visit many males and do not rush into a paired relationship, suggesting that when they choose an already mated male, they do so voluntarily and with full knowledge (Dale et al. 1992). Perhaps they choose these males anyway because of the high costs of finding other potential partners (Stenmark et al. 1988) or because the remaining unmated males have extremely poor territories (Temrin and Arak 1989). If this is true, the options for some females may be either to accept secondary status or to not breed at all, just as late-arriving female red-winged blackbirds may have to make do with what is available in their neighborhood, even if it means making the best of a bad situation.

Lek Polygyny

In some species, males do not defend groups of females or resources that several females come to exploit. Instead, they wait for females to come to them. These males fight to control a very small area that is used only as a display arena. These mini-territories may be clustered in a traditional display ground called a **lek**, or they may be somewhat scattered, forming a dispersed lek (sometimes called a hidden lek), as is the case for satin bowerbirds (see Chapter 9) (Bradbury 1981). Despite the fact that leks do not contain food, nesting sites, or anything else of practical utility, females come to them anyway. They come to choose a mate, receive his sperm, and then depart to rear the offspring alone. Leks are often established in places where females need to pass through, as in the rocky intertidal areas that provide access for feeding in Galápagos marine iguanas (*Amblyrhynchus cristatus*) (Partecke et al. 2002, Vitousek et al. 2008), or they may occur in places that are convenient for displaying to females. Take, for example, the white-bearded manakin (*Manacus manacus*) which displays to females on the rainforest floor. When female manakins arrive at a lek in a Trinidadian forest, they may find as many as 70 sparrow-sized males in an area only 150 meters square (**VIDEO 10.2**). Much like the golden-collared manakins (*Manacus vitellinus*) we introduced in Chapter 9, each male white-bearded manakin will have cleared the ground around a little sapling rising from the forest floor. The sapling and cleared court are the props for his display routine, which consists of rapid jumps between perches accompanied by loud sounds produced by snapping his clubbed primary wing feathers together. When a female is around, the male jumps to the ground with a snap and immediately back to the perch with a buzz, and then back and forth "so fast he seems to be bouncing and exploding like a firecracker" (Snow 1956). If the female is receptive and chooses a partner, she flies to his perch for a series of mutual displays followed by copulation. Afterward, she leaves to begin nesting, and the male remains at the lek to court newcomers. Alan Lill found that in a lek with 10 male manakins, where he recorded 438 copulations, 1 male achieved nearly 75 percent of the total copulations; a second male mated 56 times (13 percent), while 6 other males together accounted for only 10 matings (Lill 1974). Preferred males tend to occupy sites near the center of the manakin lek, and they engage in more aggressive displays than less successful males (Shorey 2002).

Huge inequalities in male mating success (high reproductive skew) are standard features of lekking species. In the topi (*Damaliscus lunatus*), a type of African antelope, older males tend to occupy the center of a lek, and therefore copulate much more often than younger rivals forced to the periphery (**FIGURE 10.30**) (Bro-Jørgensen and Durant 2003). Likewise, just 6 percent of the males in a lek of the West African hammer-headed bat (*Hypsignathus monstrosus*) were responsible for 80 percent of the matings recorded by Jack Bradbury (**FIGURE 10.31**). In this species, males gather in groups along riverbanks, each bat defending a display territory high in a tree, where he produces loud cries that sound like "a glass being rapped hard on a porcelain sink" (**AUDIO 10.1**) (Bradbury 1977). Receptive females fly to the lek

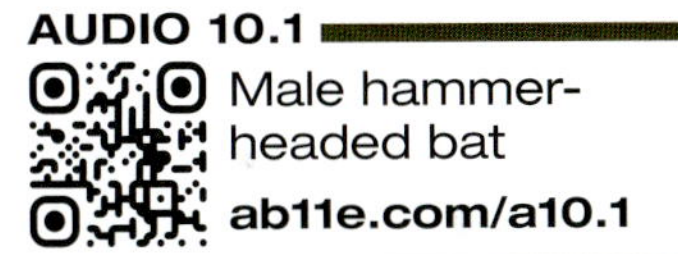

Male Topi

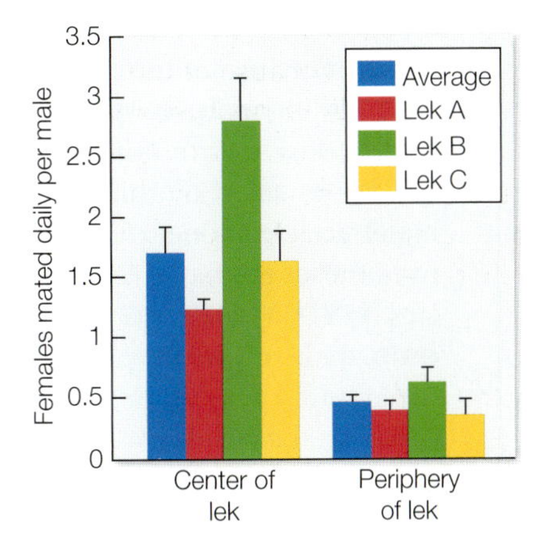

FIGURE 10.30 Mating success at topi leks. Each male topi in a central position at his lek mates with more females per capita than do males forced to peripheral sites. The blue bars represent the averages for the three leks shown. Bars depict mean +/– SE. (After Bro-Jørgensen and Durant 2003; photograph © J. Norman Reid/ShutterStock.com.)

FIGURE 10.31 A lek polygynous mammal: the hammer-headed bat. Males cluster in dense groups at mating arenas where they defend display territories in trees overhanging riverbanks. (Photograph © Ivan Kuzmin/Alamy Stock Photo.)

and visit several males, each of which responds with a paroxysm of wing-flapping displays and strange vocalizations.

Why do male hammer-headed bats, white-bearded manakins, and topi antelope behave the way they do? Bradbury has argued that lekking evolves only when other mating tactics do not pay off for males, largely due to a wide and even distribution of females (Bradbury 1981). Hammer-headed bats do not live in permanent groups but instead travel over great distances in search of widely scattered sources of food of unpredictable availability, especially figs and other tropical fruits. A male that tried to defend one tree might have a long wait before it began to bear attractive fruit, and when it did, the large amount of food would attract hordes of consumers, which could overwhelm the territorial capacity of a single defender. Thus, the feeding ecology of females of these species makes it hard for males to monopolize them, directly or indirectly. Instead, males display their merits to choosy females that come to leks to inspect them.

Just as the defense of resources or females was invoked to explain two types of defense polygyny (female versus resource), the absence of defensible clusters of females or resources has been invoked as an explanation for both nonterritorial scramble competition polygyny (see below) as well as for leks. Why some species subject to these conditions simply search for mates while others form elaborate leks is not completely clear. Nor is it known for certain why males of some lekking species congregate in the small areas of traditional leks as opposed to displaying solitarily in a dispersed lek (as in the satin bowerbird). One possibility is that congregated males may be relatives that assist one another in the attraction of mates (Concannon et al. 2012). Yet although we have long recognized that male relatives come together on leks in a variety of avian species (Höglund et al. 1999, Shorey et al. 2000), there appears to be little evidence that male relatedness drives lekking behavior in manakins (Loiselle et al. 2006), that male–male competition on a lek is correlated with relatedness in grouse (Lebigre et al. 2016), or even that female mate choice is based on relatedness to males in bowerbirds (Reynolds et al. 2014). Another possibility is that predation favors the adoption of a safety-in-numbers tactic (Gibson et al. 2002).

Ultimately, any reductions in the cost of displaying help explain why lekking males congregate, but we still need to know what benefit a male derives from defending a small display territory at a group lek. We touched on this idea earlier when we noted that only a few males on leks monopolize most of the matings. Here we review three hypotheses related to female choice of these top males in lekking species (**HYPOTHESES 10.5**). According to the **hotspot hypothesis**, males cluster in places ("hotspots") where the routes frequently traveled by receptive females intersect (Bradbury et al. 1989). In contrast, the **hotshot hypothesis** argues that subordinate males cluster around highly attractive males to have a chance to interact with females drawn to these "hotshots" (Bradbury and Gibson 1983). Finally, the **female preference hypothesis** suggests that males cluster because females prefer sites with large groups of males, where they can more quickly, or more safely, compare the quality of many potential mates (Bradbury 1981).

To test these alterative hypotheses for why males congregate at leks, Frédéric Jiguet and Vincent Bretagnolle created artificial leks

HYPOTHESES 10.5

Alternative hypotheses to explain why males congregate at leks

hotspot hypothesis Males cluster in places (called "hotspots") where the routes frequently traveled by receptive females intersect.

hotshot hypothesis Subordinate males cluster around highly attractive males to have a chance to interact with females drawn to these "hotshots."

female preference hypothesis Males cluster because females prefer sites with large groups of males, where they can more quickly, or more safely, compare the quality of many potential mates.

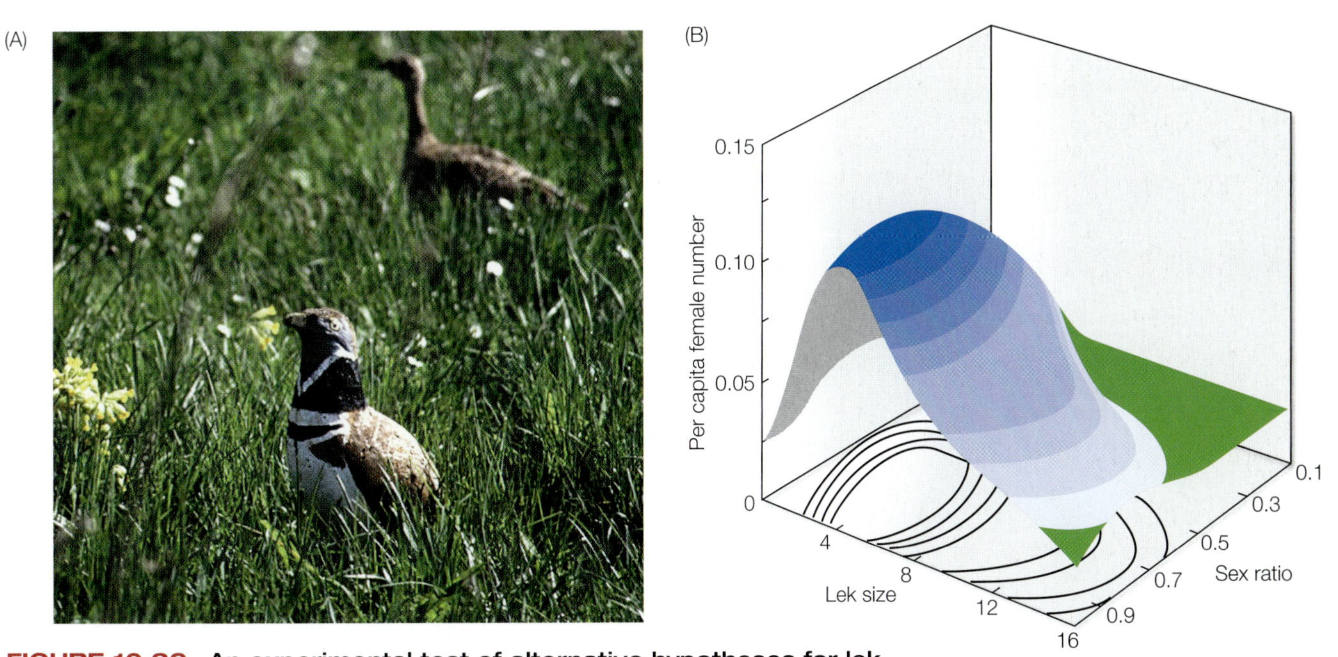

FIGURE 10.32 An experimental test of alternative hypotheses for lek formation. (A) A female little bustard visiting a decoy made to look like a male of her species. (B) More females visit groups of four decoys than smaller (or larger) groups, as shown by the peak in the graph of the number of females visiting per decoy when there were four decoys in the lek. The sex ratio represents the number of male decoys divided by the total number of decoys. (A, photograph courtesy of Frédéric Jiguet; B after Jiguet and Bretagnolle 2006.)

populated by plastic decoys painted to resemble males and females of the little bustard (*Tetrax tetrax*), a bird that exhibits a dispersed-lek mating system. The researchers placed different numbers of decoys of each sex in different fields and then, over time, counted the number of living little bustards attracted to their experimental leks. Female decoys failed to draw in males, leading to the rejection of the hotspot hypothesis. In contrast, male decoys regularly attracted both females and males, particularly if those decoys had been painted to resemble individuals with highly symmetrical plumage patterns. Therefore, the hotshot hypothesis may apply to little bustards. However, the fact that more females per decoy were attracted to clusters of four decoys than to smaller groups (**FIGURE 10.32**) is consistent with the female preference hypothesis as well, though more than four decoys did not draw in additional females (Jiguet and Bretagnolle 2006). Thus, this experiment ruled out the hotspot hypothesis but could not distinguish between the hotshot and female preference hypotheses.

Another way of testing between the hotspot and hotshot hypotheses is by temporarily removing males that have been successful in attracting females. If the hotspot hypothesis is correct, then removal of these successful males from their territories will enable other males to move into the favored sites. But if the hotshot hypothesis is correct, removal of attractive males will cause the cluster of subordinate males to disperse to other popular males or to leave the site altogether. In a study of the great snipe (*Gallinago media*), an Old World sandpiper that displays at night, removal of central dominant males caused their neighboring subordinates to leave their territories. In contrast, removal of a subordinate while the alpha snipe was in place resulted in his quick replacement on the vacant territory by another subordinate. At least in this species, the presence of attractive hotshots, not the real estate per se, determines where clusters of males form (Höglund and Lundberg 1987). Likewise,

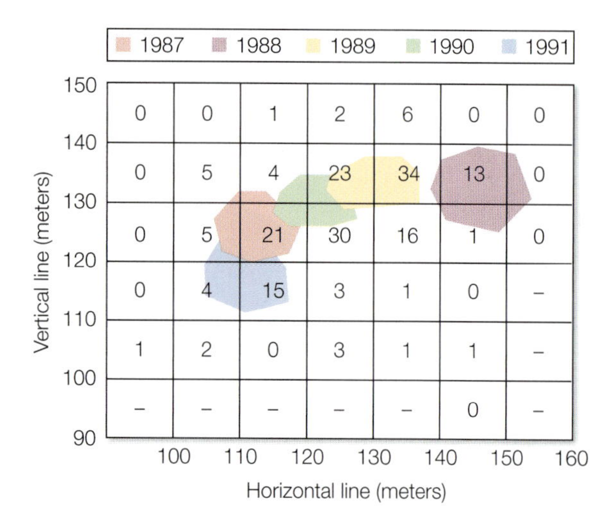

Male black grouse

FIGURE 10.33 Hotspots or hotshots? Researchers divided a black grouse lek into 100-square-meter sectors and recorded the total number of copulations in each sector over a 5-year period from 1987 to 1991. The irregular polygons show the location of the top territory for each of the 5 years. The shifts in the preferred territory suggest that male attractiveness, rather than the territory itself, plays the key role in reproductive success in this species, as required by the hotshot hypothesis. (After Rintamäki et al. 1995; photograph © iStock.com/Chiara_naturecolors.)

in the unrelated black grouse (*Lyrurus tetrix*), although relatively large, centrally located display sites are associated with higher mating success (Rintamäki et al. 1995), the exact location of the most successful territory can change somewhat from year to year, suggesting that a popular male, rather than a particular spot, most influences the behavior of other males (**FIGURE 10.33**) (Rintamäki et al. 2001).

Although the hotshot hypothesis seems likely to explain male lekking behavior in a great many cases, the hotspot hypothesis has received support in others (**FIGURE 10.34**) (Gibson 1996). For example, a site at which male fallow deer (*Dama dama*) gathered to display shifted when logging activity altered the paths regularly followed by fallow deer of both sexes (Apollonio et al. 1998). Likewise, Indian peacocks (*Pavo cristatus*) tend to gather near areas where potential mates are feeding (hotspots); the removal of some males has no effect on the number of females visiting leks of this species, as one would predict if females are inspecting those areas because of the food they provide, not because of the males there (Loyau et al. 2007). In addition, the leks of certain tropical manakins are located in areas relatively rich in fruits, the food of these birds, which is consistent with the idea that males gather near places with resources attractive to females (Ryder et al. 2006).

The hotspot hypothesis cannot, however, apply to those ungulates such as the Kafue lechwe (*Kobus lechwe kafuensis*) in which ovulating females leave their customary foraging ranges to visit groups of males some distance away, perhaps to compare the performance of many males simultaneously (**FIGURE 10.35**) (Balmford et al. 1993). According to the female preference hypothesis, a female preference for quick and easy comparisons might make it advantageous for males of the Uganda kob (*K. kob thomasi*), an African antelope related to the Kafue lechwe, to form large groups. If so, those leks with a relatively large number of males should attract disproportionately more females than leks with fewer males. Contrary to this prediction, however, the operational sex ratio is the same for leks across a spectrum of sizes, so males are no better off in large groups than in small ones (**FIGURE 10.36**) (Deutsch 1994). For this species at least, the female preference hypothesis can be rejected. The same

(A)

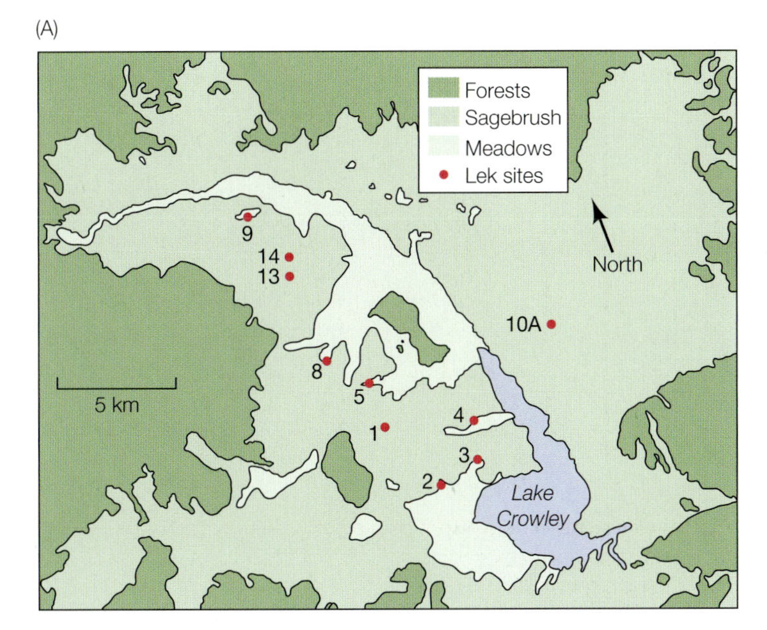

(B)

Male greater sage grouse

FIGURE 10.34 A test of the hotspot hypothesis. (A) The location of greater sage-grouse (*Centrocercus urophasianus*) leks (numbered red circles) in relation to sagebrush, meadows, forests, and a lake. (B) The distribution of nesting females in relation to the leks where males gather to display. The darker the shading, the more females present. (After Gibson 1996; photograph © Tom Reichner/Shutterstock.com.)

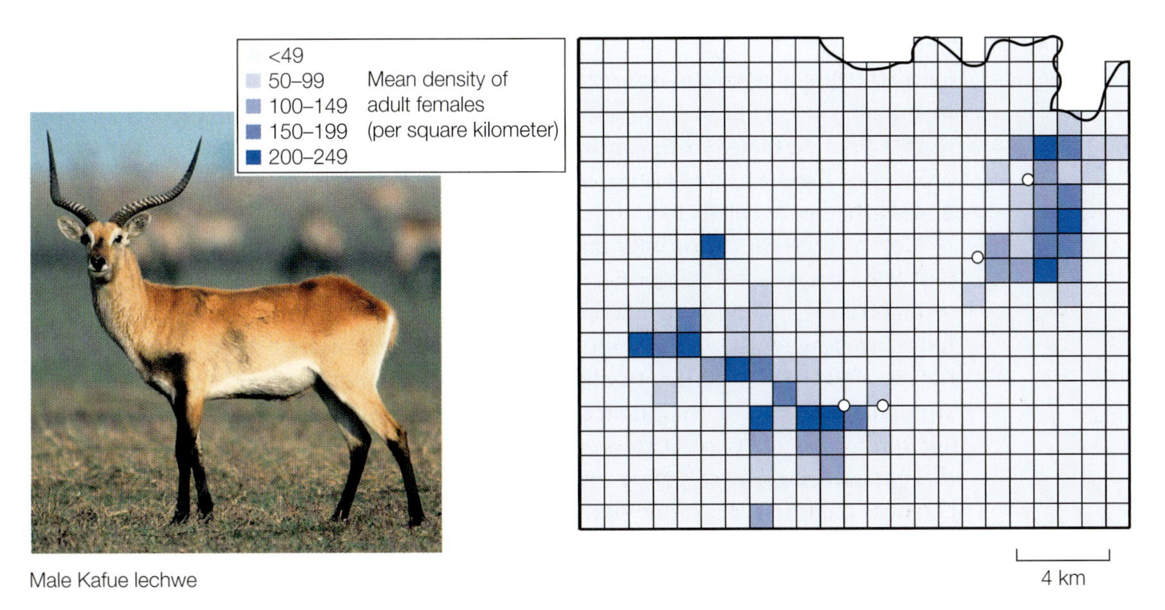

Male Kafue lechwe

FIGURE 10.35 Female density in the Kafue lechwe is not correlated with lek formation. Four leks of the antelope (open circles) are not located in the areas of highest female density, providing evidence against the hotspot hypothesis for this species. (After Balmford et al. 1993; photograph © Stockbyte/Alamy Stock Photo.)

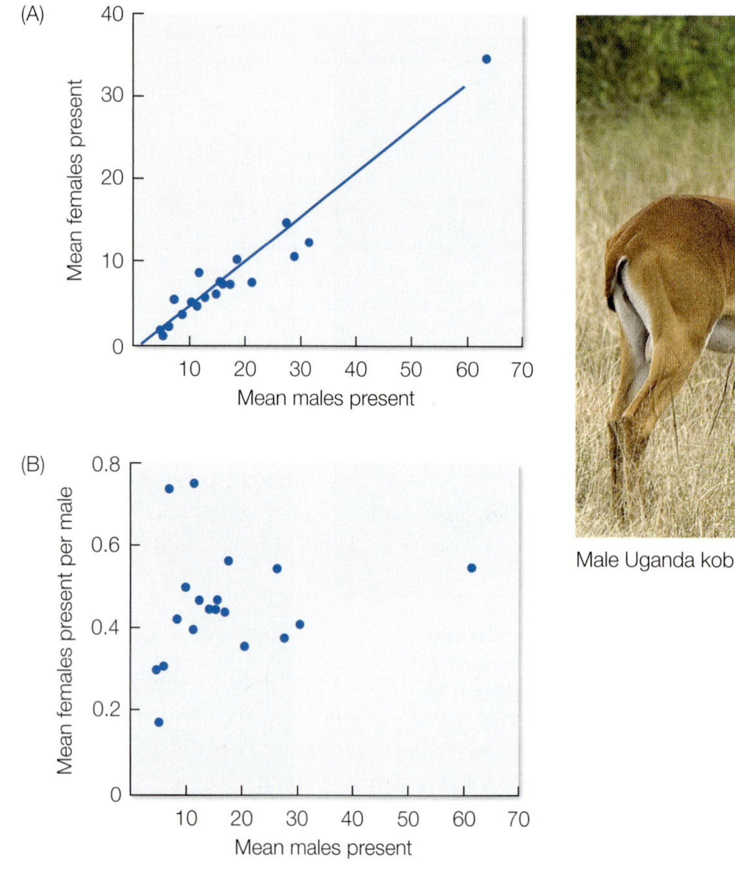

Male Uganda kob

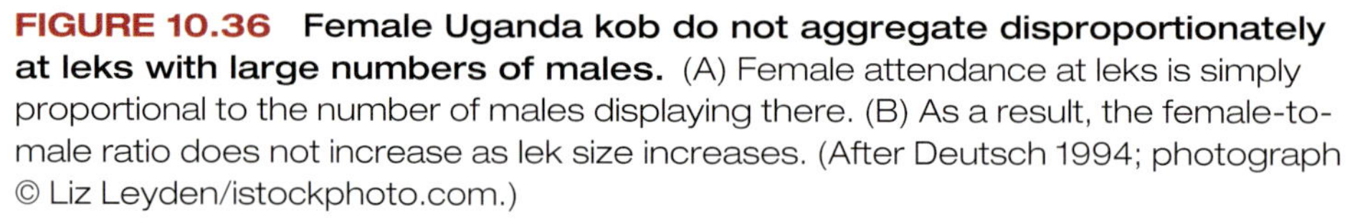

FIGURE 10.36 Female Uganda kob do not aggregate disproportionately at leks with large numbers of males. (A) Female attendance at leks is simply proportional to the number of males displaying there. (B) As a result, the female-to-male ratio does not increase as lek size increases. (After Deutsch 1994; photograph © Liz Leyden/istockphoto.com.)

is true for the barking treefrog (*Hyla gratiosa*). Here too, the more males chorusing in a pond, the more receptive females show up on a given night (**FIGURE 10.37**). But preventing males from coming to the pond to call does not reduce the number of females arriving, which is not what one would expect if a large number of calling males is needed to attract a large number of choosy females (Murphy 2003).

No one hypothesis for why lekking males form groups holds for every species. Nevertheless, the interactions among males on a lek usually seem to enable individuals of high physiological competence to demonstrate their superior condition to their fellow males and to visiting females (Fiske et al. 1998). Whatever the basis for lek formation is, lekking males are forced to compete in ways that separate the strong from the weak, making it potentially advantageous for females to come to a lek to compare and choose, the better to select a partner of the highest quality. Could the same rules apply to showy females in some species? This is something that has been explored in at least one sex-role reversed species (**BOX 10.3**).

Scramble Competition Polygyny

Female defense and resource defense tactics by competitive males make intuitive sense when either females or resources are clumped. However, in many species, receptive females and the resources they need are widely dispersed. And if females

FIGURE 10.37 The relationship between mating females and calling males of the barking treefrog. The more female barking treefrogs mating at a pond on a given night, the more males found chorusing there. However, the relationship stems not from the ability of large numbers of males to attract more females with their calls but from the fact that both sexes respond similarly to a set of environmental variables, including temperature and rainfall. (After Murphy 2003; photograph © Jay Ondreicka/ Shutterstock.com.)

are unlikely to go in search of groups of males or frequent places such as a potential lek site regularly, what is a male to do? When the costs of establishing territories (either with or without resources) exceed the benefits, males may simply try to find scarce receptive females before other males do. Flightless females of a *Photinus* firefly, for example, can appear almost anywhere over wide swaths of Florida woodland. Searching males of this species make no effort to be territorial; instead, they fly, and fly, and fly some more. In other words, they scramble to find females as quickly as they can to outcompete other males. When Jim Lloyd tracked flashing males, he walked 10.9 miles in total, following 199 signaling males, and saw exactly two matings. Whenever Lloyd spotted a signaling female, a firefly male also found her in a few minutes (Lloyd 1980). Mating success in this species almost certainly goes to those searchers that are the most persistent, durable, and perceptive, not the most aggressive.

Male thirteen-lined ground squirrels (*Spermophilus tridecemlineatus*) behave just like fireflies, searching widely for females, which become receptive for a mere 4 to 5 hours during the breeding season. The first male to find an estrous female and copulate with her will fertilize about 75 percent of her ova, even if she mates again. Given the widely scattered distribution of females and the first-male fertilization advantage, the ability to keep searching should greatly affect a male's reproductive success. In addition, male fitness may depend on a special kind of intelligence, namely the ability to remember where potential mates can be located (Foltz and Schwagmeyer 1989). After visiting several females near their widely scattered burrows, searching males often return to those places the following day. When researchers experimentally removed several females from their home sites, returning males spent more time searching for those missing females that had been on the verge of estrus. Moreover, the males did not simply inspect places that the females had used heavily, such as their burrows, but instead biased their searches in favor of the spots where they had actually interacted with about-to-become-receptive females (Schwagmeyer 1994, 1995).

If scramble competition explains why some males mate multiply, then when the distribution of females changes, so too should the mating system. Andrew Sih and colleagues tested this prediction in water striders by experimentally altering pool and group size and following the same males in both treatments (Sih et al.

BOX 10.3 **DARWINIAN PUZZLE**

Lekking females in a sex-role reversed pipefish

Given what we have already discussed in this chapter and in Chapter 9, you might start to wonder if leks exist in sex-role reversed species. After all, the same principles that apply to sexual selection and parental investment theories in traditional species also apply to those in sex-role reversed species, just in reverse. There are generally four criteria to determine if a species forms a lek: (1) displaying individuals are spatially aggregated; (2) displaying individuals contribute nothing to offspring or their mates beyond gametes; (3) the choosy sex visits the lek only to mate; and (4) the choosy sex has unrestrained access to mating partners (Bradbury 1981). Given these criteria, it should be theoretically possible for lekking to evolve among females in sex-role reversed species. But does it?

There is recent evidence of lekking in the worm pipefish (*Nerophis lumbriciformis*), a sex-role reversed species in which females display on patches on the ocean floor to attract males to which they provide their eggs and little else (Monteiro et al. 2017). Researchers found that the most highly ornamented females occupy central positions on the lek, and males mating with these females receive larger broods with bigger eggs compared with males that mate with less ornamented females. Although it remains to be seen whether displaying females congregate at leks for the same reason that males do, it is likely that the same principles hold in this sex-role reversed species.

2017). The researchers found that males in small groups (3 females and 3 males) in large pools exhibited scramble competition, but when placed in small pools with large groups (12 females and 12 males), the same animals often exhibited female defense polygyny whereby the largest male drove other males into hiding and off the water but allowed females to be relatively active. In addition to there being a change in the mating system, matings themselves also changed. In the large pools/small groups, copulations typically lasted for more than 100 minutes, but in the small pools/large groups they were less frequent and much shorter in duration. Since it typically takes only 30 minutes for males to transfer sperm, the researchers hypothesized that these shorter duration matings in small pools/large groups occurred because the dominant male had less need to mate repeatedly, as well as little need to guard females after transferring sperm because the male drove off all of his competitors.

Another form of scramble competition polygyny is influenced not by the distribution of females but by the length of the breeding period. An **explosive breeding assemblage** occurs in species with a highly compressed breeding season in which females may be receptive only during a short window of time. One such species is the horseshoe crab *Limulus polyphemus*, in which females lay their eggs on just a few nights each spring and summer. Males are under the gun to be near the egg-laying beaches at the right times and to accompany females to the shore, where egg laying and fertilization occur (Brockmann and Penn 1992). Explosive breeding assemblages are also common in amphibians that typically breed on only one or a few nights each year, such as the wood frog (*Rana sylvaticus*). On that night or nights in spring in the temperate Northern Hemisphere, most of the adult male wood frogs in a population are present at the ponds that females visit to mate and

BOX 10.3 DARWINIAN PUZZLE (*continued*)

Thinking Outside the Box

Why do you think female pipefish gather at leks? Look at the **FIGURE** below, which shows a heat map based on 239 pipefish captures at a lek. Do you think these data are consistent with the hotshot or hotspot hypotheses? These data are purely correlational. Design an experiment to further test whichever hypothesis you think these data support.

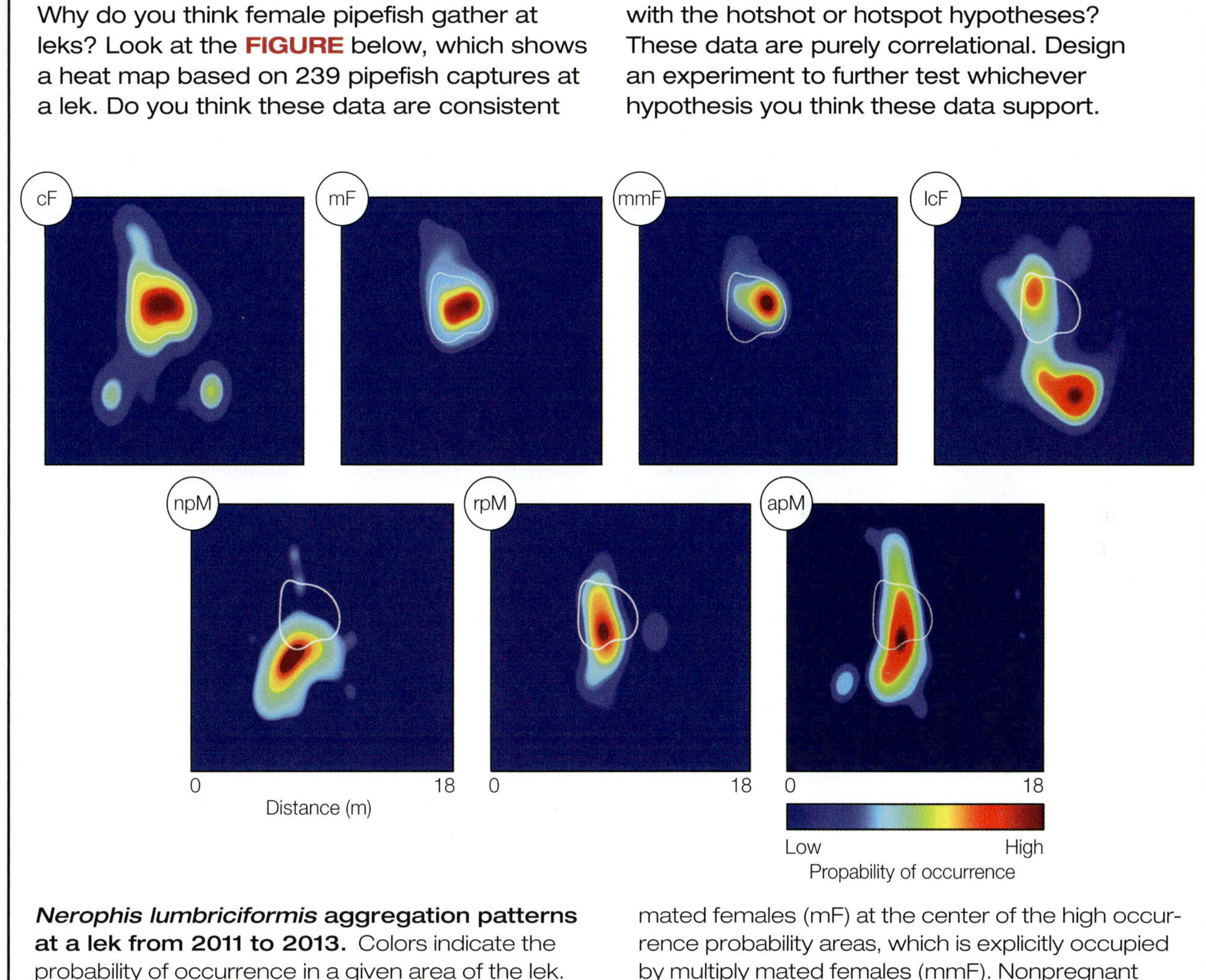

***Nerophis lumbriciformis* aggregation patterns at a lek from 2011 to 2013.** Colors indicate the probability of occurrence in a given area of the lek. Each panel corresponds to a different category of fish. Less colorful females (lcF) are preferentially observed outside the high occurrence probability areas. The more colored females (cF) are distributed through concentric layers, with genetically confirmed mated females (mF) at the center of the high occurrence probability areas, which is explicitly occupied by multiply mated females (mmF). Nonpregnant males (npM) occupy an area adjacent to the female lek. Recently pregnant males (rpM) are observed within the lek, and as pregnancies advance further (apM) the male distribution diffuses from the lek. (From Monteiro et al. 2017.)

lay their eggs. Just as in horseshoe crabs, the high density of rival males raises the cost of repelling them from a defended area. And because females are available only on this one night, a few highly aggressive territorial males cannot monopolize a disproportionate number of mates. Therefore, male wood frogs eschew territorial behavior and instead hurry about trying to encounter as many egg-laden females as possible before the one-night orgy ends (**FIGURE 10.38**) (Breven 1981). They then have everything to gain by moving far from the breeding pond where predators come to kill and eat the frogs (Rittenhouse et al. 2009).

FIGURE 10.38 An explosive breeding assemblage. A male wood frog grasps a female (upper left) that he has found before rival males, two of which are near the mating pair. Numerous fertilized egg masses float in the water around the frogs. (Photograph by Rick Howard.)

Polygynandry and Promiscuity: Multiple Mating by Both Sexes

Throughout this chapter, we have discussed the reasons that one sex has multiple partners while the other sex tends to just have one. However, there are several species in which both sexes mate multiply, either polygyandrously when there is a pair-bond or promiscuously when there is not. Although polygynandry is extremely rare, promiscuity is more common, particularly in species that do not provide parental care and those with external fertilization (such as many fishes).

Polygynandry

Because polygynandrous mating systems require males and females to form pair-bonds, this type of system tends to occur in species that form social groups (see Chapter 13). One such species is the dunnock, which we introduced earlier in this chapter. You may have noticed that we discussed female mating behavior in the context of direct benefits to polyandry, as well as male mating behavior in the context of resource defense polygyny. As we mentioned at the outset of this chapter, mating systems are flexible, and within a species they often vary according to social or ecological circumstances. The dunnock is perhaps one of the prime examples of this plasticity in mating behavior. Dunnocks exhibit not just polyandry and polygyny, but also monogamy and polygynandry. For example, in a population in England where 227 breeding attempts were monitored over 4 successive years from 1981 to 1984, 22 percent of the groups were monogamous, 38 percent were polyandrous, 10 percent were polygynous, and 30 percent were polygynandrous (Davies 1986). The majority of the polygynandrous groups consisted of two males and two females, but there were some cases of groups with two males and three females, and even some larger groups. In an introduced dunnock population in New Zealand where 80 social groups were monitored over 3 successive years from 2009 to 2012, 45 percent of the groups were monogamous, 54 percent were polyandrous, and only 1 percent were polygynandrous (Santos and Nakagawa 2013). This variation in dunnock mating systems is thought to be a result of sexual conflict, and both monogamy and polygynandry are cases where neither sex is able to gain an advantage at the expense of the other (Davies and Houston 1986).

A close relative of the dunnock, the alpine accentor (*Prunella collaris*), also forms polygynandrous groups, typically with one dominant male, one to three subordinate males, and two to four females (Davies et al. 1995). The fact that alpine accentors live at high elevation and feed on insects that are widely dispersed and patchy leads to overlapping ranges in both males and females. Individuals form groups partly to have multiple individuals available to help rear young. As in dunnocks, reproduction is shared among both males and females. Although dominant males typically gain a higher share of the reproduction than subordinates, both types of males actively engage in parental care at nests (Hartley et al. 1995). Thus, in the dunnock and the alpine accentor, the density of both birds and resources drives grouping patterns. The need for additional parental care facilitates this grouping and influences multiple mating in both sexes.

Promiscuity

Although promiscuity is most common in species lacking parental care, it does occur in those in which parents play at least some role in caring for young. One bird that exhibits promiscuity is the great tinamou (*Tinamus major*). Unlike in dunnocks and alpine accentors, multiparental care is unnecessary in the great tinamou because young are precocial and can largely take care of themselves after hatching. This freedom from extended offspring care enables females to roam the forest floor in search of males. When a female finds a male, she may mate with him, leave him with a clutch of eggs, and then be on her way in search of another mate. Males can mate with up to three females, and it is these fathers that incubate the combined clutches of eggs in this promiscuous species (Brennan 2009). While both males and females have multiple partners in this species, only the father provides care.

One of the most extreme cases of promiscuity occurs in the marine snail *Littorina saxatilis*, in which females provide care by retaining fertilized eggs inside a brood pouch until they hatch (Panova et al. 2010). There is no evidence of female mate choice and no male–male competition in this species. Males encounter dozens of receptive females each day during the mating season, and 15 to 23 males can sire clutches of 53 to 79 offspring. Whereas both sexes in this marine snail appear to mate indiscriminately and frequently, only the mother takes care of the young. Thus, unlike in dunnocks and alpine accentors where groups are needed to provision food to young, in promiscuous great tinamous and *L. saxatilis*, offspring care (by one sex) extends only through the time eggs hatch.

Promiscuity is even more common in species in which no parental care is provided by either the mother or the father. Take for example, the seaweed fly *Coelopa frigida*, which gets its name from the fact that it lives and breeds in seaweed washed up on beaches in the temperate Northern Hemisphere (**FIGURE 10.39**). Females lay eggs on the seaweed, and the larvae feed on the bacteria that decompose the seaweed. Males and females in this species both desire larger mates (Pitafi et al. 1990), a preference that is controlled by a chromosomal inversion (Gilburn and Day 1994). However, this species has no courtship behavior, and males simply attempt to mount females when they find them. Females will struggle with males, but females

FIGURE 10.39 Seaweed flies exhibit extreme promiscuity. When population densities are high, both males and females may copulate hundreds of times during their short, 3-week lives. (Photograph © Ashley Cooper/Alamy Stock Photo.)

often acquiesce to male mating attempts in a strategy that has been called **convenience polyandry**. One male was reported to have mated 37 times in a 5-minute period, and males tend on average to mount a female every 8 minutes in dense populations. Jennifer Blyth and André Gilburn calculated that females may mate two to three times every hour, and therefore hundreds of times over their 3-week lifetime (Blyth and Gilburn 2006). In species such as this where male harassment is common, promiscuity resulting from convenience polyandry may be the result of sexual conflict between males and females.

SUMMARY

1. Mating systems can be defined in terms of the number of sexual partners an individual acquires during a breeding season. Monogamy occurs when males and females both have only a single mate. Polygyny occurs when males have multiple partners but females have just one, whereas polyandry occurs when females have multiple partners and males have just one. When both sexes mate multiply, it is termed polygynandry if there is a pair-bond between males and females, and promiscuity if there is no association between mates beyond sperm transfer.
2. Monogamy is likely to evolve when the costs to mating multiply are prohibitively high for both sexes and neither sex can monopolize access to mates or resources. This can occur because potential mates do not form groups and roam widely, making them potentially costly to locate, when individuals have the ability to restrict mating behavior in their partner, or when resources are so critical to successful reproduction that both parents are necessary to rear young.
3. Polyandry is likely to evolve when the benefits of mating multiply are higher for females than remaining monogamous. These potential benefits include indirect benefits in the form of genetic advantages for offspring or direct benefits for the female. Potential genetic benefits include higher individual offspring quality or viability, greater genetic compatibility between the female and her mate, and greater genetic diversity among a female's offspring. Females also may mate with many partners to avoid the risk of inbreeding. Potential direct benefits include access to additional resources, access to additional parental care, better protection from sexually harassing males, and reduced risk of infanticide.
4. The diversity of polygynous mating systems has evolved in response to different patterns of female distribution, which are often driven by the distribution of resources or the risk of predation. These female decisions in turn affect the profitability of different kinds of mating tactics that males use to gain access to females. When females or the resources they need are clumped in space, female defense or resource defense polygyny becomes more likely. If, however, females are widely dispersed or male density is high, males may engage in nonterritorial scramble competition for mates, or they may acquire mates by displaying at a lek where females come specifically to seek out mates and nothing else.
5. Multiple mating by both sexes is likely to evolve when neither sex is able to gain an advantage over the other. Species living in social groups, often because of resource or offspring-care needs, are likely to form long-term pair-bonds and to be polygynandrous when multiple caregivers are required to successfully raise young. In contrast, when there is no pair-bond and often uniparental or no form of parental care, both sexes may mate promiscuously, or indiscriminately. Promiscuous mating systems are often the result of sexual conflict between males and females.

COMPANION WEBSITE

Go to **ab11e.com** for discussion questions and all of the audio and video clips.

CHAPTER

11

Parental Care

As we discussed in Chapter 10, mating systems are defined by the number of mates that individuals of each sex have, which—particularly in males—can influence reproductive success. Yet mating can only increase fitness if the offspring survive long enough to reproduce. In many species, parental care is critical for offspring survival, though it is often constrained by the amount of time the mates spend together. For example, in socially monogamous systems, pairs stay together throughout the breeding season to care for their offspring. In contrast, polygynous systems are often categorized by either fleeting interactions between the sexes and subsequent maternal care of offspring or multiple females rearing offspring in a territory held by a single male. In polyandrous systems, multiple males often spend the breeding season with one female. Of course,

Chapter 11 opening photograph © Hermann Brehm/Minden Pictures.

many species exhibit no parental care at all, but for those that do, the key to explaining this diversity in parental behavior lies with the cost–benefit approach. The benefits of caring for one's offspring typically include the improved survival of the assisted progeny. Yet individuals only have a limited amount of time and energy that they can invest in or devote to any given offspring (referred to as parental investment), which often requires them to make difficult choices about whom to care for or how much care to provide. Although caring for offspring can help parents pass on their genes to the next generation in these offspring, taking care of young may also prevent parents from investing in other activities such as self-maintenance or the attraction of additional mates, which also have reproductive benefits for the parent. Moreover, caring for young can also entail a future cost in terms of reduced investment in offspring in later breeding bouts (Trivers 1972). In this chapter we consider both sides of the parental equation, particularly the costs of parental care, which are often less obvious than the benefits. First we focus on the value of offspring and the decisions parents must make about which offspring to care for and how much care to provide. Clearly these decisions have the potential to generate conflict between parents and offspring, as well as among siblings. We then discuss the potential downsides of providing care and how these costs—along with the benefits—influence which sex or sexes perform the care duties. Finally, we discuss the curious cases of species that care for offspring that are not their own, exploring how brood parasitism evolves and leads to coevolutionary arms races.

Offspring Value and Parental Investment

If you have a sibling, you might have heard your parents say that they love you both equally. Although that may be comforting to hear, it turns out that not all offspring are created equal. The fields of population biology and population genetics give us the term **reproductive value**, which is a measure of the potential of an individual to leave surviving descendants in the future. Offspring with high reproductive value are likely to go on to be successful breeders, whereas those with low reproductive value are unlikely to reproduce much, if it all, during their lifetimes. As we will explore in the following section, the reproductive value of any given offspring will depend on the ecological or social circumstances into which it is born, and this value often influences the amount of care that parents are willing to provide. Indeed, parents are often able to assess offspring value when making care decisions, and therefore do not always treat each offspring equally. Variation in offspring care can be due to the behavior of the offspring (advertising their need or quality) or of the parents (who may favor the offspring most likely to succeed by committing infanticide or permitting siblicide, which removes unwanted offspring), and as you will see later, this tension sets the stage for another form of evolutionary conflict: parent–offspring conflict.

HYPOTHESES 11.1

Non-mutually exclusive hypotheses to explain how offspring signal to parents or other caregivers

signal of need hypothesis
Signals that advertise an offspring's level of need in order to maximize their chance of being fed by their parents.

signal of quality hypothesis
Signals that advertise an offspring's quality or merit in order to maximize their chance of being fed by their parents.

Parental Care Decisions

If parents have limited food resources at their disposal and cannot feed all of their offspring equally, they must decide which young should be fed at the expense of the others. How do parents make this seemingly difficult decision of valuing their young? In many species, parents use their offsprings' behavior to decide how to allocate food most effectively (**HYPOTHESES 11.1**). For example, a parent may judge the physiological state of its offspring by their appearance or by their begging behavior, the better to give more to individuals more

FIGURE 11.1 An honest signal of condition? The red mouth gape of nestling lark buntings (*Calamospiza melanocorys*) is exposed when the birds beg for food from their parents. The brightness of a baby bird's gape may reveal something about the strength of the chick's immune system or general health. (Photograph by Anthony Mercieca/Getty Images.)

likely to become reproductively successful adults over their lifetimes (in other words, those with higher reproductive value). In many species, offspring have evolved signals designed to advertise their need (hunger level) and maximize their chance of being fed by their parents (**signal of need hypothesis**) (Godfray 1995). In a bird's nest, for example, parents may use the chick's behavior as a signal, preferentially feeding the individual that begs more loudly or pushes its mouth higher than that of its siblings. Yet in other avian species, parents ignore the neediest offspring, instead feeding those that signal quality or merit (**signal of quality hypothesis**) (Grafen 1990). One informative aspect of a nestling's appearance that might signal quality is the bright red lining of its mouth, which is conspicuously displayed by many nestling songbirds as they stretch up to solicit food from a returning parent (**FIGURE 11.1**). Because the red color of the mouth lining is generated by carotenoid pigments in the blood (which cannot be synthesized and can only be obtained from the diet), and because, as we saw in Chapter 8, carotenoids are believed to contribute to immune function by scavenging free radicals, a bright red gape could signal a healthy nestling capable of making the most of any food it receives. Barn swallow (*Hirundo rustica*) chicks, for example, with redder mouths weighed more 6 days after hatching and had greater feather growth at 12 days of age than chicks with paler gapes (de Ayala et al. 2007). Parents that preferentially fed those members of their brood that had bright red mouths might be investing in nestlings of high reproductive value, that is, those young that were healthy and on the road to becoming successful breeders. If parents do indeed make adaptive parental decisions of this sort, then nestlings made ill by injection of a foreign material should have paler mouth linings than their healthier nestmates. Furthermore, parents should feed offspring with artificially reddened gapes more than they feed offspring with unaltered mouth coloration. And that is exactly what Nicola Saino's research team found when they tested both predictions in the barn swallow (**FIGURE 11.2**) (Saino et al. 2000).

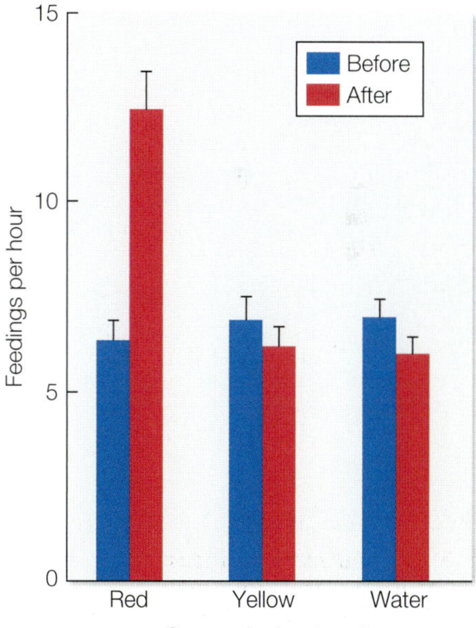

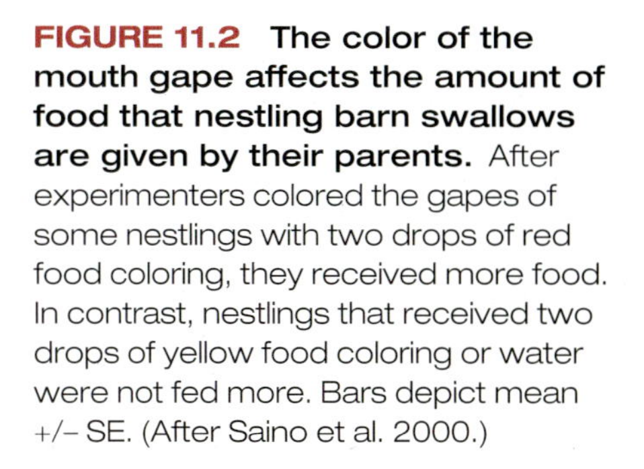

FIGURE 11.2 The color of the mouth gape affects the amount of food that nestling barn swallows are given by their parents. After experimenters colored the gapes of some nestlings with two drops of red food coloring, they received more food. In contrast, nestlings that received two drops of yellow food coloring or water were not fed more. Bars depict mean +/– SE. (After Saino et al. 2000.)

Shana Caro and colleagues hypothesized that the reason parents in some species feed the neediest offspring, whereas those in other species feed the highest-quality offspring (and in some species ignore chicks altogether), is because of the different ecological conditions that parents experience while raising their young (Caro et al. 2016a). The researchers predicted that when food is plentiful, parents should preferentially feed the offspring signaling the greatest need, whereas when food is scarce, parents should ignore begging and preferentially feed bigger chicks that are more likely to survive. Using comparative data on parent–offspring communication from 143 species of birds, the researchers quantified how offspring signaling relates to nestling body condition, and how parents respond to signals and cues of offspring condition. They then related these variables to measures of environmental quality and predictability, finding that offspring and parental strategies depend on both. In predictable and high-quality environments, chicks in poor condition beg more and parents respond to these signals of need and preferentially feed those chicks. However, in unpredictable and low-quality environments, parents pay less attention to begging, instead relying on size cues or structural signals of quality (traits such as gape coloration). In other words, ecological variation shapes the evolution of parent–offspring signaling systems and explains why some parents preferentially feed the neediest chicks, some feed the highest-quality chicks, and some ignore their chicks' begging altogether.

Although this and many other studies suggest that gape coloration in birds has evolved to signal offspring quality, this may not be the case in all avian species. Among the alternative explanations for bright gape coloration (Clotfelter et al. 2003) is the possibility that young birds gain by having colored gapes simply because these make a begging bird's mouth more visible to its parents, especially when the nest is placed in a dark tree cavity (Heeb et al. 2003). Philipp Heeb and his coworkers found that great tit (*Parus major*) nestlings with yellow-painted gapes were fed more often in relatively dark nest boxes than were red-painted birds, whose mouths were less visible under low-light conditions. However, in nest boxes with clear Plexiglas windows on top, the red-mouthed nestlings suffered no begging handicap, as shown by their ability to achieve the same weight as their yellow-mouthed companions (Heeb et al. 2003). Thus, gape coloration in at least some avian species may serve to help parents find hungry mouths rather than signal quality.

Not all offspring signals are located around the mouth area, particularly in species whose chicks spend little time in the nest. For example, the orange-tipped plumes on the heads and throats of baby American coots (*Fulica americana*) can hardly help food-bearing adults stuff edibles into the gapes of the young birds (**FIGURE 11.3A**). Unlike swallows, which have **altricial** young that reside in the nest for an extended period of time, coots have **precocial** young that are mobile soon after hatching. Therefore, baby coots do not spend much time in the nest and parents are likely to see more than just their outstretched gapes. Bruce Lyon and his colleagues decided to test the hypothesis that the brightly colored throat and head feathers of young coots might provide cues used by the parents to determine which individuals to feed and which to ignore (Lyon et al. 1994). Coots produce large clutches of eggs, but soon after the young begin to hatch, the adult birds often start the process of brood reduction. As some babies swim up to beg for food from a parent, the parent may not only refuse to provide something to eat but may also aggressively peck the head of its chick (**VIDEO 11.1**). Eventually these chicks stop begging and expire face-down in the water, pecked to death by their own parents in a case of **infanticide**. Apparently these chicks are needy, but not of as high reproductive value as their siblings.

VIDEO 11.1

Mother coot pecking at her chick
ab11e.com/v11.1

To test the link between chick ornamentation and parental care, Lyon and colleagues trimmed the thin orange tips from these special feathers on half of the chicks in a brood, while leaving the other members of the brood untouched. The unaltered orange-plumed chicks were fed more frequently by their parents, and as a result they grew more rapidly (**FIGURE 11.3B–E**) (Lyon et al. 1994). In control broods in which all of the chicks had had their orange feathers trimmed, the young were fed

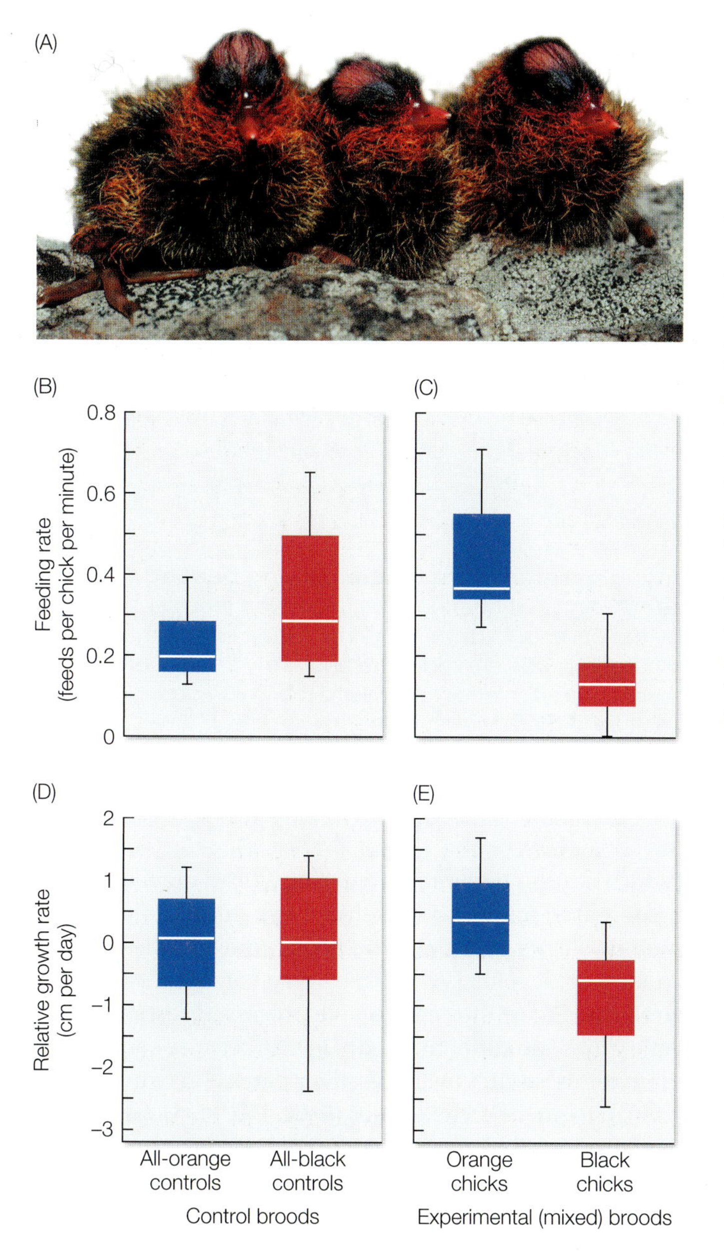

FIGURE 11.3 The effect of orange feather ornaments of baby coots on parental care. (A) Baby coots have unusually colorful feathers on their head and throat. (B) Control broods composed entirely of either unaltered (orange chicks, which are shown in blue) chicks or chicks that had had the orange tips trimmed from their ornamental feathers to make them look darker (black chicks, which are shown in red) were fed at the same mean rate (mean rates for each group are shown as white lines in the red and blue rectangles). Although some black chicks were fed at a higher rate, the difference between the mean values for the two groups was not statistically significant. (C) In experimental broods in which half the chicks were orange and half were black, the ornamented individuals received more frequent feedings from parent birds. (D) The relative growth rates of chicks in both types of control broods were the same, but (E) in mixed broods ornamented (orange chicks, which are shown in blue) chicks grew faster than the experimentally altered (black chicks, which are shown in blue) chicks. Values shown are medians (white lines), interquartile ranges (boxes), and 10–90th percentiles (whisker lines). (A, photograph © iStock.com/sandergroffen; B–E after Lyon et al. 1994.)

as often as control broods consisting only of untouched orange-feathered chicks. This result shows that the parents of the experimental mixed broods discriminated against the chicks without orange feathers because they were not as well ornamented as their feather-intact brood mates, not because the parents failed to recognize them as their offspring (Lyon et al. 1994). The fundamental message provided by American coots and many other animals is that parents do not necessarily treat each offspring the same. Cases in which parents help some of their brood survive at the expense of others remind us that natural selection acts not on variation in the number of offspring produced but on the number that survive to reproduce and pass on the parents' genes. Because of this, providing offspring care—either by parents or alloparents (see Chapter 13)—is often critical to an animal's fitness.

Although honest signaling systems between parents and offspring occur commonly in birds (because one or both parents often provide extended care to young), they exist in other taxonomic groups as well. Take, for example, the burying beetle

FIGURE 11.4 Discriminating parental care by the burying beetle *Nicrophorus vespilloides*. (A) An adult beetle inspecting larvae feeding on a ball of carrion prepared by the parents for their offspring. (B) When the mother is present, the senior (older, larger) grubs are fed more and grow to a larger size than when the mother is absent. No such effect applies to the junior (younger, smaller) grubs. (A, photograph by © Mark Moffett/Minden Pictures; B after Smiseth et al. 2007.)

Nicrophorus vespilloides, in which adults cooperate by burying a dead mouse or vole, removing the hair from the deceased animal, and fashioning a lump of flesh from the remains. The female then lays eggs near this brood ball, which will serve as a source of nourishment as offspring develop. When the larvae hatch, they can feed themselves from the prepared carcass, or they can receive processed carrion regurgitated by a parent. The beetle grubs can differ markedly in size because some hatch out sooner than others, and under these circumstances, their parents give more food to the earlier-hatched grubs than to later-hatched grubs (**FIGURE 11.4**). Given that only so many progeny can be supported by one mouse or vole carcass, parents may gain by helping those offspring most likely to achieve adulthood (those with higher reproductive value), especially since the absolute amount of food needed to reach maturity is less for grubs that are farther along the road to adult metamorphosis (Smiseth et al. 2007).

You have just seen how apparently honest signaling systems between parents and offspring can evolve, and how parents can use offspring signals to adaptively adjust their care behavior according to offspring need or quality. But can offspring manipulate parental care behavior, perhaps dishonestly signaling their level of need? In a comparative analysis of 60 avian species, Caro and colleagues tested the hypothesis that offspring will be more likely to exaggerate signals of need when they compete with either less-related siblings or a greater number of siblings (Caro et al. 2016b). In other words, decreased sibling relatedness or increased sibling competition could drive chicks to exaggerate their need and dishonestly signal to their parents. In support of this prediction, the researchers found that begging was less reliable in species where parents produced larger or more broods over their lifetime, as well as where parents were unlikely to breed together again, either because of the death of one member of a pair or because of divorce. Thus, chicks in some avian species are capable of exploiting parents that are unlikely to produce additional full siblings in the future.

Parental Favoritism in Offspring Care and Production

As you have just seen, parents rarely distribute their care in a completely equitable fashion, even when they are equally related to all of their offspring. Moreover, just as offspring body condition can sometimes influence levels of parental care, so too can parental condition affect how adults care for their young (and even which sex of progeny they produce, as you will see below). Consider the biased parental decisions made by the red mason bee (*Osmia rufa*), in which females nest in hollow stems and supply pollen and nectar for a series of brood cells, provisioning one after another. Initially, when the adult females are young and in good condition, they tend to give the first few offspring large amounts of food (Seidelmann et al. 2010). These initial offspring are the products of fertilized eggs and so develop into daughters of the red mason bee mothers. But as the season progresses and the females get older, their body condition deteriorates, making it more and more difficult for them to forage efficiently. As this happens, females provide less food per brood cell and they lay unfertilized eggs in these cells, which develop into sons that weigh much less than their sisters (see Chapter 12 for a discussion of sex determination in bees and other insects) (**FIGURE 11.5**). Because females of this and other bee species are able to control both the sex of an egg and the amount

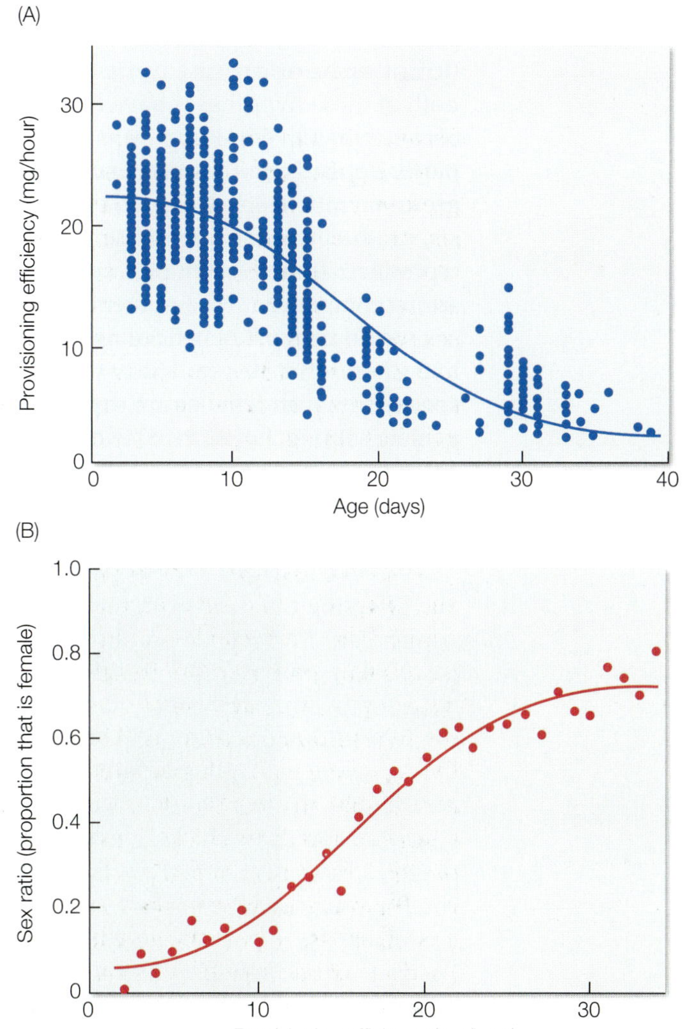

FIGURE 11.5 Adjustment of investment in sons and daughters by the red mason bee. (A) When females are young at the start of the breeding season, their provisioning efficiency is high, and (B) under these conditions, the sex ratio of their offspring is biased toward daughters. Small sons are produced when the females are older and their efficiency in filling their brood cells is low. Provisioning efficiency is measured in terms of the average increase in the mass of a larva per hour of bringing food to the nest. (A, B after Seidelmann 2006; photograph © Ant Cooper/Shutterstock.)

HYPOTHESES 11.2

Non-mutually exclusive hypotheses to explain population sex ratios that deviate from a ratio of 1:1 males to female

local competition hypothesis
When related individuals compete for resources or mates, then one sex is more costly to produce.

local enhancement hypothesis
When one sex provides resources or enhances the mating success of its relatives, then that sex is cheaper to produce.

of brood provisions the offspring receive, they can invest more in a daughter than a son. In this way, mothers give their daughters the resources needed to make their energy-demanding eggs and to do the hard work of foraging for their offspring. Sons can afford to be smaller, since they are short-lived, make tiny sperm, and spend their time searching for receptive females, presumably less demanding endeavors than those tackled by females.

Red mason bees, along with many other species of animals, can manipulate the sex of their offspring and invest in the sex with the higher reproductive value. How common is it for parents to manipulate the sex of their offspring? Ronald Fisher argued that, on average, parents should invest equally in sons and daughters (Fisher 1934). Assuming that males and females are equally costly to produce, he reasoned that as females become more abundant in a population, males have better mating prospects than additional females (and vice versa when males are more abundant). Over time, the sex ratio of the population should equilibrate and reach a ratio of 1:1 males to females. However, there are cases in which the population sex ratio should deviate from a Fisherian ratio of 1:1 (**HYPOTHESES 11.2**). For example, when related individuals compete for resources or mates (**local competition hypothesis**), then sex ratios can become skewed in one direction or the other because one sex is more costly to produce (Hamilton 1967, Clark 1978). There are also situations when one sex is cheaper to produce because it provides resources or enhances mating success of its relatives (**local enhancement hypothesis**) (Gowaty and Lennartz 1985, Sieff 1990). Both of these hypotheses have been well studied in cooperatively breeding birds because they live in kin groups and because the reproductive value of the sexes partly depends on the costs and benefits to parents of receiving help from other group members (see Chapter 13). According to the local enhancement hypothesis, producing offspring of the primary helping sex could enhance the parent's reproductive success, whereas according to the local competition hypothesis, if offspring compete with their parents for food, then producing the primary dispersing sex would be most beneficial to parents. Jan Komdeur tested predictions of these hypotheses in a 10-year study of Seychelles warbler (*Acrocephalus sechellensis*), a species in which females are the primary helping sex and males the primary dispersing sex) (**FIGURE 11.6**) (Komdeur 1996). Komdeur found that on low-quality

FIGURE 11.6 Photograph of Seychelles warbler. (Photograph © Brent Stephenson/Minden Pictures.)

territories where food was scarce, parents produced male offspring that tended to disperse from the territory. On high-quality territories where food was plentiful, however, parents produced female offspring that tended to remain on the territory as helpers. Thus, the quality of the Seychelles warbler's territory affects both competition for food and the likelihood of productive inheritance of territory, both of which influence the sex of the offspring that mothers produce.

The manipulation of offspring sex ratio observed in Seychelles warblers turns out to be surprisingly common in the animal world. Robert Trivers and Dan Willard have argued that a parental bias toward one sex or another is expected to evolve either in response to variation in parental body condition (which can be related to territory quality and resource availability) or to variation in how well one sex converts extra parental investment into more of its own offspring (Trivers and Willard 1973). According to the **Trivers–Willard hypothesis**, maternal condition can influence not only how much care a mother invests in each sex, but which sex she actually produces in the first place. This hypothesis predicts that when one sex exhibits greater variation in reproductive value (in other words, higher reproductive variance), then mothers in good condition should produce offspring of that sex, whereas mothers in poor condition should produce offspring of the opposite sex. In most polygynous species, the sex with higher reproductive variance is male because only one or a few males in the population monopolize most of the reproduction (which results in high reproductive skew among males); most males will fail to reproduce. In contrast, since most females in polygynous species will reproduce, the reproductive variance in females is much lower than that in males. Therefore, mothers in good body condition should produce more male than female offspring so that those high-quality sons will produce more of their own offspring, but mothers in poor body condition should produce more female than male offspring because a daughter is likely to produce more offspring over the course of her lifetime than a low-quality son.

Trivers–Willard hypothesis Mothers can adjust offspring sex ratio according to their own body condition.

How common is maternal sex-ratio adjustment in offspring? Although there is good evidence of sex-ratio adjustment in Seychelles warbler and other birds, the best evidence comes from mammals. For example, although a meta-analysis of more than 400 studies of sex ratio in mammals found that only 34 percent of the studies supported the Trivers–Willard hypothesis, 74 percent of the studies where weight or food were measured or manipulated around the time of conception supported the hypothesis (Cameron 2004). But do mothers of mammal species that bias the offspring sex ratio receive a fitness benefit? Using 90 years of captive breeding data from 198 mammalian species, Thogerson and colleagues showed that mothers tend to bias the sex ratio of their offspring toward the sex that maximizes the number of grand-offspring produced (Thogerson et al. 2013). That is, compared with non-skewing conspecifics, mothers that bias their offspring sex ratio toward daughters produce 29 percent more grand-offspring, and those that bias toward sons produce 25 percent more grand-offspring. Clearly, biasing the sex of offspring can be adaptive for some species under certain circumstances.

Parental Favoritism in Humans

If maternal sex-ratio manipulation occurs in mammals, does it also occur in our own species? We know, for example, that women with access to abundant food resources are slightly more likely to have sons than women with less to eat. This outcome reflects the fact that sons have the potential in some human societies to have many children by securing multiple wives (**polygamy**) or many sexual partners (see Chapter 14). In other words, the potential to be polygamous gives sons a higher reproductive value than daughters, at least under some conditions. Moreover, sons are likely to have a higher reproductive value if they have been developmentally advantaged and so are capable of competing successfully with rival males for women. Mothers with high-quality diets can give their male embryos the good start in life that they will need to become potent competitors as adults. In keeping with this argument, high-ranking women in polygamous marriages in Rwanda are more likely to have sons than are subordinate low-ranking, later-arriving wives, who appear to experience reduced access to resources (and more stress) (Pollet et al. 2009).

Additional evidence in support of the idea that human parents bias their investment in sons when they can help sons achieve especially high reproductive success comes from data on inheritance rules in relation to the potential for polygamy. John Hartung showed that in societies in which some men had many wives because of culturally sanctioned polygamy, sons were much more likely to receive a disproportionate share of the family inheritance than were daughters. In these groups, sons can potentially convert inherited wealth into multiple wives, providing many more grandchildren for their (usually deceased) parents than can daughters, whose reproductive success will of course generally be less than that of men with more than one wife (Hartung 1982, Smith et al. 1987). Importantly, this is only true for sons of high reproductive value, since those of low reproductive value may have zero chance at reproduction, whereas females are generally guaranteed to have some offspring.

Does this mean that parents in most human societies favor sons over daughters when they have the means to do so? Of course not. Take, for example, the Mukogodo tribe of Kenya (Cronk 1993, 2000). The forest-dwelling Mukogodo were traditionally an isolated tribe of beekeepers and hunter-gatherers who at the beginning of the twentieth century transitioned to livestock pastoralism (the typical lifestyle of most of their neighboring, savanna-dwelling tribes), largely because of population growth and British colonial policies that led to population movement. As the Mukogodo began to intermarry with their tribal neighbors, including the historically pastoralist Samburu and Maasai, young Mukogodo males needed livestock for bridewealth payments. As hunter-gathers without much livestock wealth, the Mukogodo were traditionally looked down on by the other tribes, residing at the bottom of a regional hierarchy of wealth, status, and marital success. In fact, the Mukogodo continue to this day to be referred to as *il-torrobo*, or "poor scum," by the Maasai. Because of the Mukogodo's low status and wealth, their polygamy rates are lower and men's ages at first marriage are higher than among neighboring tribes. Yet despite the challenges that Mukogodo men face in finding a bride, Mukogodo women all get married, often to men from wealthier neighboring tribes. Ultimately, this means that Mukogodo women tend to have more children than the average Mukogodo man and generally live a better life.

Since Mukogodo women appear to have a higher reproductive value than Mukogodo men, Lee Cronk predicted that, according to the Trivers–Willard hypothesis, the Mukogodo would invest more in a daughter than in a son (Cronk 1993). Indeed, the childhood (ages 0–4) sex ratio is heavily biased, with only 67 boys for every 100 girls. What causes this biased sex ratio? Cronk discovered that mothers tend to nurse daughters for longer than sons, and they are also more likely to take their daughters for medical care than their sons. Cronk also found that parents remain closer to daughters than to sons and hold daughters more than sons. Clearly Mukogodo parents tend to favor their daughters over their sons, which likely explains the female-biased childhood population sex ratio. But are parents aware of their favoritism to daughters? Surprisingly, when Cronk asked parents whether they favored sons or daughters, they overwhelmingly said they favor sons, not daughters! Thus, despite conscious feelings to the contrary, the data from the Mukogodo are consistent with the Trivers–Willard hypothesis and demonstrate how human parents will invest in the sex with the highest reproductive value, be it male or female.

Family Conflict

Not all animals appear to directly favor some of their offspring over others. For example, nesting great egrets (*Ardea alba*) bring small fish back to the nest and drop them in front of their brood, allowing their sons and daughters to fight for possession of their meals (**FIGURE 11.7**) (Mock 1984). Fighting among siblings, however, can escalate, with a dominant nestling bludgeoning a brother or sister to death or pushing it out of the nest to fall to its death or die of starvation so that the dominant individual can get more food (**VIDEO 11.2**). You might think that the parent should intervene to reduce **sibling conflict** and prevent the loss of even one chick from a brood of three or four, but they never do in this species. It is conceivable that this **siblicidal behavior** may have evolved only because of the fitness advantages enjoyed by offspring able to dispose of siblings that were competing for the same supply of food. This situation is predicted to lead to **parent–offspring conflict**, a concept also developed by Robert Trivers after he realized that some actions can advance the fitness of an offspring while reducing the reproductive success of its parent, and vice versa (Trivers 1974).

FIGURE 11.7 Sibling aggression in the great egret. Two chicks fight viciously in a battle that may eventually result in the death of one of them. (From Mock et al. 1990; photograph © DMS Foto/Shutterstock.)

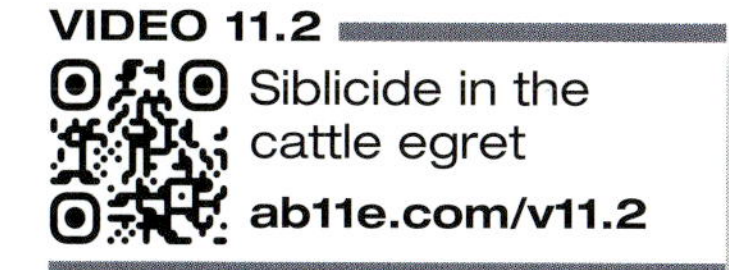

VIDEO 11.2 Siblicide in the cattle egret ab11e.com/v11.2

In some animals in which siblicide occurs, parents can—and apparently do—resist their offspring's selfish behavior. Evidence for this claim comes from studies of seabirds called boobies, in which an older chick of some species disposes of a younger chick in the first few days of life (Lougheed and Anderson 1999). The older chick's ability to kill its younger brother or sister stems in part from the pattern of asynchronous egg laying and incubation in boobies. In these birds, females lay one egg, begin incubating it at once, and then some days later lay a second egg. Because the first egg hatches sooner than the second, the first hatched chick is relatively large by the time the second egg hatches. In booby species in which siblicide occurs, the older chick typically forces the younger chick out of the nest scrape, where it dies of exposure and starvation (**FIGURE 11.8**).

In the Nazca booby (*Sula granti*), siblicide is standard practice (referred to as **obligate siblicide**), but this is not the case in the blue-footed booby (*Sula nebouxii*), in which chicks engage in siblicide less often and generally later in the nesting period (referred to as **facultative siblicide**). If, however, you give a pair of blue-footed booby chicks to Nazca booby parents, which tolerate sibling aggression, the older chick often quickly kills the younger chick under the vacant gaze of one of its substitute parents. In contrast, blue-footed booby parents appear to keep their first chick under control during its initial days with its sibling. What happens in the opposite experiment, when Nazca booby chicks are given to blue-footed booby parents? As you

FIGURE 11.8 Early siblicide in the brown booby (*Sula leucogaster*). A very young chick is dying in front of its parent, which continues to brood the larger, siblicidal chick that has forced its younger sibling out of the nest and into the sun. (Photograph by John Alcock.)

might predict, the foster parents are sometimes able to prevent the older chick from immediately killing its sibling (**FIGURE 11.9**) (Lougheed and Anderson 1999), providing evidence that parents can interfere with lethal sibling rivalries, should it be in their interest to do so.

Given the differences between Nazca and blue-footed boobies in how parents interfere with siblicide, could siblicidal behavior actually benefit parents (in addition to the surviving offspring through increased access to food)? One way to address this question might be to determine why Nazca boobies bother to lay two eggs when they always allow the first chick to kill the second. To address the first question, David Anderson proposed the **reproductive insurance hypothesis**, which argues that the second egg is laid as insurance against hatching failure (Anderson 1990). Consistent with predictions of this hypothesis, only 60 percent of first-laid Nazca booby eggs hatch, and laying two eggs reduces complete nest failure by 16 percent. By experimentally manipulating clutch sizes, Anderson's team also found that adding an additional egg to naturally occurring one-egg clutches increased the number of young fledged on average (Clifford and Anderson 2001a). So why were some female Nazca boobies laying only one egg when there was clearly a benefit to laying two? Anderson's team proposed that it was because they were food-limited. To test this idea, the researchers provided some females with extra food twice a day during the early breeding season. They found that a higher proportion of food-supplemented females produced two-egg clutches relative to controls (Clifford and Anderson 2001b). Thus, laying a second egg provides insurance in case the first fails to hatch, but food limitation may prevent some females from laying a second egg and caring for a second chick.

reproductive insurance hypothesis Mothers in siblicidal species lay a second egg as insurance against hatching failure.

Could resource limitation also explain why blue-footed boobies are facultatively siblicidal, with parents allowing siblicide to occur in some years but not others? An older blue-footed booby chick seems to kill the younger chick only in years when food is scarce. By experimentally depriving blue-footed booby chicks of food temporarily, Hugh Drummond and Cecilia Chavelas showed that older chicks attacked younger ones when they were food-restricted, but not when food deprivation ended (Drummond and Chavelas 1989). Aggressive behavior in blue-footed booby chicks was therefore mediated by the chick's nutritional status, which was in turn a result of how much the parents fed it. Thus, ecological factors in the form of food availability influence not only a mother's ability to lay eggs, but also the chick's siblicidal behavior.

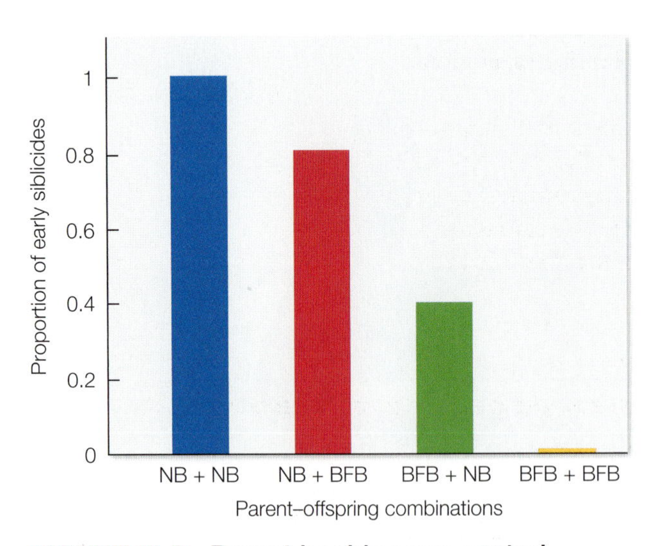

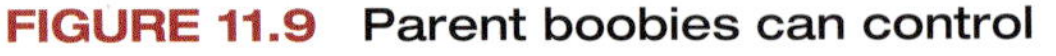

FIGURE 11.9 Parent boobies can control siblicide to some extent. The rate of early siblicide by Nazca booby (NB) chicks declines when they are placed in nests with intervention-prone blue-footed booby (BFB) foster parents. Conversely, the rate of early siblicide by blue-footed booby chicks rises when they are given Nazca boobies as foster parents. (After Lougheed and Anderson 1999.)

As we discussed earlier, egrets differ from blue-footed boobies—but are similar to Nazca boobies—in that parental intervention in egrets does not occur when two offspring fight. Indeed, lethal sibling battles are actually promoted by earlier parental decisions about when to begin incubating eggs. Thus, as soon as a female egret lays her first egg, incubation begins, just as is true for boobies. Because there are 1 or 2 days between egg layings in a three-egg clutch, the young hatch asynchronously, with the firstborn getting a head start in growth. As a result, this chick is much larger than the third-born chick, which helps ensure that the senior chick monopolizes the small fish its parents bring to the nest. The first-born chick is not only bigger but also more aggressive and outcompetes its younger siblings for access to food. At least in egrets, boobies, and many other avian species, the first-laid egg contains relatively large quantities of androgens—hormonal aggression facilitators given to the first-hatched offspring by its mother. The unequal feeding rates that result from this maternal effect further exaggerate the size differences in the siblings, creating a runt of the litter that often dies from the combined effects of starvation and assault (Mock and Parker 1997, Schwabl et al. 1997).

These results suggest that egret parents not only tolerate siblicide, they actually promote it. Why would they do this? Perhaps because, just as in the blue-footed booby, the parental interests of egrets are served by having the chicks themselves eliminate those members of the brood that are unlikely to survive to reproduce. Although in good years parents can supply a large brood with enough for all to eat, in most years food will be moderately scarce, making it impossible for the adults to rear all three offspring. When there is not enough food to go around, a reduction in the brood, accomplished by siblicide, saves the parents the time and energy that would otherwise be wasted in feeding offspring with little or no chance of reaching adulthood even if they were not attacked by their siblings. Thus, we can hypothesize that this form of occasional, or facultative, siblicide exhibited by egrets is driven by resource availability, as it also seems to be in blue-footed boobies.

One way to test this **facultative siblicide hypothesis** is to create unnaturally synchronous broods of egrets (Mock and Ploger 1987), which was done by taking cattle egret (*Bubulcus ibis*) chicks that had hatched in several nests on the same day and putting them in the same nest. As a control, normal asynchronous broods were assembled by bringing chicks together that differed in age by the typical 1.5-day interval. A category of exaggeratedly asynchronous broods was also created by putting chicks that had hatched 3 days apart into the same nests. If the normal hatching interval is optimal in promoting efficient brood reduction, then the number of offspring fledged per unit of parental effort should be highest for the normal asynchronous broods. This prediction was confirmed. Members of synchronous broods not only fought more and had lower survivorship but also required more food per day than normal broods, resulting in low parental efficiency (**TABLE 11.1**). The same result occurred in similar experiments with the blue-footed booby (Osorno and Drummond 1995), confirming the role of resource availability in driving facultative siblicide in these species.

facultative siblicide hypothesis Parents permit siblicidal behavior only when resource availability is low.

Natural selection has led to cattle egret parents and others like them that manipulate the hormone content of their eggs and incubate them in ways that lead to differences in size and fighting ability among their chicks. Committing infanticide or simply allowing siblicidal behavior to occur helps parents deliver their care only to offspring that have a good chance of eventually reproducing, while keeping their

TABLE 11.1 The effect of hatching asynchrony on parental efficiency in cattle egrets

	Mean survivors per nest	Food brought to nest per day (ml)	Parental efficiency[a]
Synchronous brood	1.9	68.3	2.8
Normal asynchronous brood	2.3	53.1	4.4
Exaggerated asynchronous brood	2.3	65.1	3.5

Source: Mock and Ploger 1987.

[a]The number of surviving chicks divided by the volume of food brought to the nest per day × 100.

food delivery costs to a minimum. Although cases of this sort represent extreme examples of parental favoritism, even those nesting birds that bias their allocation of food resources toward vigorous offspring are really practicing infanticide by increasing the likelihood of death for those progeny unlikely to reproduce even if well fed. In other words, when making parental care decisions, parents favor offspring with high reproductive value, something that the young in many species attempt to signal to caregivers with their behavior or physical traits.

To Care or Not to Care

Up until this point, we have considered why parents care more for some of their offspring than others, as well as how this has the potential to generate conflict among siblings and among parents and their offspring. We discussed how the reproductive value of young born under certain environmental or social conditions influences how much parents are willing to invest in care, as well as how environmental differences influence parent–offspring conflict. Yet in all of these cases, we only considered the benefits to the parents of caring for young. If we are to understand how parental care systems have evolved, we must also consider the costs to parents of caring for young. Costs influence not only whether parents should invest at all in offspring (versus saving their investment for offspring that might be produced in the future), but also which and how many parents should perform most of the care duties. After all, taking care of young is costly in terms of energy expenditure, and it is not always in a parent's best interest to do so. Recall from Chapter 9 the concept of sexual conflict, which describes conflict between males and females over mating decisions. As you will now see, conflict between males and females extends beyond the period of mate choice into the period of parental care because caring for young is so costly, yet is often essential to achieving reproductive success.

The Costs and Benefits of Parental Care

The decision to invest in some offspring over others is not always about deciding among young in the current batch, but instead between current and future batches of offspring. Although feeding young can increase the survival chances of offspring (and therefore the reproductive success of the parent), it can also reduce survival of both offspring and parents in some situations. For example, predators can use parental comings and goings to find the nest or burrow and not only feast on its occupants, but also lurk near the nest to intercept and eat the parents as they return to their offspring with food. How does a parent then balance the survival and reproductive costs and benefits of its parental care duties and activities? Cameron Ghalambor and Tom Martin predicted that parent birds, for example, should adjust their provisioning behavior adaptively in accordance with two key factors: the nature of the predator (whether it consumes nestlings or adults) and the annual mortality rate for breeding adults (Ghalambor and Martin 2001). In birds with low adult mortality rates, parents should minimize the risk of getting themselves killed by a predator because they will probably have many more chances to reproduce in the future. However, in species with high adult mortality rates, parents should be less concerned with their own safety and more sensitive to the risks that their nestlings may confront from nest-raiding predators. Since these parents will have relatively few chances to reproduce over their lifetime, they can gain by doing more for their current brood than waiting for a future brood, which may never come. This decision represents a classic life history trade-off, where parents must decide how much to invest in current versus future reproduction.

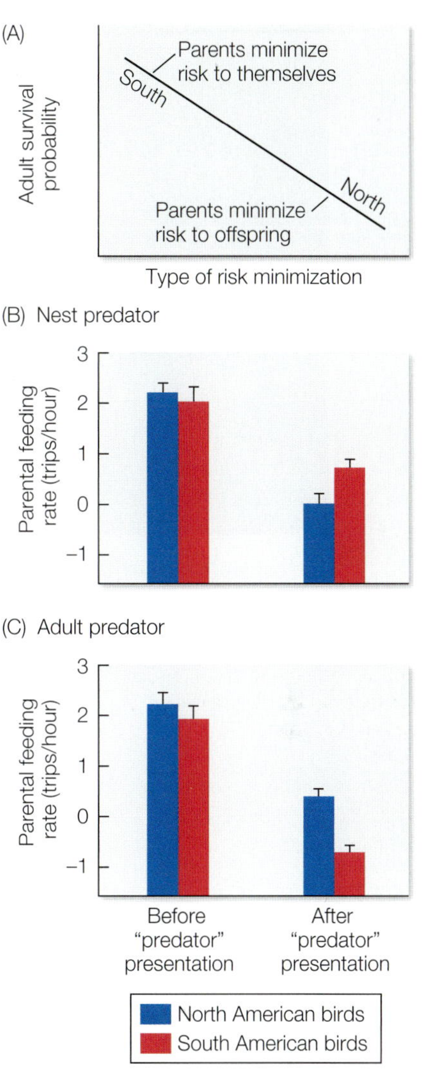

FIGURE 11.10 Parental care has fitness costs and benefits. (A) The reactions of parent birds to predatory threats to themselves and their offspring should vary in relation to the annual mortality rates of adults, which differ between South and North America. Shorter-lived North American birds are predicted to do more to reduce risks to their offspring; longer-lived South American birds are predicted to do more to promote their own survival. (B) The rate at which parents feed their nestlings falls more sharply in North American birds than in South American birds in response to an apparent risk of nest predation by jays, which can find nests by watching where parent birds go. (C) The rate at which parents feed their nestlings falls more sharply in South American birds than in North American birds after the adults see a hawk capable of killing them. Bars depict mean +/– SE. (After Ghalambor and Martin 2001.)

Ghalambor and Martin knew that birds that breed in North America tend to have shorter lives than their close relatives that breed in South America. Taking advantage of this latitudinal difference in life span, they matched up five pairs of closely related bird species, including, for example, two members of the same genus, the short-lived American robin (*Turdus migratorius*) from North America and the longer-lived rufous-bellied thrush (*T. rufiventris*) from South America (Ghalambor and Martin 2001). They then compared parental care behavior in these matched pairs of bird species. When the researchers played recordings of Steller's jays (*Cyanocitta stelleri*) to American robins in Arizona and recordings of plush-capped jays (*Cyanocorax chrysops*) to rufous-bellied thrushes in Argentina (both jays are nest predators but do not prey on adults), the robins and thrushes both greatly reduced their visits to their nests for some time, which under natural conditions would help hide their nests from these predators. However, the robins curtailed their activity around the nest more strongly than the thrushes did, presumably because the robins had more to gain by protecting their current crop of nestlings from keen-eyed jays, given their relatively low probability of reproducing in subsequent years.

Next, to further test the prediction that parents adjust their provisioning behavior adaptively depending on the predation risk to their offspring versus the risk to themselves, the researchers did the same experiment, but with an adult rather than a nest predator. When they placed a stuffed sharp-shinned hawk (*Accipiter striatus*) (a killer of adult birds) near active nests and played recorded sharp-shin calls, the parent birds in the sampled species again reduced their visits to their nests for some time. In this round of tests, however, the potentially long-lived thrushes in Argentina delayed their return to the nest longer than the corresponding robins in Arizona did (**FIGURE 11.10**) (Ghalambor and Martin 2001). This case is merely one of many in which the parental strategies of North American birds appear to differ from those of related South American species, likely because of differences in predation pressure on nesting adults in the two regions (Martin and Schwabl 2008). Thus, both the costs and benefits of taking care of offspring—as well as the reproductive value of current versus future offspring—have fine-tuned the evolution of parental behavior in these birds.

Sexual Conflict and Parental Care: Who Cares?

Although in many species of birds, including American robins and rufous-bellied thrushes, both fathers and mothers help their progeny survive, care by just one adult is extremely common in the animal kingdom. Because the creatures we live with

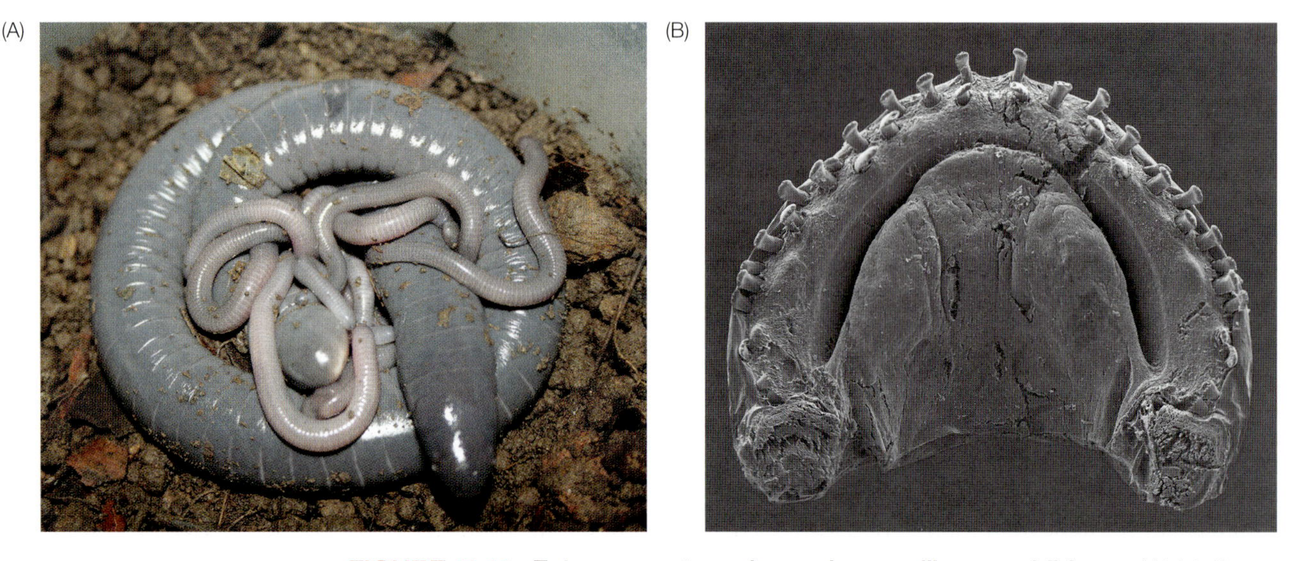

FIGURE 11.11 Extreme maternal care in caecilian amphibians. (A) Mother caecilians live with their offspring and permit them to remove and feed on their nutritious skin. (B) Nestling caecilians have curved teeth for the very purpose of stripping the edible skin from their mothers. (A, photograph © Uncatalogued/Minden Pictures; B, courtesy of Carlos Jared.)

and pay most attention to are mammals, you may think that all uniparental care is performed by mothers. Indeed, some maternal sacrifices are quite striking, as in the readiness of certain mother caecilians (burrowing wormlike amphibians) to let their youngsters feed on their lipid-rich epidermis, which the infant caecilians strip from their mother's body (**FIGURE 11.11**) (Wilkinson et al. 2008). Then there are the females of a spider, *Stegodyphus lineatus*, that not only feed their spiderlings regurgitated food but eventually permit their brood to cannibalize them completely (Salomon et al. 2005), a phenomenon also observed in some pseudoscorpions (Del-Claro and Tizo-Pedroso 2009).

One intuitively appealing explanation for the female domination of parental care-giving, as illustrated by the Membracinae treehoppers (**BOX 11.1**), is that because females (unlike males) have already invested so much energy in making eggs, they have a greater incentive to make sure that their large initial gametic investment is not wasted, and so they come to the aid of their offspring more often than males. However, this argument starts to break down when we observe that females of a substantial number of species, including the spotted sandpiper (*Actitis macularius*) and many fishes, abruptly end their parental activities after laying their large and costly eggs, which they leave totally in the care of their male partners (**FIGURE 11.12**). Although maternal-only care may be the majority form of parental care, exclusive paternal care occurs in some taxonomic groups. And of course some species of animals provide no parental care to their offspring, either allowing them to care for themselves, as in many fishes, or allowing other species to do it for them, as in brood parasites.

To understand why various forms of parental care systems have evolved, we need to consider both the costs and benefits to each sex in providing care. Birds are an excellent group in which to examine this because unlike in mammals, where females are the only sex that lactates, in most species of birds both sexes can share equally in care duties from the time eggs are laid until when chicks fledge the nest. To illustrate this cost–benefit approach to understanding how distinctive parental care systems have evolved, we begin with a single polygynous species—the Eurasian penduline tit

(A)

(B)

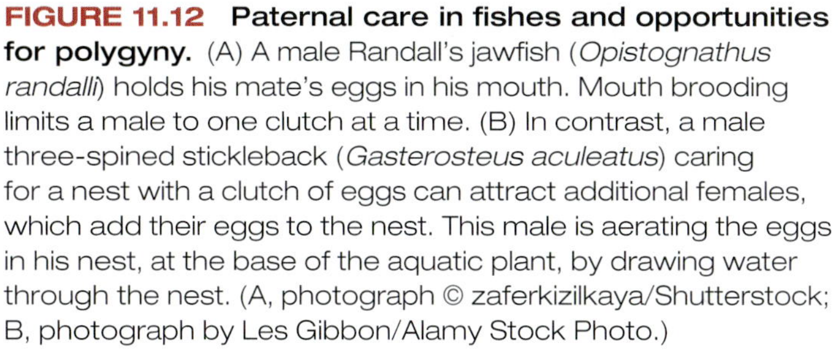

FIGURE 11.12 Paternal care in fishes and opportunities for polygyny. (A) A male Randall's jawfish (*Opistognathus randalli*) holds his mate's eggs in his mouth. Mouth brooding limits a male to one clutch at a time. (B) In contrast, a male three-spined stickleback (*Gasterosteus aculeatus*) caring for a nest with a clutch of eggs can attract additional females, which add their eggs to the nest. This male is aerating the eggs in his nest, at the base of the aquatic plant, by drawing water through the nest. (A, photograph © zaferkizilkaya/Shutterstock; B, photograph by Les Gibbon/Alamy Stock Photo.)

(*Remiz pendulinus*)—that exhibits a combination of male-only, female-only, and biparental care within the same population (van Dijk et al. 2007). Understanding how care decisions are made in this behaviorally plastic species may inform us more generally about why different parental care systems evolve in different species.

In Eurasian penduline tits, unmated males start building nests and sing to attract a female, and then both sexes finish the building of their nest together. Once the female has started egg laying, either the male or the female, or both, desert the nest during the egg-laying period to renest with a new mate. When both parents desert, the chicks starve. But when only one parent deserts, the other parent will do all of the care (unless they too decide to desert). Between 30 and 40 percent of nests are deserted by both parents, typically on the same day, meaning that one parent is left to rear the young in the majority of cases (van Dijk et al. 2007, 2011). Since deserting parents may remate up to six times within a single breeding season, the fitness benefits of desertion are clear, as long as the partner takes care of the young the deserter left behind. Thus, despite a shared interest in their current brood, there is typically conflict between male and female penduline tits over who should care for the young and who should attempt another brood with a new mate. The decision to desert appears to be strongly influenced by individual differences in the expected reproductive output for each male–female pair (van Dijk et al. 2011). Both sexes must weigh their own current versus future reproductive options and make a decision to care for young alone, together, or not at all. Although it remains unclear which factors actually lead an individual to desert (van Dijk et al. 2011), evidence from another population of Eurasian penduline tits suggests that individuals in good body condition are more likely to desert (Bleeker et al. 2005).

As you can see from the Eurasian penduline tit example, parental care is simply another form of parental investment, and as we discussed in Chapter 9, the difference in parental investment between males and females begins at the earliest stage of investment, when a tiny sperm fuses with a giant egg (the fundamental asymmetry of sex). When the interests of mother and fathers are dissimilar, then which

BOX 11.1 **EXPLORING BEHAVIOR** BY **INTERPRETING DATA**

Why do females provide all of the care in treehoppers?

Only females provide parental care in the Membracinae treehoppers. One approach for determining why females exclusively care for young is to use ancestral state reconstruction to determine when this behavior evolved in the group. Lin and colleagues placed the genera of Membracinae that exhibit maternal care on a molecular phylogeny of this subfamily of treehoppers (**FIGURE**) (Lin et al. 2004). The circles at the bases of the lineages represent ancestral species; the proportion of a given circle that is blue reflects the probability that females of this ancestral species cared for their eggs. The numbers mark the three probable origins of maternal care.

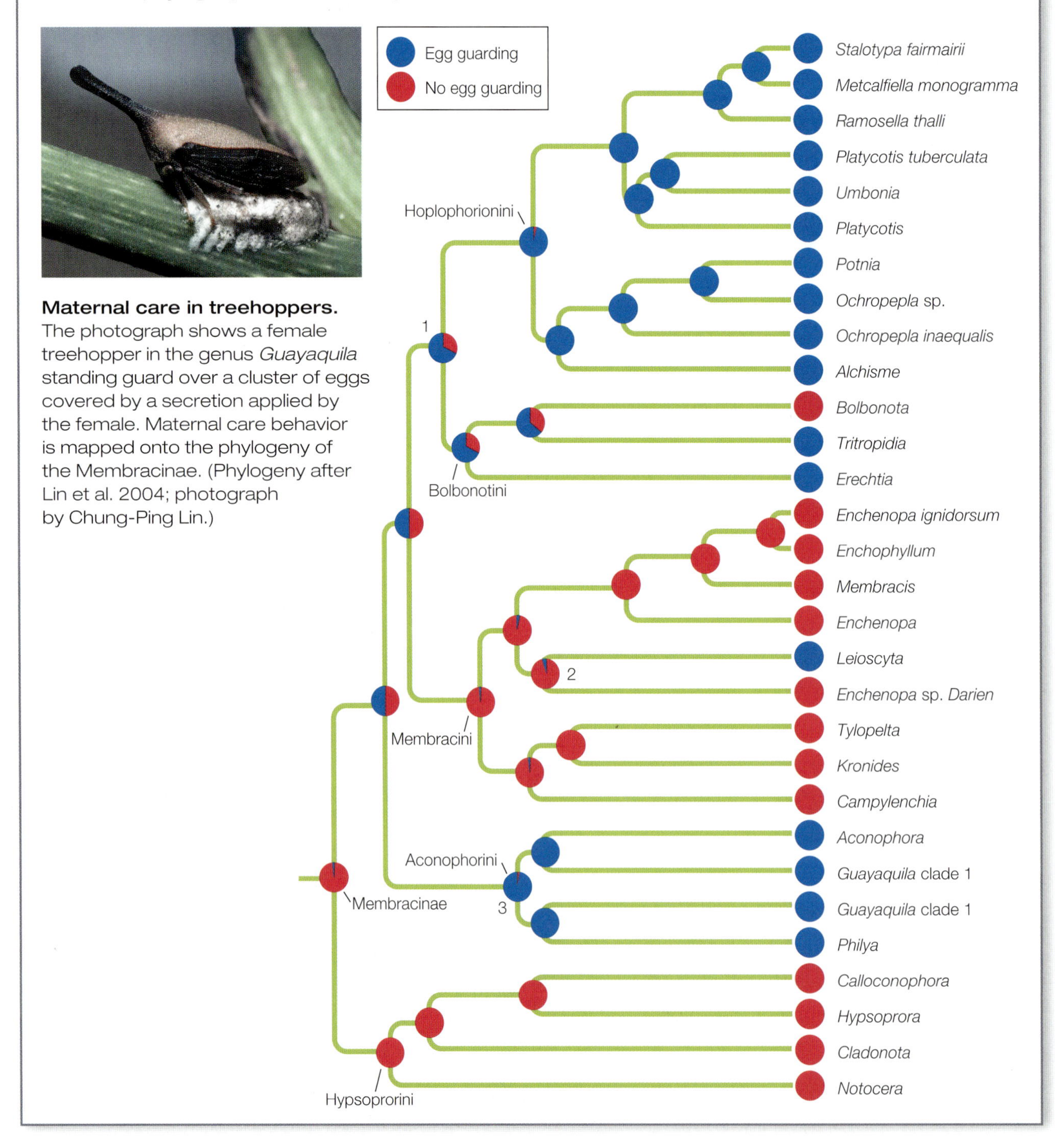

Maternal care in treehoppers. The photograph shows a female treehopper in the genus *Guayaquila* standing guard over a cluster of eggs covered by a secretion applied by the female. Maternal care behavior is mapped onto the phylogeny of the Membracinae. (Phylogeny after Lin et al. 2004; photograph by Chung-Ping Lin.)

BOX 11.1 EXPLORING BEHAVIOR BY INTERPRETING DATA *(continued)*

Thinking Outside the Box

On the basis of the phylogeny, how many times has maternal care been lost after evolving in the Membracinae? If you were to employ the comparative approach to examine the evolved function of maternal care in this group, what genera would be of special interest to you, and why?

sex takes care of the young becomes a source of conflict between the parents. Over time, this conflict can lead to the evolution of male-only, female-only, or biparental care systems. To understand how these systems evolve, let's consider the reasons why females and males care for young.

Why Do Females Care?

Let's begin by considering the costs of maternal care, which can be illustrated by examining the effects of brood tending on female European earwigs (*Forficula auricularia*). In this species, mothers often stay with clutches of eggs laid in burrows to feed their larval offspring after the eggs have hatched (**FIGURE 11.13**). Females can help their young survive, an obvious benefit of their tending behavior, but they pay a price as well. For maternal females, the interval between laying one clutch of eggs and the next is a week longer than for females that do not stay around to help one group of offspring get a good start in life (Kölliker 2007). The same cost applies to females of the harvestman *Serracutisoma proximum* (an arachnid), which lay more clutches and have nearly a 20 percent higher lifetime fecundity if they are prevented from standing guard over their eggs, compared with those females that tend their clutches (Buzatto et al. 2007). Thus, females often pay a reproductive price in terms of reduced future reproductive success when they care for young.

Because there is no guarantee that females will always derive a net benefit from an extra period of parental care, how can we explain the general pattern of female-only parental care? David Queller has argued that if the costs of parental care are lower for females than for males, then this could tip the scales so that females would be expected to provide more care than males (Queller 1997). Let's assume for the sake of simplicity that one standard unit of parental care invested in a current offspring reduces the future reproductive output of a male and a female by the same amount. Let's also assume that we are looking at a polyandrous species in which females sometimes mate with more than one male in a breeding season. Mating multiple times can increase the value of the offspring to the female (by, for example, increasing genetic diversity within the brood) and therefore reducing the cost to her of losing out on mating opportunities while parenting. In this case, the average benefit to a male from caring for a brood of offspring will be reduced to

FIGURE 11.13 A maternal female earwig guards her clutch of eggs, which she protects against predators. In nature, most females lay their eggs under leaf litter or some other shelter. (Photograph © Bert Pijsr/Minden Pictures.)

the extent that some of "his" offspring were actually fathered by other males. For example, if his paternity is, on average, 80 percent (he loses 20 percent to another male), then for every five offspring assisted, the male's investment can yield at most only four descendants. In contrast, all five offspring could advance the female's reproductive success. In other words, when a paternal male provides parental care to offspring that are not his own, the benefits of being a good parent decrease. Therefore, according to the **paternity assurance hypothesis**, males should only provide parental care when they are certain to be the fathers of the individuals they are feeding.

paternity assurance hypothesis Males are more likely to provide parental care when they are certain to be the fathers of those offspring.

One key prediction of the paternity assurance hypothesis is that when a male knows that his mate has mated with someone else and the likelihood of him fathering all of the offspring declines, he should reduce his level of parental care. Eduardo Santos and Shinichi Nakagawa tested the paternity assurance hypothesis in the dunnock (*Prunella modularis*), a species that, as we discussed in Chapter 10, forms both monogamous and polyandrous groups (Santos and Nakagawa 2017). In polyandrous groups, a male is likely to share paternity. Although in monogamous groups a male should be more likely to father more offspring, females can and do seek extra-pair fertilizations. In support of the paternity assurance hypothesis, the researchers found that socially monogamous males reduce their level of paternal care when they lose sires to extra-pair mates, an outcome that the males likely determine by assessing the behavior of their mates. However, in polyandrous groups in which males are already resigned to sharing paternity, there is no relationship between levels of paternal care and actual levels of paternity. Females in monogamous groups might not worry about this loss of paternal care if they are able to make up the difference themselves. However, the researchers also demonstrated that in monogamous—but not in polyandrous—broods, increased male feeding leads to higher fledgling success. Thus, since socially monogamous pair-bonded males reduce their level of paternal care when they suspect they lose paternity, females must weigh the genetic or other benefits of extra-pair copulations against the possibility of reduced parental care from their mates.

Even when a pair is faithful, there can still be conflict over investment and care. Fathers can continue to adjust the care they provide, and in some species mothers may actually manipulate the amount of paternal care their offspring receive. For example, in the burying beetle *Nicrophorus vespilloides*, Matthieu Paqueta and Per Smiseth tested the idea that mothers might be able to manipulate male parental care behavior via maternal effects at egg laying that alter offspring phenotype (Paquet and Smiseth 2017). The researchers performed a cross-fostering experiment in which they exposed females to males (or not) at the time of egg laying. They found that offspring were smaller at hatching when females laid eggs in the presence of a male, suggesting that females invest less in eggs when they expect males to help care for the offspring. Although offspring laid in the presence of a male gained more weight after hatching, this increase in offspring body mass was not because males cared for the offspring more (males provided less for the offspring). However, since males ate less of the dead mouse provisions themselves, there was more food available for the young. Thus, even though males are able to adjust their level of paternal care based on perceived levels of paternity or quality of the offspring, female burying beetles, as well as those of some other species, have the ability to manipulate males into caring more by altering offspring phenotype.

The relative proportion of parental care performed by mothers and fathers can also be influenced by ecological factors. For example, in a study of 36 populations from 12 plover species covering 6 continents, Vincze and colleagues examined how temperature fluctuations influenced paternal care during incubation (Vincze et al. 2016). The researchers found that both average ambient temperature and its fluctuation influenced parental cooperation during incubation, such that the share of parental duties performed by males increased with mean temperature and between-year variation in temperature during daylight hours. That is, when conditions became harsher (the mean temperature and/or the between-year unpredictability of temperature was high), males increased their incubation effort relative to females. The authors argued that this pattern of greater male care in harsher environments was a result of the need to protect the embryo when extreme weather events were more likely. Thus, males were willing to pay a greater cost and help females incubate more in environments where temperature-induced nest failure is more probable, likely leading to higher reproductive success.

Why Do Males Care?

As we discussed above, females reap higher benefits from parental care due to their assurance of genetic relatedness and because they can defray some of the cost of lost mating opportunities by producing mixed paternity broods. But what about males? What conditions lead to a better cost–benefit ratio for males than for females? As we noted when discussing sexual selection theory in Chapter 9, males that acquire many mates generally leave many descendants. Males that are successful in mating with many females would pay a steep price if they were to care for a few offspring at the cost of missing some additional mates. Imagine a lek of black grouse (*Lyrurus tetrix*) in which the top male fertilizes most of the eggs of the 20 or so females that come to the lek to mate. Because regular attendance at the lek is one of the main correlates of male mating success, a grouse with a reasonable chance of becoming an alpha male (a high-quality male with the necessary attributes to repel rival males and attract mates) would lose fitness if he took time off from lekking to incubate a batch of eggs (Queller 1997). The same rule surely applies to sexually attractive males of many other species, where males must often make a trade-off between parental care and mate attraction.

Given the logic of Queller's analysis of the costs of parental care and the fact that males have so much to lose by investing in care rather than seeking additional mating opportunities, you might think that paternal care should be exceptionally rare. In reality, male-only parental has evolved in a number of animals, including among fishes where it is actually quite common (see Figure 11.12). Because male fish are loaded with sperm and could therefore potentially have many more offspring than the most fecund female of their species, they would seem, at first glance, to have much to lose by taking time and energy away from mating effort to be good parents. Upon reflection, however, we can see that the trade-off between parental care and mate attraction is greatly reduced in mating systems in which females are drawn to egg-guarding, parentally committed males. Although the top males could still potentially achieve more fertilizations than paternal males that rear eggs from a handful of females, it equalizes the situation between males and females in terms of potential reproductive success when males can attract many egg-laying females *and* care for a large number of offspring.

In birds, the number of offspring a parent can care for is often constrained by the number of eggs it can incubate. Fish, however, do not face this constraint (because they do not have to warm the eggs, which are also much smaller than bird eggs) and can therefore care for many more eggs than can birds (**BOX 11.2**). In fact, male three-spined sticklebacks can care for up to ten clutches of eggs in one nest over the 2 weeks or so that it takes the eggs to hatch. But why do males perform the care rather than the females? It turns out that an average female stickleback can produce only seven clutches of eggs during this period, even without taking time out to guard her eggs (Clutton-Brock and Parker 1992). Thus, parenting would be far less advantageous for female sticklebacks than for males, given that a female would be

BOX 11.2 EXPLORING BEHAVIOR BY INTERPRETING DATA

Reactions of nest-defending bluegill males to potential egg and fry predators under two conditions

Territorial male bluegill sunfish (*Lepomis macrochirus*) defend the eggs and fry in their nests against predatory fish such as largemouth bass (*Micropterus salmoides*). The **FIGURE** shows how intensely males defended their nests in an experiment in which some territorial bluegills were exposed to potential cuckolders during the spawning season. Bryan Neff put sneaker males in plastic jars near the nests of his experimental subjects to provide the cues associated with a high risk of cuckoldry. He then measured male brood defense by quantifying how intensely bluegill dads charged and threatened a pumpkinseed sunfish (*Lepomis gibbosus*), which eats bluegill eggs and fry. Neff placed the sunfish predators in clear plastic bags before introducing them next to bluegill nests.

Reactions of nest-defending bluegill males to potential egg and fry predators under two conditions. Experimental males had been exposed to clear containers holding smaller male bluegills, mimicking the presence of rivals that might fertilize some of the eggs in the defenders' nests, whereas control males were not subjected to this treatment. Parental care was quantified using a formula based on the number of displays and bites directed at a plastic bag holding a predatory pumpkinseed sunfish. Bars depict mean +/– SE. (After Neff 2003.)

Thinking Outside the Box

How do you interpret the results shown in the figure? What is puzzling about them? Does it help to know that bluegill males can apparently evaluate the paternity of fry, but not of eggs, by the olfactory cues they offer?

brooding only one clutch at a time. Furthermore, while the female was so engaged, she would not be able to forage freely, and so would not grow as rapidly as she might otherwise, an especially damaging outcome for those species in which female fecundity increases exponentially with increasing body size. Males that are parental also grow more slowly than they would otherwise, but since they must remain in a territory anyway if they are to attract mates, the decrease in their growth resulting exclusively from parental care is negligible.

The costs of parental behavior for the two sexes have been directly measured for a mouth-brooding cichlid, the St. Peter's fish (*Sarotherodon galilaeus*), in which either the male or the female may care for their young by orally incubating the fertilized eggs. Both sexes lose weight while mouth brooding, not surprisingly since it is difficult to eat with a mouth full of eggs or baby fish. Furthermore, the interval between spawnings increases for parental fish of both sexes compared with individuals whose clutches are experimentally removed (**FIGURE 11.14**). However, parental females wait 11 more days between spawnings than nonparental females, whereas brooding males pay a smaller price of only 7 extra days between spawnings compared with nonparental males. Moreover, parental females produce fewer young in the next clutch than do nonparental females, whereas parental males are

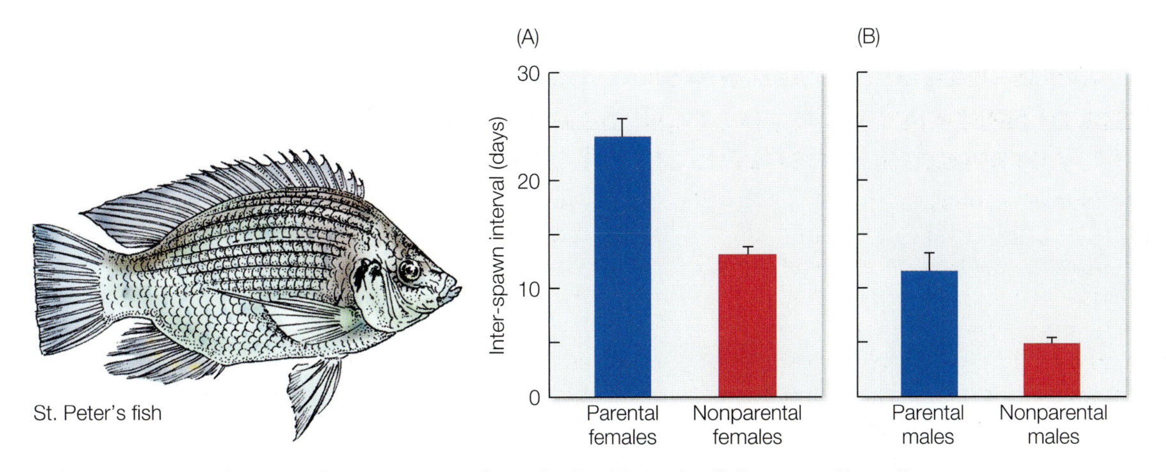

FIGURE 11.14 Parental care costs female St. Peter's fish more than it costs males. (A) Females that have cared for a clutch of eggs are much slower to produce a new batch of mature eggs than are females that have not provided care to their previous clutch. (B) Parental males also spawn less often than nonparental males, but the difference between the two groups is less than for females. Bars depict mean +/– SE. (After Balshine-Earn 1995.)

just as able to fertilize complete clutches of eggs in their next spawning as are nonparental males. Therefore, although both sexes pay a price for being parental, the costs of brood care are greater for females than for males (Balshine-Earn 1995). Thus, it is more likely that paternal males will have higher fitness than maternal females, leading to male parental care (Gross and Sargent 1985).

Another group in which exclusive paternal care is relatively common is the neotropical poison frogs (families Aromobatidae and Dendrobatidae) (**FIGURE 11.15**). Parental care is actually quite rare in frogs, occurring in less than 20 percent of species. However, in the poison frogs nearly all species exhibit some form of care, ranging from the transport of tadpoles from terrestrial egg clutches to bodies of water, to extended care by females, males, or both sexes. Male parental care seems to be the ancestral state in poison frogs, with biparental and female-only care having evolved later (Summers and Tumulty 2014). As in fishes, paternal care in frogs may be an outgrowth of territorial defense of oviposition sites for females. Many species transport and care for eggs in arboreal pools of water that collect in leaf axils (such as those of bromeliads, *Heliconia*, tree holes, bamboo stalks, palm fronds, and so forth) where young are safe from predators. Defense of these resources is generally done by males, partly as a way to ensure higher confidence in paternity. Thus, in frogs, male territoriality and the guarding of safe places to lay eggs, as well as the need to ensure paternity, promote paternal care in several species.

Although exclusive paternal care of young is common among fishes and some frogs, it is rarer in other animals (Clutton-Brock 1991, Tallamy 2001). Yet studying these cases of sex role reversal can tell us a great deal about how parental care more broadly has evolved. Polyandry, which is rare in birds (occurring in less than 1 percent of species, primarily waders and shorebirds), is often accompanied by sex role reversal and male-only parental care. For example, female northern jacanas (*Jacana spinosa*) guard territories in ponds or lakes that contain floating vegetation, on which males build nests (Emlen and Oring 1977). Within a single territory, a female will lay eggs in the nests of multiple males, which perform most parental duties. Female jacanas are larger than males, as is the case in many role-reversed species due to sexual selection, and they help defend the territory and nests from predators and other jacanas. In another example, males of the harvestman *Iporangaia pustulosa*

FIGURE 11.15 Poison dart frogs. (A) Strawberry poison-dart frog (*Oophaga pumilio*) carrying tadpoles; (B) Strawberry poison-dart frog (*Oophaga pumilio*); (C) Amazonian poison-dart frog with tadpoles (*Ranitomeya ventrimaculata*); (D) Blue poison-dart frog (*Dendrobates azureus*); (E) Harlequin poison-dart frog (*Oophaga histrionica*); (F) Yellow banded poison-dart frog on a leaf (*Dendrobates leucomelas*). (Photographs: A, adrian hepworth/Alamy Stock Photo; B, © iStock.com/NTCo; C, © iStock.com/kikkerdirk; D, © iStock.com/Azureus70; E, © iStock.com/Studio-Annika; F, © iStock.com/SHAWSHANK61.)

sometimes spend months standing near eggs laid on the underside of leaves (**FIGURE 11.16**). In a male-removal experiment lasting 12 days, clutches that were left unattended were attacked by other harvestmen in almost every instance, whereas half of the control clutches with a guarding male were intact at the end of the study (Requena et al. 2009). In other species of harvestmen with paternal care, females appear to be attracted to partners with eggs in their care, suggesting that the males'

FIGURE 11.16 A male harvestman (*Iporangaia pustulosa*) protecting eggs that he has surely fertilized. In this neotropical species, males are the guardians of offspring; without paternal care, egg survival is seriously compromised. (Photograph by Andrew M. Snyder/Moment Open/Getty Images.)

unusual behavior has evolved via sexual selection, just as is often true for fishes, frogs, and birds with highly paternal males (Machado et al. 2004, Nazareth and Machado 2010, Summers and Tumulty 2014).

Among the paternal insects are large male water bugs in the genus *Lethocerus*, which guard and moisten clutches of eggs that females glue onto the stems of aquatic vegetation above the waterline (**FIGURE 11.17A**) (Smith and Larsen 1993). In some other water bug genera (such as *Abedus* and *Belostoma*), care goes a step further, as males permit their mates to lay eggs directly on their backs (**FIGURE 11.17B**). A brooding male *Abedus* spends hours perched near the water surface, pumping his body up and down to keep well-aerated water moving over the eggs. Clutches that are experimentally separated from male attendants do not develop, demonstrating that male parental care is essential for offspring survival in this case.

Bob Smith has explored both the history and the adaptive value of these unusual paternal behaviors in water bugs (Smith 1997). Since most insects, including some close relatives of the Belostomatidae, lack male parental care, it is probable that species with egg-brooding males evolved from nonpaternal ancestors (**FIGURE 11.18**). Whether out-of-water brooding and back brooding evolved independently from such an ancestor, or whether one preceded the other, remains unknown. However, when *Lethocerus* females cannot find suitable exposed vegetation for their eggs, they sometimes lay their eggs on the backs of other individuals, male or female. This unusual behavior indicates how the transition from out-of-water brooding to back brooding might have occurred.

(A)

(B)

FIGURE 11.17 Male water bugs provide uniparental care. (A) In the genus *Lethocerus*, males stand watch over eggs their mates lay on vegetation out of water. (B) In other genera of belostomatids, males remain in water to brood eggs glued onto their backs by their mates. (A, photograph © CSP_feathercollector/Fotosearch LBRF/AGE Fotostock; B, photograph by Greg Hume/CC BY-SA 3.0.)

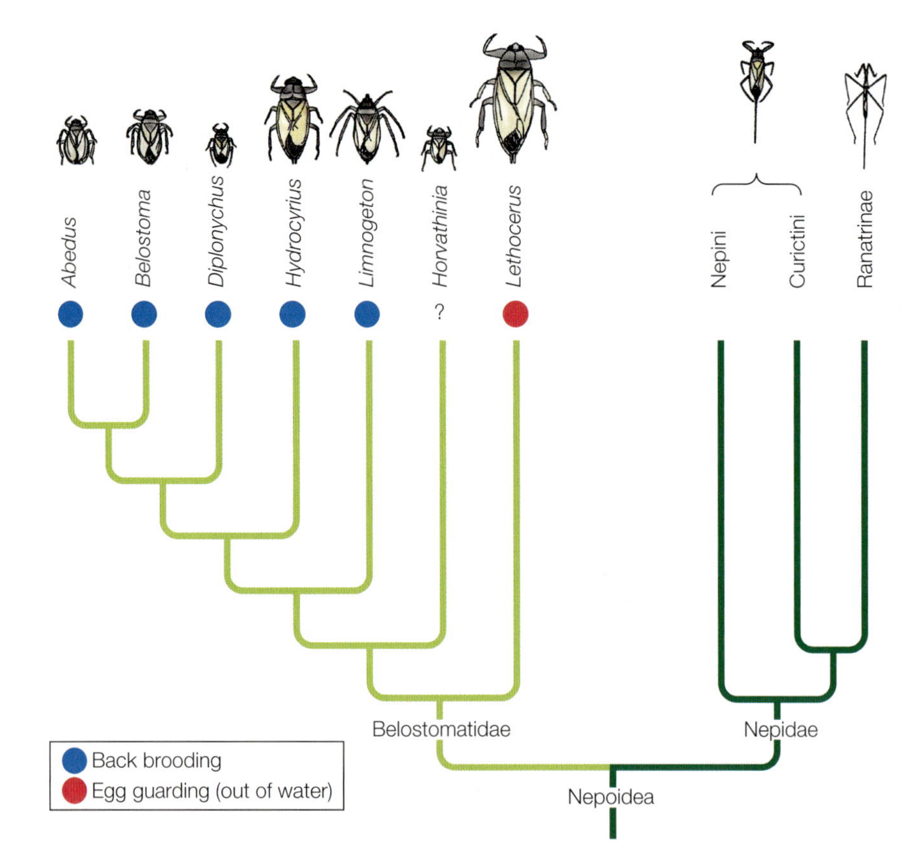

FIGURE 11.18 Evolution of brood care by males in the Nepoidea (the group that includes the belostomatid water bugs). The illustrations, drawn to scale, show the largest representatives of each group. Paternal care is widespread throughout the family Belostomatidae, but no species in the family Nepidae exhibits this trait. (After Smith 1997.)

But why do the eggs of water bugs require costly brooding? Huge numbers of aquatic insects lay eggs that do just fine without a caretaker of either sex. Smith notes that belostomatid eggs are much larger than the standard aquatic insect egg, with a correspondingly large requirement for oxygen needed to sustain the high metabolic rates underlying embryonic development. But the relatively low surface-to-volume ratio of a large aquatic egg leads to an oxygen deficit inside the egg. Since oxygen diffuses through air much more easily than through water, laying eggs out of water can solve that problem. But this solution creates another problem, which is the risk of desiccation that the eggs face when they are high and dry. The solution, brooding by males that moisten the eggs repeatedly, sets the stage for the evolutionary transition to back brooding at the air–water interface.

Wouldn't things be simpler if belostomatids simply laid small eggs with large surface-to-volume ratios? To explain why some water bugs produce eggs so large that they need to be brooded, Smith points out that water bugs are among the world's largest insects, an advantage when it comes to grabbing and stabbing fish, frogs, and tadpoles. Water bugs, and all other insects, grow in size only during the immature stages. After the final molt to adulthood, no additional growth occurs. As an immature insect molts from one stage to the next, it acquires a new flexible cuticular skin that permits an expansion of size, but no more than a 50 to 60 percent increase per molt. One way for an insect to grow large, therefore, would be to increase the number of molts before making the final transition to adulthood. However, no member of the belostomatid family molts more than six times. This observation suggests that these insects are locked into a five- or six-molt sequence, just as spotted sandpipers evidently cannot lay more than four eggs per clutch. If a water bug is to grow large enough to kill an adult frog, then the first instar (the nymph that hatches from the egg) must be large, because it will get to undergo only five or so 50-percent expansions. And for the first-instar nymph to be large, the egg must be large. For large eggs to develop quickly, they must have access to oxygen, which is where male brooding comes into play. Thus, male brooding is an ancillary evolutionary development whose foundation lies in selection for a body size that enables a bug to take down relatively large prey (Smith and Larsen 1993). The water bug story revolves around the "panda principle" (see Chapter 8) in that evolutionary modifications of body size were layered on what had already evolved in this lineage.

Female water bugs, however, could conceivably provide care for their own eggs after laying them on exposed aquatic vegetation. Why is it that the males do the brooding, never the females? Here we have a story that parallels the fish case closely. First, male water bugs with one clutch of eggs sometimes attract a second female, perhaps because a male bug with a partial load of eggs is effectively advertising his capacity for parental care, just like a male stickleback at his nest (Tallamy 2001). Second, just as is true for some fishes, the costs of parental care may be disproportionately great for females in terms of lost fecundity. To produce large clutches of large eggs, female belostomatids require far more prey than males do. And because brooding limits mobility and thus access to prey, parental care probably has greater fitness costs for females than for males, biasing selection in favor of male parental care.

Discriminating Parental Care

No matter which parent provides care, one would not expect the caring parent (or parents) to provide assistance freely to young animals that are not their own genetic offspring. After all, why would a parent expend energy to care for an individual that is not related to itself (see Chapter 12)? As you will see below, there are several cases where parents take care of others' offspring, including examples where parents are unable to discriminate between the young of their own species, or even those of other species.

Recognizing One's Own Offspring

Sometimes parents might have difficulty identifying their own offspring. Consider the Mexican free-tailed bat (*Tadarida brasiliensis*), which migrates to certain caves in the American Southwest, where pregnant females form colonies in the millions. After giving birth to a single pup, a mother bat leaves her offspring clinging to the roof of the cave in a crèche that may contain 4000 pups per square meter (**FIGURE 11.19**). When the female returns to nurse her infant, she flies back to the spot where she last

FIGURE 11.19 Mexican free-tailed bat mothers recognize their pups in the crowded colony. Mothers are able to do this despite the fact that they leave their infants in a dense mass of baby bats when they leave their caves to forage outside. When a female returns to nurse her pup, she almost always relocates her son or daughter among thousands of other individuals. (Photograph by Johner Images/Alamy Stock Photo.)

nursed her pup and is promptly besieged by a host of hungry baby bats (McCracken 1984). Given the swarms of pups, early observers believed that mothers simply provided milk on a first-come, first-served basis. But are Mexican free-tailed bat mothers really indiscriminate parents? To find out, Gary McCracken captured female bats and the pups nursing from them and took blood samples from both (McCracken 1984). He found that females located their own pups at least 80 percent of the time, a conclusion confirmed and extended by more recent evidence that females use the vocal and olfactory signals of pups to recognize their own pups before letting them nurse (Balcombe 1990, McCracken and Gustin 1991).

If the ability to recognize one's genetic offspring (a type of kin recognition; see Chapter 13) is advantageous in proportion to the risk of misdirected parental care, then we would expect convergent evolution of parent–offspring recognition in colonial mammals other than Mexican free-tailed bats. Tests of this prediction have been supported in species as different as the common degu (*Octodon degus*), a small rodent in which females rear young together in communal burrows (Jesseau et al. 2008), and the subantarctic fur seal (*Arctocephalus tropicalis*) in which females give birth on crowded island beaches. Female seals remain with their pups for about a week before leaving for an oceanic fishing trip that can last as long as 3 weeks. Playback experiments demonstrate that a baby seal takes no more than 5 days to learn its mother's call, while the females are also quick learners. When a mother seal returns to the beach, she calls out, and her infant calls back, usually leading to their reunion in less than 15 minutes (Charrier et al. 2003).

A comparative test of the hypothesis that offspring recognition functions to prevent misdirected care takes advantage of variation among swallow species in the risk of making parental mistakes. Although both the bank swallow (*Riparia riparia*) and the northern rough-winged swallow (*Stelgidopteryx serripennis*) nest in clay banks, the bank swallow is colonial, whereas the rough-winged swallow nests by itself. Individual fledglings of the colonial bank swallow produce highly distinctive vocalizations, giving their parents a reliable cue to use when making decisions about which individuals to feed and which to repel (such as the fledglings that sometimes wind up in the wrong nests begging for food). Bank swallow parents rarely make mistakes, despite the high density of nests in their colonies (Beecher et al. 1981, Medvin and Beecher 1986). In contrast, the solitary rough-winged swallow never has a chance in nature to feed another's fledglings and so would not be expected to have evolved sophisticated offspring recognition mechanisms, which surely entail an energetic cost. Indeed, rough-winged swallow chicks produce calls that sound much more alike than those of bank swallows, a reflection of the fact that young rough-wings need not communicate their identity to their parents.

Two other swallow species, the highly colonial cliff swallow (*Petrochelidon pyrrhonota*) and the less social barn swallow, also differ in their chick recognition attributes. As expected, cliff swallow chicks produce calls containing about 16 times as much variation as the corresponding calls of barn swallow chicks (**FIGURE 11.20**) (Medvin et al. 1993). Therefore, it should be easier for cliff swallows to recognize their young than for barn swallows to discriminate among barn swallow chicks based on their vocalizations. In operant conditioning experiments that required adults of both species to discriminate between pairs of chick calls, cliff swallows reached 85 percent accuracy significantly faster than barn swallows. These results suggest that the acoustic perception systems of the cliff swallow, as well as its calls, have evolved to promote accurate offspring recognition (Medvin et al. 1993), just as is true of Mexican free-tailed bats and subantarctic fur seals.

Even though many swallows, bats, and other animals that live in colonies have evolved specialized ways to recognize their own offspring, parents in some species occasionally make mistakes that lead them to adopt genetic strangers (**BOX 11.3**).

(A)

Cliff swallow

(B)

Barn swallow

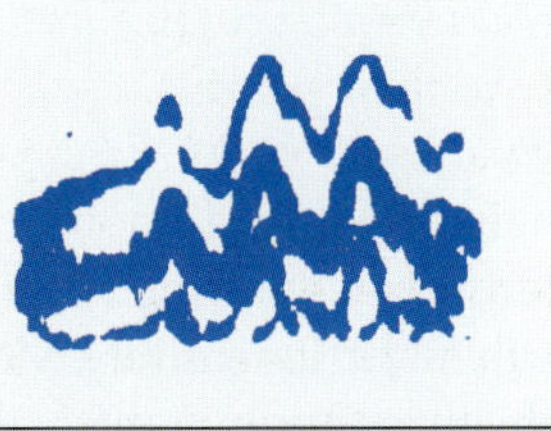

FIGURE 11.20 Recognition of offspring evolves when the risk of misdirected parental care is high. (A) Chicks of cliff swallows, a colonial species, produce highly structured and distinctive calls, helping their parents recognize them as individuals. (B) The calls of barn swallow chicks, a less colonial species, are much less structured and more similar. Sonograms of the sort shown here (and elsewhere) provide a visual record of the sound frequencies produced by the signalers over time. The call frequencies of both species lie between 1 and 6 kHz; the durations of the calls range from 0.7 to 1.3 seconds for the cliff swallows and from 0.4 to 0.8 seconds for the barn swallows. (After Medvin et al. 1993; A, photograph © Natalia Kuzmina/Shutterstock; B, photograph © iStock.com/Kaphoto.)

BOX 11.3 **DARWINIAN PUZZLE**

Why do parents in some species adopt genetic strangers of their own species?

The cases described thus far support the prediction that parents should recognize their own young and discriminate against others of their own species when the probability of being exploited by someone else's offspring is high. And yet some colonial, ground-nesting gulls occasionally adopt unrelated chicks of their own species. Although researchers initially reported that adults consistently rejected older, mobile chicks when they were experimentally transplanted between nests (Miller and Emlen 1975), attacks by adults on these transferred youngsters apparently occurred because of the frightened behavior of the displaced chicks (Graves and Whiten 1980). When juveniles voluntarily leave their natal nests—which they sometimes do if they have been poorly fed by their parents (**FIGURE**)—they do not flee from potential adopters but instead beg for food and crouch submissively when threatened. These young chicks have a good chance of being adopted, even at the ripe old age of 35 days (Holley 1984). If they are taken in by a set of foster parents, they are more likely to survive than if they had remained with the genetic parents that were failing to supply them with enough food (Brown 1998).

When parents apparently fail to act in the best interests of their genes by taking care of young that are not their own, we must consider the costs, not just the benefits, of what might appear to be better for those genes. And learned recognition of offspring carries costs as well as benefits, notably the risk of making a mistake by not feeding, or even attacking and killing, one's own offspring. Rather than erring on the side of harming their genetic offspring, gulls have evolved a cautious readiness to feed any chicks in their nest that beg confidently when approached by an adult (Pierotti and Murphy 1987). Sometimes this rule of thumb permits a genetic outsider to steal food from a set of foster parents by slipping into a nest with other youngsters of its age and size (Knudsen and Evans 1986). When adoption occurs, the adoptive parents lose about 0.5 chicks of their own on average. However, adoption is rare, with fewer than 10 percent of adult ring-billed gulls (*Larus delawarensis*) taking in a stranger in any year (Brown 1998). The modest average annual fitness cost of a rule of thumb that results in occasional adoptions has to be weighed against the cost of rejecting one's own genetic offspring that would arise if parent gulls were more reluctant to feed chicks in their nests. This factor could help explain the gullibility of adoptive parent gulls.

Interspecific Brood Parasitism

Similar rules of thumb are employed by solitary living birds as well, which unlike their colonial counterparts, are at risk for exploitation by other species entirely. Most solitary living species have never needed to develop sophisticated egg or offspring recognition systems, and so they (like some colonial species as well) are easy prey for specialized **interspecific brood parasites**, which can insinuate themselves into a family of a completely different species by employing deceptive signals that trigger parental care by their hosts. For example, after killing all of the host's young (**VIDEO 11.3**), a parasitic common cuckoo (*Cuculus canorus*) chick mimics an entire brood of reed warbler (*Acrocephalus scirpaceus*) chicks, stimulating their caregivers to bring it as much food as they would provide to several of their own chicks (**FIGURE 11.21**). In contrast to the homicidal tendencies of common cuckoo chicks, brown-headed cowbird (*Molothrus ater*) chicks tolerate host young, whose presence in the nest actually procures the larger cowbird chick a higher provisioning rate than if it

VIDEO 11.3

Common cuckoo tossing great reed warbler eggs

ab11e.com/v11.3

BOX 11.3 **DARWINIAN PUZZLE** *(continued)*

Ring-billed gull and chick

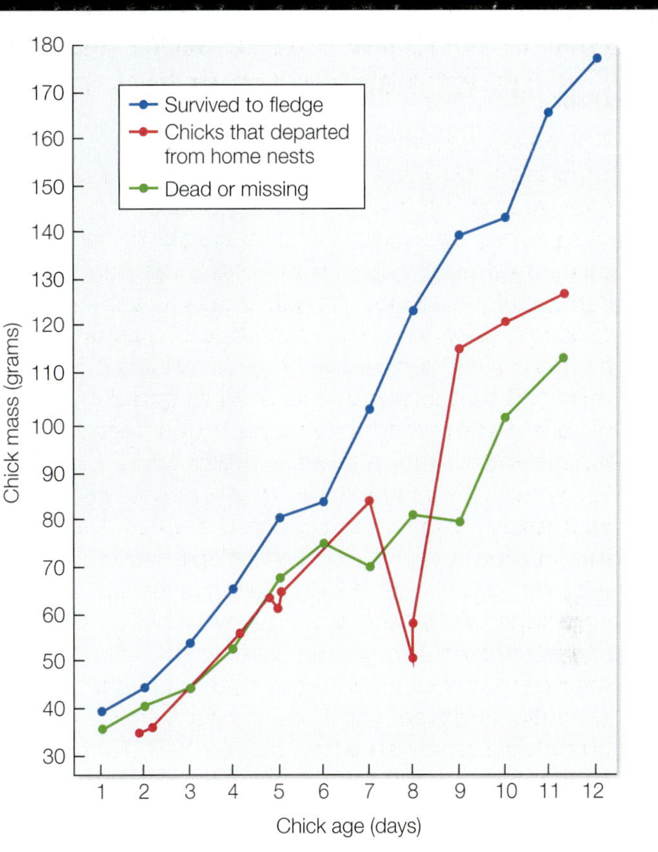

Why seek adoptive parents? Gull chicks that abandoned their natal nests in search of foster parents weighed much less than average for chicks their age. Yet these potential adoptees were sometimes adopted by nonrelatives, and as a result they weighed more on average at day 11 than did the subset of unadopted (poorly fed) gull nestlings that later died or disappeared. (After Brown 1998; photograph © Reimar/Alamy.)

Thinking Outside the Box

Extend this cost-benefit approach and consider some scenarios when foster parents may be more or less likely to adopt unrelated gull chicks. Consider both the social and the ecological environment as you think outside the box.

were alone (Kilner et al. 2004). Interspecific brood parasitism is not limited to birds; it also occurs in bees and other insects. Take, for example, the cuckoo bees, which get their name from avian cuckoos and which lay their eggs in the brood chambers of other solitary bee species. Since both cuckoo bees and wasps get their name from cuckoo birds, the remainder of this section will explore brood parasitism in birds, the group in which this parental care system was first described.

The energetic benefits (to the parasite) and costs (to the host) of interspecific brood parasitism should be obvious. Yet for a parasite nestling to take advantage of a host species' parental decisions, the egg containing the parasite has to hatch. Because some host species take immediate action against the eggs of parasites (for example, by burying a foreign egg, removing it from the nest, or abandoning the

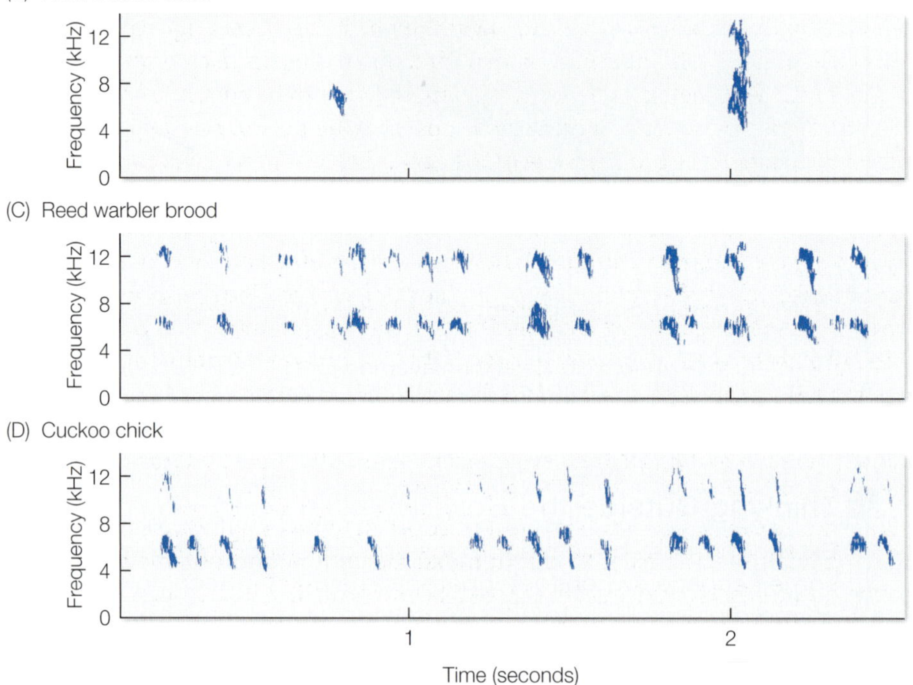

FIGURE 11.21 The common cuckoo chick's begging call matches that of four baby reed warblers. (A) A cuckoo chick begging for food from its reed warbler foster parent. The calls shown below the photograph are those of (B) a single reed warbler chick, (C) a brood of four reed warblers, and (D) a single cuckoo chick. (A, photograph © iStock.com/Maurizio Bonora; B–D after Davies et al. 1998.)

nest altogether), we can ask why don't all parasitized bird species reject unwanted eggs (**FIGURE 11.22**) (Winfree 1999)? One reason may be that parents that make incubation of eggs dependent on their learned recognition of the eggs they lay might sometimes abandon or destroy some of their own eggs by mistake, treating them as if they were foreign eggs. Indeed, reed warblers sometimes do throw out some of their own eggs while trying to get rid of cuckoo eggs (Davies and de L. Brooke 1988). If brood parasites victimize only a very small minority of the host population, then even a small chance of recognition errors by the host could make accepting parasite eggs the adaptive option (Lotem et al. 1995). This prediction has been tested with reed warblers in Britain, where the risk of parasitic exploitation is low while the chance of recognition errors is modest but not zero. Unless the

FIGURE 11.22 Why do many birds accept a cowbird's egg? This nest contains two eggs of the indigo bunting (*Passerina cyanea*) host and one very different egg laid by a cowbird. The puzzle is, why doesn't the bunting recognize that its nest has been parasitized? (Photograph © Agnieszka Bacal/Shutterstock.com.)

warblers had seen a (stuffed) cuckoo near their nest, they generally accepted model cuckoo eggs placed in their nest by experimental biologists (Davies et al. 1996).

Acceptance of the parasite egg is also more likely to be adaptive when the host is a small species, perhaps because the host may be unable to grasp and remove the larger parasitic egg, or because of the general rule of parental favoritism to feed the biggest chick more, thus giving the parasite chick an advantage (Rohwer and Spaw 1988). Such small-billed birds have two choices: abandon the clutch, either by leaving the site or by building a new nest on top of the old one, or stay put and continue brooding the clutch along with the parasite egg. The abandonment option imposes heavy penalties on the hosts, which must at a minimum build a new nest and lay a new clutch of eggs. The costs of this option are especially high for cavity-nesting species because acceptable tree cavities are generally rare. Likewise, yellow warblers (*Setophaga petechia*) tend to accept foreign eggs when their nests are parasitized near the end of the breeding season, when little time remains to rear a new brood from scratch (Sealy 1995).

Even if host birds could throw out or cover up a brood parasite's eggs without making mistakes, the parasite might make this option unprofitable by returning to the nest to check whether its egg had been harmed; if it had been, the parasite might retaliate against the host by destroying its eggs (**FIGURE 11.23**). This **mafia hypothesis** (Zahavi 1979) has been tested by examining the interactions between parasitic great spotted cuckoos (*Clamator glandarius*) and a host species, the European magpie (*Pica pica*) (Soler et al. 1995). Magpie nests from which cuckoo eggs had been ejected suffered a significantly higher rate of predation than did nests with accepted cuckoo eggs (87 percent vs. 12 percent). Furthermore, when researchers removed the cuckoo egg from a nest that was apparently being checked by a cuckoo and replaced the magpie's eggs with Plasticine imitations, the cuckoo approached the nest after the researchers had finished and left its beak marks on the false magpie eggs. Magpies that lose their clutches have to renest, which exposes them to all the negative effects that delays in breeding have in a seasonal environment. In this light, it is not surprising that acceptor magpies actually have somewhat higher reproductive success, as measured in terms of fledglings produced, than do ejectors of cuckoo eggs (Soler et al. 1995).

mafia hypothesis Hosts accept brood parasitic eggs out of fear of retaliation by the brood parasite for destroying its eggs.

Brown-headed cowbirds are also avian mafiosi, as Jeffrey Hoover and Scott Robinson demonstrated in a study of prothonotary warblers (*Protonotaria citrea*)

FIGURE 11.23 Cowbirds, such as this female visiting a host's nest, are among the parasitic birds that may punish hosts that fail to incubate the parasite's eggs. (Photograph © S & D & K Maslowski/Minden Pictures, Inc.)

(Hoover and Robinson 2007). The researchers worked with a large sample of warblers that had built their nests in boxes on top of greased poles, which made them immune to snakes and small mammals that often prey on the eggs and young of prothonotary warblers. But these nests were still vulnerable, at least initially, to cowbirds, which could enter a nest and deposit an egg there. Nests parasitized in this fashion were then divided into three groups: (1) the cowbird egg was removed but the nest opening was left as it was—wide enough to permit the reentry of a cowbird; (2) the cowbird egg was not touched and the nest entrance was not modified; and (3) the cowbird egg was removed and the nest entrance was made smaller so that only warblers, not cowbirds, could then enter the nest box. When the cowbird egg was removed, often the nest was subsequently visited by an avian predator, almost certainly the cowbird whose egg had been taken away, and the warbler eggs were destroyed. However, when the parasite's egg was not ejected by the experimenters, the warbler's eggs were much less likely to disappear, even though adult cowbirds still had free access to the nests in this category. Finally, when cowbirds were prevented from revisiting nests they had parasitized, the loss of warbler eggs did not occur (**FIGURE 11.24**). These results strongly suggest that cowbird females often come back to nests they have parasitized, which enables them to destroy the eggs of any host bold enough to get rid of an unwanted egg. This tactic may cause the host to renest, a decision that may enable the parasite to strike again with greater success (Hoover and Robinson 2007).

Choosing the Correct Host

Perhaps the key decision that every parasite mother must make is choosing which nest to lay her egg in. In brown-headed cowbirds, the size of the nest (Mclaren and Sealey 2003) as well as the size of the host eggs can affect a parasite's decision to lay (Merrill et al. 2017). This decision, however, means not only choosing a specific nest structure, but in some cases choosing a host species. That is, some parasitic species are host specialists that lay their eggs in the nest of only a single host species, but others are host generalists that can lay their eggs in the nests of many host species. Although host generalists seem to have many options of where to lay their eggs, they must deal with nests of different shapes and sizes, eggs of different colors and

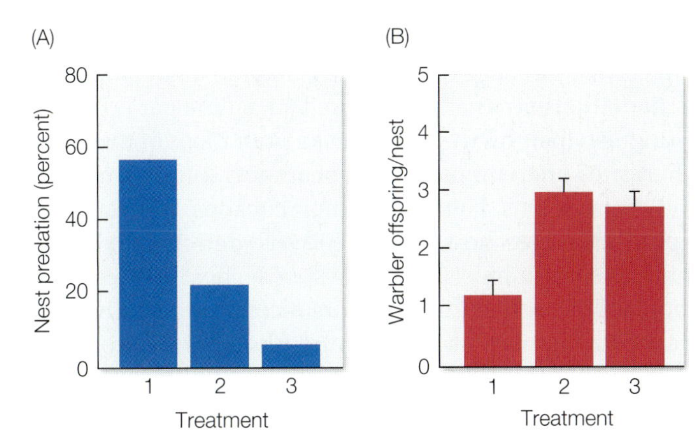

Female brown-headed cowbird

FIGURE 11.24 The mafia hypothesis as tested with parasitic cowbirds and prothonotary warblers. (A) In treatment 1, a cowbird laid an egg in the nest, which was then removed by the experimenter. Subsequently, the warbler eggs in most nests in this treatment were destroyed, presumably by the thwarted cowbird. In treatment 2, all nests were parasitized but the cowbird eggs were left in the nests, which were largely untouched by predators thereafter. In treatment 3, the cowbird eggs were removed from the parasitized nests, which were then made inaccessible to cowbirds; none of these nests was harmed after removal of the parasite's egg. (B) The warblers produced more offspring under treatments 2 and 3 than under treatment 1. Bars depict mean +/– SE. (After Hoover and Robinson 2007, © 2007 National Academy of Sciences, USA; photograph © iStock.com/ziggy7.)

patterns, and different behavioral responses to parasitic eggs and chicks. Thus, while it may be easier for host generalists to locate a host nest compared with host specialists, generalists often cannot produce particularly good mimetic eggs, which are perhaps at greater risk of being identified by the host. Perhaps then, while a species may be a generalist brood parasite, some individuals or populations may be more specialized on one or a few hosts. Indeed, it turns out that in some brood parasitic species, such as the common cuckoo, there are distinct races that parasitize select target host species. Using mitochondrial DNA, which is passed down maternally from mother to daughter, Claire Spottiswoode and colleagues have shown that the greater honeyguide (*Indicator indicator*) (see Chapter 8), which parasitizes several African birds (**VIDEO 11.4**), exhibits two highly divergent lineages that are associated with ground- versus tree-nesting host species (Spottiswoode et al. 2011). In other words, some honeyguides are specialized parasites of ground-nesting bird species and others of tree-nesting species; few honeyguides parasitize both types of species. A lack of genetic differentiation in nuclear DNA—which is passed down both maternally and paternally—further suggests that these distinct races interbreed normally. These results suggest that mothers pass their host preference on to their daughters, as has also been suggested from studies of common cuckoos (Gibbs et al. 2000) and shiny cowbirds (*Molothrus bonariensis*) (Mahler et al. 2007).

VIDEO 11.4
Parasitic honeyguides killing host chicks
ab11e.com/v11.4

Coevolutionary Arms Races

Whenever there are two parties in conflict with each other, as there are in the case of a brood parasite and its host, the two sides exert reciprocal selection pressure on each other, with an adaptive advance made by one often leading eventually to an adaptive counterresponse by the other. This phenomenon is called a **coevolutionary arms race**, something that has also been seen in host–parasite interactions (Kilner and Langmore 2011). The adaptations and counteradaptations that avian hosts and parasites employ in the arms race can occur throughout the nesting cycle.

Although these can begin prior to egg laying (for example, hosts mobbing parasites, and parasites evolving cryptic plumage or secretive behaviors to avoid angry hosts), most of the counteradaptations occur after eggs are laid, a time period when hosts need to be able to distinguish their own eggs or chicks from those of the parasite (Feeney et al. 2014b). Parasites often mimic host appearance, sound, smell, or chemical makeup to exploit their victims. Female common cuckoos, for example, belonging to distinct races parasitize select host species. To avoid detection by hosts, cuckoos have evolved remarkable egg mimicry, and in response, host species have evolved better discrimination abilities, often through more recognizable individual egg signatures. Using advanced computer vision tools and photographs from 689 host eggs from 206 cuckoo-parasitized clutches belonging to eight of the cuckoo's preferred European hosts, Cassie Stoddard and colleagues asked whether hosts of the common cuckoo have evolved eggs with individual pattern signatures (Stoddard et al. 2014). The researchers showed that, consistent with the idea of a coevolutionary arms race, those hosts subjected to the best cuckoo egg mimicry have evolved the most distinctive egg signatures.

Unique host egg signatures not only make it easier for hosts to recognize parasite eggs, but they also make it harder for parasites to produce mimetic eggs. After all, a single parasite may not be able to match all of the egg patterns in different host nests. One way that parasites might be able to circumvent these unique host egg signatures is to repeatedly parasitize the same host individual with multiple eggs, as doing so should make accurate rejection decisions harder for the host, regardless of the mechanism it uses to identify foreign eggs. Martin Stevens and colleagues tested this prediction experimentally in the parasitic cuckoo finch (*Anomalospiza imberbis*) and its most common host, the African tawny-flanked prinia (*Prinia subflava*) (Stevens et al. 2013). In both the parasite and host species, egg coloration and patterning varies greatly among individuals, but each individual female always lays a single egg type throughout her life. By switching prinia eggs from other prinia nests (to simulate brood parasitism), the researchers showed that prinia females can recognize their own eggs and reject others' eggs (**FIGURE 11.25A**). However, the researchers found that as the proportion of "parasitic" eggs in a nest increases, hosts are less likely to reject them and require greater differences in appearance to do so (**FIGURE 11.25B**). Thus, if cuckoo finches were to lay multiple eggs in a prinia nest, they would have a better chance of hatching because the prinia would not be able to distinguish its own eggs from those of the parasite.

The arms race approach helps us make sense of another well-studied interaction, between the parasitic Horsfield's bronze-cuckoo (*Chrysococcyx basalis*) and the superb fairy-wren (*Malurus cyaneus*) (Langmore and Kilner 2007). Although the Horsfield's bronze-cuckoo is a generalist parasite that can lay its eggs in the nests of a variety of species, its preferred host is the cooperatively breeding superb fairy-wren, which you might recall from Chapter 10 exhibits the highest degree of extra-pair paternity of any bird studied. Unlike the common cuckoo from Europe, the Horsfield's bronze-cuckoo does not mimic its host eggs very well, presumably because it can use multiple hosts rather than specializing on a single one (Feeney et al. 2014a). If breeding fairy-wrens find an egg in their nest before they have begun to deposit their own eggs there, they almost always build over the intruder's egg, and they abandon nests altogether if a cuckoo drops an egg in after the wrens have begun to incubate their own complete clutch. However, cuckoos turn out to be very good at slipping in to lay an egg when the host nest contains only a partial clutch of fairy-wren eggs. In such a case, the cuckoo egg is almost always accepted and incubated along with the hosts' eggs.

The arms race between the Horsfield's bronze-cuckoo and the superb fairy-wren continues into the nestling stage. When the cuckoo chick hatches and kills its wren

(A)

Host
Foreign
Host
Foreign

(B)

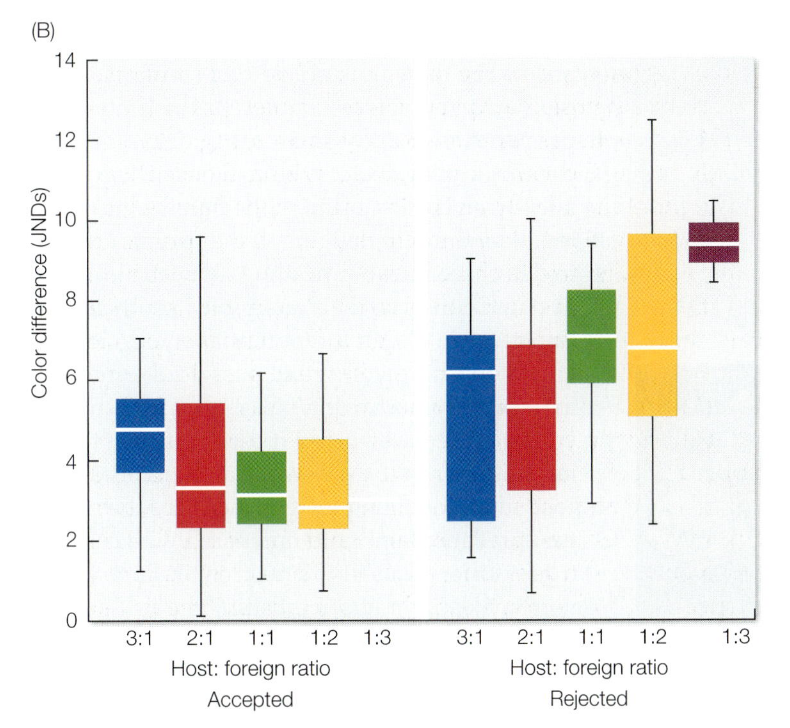

FIGURE 11.25 Prinia produce eggs that vary in color and patterning, which influences their ability to detect and reject parasitic cuckoo finch eggs. (A) Examples of eggs from experimental clutches and naturally parasitized nests. The images in the top three rows show the range of colors and patterns found in prinia eggs. Lines under the eggs indicate groups originating from the same clutch. The two images in the bottom row show naturally parasitized clutches with one cuckoo finch egg (left) or two cuckoo finch eggs (right). Hosts do not seem to use marking variations in egg rejection decisions. (B) Foreign eggs were more likely to be rejected the more they differed in color from the host's own eggs. As foreign eggs increased in number, hosts required a greater difference in color to reject them. Color differences are discrimination units or "just noticeable differences" (JNDs), where a JND of less than 1 means that two objects cannot be discriminated, and values between 1 and 3 should be difficult to discriminate. Box plots show median bars and interquartile ranges; whisker lines show outer quartiles. (From Stevens et al. 2013.)

nest mates by pushing them out of the nest, the adult fairy-wrens abandon the nest about 40 percent of the time, leaving the baby cuckoo to die as well. In the remainder of cases, the fairy-wrens continue to care for the sole occupant of their nest, a baby cuckoo, and so waste their time and energy rearing the killer of their offspring. But how does an adult fairy-wren decide when to abandon the nest or not? Horsfield's bronze-cuckoo chicks produce begging calls very much like those of their host nestlings (**FIGURE 11.26**). Presumably this call mimicry is an evolved response to their hosts' ability to discriminate against illegitimate signalers. In support of this conclusion, Naomi Langmore and her colleagues point to the fact that superb fairy-wrens always abandon nests that have been parasitized by a different species, the shining bronze-cuckoo (*Chrysococcyx lucidus*), whose nestlings produce vocalizations very different from those of the host chicks (Langmore et al. 2003).

But are superb fairy-wrens able to recognize their own offspring and distinguish them from a Horsfield's bronze-cuckoo chick? Diane Colombelli-Negrel and colleagues demonstrated using a cross-fostering experiment that the highest mother–nestling call similarity occurred between foster mothers and chicks, rather than between genetic mothers and chicks or between mothers and cuckoo nestlings (Colombelli-Negrel et al. 2012). Using playback experiments, the team then showed that adults respond more to the begging calls of offspring hatched in their own nest than to calls of other wren or cuckoo nestlings. The researchers concluded that fairy-wrens use a parent-specific password that is learned while their offspring are still in the egg to shape call similarity with their own young (see Chapter 3). This embryonic password then allows parents to detect bronze-cuckoo nestlings before the cuckoo chicks kill the fairy-wren chicks and trick the parents into feeding them through fledging. If the parents are able to detect the intruder nestling, they abandon the nest and conserve their energy for another nesting attempt. Given the potential energetic benefits of teaching young the password while still in the embryo, why don't all mothers teach their young these passwords? It turns out that teaching the embryo the password also entails a high cost: mothers calling to eggs increases the risk of nest predation eightfold compared with controls (Kleindorfer et al. 2014). Thus, female fairy-wrens must make trade-offs in escaping the risks of nest

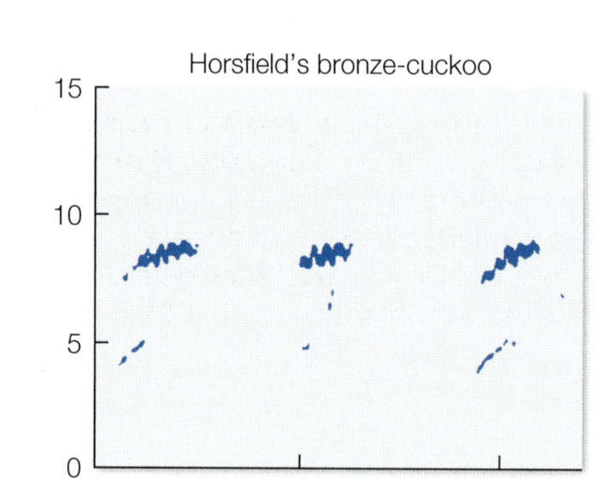

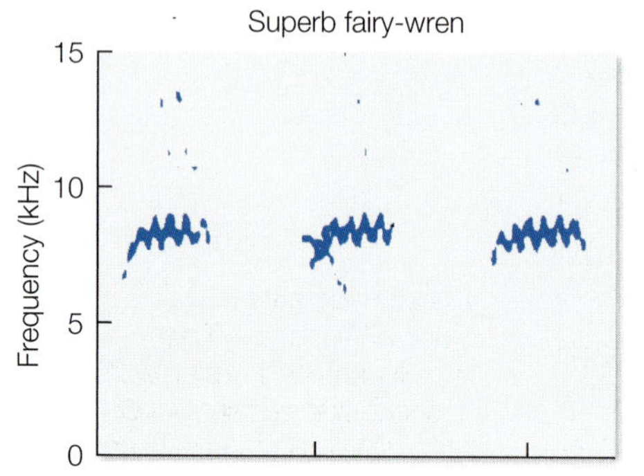

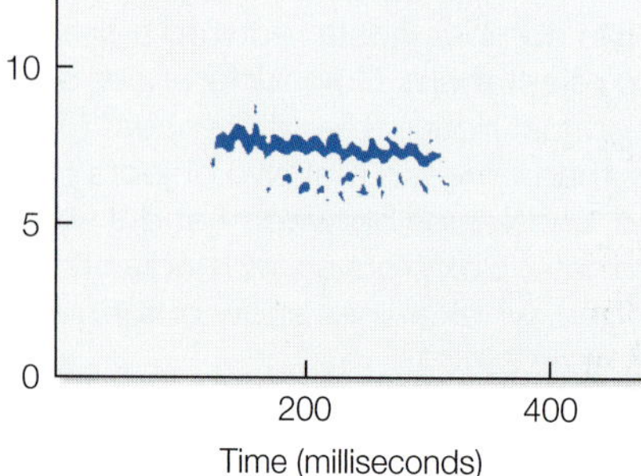

FIGURE 11.26 A product of an evolutionary arms race? Chicks of Horsfield's bronze-cuckoo, a specialist brood parasite on superb fairy-wrens, mimic the calls of their hosts' chicks very closely, which may help them overcome the defenses of fairy-wren host parents. In contrast, the chicks of another species, the shining bronze-cuckoo, which rarely parasitizes superb fairy-wren nests, not only do not look like fairy-wren chicks but lack a good facsimile of the begging calls of superb fairy-wren nestlings. (After Langmore et al. 2003; photographs © 2003 Nature Publishing Group.)

predation and brood parasitism, something that is likely to help facilitate the arms race with cuckoos.

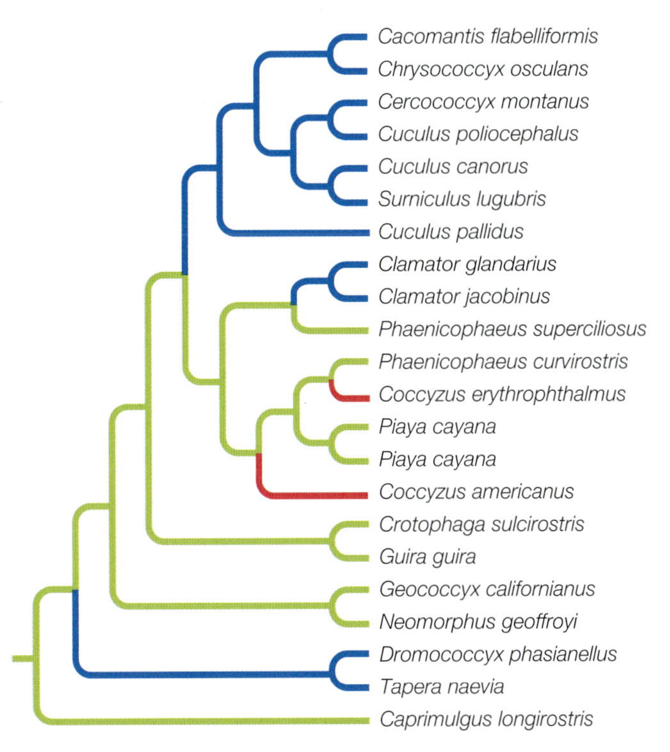

FIGURE 11.27 Multiple origins of brood parasitism by cuckoos. In this family of birds, brood parasitism has evolved three times. The blue branches indicate those lineages whose modern descendants are specialist parasites. The red branches are for two species that are occasional parasites. (After Aragón et al. 1999.)

The Evolution of Interspecific Brood Parasitism

How did cuckoos, cowbirds, and other avian brood parasites evolve the capacity to parasitize the parental care behavior of other bird species? We turn again to Darwin's theory of descent with modification for an answer. In the case of cuckoos, one phylogenetic reconstruction based on molecular data indicates that specialized parasitism has arisen three times during the evolutionary history of this group (**FIGURE 11.27**) (Sorenson and Payne 2005). Given this phylogeny and the overall rarity of specialist brood parasitic bird species, Oliver Krüger and Nick Davies hypothesized that the ancestor of the current parasitic cuckoos was a nonparasitic bird in which adults cared for their own offspring. By taking advantage of detailed information on the natural history of many of the 136 species of cuckoos worldwide, of which 83 are nonparasitic and 53 are parasitic, Krüger and Davies were able to demonstrate that the ancestral state was represented not only by parental care but also by the occupation of small home ranges and the absence of migration. Subsequently, species evolved that provided parental care for their offspring but that possessed relatively large ranges and a tendency to migrate. From lineages of this sort came the modern brood parasites, which also have large home ranges and are generally migratory (Krüger and Davies 2002).

But what about the transition between parental behavior and specialized brood parasitism? This shift could also have taken place in stages, with an intermediate phase when the parasites of the day targeted nesting adults of their own species (termed **conspecific**, or **intraspecific**, **brood parasitism**), with the shift to members of one or more other species occurring later. Alternatively, specialized interspecific parasitism may have arisen abruptly with the exploitation of adults of another species right from the start. The first scenario, the **gradualist shift hypothesis**, predicts that females of some of today's birds should lay their eggs in nests of their own species, and indeed, parasitism of this sort has been documented in about 200 species (Lyon and Eadie 2008), including birds as different as the zebra finch (*Taeniopygia guttata*) (Schielzeth and Bolund 2010) and the American coot (Lyon 1993). Examining intraspecific brood parasitism in some of the extant species that commonly exhibit the behavior might provide insights into the evolution of interspecific brood parasitism.

gradualist shift hypothesis The evolution of interspecific brood parasitism involved intraspecific parasitism as an intermediate stage.

Wood ducks (*Aix sponsa*) commonly exhibit intraspecific brood parasitism. In this species, suitable nest cavities in trees are scarce, with the result that two females sometimes lay eggs in the same nest before one duck evicts the other. The "winner" then cares for the eggs of the other female along with her own, having made the "loser" an involuntary parasite in the process (Semel and Sherman 2001). If the losers were to produce more offspring than they would have if they had retained the nest site, then selection could favor variants that voluntarily deposit and abandon their eggs in the nests of others—a behavior that is fairly common in another hole-nesting duck, the Barrow's goldeneye (*Bucephala islandica*) (Eadie and Fryxell 1992).

In the American coot, "floater" females without nests or territories of their own lay their eggs in the nests of other coots, apparently in an effort to make the best of a bad situation, since they cannot brood their own eggs themselves. But some fully territorial females with nests of their own also regularly pop surplus eggs into the nests of unwitting neighbors, indicating that several hypotheses can apply to intraspecific brood parasitism (Lyon and Eadie 2008). Since there are limits to how many young one female coot can incubate and rear with her partner, even a territorial female can boost her fitness a little by surreptitiously enlisting the parental care of other pairs (Lyon 1993). The exploitative nature of this behavior is revealed by the finding that older, larger females select younger, smaller ones to receive their eggs, presumably because this kind of host cannot easily prevent a larger female from gaining access to her nest (Lyon 2003). Pressure of this sort has apparently shaped the evolution of coot behavior, judging from the fact that parasitized females tend either to bury the eggs of others or to keep their own eggs in the better brooding position in the center of the nest (Lyon 2007). Eggs in the outer part of the nest tend to hatch later than the eggs in the nest center, possibly giving these young, as we discussed earlier, a disadvantage when it comes to begging for food from parents. Coots can discriminate between their own chicks and those of a parasite if members of their own brood hatch first (Shizuka and Lyon 2010). By increasing the odds that parasitic chicks hatch later, the adult coots have a chance to learn the cues associated with their own offspring, which enables them to use this information to reject later-hatched chicks if these are not their own—a nice example of how animals avoid misdirecting their valuable parental care (Shizuka and Lyon 2011).

The gradualist shift hypothesis for the evolution of brood parasitism among species also predicts that when intraspecific brood parasites first began to exploit other species as hosts, they should have selected other related species with similar nestling food requirements. Currently, most specialized brood parasites take advantage of species that are not closely related to them, but perhaps most brood parasites have been evolving for many millions of years since the onset of their interspecific parasitic behavior. To test this prediction, we need to find brood parasites that have a relatively recent origin. Female whydahs and indigobirds, in the family Viduidae, parasitize finch species that belong to a closely related family, Estrildidae (**FIGURE 11.28**). This close relatedness may be why both parasites and hosts share several important features, in particular bright white eggs and an unusual form of begging behavior by nestlings, in which the baby birds turn their head nearly upside down and shake it from side to side, rather than stretching upward in the manner of most other nestling songbirds. Assuming that the ancestral parasitic species also possessed these attributes, sensory exploitation (see Chapter 8) could account for the success the offspring of the original brood parasite had after hatching in the nest of an estrildid finch host (Sorenson and Payne 2001).

Despite some evidence that parasites target closely related host species, the large majority of living brood parasites take advantage of unrelated species much smaller than they are (Slagsvold 1998, Hauber 2003), a finding that could be explained if the ancestral parasites made an abrupt shift from normal parental care to exploiting one or more unrelated smaller host species. Such a shift might well have been more likely to succeed, given that, as already noted, brood parasite nestlings that become larger than their hosts' offspring are more likely to be fed, another form of sensory exploitation that works to a parasite's advantage. The importance of the size disparity between host and parasite has been demonstrated experimentally. When Tore Slagsvold shifted eggs of blue tit (*Cyanistes caeruleus*) into great tit nests, a manipulation discussed in Chapter 3, the experimentally produced brood-parasitic

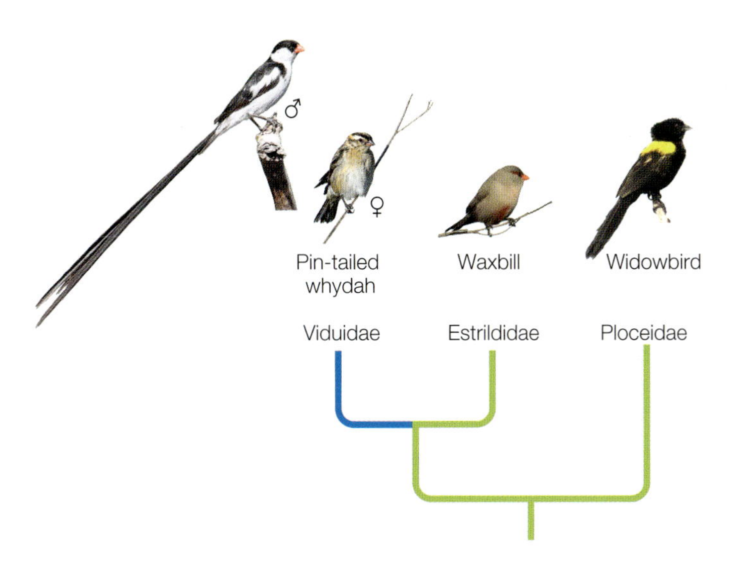

FIGURE 11.28 Whydahs parasitize closely related species. Whydahs (in the family Viduidae) parasitize nests of finches in the family Estrildidae (the family most closely related to Viduidae). Adult male wydahs, like the male pin-tailed whydah (*Vidua macroura*) shown here, look nothing like their hosts, but female pin-tailed whydahs and their nestlings do resemble adult and nestling waxbills (*Estrilda astrild*), an estrildid finch parasitized by this whydah. (Photographs from left to right: © Neal Cooper/Shutterstock.com; © Simon_g/Shutterstock.com; © iStock.com/Leopardinatree; Nigel Voaden/CC BY-SA 2.0.)

blue tit nestlings, which are smaller than those of great tits, did very poorly. In the reciprocal experiment, however, most of the great tit chicks cared for by blue tit parents survived to fledge (**FIGURE 11.29**) (Slagsvold 1998). These findings suggest that unless the original mutant interspecific brood parasites happened to deposit their eggs in the nests of a smaller host species, the likelihood of success (from the parasite's perspective) was not great.

Finally, Yoram Yom-Tov and Eli Geffen have employed the comparative approach to determine which historical scenario for the evolution of obligatory avian parasites is more likely—the indirect, or gradual, pathway with intraspecific parasitism as an intermediate stage (gradualist shift hypothesis), or the direct pathway in which standard parental behavior was quickly replaced by obligate parasitism (**FIGURE 11.30**). Their analysis indicates that for a large group of altricial birds, the probability was much greater that the ancestor of today's obligate parasites was a bird that did not engage in intraspecific parasitism and instead took advantage of members of an entirely different species (Yom-Tov and

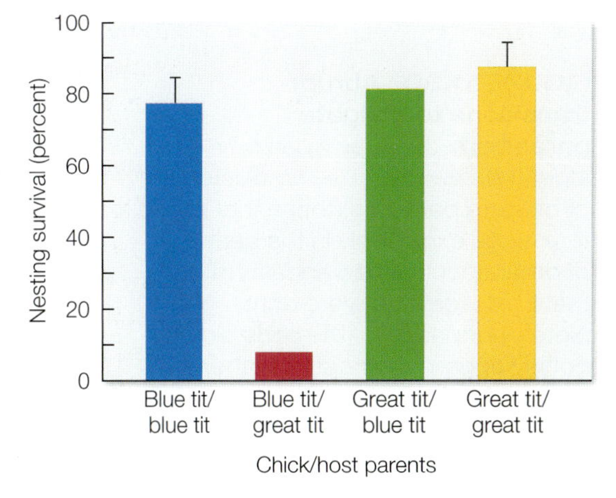

FIGURE 11.29 The size of an experimental "brood parasite" nestling relative to its host species determines its survival chances. Larger great tit nestlings survived well when transferred to the nests of smaller blue tits, whereas blue tits did poorly in great tit nests. Bars depict mean +/– SE. (After Slagsvold 1998; left photograph © David Dohnal/ShutterStock; right photograph © iStock.com/nieudacza.)

Geffen 2006). Thus, proponents of both evolutionary scenarios for interspecific brood parasitism have supportive evidence to which they can point, which leaves the resolution of this issue up in the air until we have a more detailed understanding of the breeding ecologies of more brood parasitic species (Feeney et al. 2014b).

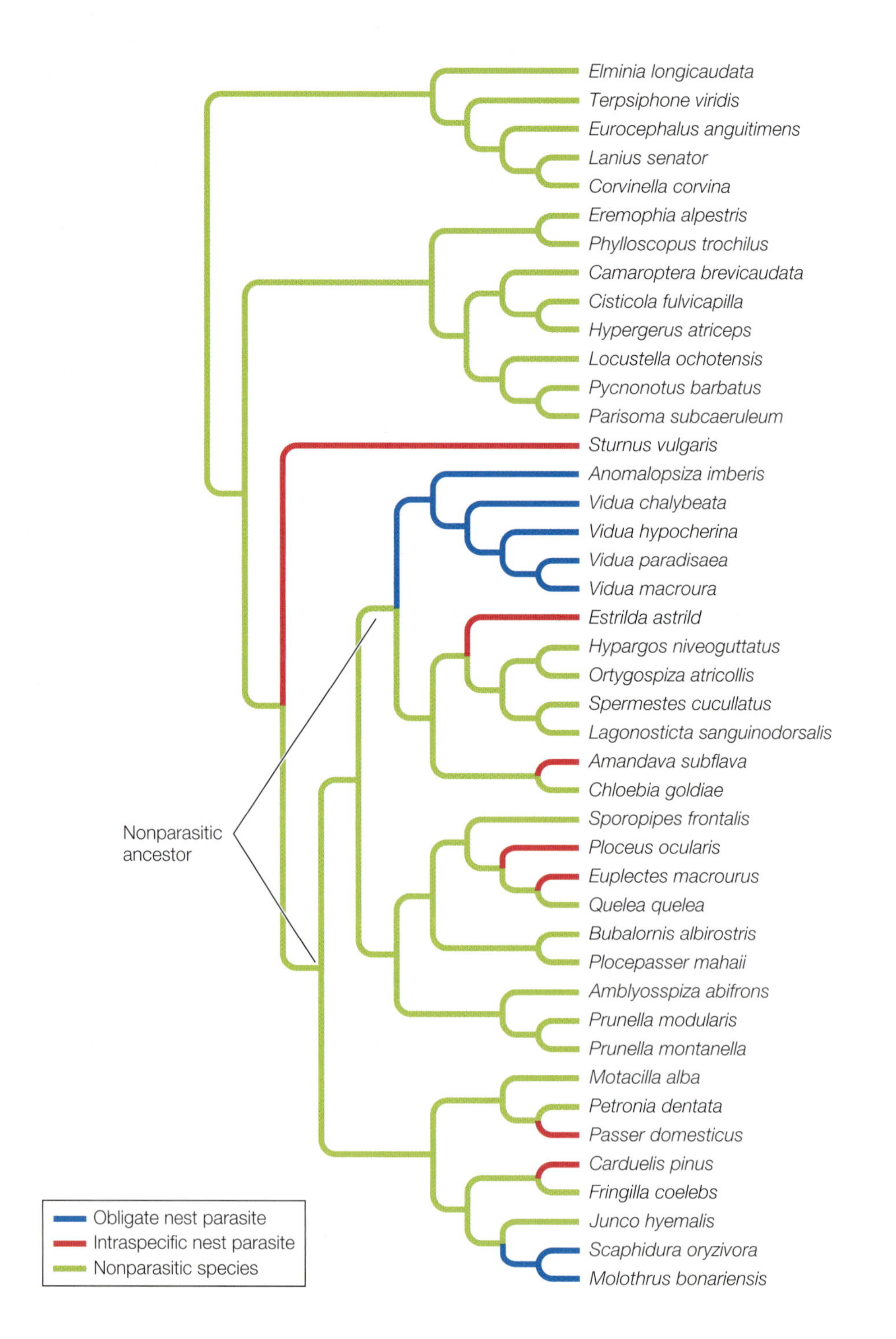

FIGURE 11.30 Abrupt transitions to obligate parasitism. Here an avian phylogeny shows that the two clusters of obligate parasites (those that lay their eggs exclusively in the nests of other species) had ancestors that in all likelihood were completely nonparasitic (the birds did not lay eggs in the nests of other members of their species). (After Yom-Tov and Geffen 2006.)

SUMMARY

1. Not all offspring are created equal, or have the same reproductive value. Offspring with high reproductive value are likely to go on to be successful breeders themselves, whereas those with low reproductive value are unlikely to reproduce much, if it all, during their lifetimes. An individual's reproductive value influences how much care a parent is willing to invest for that individual. To help communicate their value to parents and ensure that they are fed, offspring have evolved complex signaling systems that advertise their need or quality. Parents not only favor some offspring in terms of care, but they also can manipulate the sex of the progeny they produce. Maternal condition, as well as the local social and ecological conditions, can influence the relative value of each sex and which one parents produce or invest in.

2. The parents of some species show complete indifference to lethal aggression among their young offspring. This may be explained as part of a parental strategy to let the offspring themselves identify which individuals are most likely to provide an eventual genetic payoff to the parents that produced them. The more general principle is that selection rarely favors completely even treatment of offspring, because some young are more likely than others to survive to reproduce.

3. Many species lack biparental care even though offspring that receive parental help are generally more likely to survive to reproduce. One explanation for this is that the time, energy, and resources that parents devote to their offspring have costs, including reduced fecundity in the future and fewer opportunities to mate in the present. Moreover, because the costs of paternal care are often greater than the costs of maternal care, particularly in polygynous species where males forego additional mating opportunities by caring for young, paternal care is generally less common in most taxonomic groups than is maternal care.

4. Examining cases of sex role reversal can be instructive for our understanding of how parental care systems evolve. Thus, uniparental care by male fishes may have evolved because males caring for eggs laid in the nests they guard can be more attractive to potential mates than if the males lacked eggs to protect. In contrast, the costs of maternal care to female fishes may include large reductions in growth rate and consequent losses of fecundity. In frogs, paternal-only care has evolved in species where males guard territories that house oviposition sites, and as a way to ensure paternity certainty. Fecundity losses for females may also be involved in the evolution of exclusive paternal care by some water bugs.

5. An evolutionary approach to parental care yields the expectation that when the risk of investing in genetic strangers is high, parents will be able to identify their own offspring. As predicted, offspring recognition is widespread, particularly in colonial species in which adults have many opportunities to misdirect their care to foreign offspring. However, adults do sometimes adopt nongenetic offspring, including those of specialist brood parasites, with consequent losses of fitness. Multiple hypotheses exist to account for these puzzling cases, including the possibility that highly discriminating host adults could lose fitness by sometimes erroneously rejecting their own offspring. Host–parasite interactions lead to coevolutionary arms races as hosts evolve ways to discriminate between their own offspring and those of the parasite, and parasites evolve ways to trick hosts into accepting their parasitic eggs and chicks.

COMPANION WEBSITE

Go to **ab11e.com** for discussion questions and all of the audio and video clips.

CHAPTER 12

Principles of Social Evolution

In many parts of the world, termites are feared as great destructors, capable of leveling a house, chewing apart the wood support structure in no time. However, termites are actually among the most impressive builders in the animal world. Many of the several thousand termite species build immense structures of packed earth that take your breath away (**FIGURE 12.1**). These castles of clay can reach more than 7 meters above the ground, providing a shelter for millions of individuals and helping regulate the temperature and protect the colony from predators. All termite groups, large or small, are composed mostly of individuals that build the colony's home, forage for the resources needed to feed their fellow termites, or care for others' eggs and the immature termites that hatch from those eggs. Amazingly, none of these workers or soldiers will ever reproduce—they are sterile. Instead, they spend their entire lives helping others,

Chapter 12 opening photograph of Portuguese man-of-war © Wild Wonders of Europe/ Lundgren/Minden Pictures.

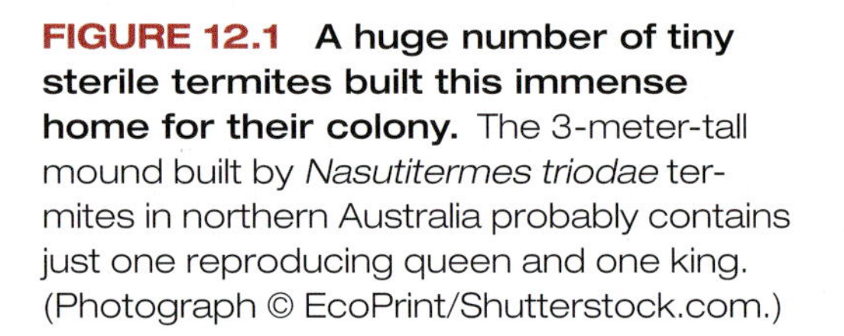

FIGURE 12.1 A huge number of tiny sterile termites built this immense home for their colony. The 3-meter-tall mound built by *Nasutitermes triodae* termites in northern Australia probably contains just one reproducing queen and one king. (Photograph © EcoPrint/Shutterstock.com.)

generally a large reproducing female and her much smaller male partner—the queen and king of the colony—as well as a new generation of brothers and sisters, some of whom may eventually try to establish colonies of their own (Wilson 1971).

The presence of sterile workers seems to defy Darwin's theory of evolution by natural selection. After all, how can individuals with genes that support foregoing their own reproduction exist if their ancestors didn't reproduce? How would such sterility genes be carried into the next generation? This puzzle applies to more than termites, because all ants, some wasps, and some bees—a group known as the Hymenoptera—also form groups with nonreproducing workers. Many of these species also exhibit complex behaviors, such as the ability of army ants (*Eciton* spp.) to build living bridges (**VIDEO 12.1**). The Hymenoptera are the most noted of the eusocial insects—species defined as having overlapping generations, cooperative care of young, and reproductive division of labor where many individuals in a group are temporarily or permanently sterile. However, **eusociality** occurs in other animals, ranging from spiders to shrimps and even a single mammal, the naked mole rat (*Heterocephalus glaber*) (Rubenstein and Abbot 2017a). In many eusocial insects, the worker caste is composed of females with atrophied ovaries, which make them incapable of reproduction. These sterile individuals help other members of their colony reproduce (such as the queen or the future kings and queens produced by the colony).

VIDEO 12.1

Army ants building a living bridge

ab11e.com/v12.1

In this chapter, we introduce the concept of social evolution and explore how cooperative behaviors can evolve. Our goal is to provide a theoretical framework from which we can begin to think about why some animals cooperate, why some form groups, and even why some live in eusocial societies where reproduction is monopolized by a single queen. We examine the implications of why, for example, in a honey bee colony containing more that 50,000 individuals, all of the daughters and nearly all of the sons are typically the offspring of just a single mother (though as you will learn later in the chapter, often of many fathers) (Ratnieks and Helanterä 2009). Although cooperation is a defining feature of eusocial societies, conflict is inherent to every social group. We explore a variety of forms of social conflict, and discuss how conflict is mediated in animal societies. Throughout this chapter, we use eusocial insects to illustrate most of our points because these social animals exhibit some of the most complex societies of any organisms on the planet. They also illustrate one of the major transitions in evolution (**BOX 12.1**). However, we also introduce several other social organisms, ranging from amoebae to flatworms. Ultimately, we build from the theoretical framework presented here to discuss—in Chapter 13—cooperation in vertebrate societies, as well as to contrast vertebrate groups with the animal societies that we introduce here. But first, we need to return to the Darwinian theory of evolution by natural selection and consider the level at which selection acts.

BOX 12.1 DARWINIAN PUZZLE

The major evolutionary transitions

The evolution of complex animal societies such as those of eusocial insects and even our own species represents a Darwinian puzzle that has fascinated behavioral biologists for centuries. Yet this puzzle is not unique to social evolution, as one of the major features of evolution is that lineages increase in complexity over time, a process referred to as major evolutionary transitions (Maynard Smith and Szathmáry 1995, Szathmáry and Maynard Smith 1995). The transition from non-eusocial to eusocial insect societies was associated with an increase in behavioral diversity and specialization. John Maynard Smith and Eörs Szathmáry saw this social transition as just one of many major evolutionary transitions (**TABLE**).

The major evolutionary transitions

Transition from	Transition to
Replicating molecules	Populations of molecules in compartments
Independent replicators	Chromosomes
RNA as both genes and enzymes	DNA and protein (the genetic code)
Prokaryotes	Eukaryotes
Asexual clones	Sexual populations
Protists	Multicellular organisms (plants, animals, fungi)
Solitary individuals	Colonies with nonreproductive castes (eusociality)
Primate societies	Human societies

Source: Maynard Smith and Szathmáry 1995, Szathmáry and Maynard Smith 1995.

Although increased complexity is often associated with changes in the way information is stored and transmitted, biological complexity is difficult to define. Since complexity can be measured as anything from genome size to behavioral variety, rather than emphasizing the evolutionary outcome, behavioral biologists have instead focused on the mechanism whereby the quantity of genetic information or behavioral diversity can increase. Andrew Bourke argued that the major evolutionary transitions should be viewed as occurring in three phases: the formation, maintenance, and transformation of cooperative groups (Bourke 2011a). In all three phases, the transition from smaller, more simple to larger, more complex units is marked by several shared properties, including: (1) smaller units combine and become differentiated to form larger units; (2) smaller units are often unable to replicate in the absence of larger units; (3) smaller units can disrupt the development of larger units; and (4) with the formation of larger units, new ways of transmitting information arise. These properties were later simplified by Szathmáry into three critical features of all major transitions (Szathmáry 2015): (1) a transition from independent replicators to higher-level evolutionary units; (2) the emergence of a division of labor (or the combination of functions) that allows higher-level units to be more efficient under certain conditions; and (3) the emergence of novel inheritance systems that allow information to spread.

When viewed in this way it is easy to see that the evolution of eusocial insect societies (as well as those of vertebrates, including humans) has occurred in much the same way as other evolutionary transitions. We can see that

(Continued)

BOX 12.1 **DARWINIAN PUZZLE** *(continued)*

information transfer (communication) is critical to social evolution, a topic we will explore in Chapter 13, as well as in Chapter 14 when we propose that the development of language has been essential to our own evolution. Ultimately, Maynard Smith and Szathmáry proposed a shared theoretical framework for the evolution of biological complexity. Although testing their ideas is challenging, doing so may provide key insights into not just social evolution, but the evolutionary process more broadly.

Thinking Outside the Box

Can you think of some other examples that fit the major evolutionary transitions model? Szathmáry (Szathmáry 2015) and others have argued that biological systems such as the nervous system or the adaptive immune system can be viewed in light of major evolutionary transitions. How might this work? The model implies that evolution is continually creating ever more complex systems. Do you think this a valid statement?

Altruism and the Levels of Selection

One of the most familiar eusocial insects is the honey bee *Apis mellifera*, a species with which most of us are more familiar than we would prefer. As you learned in Chapter 1, when a honey bee worker stings a person in defense of her hive mates, she dies. This suicidal sting is a classic example of a self-sacrificing behavior that appears to reduce the stinging bee's own fitness (**FIGURE 12.2**) (Ratnieks and Helanterä 2009). There are many other cases of self-sacrificial behavior among eusocial insects, such as worker ants that stay outside the nest to seal off the entrance in the evening, even though this activity means they will die before morning (Bourke 2008, Tofilski et al. 2008). Likewise, when colonies of the ant *Colobopsis saundersi* are under attack, the workers break open their abdomens, spilling a gluey substance onto the intruders, a suicidal line of defense for the colony (Maschwitz and Maschwitz 1974). Self-sacrifice of this sort is what behavioral biologists refer to as **altruism**, a behavior

(A)

(B)

FIGURE 12.2 Sacrifices by eusocial insect workers. (A) In a nasute (pointed-nose) termite colony, soldiers that are incapable of reproducing attack intruders to their colony at great personal risk and spray these enemies with sticky repellents stored in glands in their heads. (B) When a honey bee stings a vertebrate (but not other insects), she dies after leaving her stinger and the associated poison sac attached to the body of the victim. (A, photograph © sunipix55/Shutterstock.com; B, photograph by Waugsberg/CC BY-SA 3.0.)

that benefits others to the detriment of the one performing the behavior. For a long time, it was difficult to see how worker sterility and reproductive suicide could spread by natural selection. Altruism turned out to be the key to explaining how these behaviors could evolve.

Altruism represents not only one of the premier Darwinian puzzles, but one that puzzled even Darwin. Darwin's theory of natural selection argues that individuals in a population exhibit heritable variation in traits that influence survival and reproduction, and that selection acts on some of these traits to influence the fitness of those that bear them. Over time, variation that results in greater reproduction and survival will increase in frequency in the population. Natural selection theory—and its corollary, sexual selection theory—does well at explaining the evolution of many of the traits we have discussed throughout this book. Yet in *On the Origin of Species*, Darwin confronted several problems for his theory of natural selection illustrated by insect societies, one of which was the evolution of sterile castes (Ratnieks et al. 2011).

Individual versus Group Selection

Arguably, it was not until 1962 when the English zoologist V. C. Wynne-Edwards published his book *Animal Dispersion in Relation to Social Behaviour* that the modern scientific analysis of altruism really began (Wynne-Edwards 1962). Unlike Darwin, who proposed that natural selection operates at the level of the individual, Wynne-Edwards proposed something radical, that social attributes such as altruism had evolved to benefit groups or species as a whole, rather than the individual (**FIGURE 12.3**). According to Wynne-Edwards's theory of **group selection**, groups or species with self-sacrificing (altruistic) individuals are more likely to survive than groups without altruists, leading to the evolution of group-benefiting altruism. In other words, selection will favor altruistic traits that favor the fitness of groups over the fitness of the individual (Wilson 1975). Consider the honey bee's stinging behavior. According to Wynne-Edwards, this self-sacrificial behavior by one individual might save the colony from destruction by a potential predator, allowing the group to survive when it otherwise might not have.

Wynne-Edwards's theory of group selection was controversial (as it still is today), and it was not long before George C. Williams challenged this for-the-benefit-of-the-group or -species type of selection in a book of his own, *Adaptation and Natural Selection* (Williams 1966). Williams showed that the persistence of a hereditary trait (including a behavior), and of the genes underlying the development of that trait, was much more likely to be determined by differential reproductive success of genetically distinctive individuals than by differential success among genetically distinctive groups. In other words, Williams countered Wynne-Edwards's group selection argument with a renewed focus on individual selection. He argued that if a group benefit requires an individual to sacrifice itself and reduce its own fitness, then natural selection will trump group selection (Williams 1966). Moreover, since many of the so-called group benefits also benefited individuals, individual selection could just as easily explain the evolution of an altruistic trait as group selection. Ultimately, Williams asserted that

FIGURE 12.3 A cartoon of presumably suicidal lemmings headed into the ocean. The idea that lemmings commit mass suicide when population density outstrips the food supply is actually a myth. Instead, when food becomes scarce, large numbers of individuals migrate in search of additional resources. This inevitably means having to cross large bodies of water, which can be dangerous for these small creatures. Note the lemming with the parachute, a cartoonish demonstration of the critique of group selection offered by George Williams, who argued that suicide would be selected against by natural selection, even if it were beneficial to the group or species as a whole. Wearing a parachute would be selected for by natural selection. (© Mark Godfrey/www.CartoonStock.com.)

Darwinian selection acting on differences among individuals in a population or species will usually have a stronger evolutionary effect than group selection acting on differences among entire groups.

As we will discuss in greater detail below, there have been numerous attempts since Wynne-Edwards to develop more convincing forms of group selection theory. Indeed, this "levels of selection" debate over the level of the biological hierarchy at which natural selection acts (individual, group, gene, species, community, and so forth) has generated much debate among behavioral biologists (Okasha 2010). Yet despite generating controversy, these attempts have generally failed to persuade the vast majority of behavioral biologists that Darwin's view of individual selection—even on altruistic traits that provide clear benefits to others—is wrong. So if we are to explain the altruism exhibited by sterile worker termites, honey bees, or other eusocial insects in Darwinian terms, we have to produce a hypothesis in which the apparent self-sacrificing behavior of these workers actually has an evolutionary benefit to them as individuals.

Altruism and the Role of Kin Selection

In *On the Origin of Species*, Darwin came tantalizingly close to solving the puzzle of worker sterility in the form of an analogy with domesticated cattle. Although it is not possible to breed the tastiest steers (because they have already been eaten), it is possible to keep the parents (or other relatives) of those individuals with desirable attributes in the breeding stock even as their offspring are shipped off to the slaughterhouse. Those young cows that are butchered are the equivalent of workers in eusocial insect societies, in that they fail to reproduce, but their hereditary attributes (nicely marbled beef, efficient conversion of grass into muscle, and the like) can persist if their parents go on to produce the next generation of breeders as well as the sacrificed animals. In this way, Darwin came close to recognizing that the propagation of genes shared by *relatives* might explain altruistic behavior in insect societies.

FIGURE 12.4 How to achieve indirect fitness. By adopting two nephews and a niece, this woman could be propagating her genes indirectly because relatives share rare family alleles with one another. (This assumes that the children would not have survived or reproduced as successfully had they not been adopted.) (Photograph © Sacramento Bee/Zuma Press.)

It would be more than 100 years before William D. Hamilton solved the puzzle of altruism in eusocial insects. Hamilton reshaped Darwin's argument using an understanding of heredity far superior to that available to Darwin (Hamilton 1964). Hamilton's theory was based on the premise that relatives tend to share genes. He realized that although personal reproduction contributes to this ultimate goal in a *direct* fashion, because parents and offspring share genes in common, helping genetically similar individuals other than one's own offspring—that is, one's nondescendant kin—survive to reproduce can provide an *indirect* route to the same end (**FIGURE 12.4**). In other words, it does not matter which bodies are carrying certain genes; a trait becomes more common if the allele that promotes the development of that attribute becomes more common, whether through propagation from direct descendants or nondescendant relatives. Thus, as you saw in Chapter 1, genes are the unit of replication (or replicators) and organisms (or even groups of organisms) are simply vehicles that help genes propagate into the next generation (Dawkins 1982).

To understand why passing genes either directly through one's own offspring or indirectly through one's relatives can have similar fitness effects, we must explore the concept of the **coefficient of relatedness** (***r***). This term refers to the probability that an allele in one individual is present in another because both individuals have inherited it from a recent common ancestor. In other words, it estimates the degree of genetic relatedness between two individuals (**BOX 12.2**). We begin with a diploid organism, such as a human, where one copy of our genetic material comes from our mother and another copy from our father (as you will see below when we discuss haplodiploidy, not all sexually reproducing organisms inherit their genetic material in this way). Imagine, for example, that a parent in a diploid species has the genotype *Aa*, which is to say that *A* and *a* are two alleles of the gene in question. Any offspring of this parent will have a 50 percent chance of inheriting the *a* allele (as well as a 50 percent chance of inheriting the *A* allele) because any egg or sperm that the parent donates to the production of an offspring has one chance in two of bearing the *a* allele. The coefficient of relatedness between parent and offspring is therefore 1/2, or 0.5. This means that, on average, a parent and its offspring in a diploid species share half of their genetic material.

The coefficient of relatedness varies for different categories of relatives. For example, let's consider two sisters. On average, two sisters share half of their genetic material, though it is important to recognize that the half that they share is different than the half that a mother and daughter share. In the case of the sisters, they share one-fourth of their genetic material with their mother and one-fourth of their genetic material with their father, whereas a daughter shares half of her genetic material with each parent. Thus, while the coefficient of relatedness is the same for two sisters as it is for a mother and a daughter, the amount of shared maternal and paternal genes differs.

What do coefficients of relatedness look like for less closely related individuals than first-order relatives (parents, offspring, and siblings)? Let's start with nieces or nephews. A man and his sister's son have one chance in four of sharing an allele by descent because the man and his sister have one chance in two of having this allele in common, and the sister has one chance in two of passing that allele on to any given offspring. Therefore, the coefficient of relatedness for a man and his nephew is $1/2 \times 1/2 = 1/4$, or 0.25. For two cousins, the r value falls to 1/8, or 0.125. In contrast, the coefficient of relatedness between an individual and an unrelated individual is 0 (and the coefficient between monozygotic twins is 1). **FIGURE 12.5** illustrates the coefficients of relatedness for diploid organisms, including humans.

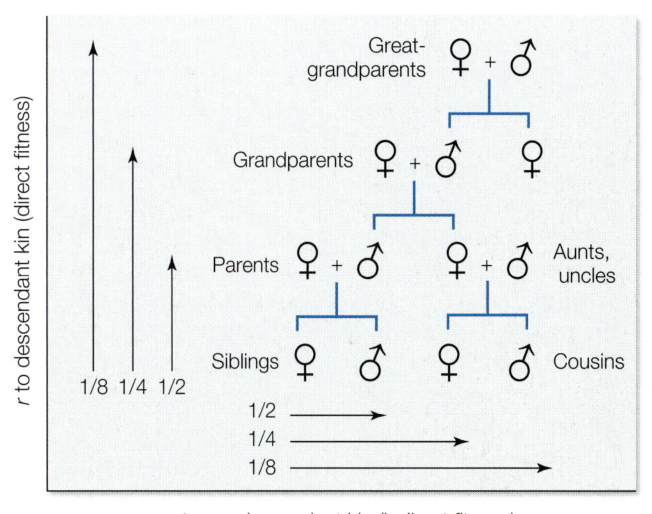

FIGURE 12.5 Coefficients of relatedness in humans and other diploid species. Relatedness to both descendant kin (direct fitness) and nondescendant kin (indirect fitness) declines from first-order relatives (parents, offspring, and siblings) to other family members.

With our knowledge of coefficients of relatedness between altruists and the individuals with whom they cooperate, we can determine the fate of a rare "altruistic" allele that is in competition with a common "selfish" allele. The key question is, will the altruistic allele become more abundant than the selfish allele if its carriers forego reproduction and instead help a relative reproduce? Let's do a thought experiment to answer this question. Imagine

BOX 12.2 INTEGRATIVE APPROACHES

Calculating genetic relatedness

Relatedness values are pairwise estimates of genetic similarity between two individuals that range from 0 (unrelated) to 1 (identical). In a colony of eusocial insects, all of the pairwise values between group mates can be averaged for a given caste type (for example, worker–worker or queen–queen relatedness) to generate estimates of kin structure.

There are many approaches to estimating relatedness values. The simplest is to use pedigrees, or family trees. Although pedigrees provide accurate information about the social relationships among parents and offspring, they fail to capture genetic relationships the way genetic markers do. The types of genetic markers used to study relatedness have changed greatly over the years (Schlotterer 2004). The first genetic markers were biochemical markers called allozymes, which are variant forms of an enzyme that differ structurally (but not functionally) and exhibit substantial polymorphism (they occur in a variety of forms) in natural populations. These protein variant markers were soon replaced by molecular markers that more directly detected DNA variation, either sequence polymorphism or repeat variation. Early molecular markers included Amplified Fragment Length Polymorphisms (AFLPs), which use restriction enzymes to digest DNA and then polymerase chain reaction (PCR) to amplify selected fragments. Unlike allozymes, AFLPs are dominant markers, which means individuals carrying two different alleles (heterozygotes) cannot be distinguished from those that carry two copies of the same allele (homozygotes). In contrast, heterozygotes can be distinguished from homozygotes with codominant markers, allowing for the determination of specific genotypes and allele frequencies at a given locus, two factors that are important for accurately estimating relatedness values. For this reason, AFLPs were soon replaced by genetic fingerprinting methods involving short tandems repeats (STRs), or repetitive regions of DNA that have high mutation rates and high within-population variation. The first of these codominant markers that measured DNA repeat variation to estimate kinship were minisatellites, which contain tandem repeats of DNA ranging in length from 10 to 60 base pairs. Minisatellites were soon replaced by their close cousin, microsatellites, which contain shorter and more variable repeat motifs. Since microsatellites are di-, tri-, or tetranucleotide tandem repeats (**FIGURE A**) that tend to occur in noncoding regions of the genome, large numbers of markers can be identified easily in a cost-effective manner and used effectively for studies of kinship (Queller et al. 1993). More recently, however, microsatellites have begun to be replaced by single nucleotide polymorphisms (SNPs), which are sequences of DNA for which two alleles at a particular locus differ by only one base pair in an otherwise identical sequence (**FIGURE B**). Although SNP-based estimates of kinship require many more loci than microsatellite-based approaches to accurately estimate relatedness, since SNPs are simpler to score than microsatellites and cheaper and easier to develop, this DNA sequence polymorphism approach will likely be the marker of choice going forward (at least until something else comes along to replace it).

(A) Microsatellite (GT repeats)

..TAC(GTGTGTGTGTGTGTGTGTGTGTGT)ACC..
..TAC(GTGTGTGTGTGTGTGTGTGTGTGTGT)ACC..
..TAC(GTGTGTGTGTGTGTGTGTGTGTGTGTGT)ACC..
..TAC(GTGTGTGTGTGTGTGTGTGTGTGTGTGTGT)ACC..
..TAC(GTGTGTGTGTGTGTGTGTGTGTGTGTGTGTGT...)ACC..

(B) Single nucleotide polymorphism (SNP)

.....GCCTCCGTCAGTGCTGCCT....
.....GCCTCCATCAGTGCTGCCT....

Examples of a microsatellite and SNP. (A) A GT dinucleotide microsatellite repeat that varies in length (repeat number) in five individuals. (B) A single nucleotide polymorphism (G vs. A) in two individuals.

that an animal could potentially have one offspring of its own, or alternatively, it could invest its efforts in the offspring of its siblings, thereby helping three nephews or nieces survive that otherwise would have died. Recall that a parent shares

BOX 12.2 INTEGRATIVE APPROACHES *(continued)*

Thinking Outside the Box

In less than 20 years, the genetic approaches for estimating pairwise genetic relatedness between two individuals have changed multiple times. For each study organism, behavioral biologists must consider not only the difficulty and cost in developing a given marker type, but also the amount of information it will provide. Since the resolving power of a set of markers depends in large part on the number of independent alleles per locus and their frequency, some types of markers are more informative than others. What do you think will happen to studies of genetic relatedness when we are able to easily and cheaply sequence entire genomes? What other types of challenges do you think researchers will face when they compare entire genomes of group members? As a student interested in this area of research, what types of skills do you think will be useful for analyzing these large datasets?

half of its genes with an offspring, and one-fourth of its genes with each nephew or niece (see Figure 12.5). Therefore, in this example, personal reproduction of one offspring yields $r \times 1 = 0.5 \times 1 = 0.5$ genetic units that contribute directly to the next generation, whereas altruism directed at three relatives yields $r \times 3 = 0.25 \times 3 = 0.75$ genetic units passed on indirectly in the bodies of relatives. In this case, the altruistic behavior is adaptive because individuals with this trait pass on more of the altruism-promoting alleles to the next generation than do individuals with the selfish allele that try to breed on their own.

An alternative way to consider this problem is to compare the genetic consequences for individuals who aid others at random compared with those who help close relatives. If aid is delivered indiscriminately to any individual, then no one form of a gene is likely to benefit the bearer more than any other, and the carrier of an altruism allele actually pays a price for the help that raises the fitness (the number of genes contributed to the next generation) of individuals with other forms of the gene. In other words, indiscriminate group benefits are not adaptive, as Williams pointed out (Williams 1966). But if close relatives aid one another preferentially, then any alleles they share with other family members may survive better, causing those alleles to increase in frequency compared with other forms of the gene in the population at large. When one thinks in these terms, it becomes clear that a kind of natural selection can occur when individuals differ in their effects on the reproductive success of relatives. Although Hamilton developed this framework of thinking about selection that favors the reproductive success of an organism's helped relative, it was actually the theoretician and geneticist John Maynard Smith who gave it a name; he called it **kin selection** (Maynard Smith 1964).

Kin Selection and Inclusive Fitness Theory

Because fitness gained through personal reproduction (termed **direct fitness**) and fitness achieved by helping nondescendant kin survive and reproduce (termed **indirect fitness**) can both be expressed in identical genetic units, we can sum up an individual's total contribution of genes passed to the next generation, creating a quantitative measure called **inclusive fitness** (**TABLE 12.1**). It is important to remember, however, that an individual's inclusive fitness is not calculated simply by adding up that animal's genetic representation in its offspring plus that in all of its other relatives. Instead, what counts is *only* an individual's own effects on gene

TABLE 12.1 Key terms used in the study of altruism

Altruism: Cooperative behavior that lowers the donor's reproductive success while increasing the reproductive success of the recipient of the altruistic act.

Kin selection: A type of natural selection that favors the reproductive success of the relatives an organism helps, even at a cost to the organism's own fitness.

Direct fitness: A measure of the reproductive (genetic) success of an individual based on the number of its offspring that live to reproduce.

Indirect fitness: A measure of the genetic success of an altruistic individual based on the number of relatives (or genetically similar individuals) that the altruist helps reproduce that would not otherwise have survived to do so.

Inclusive fitness: A total measure of an individual's contribution of genes to the next generation generated by both the direct fitness (derived from reproduction) and the indirect fitness (which depends on social interactions with relatives).

propagation *directly* in the bodies of its surviving offspring that owe their existence to the parent's actions, not to the efforts of others, and *indirectly* via nondescendant kin that would not have existed except for the individual's assistance (for example, in rearing or protecting these relatives).

Let's do another thought experiment to better understand the concept of inclusive fitness. Imagine that a typical diploid organism managed to rear two offspring to reproductive maturity while at the same time helping three siblings each raise an offspring that otherwise would not have survived to reproduce. In such a case, the direct fitness gained by the individual would be $2 \times 0.5 = 1.0$, the indirect fitness gained by that same individual would be $3 \times 0.5 = 1.5$, and therefore the total, or inclusive, fitness of this individual would be $1.0 + 1.5 = 2.5$.

Despite our thought experiments, the concept of inclusive fitness is rarely used to secure absolute measures of the lifetime genetic contributions of individuals. Instead, inclusive fitness theory is typically used to help compare the evolutionary (genetic) consequences of two alternative hereditary traits (Queller 1996). In other words, inclusive fitness becomes important as a means to determine the *relative* genetic success of two or more competing hereditary behavioral traits or strategies. For example, if we wish to know whether an altruistic strategy that helps another individual reproduce is superior to one that promotes only personal reproduction, we can compare the inclusive fitness consequences of the two strategies. For an altruistic trait to be adaptive, the inclusive fitness of altruistic individuals has to be greater than it would have been if those individuals had tried to reproduce personally. In other words, a rare allele "for" altruism will become more common only if the indirect fitness gained by the altruist is greater than the direct fitness it loses as a result of its self-sacrificing behavior. This statement is often presented as **Hamilton's rule**: a gene for altruism will spread only if $rB > C$, where r is the genetic relatedness between two individuals, B is the benefit (in terms of reproductive units) gained by the recipient of the altruistic act, and C is the cost (in terms of reproductive units lost) to the individual performing the altruistic act. We actually used this same approach in the thought experiment above when we concluded that altruism would be adaptive in cases in which the altruist gave up two offspring ($2 \times 0.5 = 1.0$ genetic units) as the cost of helping siblings survive that otherwise would have perished ($3 \times 0.5 = 1.5$ genetic units). Importantly, Hamilton's rule is a general equation that can be used to explain not only the evolution of altruistic behaviors, but also spiteful behaviors in which an individual damages the fitness of another at cost to itself. Moreover, as you will see in Chapter 13, Hamilton's rule does more than simply consider the role of genetic relatedness among potential altruist; it can also be used to explore how things such as ecological factors influence social evolution.

Challenges to Kin Selection and Inclusive Fitness Theory

Hamilton's rule—and inclusive fitness theory more generally—is an elegant framework for studying altruistic behaviors, not just in insects but in all animals (**BOX 12.3**). Its elegance lies in its simplicity. Yet ever since the debate between Wynne-Edwards and Williams over group selection, some scientists have criticized kin selection theory and argued that it is still possible that group selection might occur under some conditions (Wilson 1975, Eldakar and Wilson 2011, Nowak et al. 2011). Although the term *group selection* has been applied to many different evolutionary processes over the past half century (West et al. 2011), there is little doubt that competition among genetically different groups can have genetic consequences over time, thereby affecting the course of evolution. The question then becomes, when, if ever, should group selection theory—instead of kin selection theory, which acts at the level of the individual—be used to explain complex social behavior?

David Sloan Wilson and others (going back to Wynne-Edwards) have argued that because social insects live in colonies, they are ideal candidates for an analysis based on some form of group selection (Wilson and Wilson 2008). Here the basic argument

BOX 12.3 **EXPLORING BEHAVIOR BY INTERPRETING DATA**

Altruism in amoebae

Altruistic behavior is not limited to insects, or even to higher-level animals. Indeed, some species of protozoa are known to cooperate with relatives. Perhaps the best studied of these organisms is the amoeba *Dictyostelium discoideum*, in which individuals occasionally aggregate into a colony (Kuzdzal et al. 2007). These social amoebae exhibit several complex behaviors, including rudimentary farming of the bacteria that they consume (Brock et al. 2011). *D. discoideum* lives most of its life as haploid amoeboid cells, feeding and reproducing asexually. When food becomes scarce, however, thousands of amoebae aggregate to form a slug. About 20 percent of these cells produce a stalk that supports a ball-shaped fruiting body composed of the remainder of the cells (**FIGURE**). The stalk-forming individuals die without reproducing, whereas those in the fruiting body give rise to spores that can become new individuals after the spores are carried away from the colony by passing insects and the like. In other words, some colony members forego their own reproduction to help others achieve the height necessary to disperse their propagules and reproduce successfully (**VIDEO 12.2**).

The social amoeba *Dictyostelium discoideum*. Notice the fruiting bodies at the top of each stalk. (Photograph by Usman Bashir, CC BY 4.0.)

VIDEO 12.2

Dictyostelium discoideum developing fruiting bodies
ab11e.com/v12.2

(Continued)

BOX 12.3 EXPLORING BEHAVIOR BY INTERPRETING DATA *(continued)*

Thinking Outside the Box

Based on what you have learned about inclusive fitness theory in eusocial insect colonies, you might assume that *D. discoideum* slugs always consist of relatives and that the non-reproducing stalk cells would be favored by kin selection. But what if we told you that more than half of the slugs studied in the lab contained chimeric mixtures of multiple amoeba strains (Strassman et al. 2000)?
Joan Strassman and David Queller have argued that these chimeric slugs set the stage for social conflict (see later in this chapter) and the evolution of freeloading or cheater behavioral strategies. In other words, because slugs are potentially a mixture of unrelated amoeba strains, some strains may cheat and form fruiting bodies, but not stalks, giving them a better chance at passing along their genes to the next generation. Given these concerns, could you still use kin selection theory to make a prediction about the genetic similarity of the strains that work together? Why or why not? You might wonder, for example, if *D. discoideum* in the lab form chimeras because they come into contact with other strains in ways that they might not in nature. It turns out that researchers led by Strassman and Queller have found that chimeras also form in nature (Fortunato et al. 2003). For kin selection to work, relatedness among cells must be high. What would the average relatedness among cells be in this asexually reproducing species if all of the amoebae were of a single strain? It turns out that the average relatedness of cells in wild-caught slugs is 0.52 (Fortunato et al. 2003), or equivalent to that of parents and offspring, or of siblings, in diploid species. However, follow-up studies using additional fruiting bodies from various locations and times of year found slightly higher relatedness values of 0.68 in chimera fruiting bodies and values of 0.98 in non-chimeric fruiting bodies (Gilbert et al. 2007). Given these results, what can you conclude about the role that kin selection plays in the lives of social amoebae?

is again that colonies with more self-sacrificing individuals will be favored by group selection if groups with more altruists outcompete rival groups and so contribute more genes to the next generation. Indeed, there are cases when competition of this sort between groups becomes so intense that selection may act at both the group and the individual level (Reeve and Hölldobler 2007). David Sloan Wilson envisioned a scenario where these layers resembled Russian matryoshka, or stacking, dolls (**FIGURE 12.6A**) (Wilson and Sober 1994). At the lowest level were genes, then cells, then organisms, and finally groups of organisms. Wilson called this view **multilevel selection** because it encompassed both individual- and group-level selection (**FIGURE 12.6B**).

David Sloan Wilson and E. O. Wilson use the siphonophores—an order of marine animals related to jellyfishes and that includes the Portuguese man-of-war (*Physalia physalis;* see opening photo for this chapter)—to illustrate multilevel selection (**FIGURE 12.7**) (Wilson and Wilson 2008). Although most siphonophores look like a single individual, they are actually made up of many multicellular individuals called zooids that exhibit specialized forms and functions. Much like eusocial insects, siphonophores exhibit a division of labor (Dunn 2009). There are more than 12 functional classes of zooids, including those that provide locomotion, ingest food, capture prey, circulate blood, or reproduce sexually. Some siphonophore species even have a rudimentary nervous system. As Wilson and Wilson have argued, siphonophores have essentially created a new kind of organism by turning simpler organisms into organs (Wilson and Wilson 2008). Bert Hölldobler and E. O. Wilson have argued much the same thing for eusocial insect colonies in which some types of individuals perform very specialized tasks (Hölldobler and Wilson 2008). In both

(A)

FIGURE 12.6 Multilevel selection describes a hierarchy of evolutionary processes involving interacting layers of competition. (A) Much like Russian matryoshka dolls, these layers are nested one within another. (B) Within a single organism, genes contend with each other for a place in the next generation; within a group of organisms, selection acts on the relative fitness of individuals; groups within a population also differ in their collective survival and reproduction. At even higher levels (not shown), populations, multispecies communities, and even whole ecosystems can be subject to selection. (A, photograph © iStock.com/efcarlos; B after Wilson and Wilson 2008.)

(B)

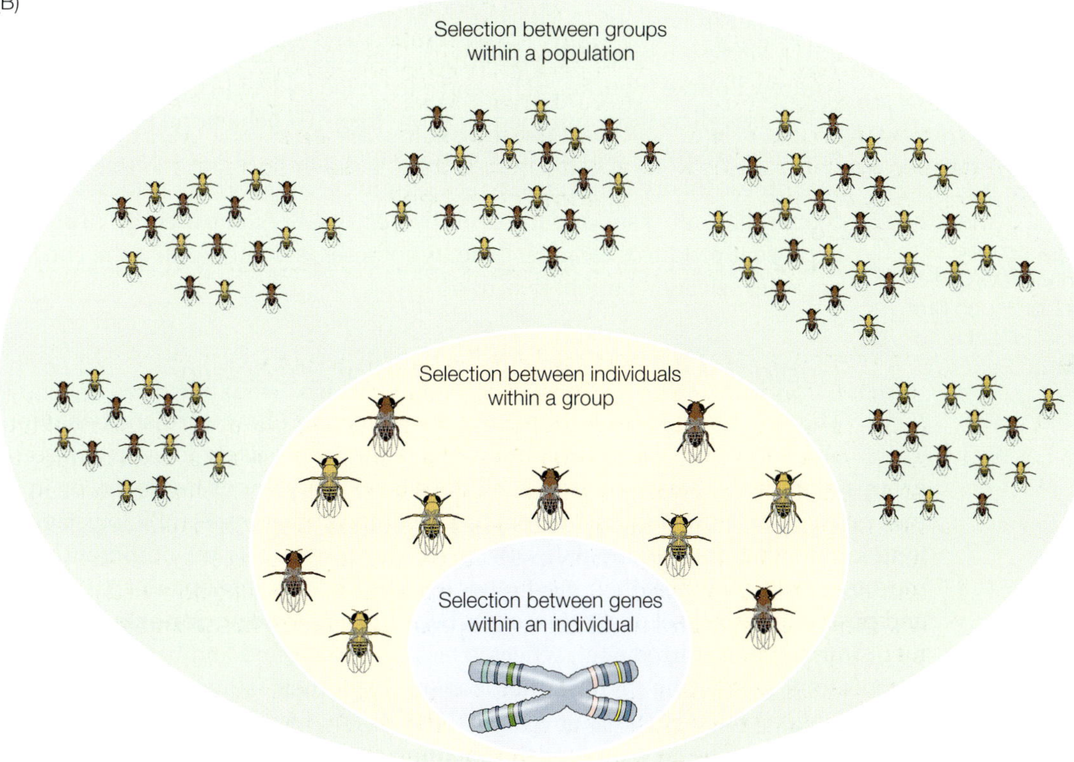

siphonophores and insect colonies, group selectionists argue that individual specializations can be interpreted as being "for the good of the colony" in the much the same way that organs can be interpreted as being "for the good of the individual."

Viewed in this way, multilevel selection is an attractive idea because it combines individual- and group-level selection into a single framework. Even so, when should you use an individual versus a group selection perspective for understanding social evolution? In other words, when is it appropriate to calculate the contribution of alleles to the next generation by quantifying the changes in allele frequencies within

FIGURE 12.7 Siphonopores are made up of many multicellular individuals that perform specialized functions. Individual zooids, including those from this individual captured in the Gulf of Mexico, perform tasks such as locomotion, prey capture, and reproduction, resulting in a division of labor much like that observed in eusocial insect colonies. (Photograph © Danté Fenolio/Science Source.)

groups rather than by adding up the genes passed on by the individual members of that species? Recall from both Chapter 1 and our discussion above that only genes replicate themselves; groups of individuals and individuals by themselves are "vehicles" that can contribute to the transmission of genes, but they are not replicators (Dawkins 1989, Bourke 2011a). As a result, it doesn't matter if you use a group-based method of gene accounting or a system based on individual-level kin selection; the two methods are mathematically equivalent (Reeve 2000, Marshall 2011).

Although most social evolutionary biologists use the individual selection perspective derived from kin selection theory, David Sloan Wilson and others have complained that the preference for kin selection theory stems from prejudice against the rejected Wynne-Edwardsian version of group selection (Wilson and Wilson 2007). But as Stuart West and his colleagues have pointed out, kin selection theory is widely accepted primarily because it has helped so many researchers develop testable hypotheses for puzzling social behaviors exhibited by creatures as different as termites, bacteria, amoebae, and humans (West et al. 2011). Nonetheless, other researchers have continued to attack kin selection theory. For example, Martin Nowak, Corina Tarnita, and E. O. Wilson claimed that kin selection theory should be discarded altogether in favor of an alternative mathematical approach (Nowak et al. 2011). Yet few researchers agree with this idea, for reasons spelled out by Andrew Bourke (Bourke 2011a) and a team led by David Queller (Liao et al. 2015), as well as those outlined by more than 100 behavioral biologists in a response to Nowak and his colleagues' paper (Abbot et al. 2011). Thus, inclusive fitness theory continues to be a guiding framework for understanding cooperation, altruism, and the evolution of sociality more broadly, primarily because it produces clear, testable hypotheses, something that the concept of group selection largely fails to do.

Haplodiploidy and the Evolution of Eusociality

The concept of kin selection helps us understand how altruistic behaviors and eusociality can evolve, but why are they more likely to evolve in some groups of animals than in others? Why does eusociality, for example, occur in so many species of Hymenoptera but remain rare among other insects? Biologists have long known that male and female ants, bees, and wasps have very different chromosomes (and genomes) than most other insects. Male Hymenoptera are **haploid** and possess only one set of chromosomes because of **parthenogenesis** (development from an unfertilized egg), whereas females are diploid and have two sets of chromosomes, one from their mother and one from their father. This type of sex-determination system, in which males are haploid and females are diploid, is called **haplodiploidy**. (The type of sex-determination system in which males are diploid is called **diploidy**, or **diplodiploidy**.) Because all male hymenopterans are haploid, all of the sperm a male makes are genetically identical. So if a female ant, bee, or wasp mates with just one male, all of the sperm she receives will have the same set of genes. When the female uses those sperm to fertilize her eggs, all of her diploid daughters will carry the exact same set of paternal genes, but only an average of 50 percent of the maternal genes. This difference occurs because females are diploid but males are only haploid. Thus, when a queen bee's eggs unite with genetically identical sperm, the resulting offspring share 50 percent of their genes thanks to their father and 25 percent on average because of their mother, for a total r of 0.75 (**FIGURE 12.8**). In other words, sisters are more related to each other ($r = 0.75$) than they are to their mother (or they would be to their own daughters; $r = 0.5$).

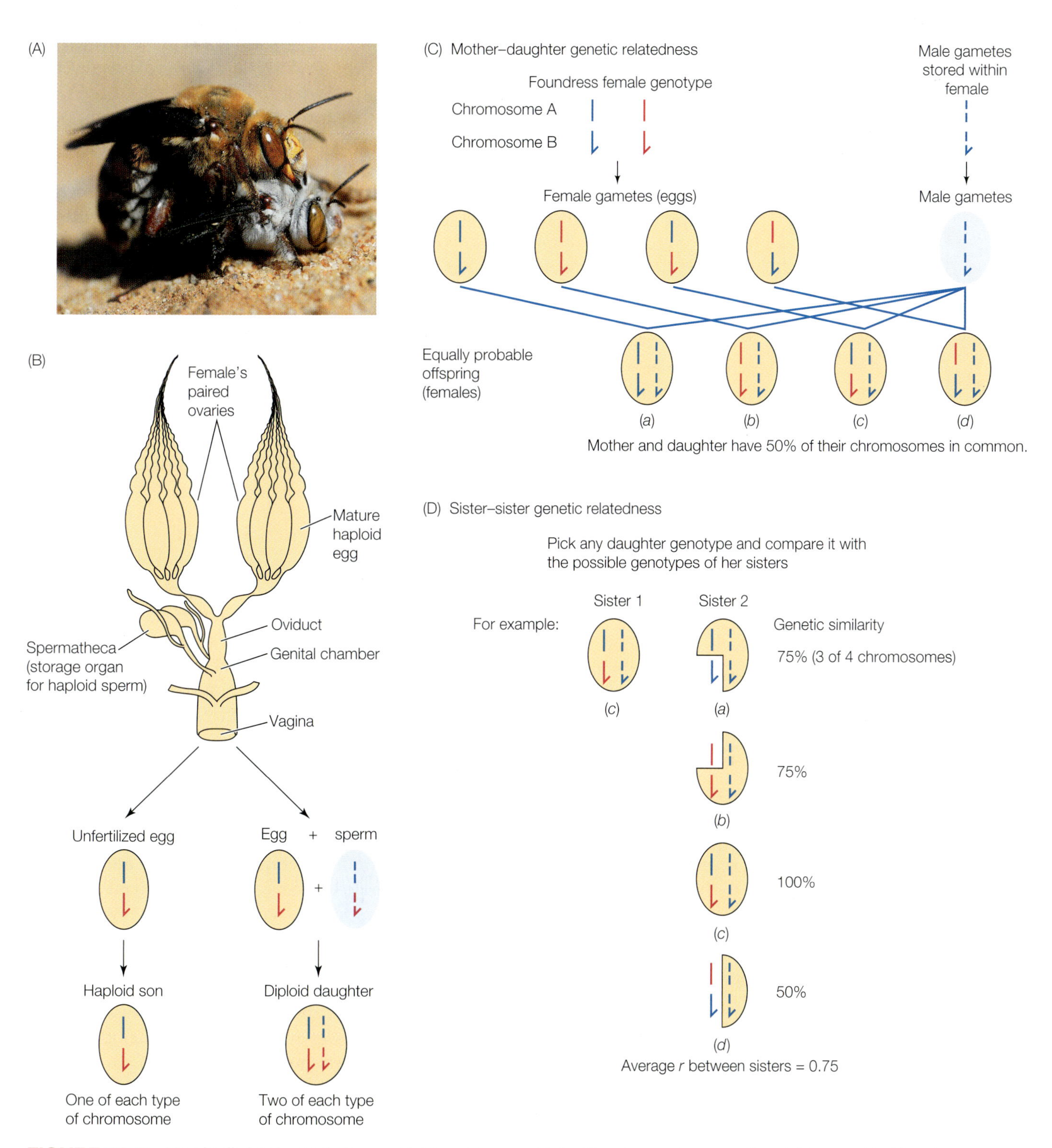

FIGURE 12.8 Haplodiploidy and the evolution of eusociality in the Hymenoptera. (A) When a haploid male hymenopteran copulates with a diploid female, all the sperm she receives are genetically identical. (B) Then when the female releases a mature haploid egg from her ovaries, it may or may not be fertilized with sperm from the spermatheca as the egg passes down the oviduct. Unfertilized eggs become haploid sons. Fertilized eggs become diploid daughters. (C) The genetic relatedness of mothers that have mated with a single male to their daughters ($r = 0.5$). (D) The average genetic relatedness between sisters ($r = 0.75$) that are daughters of a monogamous female. (Photograph by John Alcock.)

haplodiploidy hypothesis The relatively higher relatedness of full sisters in haplodiploid populations promotes altruism among siblings and, consequently, the evolution of eusociality.

Hamilton realized the significance of the fact that the individuals that make up a typical eusocial colony are primarily closely related sisters that share on average three-fourths of their genetic material, something he formalized in the **haplodiploidy hypothesis** (Hamilton 1964). Remember that Hamilton's rule argues that the benefit of helping genetically similar individuals must exceed the cost of sacrificing personal reproduction in order for the altruism to spread through a population. Hamilton had the keen insight that the coefficient of relatedness between full sisters in hymenopteran species could be higher than the 0.5 figure that applies to full siblings in most other organisms, including humans, because they are haplodiploid. As a consequence of this genetic fact, altruistic behaviors (according to Hamilton's rule) should be more likely to evolve in haplodiploid Hymenoptera than in diplodiploid animals. If sisters really are especially closely related, kin selection could more easily favor female hymenopterans that, so to speak, put all their eggs (alleles) in a sister's basket rather than reproducing themselves. In other words, selection would favor most females helping to raise their sisters (their mother's daughters) rather than trying to breed on their own (to produce daughters). This scenario is exactly what we observe in eusocial insect societies: workers (which are female) help rear their mother's daughters (their sisters) rather than trying to produce their own daughters. Perhaps not coincidentally, the Hymenoptera have the greatest number of highly altruistic insect species, and these species have worker castes that are composed of only females.

Testing the Haplodiploidy Hypothesis

It seems quite logical that a higher coefficient of relatedness among sisters because of their haplodiploid genetic system could make altruism among sisters more likely to evolve, potentially explaining why sterile female workers are especially well represented among the Hymenoptera (Hamilton 1964). Yet just because many haplodiploid species are eusocial does not mean that haplodiploidy is the reason that eusociality evolves. After all, while all of the more than 15,000 species of ants are eusocial, most of the nearly 20,000 species of bees are solitary and do not live in groups. Additionally, there are many diploid species—including all termites—that are eusocial. Clearly, further evidence is needed to make the causal link between haplodiploidy and eusociality.

The haplodiploidy hypothesis generates additional predictions. For example, if the haplodiploid system of sex determination contributed to the evolution of eusociality in the Hymenoptera, then sterile worker castes would be predicted to bias their help toward reproductively competent sisters rather than toward male siblings. Why? Because two hymenopteran sisters share, on average, 75 percent of their genes whereas a sister shares, on average, only 25 percent of her genes with her haploid brother (see Figure 12.8). Recall that half of the genes a sister possesses are paternal, of which males do not share (because they do not have fathers), whereas the remaining half of the genes a sister possesses are maternal and shared on average 50 percent of the time with her brother. This means that a sister shares, on average, only one-fourth of her genes with her brothers ($r = 0.25$).

Bob Trivers and Hope Hare expanded on this logic of the haplodiploid genetic system and argued that since sisters are three times more closely related to one another than they are to their brothers (in species whose queen mates only once), then worker hymenopterans should invest three times as much in sisters as in brothers (Trivers and Hare 1976). They predicted that if female workers employ a behavioral strategy that benefits their reproducing sisters and maximizes their indirect fitness, then the combined weight of all the adult female reproductives (a measure of the total resources devoted to the production of females) raised by the colony's workers should be three times as much as the combined weight of all the adult male reproductives. When Trivers and Hare surveyed the literature on the ratio of total weights of the two sexes produced in colonies of different species of ants, they found a 3:1 investment ratio, as predicted by Hamilton's haplodiploidy hypothesis (**FIGURE 12.9**) (Trivers and Hare 1976).

FIGURE 12.9 Ants attending eggs in an anthill. These workers are tending the eggs laid by the queen. (Photograph © iStock.com/benedamiroslav.)

The results of Trivers and Hare's study suggest that workers maintain some level of control over offspring production, because if queens were in complete control of offspring production, they should ensure that the investment ratio for the two sexes was 1:1 since a queen donates 50 percent of her genes to each offspring, whether male or female. Queens therefore gain no genetic advantage by making a larger total investment in daughters than sons, or vice versa (Williams 1966). Imagine a hypothetical population of an ant species in which queens do tend to produce more of one sex than the other. In this situation, any mutant queens that did the opposite and had more offspring belonging to the rarer sex would be handsomely repaid in grand-offspring. If males were scarce, for example, then a queen that used her parental capital to generate sons would create offspring with an abundance of potential mates and thus many more opportunities to reproduce than a comparable number of daughters would have. The greater fitness of son-producing queens would effectively add more males to the next generation, moving the sex ratio back toward equality. If, over time, the sex ratio overshot and became male biased, then daughter-producing queens would gain the upper hand, shifting the sex ratio back the other way. When the investment ratio for sons and daughters is 1:1, neither son-producing specialist queens nor daughter-producing queens have a fitness advantage. The fact that most (but not all) studies have found an investment ratio skewed toward females in ant, bee, and wasp colonies suggests that when there is a conflict between the fitness interests of queens and workers, the workers win (Meunier et al. 2008). However, if all colonies of eusocial Hymenoptera invest three times as much in females as in males, then the overall investment ratio will be biased 3:1 in favor of females. Because they are relatively rare (investment wise), brothers will therefore provide three times the fitness return per unit of investment relative to females, canceling any indirect fitness gain for workers that put more into making sisters than brothers (Gardner et al. 2012). For this and several other reasons (including the fact that eusocial termites are diplodiploid organisms), haplodiploidy alone is inadequate to explain the evolution of eusociality (**BOX 12.4**) (Alpendrinha et al. 2013).

Inclusive Fitness and Monogamy in Eusocial Insects

If the haplodiploidy hypothesis is unlikely to explain the evolutionary origin of eusociality in the Hymenoptera—let alone in non-haplodiploid species—then we need a more general hypothesis to explain the evolution of complex societies. As we mentioned earlier, Darwin provided a tentative explanation for the evolution of sterile workers in eusocial insects, namely that selection acting on other (reproductively competent) family members could result in the spread of self-sacrificing traits in the

FIGURE 12.10 Family living is the norm in eurosocial Hymenoptera. (A) Although all ants are eusocial, eusociality is less common in (B) bees (like these honey bees) and (C) wasps (like these *Polistes* wasps). In all of these hymenopterans, groups are composed primarily of parents and offspring. (A, photograph © iStock.com/Adisak Mitrprayoon; B, © iStock.com/fpwing; C, © Leonid Eremeychuk/Shutterstock.com.)

relatives of those reproducers. If we use the modern form of Darwin's argument, namely the Hamiltonian kin selection approach and the framework of inclusive fitness theory, we can make a prediction about the makeup of eusocial insect societies, which is that these groups will generally be composed almost entirely of parents and offspring. This prediction is essentially correct (Boomsma 2007, Strassmann and Queller 2007): most eusocial insect societies are families that consist of related individuals (**FIGURE 12.10**). Even in insect colonies with more than one queen, groups are typically composed of multiple families. And even in these polygynous (multi-queen) societies, if the sterile workers of one family are helping in ways that benefit the entire colony, the fact that their reproducing relatives are present means that sterile individuals can also gain fitness benefits from the enhanced output of their reproducing relatives. Moreover, queens that join others in forming a single colony are often relatives, a factor that increases the coefficient of relatedness among their families of offspring.

Although most eusocial insect colonies consist of relatives, the average relatedness among group members can vary from quite high in some colonies (or species) to quite low in others. What might cause this variation in kin structure? We can see that when colonies are headed by a single queen that mates with only one male in her lifetime, the offspring of this monogamous mating pair will be full siblings and share the same paternal DNA. But in many Hymenoptera and other eusocial insect species, queens are polyandrous and mate with multiple males, sometimes dozens or more, as we saw in *Apis* bees in Chapter 10. In fact, polyandry is much more common than monogamy in insects, even eusocial ones. Jacobus Boomsma realized that while polyandry can erode kin structure in eusocial societies by creating siblings that are less related than they would be if they shared the same father, monogamy might actually help promote it by maintaining high relatedness among siblings. In other words, when queens mate monogamously, workers need gain only a tiny indirect fitness advantage from helping their siblings survive to reproduce in order to make helping the adaptive option. According to this **monogamy hypothesis**, lifetime monogamy must be tightly linked to the evolution of obligately sterile workers (those individuals completely incapable of reproducing) in Hymenoptera, termites, and other eusocial insects (Boomsma 2007, 2009). If Boomsma is correct in his hypothesis that identifies a mechanism (monogamy) for promoting kin selection, then a species' mating system should directly affect the likelihood of that species evolving eusociality.

Since the monogamy hypothesis is an evolutionary one, we can use the comparative approach to test the key prediction that in ancestral hymenopterans that gave rise to today's eusocial bees, ants, and wasps, females should have mated with just one male (or have been monogamous). A team led by William Hughes created a phylogeny of the Hymenoptera based on the molecular and structural similarity of 267 species of ants, bees, and wasps and used ancestral state reconstruction to test the monogamy hypothesis (Hughes et al. 2008). By mapping onto the tree data on whether female colony foundresses mate with just one male (monogamy) or with more than one male (polyandry), it was possible to determine whether the ancestral species of different lineages were more likely to have been monogamous

monogamy hypothesis Lifetime monogamy ensures that siblings are highly related, making obligately sterile workers (those individuals completely incapable of reproducing) and ultimately eusociality more likely to evolve.

BOX 12.4 **EXPLORING BEHAVIOR BY INTERPRETING DATA**

Division of labor in clonal trematode flatworms

Trematodes are a class of parasitic flatworms that have a complex lifestyle involving at least two hosts. These worms reproduce sexually in their primary vertebrate host but asexually in their intermediate host (typically a snail). During their asexual phase, trematodes produce large numbers of clonal offspring called parthenitae. In species in which the parthenitae have mouthparts, they prey on the young of less dominant species, which often have smaller mouthparts or lack mouthparts altogether. A team led by Ryan Hechinger documented morphological and behavioral specializations among parthenitae to form distinct soldier and reproductive castes in an undescribed trematode species called *Himasthla* sp. B collected from the Pacific Ocean off the coast of California (**FIGURES A** and **B**) (Hechinger et al. 2010).

(Continued)

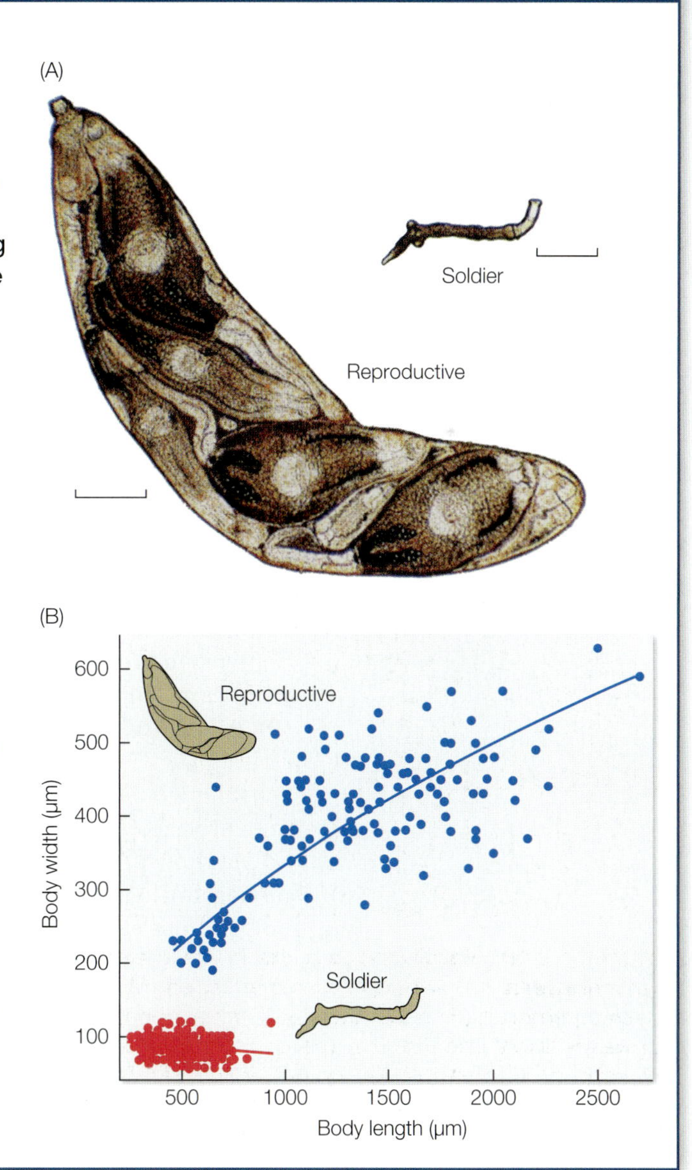

Soldier and reproductive trematodes differ in body size and shape. (A) A reproductive morph and a soldier morph of the trematode *Himasthla* sp. B. The soldier morph is visibly smaller and differently shaped. Scale bar = 0.2 mm. (B) Soldier morphs are shorter and thinner than reproductive morphs and exhibit a different length-to-width ratio. (After Hechinger et al. 2010.)

BOX 12.4 EXPLORING BEHAVIOR BY INTERPRETING DATA *(continued)*

Unlike reproductive individuals, soldiers have large mouthparts, are small and thin, and readily attack both heterospecifics and conspecifics from other colonies. These behavioral differences between soldiers and reproductives were also demonstrated experimentally in the related trematode species *Himasthla elongata* from Denmark (Mouritsen and Halvorsen 2015). Soldiers in this species were more likely to attack both prey items and competing parasites (**FIGURES C** and **D**).

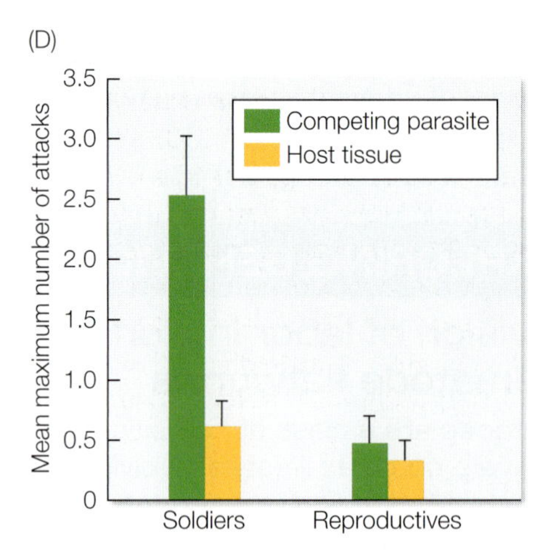

Attack rates for soldier and reproductive trematodes. (C) In the species *Himasthla elongata*, soldiers attacked prey at higher rates than reproductives did. Values show mean +/– SE. (D) *H. elongata* soldiers also attacked parasite competitors more than reproductives did, but there was no difference in attack rates on snail host tissue used as a control prey choice. Bars depict mean +/– SE. (After Mouritsen and Halvorsen 2015.)

or polyandrous. Eusociality has evolved independently several times in the Hymenoptera, and for all eight lineages for which mating frequency data are available, it appears that the ancestral species at the base of each lineage was monogamous (**FIGURE 12.11**), which would have facilitated altruism and the evolution of eusociality (Hughes et al. 2008).

Although monogamy occurred prior to eusociality in all of the lineages containing eusocial species, many of those lineages today contain species that are in fact polyandrous. At first glance, this pattern would seem to be inconsistent with the monogamy hypothesis. However, in all of those lineages that eventually gave rise to polyandrous eusocial species, the workers appear to have lost the capacity to mate. This means that when foundress females mate with several males (and produce offspring that are less related to their siblings than they would be if they had the same father), the queens are still largely guaranteed that a sterile workforce will be there to assist in the production of their reproductively competent sisters. In other words, workers are physiologically incapable of defecting and trying to breed elsewhere should their siblings be less related to them.

Why would females mate polyandrously when released from having to be monogamous in order to ensure a happy and helpful workforce? As you read in

BOX 12.4 **EXPLORING BEHAVIOR BY INTERPRETING DATA**

Thinking Outside the Box

The discovery of caste differentiation in flatworms was the first observation of its type for an entire phylum of animals. How widespread are castes within trematodes? Ana García-Vendrenne and colleagues explored this question in five additional species, four of which infect a marine snail and one (*Echinostoma liei*) that infects a freshwater snail (**FIGURE E**) (García-Vedrenne et al. 2016). How many of these five species appear to exhibit a soldier caste? How did you come to this conclusion? Do you think trematode species that exhibit a reproductive division of labor are actually eusocial? Consider the three defining characteristics of eusociality that we introduced earlier. Do you think kin selection is likely to play a role in social evolution in trematodes? Finally, some researchers have criticized these studies as not providing convincing evidence of caste differentiation, instead arguing that in some related trematode species, the bimodal distribution of body sizes (see Figure E) turns out be the result of parthenitae at various stages of maturity (Galaktionov et al. 2015). Moreover, these researchers contend that the behavioral differences are probably the result of age-related feeding preferences and not caste-specific behaviors. Do you think these observations invalidate the other work on castes in trematodes? What other information might you need to conclude that a species is eusocial?

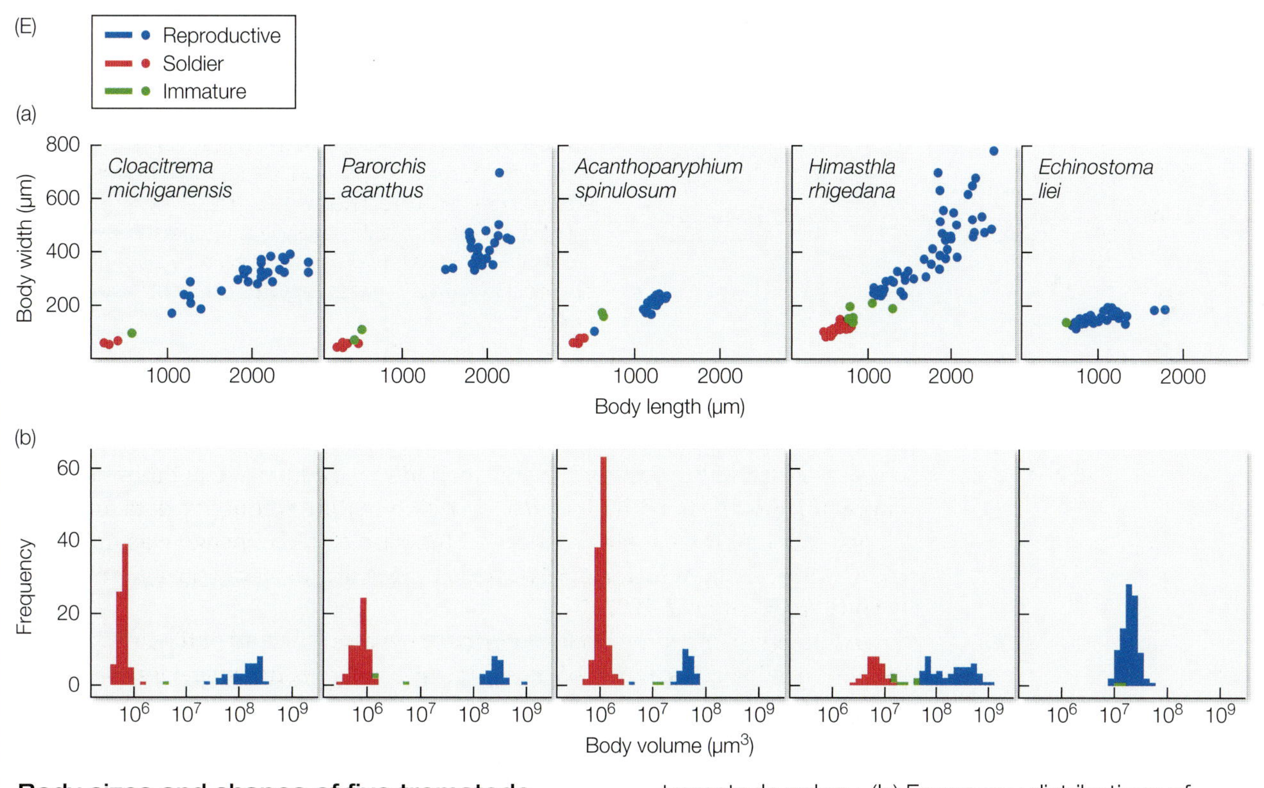

Body sizes and shapes of five trematode species. (a) Body width to body length relationships of soldiers, immatures, and reproductives (only the last are shown for *Echinostoma liei*). Each point represents a randomly sampled individual from a single trematode colony. (b) Frequency distributions of body volume for randomly sampled individual soldiers, immatures, and reproductives. (After García-Vedrenne et al. 2016.)

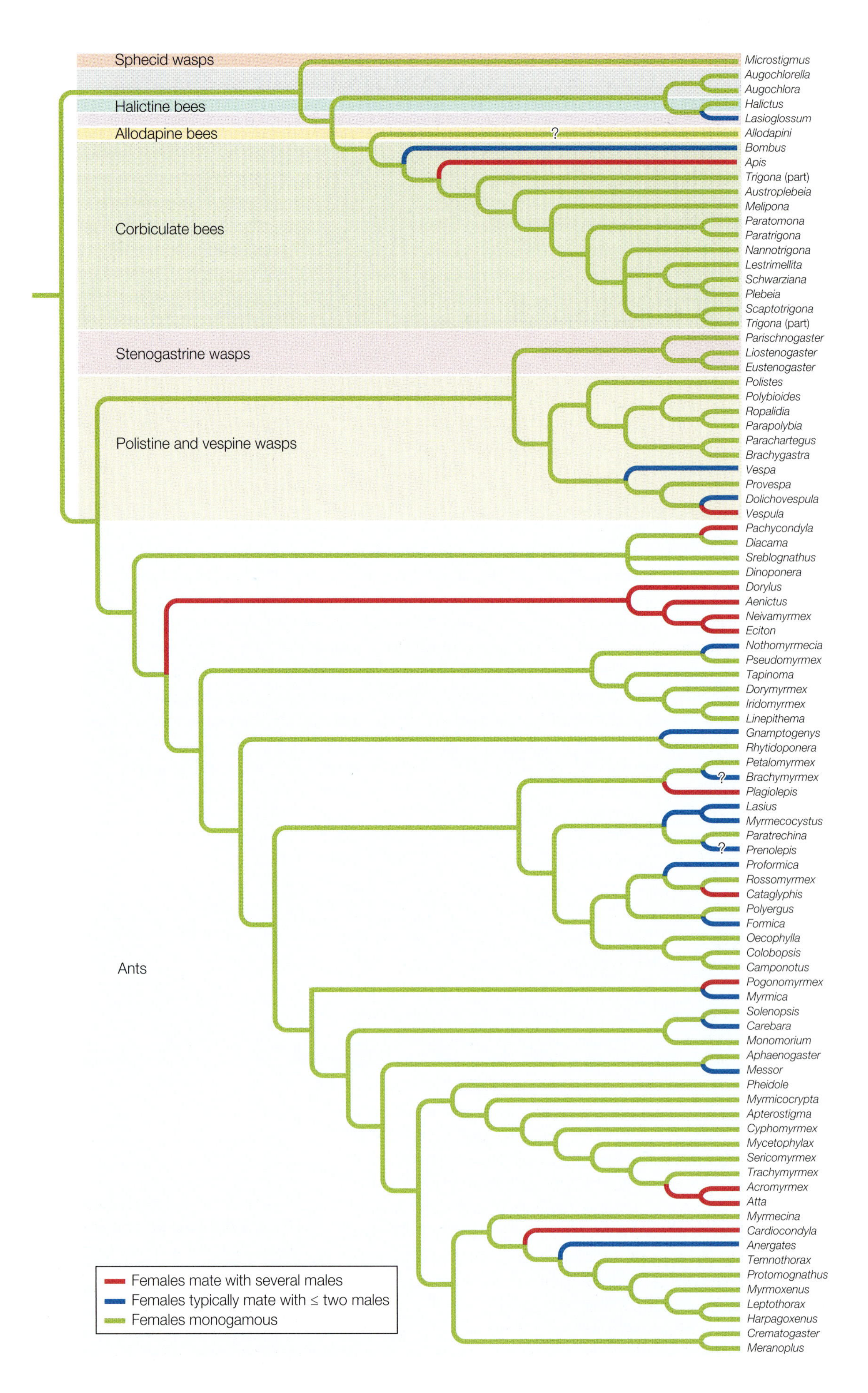

Sphecid wasps
Halictine bees
Allodapine bees
Corbiculate bees
Stenogastrine wasps
Polistine and vespine wasps
Ants
Microstigmus
Augochlorella
Augochlora
Halictus
Lasioglossum
Allodapini
Bombus
Apis
Trigona (part)
Austroplebeia
Melipona
Paratomona
Paratrigona
Nannotrigona
Lestrimellita
Schwarziana
Plebeia
Scaptotrigona
Trigona (part)
Parischnogaster
Liostenogaster
Eustenogaster
Polistes
Polybioides
Ropalidia
Parapolybia
Parachartegus
Brachygastra
Vespa
Provespa
Dolichovespula
Vespula
Pachycondyla
Diacama
Sreblognathus
Dinoponera
Dorylus
Aenictus
Neivamyrmex
Eciton
Nothomyrmecia
Pseudomyrmex
Tapinoma
Dorymyrmex
Iridomyrmex
Linepithema
Gnamptogenys
Rhytidoponera
Petalomyrmex
Brachymyrmex
Plagiolepis
Lasius
Myrmecocystus
Paratrechina
Prenolepis
Proformica
Rossomyrmex
Cataglyphis
Polyergus
Formica
Oecophylla
Colobopsis
Camponotus
Pogonomyrmex
Myrmica
Solenopsis
Carebara
Monomorium
Aphaenogaster
Messor
Pheidole
Myrmicocrypta
Apterostigma
Cyphomyrmex
Mycetophylax
Sericomyrmex
Trachymyrmex
Acromyrmex
Atta
Myrmecina
Cardiocondyla
Anergates
Temnothorax
Protomognathus
Myrmoxenus
Leptothorax
Harpagoxenus
Crematogaster
Meranoplus
Females mate with several males
Females typically mate with ≤ two males
Females monogamous

FIGURE 12.11 Monogamy and the origin of eusociality by kin selection in the Hymenoptera. In this group, there are many different eusocial species, and among these, polyandry has often evolved independently. However, the phylogeny reveals that the ancestral species of the different eusocial groups (each lineage is presented with a different background color) were in every case monogamous, the condition necessary for sister hymenopterans to be unusually closely related. (After Hughes et al. 2008.)

Chapter 10, there are many potential benefits available to females from mating polyandrously. In the honey bee, for example, polyandry may have spread secondarily because of the benefits of having high genetic diversity within a colony to promote disease resistance (Tarpy 2003) or to make worker specialization and task efficiency more likely (Page et al. 1995). Thus, the highly eusocial lifestyle of the now polyandrous honey bee may currently be maintained by selection pressures that differ from those that were responsible for the origin of eusociality in this species. The key point, however, is that monogamy appears to have been essential for the evolution of sterile workers in the Hymenoptera (Hughes et al. 2008).

And as you will see in Chapter 13, these same ideas about monogamy and offspring relatedness are important for explaining the evolution not only of insect societies, but also of bird and mammal societies. Ultimately, this comparative research reinforces the point that to understand the evolution of altruism (and complex animal societies), we need to take into account mating systems, which affect the distribution of relatedness in family groups—the r term in Hamilton's rule. Of course, the B and C terms in Hamilton's rule are equally important. It is fair to say that going all the way back to Hamilton, behavioral biologists studying eusocial insects have emphasized the importance of genetic relatedness in social evolution (Elgar 2015, Rubenstein and Abbot 2017b). In contrast, researchers studying sociality in vertebrates have tended to emphasize the ratio of benefits to costs. In Chapter 13 we will turn our attention to focus more on these costs and benefits of cooperative behavior and social living.

Sterility and Caste Differentiation

It should now be clear that the evolution of worker sterility has been critical to the functioning of eusocial societies. Indeed, one of the defining characteristics of eusocial insect societies is a reproductive division of labor, or the presence of castes. In all eusocial species there is a reproductive caste—the queen (or queens), and sometimes a king, as in the mound-building termites. In many species there is also a worker caste whose members forage for the colony, take care of young, and perform other work around the nest (**FIGURE 12.12**). Finally, there is sometimes a soldier caste that, as we saw in trematodes, helps defend the colony from attack by predators and others of the same species. In some eusocial insect species there can be many thousands of workers and/or soldiers, none of which can produce reproductively capable offspring. For example, leaf cutter ant colonies can contain tens of thousands of workers but only a single queen. As we mentioned earlier in this chapter, Darwin was puzzled by worker sterility in ants. But you should now be able to understand how sterile castes can evolve. Indeed, kin selection, and inclusive fitness theory

FIGURE 12.12 Morphological castes in the Florida harvester ant *Pogonomyrmex badius*. The queen (left), a major worker (right), and a minor worker (bottom) are not only morphologically distinct, they also exhibit different behavioral roles. The major worker, for example, has an enormous head with large mandibles that it uses to crack seeds. (Photograph by Alex Wild/alexanderwild.com.)

more generally, has provided us with a shared framework for understanding how eusociality and worker sterility can evolve in insects as different as haplodiploid ants and diplodiploid termites (and as you will see in Chapter 13, in vertebrates as diverse as birds and primates). In other words, we are beginning to develop an understanding of why castes have evolved and the demographic conditions in which social evolution is more likely.

But in a eusocial species, are queens and workers really that different beyond their behavioral roles? After all, in many eusocial insects, a single egg can give rise to either a queen or a worker depending on how the egg is provisioned. In Chapter 3 we introduced the honey bee *Apis mellifera* and discussed how larvae provisioned with copious amounts of a substance called royal jelly become queens. Workers produce royal jelly in a special gland and feed it to all developing larvae, but for larvae identified to become new queens (those laid in specially constructed queen brood cells), large quantities of royal jelly trigger queen-specific morphological development, such as the growth of ovaries that are needed to someday lay fertilized eggs. We also introduced in Chapter 3 the interactive theory of development, the idea that genes and the environment interact to influence the development of behavioral traits. So how do genes contribute to the differences we observe among different castes both within and among eusocial species?

Modern molecular tools and reverse genetics approaches are providing a way to not only probe the genetic architecture that underlies sociality in animals, but to do so comparatively to determine whether castes in disparate social species share similar molecular architectures (Robinson 2005). Gene Robinson calls this emerging field "sociogenomics" and argues that studies should identify the genes that influence social behavior, explore the effects of the environment (both social and ecological) on gene action, and then use these genes to study the evolution of behavioral diversity in a comparative way (Robinson 1999). Sociogenomics has matured as a field over the past two decades, and studies in eusocial insects in particular are beginning to reveal general patterns in the genetic architecture underlying caste differentiation. For example, caste determination in many species is environmentally mediated by both the ecological environment (such as nutritional state) and the social environment (Robinson 2005). Early work in this area, first with microarray studies and later with transcriptomic studies, showed that in several eusocial insect species, most notably ants, bees, wasps, and termites, individuals of different castes show many differentially expressed genes and gene networks. For example, a comparative transcriptomic study in ten species of bees that varied in social structure from non-eusocial to highly eusocial found a shared set of more than 200 genes with molecular signatures of accelerated evolution in the social species (**FIGURE 12.13**) (Woodard et al. 2011). A subsequent comparative study using full genome sequences of ten species of bees that also varied in social structure suggested that differences in social behavior were the result of differences in gene regulation, but that there was no single genomic path to eusociality (Kapheim et al. 2015). Thus, eusociality may arise through different mechanisms, but always involves an increase in the complexity of gene networks (see Box 12.1).

Another emergent theme from sociogenomic research is that many of the genes that appear to play a role in social behavior and caste differentiation were co-opted from those used in solitary behaviors (Robinson 2005). That is, shared genetic "toolkits" may be responsible not only for more simple social behaviors such as responses to social challenges in diverse organisms (Rittschof et al. 2014), but also for more complex social behaviors such as the behavioral roles associated with different eusocial insect castes (Toth and Robinson 2007, 2009). As Kapheim and colleagues' comparative study in bees demonstrated, many of these genes are linked

(A)

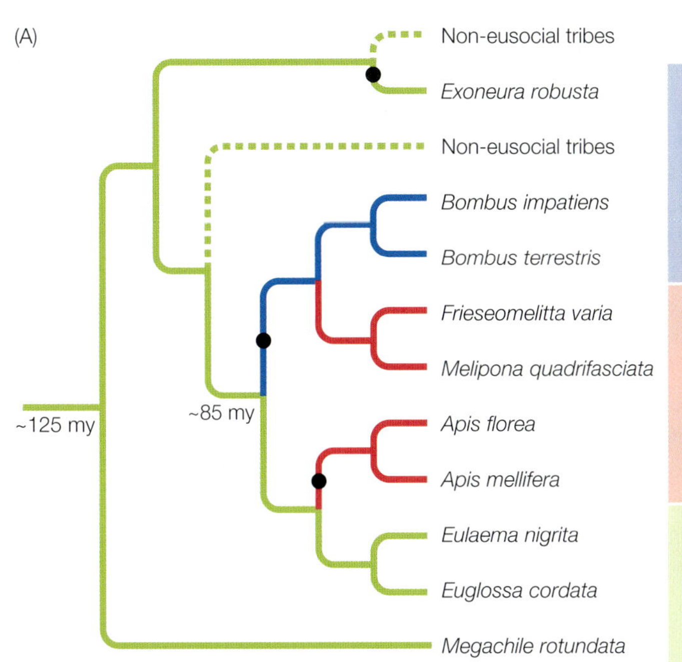

Primitively eusocial

Small colonies (10–100s)
Annual colony cycle
Less specialized queen and worker castes
Nutritional influence on caste
Dominance hierarchies
Diverse exocrine gland functions

Highly eusocial

Large colonies (1000s–10,000s)
Perennial colony cycle
Highly specialized queen and worker castes
Long queen life span (10× worker life span)
Nutritional influence on caste
Year-round nest thermoregulation
Diverse exocrine gland functions

Non-eusocial

Communal (individuals share a nest) or solitary
Annual life cycle
All individuals reproductive
Females only provide care for their own offspring

(B)

Primitively eusocial

Biosynthesis
Chromatin-related
Neuron differentiation
Nitrogen compound metabolism
Growth and development
Response to stimulus

Macromolecular complex assembly
Transcription

All eusocial

Signal transduction
Gland development

Carbohydrate metabolism
Phosphorylation

Highly eusocial

General metabolic processes
Acid metabolism

FIGURE 12.13 Bee species used to identify genes and biological properties evolving rapidly in eusocial lineages. (A) Phylogeny of bee species with lineages color-coded by lifestyle. The boxes list key characteristics of each lifestyle. Circles represent independent origins of eusociality; (my) million years ago. (B) Biological processes that have undergone accelerated evolution in all eusocial lineages, in only highly eusocial lineages, and in only primitively eusocial lineages. (After Woodard et al. 2011.)

in networks, and not all of them are evolutionarily conserved. For example, in the pharaoh ant (*Monomorium pharaonis*), the genes of older foraging workers are highly connected and more evolutionarily conserved, but those of younger nurse workers are more loosely connected and rapidly evolving (Mikheyev and Linksvayer 2015). In another ant species, the clonal raider ant (*Ooceraea biroi*), genes underlying odorant receptors used in chemical communication have undergone a massive expansion, suggesting that novel genes related to communication have played a key role in the evolution of eusociality (McKenzie et al. 2017). Thus, despite much research effort in a range of species, there does not appear to be a shared molecular pathway underlying caste differentiation in closely related species, let alone among eusocial insects. Instead, there may be some shared genetic toolkits and building blocks that set the stage for social evolution and the emergence of new social genes. Additionally, regulatory elements that control gene expression may explain how some individuals become queens and others workers. Thus, caste differentiation in eusocial insects appears to have evolved in different ways in different species. Yet the behavioral similarities among workers in taxa as diverse as Hymenoptera and termites, as well as the loss of the ability to reproduce in the workers of nearly all eusocial species, suggest that natural selection has worked in similar ways across disparate taxa.

Social Conflict in Animal Societies

As we illustrated above, even though eusocial insect societies are essentially family groups consisting of relatives, their genetic interests are not always identical. The fact that daughters prefer their mothers to produce more daughters, while the mothers are essentially indifferent as to which sex they produce, creates the potential for social conflict. In fact, since in most Hymenoptera, daughters can produce unfertilized eggs that become males, there is the potential to create additional conflict, not just between mothers and daughters but also among sisters. And even though these societies are defined by altruism and cooperative behaviors, the potential for conflict, as Hamilton noted, is an inherent part of every animal group:

> Every schoolchild, perhaps as part of religious training, ought to sit watching a *Polistes* wasp nest for just an hour.... I think few will be unaffected by what they see. It is a world human in its seeming motivations and activities far beyond all that seems reasonable to expect from an insect: constructive activity, duty, rebellion, mother care, violence, cheating, cowardice, unity in the face of a threat—all these are there.
>
> *(Hamilton 1996)*

Reproductive Conflict

Consider, for example, paper wasps, in which rebellion and violence take any number of forms. One female lunging at another or biting or climbing on top of her opponent are obvious forms of social conflict. But there are more subtle forms of conflict as well. For example, should a worker wasp lay an egg (**FIGURE 12.14**), then another member of the group, most often another worker, may well discover the haploid egg and eat it (Ratnieks and Helanterä 2009). Egg destruction is a common form of **worker policing** in eusocial insects. Negative interactions can also be more elaborate, as in the ant *Dinoponera quadriceps*, in which the dominant reproducing female (the queen) smears a potential competitor with a chemical from her stinger, after which lower-ranking workers immobilize the unlucky pretender queen for

FIGURE 12.14 Workers and the queen monitor the reproductive behavior of others. Eggs laid by individuals other than the queen wasp will often be eaten by other colony members. Eggs can be seen in several of the cells in the upper left side of this nest; larvae occupy some of the other cells and are being fed by the workers. (Photograph © Bartomeu Borrell/age fotostock.)

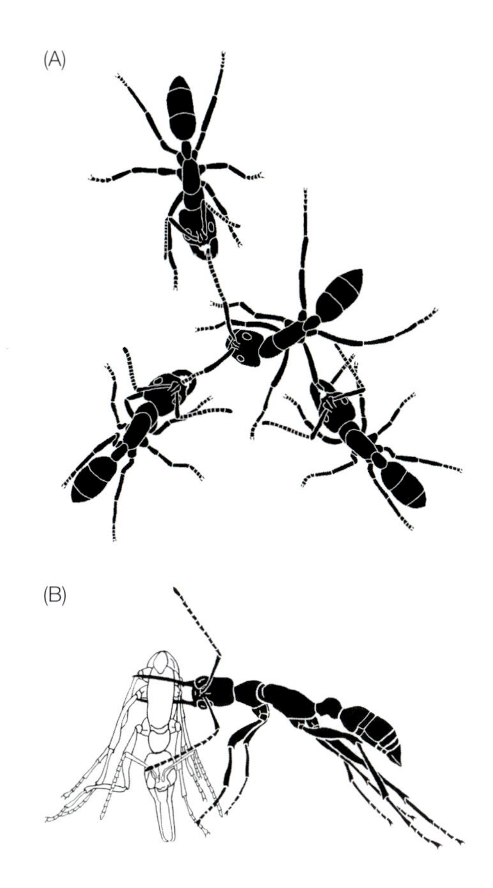

FIGURE 12.15 Conflict within ant colonies. (A) In the ant *Dinoponera quadriceps*, three workers grasp the would-be reproductive that has been marked by the queen, preventing her from moving. (B) In the ant *Harpegnathos saltator*, the individual in black has grabbed and immobilized a nest mate whose ovaries have begun to develop. After holding her nest mate captive for 3 or 4 days, that worker may turn her prisoner over to another worker to continue the imprisonment. (A, from Monnin et al. 2002, © 2002 by the Nature Publishing Group; B, drawing by Malu Obermayer, from Liebig et al. 1999, © 1999 by The Royal Society of London.)

days at a time (**FIGURE 12.15A**) (Monnin et al. 2002). Likewise, workers of the ant *Harpegnathos saltator* punish nest mates that are developing their ovaries by holding the offenders in a firm grip (**FIGURE 12.15B**), preventing them from doing anything and thereby inhibiting further development of the immobilized ant's ovaries (Liebig et al. 1999).

These examples from wasp and ant groups illustrate one of the primary forms of social discord in animal societies: **reproductive conflict** (the other is sexual conflict, a topic we introduced in Chapter 9). Reproductive conflict encompasses not only conflict over which individuals get to breed, but also the battle over the resources that individuals need to breed. Importantly, reproductive conflict in eusocial species occurs not just in societies with multiple queens, but also between queens and workers (and among workers) over the production of unfertilized eggs. In colonies with multiple queens, reproduction is unlikely to be shared equitably, resulting in reproductive skew. When reproduction becomes skewed, kin structure within a society is also affected. As more females reproduce and reproductive skew decreases, the genetic relatedness among offspring also decreases. This decrease in relatedness with an increase in the number of breeding females in the group is analogous to the reduction in relatedness among siblings that occurs as the number of males with which a female mates (polyandry) increases (Rubenstein 2012). Daniel Kronauer and Jacobus Boomsma showed in army ants that there is a negative association between multi-queen colonies and multiple matings by queens (Kronauer and Boomsma 2007), suggesting that these are independent mechanisms that alter kin structure within eusocial societies. Thus, even in family groups characterized by altruism and relatively high genetic relatedness among group mates, the potential for reproductive conflict over who gets to reproduce—as well as sexual conflict over with whom the queen mates in some species—is likely to exist in many eusocial societies.

Social conflict clearly occurs in many forms in eusocial insect societies, but does this conflict actually influence the evolution of altruism? Tom Wenseleers and Francis Ratnieks tested whether sanctions imposed on workers that try to reproduce can make it more profitable for the sanctioned workers to behave altruistically. Using data collected from 20 species of eusocial insects on the effectiveness of policing efforts within colonies, as well as the proportion of workers that laid unfertilized eggs in their colonies, the authors were able to show that when the eggs of workers were reliably destroyed, a smaller proportion of workers developed ovaries in the colony (**FIGURE 12.16**) (Wenseleers and Ratnieks 2006b). In other words, the policing behavior of colony mates can mean that workers have little chance of boosting their inclusive fitness directly—thus, the indirect route to inclusive fitness is superior.

Another way of looking at this issue is to consider what percentage of the males in a colony of eusocial insects are the sons of workers, a figure that varies from 0 to 100

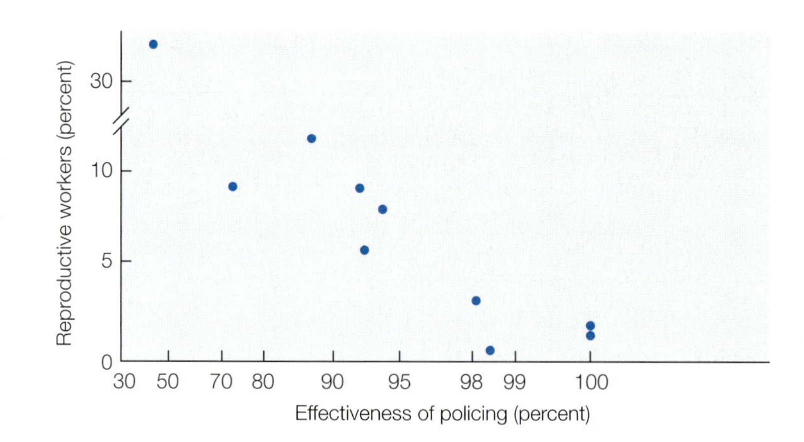

FIGURE 12.16 A test of the effectiveness of policing on the likelihood that workers will reproduce in social insect colonies. The better that workers are at destroying the eggs of their fellow workers, the less likely workers will try to reproduce. Each point represents values from different wasp or bee species. (After Wenseleers and Ratnieks 2006b.)

percent. If worker policing is responsible for reducing the number of workers' sons, then in cases with low numbers of worker sons, the policing workers are predicted to be more closely related to the queen's sons than to the other workers' sons. This prediction is correct (**FIGURE 12.17**). In *Melipona* bees, female larvae are sealed within brood cells that are the same size for both workers and queens and so are fed equally by other workers. Under these circumstances, female larvae control their own fate and can become either workers or queens, depending on what option confers the higher inclusive fitness. The excess mature queens (only one is needed to head up a colony) are quickly killed by the workers, but even so, the fact that they are produced at all lowers the reproductive output of the colony as a whole. In different species, different proportions of the female larvae opt to develop into queens, with as many as 25 percent of all immature larvae becoming queens (Wenseleers and Ratnieks 2004). Evidently it pays to help, but it can pay even more to *be helped*. The production of so many queens ready to take advantage of worker assistance reduces the size of the workforce because these individuals lack the modifications of the hindlegs that workers use to collect pollen for brood pots.

Wenseleers and Ratnieks predicted that the extent of queen overproduction in *Melipona* bees would be a function of how many reproductively competent males

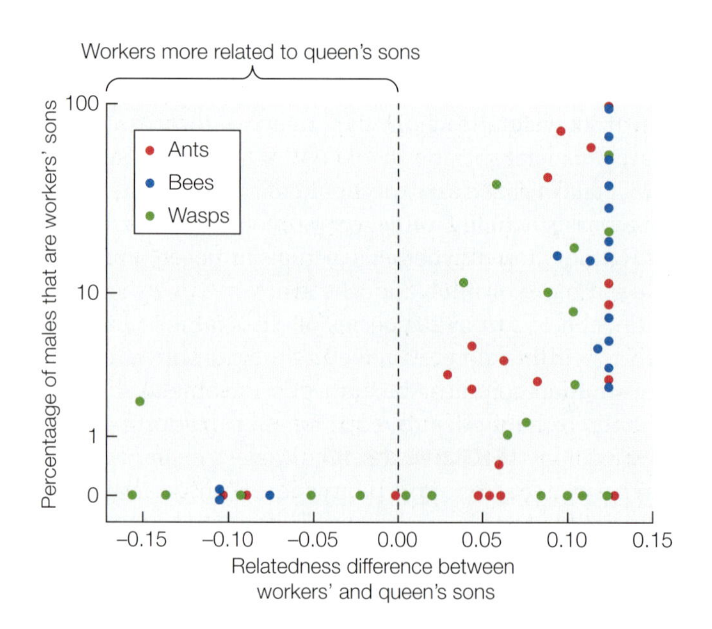

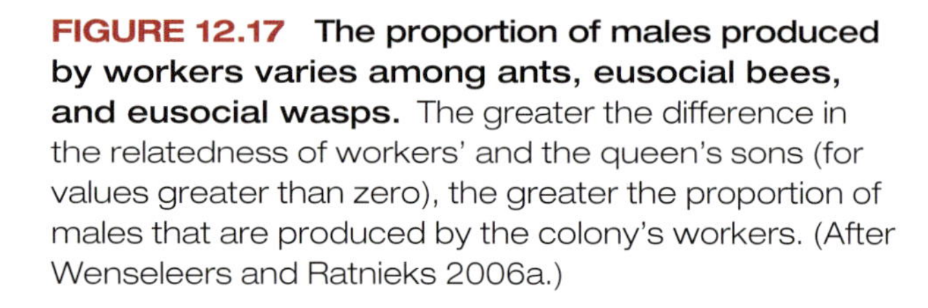
FIGURE 12.17 The proportion of males produced by workers varies among ants, eusocial bees, and eusocial wasps. The greater the difference in the relatedness of workers' and the queen's sons (for values greater than zero), the greater the proportion of males that are produced by the colony's workers. (After Wenseleers and Ratnieks 2006a.)

(A)

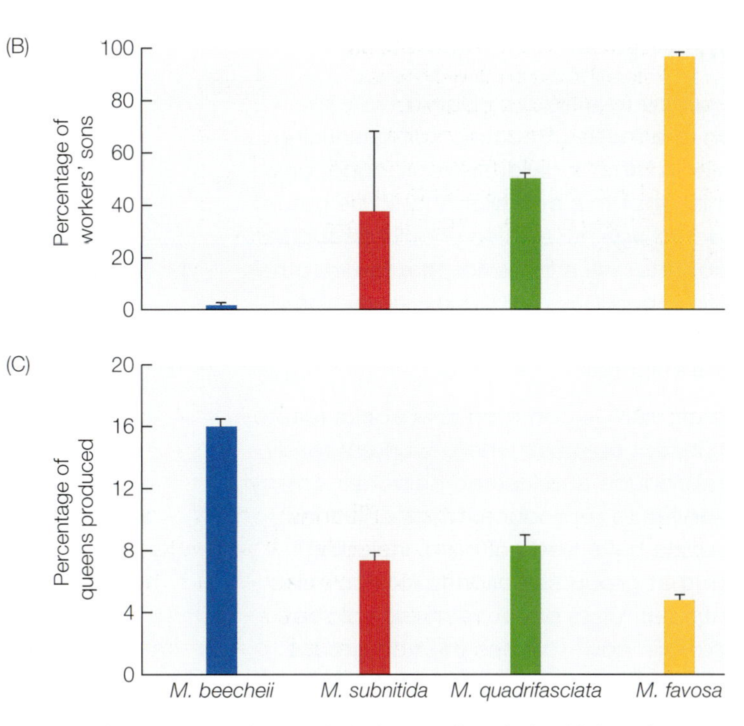

FIGURE 12.18 Colony kin structure is linked to queen production in eusocial *Melipona* bees. (A) *Melipona beecheii*. (B) The percentage of males produced by workers varies enormously among four species of *Melipona*. (C) To the extent to which workers are reproducing, creating nephews (r = 0.375) of their sisters, females should be less likely to become queens (which cannot carry out worker tasks). As predicted, the higher the percentage of workers' sons in the colonies of different species, the less likely immature females are to try to become queens. Bars depict mean +/– SE. (A, photograph © Eric Tourneret/Visuals Unlimited, Inc.; B, C after Wenseleers and Ratnieks 2004.)

in the colony were the sons of workers (Wenseleers and Ratnieks 2004). The greater the proportion of workers' sons, the greater the kin-selected cost to a female larva for opting to develop into a queen. The coefficient of relatedness between a female and her sisters' sons is 0.75 × 0.5 = 0.375 (the *Melipona* mother of these females mates only once, resulting in the high r for sisters). In contrast, the coefficient of relatedness for a female and her brothers, the sons of the queen, is only 0.25, as discussed earlier. If worker sisters are producing many sons, a new queen harms these nephews by removing herself from the colony workforce. Harm done to relatives reduces an individual's inclusive fitness in proportion to the similarity between these kin. And in fact, as predicted, in a species in which many workers are laying haploid eggs (which develop into their sons), most females "choose" not to develop into queens. In a species in which workers' sons are relatively rare, larval females are more likely to opt to develop into queens (**FIGURE 12.18**) (Wenseleers and Ratnieks 2004).

The fact that kin selection theory (or inclusive fitness theory more generally) has been used to make testable predictions about the kinds of conflicts that occur within colonies explains why this approach is so strongly favored by most scientists who study the social behavior of eusocial insects (and other highly social animals) (Strassmann and Queller 2007, Bourke 2011b). Moreover, group selection does not really account for the aggressive interactions that occur within animal societies, which according to the group selectionists are predicted to exhibit group-promoting adaptations, not individual-level conflicts that arise from natural selection.

SUMMARY

1. A great puzzle identified by Darwin was the evolution of altruism. Because self-sacrificing altruists help other individuals reproduce, one would think that this behavior should be naturally selected against and so should disappear over time. After all, a behavior that favors other individuals reproducing at the expensive of one's self seems to defy the theory of evolution by natural selection.
2. Extreme altruism is common in eusocial (caste-forming) insect colonies where workers rarely, if ever, reproduce and instead help their colony mates survive to reproduce. Many solutions to this puzzle have been offered, including the claim that group selection favors colonies that contain altruists because these colonies produce more new colonies than do groups without altruists.
3. Most students of animal social behavior employ inclusive fitness theory and the concept of kin selection, both of which operate at the level of the individual, because they have repeatedly proven to be useful for scientific investigation of sociality. Specifically, inclusive fitness theory has been used to show that altruism can spread through a population if the cost to the altruist in terms of a reduction in the number of offspring produced (multiplied by the coefficient of relatedness between the altruist and those offspring) is less than the increase in the number of related individuals helped by the altruist (multiplied by the average coefficient of relatedness between the altruist and the helped relatives).
4. Kin selection can result in an increase in the number of genes transmitted indirectly by an individual to the next generation in the bodies of relatives that exist because of the altruist's help. An individual's total genetic contribution to the next generation is called its inclusive fitness.
5. Inclusive fitness theory underlies modern "gene-centered thinking," with researchers now aware that individuals should behave in ways that boost their inclusive fitness, whether this is achieved through self-sacrificing cooperation or through self-serving conflict with others, even close relatives.
6. Sociogenomics, the study of the genes that influence social behavior and social diversity, allows researchers to probe the genetic architecture of sociality and caste differentiation. Although many socially relevant genes have been co-opted from those underlying solitary behaviors, others are rapidly evolving and appear to play a key role in social evolution.
7. Although cooperation broadly and altruism specifically form the basis of all eusocial insect societies, social conflict in the form of opposition over who breeds (reproductive conflict) and occasionally over who mates with whom (sexual conflict) is an inherent part of social living. In many eusocial societies, social conflict occurs not just among queens and/or kings, but also among workers, which in some Hymenoptera can produce unfertilized eggs that become males.

COMPANION WEBSITE

Go to **ab11e.com** for discussion questions and all of the audio and video clips.

Social Behavior and Sociality

There are few places on Earth that draw more people interested in animal behavior than the Serengeti, a region of East Africa that extends from northern Tanzania to southern Kenya. In addition to hosting the greatest terrestrial mammal migration in the world, it is home to perhaps the largest remaining populations of megafauna on Earth. If you are fortunate to travel to the Serengeti, you will likely see large herds of ungulates (including zebras, wildebeests, and more), parades of elephants, prides of lions, towers of giraffes, and troops of baboons (**FIGURE 13.1**). The common denominator for all of these mammals, as well as for many other species you may encounter on the African savanna, is that they spend much of their lives living in groups. In other words, most of these species are social. Yet nearly all of these vertebrate groups are quite different from the eusocial

Chapter 13 opening photograph © Anthony Bannister/Science Source.

(A)

(B)

(C)

FIGURE 13.1 From (A) zebras to (B) elephants to (C) baboons, many species of African mammals live in groups. (A, photograph © meunierd/Shutterstock.com; B, © iStock.com/nicolamargaret; C, © iStock.com/geneward2.)

insect societies that we discussed in Chapter 12. Although many of these Serengeti mammals, such as elephants and lions, live in family groups, some, such as the wildebeests, form groups that largely lack kin structure. Whereas some of these species live in permanent social groups like the Hymenoptera, other species, such as zebras, form fission–fusion societies in which group membership is constantly in flux.

To understand the evolution of sociality in animals as diverse as termites and baboons, we must first discuss social behavior more broadly and how it relates to living in groups. We must consider what a group is and understand the dynamics of group living, including how individuals interact to form these collectives. This requires an explanation of the many types of social behaviors, including ones that benefit all, some, or none of the interacting individuals. In this chapter we explore altruism in greater detail, and define what a cooperative behavior really is. We then explore individual differences in social behaviors, followed by a discussion of the reasons that animals come together to form cooperative groups in the first place. Included in our analysis of animal groups is a discussion of kin selection, a topic that we detailed in Chapter 12 as influencing the evolution of altruism and eusociality. Recall that Hamilton's rule encompassed not just genetic relatedness among group-living individuals, but also the costs and benefits of cooperative acts. In discussing kin groups and altruism, we will begin with what you learned in Chapter 12 in the context of eusocial societies. From there, we will begin to think about vertebrate societies in more detail by extending Hamilton's rule to consider the role of ecology in shaping animal groups. Throughout this chapter, we contrast vertebrate and invertebrate societies, discussing some of the similarities as well as the major differences.

The Evolution of Social Behavior

Some of the most mesmerizing behaviors are the **collective behaviors** of animals that form groups. Whether it be the murmurations of European starlings (*Sturnus vulgaris*) at dusk looking for a site to roost (**VIDEO 13.1**), the swirling spheres of anchovies attempting to escape or confuse predators, or the waves of marching army ants in search of food, the behavior of these collectives is really the synchronized movement of individuals following a series of basic interaction rules. In this context of collective action, we need to think about groups only as a series of interacting individuals; we do not need to worry about whether those individuals are related or not—we'll worry about that later. After all, the fish in a school of anchovies are almost certainly not related, but the individuals in a column of army ants are all likely to be close kin. Yet both groups exhibit a similar set of coordinated behaviors. We use the term *interaction* to describe the non-independence between individuals' movements, and although individuals are behaving independently, they typically follow the same set of interaction rules, giving the appearance of highly coordinated movement. These decision rules can be used to understand how, for example, individual fish orient themselves within a school and respond to the threat of a predator, or how group-level decisions are made when members have divergent interests (**BOX 13.1**). Interactions among conspecifics form the basis of social behavior, and as you will see, these social interactions can be either positive or negative (Hofmann et al. 2014).

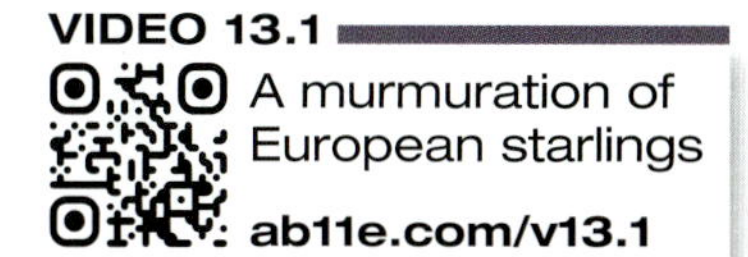

Forms of Social Behavior

In Chapter 12, we argued that cooperation is critical to the evolution of eusociality and other types of complex animal societies. While this is true, it is important to realize that cooperation is just one form of social behavior. Before we discuss cooperative behavior in vertebrates in more detail, we must first more broadly examine the four general kinds of social behavior that can take place between two individuals of the same species (or even among individuals of different species), whether these are amoebae or zebras. All social behaviors have fitness consequences (gains or losses termed **fitness payoffs**) for both the individual that performs the behavior (the donor) and another individual (the recipient) (**FIGURE 13.2**). When a social behavior results in both individuals receiving a fitness benefit (+/+), it is referred to as **mutual benefit** or a mutualism. In contrast, when the recipient receives a fitness benefit but the donor pays a fitness cost (–/+), it is called altruism, as we discussed in Chapter 12. Cooperation therefore includes both altruistic and mutually beneficial behaviors (West et al. 2007). Whereas both mutually beneficial and altruistic behaviors benefit the recipient, there are two forms of social behavior that do not: **selfishness**, sometimes called deceit or manipulation, occurs when the donor benefits but the recipient does not (+/–), and **spite** occurs when both the donor and the recipient fail to benefit from a behavior and instead both pay a cost (–/–). Below we discuss each of these forms of social behavior in greater detail.

Mutual Benefit

Cooperation in the form of mutual benefit is one of the best-studied animal social behaviors, but as you have already seen, it can be challenging to identify whether benefits are available to both participants in a social interaction. It can also be particularly challenging to distinguish a mutual benefit from a kin-selected benefit for species that live with relatives. Take, for example, the paper wasp *Polistes dominula*, in which several females may band together to build and provision a nest in the spring. At any one moment, only one wasp in a group of females acts as a queen, while the

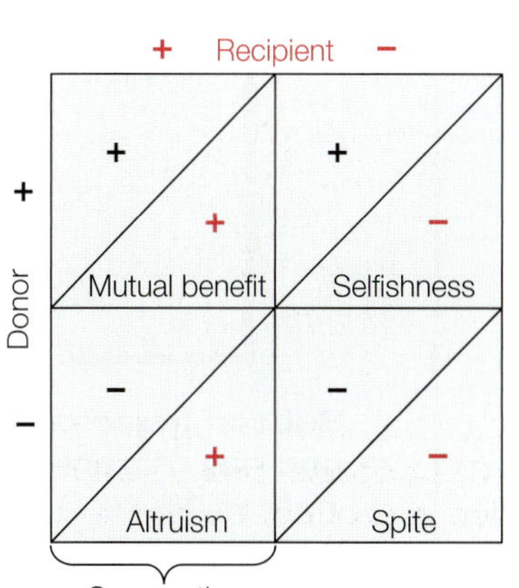

FIGURE 13.2 Fitness payoffs for social behaviors. Mutual benefit occurs when both individuals receive a fitness benefit (+/+). Altruism occurs when the recipient receives a fitness benefit but the donor pays a fitness cost (–/+). Cooperation includes both altruistic and mutually beneficial behaviors. Selfishness occurs when the donor benefits but the recipient does not (+/–), and spite occurs when both the donor and the recipient fail to benefit from a behavior and instead both pay a cost (–/–).

BOX 13.1 **DARWINIAN PUZZLE**

How do groups of animals decide where to go?

If a group is simply a collection of individuals following simple decision rules, how do decisions get made when individuals might have different interests or desires? For example, when animals move as part of a collective, it is often unclear how the group decides where to go, especially when individuals might have different opinions on the best route to take or even which direction to take in the first place. Take a troop of olive baboons (*Papio anubis*) living on the Kenyan savanna. Each morning when individuals wake up, some must decide where the group should go to forage that day. To determine how individual interests influence where groups decide to travel, a team led by Margaret Crofoot used global positioning system (GPS) collars to record the movement of individual baboons every second from sunrise when they awoke until sunset when they settled down to sleep (Strandburg-Peshkin et al. 2015). In this way, researchers were able to follow the detailed movements of individual baboons across the landscape (**FIGURE**).

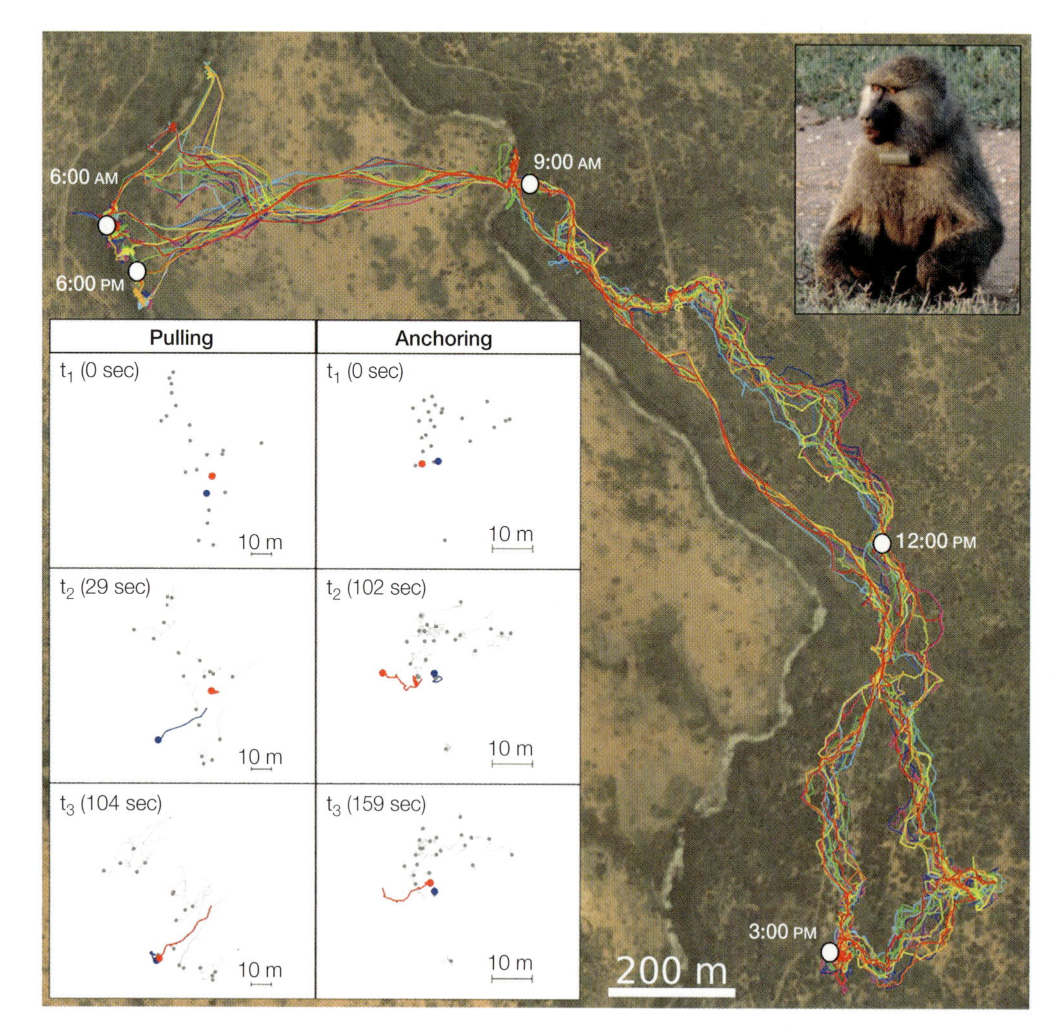

Baboon trajectories (25 individuals) during one day of tracking. (Inset, left) Successful initiation ("Pulling"), where the initiator (red) recruits the follower (blue). (Inset, right) Failed initiation ("Anchoring"), where the initiator (red) fails to recruit the potential follower (blue). Other individuals' trajectories are in gray. (From Strandburg-Peshkin et al. 2015.)

BOX 13.1 DARWINIAN PUZZLE *(continued)*

Surprisingly, baboons did not simply follow more dominant individuals, but instead followed individuals that moved in a highly directed manner. The likelihood of following depended on the number of initiators and their agreement on which direction to go. But what influenced where those initiators went? By combining the high-resolution GPS tracking with three-dimensional modeling from imagery captured by a drone (**VIDEO 13.2**), the authors showed that movement decisions were also influenced by large-scale habitat features such as roads and sleeping sites, as well as fine-scale features such as vegetation structure (Strandburg-Peshkin et al. 2017). Thus, the decision of where to go each morning was really a group-level decision for the baboons based on the interactions among individual group members, which were influenced by both large- and fine-scale features of the surrounding environment.

VIDEO 13.2

Thinking Outside the Box

Troops of baboons consist of a mixture of related and unrelated individuals. Crofoot's team did not know much about the baboons they studied beyond an individual's sex. For their study, they purposely chose a group of naïve, previously unstudied baboons. Given what you learned in Chapter 12 about cooperation and conflict in family groups, how do you think genetic relatedness might have influenced behavioral interactions and group movement decisions? Design a study similar to that of Crofoot team's that would allow you to determine how kinship influences the likelihood of being an initiator or a follower.

others are subordinate to the leader. Based on what we discussed in Chapter 12, you might think that these subordinate workers would all be sisters or close relatives of the queen wasp (thereby gaining indirect fitness from their helpful behavior). Instead, up to 35 percent of the helper wasps in the typical *P. dominula* nest are unrelated to the dominant female (Leadbeater et al. 2011). Under these circumstances, there is no possibility that these subordinates are engaged in kin-directed altruism (see below), in which their sacrifices—in terms of lowered production of their own offspring—are compensated for by increases in the number of relatives that exist because of their helpful behavior.

Just because these *Polistes* wasps do not receive kin-directed (indirect) benefits from working together doesn't mean they don't receive other benefits from grouping. Instead of the wasps receiving indirect fitness benefits, we might account for their apparent self-sacrifice by demonstrating that these unrelated subordinates derive a future direct reproductive benefit by helping. When an individual eventually gains access to a resource controlled by another individual because of its prior help, it is called postponed cooperation. A team led by Ellouise Leadbeater showed that social subordinates had a chance of inheriting a well-protected nest from the dominant queen upon the queen's demise. When this happens, the successor to the initial queen has the assistance of the remaining subordinates at the nest. Because of the considerable reproductive success of the inheritors, subordinates on average have higher direct fitness than solitary nesting females, which run a high risk that they will be killed and/or have their nests destroyed by the predatory birds and mammals that plague weakly-defended nests of paper wasps (**FIGURE 13.3**) (Leadbeater et al. 2011). Thus, by helping a dominant female, the subordinate

FIGURE 13.3 A direct fitness benefit for helping in a paper wasp. Females that help an unrelated foundress female have a chance of inheriting a colony from the dominant foundress. (A) The photograph shows an early-stage colony with three foundress females. One is dominant to the others, but one of the subordinate females has a chance to inherit the nest should the dominant female disappear. (B) Inheritors often acquire a well-defended colony with helpers, and as a result, they generally reproduce more successfully than solitary nest foundresses in *Polistes dominula*. (A, photograph © Oxana Bernatskaya/123RF; B after Leadbeater et al. 2011.)

receives a "golden ticket" from the dominant that permits the helper to stay with the colony and potentially inherit the nest at some later date. As you will see below, this idea of territorial inheritance might also influence the evolution of cooperative breeding behavior in vertebrates.

Behavioral biologists have uncovered many other cases of individuals helped by unrelated members of their species in ways that, on average, eventually result in heightened personal reproductive success for the helper. Consider, for example, the strange cooperation that links subordinate yearling male lazuli buntings (*Passerina amoena*), which have dull brown plumage, with brightly colored, dominant yearling males (**FIGURE 13.4**). The dominant males tolerate dull-plumaged birds as neighbors, letting them settle next door while aggressively driving away other males that are more brightly colored. One reproductive benefit of this tactic for the brightly colored males is that they get to mate with the subordinate males' females that presumably find the brightly colored males more appealing than their own dull-colored social partners, as is true for some other birds in which both dull and brightly colored males compete for mates (Webster et al. 2008). Erick Greene and colleagues found that dull-plumaged males regularly care for one or two extra-pair young, which are presumed to be the genetic offspring of their more attractive neighbors (Greene et al. 2000). The question in lazuli buntings is what, if any, benefit subordinate males secure from living next to such sexually attractive males. Greene and colleagues found that subordinate buntings generally rear a few offspring of their own as a result of possessing high-quality territories that appeal to females. In contrast, males of intermediate plumage brightness are often pushed by dominant rivals into habitat so poor that no female will join them, with the result that they must wait an entire year before trying to reproduce again (Greene et al. 2000).

Because both dull-colored male buntings and their brightly plumaged neighbors gain some fitness from their interactions, this social arrangement can be categorized

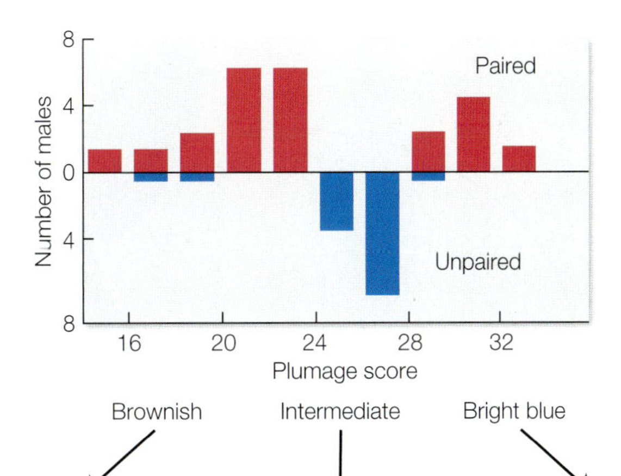

FIGURE 13.4 Cooperation among competitors. Yearling male lazuli buntings range in color from dull brownish to bright blue and orange. (Their plumage scores range from less than 16 to more than 32.) Bright yearling males permit dull males, but not males of intermediate brightness, to settle on territories neighboring their own. As a result, brownish males often pair off with females in their first year, whereas yearling males of intermediate plumage typically remain unpaired. (After Greene et al. 2000; photographs courtesy of Erick Greene.)

as mutual benefit. But is the same true for male partnerships among long-tailed manakins (*Chiroxiphia linearis*), a species in which unrelated males cooperate in an attempt to attract females with extraordinary coordinated displays (**VIDEO 13.3**) (Foster 1977, McDonald 2010)? Finding the right male partner to aid them in their mating quest can be challenging for male long-tailed manakins. Research led by David McDonald used social network analysis (**BOX 13.2**) to study manakin cooperation networks and determine how male–male associations such as these form (Edelman and McDonald 2014). The authors showed that males were more likely to cooperate if they primarily displayed at the same or neighboring leks, and if they did so with a "friend of a friend" rather than with males with which they did not share a mutual partner. Ultimately, social network analysis demonstrated that long-tailed manakin networks were highly structured, with males interacting only with a subset of the population, and so a series of local processes predicted male–male cooperative associations.

VIDEO 13.3

Male long-tailed manakins displaying

ab11e.com/v13.3

But why do male manakins cooperate in the first place? Female manakins that respond to the cooperative callers' astonishing cartwheel display may land on the males' display perch, often a horizontal section of liana that lies a foot or so above the ground (**FIGURE 13.5**). Should a female visitor start jumping excitedly on the perch in response to these "butterfly flights" that show off the males' beautiful plumage, the beta male discreetly departs, while the alpha male stays to copulate with the visitor. By marking the males at display sites, David McDonald and colleagues found that the alpha male did all of the mating, leaving nothing for the beta male (McDonald and Potts 1994).

FIGURE 13.5 Cooperative courtship of the long-tailed manakin. The two males are in the cartwheeling portion of their dual display to a female, which is perched on the vine to the right.

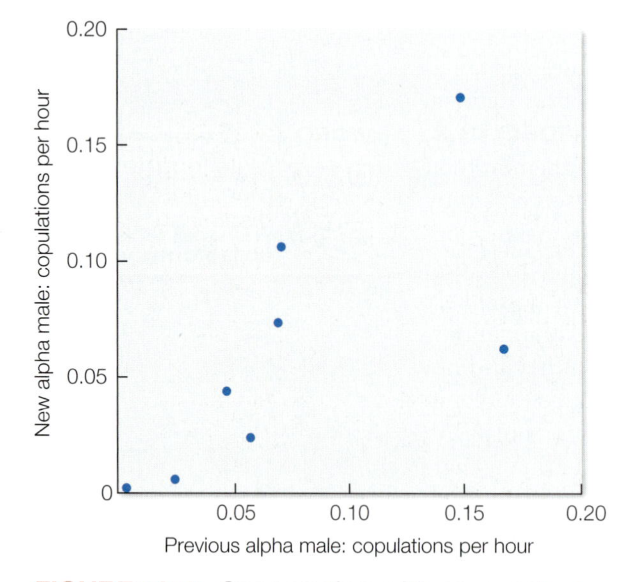

FIGURE 13.6 Cooperation with an eventual payoff. After the death of his alpha male partner, the beta male long-tailed manakin (now an alpha) copulates about as frequently as his predecessor did, presumably because the females attracted to the duo in the past continue to visit the display arena when receptive. (After McDonald and Potts 1994.)

How can it be adaptive for a celibate beta subordinate to help a sexually monopolistic alpha male year after year? McDonald proposed that by conceding all receptive females to the alpha male, a beta manakin establishes his claim to be next in line, if and when the alpha male disappears, by keeping other (mostly somewhat younger) birds at bay. After many years of careful observation, McDonald showed that when an alpha male disappears, the beta male does in fact move up in status and then usually gets to mate with many of the same long-lived females that copulated with the previous alpha male (**FIGURE 13.6**) (McDonald and Potts 1994). Thus, beta males work for unrelated partners because this is the only way to join a queue to have any chance of eventually becoming a reproducing alpha male. And beta males only have a *chance* of becoming an alpha; there is no guarantee that a beta male will become an alpha, only a better than average chance of doing so at some point in his life. However, in a closely related species, the lance-tailed manakin (*Chiroxiphia lanceolata*), in which two males also perform a display, Emily DuVal showed that because beta males do not always inherit breeding positions when they should, queuing does not fully explain cooperation in this species (DuVal 2007). Instead, she argued that beta males may be apprenticing to learn effective and appropriate displays that enhance their subsequent success as alphas.

This same queuing process has been observed in a variety of group-living coral reef fishes, including the clownfish

BOX 13.2 INTEGRATIVE APPROACHES

Social network analysis

Be it a troop of baboons, a lek of manakins, or even the class you are taking on animal behavior, individuals in a group form social relationships, and those relationships vary among individuals and over time. As social creatures ourselves, we are well aware that not all individuals interact equally with others. Some individuals interact with many others, but some only interact with a few. Still other individuals might interact strongly and frequently with specific group members but rarely with others. To illustrate these complex social dynamics, think about your favorite social media site. Be it Facebook, Twitter, Instagram, or the next big thing, you and your friends and acquaintances share information in the form of updates or photos that can then be passed to and among other individuals in your or their networks. Social network analysis has become a popular tool for studying these interactions, be they in animal or human groups. Each individual in a network can be represented as a node, and each connection or link between individuals is called an edge. More popular individuals that are connected to more nodes will have more edges, and individuals that interact more frequently together can be illustrated by weighting the edges (simply by making them bigger).

Researchers can use the information in the social network relationships to study how information (such as "fake news" on Facebook) spreads among users. While these techniques are useful for studying humans, they are also important for studying social behavior in other animal species. Take for example, these network diagrams showing how information diffuses through populations of wild birds (**FIGURE**). Lucy Alpin and colleagues introduced alternative novel foraging techniques into populations of great tits (*Parus major*) using puzzle boxes that were necessary to access food at radio-frequency identification (RFID) feeders (see Video 3.1 for a similar feeder design) (Alpin et al. 2015). In each population, one pair of birds was trained on the novel foraging technique by being subjected to one of three training regimens: (1) given no training and left in the

(Continued)

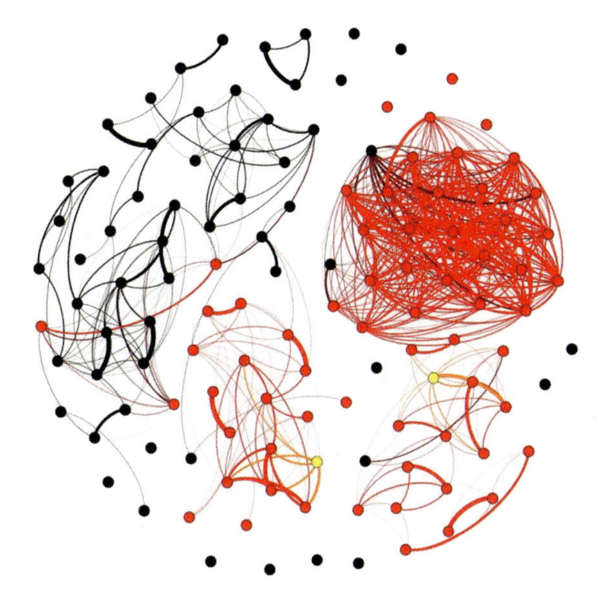

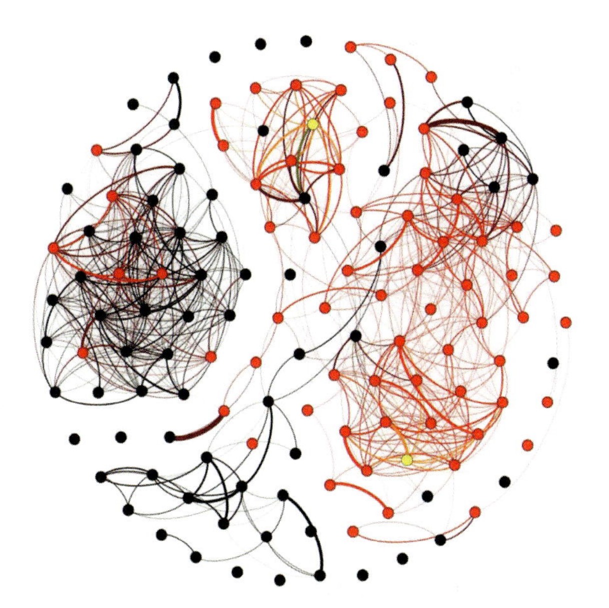

Social networks show the diffusion of innovation through great tit populations. Yellow nodes indicate trained demonstrators. Red nodes indicate individuals that acquired the novel behavior after 20 days of exposure. Black nodes indicate naïve individuals that never learned the behavior. The network on the left had 123 individuals, whereas the one on the right had 137 individuals. (From Alpin et al. 2015.)

BOX 13.2 INTEGRATIVE APPROACHES (*continued*)

cage with ad libitum food (control); (2) trained to solve the novel puzzle box by pushing the blue side of the door to the right; or (3) trained to solve the novel puzzle box by pushing the red side of the door to the left. Every individual in the population was identified and given a PIT (passive integrated transponder) tag that was logged automatically at feeders. From only two trained birds in each population, information about how to access the food at feeders spread rapidly through social network ties to reach on average 75 percent of the individuals in each population. Thus, although only two birds were taught how to open the feeders, the majority of birds learned how to solve the puzzle from these birds in less than 3 weeks. Since each population was given a different puzzle, network analysis allowed Alpin's team to show not only how information spread among the birds, but that each population was heavily biased toward using the technique that was originally introduced, resulting in established local traditions. The network diagrams also show visually how the information could spread to birds with whom the trained individuals never interacted directly.

Thinking Outside the Box

How do you think naïve individuals learned to open the puzzle boxes? Come up with a few potential mechanisms to explain the transfer of information in great tits. Design an experiment to isolate these mechanisms and determine which ones the birds used. Look closely at the network diagrams. Some individuals never interacted with others, and therefore did not learn how to access the feeders. However, some individuals interacted frequently with those that learned the behavior and yet did not learn it themselves. Look at the larger cluster of mixed red and black nodes in the network diagram on the right. These individuals are connected by many edges, yet many did not learn the behavior. Why do you think that was?

Amphiprion percula, in which subordinates wait as part of a group living in an anemone for reproductive positions to become vacant (**FIGURE 13.7**). But clownfish queue with an added twist. Most clownfish groups are made up of a few individuals that form a size-based hierarchy in which the female is largest, the breeding male is the next largest, and the other individuals—all males—are progressively smaller (Fricke 1979). If the female dies, the breeding male changes its sex to become the breeding female, and the largest nonbreeding male increases in size to become the new breeding male. As in the manakins, subordinates benefit from queueing and waiting for a vacant breeding position. However, unlike in the manakins, dominants gain no apparent benefits from having subordinates around (Buston 2003). Ultimately, the presence of freeloading subordinates waiting for the dominant to die creates a greater potential for social conflict among group mates than in the manakins. This potential conflict is controlled by precise regulation of subordinate growth to make sure that smaller individuals never approach the dominant's size. Why does the dominant male permit the subordinate males to remain in the group if they provide no fitness benefit? It is actually more costly for the dominant male to chase off his subordinate group mates than it is to permit them to remain in the anemone, which provides safe haven for all fish in the group. Thus, while the dominant male does not receive a benefit from having the subordinates around, he also does not pay a cost unless he expends energy to chase them away. By tolerating these smaller males in the anemone, he avoids paying that cost, which may be as advantageous to him as receiving a benefit.

FIGURE 13.7 Subordinate clownfish living in an anemone wait for reproductive positions to become vacant. Each anemone has just one breeding pair, and subordinate males wait for an opportunity to become breeders, and possibly change sex. (Photograph © iStock.com/crisod.)

Altruism and Reciprocity

Altruism—unselfish or self-sacrificial behavior—among non-kin is quite rare in the animal world. In Chapter 12 we introduced several examples of self-sacrificial behavior in eusocial insects, but in all of those cases, these behaviors benefit an individual's relatives, with whom it shares genes (in other words, they are cases of kin-directed altruism). Examples of altruism among non-kin are very unusual, except when the benefits are postponed. The study of long-tailed manakins revealed that some superficially self-sacrificing actions actually advance the reproductive chances of at least some patiently helpful individuals that survive long enough to graduate to alpha (mating) status. Cases of postponed cooperation are very similar to instances in which helpful individuals also receive delayed compensatory repayment *directly* from the helped individual, rather than waiting to inherit a resource. Robert Trivers called this social arrangement **reciprocal altruism** because individuals that are helped generally return the favors they receive—eventually (Trivers 1971). But because the helpful indivduals are not really sacrificing direct fitness over the long haul, reciprocal altruism is often called just plain reciprocity. Whatever the label, if the initial direct fitness cost of helping is modest but the delayed direct fitness benefit from receiving the returned favor is on average greater, then selection can favor being helpful in the first case. Reciprocity will occur only among individuals that remain together long enough for a helpful act to be repaid, something typical of many group-living species. A classic example is the grooming behavior of assorted group-living primates in which individuals pick through the fur of a companion, a helpful act that may be repaid at a later date when the groomee becomes the groomer or performs some other useful behavior for the animal that had been its helper (**FIGURE 13.8**) (Schino and Aureli 2010). Occasionally, third-party observers that witness a helpful act are more likely to help those that have helped others, a concept called **indirect reciprocity**. Indirect reciprocity has been used to explain why some individuals (particularly humans) develop cooperative reputations (Nowak and Sigmund 2005).

FIGURE 13.8 Reciprocity occurs in primates that groom one another. Here, a female hamadryas baboon (*Papio hamadryas*) is cleaning the fur of her partner and protector, a male that will keep other males from sexually harassing his groomer in the future. (Photograph © Dirk M. de Bar/ Shutterstock.com.)

Distinguishing reciprocity from other mechanisms of cooperation such as kin selection can be particularly challenging in cases such as primate grooming, in which both relatives and nonrelatives interact. The clearest cases of reciprocity actually involve different species cooperating, since kinship is excluded from playing a role (Sachs et al. 2004). Take, for example, cleaner fish that remove ectoparasites and dead skin from other species of reef fish. A cleaner benefits from getting safe access to food (because the host does not eat the cleaner), while the host benefits from the removal of parasites that have been shown to otherwise injure it (Grutter 1999). Another classic example of interspecific cooperation involves humans and an African bird called the greater honeyguide (*Indicator indicator*). Contrary to what its name would suggest, the honeyguide does not eat honey. Instead, it prefers the wax honeycombs that bees produce to store their honey. However, accessing wax inside a beehive is not an easy task for a honeyguide, so the species has evolved an ingenious method: it lets humans do the dangerous work, and then takes their spoils as a reward for guiding them to the honey (hence the bird's name). Claire Spottiswoode and colleagues followed honeyguides in northern Mozambique and demonstrated that at least 75 percent of the time, the birds succeeded in taking their human followers to a hive using a distinctive chattering call (**AUDIO 13.1**) (Spottiswoode et al. 2016). The researchers further showed that this association is no accident, as local human honey-hunters produce a specific call of their own, a loud trill followed by a grunt, that is passed down from father to son. Using a set of playback calls, including the human honey-hunting sound, a control animal sound, and a control human sound, the authors found that honeyguides do indeed react to the special call.

AUDIO 13.1

Greater honeyguide chatter call

ab11e.com/a13.1

Although some primates, fishes, and birds appear to be capable of reciprocity, direct payback behavior is not particularly common (Connor 2010, West et al. 2011), perhaps in part because a population composed of reciprocal helpers would generally be vulnerable to invasion by individuals happy to accept help but likely to "forget" about the payback. "Defectors" reduce the fitness of helpers in such a system, which ought to make reciprocity less likely to evolve. The problem of defection

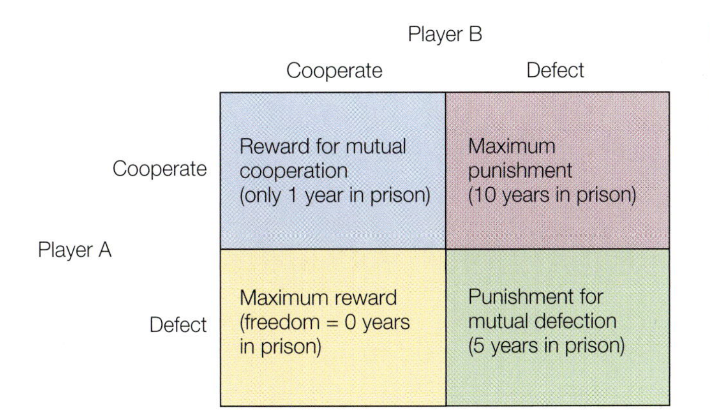

FIGURE 13.9 The prisoner's dilemma. The diagram lays out the payoffs for player A that are associated with cooperating or not cooperating with player B. Defection is the adaptive choice for player A given the conditions specified here (if the two individuals will interact only once).

can be illustrated with a game theoretical model called the **prisoner's dilemma** (**FIGURE 13.9**). Imagine that a crime has been committed by two people who have agreed not to squeal on each other if caught. The police have brought them in for interrogation and have put them in separate rooms. The cops have enough evidence to convict them both on lesser charges but need to have the criminals implicate each other in order to jail them for a more serious crime. In the case of the prisoner's dilemma, the charges can be modeled as payoffs, and each payoff will vary depending on what the alleged criminals decide to do. The police therefore offer each suspect freedom if they will squeal on their partner. If suspect A accepts the offer ("defect") while B maintains their agreed-upon story ("cooperate"), A gets freedom (the maximum payoff) while B gets hit with the maximum punishment—say, 10 years in prison (the "sucker's payoff," shown in the upper right panel of Figure 13.9). If together they maintain their agreement (cooperate + cooperate), then the police will have to settle for conviction of both on the lesser charges, leading to, say, a 1-year prison term for each suspect. And if each one fingers the other, the police will use this evidence against both and renege on their offer of freedom for the snitch, so both A and B will be punished quite severely with, say, a 5-year prison sentence each.

In a setting in which the payoffs for the various responses are ranked "defect while other player cooperates" > "both cooperate" > "both defect" > "cooperate while other player defects," the optimal response for suspect A is always to defect, never to cooperate. Under these circumstances, if suspect B maintains their joint innocence, A gets a payoff that exceeds the reward that could be achieved by cooperating with a cooperative B; if suspect B squeals on A, defection is still the superior tactic for A, because A suffers less punishment when both players defect than if one cooperates while the other squeals. By the same token, suspect B will always come out ahead, on average, if B defects and points the finger at A.

In our example of the prisoner's dilemma, cooperation was not the best strategy. How, then, can we account for the cases of reciprocity that have been observed in nature (and that are also common features of human social behavior; West et al. 2011)? In our example, the criminals interacted only one time. However, when the game is modified so that individuals can interact repeatedly, what we call the multiplay or iterated prisoner's dilemma, individuals who use the simple decision rule called tit-for-tat, or "do unto individual X as X did unto you the last time you met," will reap greater overall gains than cheaters who accept assistance but never return the favor (Axelrod and Hamilton 1981). When multiple interactions are possible, the rewards for back-and-forth cooperation add up, exceeding the short-term payoff from a single defection. In fact, the potential accumulation of rewards can even favor individuals who "forgive" a fellow player for an occasional defection,

FIGURE 13.10 Blood meal sharing in vampire bats is explained by reciprocity. Individuals ensure against future starvation if they share blood with a neighbor after a successful night of feeding, assuming that another individual in the colony reciprocates with them after they have had an unsuccessful night of feeding at some point in the future. (Photograph © Nick Hawkins/Minden Pictures.)

because that tactic can encourage maintenance of a long-term relationship with its additional payoffs (Wedekind and Milinski 1996, St-Pierre et al. 2009).

Vampire bats appear to meet the required conditions for multiplay reciprocity. These animals must find scarce vertebrate victims from which to draw the blood meals that are their only food source (**FIGURE 13.10**). (Usually they feed on birds or livestock, though recent genetic analysis has confirmed that a population in a highly human-modified habitat in Brazil has begun feeding on people; Ito et al. 2016.) After an evening of foraging, the bats return to a roost where individuals known to one another regularly assemble. A bat that has had success on a given evening can collect a large amount of blood, so much that it can afford to regurgitate a life-sustaining amount to a companion that has not been so lucky. Under these circumstances, the cost of the gift to the donor is modest, but the potential benefit to the recipient is high, since vampire bats die if they fail to get food three nights in a row. Thus, a cooperative blood donor is really buying insurance against starvation down the road. Individuals that establish durable give-and-take relationships with one another are better off over the long run than those cheaters that accept one blood gift but then renege on repayment, thereby ending a potentially durable cooperative arrangement that could involve many more meal exchanges (Wilkinson 1984).

Without knowing the relatedness among bats, it is difficult to rule out kin selection as at least partially explaining the blood-sharing behavior of vampire bats. Although common vampire bats (*Desmodus rotundus*) cooperate with both kin and non-kin alike, an experiment by Gerald Carter and Gerald Wilkinson that involved simultaneously fasting and removing bats from an experimental colony demonstrated that bats that had fed more non-kin in previous years had more donors and received more food (Carter and Wilkinson 2015). To rule out the idea that better-connected bats always receive more food, Carter and colleagues expanded this work to show that helping more non-kin did not increase food received when key donors were available, but it did reduce the negative impact on food received when a key donor was removed (Carter et al. 2017). Together, these results confirmed experimentally that although kin selection plays a role in some cases, reciprocity can at least partially explain cooperative food sharing in vampire bats.

Non-cooperative Social Behaviors: Selfishness and Spite

There are many examples of selfish behavior in nature, including animals stealing food from group mates or failing to do their part in foraging or defense. Stealing food from others, or **kleptoparasitism**, is a common form of selfishness and occurs in many animals, including insects, mammals, and birds. The key to selfishness in animal societies is that individuals often use coercive tactics to manipulate others to perform actions likely to increase their fitness. In such manipulative situations, the behavior of both individuals need not be adapted to providing assistance to their partner. Additionally, the costs and benefits of interactions often differ substantially between partners so that there is no alternation of assistance given and received, as there is in reciprocity (Clutton-Brock 2009a).

Although selfish behaviors are quite common, spite is rare and has been hard to demonstrate, outside of humans of course. It is not surprising that spite appears to be rare in nature, as it is not something that will typically be favored by natural selection, since neither party receives a fitness benefit from a spiteful act. Spite could be adaptive if it leads to modification of the recipient's behavior (or that of others observing the punishment), which could then benefit the spiteful individual in the long run. There is some experimental evidence from brown capuchin monkeys (*Cebus apella*) that when captive individuals observe others getting more of a food resource, they punish those individuals (Leimgruber et al. 2016). This has been interpreted as spite because individuals that punish others also end up paying a cost in the process.

Another potential example of spite occurs in the polyembryonic parasitoid wasp *Copidosoma floridanum* (Gardner and West 2004). Eggs of this species are deposited by multiple females into moth eggs, where the wasps develop inside the developing caterpillar. One wasp egg proliferates clonally to produce multiple larvae, some of which develop normally while others develop into sterile soldiers that, after hatching, attack and kill some of the other larvae before dying themselves. Which larvae do the soldiers attack? Experimental work by Giron and colleagues demonstrated that soldiers can recognize kin and preferentially attack non-kin (Giron et al. 2004, 2007). Such **kin recognition** is central to the evolution of cooperative behaviors among relatives, not just in wasps but in birds (where vocalizations often play a role) and in mammals (where olfactory cues often play a role). So while the data from the embryonic wasp studies are consistent with the concept of spite, kin-directed altruism could at least partially explain the results, since the soldiers that fail to reproduce directly might benefit indirectly by helping their relatives.

Individual Differences in Social Behavior

If we are to understand the adaptive basis of social behavior and the role that natural selection plays in behavioral evolution more broadly, we must consider the factors that shape behavioral variation within and among populations and species. Individual variation in animal behavior is typically studied under the theme of consistent differences in behavior (referred to as **animal personality**) or of behavioral correlations among different contexts (referred to as **behavioral syndromes**) (Sih et al. 2004, Reale et al. 2007). Animal personalities typically refer to behaviors that are repeatable through time, whereas behavioral syndromes are defined more broadly as behavioral consistency within individuals across contexts. For example, animals that tend to be more bold than others in the face of predation risk also tend to be more aggressive toward conspecifics in a reproductive or other context (Sih and Bell 2008). Indeed, a syndrome that involves exploratory behavior, fearfulness, aggression, and response to environmental change (a "proactive reaction axis") has been identified in several species (Sih et al. 2004). Whether or not studying individual behavioral variation in the context of animal personalities or behavioral syndromes adds anything new to the field of animal behavior beyond what behavioral biologists have already been doing for decades is a topic of great debate (Beekman and Jordan 2017), but here we explore animal personalities in social species.

Personalities in Social Species

Evidence from numerous species suggests that both individual repeatability and correlations among different behaviors as varied as boldness, aggressiveness, activity, exploration, and sociability are common (Sih et al. 2015). Indeed, personalities have been described in taxa as diverse as fishes, reptiles, birds, mammals, and insects. Why do animals have personalities? From a proximate perspective, differences in adult personalities could simply be the result of developmental differences. Early life experiences have been shown to influence personalities later in life in several species (Sih et al. 2015). Yet from an ultimate perspective, it is important to know if personality differences are correlated with differences in fitness. In a meta-analysis of more than 30 studies from a diversity of animals, Brian Smith and Daniel Blumstein found that bolder individuals had increased reproductive success but lower survival (Smith and Blumstein 2008). In contrast, exploratory behavior did not influence reproductive success, and aggression had only a small positive effect. Thus, there is some evidence that similar types of personality traits influence fitness in similar ways in different species.

Cooperative breeding species, in which more than two individuals care for young, are ideal for studying the fitness effects of personalities because, like eusocial insects, individuals adopt different reproductive roles that vary in the likelihood of gaining direct reproductive success. Unlike in eusocial insects, however, these reproductive roles are rarely permanent and individuals have the flexibility to switch roles during their lives. This means, for example, that an individual with a specific type of personality might be more (or less) likely to become a breeder than an individual with another personality. A study in cooperatively breeding meerkats (*Suricata suricatta*) asked whether individuals with different behavioral personalities were associated with the adoption of a dominant breeder versus a subordinate alloparent or helper social role (a nonbreeder that aids others in raising young) (Carter et al. 2014). Behaviors such as babysitting, provisioning, and raised guarding (sentinel behavior) are repeatable in meerkats, suggesting that they are personality traits. However, the researchers found no relationship between early life behavior for these personality traits and the acquisition of dominance status as adults. In other words, individuals with specific personality traits were not more or less likely to become breeders than any other individual.

Another cooperative breeder, the Seychelles warbler (*Acrocephalus sechellensis*), also exhibits personality traits in the form of novel environment and novel object exploration, both of which are at least partly heritable in this species (Edwards et al. 2017a). Since exploratory behavior correlates positively with traits such as risk-taking behavior and activity in other wild bird species, Hannah Edwards and colleagues predicted that exploratory behavior might promote extra-pair mating by increasing the rate at which potential extra-pair partners are encountered (Edwards et al. 2017b). However, the researchers found that neither the total number of offspring nor the number of extra-pair offspring produced were associated with a male's or female's exploratory behavior. Thus, while personalities have been identified in numerous animal species and shown to influence fitness in some, evidence that they are related to the adoption of different social roles or mating tactics in cooperatively breeding species is scarce. Instead, these studies suggest that reproductive behaviors in cooperatively breeding species may be more plastic than heritable.

The Evolution of Cooperative Breeding

Although the idea that sociality is always superior to solitary living is popular, this view is rarely held by those aware of the cost–benefit approach to behavioral biology. While living with others can have advantages, it also comes with substantial fitness costs to individuals. As Richard Alexander pointed out, there are no automatic benefits of living in a group, but there are some automatic costs of living with others (Alexander 1974). For example, because cliff swallows (*Petrochelidon pyrrhonota*) nest side by side in colonies, these birds are subject to parasitic bugs that make their way from nest to nest. The larger the swallow colony, the more it attracts transient birds, and with these visitors come more bugs, some of which slip off their carriers and onto resident swallows and their offspring. As a result, the larger the colony, the higher the infestation rate of the parasite (Brown and Brown 2004). When the blood-sucking parasites attack swallow chicks, the price for being social becomes evident in the form of stunted nestlings (**FIGURE 13.11**) (Brown and Brown 1986). For social life to be adaptive for cliff swallows, some benefit must exist that more than compensates the birds for the risk of having bug-infested offspring. For the cliff swallow, a major benefit appears to be improved foraging success for adult birds searching in groups for flying insects (Brown 1988).

There are other costs to group living, including greater competition for resources such as food, breeding sites, or mates, and greater conspicuousness, either for groups of prey becoming more vulnerable or for groups of predators becoming less effective at hunting. Yet some of these same costs can be countered by living in groups. For example, although competition for access to food resources may increase in larger groups, there are more eyes to locate food. Similarly, more eyes also mean it may be easier to spot predators, even if it is also easier for the predators to spot the prey. Groups of males, such as those that form in lekking species, are better able to attract and gain access to females. Other benefits to grouping include the ability to maintain homeostasis, or body temperature, which is important not just in eusocial

FIGURE 13.11 Effect of parasites on cliff swallow nestlings. The much larger nestling on the right came from an insecticide-treated nest; the stunted baby of the same age on the left occupied a nest infested with swallow bugs. (From Brown and Brown 1986, © 1986 by the Ecological Society of America.)

insects but also in several social species of primates, birds, and even lizards (Rubenstein and Abbot 2017b). And of course, one of the primary benefits of grouping is access to offspring care, often from the father or social partner, and sometimes from kin that help a mother rear her brood. As you saw in Chapter 9, these types of benefits are often direct benefits, which are distinct from indirect benefits, those that are typically genetic benefits for offspring that, in social species in particular, are associated with kin selection and living in family groups. The relative importance of direct versus indirect benefits for social living in vertebrates is a source of great debate and clearly varies among species (Clutton-Brock 2002). Whatever the type of benefit that leads to grouping, when social individuals do not gain sufficient benefits to replace solitary types over evolutionary time, group living will not evolve (Alexander 1974). And of course, when the benefits of living together outweigh those of living alone, groups will form. Thus, considering both the benefits (direct and indirect) as well as the costs to living in groups is essential for understanding why cooperation—and cooperative breeding behavior in particular—evolves, a topic that has generated many hypotheses (**HYPOTHESES 13.1**).

HYPOTHESES 13.1

Non-mutually exclusive hypotheses to explain the evolution of cooperative breeding behavior

kin selection hypothesis Indirect fitness benefits explain why some individuals delay independent breeding and become nonbreeding helpers that aid others to reproduce.

group augmentation hypothesis Individuals survive or reproduce better by living in larger groups.

ecological constraints hypothesis When resources required to breed successfully are limiting, individuals will delay dispersal and remain in their natal territory to help raise their relatives.

life history hypothesis Specific life history traits, such as high juvenile and adult survival, play a role in the evolution of cooperative breeding by creating a surplus of individuals in a given habitat.

benefits-of-philopatry hypothesis The benefits of delaying dispersal outweigh the costs associated with attempting to disperse and breed independently.

temporal variability hypothesis Environmental uncertainty promotes cooperative breeding because having helpers at the nest allows birds to breed successfully under both good and bad conditions.

bet-hedging hypothesis Having helpers at the nest reduces environmentally-induced reproductive variance.

Reproductive Cooperation and Kin Selection

As you saw in Chapter 12, kin selection has been valuable in analyzing cases in which helpful behavior seems unlikely to ever provide *direct* fitness benefits for helpers, as in the case of sterile worker insects that will never have the chance to pass on their genes directly. One of the first people to consider the possible effects of kin selection on social interactions in a vertebrate species was Paul Sherman, who studied a medium-size North American rodent that produces a staccato whistle when a predator approaches (**FIGURE 13.12**). The sound of one Belding's ground squirrel (*Urocitellus beldingi*) whistling sends other nearby ground squirrels scurrying to safety (**VIDEO 13.4**) (Sherman 1977). Sherman discovered that calling squirrels are tracked down and killed by weasels, badgers, and coyotes at a higher rate than noncallers, which indicates that callers pay a survival—and thus reproductive—cost for sounding the alarm. However, this cost might be more than offset by an increase in the survival of the caller's offspring, which if true would boost the caller's direct fitness. The fact that adult females generally live near their offspring and are more than twice as likely to give the alarm than adult males, which move away from their progeny and so cannot alert offspring to danger, is consistent with a direct fitness explanation for alarm calling.

But could there also be an indirect fitness benefit for alarm calling in Belding's ground squirrels? Sherman observed that adult female squirrels with nondescendant kin (relatives other than offspring) living nearby were significantly more likely to yell when a predator appeared than were adult females without sisters, aunts, and nieces as neighbors. This result is what we would predict if alarm callers sometimes help relatives other than offspring survive to pass on shared family genes, resulting in indirect fitness gains for those that warn their relatives of danger (**BOX 13.3**). Sherman's work therefore indicates that both direct and indirect selection contribute to the maintenance of alarm-calling behavior in this species (Sherman 1977).

VIDEO 13.4
Belding's ground squirrel alarm call
ab11e.com/v13.4

The Belding's ground squirrel example demonstrated how kin selection can influence behavior in a survival context. But kin selection can also explain cooperative behavior in a mating context. For example, much like the long-tailed manakins discussed earlier, male wild turkeys (*Meleagris gallopavo*) also engage in cooperative courtship to attract mates. However, Alan Krakauer demonstrated that, unlike in the manakins, in which cooperative males are unrelated, male turkeys that cooperate to attract mates tend to be highly related with an average relatedness of 0.42, which is close to the value of 0.5 for full siblings (**FIGURE 13.13**) (Krakauer 2005). In other words, brothers often cooperate to find mates. Thus, Krakauer demonstrated that cooperation in wild turkeys is consistent with Hamilton's rule, because if a brother helps a male become dominant, and the dominant male reproduces more, then the helper gains genetic representation in the next generation.

FIGURE 13.12 An alarmed Belding's ground squirrel gives a warning call after spotting a terrestrial predator. (Photograph © Richard R. Hansen/Science Source.)

Reproductive Benefits and Cooperative Breeding

Although eusocial societies are often said to represent the pinnacle of social evolution, eusociality is actually quite rare. Only about 2 percent of insect species are eusocial. In contrast, a greater proportion of vertebrate species live in groups with overlapping generations, cooperative care of young, and at least a temporary reproductive division of labor whereby some individuals in the group forego their own reproduction to help raise others' offspring (Rubenstein and Abbot 2017a). This type of social system in which more than two individuals care for young is called cooperative breeding, and in many ways it parallels eusociality in insects. Cooperative breeding occurs in about 5 percent of all mammals, nearly 10 percent of all birds, and a handful of fish species. The hallmark of cooperatively breeding societies is the presence of nonbreeding helpers that aid in rearing others' young. In most cooperatively breeding birds, helpers tend to be male, whereas in mammals helpers tend to be female. This pattern of sex-biased helping is the result of sex-biased dispersal, with female-biased dispersal occurring in birds and male-biased dispersal occurring in mammals (see Box 7.2) (Greenwood and Harvey 1982), largely as a result of inbreeding avoidance. In most species, these helpers are capable of breeding, yet they don't. And in many—but not all—species, these helpers are offspring of the breeding pair from previous years. As in the eusocial insects, is cooperative breeding behavior simply the result of indirect selection to help raise relatives (**kin selection hypothesis**), or is there more going on?

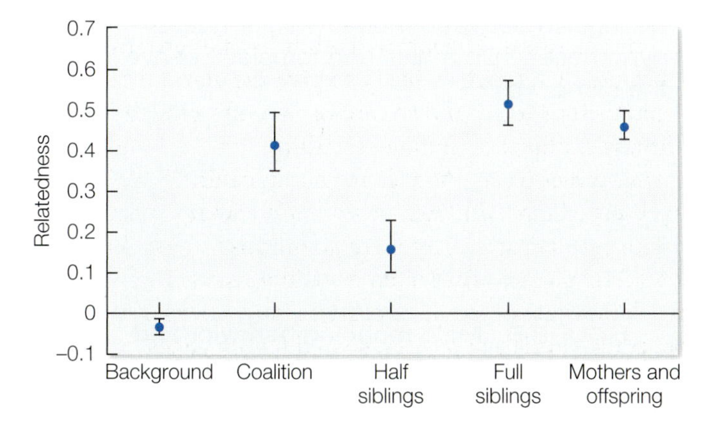

FIGURE 13.13 Male turkeys cooperate with kin to form coalitions in order to attract mates. The graph shows the background genetic relatedness (calculated from microsatellite genotypes) for all adult males in the population, the relatedness between subordinate males and their dominant partner ("Coalition"), and the relatedness between half siblings, full siblings, and mothers and their offspring. Values are mean +/– SE. (After Krakauer 2005; photograph © Monte Loomis/Shutterstock.com.)

BOX 13.3 EXPLORING BEHAVIOR BY INTERPRETING DATA

Mobbing and kinship in groups of Siberian jays

Many species of jays breed cooperatively in kin-based social groups with helpers at the nest. Siberian jays (*Perisoreus infaustus*) (**FIGURE A**) live in similar social groups, but no birds other than the parents take care of young. Why do Siberian jays live in family groups but not have helpers at the nest? As in Belding's ground squirrels, kin may receive a benefit from living together in the form of predator detection. Michael Griesser and Jan Ekman tested experimentally whether dominant Siberian jays protect their independent, retained offspring by giving alarm calls during simulated predator attacks (Griesser and Ekman 2004). The researchers found that dominant females were nepotistic in their alarm calling, in that they called more frequently when accompanied by their retained offspring than by unrelated immigrants, whereas dominant males called indiscriminately. Furthermore, playback experiments suggested that alarm calls convey information about danger and incite an immediate escape reaction. These experiments suggest that offspring gain direct benefits in the form of antipredator defense, but only when they live in groups with kin.

(A)

FIGURE A Siberian jays (Photograph © Franz Christoph Robiller/imagebroker/Alamy.)

Thinking Outside the Box

When a Siberian jay identifies a predator, usually an owl or hawk, in the forest where it lives, it often calls to others to join it in harassing the predator with noisy vocalizations and swooping attacks. This behavior could have direct fitness benefits for a mobber if its actions induce a dangerous predator to move out of the home range of the jay and its offspring. But the behavior could also have indirect fitness benefits for a bird whose behavior helps nondescendant kin by driving away a predator. **FIGURE B** presents data on the mean number of all mobbing calls given by members of groups of Siberian jays, some of which were composed of kin and others of which were made up of non-kin. How do these data help distinguish between these two alternative hypotheses? Do males and females behave in the same way?

(B)

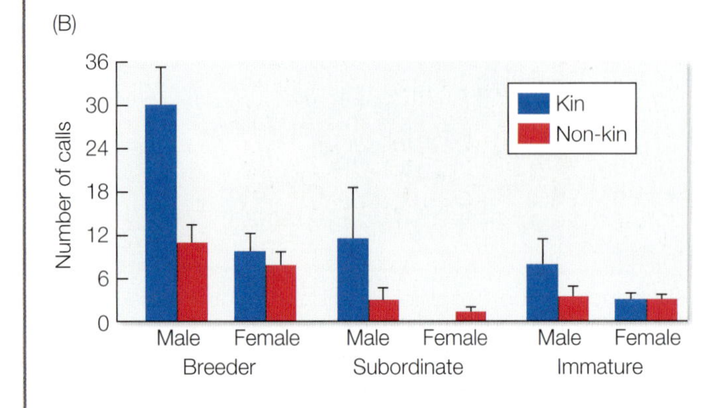

FIGURE B Male mobbing behavior and kinship. Male members of kin groups gave many more mobbing calls upon discovering a predator in their territory than did males in groups composed of non-kin. Bars depict mean +/– SE. (After Griesser and Ekman 2005.)

To understand why some individuals in cooperatively breeding groups consisting of kin become nonbreeding helpers, we need to consider not only the potential indirect benefits of helping, but also the potential direct benefits. In some of the earliest work studying cooperative breeding behavior in birds, Uli Reyer showed that both direct and indirect benefits played a role in the evolution of helping by male pied kingfishers (*Ceryle rudis*) (Reyer 1984). These African birds nest in tunnels in the banks of large lakes and rivers. Year-old males that are unable to find mates become primary helpers that bring fish to their mothers and her nestlings while attacking predatory snakes and other nest enemies. But these primary helpers are not the only type of helper in pied kingfishers. Some birds, termed secondary helpers, help unrelated nesting pairs in the manner of primary helpers. And rather than trying to breed or help, some birds, called delayers, simply sit out the breeding season, waiting for next year.

To understand why helpers help, we need to measure the costs and benefits of their actions. Primary helpers work harder than both secondary helpers and delayers (**FIGURE 13.14**), which translates into a lower probability of their surviving to return to the breeding grounds the next year (54 percent return) compared with secondary helpers (74 percent return) and delayers (70 percent return). Furthermore, only 67 percent of surviving primary helpers find mates in their second year and reproduce, whereas 91 percent of returning secondary helpers go on to breed. Many one-time secondary helpers bond with the females they helped the preceding year (10 of 27 in Reyer's sample), suggesting that improved access to a potential mate, rather than the inclusive fitness benefits of raising kin, is the ultimate payoff for a secondary helper's behavior.

These data enable us to use Hamilton's rule to calculate the direct fitness cost to the altruistic primary helpers in terms of reducing their own reproduction in their second year of life. For simplicity, we will restrict our comparison to solo primary helpers that help their parents rear siblings in the first year and then breed on their own in the second year (if they survive and find a mate) versus secondary helpers

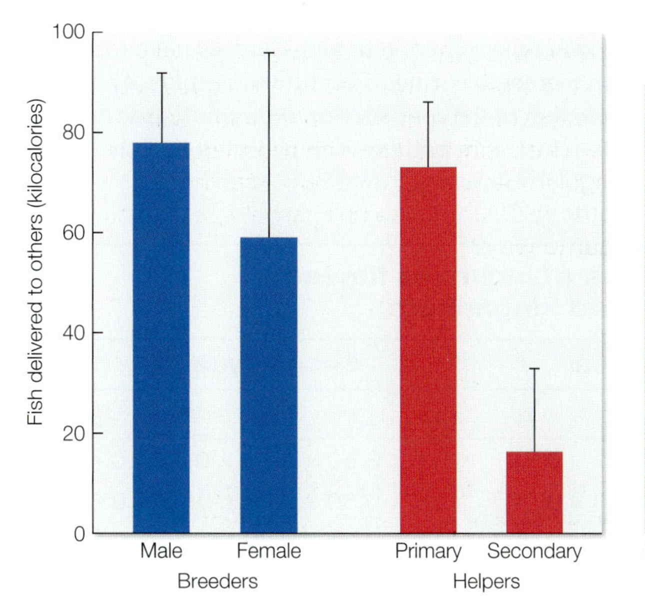

Pied kingfisher

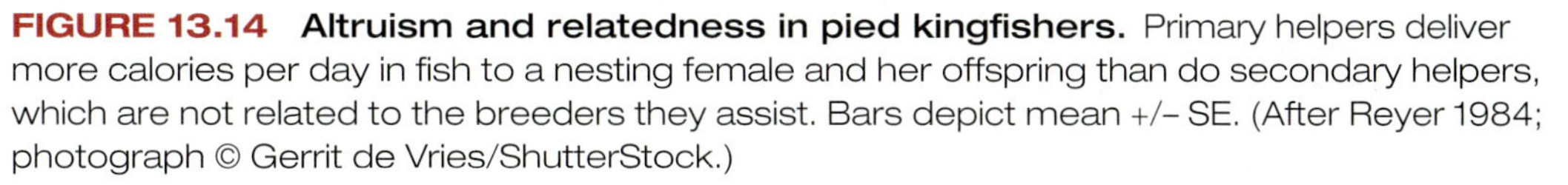

FIGURE 13.14 Altruism and relatedness in pied kingfishers. Primary helpers deliver more calories per day in fish to a nesting female and her offspring than do secondary helpers, which are not related to the breeders they assist. Bars depict mean +/− SE. (After Reyer 1984; photograph © Gerrit de Vries/ShutterStock.)

that help nonrelatives with no other helpers present in the first year and then reproduce on their own in the second year (if they too survive and find a mate).

Primary helpers that throw themselves into helping their parents produce offspring at the cost of being less likely to survive and thus to reproduce themselves in the next year. Although primary helpers do better than delayers in the second year (0.41 versus 0.29 units of direct fitness), secondary helpers do better still in year 2 (0.84 units of direct fitness) because they are more likely to survive to year 2, at which time they usually secure a partner (**TABLE 13.1**). But is the cost to primary helpers of 0.43 lost units of direct fitness (0.84 – 0.41 = 0.43) in the second year offset by a gain in indirect fitness during the first year? To the extent that these males increase their parents' reproductive success, they create siblings that otherwise would not exist, indirectly propagating their genes in this fashion. In Reyer's study, the parents of a primary helper gained an extra 1.8 offspring, on average, when their son was present. Some primary helpers assisted their genetic mothers and fathers, in which case the extra 1.8 siblings were full brothers and sisters, with a coefficient of relatedness of 0.5. But in other cases, one parent had died and the other had remated, so the offspring produced were only half siblings ($r = 0.25$). The average coefficient of relatedness for sons helping a breeding pair was thus between 0.25 and 0.5 ($r = 0.32$). Therefore, the average gain for primary helper sons was 1.8 extra siblings × 0.32 = 0.58 units of indirect fitness in the first year, a figure higher than the mean direct fitness loss experienced in their second year of life. Note that Reyer used Hamilton's rule to establish that primary helpers sacrifice future personal reproduction in year 2 in exchange for increased numbers of nondescendant kin in year 1 (Reyer 1984). Because these added siblings carry some of the helpers' alleles, they provide indirect fitness gains that more than cover the loss in direct fitness that primary helpers experience in their second year relative to secondary helpers.

This analysis shows that primary pied kingfisher helpers raise their fitness *indirectly* through their increased production of nondescendant kin, whereas secondary helpers raise their fitness *directly* by increasing their future chances of reproducing personally. Primary helpers reproduce less over a lifetime but are compensated genetically for their helpful behavior via a gain in indirect fitness; secondary helpers temporarily reproduce less but sometimes inherit territories and sexual partners as a result of their helpful behavior, thereby securing direct fitness benefits. Thus, this one species offers support for two different hypotheses on the evolution of helping at the nest. The indirect and direct fitness hypotheses can be tested for other cases

TABLE 13.1 Calculations of inclusive fitness for male pied kingfishers

	First year			Second year				
Behavioral tactic	y	r	f_1	o	r	s	m	f_2
Primary helper	1.8 ×	0.32 =	0.58	2.5 ×	0.50 ×	0.54 ×	0.60 =	0.41
Secondary helper	1.3 ×	0.00 =	0.00	2.5 ×	0.50 ×	0.74 ×	0.91 =	0.84
Delayer	0.3 ×	0.00 =	0.00	2.5 ×	0.50 ×	0.70 ×	0.33 =	0.29

Source: Reyer 1984.

Symbols: y = extra young produced by helped parents; o = offspring produced by breeding ex-helpers and delayers; r = coefficient of relatedness between the male and y, and between the male and o; f_1 = fitness in the first year (indirect fitness for the primary helper); f_2 = direct fitness in the second year; s = probability of surviving into the second year; m = probability of finding a mate in the second year.

of helpers, which are found in some other birds as well as certain other vertebrates (Brown 1987, Jennions and Macdonald 1994, Taborsky 1994).

Another direct benefit of breeding cooperatively is that individuals survive or reproduce better simply by living in larger groups, a process referred to as **group augmentation** (Clutton-Brock 2009a). According to the **group augmentation hypothesis**, delayed reciprocity rather than kin selection might explain cooperative breeding behavior in some species. Although the group augmentation hypothesis has been challenging to test, Hanna Kokko and colleagues demonstrated in a model that group augmentation can explain cooperative breeding in species in which kin structure is low and yet helping behavior persists (Kokko et al. 2001). Group augmentation could act through the recruitment of new group mates, even those unrelated to the helpers, as well as through the retention of offspring. As you saw in the example of pied kingfishers, the direct benefit of territorial inheritance is important for explaining why secondary helpers help. Sjouke Kingma used a comparative analysis of 44 cooperatively breeding bird species to show that variation in the likelihood of territory inheritance explains why helping effort is so variable among species, as well as why helpers preferentially direct care to related individuals in some species but not in others (**FIGURE 13.15**) (Kingma 2017). In other words, the delayed direct benefit of inheriting a territory not only influences why some individuals help, but also influences whom those individuals help and how much help they give.

group augmentation hypothesis Individuals survive or reproduce better by living in larger groups.

pay-to-stay hypothesis Parents require individuals to help as a form of payment for remaining in the group.

Direct benefits to the breeders—not just the helpers—can also explain how cooperative breeding systems can evolve. In many avian species, helpers reduce the workload of parents in offspring care, a phenomenon called **load-lightening** (Kingma et al. 2010). The presence of helpers allows parents to reduce their workload, and thus energy expenditure, because the helpers share in the burden of parental care. According to the **pay-to-stay hypothesis**, parents in some species require individuals to help as a form of payment for remaining in the group (Gaston 1978). Evidence of pay-to-stay comes primarily from fishes. Cooperatively breeding cichlids tend to form non-kin social groups, and paying via helping may enable helpers not just to remain in the group, but also to inherit the territory (**FIGURE 13.16**) (Balshine-Earn et al. 1998, Bergmuller et al. 2005). In some species, however, reducing the costs of parental care manifests in other ways than simply reducing feeding effort. In the superb fairy-wren (*Malurus cyaneus*), for example, the presence of helpers in the group at the time egg laying begins enables the breeding female to make lighter eggs that contain significantly less fat and protein than those laid by a breeding female without available helpers in her group (Russell et al. 2007a). By cutting back on their egg investments in each breeding attempt,

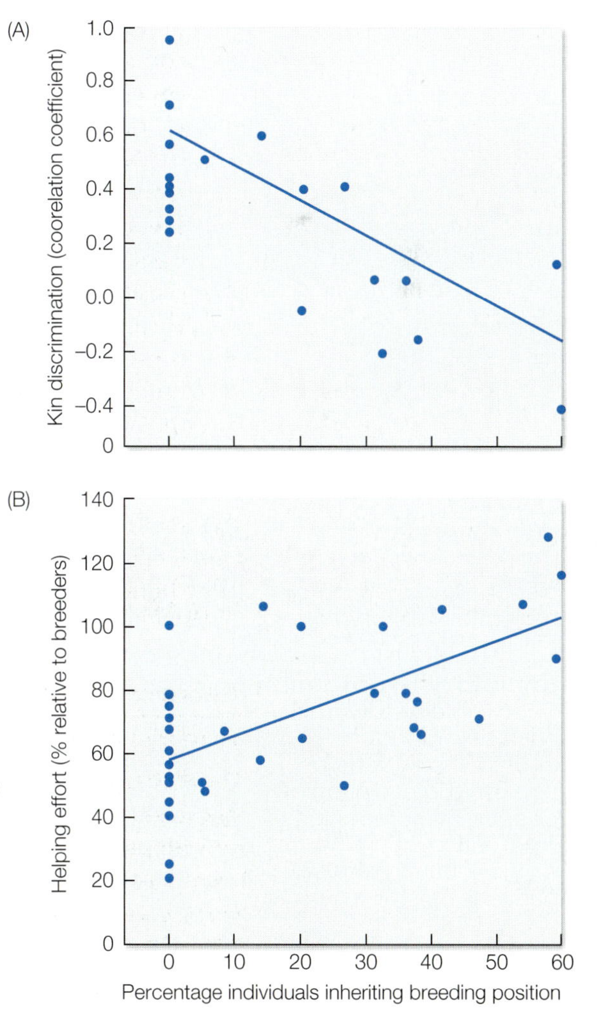

FIGURE 13.15 The likelihood of territory inheritance drives helping decisions in cooperatively breeding birds. (A) Helpers with a high likelihood of inheriting their resident territory do not invest more in more related offspring, but when prospects of territory inheritance are limited, subordinates mainly direct help toward related offspring. (B) Helpers provision offspring more when the probability of inheriting their resident territory is larger. Dots reflect species averages. (After Kingma 2017, CC BY 4.0.)

FIGURE 13.16 Cooperative breeding in the cichlid *Neolamprologus pulcher*. This species forms primarily non-kin social groups in which subordinates "pay" to stay in the group by helping protect and rear young. (Photograph by Michael Taborsky.)

females with helpers may live longer and thus have more lifetime opportunities to reproduce. If this is true, helpers may boost the number of siblings created by their mother, adding to the helpers' indirect fitness if they are related to the adult female they are assisting (Cockburn et al. 2008).

Reproductive Costs and Cooperative Breeding

Up until this point, we have discussed exclusively the benefits of breeding cooperatively. Think back to Hamilton's rule, which states that a gene for altruism will spread only if $rB > C$, where r is the genetic relatedness between two individuals, B is the benefit (in terms of reproductive units) gained by the recipient of the altruistic act, and C is the cost (in terms of reproductive units) to the individual performing the altruistic act. To understand why cooperative breeding behavior evolves, we must consider not only the benefit term but also the cost term in the equation. Stephen Emlen argued that the costs of breeding independently may directly contribute to the evolution of cooperative breeding behavior (Emlen 1982). He called this idea the **ecological constraints hypothesis**, which argues that when resources required to breed successfully are limiting, individuals will delay dispersal and remain in the territory in which they were born (their natal territory) to help raise their relatives. Limiting resources typically refer to the availability of suitable territories with enough food and nesting or denning sites to successfully raise a family. Territories or breeding sites are often limiting to a population because most of the best sites are already occupied by other members of that species, a phenomenon called **habitat saturation**. Habitat saturation occurs in many species of animals, particularly ones that are long-lived because there is little territorial turnover. The fact that many cooperative breeders also happen to be long-lived, which leads to habitat saturation, has been captured in the **life history hypothesis**, an evolutionary hypothesis to explain cooperative breeding behavior in birds (Arnold and Owens 1998).

ecological constraints hypothesis When resources required to breed successfully are limiting, individuals will delay dispersal and remain in their natal territory to help raise their relatives.

life history hypothesis Specific life history traits, such as high juvenile and adult survival, play a role in the evolution of cooperative breeding by creating a surplus of individuals in a given habitat.

Delayed dispersal is critical to the evolution of cooperative breeding behavior and central to the ecological constraints hypothesis. As available territories or other resources become limited on the landscape, then the costs to dispersing should drive some individuals to become helpers. Whether saturated breeding habitats

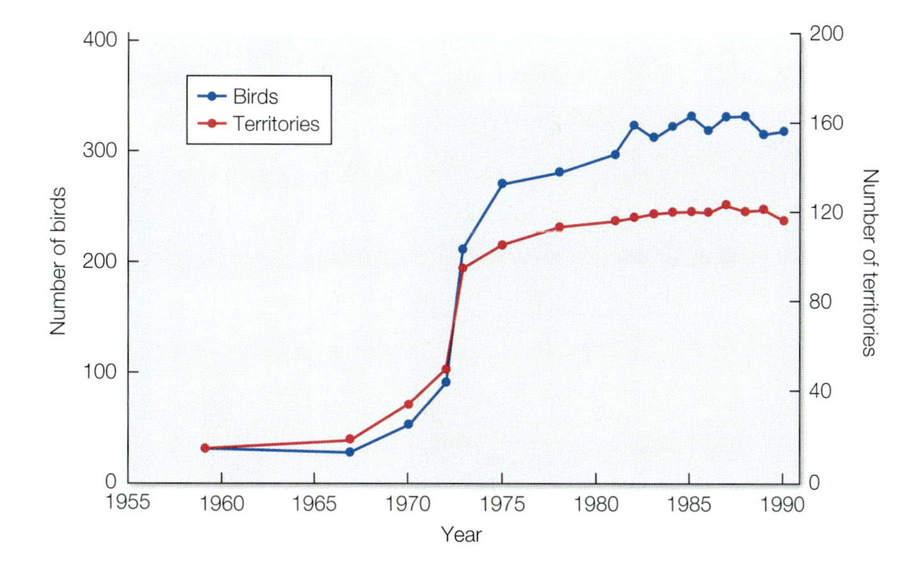

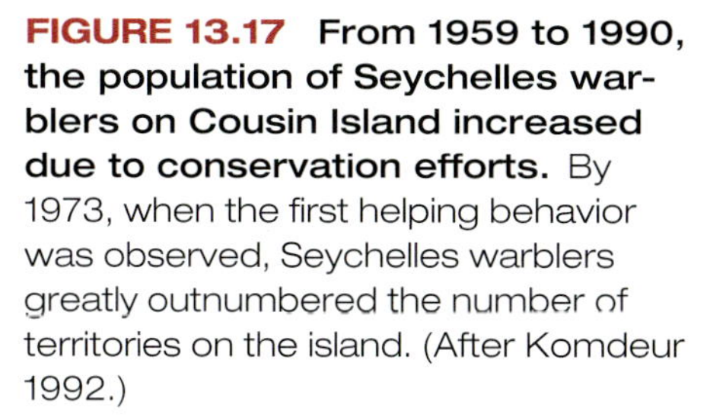
FIGURE 13.17 From 1959 to 1990, the population of Seychelles warblers on Cousin Island increased due to conservation efforts. By 1973, when the first helping behavior was observed, Seychelles warblers greatly outnumbered the number of territories on the island. (After Komdeur 1992.)

contribute to the maintenance of facultative (optional) helping is testable. If young birds remain on their natal territories because they cannot find suitable nesting habitat, then yearlings given an opportunity to claim unoccupied territories of high quality should promptly exercise the option to become breeders. Jan Komdeur did the necessary experiment with the Seychelles warbler, a small bird that has played a big role in testing evolutionary hypotheses about helping at the nest. The Seychelles warbler was on the verge of extinction in the 1960s before a massive conservation effort saw its numbers rebound on the only island in the Seychelles archipelago where it remained, Cousin Island. As the population increased on Cousin Island, Komdeur observed helping behavior at the nest, something that had never before been seen in this species (**FIGURE 13.17**). Later, to test the hypothesis that this behavior was being influenced by habitat saturation as well as efforts to conserve the species, Komdeur transplanted 58 birds from Cousin to two other nearby islands with no warblers. This translocation created vacant territories on Cousin. Almost immediately, helpers at the nest there stopped helping and moved into the open spots to begin breeding. Since the islands that received the transplants initially had many more suitable territorial sites than warblers, Komdeur expected that the offspring of the transplanted adults would also leave home promptly in order to breed on their own. They did, providing further evidence that young birds help only when they have little chance of making direct fitness gains by dispersing and breeding on their own (Komdeur et al. 1995).

As Komdeur continued to study the warblers, he realized that the dispersal decision rules made by young birds were a bit more complex than simply finding an open territory on which to breed. He predicted that young Seychelles warblers should also be sensitive to the quality of their natal territory. Breeding birds occupy sites that vary in size, plant cover, and edible insects. By using these variables to divide warbler territories into categories of low, medium, and high quality, Komdeur showed that young helpers on high-quality territories were likely to survive there while also increasing the odds that their parents would reproduce successfully. Young birds whose parents had prime sites often stayed put, securing both direct and indirect fitness gains in the process. In contrast, young birds on low-quality natal territories had little chance of helping to boost the reproductive success of their parents. Instead, they left home and tried to find breeding opportunities of their own (Komdeur 1992). The probability that a warbler would disperse was also influenced by whether both of its genetic parents were alive and in control of

FIGURE 13.18 Dispersal by helper Seychelles warblers. Helpers are more likely to leave their home territory if they lose one or both of the parents they have been helping. Note the colored leg bands that enable researchers to track individuals over time. (After Eikenaar et al. 2007; photograph by Cas Eikenaar.)

the family territory or whether one or both had been replaced by new stepparents (**FIGURE 13.18**) (Eikenaar et al. 2007). Thus, the dispersal of helpers became more likely if opportunities to help close kin were low.

Although the Seychelle's warbler example supports predictions of the ecological constraints hypothesis and the idea that habitat saturation influences delayed dispersal and the retention of helpers in their natal group, Komdeur actually argued that the example was consistent with another hypothesis for the evolution of cooperative breeding, the **benefits-of-philopatry hypothesis**. According to this hypothesis, the benefits of delaying dispersal outweigh the costs associated with attempting to disperse and breed independently. An astute reader will see that the ecological constraints and benefits-of-philopatry hypotheses are really just two sides of the same coin. According to Hamilton's rule, one must consider both the costs and benefits of an altruistic behavior. The ecological constraints hypothesis emphasizes the costs in terms of lost direct reproductive benefits when attempting to breed in an environment where territories or other resources are limiting, whereas the benefits-of-philopatry hypothesis emphasizes the indirect reproductive benefits of helping to raise relatives.

benefits-of-philopatry hypothesis The benefits of delaying dispersal outweigh the costs associated with attempting to disperse and breed independently.

The behavioral flexibility exhibited by Seychelles warblers is not unique to that species. Consider how young female white-fronted bee-eaters (*Merops bullockoides*) make adaptive decisions about reproducing (**FIGURE 13.19**). This African bird nests in loose colonies in clay banks. Like male pied kingfishers, young female white-fronted bee-eaters can choose to breed, to help a breeding pair at their nest burrow, or to sit out the breeding season altogether. If an unpaired, dominant, older male courts a young female, she almost always leaves her family and natal territory to nest in a different part of the colony, particularly if her mate has a group of helpers to assist in feeding the offspring they will produce. Her choice usually results in high direct fitness payoffs. But if young subordinate males are the only potential mates available to her, the young female will usually refuse to pair with them. Young males come with few or no helpers, and when they try to breed, their fathers often harass them, trying to force their sons to abandon their mates and return home to help rear their siblings, a phenomenon called redirected helping (Emlen et al. 1995).

Having available helpers is important in the mating decisions of another African cooperative breeder, the superb starling (*Lamprotornis superbus*). First-time breeding superb starling females, which lack offspring from previous years to act as helpers, will mate with extra-pair males from inside their social group that then aid the

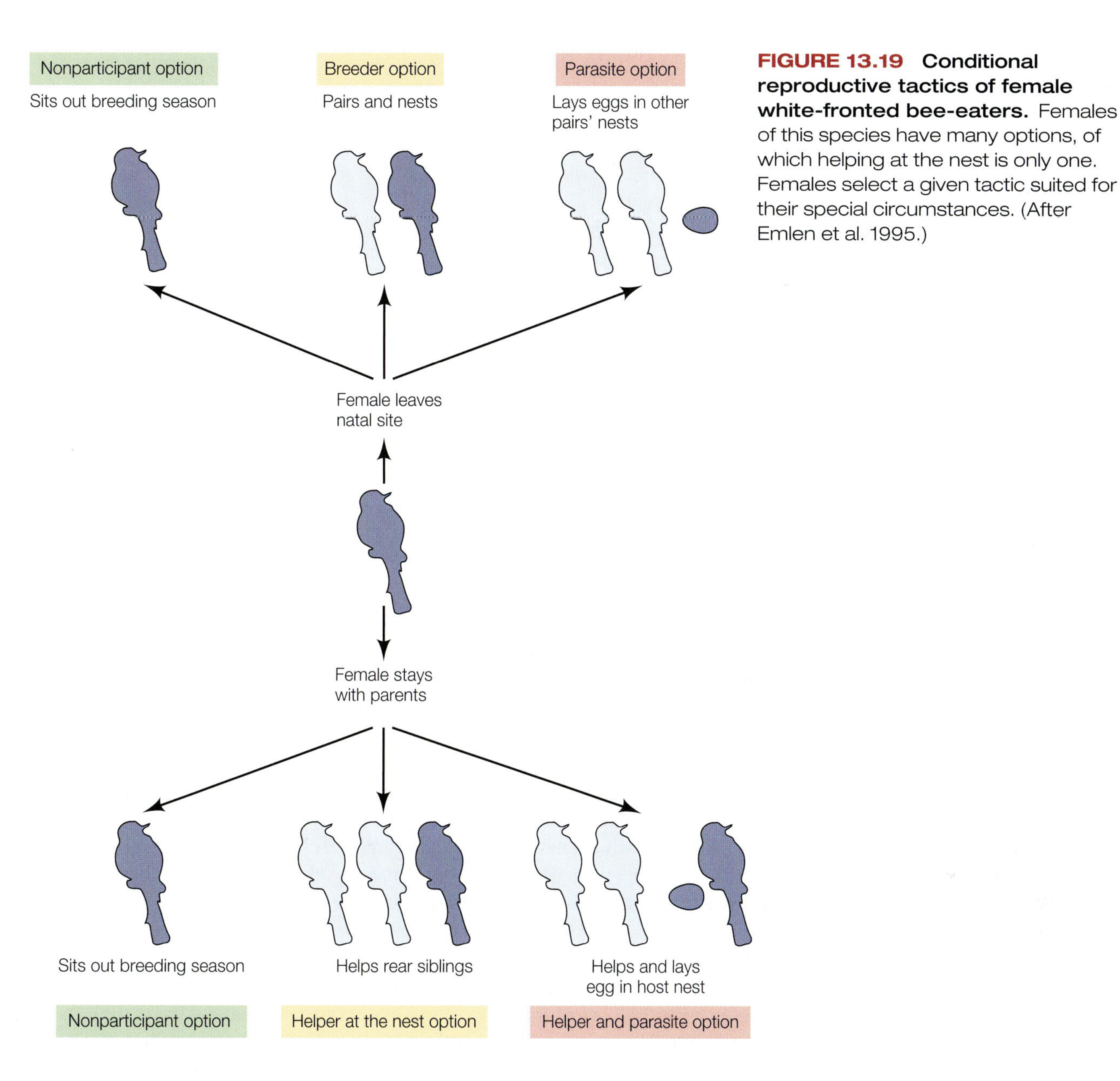

FIGURE 13.19 Conditional reproductive tactics of female white-fronted bee-eaters. Females of this species have many options, of which helping at the nest is only one. Females select a given tactic suited for their special circumstances. (After Emlen et al. 1995.)

females in rearing the young, some of which are likely to have been sired by those recruited helper males (Rubenstein 2007). Female white-fronted bee-eaters can take it even a step further. A female that opts not to pair off under unfavorable conditions may choose to slip an egg into someone else's nest (a case of intraspecific brood parasitism; see Chapter 11) or to become a helper at the nest in her natal territory—provided that the breeding pair there are her parents (see Figure 13.19). If one or both of her parents have died or moved away, she is unlikely to help rear the chicks there, which are at best half siblings, and instead will simply wait, conserving her energy for a better time in which to reproduce. Thus, although daughter bee-eaters have the potential to become helpers at the nest, they choose this option only when the indirect fitness benefits of helping are likely to be substantial (Emlen et al. 1995).

Although tests of Emlen's ecological constraints hypothesis have largely focused on habitat saturation and territory limitation, more recently behavioral biologists have also tested the idea that helping evolves because rearing young in harsh environments—like the African savannas where pied kingfishers, white-fronted bee-eaters, and superb starlings live—is challenging. Phylogenetic comparative analyses in birds (Rubenstein and Lovette 2007, Jetz and Rubenstein 2011) and mammals (Lukas and Clutton-Brock 2017) have shown that cooperative breeders tend to occur in arid and semiarid environments where rainfall is low, variable, and unpredictable. In other words, cooperative breeders tend to live in environments where life is challenging. Dustin Rubenstein and Irby Lovette termed this idea the **temporal variability hypothesis**, arguing that environmental uncertainty promotes cooperative breeding because having helpers at the nest allows birds to breed successfully under both good and bad conditions (Rubenstein and Lovette 2007). Indeed, many, although not all, studies of cooperatively breeding birds in arid and semiarid environments around the world have shown that having helpers increases reproductive success in poor years when rainfall and food availability are low (Shen et al. 2017). Another benefit of having helpers in these harsh environments has been formalized in the **bet-hedging hypothesis**, the idea that helpers reduce complete nest failure, thereby leading to a reduction in reproductive variance (Rubenstein 2011), another prediction of Hamilton's rule (Kennedy et al. 2018).

temporal variability hypothesis Environmental uncertainty promotes cooperative breeding because having helpers at the nest allows birds to breed successfully under both good and bad conditions.

bet-hedging hypothesis Having helpers at the nest reduces environmentally-induced reproductive variance.

Reproductive Conflict in Cooperative Breeders

As you have seen throughout this chapter, self-sacrificing and cooperative behaviors clearly occur in animals other than the eusocial bees, wasps, ants, and termites (Rubenstein and Abbot 2017a). But unlike workers in eusocial species, almost all vertebrate helpers retain the capacity to reproduce. Instead of working selflessly for others for an entire lifetime, they typically assist their kin only for a limited time and then attempt to reproduce personally. In other words, social vertebrates engage in facultative altruism, not obligate altruism, and appear to maximize their lifetime inclusive fitness, both directly and indirectly. With respect to that period in a social mammal or helper bird's life when it is assisting others, kin selection theory and Hamilton's rule provide a powerful framework for understanding the adaptive significance of the helper's behavior. As you have seen, kin selection theory is the basis for predictions that altruists will direct their assistance to relatives, usually close ones, and that such assistance will boost the reproductive success of the individuals they help. Avian helpers at the nest do indeed tend to help others in relation to the degree to which they are related (Griffin and West 2003). Furthermore, Hamilton's rule predicts that altruism is more likely to evolve when helpers give up relatively little in the way of direct fitness, since their chances of reproducing personally are very low at that time. Among the factors that reduce the cost of foregoing reproduction is an environment saturated by competitors so that few territories or other resources are available to a would-be disperser. If the cost of not reproducing is low for this reason or simply because leaving a relatively safe home base is very dangerous, then the downside of giving up a small chance to reproduce is likely to be outweighed by the indirect fitness benefits derived from boosting the fitness of close relatives, such as siblings. These conditions apply with equal strength to termites, Seychelles warblers, and meerkats (Clutton-Brock 2009b).

Reproductive Suppression

Perhaps the best link between social vertebrates and insects came with the discovery of one of the strangest social mammals, the naked mole rat (*Heterocephalus glaber*) (Sherman et al. 1991, Burda et al. 2000). This little hairless, sausage-shaped creature (**FIGURE 13.20**) forms colonies of up to 200 individuals that live in a complex

FIGURE 13.20 A mammal with an effectively sterile caste. Naked mole rats live in large colonies made up of many workers that serve a queen and one or a few breeding males. Three pups are shown here nursing from their mother, the queen mole rat. (Photograph © J. Jarvis/Visuals Unlimited, Inc.)

maze of burrows under the African plains. The impressive size of their subterranean home stems from extraordinary cooperation among chain gangs of colony members, which work together to move tons of earth to the surface each year while burrowing about in search of edible tubers. Yet when it comes to reproducing, breeding is restricted to a single large "queen" and several "kings" that live in a centrally located nest chamber. Females other than the queen do not even ovulate. Instead, they serve as functionally sterile helpers at the nest, consigned to specialized support roles for the queen and kings, as is also the case for most of the males in the colony (Lacey and Sherman 1991).

The altruism seen in this eusocial, caste-forming vertebrate appears to stem at least in part from policing by the queen mole rat in particular. The chief policewoman shoves other members of the colony around, inducing high levels of stress in subordinate females and males. Aggressive interactions such as these also occur in groups of birds where there are helpers at the nest and involve such things as destruction of the eggs of joint-nesting, related females (Mumme et al. 1983, Koenig et al. 1995), and as we will discuss in more detail later, the forced eviction of potential reproductive competitors from the group's territory in other mammalian species. Bullying by the queen mole rat suppresses the production of sex hormones in her underlings, so they become incapable of reproducing. At a proximate level of analysis, these workers are therefore functionally sterile like the workers in an insect colony. At the ultimate level, however, the altruism shown by subordinate naked mole rats occurs in part because they are forced to forego reproduction. At that juncture, their only options are to leave the colony to try to reproduce elsewhere (a very risky proposition because they are vulnerable to predation on the surface) or to accept their non-reproductive status and assist the queen mother sufficiently to be permitted to remain within the safety of the group. Nonetheless, the decision to stay with the bullying queen may well be adaptive because naked mole rat workers help produce the occasional reproductively capable sibling, a plump brother or sister that leaves home to found a new colony elsewhere, presumably with an unrelated individual of the opposite sex from another group (Braude 2000). If the mole rats always failed to increase their kin-based genetic success, they would surely leave the queen behind.

The type of reproductive suppression of potential breeders observed in naked mole rats is also seen in other mammalian cooperative breeders. In meerkats, for example, a team led by Tim Clutton-Brock demonstrated that dominant females evict subordinate females from the group to suppress reproduction among potential competitors (**FIGURE 13.21**) (Young et al. 2006). Eviction results in a decrease in sex steroids in the subordinates and in an increase in stress hormones, suggesting that reproductive suppression in this species is directly related to dominant-enforced stress. In a close relative of the meerkat, the banded mongoose (*Mungos mungo*), dominant females also suffer reproductive costs when subordinate females breed, to which they respond by evicting breeding subordinates from the group en masse (Cant et al. 2010). Michael Cant and colleagues predicted that, as in insects (see Chapter 12), banded mongooses should police the threat of subordinate reproduction before accepting the potential costs (fighting, injury, etc.) of eviction (Cant et al. 2014). Using a 7-year contraception-injection experiment to manipulate maternity within the social group, Cant and his team showed that dominant females use the threat of infanticide to deter subordinate reproduction. A similar set of experiments in meerkats also showed that when subordinate breeding is suppressed, dominant breeders produce higher quality offspring because there are more helpers to feed the pups (Bell et al. 2014). Additionally, when subordinate meerkats are prevented from breeding, dominant breeders are able to feed more, gain more weight during pregnancy, and produce heavier pups (**FIGURE 13.22**). Thus, much as in eusocial insect societies where conflict between breeders and workers exists, hidden conflict over reproduction also exists in vertebrate societies. This potential for conflict between dominant and subordinates can be reduced

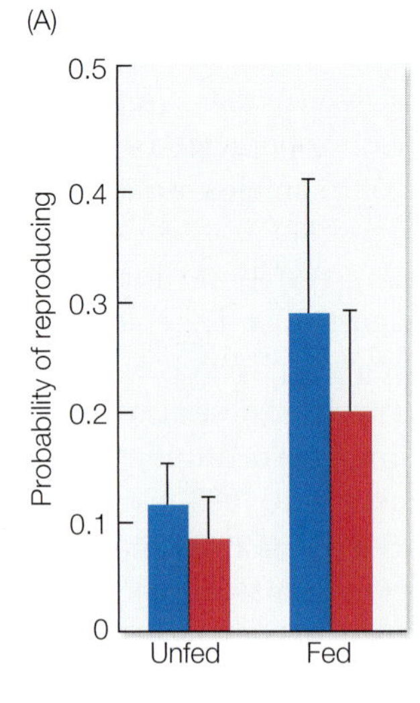

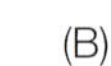

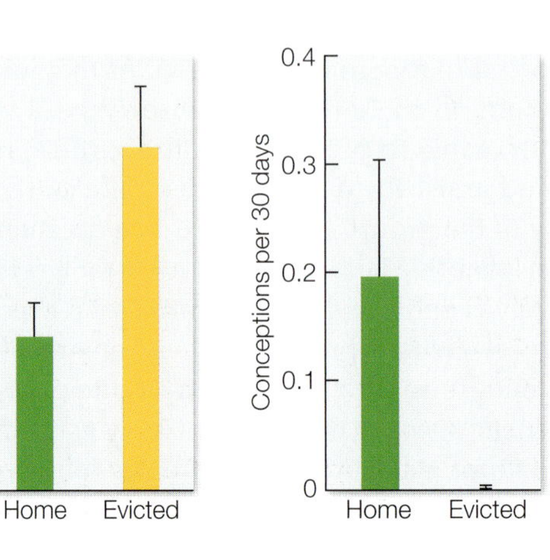

FIGURE 13.21 Reproductive suppression in meerkats. (A) The effect of supplemental food on the probability that young females (blue) and young males (red) would reproduce during their lifetimes. (B) Subordinate females that were temporarily driven from their group were more likely to abort their fetuses and less likely to conceive during the period of eviction than were females that avoided being evicted. Bars depict mean +/– SE. (A after Russell et al. 2007b; B after Young et al. 2006, © 2006 National Academy of Sciences, USA; photograph courtesy of Andrew D. Sinauer.)

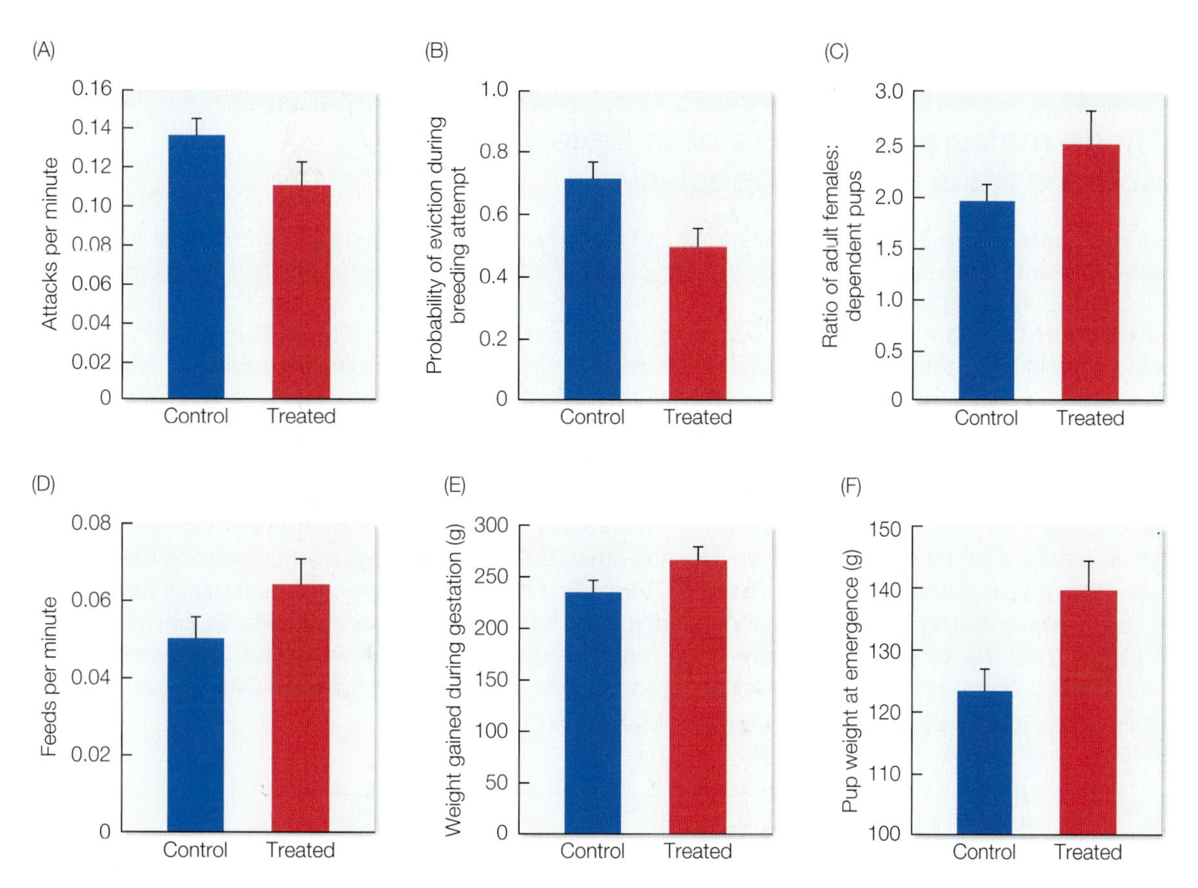

FIGURE 13.22 The effects of experimental suppression of subordinate female reproduction in meerkats. Both treated and control groups consisted of subordinates who were injected with contraceptives (treated) or saline (control). Suppression of reproduction in subordinate females led to (A) a decrease in attack rates by female dominants on subordinates, (B) a decrease in the probability that a subordinate female would be evicted from the group, (C) an increase in the number of female helpers, (D) a greater pup provisioning rate by female subordinates, (E) greater weight gain during gestation by dominant females, and (F) greater pup weight at emergence. Bars depict mean +/– SE. (After Bell et al. 2014, CC BY 4.0.)

simply by the threat of eviction, and in some cases, the act of eviction itself. It appears, therefore, that most social vertebrate and insect species defined by altruism and cooperation also experience regular conflict over reproduction, contrary to the argument promoted by advocates of multilevel selection.

Reproductive Skew, Extra-pair Paternity, and Social Structure

Although closely related, meerkats and banded mongooses differ fundamentally in their social organization and patterns of reproductive skew. Meerkats engage in **singular breeding**, meaning social groups typically contain only a single breeding female and reproductive skew is high. In contrast, banded mongooses engage in **plural breeding**, meaning there are multiple breeding females in each group so that reproductive skew tends to be lower. In fact, in most banded mongoose groups, all of the adult females breed, and they do so synchronously. This difference in skew and social organization has major implications for both genetic structure and the degree of conflict within the group, as well as for the evolution of traits used in competition for access to mates or other resources related to reproduction (**BOX 13.4**). Because multiple females breed with multiple males in plural breeding groups, the offspring

BOX 13.4 DARWINIAN PUZZLE

Why do males and females both have elaborate traits in social species?

Recall that in Chapter 8 we introduced the pseudopenis of the female spotted hyena (*Crocuta crocuta*). We suggested that one reason for this odd structure is that females in this cooperatively breeding species are particularly aggressive and use their enlarged clitoris for signaling dominance rank. It turns out that females in many cooperatively breeding species aggressively attack other females, fighting for access to mates, other resources needed for reproduction, or dominance rank (Rubenstein 2012). For example, female meerkats gain greater benefits from acquiring dominant status than males, and traits that increase a female's competitive ability exert a stronger influence on female breeding success (Clutton-Brock et al. 2006). These types of studies suggest that sexual selection could operate differently on females in cooperative versus non-cooperative species. Using the comparative approach, Dustin Rubenstein and Irby Lovette showed in all 45 species of African starlings (Sturnidae) that males and females in cooperatively breeding species tend to have more similar plumage and wing sizes (a proxy for body size in birds) than do males and females in non-cooperatively breeding species (**FIGURE A**) (Rubenstein and Lovette 2009). Plumage in this study was scored using field guides, but in subsequent studies Rubenstein's team demonstrated the same pattern when looking at plumage reflectance (Maia et al. 2013, 2016).

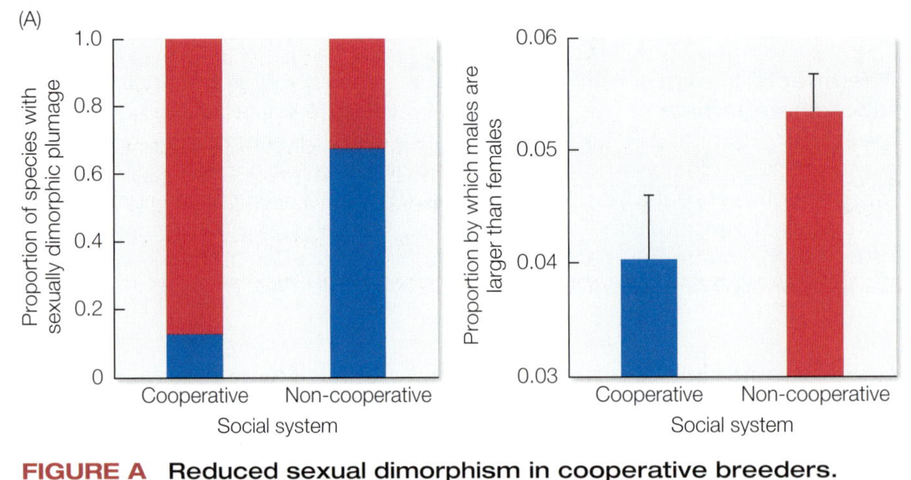

FIGURE A Reduced sexual dimorphism in cooperative breeders. Among African starlings, species that breed cooperatively show less sexual dimorphism in plumage (left) and body size (right) than those that do not breed cooperatively. On the left, blue is dimorphic and red is monomorphic. On the right, bars depict mean +/– SE. (After Rubenstein and Lovette 2009.)

tend to be less related to each other than they are in singular breeding groups, resulting in lower overall kin structure in plural breeders. Of course, how related the offspring are also depends on how related the breeders are, since when breeders are related their offspring will also be more related (Rubenstein 2012). Many plural breeding groups consist of a mixture of related and unrelated individuals. While 55 percent of avian cooperative breeders nest in family groups, 30 percent nest in mixed groups of relatives and nonrelatives, and 15 percent nest largely with nonrelatives (Riehl

BOX 13.4 DARWINIAN PUZZLE *(continued)*

Thinking Outside the Box

The pattern of reduced sexual dimorphism in plumage and body size in cooperatively breeding African starlings suggests that female traits used in female–female competition in these species are under stronger sexual selection than in females of non-cooperatively breeding species. How would you test this hypothesis in African starlings? Rubenstein and Lovette did this in two ways. In one test, they compared the variance in wing length between males and females in cooperative and non-cooperative species. What would looking at sex-specific variation in a trait tell you? What would you predict the researchers should find? Does **FIGURE B** support your prediction? Rubenstein's team also found that relative to males, females in cooperatively breeding species tend to have longer wings than females in non-cooperatively breeding species. Together, these results suggest that females in cooperatively breeding species are more ornamented that those in non-cooperatively breeding species, a result that Jim Dale and colleagues later confirmed in an analysis of nearly all of the 6000 species of passerine birds (Dale et al. 2015).

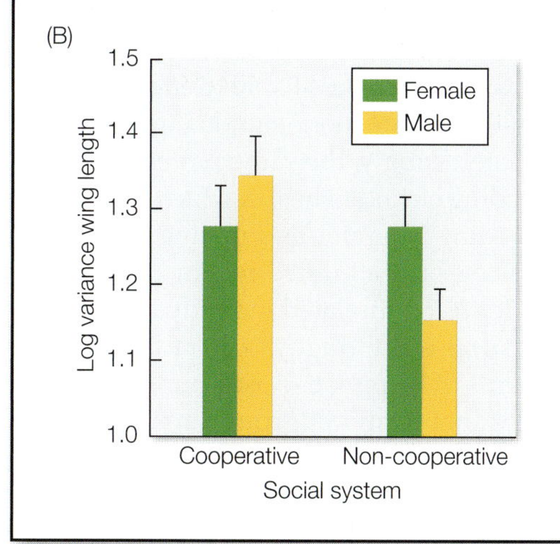

FIGURE B Among African starlings, females exhibit greater variance in wing length than males in non-cooperative species, but not in cooperative species. This supports the prediction that males and females of cooperatively breeding species should have similar variances in body size, whereas males and females of non-cooperatively breeding species should have dissimilar variances in this trait, since the relative intensity of selection is more similar between the sexes in the cooperatively breeding species. Bars depict mean +/– SE. (After Rubenstein and Lovette 2009.)

2013). Because there are more breeding opportunities in plural breeding groups than in singular breeding groups, reproductive conflict for breeding positions can be high in many plural breeding species. Of course, because the potential payoff for breeding in a singular breeding group is higher, conflict can still be strong in many singular breeders when breeding opportunities arrive, say as a result of the death of a parent. Ultimately, differences in the patterns of reproductive skew and social structure can influence kin structure within cooperatively breeding groups.

Recall that in Chapter 12 we introduced the monogamy hypothesis and the idea that mating behavior can also influence kin structure. If females mate polyandrously and have extra-pair fertilizations with other males, the offspring will be less related (Boomsma 2007). Insect queens, whether hymenopterans or termites, often produce offspring with a relatively high coefficient of relatedness. Indeed, as noted in Chapter 12, eusociality is thought to have evolved in the Hymenoptera (ants, bees, and wasps) only in lineages that were originally monogamous (Hughes et al. 2008). These same ideas appear to hold in vertebrates as well. Dieter Lukas and Tim Clutton-Brock showed that in mammals, the evolution of cooperative breeding was restricted to socially monogamous species (Lukas and Clutton-Brock 2012). But what about in birds, in which mating systems are much more fluid and promiscuity

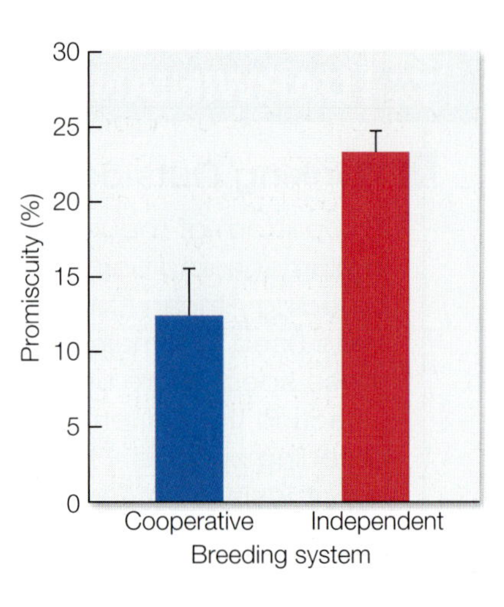

FIGURE 13.23 Cooperative breeding in birds is linked to monogamy. In 267 species of birds, female promiscuity (measured as the number of broods with extra-group chicks versus the number of broods with only within-group chicks) was lower in cooperatively breeding species than in non-cooperatively breeding species. Thus, pairs in cooperatively breeding species were more monogamous than those in non-cooperatively breeding species. Bars depict mean +/– SE. (After Cornwallis et al. 2010.)

(frequent extra-pair copulations) almost seems to be the norm (see Chapter 10)? Although there are definitely exceptions to the rule, promiscuous avian species (those in which a relatively large percentage of nests contain young sired by more than one male) are, as predicted, less likely to engage in cooperative breeding with helpers at the nest than are monogamous species (**FIGURE 13.23**) (Cornwallis et al. 2010). Interestingly, the same pattern has not been not observed in cooperatively breeding cichlid fishes, in which there is no relationship between cooperative breeding behavior and mating system (Day et al. 2017). However, kin selection plays only a limited role in group formation in cichlids, as most species form groups of unrelated individuals. The primary benefit of grouping in these fish species is thought to be reduced predation risk, and not the kin-selected benefits of helping to raise young (Groenewoud et al. 2016). Thus, in social species that form kin groups, mating and social systems appear to go hand in hand, whereas in those species that form groups of largely unrelated individuals, the mating system may be less important to the evolution of sociality.

Reproductive conflict among cooperatively breeding vertebrate societies varies as a function of the social structure and the pattern of reproductive skew within a social group. These same differences in reproductive structure are observed not only in birds and mammals, where both singular and plural breeding species are common, but also in insect societies. Monogynous insect colonies contain a single reproductive queen, whereas polygynous colonies contain multiple queens. There are of course many differences in the structure and makeup of insect and vertebrate societies, not the least of which is the presence of sterile castes in most eusocial insects, but the parallels in social structure between cooperatively breeding and eusocial societies suggest that sociality may be under similar selection pressures in these very different animals (Rubenstein et al. 2016). Ultimately, the parallels between social vertebrates and insects suggest that many of the same principles—kin selection, ecological constraints, reproductive suppression, and so on—are likely to underlie the evolution of sociality in these disparate animal groups (Rubenstein and Abbot 2017a), as well as the cooperative and collective behaviors of organisms as varied as microbes and amoebae.

SUMMARY

1. Social behavior includes interactions that benefit one party (altruism, selfishness), both parties (mutual benefit, or mutualism), or neither party (spite). Identifying which individuals receive a benefit that overcomes the cost of a social interaction is essential for demonstrating that a behavior is truly cooperative. Cooperative behaviors are those in which the recipient of an interaction receives a fitness benefit (mutualism, altruism). In contrast, collective behaviors are the synchronized movements of individuals following a series of basic interaction rules.
2. Altruism is rare in nature, but reciprocal altruism (or reciprocity) between members of the same or even different species is more common and explains many of the cooperative interactions that behavioral biologists have observed, particularly among non-kin. In these cases, an individual (the donor) provides assistance to another animal from which it later receives repayment that on average increases the donor's reproductive success enough to overcome the cost of its earlier helpfulness.
3. Individuals exhibit behavioral variation, including consistent differences over time (termed animal personalities) or across different contexts (termed behavioral syndromes). These behavioral differences can have positive effects on fitness.
4. Animals other than eusocial insects engage in various kinds of social interactions, including cooperation from which both parties derive immediate gains in direct fitness. In vertebrates, species that live in kin groups in which some individuals forego reproduction and help others rear young are called cooperative breeders.
5. Helpers in cooperatively breeding birds and mammals (and even in some insects, such as *Polistes* wasps) can gain direct and/or indirect benefits from aiding others in rearing their young. For example, in addition to gaining indirect fitness benefits, helpers may increase the likelihood of inheriting their territory.
6. In addition to these benefits, costs—often in the form of ecological constraints that limit dispersal and independent breeding—can influence the evolution of cooperative breeding behavior. Habitat saturation, which may be more likely to occur in long-lived species, can promote helping behavior, as can living in harsh and unpredictable environments where many individuals are needed to successfully raise young.
7. Reproductive conflict occurs in social vertebrates, just as it does in eusocial insects. In vertebrates, the fact that subordinate individuals are not sterile means that conflict over breeding can be intense and lead to reproductive suppression by dominant individuals. Suppression of subordinate reproduction increases reproductive skew within social groups, which can intensify social conflict and drive the evolution of traits used to gain access to reproductive roles, not just in males but in both sexes.

COMPANION WEBSITE

Go to **ab11e.com** for discussion questions and all of the audio and video clips.

CHAPTER 14

Human Behavior

We are all descendants of a long line of successful reproducers. If any of our direct ancestors, going back to the single-celled protobacterium in the primordial soup, had failed to reproduce, we would not be here today. Consequently, the same evolutionary processes that shape proximate mechanisms for the capacity and inclination to behave that we have discussed thus far in this text apply equally to the evolution of human behavior. In other words, humans are just like other animal species when it comes to understanding why we behave the way we do. Although we are unusual and quite wonderful in our own way, so too are fruit flies, kittiwakes, and rhinoceroses. Having applied the levels-of-analysis approach to other animals throughout this book, we do the same for *Homo sapiens* in this final chapter. After all, our behavior owes its existence to our

Chapter 14 opening photograph © Eric Baccega/Minden Pictures.

genetic–developmental systems, which make it possible for each of us to grow from a single cell into a creature with a brain, spinal cord, and endocrine system. These neural–hormonal mechanisms in turn make behaving possible for us in the same way that they do for other species. Since what we do affects how many copies of our genes we transmit to the next generation, studying our own behavior has implications for understanding our own fitness. There have surely been hereditary differences among people living in preceding generations that influenced their inclusive fitness, which determined the frequency of competing alleles in succeeding generations. Thus, natural selection has shaped our evolutionary history, as it has for all other organisms. Moreover, our species is derived from other, still older species in an evolutionary lineage that ultimately goes back to a primordial single-celled protobacterium of some sort. Understanding even a part of this still longer history will tell us when changes were layered onto older traits and what these changes were. We are a species with an evolutionary history, and so we have proximate mechanisms inherited from ancestral species but modified in ways that help us behave in an adaptive fashion.

Of course controversies abound with respect to any sort of research on humans, but here we intend simply to illustrate how behavioral biologists and evolutionary psychologists have gone about studying the behavior of our species from an ultimate and proximate perspective. Since we could easily write an entire book just on human behavior, we have decided to focus this chapter on only two aspects of behavioral theory as it relates to humans: communication and reproduction. We begin with a behavior that is uniquely human: spoken language. Using the tools that we discussed in previous chapters—comparative analysis, developmental biology, and genetics—we explore the evolution and adaptation of complex speech. From there we move on to discuss a topic shared across nearly all animal species: reproductive behavior. We emphasize mate choice (in both men and women) and sexual conflict. Finally, we end the chapter—and indeed the book—by discussing ways that behavioral and evolutionary theory may be able to help improve human lives. Specifically, we discuss evolutionary medicine, emphasizing how taking an adaptive approach may help us better understand—and eventually treat—a variety of diseases and disorders. We focus on just two topics in evolutionary medicine: obesity and autism. We end by reinforcing the idea that claiming that a human behavior (or for that matter, a behavior of any species) is adaptive or natural does not mean that it is desirable, moral, or necessary—only that it tends to propagate our genes.

Communication

Humans are a remarkably social species. As you learned in previous chapters, social species often need complex communication systems to transfer information or to recognize group mates, kin, or individuals. Although organisms of all sorts, even bacteria, communicate (see Chapter 8), our species is the only one that uses exceedingly complex languages to do the job. When we speak, whether in Urdu or English, we can produce a huge array of meaningful sentences and be understood because both signalers and receivers have acquired unconsciously at an early age the operating rules (the grammar) of the language of their culture. Yet where did our stunning language ability come from? Answering this question is challenging because we first need to consider *when* human language first evolved (Morgan et al. 2015). Determining when human language first evolved has become a controversial topic, as estimates range from 2.5 million to only 50,000 years ago, quite a wide range in our evolutionary history. Determining when humans first began to speak involves using many of the approaches that you learned about in previous chapters, including comparative analysis, experimental manipulations, brain imaging, and evolutionary genomics. We detail some of these approaches below.

The Development and Evolutionary History of Human Speech

Although humans are the only primates that develop exceedingly complex language, the antecedents of modern speech were likely present in now-extinct species in the lineage that gave rise to modern *Homo sapiens*. One way in which some scientists have explored this possibility is by investigating if modern chimpanzees (*Pan troglodytes*), our closest living relatives, have retained some elements of these protolinguistic abilities. The search for these abilities led early researchers to rear chimpanzees with humans. In one such experiment, Winthrop and Luella Kellogg raised a female chimpanzee named Gua along with their own young son. Their research helped demonstrate that the vocal tracts of chimpanzees are physiologically unsuited to producing speech, as Gua never produced a recognizable spoken word (Kellogg and Kellogg 1933). In fact, Charles Darwin had noted much earlier in *On the Origin of Species* that human vocal tracts differ from those of other primates. We now know that the shape of the human vocal tract allows the tongue to maneuver and produce the necessary frequencies of the most common, or "quantal," vowels, whereas other primates are physically incapable of producing these sounds because they have longer tongues and differently shaped vocal tracts (**FIGURE 14.1**) (Lieberman and McCarthy 2001).

If chimpanzees are incapable of speaking due to constraints in their morphology, can they produce language in a different manner? Trying to get chimpanzees to use language without speaking has been more successful, suggesting that they

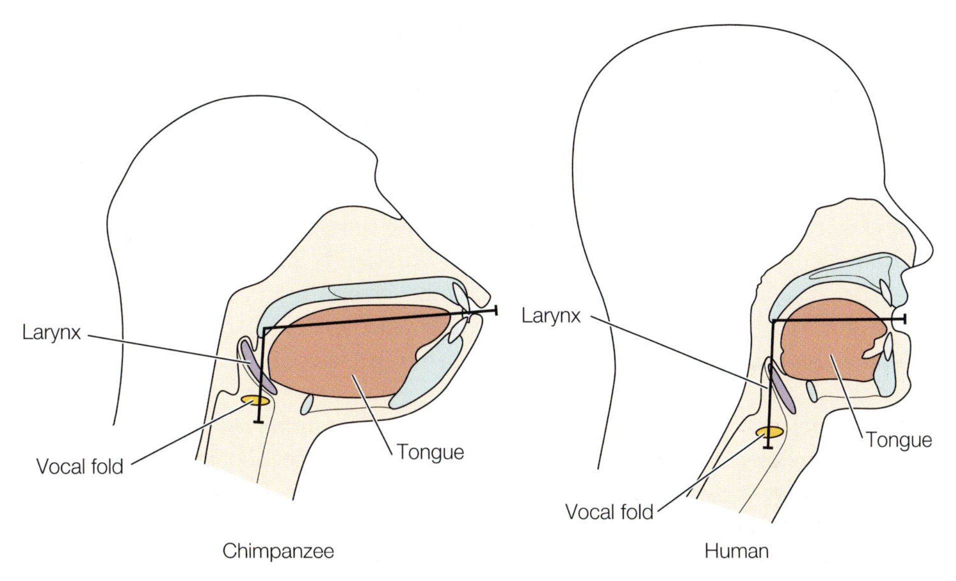

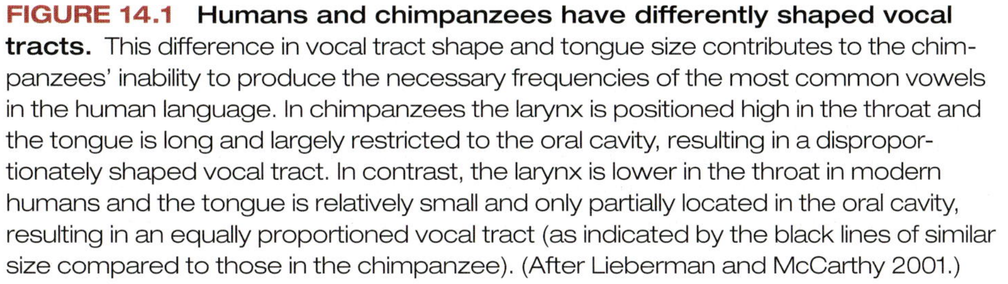

FIGURE 14.1 Humans and chimpanzees have differently shaped vocal tracts. This difference in vocal tract shape and tongue size contributes to the chimpanzees' inability to produce the necessary frequencies of the most common vowels in the human language. In chimpanzees the larynx is positioned high in the throat and the tongue is long and largely restricted to the oral cavity, resulting in a disproportionately shaped vocal tract. In contrast, the larynx is lower in the throat in modern humans and the tongue is relatively small and only partially located in the oral cavity, resulting in an equally proportioned vocal tract (as indicated by the black lines of similar size compared to those in the chimpanzee). (After Lieberman and McCarthy 2001.)

FIGURE 14.2 Chimpanzee trained to use "language." A young chimpanzee being taught sign language by its caretaker. (Photograph by © Susan Kuklin/ Science Source.)

possess the cognitive abilities to produce language, just not speech. The young chimpanzees that have taken part in nonverbal language experiments have been able to learn to associate a host of visual symbols with objects and to respond to spoken commands and requests by, for example, touching certain symbols on a computer screen or keyboard. Still other chimpanzees have been taught elements of sign language (**FIGURE 14.2**) (Rivas 2005), a task that takes advantage of the fact that these animals use a modest number of communicative gestures in their natural environment (Hobaiter and Byrne 2011). Some researchers believe that these experiments reveal that chimpanzees have evolved abilities that enable them to employ rudimentary elements of human speech or language (Ujhelyi 1996), but there are many problems and uncertainties associated with this claim (Shettleworth 2010). In addition, there is considerable debate about whether chimpanzees can acquire any grammatical rules on how to string words together to make even simple short sentences (Pinker 1994, Rivas 2005). Nonetheless, there is no doubt that there are big differences between our close relatives and us on the language front, which is not surprising given that chimpanzees have been separated evolutionarily from humans for roughly 6 million years (Wood and Harrison 2011), during which time humans have clearly evolved under very different selection pressures than our closest living anthropoid relatives.

To understand some of the selection pressures that may have shaped the evolution of human language, researchers have conducted experiments with contemporary humans (**BOX 14.1**). Early hominids were skilled at building stone hand-axes and other tools at least 2.5 million years ago, and some researchers have hypothesized that language may have facilitated effective tool making. To test this **technical hypothesis**, Morgan and colleagues used nearly 200 humans to test five mechanisms of social information transfer: reverse engineering, imitation/emulation, basic teaching, gestural teaching, and verbal teaching (Morgan et al. 2015). The researchers found that social transmission of information through rudimentary language (such as gestural or verbal teaching) greatly improved tool-making ability, concluding that tool making would have created an ideal selective environment leading from observational learning to much more complex verbal teaching. In another test of this hypothesis, Natalie Uomini and Georg Meyer compared patterns

technical hypothesis Language may have facilitated effective tool making.

of brain activity in modern humans making tools and generating language tasks (Uomini and Meyer 2013). Using functional transcranial Doppler ultrasonography (fTCD) to measure brain blood flow, the researchers found that stone tool making and cued word generation cause common cerebral blood flow signatures, which is consistent with the idea that both tasks have a shared neural substrate in the brain.

BOX 14.1 INTEGRATIVE APPROACHES

Ethical studies of humans and other animals

Throughout this book, you have read about numerous studies involving a diversity of animal species, including humans. At times, you may have wondered about the ethical decisions behind a given study or experiment. Perhaps, for example, you wondered why it was necessary to track the migration of so many free-living birds with GPS tags. Or you may have been concerned to know who decided the level of food to give captive mice in a food-restriction experiment. Or you may be interested later in the chapter to know who approved sending a college student out on campus to ask men and women questions about their sexual preferences.

Over the past few decades oversight of animal research has increased drastically, and today all studies of vertebrates in the United States—and in most developed countries—must be reviewed by a committee that applies ethical guidelines. For example, it seems unlikely that researchers in this era would be allowed to raise their own children with a young chimpanzee, as the Kelloggs did in the 1930s (Kellogg and Kellogg 1933). In fact, researchers in the United States today cannot receive funding from the National Science Foundation, the National Institutes of Health, or other federal agencies without first having their studies approved by an institutional ethics committee. For studies of humans, this committee is the **Institutional Review Board** (**IRB**), and for nonhuman vertebrates it is the **Institutional Animal Care and Use Committee** (**IACUC**). These committees are made up of scientific experts, including doctors, veterinarians, and other scientists, as well as nonscientists and at least one person not affiliated with the institution (other than being on the committee), who may be a lawyer, a member of the clergy, or a concerned citizen. Each institution's IRB and IACUC review research protocols, evaluate the potential for harm or distress, and perform cost-benefit analyses to determine whether the research should be conducted as planned. Ultimately, these committees take steps to protect the rights and welfare of the individuals in a research study, whether they are humans or other vertebrates. In this way, research that behavioral biologists and other scientists perform today is required to meet ethical standards before it is performed.

An issue closely related to the ethical study of humans and other animals is the legality of conducting the research. After all, scientists can't just show up in a country and start disrupting an animal species or its habitat. The specific ways in which research of both wildlife and humans are approved and regulated varies greatly by country. Most countries today have very strict laws about the type of work that can be conducted, as well as the type of samples that can be collected, including answers to questionnaires given to humans. Permissions vary depending on the species of study, and those that are listed by CITES (the Convention on International Trade in Endangered Species of Wild Fauna and Flora), an international agreement among governments, are under greatest scrutiny. Most behavioral biologists independently obtain research permits for conducting their studies, as well as permits for importing or exporting their samples to their home labs. Although these permissions are not as tightly regulated as the ethical issues related to doing animal research, many scientific journals will not publish a study's results unless the authors demonstrate that they obtained the necessary permissions, though there is no formal or consistent process as there is for IRB or IACUC approval. Ultimately, it is the scientists' responsibility to get legal and ethical permission to conduct their research. Most scientists today learn this during their training and take this responsibility very seriously.

(Continued)

BOX 14.1 INTEGRATIVE APPROACHES (*continued*)

Thinking Outside the Box

Are there any studies you read about in this book that you think would not be allowed today? Do you think the knowledge that they generated would be lost without such a study, or can you think of a new experimental design that might be better or more ethical and generate similar results? You may have noticed above that we referred specifically to vertebrate studies, despite the fact that in every chapter of this book we have given examples of vertebrates and invertebrates side by side. Although IRB or IACUC approval is required for all studies of vertebrate behavior, studies of invertebrate behavior require no formal approval, at least not in the United States and most other countries. Why do you think this is? Finally, one of the main issues surrounding the legality of conducting scientific research today comes down to the potential monetization of results or findings from a study. For example, who should benefit if a study of chemical communication in a rare insect in a tropical forest in a developing nation indirectly led to a cure for cancer? Should the country where that research was conducted receive some or all of the financial gain from that discovery? For most basic science, the odds of something like this happening are quite slim. Nonetheless, many countries now require signed sharing agreements to protect their legal rights should it occur.

Although experimental studies with modern humans are consistent with the technical hypothesis and the idea that human language could have coevolved with tool making at least 2.5 million years ago, it is important to remember that language does not require speech, as experiments with chimpanzees have demonstrated. In fact, comparative genomic studies suggest that human speech evolved much more recently in our evolutionary history than the experimental work would suggest. The study of the hereditary basis of speech began with the discovery of a gene called *FOXP2*, which became linked to language ability because individuals in one British family with a mutant allele of this gene have severe speech deficits, including difficulty in speaking words and in using words grammatically (Hurst et al. 1990). The protein that *FOXP2* codes for is a transcription factor, which means that when the gene is expressed, the resulting protein controls the operation of additional genes, whose protein products affect the development of neural circuitry in the brain that is necessary for adept speech.

Although there are almost certainly other regulatory genes involved in the development of speech and language centers in the human brain, *FOXP2* clearly has something to do with our capacity for language (Scharff and Petri 2011). What is striking about this gene is its occurrence in chimpanzees and gorillas, as well as in frogs, fish, rodents, and especially certain songbirds that learn their species' vocalizations. There is no question therefore that *FOXP2* originated in the distant past in an ancestor of many modern species. This gene may have been independently modified in ways that enable birds such as the white-crowned sparrow (*Zonotrichia leucophrys*) to learn their communication signals by listening to others of their species vocalize (see Chapter 2). Humans and their close relatives have retained the gene in nearly the same form from a more recent mammalian ancestor. In humans, the modifications of *FOXP2* presumably affect how the gene's transcription regulates other genes involved in the development of the neural substrate for the acquisition of learned speech (Scharff and Petri 2011). In fact, the *FOXP2* variant in humans that is thought to relate to our complex speaking abilities was shared by Neanderthals (*Homo neanderthalensis*), which are now extinct. This shared genetic variant between humans and our closest ancestor suggests that the hereditary base for human speech arose at least 300,000–400,000 years ago (Krause et al. 2007), much later than the experimental studies with modern humans suggest. However, genetic differences between Neanderthals and modern humans in the binding site for a related transcription factor (POU3F2) suggest that natural selection has continued to shape the hereditary basis of human speech since we diverged from Neanderthals (Maricic et al. 2012). Thus, as is so often the case, modern species carry genes that appeared in the distant past and that have taken on new adaptive functions as mutations occurred that affected how these genes are regulated (Carroll 2005), as is the case for *FOXP2*.

Finally, it is important to remember that language—like all behavioral traits—is not just genetically influenced. The parts of the brain that are essential for language acquisition, production, and comprehension develop as a result of an interaction between *FOXP2* and many other genes expressed in brain cells, as well as the chemical, social, and ecological environments in and around those cells. For example, the cultural environment that a baby experiences obviously has a huge effect on the interactive events that take place as the physiological foundation for language use is built. A child born and reared in Pakistan will probably speak Urdu, whereas a child born and reared in Mexico will probably speak Spanish. Yet this does not mean that either child's language is environmentally determined, only that the *differences* between the children are caused by an environmental difference that affects the gene–environment interactions that underlie their ability to speak Urdu or Spanish.

The Neurophysiology of Speech

The gene *FOXP2* exerts its developmental effect on our ability to speak by regulating a battery of other genes. The products of these genes contribute in some way to the development of particular parts of the brain that are essential for language learning or speech itself. It has long been known that damage to what is called Broca's area in the cerebral cortex of the human brain is associated with the loss of the ability to speak, while a different anatomical brain region, Wernicke's area, is responsible for the ability to comprehend spoken language (**FIGURE 14.3**). Yet neither of these regions is independent of other anatomical components of the brain, as these circuits and structures certainly work together with Broca's and Wernicke's areas in the production and comprehension of speech (Lieberman 2007). Therefore, if we wish to understand the neurophysiological mechanisms of language and speech, we must discover how various structural and connective elements in the brain operate in an integrated fashion, a task that is still far from complete. However, because speech mechanisms of adults are located on one side of the brain and include well-defined parts of the cerebral cortex, human brain development has likely evolved to make language and speech possible. In infants, exposure to speech during social interactions with adults (typically their parents) may provide the environmental input that shapes the development of those parts of the brain that gradually become dedicated to speech comprehension and production. The ability to learn a second language and speak it grammatically declines markedly by the time a child has become a teenager. This so-called critical period for language learning early in life suggests that once the neural

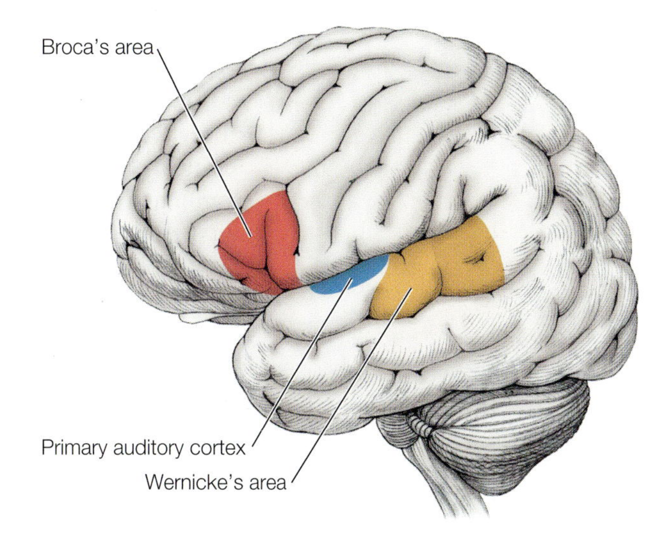

FIGURE 14.3 Regions of the brain implicated in language abilities. Broca's and Wernicke's areas are believed to play important roles in the ability of people to produce and understand speech, respectively. The primary auditory cortex also plays a role in both of these functions.

architecture of the brain has been shaped by early exposure to one language, a person's developmentally committed brain structures do not easily absorb information from the novel patterns of sound associated with other languages (Kuhl 2010). This is one reason why researchers and teachers alike suggest that children should start learning a second language when they are young.

Genetic information of a restricted and well-defined sort must underlie the ability of the brain to change in response to particular kinds of acoustic and social inputs during early childhood development. Selection presumably has eliminated all but the forms of the key genes most likely to contribute positively to the development and operation of the neural components involved in language learning. Indeed, only a tiny fraction of the human population exhibits a variant form of *FOXP2*. Those members of the one family known to possess this rare allele are unable to control the muscles of the face, lips, and mouth in the manner necessary for speech. Functional magnetic resonance images taken of these individuals reveal abnormalities in Broca's area and other components of the neural circuitry related to speech, further evidence that specific regions of the brain are essential for this very important human activity. Studies in these people of the form of the gene found in other mammals have revealed that probably *FOXP2* is usually expressed in cells in several parts of the brain. Without the normal form of the gene, the cells in the speech circuits cannot carry out tasks that are standard operating procedures for the rest of us (Vargha-Khadem et al. 2005). Vocal learning and production are impaired in mice by *FOXP2* mutations matching those that cause human speech disorders (Groszer et al. 2008, Chabout et al. 2016), as well as in zebra finches (*Taeniopygia guttata*) by experimental manipulation of *FOXP2* in the brain (Haesler et al. 2007, Murugan et al. 2013).

In addition to *FOXP2* and other genes that influence speech, many other elements of the human brain provide "skills" that contribute to our facility with language. For example, activity in our visual cortex complements our acoustic analysis of language in a most interesting way, a point discovered when researchers created videos in which a person was filmed saying one thing but the audio track played a different sound. For example, if the film showed a person saying "ga" but the audio track played "ba," an observer both watching and listening would hear "da" (McGurk and Macdonald 1976). This means that when someone sees a video of a person mouthing the nonsense sentence "my gag kok me koo grive" while an audiotape synchronously plays the nonsense sentence "my bab pop me poo brive," the test individual will hear quite clearly "my dad taught me to drive." This result demonstrates that our brains have circuits that integrate the visual and auditory stimuli associated with speech, using the visual component linked to lipreading to adjust our perception of the auditory channel (Massaro and Stork 1998). This ability increases the odds that we will understand what others say to us.

Indeed, babies about 6 months old switch their attention from the eyes of speaking adults to their lips. This happens at just about the same time that the infants begin babbling as part of the process of integrating the sounds they are hearing with the movements of a speaker's lips (Lewkowicz and Hansen-Tift 2012). When babies begin to speak their first words, they switch back to looking at the eyes of adults who are speaking. However, the importance of lipreading for verbal communication has apparently already been established in young children. The neurophysiological basis of lipreading has been traced to a region of the brain called the superior temporal sulcus, which becomes especially active when we view moving mouths, as well as hands and eyes (**FIGURE 14.4**) (Allison et al. 1994). Cells in this part of the brain are responsive to subtle movements of the lips, which presumably contributes to our great ability to comprehend the complex vocalizations of our fellow humans.

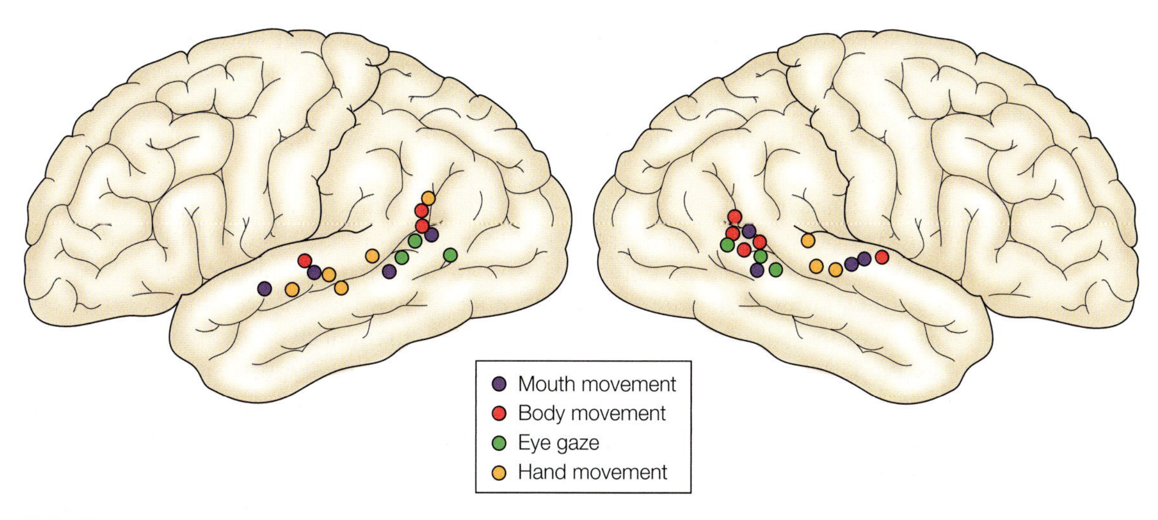

FIGURE 14.4 The function of the superior temporal sulcus of the brain. Socially relevant movements of the eyes, mouth, hands, and body activate neurons in different parts of this region of the cerebral cortex. The left and right hemispheres are shown on the left and right, respectively. Each circle represents findings by a particular research team. (After Allison et al. 2000.)

The Adaptive Value of Speech

Having introduced a few aspects of the proximate causes of speech and language in humans, let's return to an ultimate cause, namely the evolved function of language, a level of analysis that we need to consider if we are to determine why natural selection led to the spread of alleles that contributed to the development of a brain capable of such complex speech and language comprehension. It is not difficult to imagine why those of our ancestors with a protolanguage experienced higher fitness than those with less speaking facility. For example, we already mentioned the potential link between tool making and language, where the ability to communicate (either via speech or gestures) may have improved the production of stone tools, thereby allowing our ancestors to capture more resources and ultimately increase their survival and reproduction. So many benefits to individuals currently flow from being able to speak that it may seem hardly worth thinking about the adaptive value of the trait. But we must do so, because the underlying neural mechanisms of speech are extremely costly, since brain tissue requires about 16 times as much energy to maintain as an equivalent weight of muscle tissue. Indeed, the human brain's metabolic demands may have had all manner of evolutionary consequences, including the need for a nutritionally dense diet high in energy, as well as the requirement that infants be provided with large amounts of fat to fuel brain development (see Box 4.4) (Leonard et al. 2003). To the extent that language learning contributed to an increase in our brain's volume, the adaptive benefits must have been substantial if they were to outweigh the fitness costs of an energetically expensive brain.

There are many potential benefits for an individual that stem from the ability to speak, including benefits related to food acquisition and sharing, mate acquisition, offspring care, and many other behaviors we have covered throughout this book. Moreover, the transmission of useful information about one's culture and environment to one's offspring and relatives, as well as to non-kin who will reciprocate the favor or help with mutually beneficial tasks, may be another key benefit to human language (Pinker 2003). Of course, there may also be several costs associated with our ability to speak. For example, people often appear particularly interested in talking about the behavior of other people (Dunbar 1996). Although some of the information received may be accurate and benefit the listener, language also facilitates the transmission of *mis*information from manipulative senders whose speech benefits them at the expense of listeners that they deceive (Trivers 2011), a primary cost of language learning.

Let's consider a few more potentially adaptive functions of human speech.

SOCIAL GROUP FORMATION Recall from Chapter 2 when we discussed song dialects in birds that we made the analogy to language dialects in humans. Just as chaffinches (*Fringilla coelebs*) and white-crowned sparrows differ in the vocalizations they produce in different geographic regions, so too do humans, and for the same reason—the learning mechanisms underlying speech promote the learned acquisition of slight variations from group to group as speakers in different locations happen to differ in their speech. Given enough time and isolation, these differences can accumulate, and at some point, dialects can become separate languages. But even in the early stages of language differentiation, the subtle or not-so-subtle differences between groups can serve as badges of group or tribal membership, thus forming a foundation for cooperative endeavors. Indeed, the dialect that one speaks can convey information about the identity of an individual, where the person lives or was raised, their social status, or the person's desire to affiliate with others (Bryant 2010). Thus, the **affiliation function hypothesis** suggests that human speech, including language dialects, is critical to the formation of social groups.

affiliation function hypothesis Human speech, including language dialects, is critical to the formation of social groups.

The drive to be part of a group is generally so strong that people will unconsciously mimic the speech of people around them, whether they normally speak the same dialect or not. In fact, we can mimic a dialect when the only cues available come from lipreading. Dialect mimicry occurred in a study in which people were exposed to a silent video showing a person saying words such as "tennis" or "cabbage" (Miller et al. 2010). The participants, who had had no experience lipreading, were then asked to say the word they had seen being spoken in the silent video. When they said out loud "tennis" or "cabbage" or whatever word they had lip-read, they spoke this word closer to the way in which the silent model had spoken it than when asked previously to read the word in their own customary voice. Thus, people evidently copy the dialect that they lip-read and hear, even if this requires a change in how they usually speak, perhaps as a way to announce unconsciously to others their desire to be part of the group composed of speakers of another dialect.

VERBAL COURTSHIP The affiliation function of speech, if it exists, provides an example of a benefit that arises from the role that learning plays in language acquisition and production, even among adults (see Chapter 2 for a discussion of vocal learning in birds). However, another possible adaptive value of language learning has been championed by Geoffrey Miller, a strong advocate for the role of sexual selection in the evolution of speech and many other features of human behavior. Miller argues that skill in language use comes into play during what he calls "verbal courtship" (Miller 2000). According to Miller and the **sexual selection hypothesis**, men compete with one another to display their verbal competence to women who evaluate potential mates at least partly on the basis of their capacity to use language creatively, amusingly, flatteringly, and so on. In other words, speech has become a focus of sexual selection, with female choice favoring males with exceptional language skills, just as selection by choosy females is responsible for the evolution of the plumes of birds of paradise, which are far more extravagant than needed for flight or any other more prosaic function. Of course, there is no evidence to suggest that men have different language skills than females, which would be one prediction of this sexual selection hypothesis. Indeed, females with the capacity to evaluate the competence of male speakers might well be under selection for language competence themselves. Adult humans of both sexes have on

sexual selection hypothesis Female choice favors males with exceptional language skills.

average a vocabulary of about 60,000 words; our fellow primates produce at most a few dozen different vocalizations (Miller 2000). Deaf humans who use American Sign Language have more than 9000 signs at their disposal; chimpanzees have a repertoire of fewer than 100 gestures (Hobaiter and Byrne 2011), their analog of sign language. The point is that most people know and use far more words, many of which are redundant, than are actually needed to convey basic information to their listeners. If the adaptive significance of speech were simply to transmit functional information from one person to another, natural selection would presumably have acted against the extravagances associated with human language.

A lack of sex differences between males and females in vocabulary size does not fully discredit the sexual selection hypothesis. After all, in many songbird species, both sexes have large song repertoires. For example, male and female superb starlings (*Lamprotornis superbus*) both produce about 50–60 distinct syllables, and the song repertoires of both sexes are thought to be under strong sexual selection (Pilowsky and Rubenstein 2013). Perhaps a better test of the sexual selection hypothesis is to determine if women prefer males who are adept speakers during verbal courtship. When young women were given the opportunity to watch videos showing young men reading out loud and talking (among other things), the men who had previously scored well on a verbal intelligence test were judged by the women to be more sexually attractive (Prokosch et al. 2009). The women independently judged male verbal intelligence accurately on the basis of what they observed from the videos. Thus, it would make sense that men attempt to be verbally proficient when interacting with women, whereas women use their equally large vocabularies and speech capabilities to judge the verbal proficiency of potential partners, perhaps as a proxy for intelligence. Moreover, women with large vocabularies and good verbal skills might also be preferred by males (suggesting a case of mutual sexual selection).

The sexual selection hypothesis also led to the finding that, although both men and women say they value a sense of humor in the opposite sex, men think that a woman's sense of humor can be judged by how she responds to *their* jokes, whereas women think that a man's sense of humor is a function of whether he says things that strike *them* as funny (Bressler et al. 2006). In other words, perhaps men are not interested in females who are humorous but instead are looking for appreciative females—unlike women, who do like men who are amusing. If this prediction is true, then men rated relatively highly as humorists should have relatively high levels of sexual success. As predicted, the men who were considered funnier (by six student judges, based on captions that the men wrote for a set of untitled *New Yorker* cartoons) self-reported having more sexual partners than those who scored lower on the humor test (Greengross and Miller 2011).

Reproductive Behavior

For many adolescents, their first introduction to human reproductive behavior is a sex education class in high school. For decades, physical education teachers have educated America's youth about the wonders of the human body, the process of sexual intercourse, and its potential hazards in the form of pregnancy and disease transmission. Yet rarely, if ever, is sex education discussed in an evolutionary or comparative context, despite the fact that sexually transmitted diseases (STDs) have been important for shaping evolution in nonhuman primates and other species, or that sexual conflict occurs in humans just as it does in most other sexually reproducing organisms where the interests of males and females are often not the same. Here we discuss a few topics about human reproductive behavior that are almost certainly not mentioned in high school sex education classes despite the fact that humans are becoming an important system within which to study reproductive behavior and to test theories and hypotheses about sexual section and sexual conflict, thanks to long-term historical records (Wilson et al. 2017). We begin with a discussion of how men and women choose mates, and end with a consideration of how this can lead to both reproductive and sexual conflict, just as it can in other species.

An Evolutionary Analysis of Human Mate Choice

Tackling an evolutionary analysis of mate choice in our species is challenging because of the diversity of cultural rules and regulations surrounding human sexual behavior. There are monogamous, polygynous (often referred to as polygamous in humans), and polyandrous societies, some in which you are not allowed to marry an unrelated person who belongs to your clan, others in which adult men can marry prepubescent girls, some in which males and their relatives provide payment for a bride, and others in which women must bring a costly dowry with them to their wedding. Yet despite this cultural variety in human mating systems, certain biological facts of life still apply to our species. For example, women are biologically similar to other female mammals in that they retain control of reproduction by virtue of their physiological investments in producing eggs, nurturing embryos, and providing infants with breast milk after they are born. Although men are also able and often willing to make large parental investments in their offspring, their reproductive decisions nevertheless take place in a setting defined by female physiology and psychology (Daly and Wilson 1983, Geary 2010). Although women are often unable to exercise complete control over their reproduction, at least under some conditions some women have been able to choose from a number of potential partners (or their relatives have done so on their behalf).

Here we start from the premise that both men and women can adopt multiple mating strategies, ranging from committed relationships (long-term mating strategies) to sexual liaisons (short-term mating strategies) (Easton et al. 2015). Our discussion focuses on the factors that influence mate choice in both men and women. Although we do not focus on parental care per se in this chapter (see Chapter 11 for a discussion of other animal species), which mating strategy individuals adopt—particularly males—is at least partly related to their minimum obligatory parental investment.

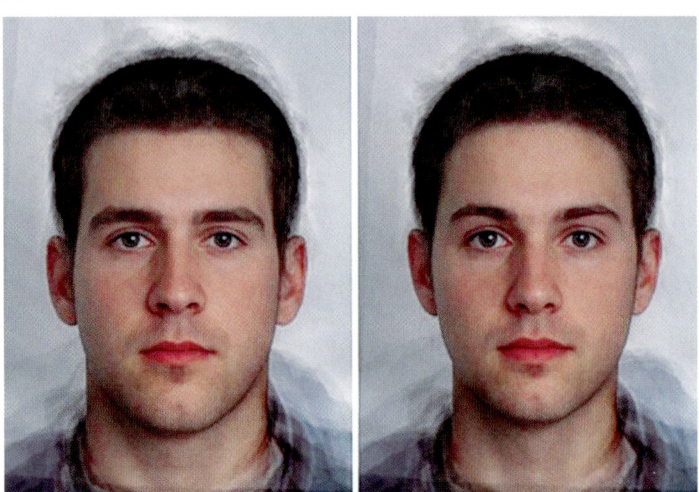

FIGURE 14.5 Mate choice based on facial appearance. Computer-modified facial images provide a way to test the preferences of women for (A) masculinized faces versus (B) feminized faces. (From Jones et al. 2008; photographs by Ben Jones and Lisa De Bruine of the Face Research Laboratory, University of Aberdeen.)

Mate Choice by Women

From an evolutionary perspective, we would expect women to be attracted to men whose physical attributes are indicative of high genetic quality or parental ability, both of which vary among males and have the potential to affect the fitness of a woman and her children. Some researchers have found that women do indeed prefer men with masculine facial features, namely a prominent chin and strong cheekbones (**FIGURE 14.5**) (Jones et al. 2008). In addition, facial symmetry has been identified as a plus (and not just in Western societies; Little et al. 2007b), as have a muscular upper body (Geary 2010), above-average height (Mueller and Mazur 2001), and a deep voice (Feinberg 2008). In the European country of Slovakia, physically attractive men are more likely to marry, and of those who are in the married cohort, the more attractive individuals (as judged by a group of young women on the basis of photographs taken of the men in their 20s) have more children (Prokop and Fedor 2011).

Why might more physically attractive males have greater reproductive success? Think back to our discussion in Chapters 9 and 10 when we addressed this topic more broadly. In humans, this relationship between male physical attractiveness and greater reproductive success might be linked to a variety of factors, including high testosterone levels, good current health, and perhaps most important, good health during juvenile development. The development of males is at special risk because of the potentially damaging side effects of testosterone. Therefore, the ability of a man to develop normally despite high levels of circulating testosterone is a possible indicator of a strong immune system capable of overcoming the handicap imposed by the sex hormone (Folstad and Karter 1992). If strong, healthy men can pass on defense against disease to their offspring, their children are more likely to survive and their mates will benefit as a result.

A study of nearly 15,000 American adults aged 25–34 years found robust associations (interestingly, in both sexes) between physical attractiveness and various measures of health, including traits such as cancer, high cholesterol, blood pressure, diabetes, heart disease, and many other health-related variables (Nedelec and Beaver 2014). Recall from Chapters 8 and 9 that males with brighter carotenoid-based ornaments (for example, feathers in birds or skin spots in fish) are often preferred by females, presumably because of the health benefits associated with higher carotenoid levels in the blood. To test this association in humans, researchers in Australia supplemented men's diets with carotenoids, which resulted in not only increased skin yellowness and redness, but also made the altered males more preferred by women than those men fed a control diet (Foo et al. 2017). However, despite increased perceived healthiness by women, there were no actual differences in oxidative stress, immune function, or semen quality between carotenoid- and control-treated men, demonstrating that although female choice was affected, the health-associated benefits of male attractiveness need further study.

In addition to being preferred by women, healthy men may also be able to compete effectively with rivals for dominance within their group. In this context, it is relevant that the men rated as having the most dominant, masculine, and attractive faces are those with the greatest handgrip strength, which correlates well with overall physical strength (Fink et al. 2007). Moreover, men with high handgrip strength develop sexual relationships sooner and have more (self-reported) sexual partners than individuals with less powerful grips (Gallup et al. 2007). Over the course of human evolution, dominant, powerful men may have been able to protect their partners from other men, as well as supply them and their offspring with the resources that usually are possessed by males of high social and political status (Grammar et al. 2003, Buss 2012). However, these types of adaptive interpretations need to be considered carefully. For example, in a study of masculinity in men's faces (a trait that is often thought to reflect male aggressiveness through the links among face shape, testosterone, and aggressive behavior; **BOX 14.2**) across 12 populations that varied in their degree of economic development, Isabel Scott and colleagues found that preferences for sex-specific facial traits were found only in highly developed societies and that the perception that masculine males look aggressive increased strongly as societies became more highly developed (Scott et al. 2014). These results challenge the idea that male facial traits linked to aggression have been under natural selection for a long time, and instead suggest that these perceptions have evolved quite recently alongside urbanization.

The importance for women of having a good provider as a mate has been established in studies showing that females in cultures without birth control who secure

BOX 14.2 EXPLORING BEHAVIOR BY INTERPRETING DATA

Female choice and the features of dominant versus attractive men

The research we have reviewed thus far tells us something about what people want in an ideal mate, but for most of us, ideal mates are in short supply. Not every man is paired off with an exceedingly fertile, gorgeous model, nor is every woman married to an extremely wealthy, highly parental, loving individual with outstanding genes. The mate choices of men and women often diverge considerably from what would seem to be best for their genes. Consider, for example, an experimental study designed to test the idea that certain features of the human face indicate hormone levels during development, and that these are the features that women use to judge the attractiveness of potential partners (**FIGURE**) (Swaddle and Reierson 2002). The results of this study indicate that women differ with respect to the facial features they associate with dominant versus attractive men.

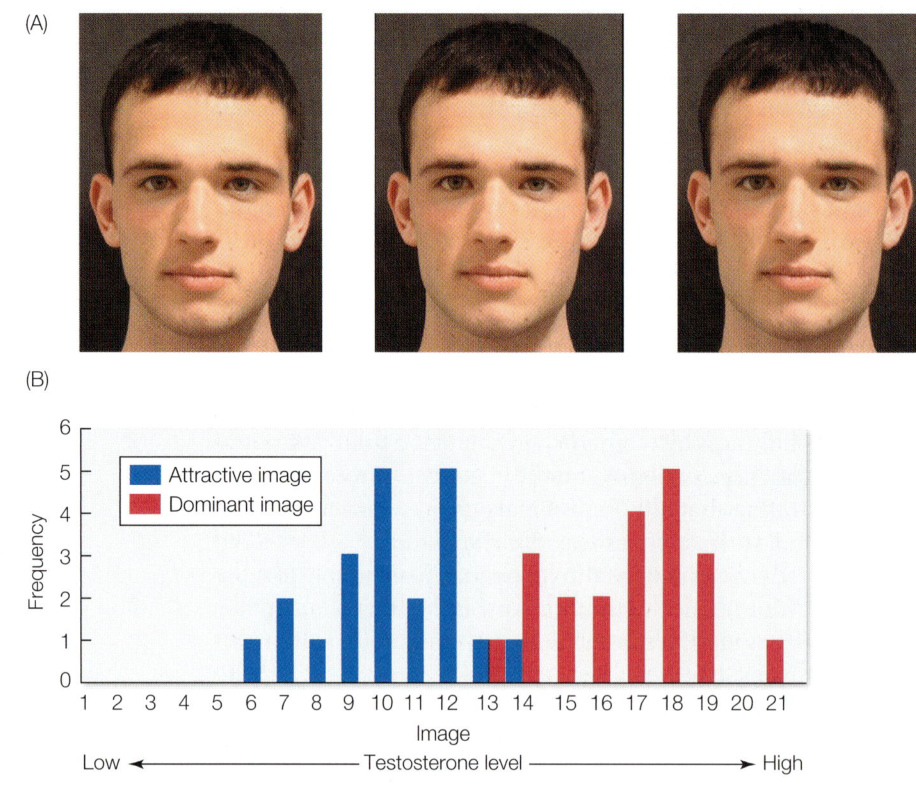

Females differ with respect to facial features they associate with dominant men versus attractive men. (A) In this study, digital images of the same male's face were altered to reflect the developmental effects of testosterone, from low to high levels (left to right). (B) When young women (ages 18–21 years) were asked to judge these photographs for physical attractiveness, they tended to pick images like the one in the middle, whereas when they were asked to rate the images in terms of social dominance, they tended to pick those similar to the right-hand image. (After Swaddle and Reierson 2002; photographs © 2002 by the Royal Society of London.)

Thinking Outside the Box

What are some of the differences that you see in these photographs? Given these differences, try to interpret the distributions in the histograms. Why do the histograms look different? Perhaps because trade-offs are involved in any pairing. Men who marry extremely attractive women may lose paternity to other men who are also attracted to their partners. Women who marry extremely strong, powerful men may lose resources to other women attracted to their partners. Do you agree with this trade-off interpretation? What do you think would happen if you replicated this experiment with older women?

relatively rich husbands tend to have higher fitness than women whose partners cannot offer many material benefits. Among the Aché of eastern Paraguay, for example, the children of men who were good hunters were more likely to survive to reproductive age than the children of less skillful hunters (Kaplan and Hill 1985). Likewise, several studies of traditional societies in Africa and Iran have revealed a positive correlation between a woman's reproductive success and her husband's wealth, as measured by land ownership or the number of domestic animals in the husband's herds (Irons 1979, Borgerhoff Mulder 1987, Mace 1998). Even in modern societies, household income is positively correlated with children's health, with the effect growing larger as children get older. Moreover, chronic illnesses in childhood can reduce the earning power of children who reach adulthood, thus perpetuating poverty across generations (Case et al. 2002), with all of the negative reproductive consequences this has for human beings.

Evidence of this sort has convinced some evolutionary psychologists to predict that females will usually put wealth, social status, and political influence ahead of good looks when it comes to selecting mates. This evolutionary prediction has been supported by questionnaire and interview studies by David Buss (Buss 1989) and others. However, even when researchers have found clear differences between the sexes in the value that they attach to "good financial prospects" versus "good looks," the absolute measures of importance given to these attributes have not necessarily been especially high for either sex. Since the men and women in these studies typically have not had to specify which items among a list of attributes are absolutely essential to their choice of mates (versus those that would simply be nice to have in a partner but are not crucial), a team of social psychologists led by Norm Li attempted to put constraints on the choices made by adults they interviewed in the United States by giving the participants a limited budget to expend on designing a hypothetical ideal mate (Li et al. 2002). Individuals were given a list of traits and a supply of "mate dollars," and then told to decide how many of their mate dollars to spend on traits such as physical attractiveness, creativity, or yearly income. Individuals could spend up to 10 mate dollars to get the highest level of traits they particularly valued, or nothing at all on traits of no importance to them. When the people interviewed were given only 20 mate dollars to work with, their investments differed greatly according to their sex. Men devoted 21 percent of the total budget to the acquisition of a physically attractive partner, whereas women spent only 10 percent of the same total amount to the same end. In contrast, women on this tight budget devoted 17 percent of their money to boost the yearly income of an ideal mate, while men invested just 3 percent of their mate dollars on this attribute. But Li's human mate choice experiment did not end here. After spending their first 20 mate dollars, the participants were given two additional 20-dollar increments. By the time they reached the third 20 mate dollars, the sexes no longer differed markedly with respect to the attributes they were buying. Having already purchased what they really valued, they could and did spend on other attributes. Confirmation of these results comes from a similar study in a Singaporean population, suggesting that this evolutionary analysis has cross-cultural validity (Li et al. 2011). These experiments tell us that the criteria for mate choice are not the same for men and women, as predicted by an evolutionary approach (Li et al. 2002).

Personal ads (which today have largely been replaced by online dating applications; **BOX 14.3**), whose cost limits the number of words used by the advertisers, also provide relevant evidence on what people consider fundamentally important in a real mating market. For example, heterosexual women seeking partners through newspapers are far more likely than men to specify that they are looking for someone who is relatively rich (Waynforth and Dunbar 1995). In keeping with this goal, women advertisers in both Arizonan and Indian newspapers also often specify an interest in someone older than they are (Kenrick and Keefe 1992),

BOX 14.3 EXPLORING BEHAVIOR BY INTERPRETING DATA

Human mate choice in an online world

As the dating world has moved beyond newspaper ads to online applications, the amount of data available to evolutionary psychologists for studying human mating preferences has grown exponentially. As of 2017, one of the leading online dating applications was Tinder, which has been downloaded more than 100 million times since 2012 and has 10 million daily users (Sevi et al. 2017). Using these new databases, researchers are able to determine not just what men and women say they prefer in a mate, but what their actual mating preferences are.

One important difference between traditional newspaper personal ads and dating applications such as Tinder (**FIGURE**) is that the latter more strongly emphasize visual traits through posting of "selfies" and other photographs that might contain information about a user's personality traits (such as whether the person tends to smile or scowl), their hobbies, their friends and social interactions, or other factors that could be useful in mate choice. For example, over one-third of all Tinder accounts have no written biography and instead contain only photographs (Tyson et al. 2016). Online profiles with photographs tend to get more views than those without (Hitsch et al. 2010), and Tinder profiles with more than one picture receive more matches than those with just a single photograph, particularly for male profiles (Tyson et al. 2016). Additionally, women are much more selective than men in their initial screening process, which suggests a greater initial time investment by females. Yet men also exhibit strong dating preferences, though these often differ from those of women. As we have seen in other studies, female preferences are more strongly related to income, education level, and occupation than are those of men, and women prefer men their own age or a few years older, whereas men prefer women their own age or a few years younger (Hitsch et al. 2010). Moreover, women prefer men with higher body mass indices and who are taller than they are, whereas men prefer the opposite. Interestingly, users may be manipulating these physical preference patterns through their choice of profile pictures, which can provide signals of physical height and impressions of power. For example, men tend to orient their Tinder profile pictures from below to appear taller and more powerful, whereas women tend to do the opposite and orient from above to appear shorter and less powerful (Sedgewick et al. 2017).

Dating applications like Tinder are changing the way humans choose mates, but are they also changing human mating preferences? (Photograph © Alex Ruhl/Shutterstock.com.)

possibly because older men usually have larger incomes than younger men (Buss 1989). Similar patterns were observed in a sample of nearly 13,000 Finnish people, in which women preferred same-aged or older men (Antfolk et al. 2015). However, the survival of infants and their subsequent reproductive success have been shown to decline with paternal age (Arslan et al. 2017), suggesting that choosing older men because of their wealth could also have negative fitness consequences for women.

BOX 14.3 EXPLORING BEHAVIOR BY INTERPRETING DATA *(continued)*

Thinking Outside the Box

What are we to make of these online dating trends and human mating preferences? The new data largely agree with previously published studies based on surveys and newspapers ads, though they often have much larger sample sizes. Can you think of some potential concerns with using these types of data? In this connected world with seemingly endless choices, it is important to remember that dating preferences are not necessarily mating preferences, though they are related. Since dating applications such as Tinder are used by some men and women for locating sexual partners (a short-term mating strategy) and others for finding potential mates with whom to marry and start a family (a long-term mating strategy) (Sevi et al. 2017), evolutionary analyses must be considered in this light. Nonetheless, many of the same principles that underlie male and female mate choice in other animals also show up in humans. And just as animals such as túngara frogs (*Engystomops pustulosus*) make irrational mate choice decisions when faced with too many choices (Lea and Ryan 2015), so too might humans, particularly when bombarded with so many online dating profiles.

Take what you have learned about mate choice and signaling in this and other chapters of the book and think about how online dating applications promote certain traits in males, as well as amplify certain preferences in females (and vice versa). For example, why are so many Tinder profiles built with only photographs? Think about the range of dating applications out there, including those designed for certain ethnic, cultural, or religious groups, those designed more for women versus men, and those made for heterosexual versus homosexual people.

If women really are highly interested in a partner's wealth and capacity to provide for offspring, then men in their 30s might be most desirable because men of this age have relatively high incomes and are likely to live long enough to invest large amounts in their children over many years. One can calculate the "market value" of men of different ages—a measure that combines both supply and demand—by using samples of personal ads and dividing the number of women requesting a particular age class of partner in their advertisements by the number of men in that age class who are advertising their availability. Men in their late 30s have the highest market value (**FIGURE 14.6**) (Pawłowski and Dunbar 1999).

Are women really that calculating when choosing potential mates? Instead, it could be that a woman's interest in the earning power of potential mates is a purely rational response to the fact that males in almost every culture control their society's economy, making it difficult for a woman to achieve material well-being on her own. If this nonevolutionary hypothesis explains why females favor wealthy males, then women who are themselves well off and not dependent on a partner's resources should place much less importance on male earning power. Contrary to this prediction, however, several surveys have shown that women with relatively high expected incomes actually put *more* emphasis—not less—on the financial status of prospective mates (Townsend 1989, Wiederman and Allgeier 1992). For example, relatively wealthy female undergraduate students seek wealth and high status in potential long-term mates (Buston and Emlen 2003). These and

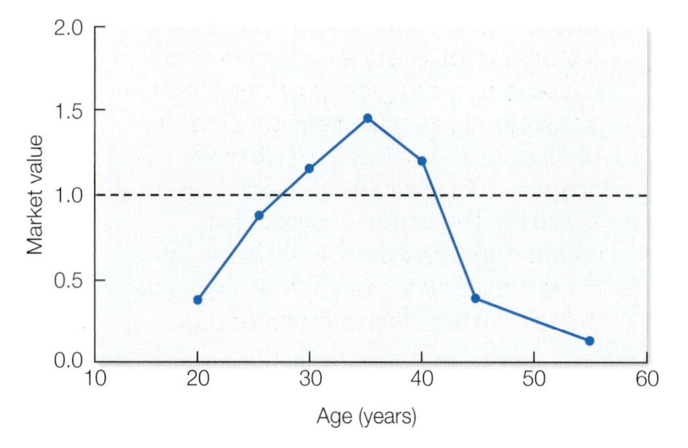

FIGURE 14.6 Age and the market value of men. Men in their 30s have the highest market value. Market value is calculated by taking the number of newspaper "personal" advertisements by women requesting men of certain age classes and dividing that by the number of men of those age classes announcing their availability. (After Pawłowski and Dunbar 1999.)

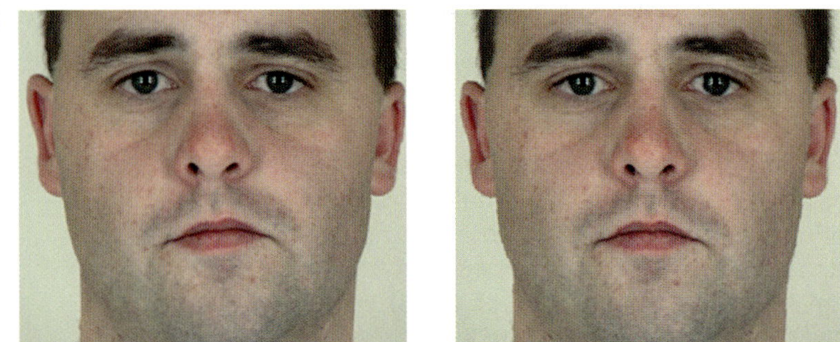

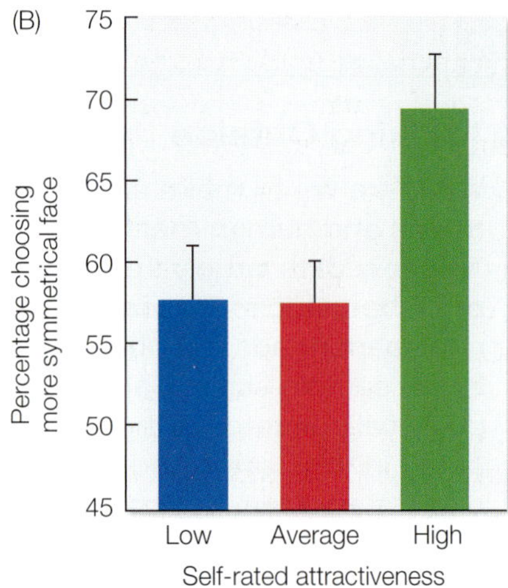

FIGURE 14.7 Women who think they are highly attractive prefer more attractive men. (A) When women were given a choice between two digitally altered photographs of men, one of which was more symmetrical than the other (right vs. left), they differed in the degree to which they said they preferred the more symmetrical face. (B) Women who rated themselves highly attractive chose the more symmetrical face nearly 70 percent of the time, whereas women of lower self-rated attractiveness did so less than 60 percent of the time. Bars depict mean +/– SE. (After Little et al. 2001; photographs © 2001 by the Royal Society of London and with permission of the Perception Lab at the University of St. Andrews.)

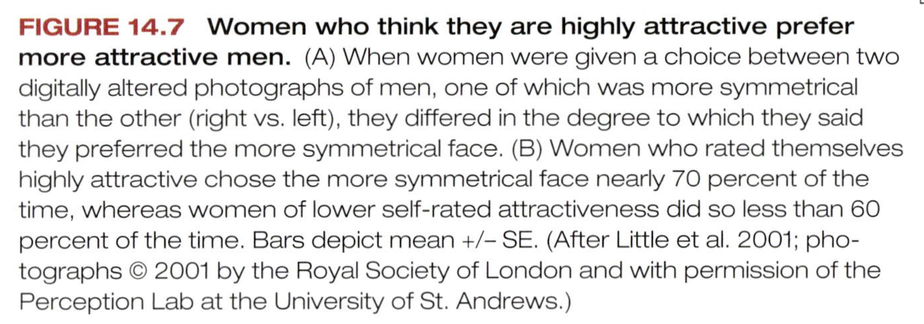

other studies suggest a general rule: that a woman's own social standing affects her mate choice criteria. Consistent with this hypothesis, women who rate themselves as "highly attractive" show a stronger preference for both relatively masculine and relatively symmetrical faces than do women who believe they are of average or low attractiveness (**FIGURE 14.7**) (Little et al. 2001). Furthermore, David Buss and Todd Shackelford report that not only are physically attractive women desirous of a highly attractive mate, they also have higher standards when it comes to partner wealth, commitment, and parental abilities (Buss and Shackelford 2008), a finding that matches the results of studies based on dating advertisements (Pawłowski and Dunbar 1999). Finally, Peter Buston and Steve Emlen found that both men and women who consider themselves high-ranking prospects for a long-term relationship expressed a preference for an equally high-ranking partner, whereas those individuals with a lower self-perception of market value were less demanding

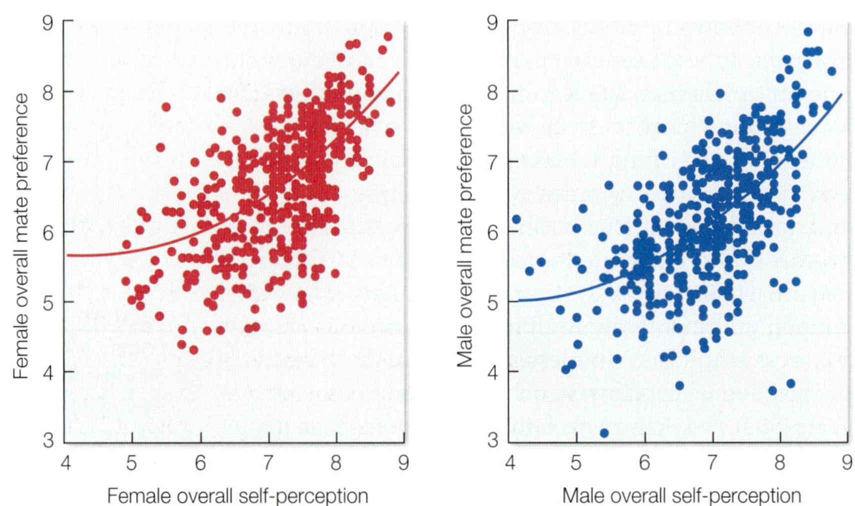

FIGURE 14.8 Self-perception of attractiveness affects mate preferences in both sexes. The degree to which women and men consider themselves attractive is correlated with their mate preferences. Less attractive individuals are willing to settle for less in a partner. The values for mate preferences are mean scores of the individuals' answers to questions about the importance of ten attributes in their dating decisions such as, on a scale of 1 to 9 how important is physical attractiveness to you? The values for self-perception are the mean scores of the respondents' own evaluations of how they would score on these ten attributes. (After Buston and Emlen 2003, © 2003 National Academy of Science, USA.)

(**FIGURE 14.8**) (Buston and Emlen 2003). The less-demanding requirements of women who are unlikely to be competitive in securing and maintaining long-term relationships may help these individuals avoid the costs of searching unrealistically for men they are unlikely to attract, as well as reducing the odds of being deserted should they enter into a relationship with a high-ranking male (Little et al. 2001).

Interestingly, results similar to those in humans have been observed in other animals. For example, in experimental studies of female zebra finches in which some were reared in small broods while others grew up in large broods, brood size affected the metabolic efficiency of the birds. That is, "privileged" small-brood females became high-quality adult finches, whereas the large-brood, low-quality females were handicapped as a result of the increased competition for food that they had experienced when young. When the two kinds of females were given an opportunity to listen to the songs of either high-quality males or low-quality males, which they could turn on by pecking a key, high-quality females preferred the songs of high-quality males, whereas low-quality females chose the songs of low-quality males. In so doing, as has been suggested for humans (Little et al. 2001), low-quality females may benefit by avoiding a competition for high-quality mates that they could not win or by ignoring high-quality males likely to reject them sooner or later (Holveck and Riebel 2011).

Sexual preferences in zebra finches and humans are one thing, but mating behavior is another. Are women actually more likely to copulate with men who possess the attributes they prefer than with men who lack them? Among the Aché of Paraguay, good hunters with high social status are more likely to have extramarital affairs and produce illegitimate children than are poor hunters, suggesting that females in this society find skillful providers sexually attractive (Kaplan and Hill 1985). Likewise, in Renaissance Portugal, noblemen were not only more likely to marry more than once, but also more likely to produce illegitimate children than men of lower social rank. These results are consistent with the prediction that women use possession of resources as a cue when selecting a father for their children (Boone 1986). However, it is also possible that noblemen or those of high status could have forced themselves on women and produced illegitimate children. After all, in many societies it is uncommon for men to provide resources for illegitimate children. Thus, one must be very careful when inferring female preferences in humans without directly studying the underlying mechanism.

If we bear the imprint of past evolution on our psyches, then women in today's Western societies should also use resource control and its correlates, such as high social status, when deciding which men to accept as sexual partners. To study the relationship between male income and copulatory success in modern Quebec, Daniel Perusse secured data from a large sample of respondents on how often they had copulated with each of their sexual partners in the preceding year (Perusse 1993). With this information, Perusse was able to estimate the number of potential conceptions (NPC) a male would have been responsible for, had he and his partner(s) abstained from birth control. Male mating success, as measured by NPC, was highly correlated with male income, especially for unmarried men (**FIGURE 14.9**). Perusse concluded that single Canadian men attempt to mate with more than one woman, but their ability to do so is affected by their wealth and social standing. These findings have been replicated in great detail with a much larger random sample of men living in the United States (Kanazawa 2003). Thus, males striving for high income and status may be the product of past selection by choosy females, which occurred in environments in which potential conceptions were likely to result in successful pregnancy (Perusse 1993). In the past, those men who secured wealth and high social standing almost certainly were likely to have had higher fitness on average than men who cared little

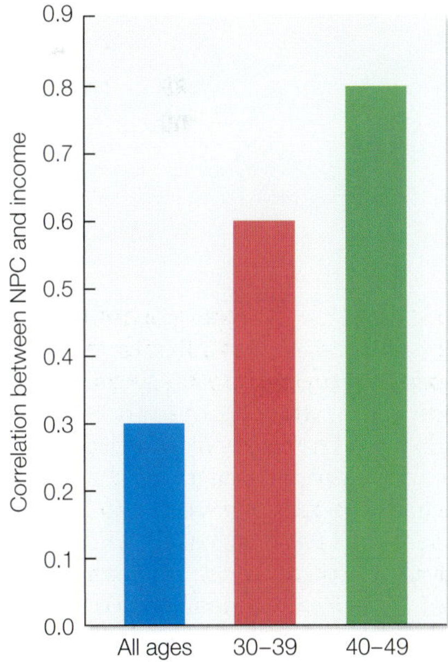

FIGURE 14.9 Higher income increases male copulatory success. Income is positively correlated with the number of potential conceptions (NPC) in the preceding year for unmarried Canadian men of various age groups, and especially for older men. (After Perusse 1993.)

about these attributes. Consistent with runaway selection theory (see Chapter 9), these data suggest selection for females to produce daughters with the preference for males with indicators of wealth and status, as well as to produce sons who strive for these goals.

Mate Choice by Men

Recall from Angus Bateman's classic experiments on *Drosophila* introduced in Chapter 9 that males in many species can increase their reproductive success by mating with additional females (Bateman 1948). It is therefore no surprise to learn that men report wanting to have more sexual partners than women, something that men can often achieve by lowering their standards for what they view as a potential mate (Easton et al. 2015). Although using a short-term mating strategy to secure multiple sexual liaisons and avoid the entanglement of a committed relationship that requires men to provide parental investment in offspring is a successful reproductive strategy, there is also evidence to suggest that many men adopt a long-term mating strategy. As you learned in Chapter 10, since seeking out mates can be costly (for humans in terms of time, energy, or money), monogamy can help reduce those costs. This is not to say, however, that this strategy is without cost. For example, high-quality women may be choosier and demand more time and investment from men before they engage in sex, and once paired, men may have to invest in mate guarding because their partners are likely to be desired by others (Buss 2003). But what about the reproductive benefits of a committed relationship? Since men who engage in short-term mating cannot be certain about the paternity of their offspring, long-term mating may increase paternity assurance (Easton et al. 2015). Moreover, if men are willing to enter into a committed, long-term relationship, they may be able to gain access to higher quality women, who typically also want the same thing. Thus, while these men might produce fewer offspring this way, they might produce higher quality offspring. Ultimately, the strategy that any man uses to attract women will depend both on his attributes (his quality) and his environment (the number of women and other men who are out there).

Which attributes do men actually prefer in their potential mates? You will recall the prediction that men will place more value on "good looks" than will women, who should instead tend to be more interested in the earning potential or resources controlled by a would-be mate. Why then might men be interested in attractive women? From an evolutionary perspective, a good-looking woman is predicted to be a fertile one, given that fertility varies considerably from woman to woman as a result of her age, health, and body weight, among other things. Preadolescent and postmenopausal women obviously cannot become pregnant. Moreover, women in their 20s are more likely to become pregnant than women in their 40s. Healthy women are more fertile than sick ones, and women who are substantially overweight or underweight are less likely to become pregnant than women of average weight (Reid and van Vugt 1987).

Men in Western cultures also tend to favor feminine features that are associated with developmental homeostasis (normal development), a strong immune system, good health, high estrogen levels, and most important, youthfulness (Thornhill and Gangestad 1999a, Geary 2010). In particular, men favor full lips, small noses, large breasts, a waist circumference that is substantially smaller than hip circumference (Singh 1993), and intermediate weight rather than extreme thinness or obesity (Tovée et al. 1999). The level of circulating estrogen in the large-breasted, narrow-waisted participants in a sample of healthy Polish women meant that these women were about three times more likely to conceive a child than the other women in the

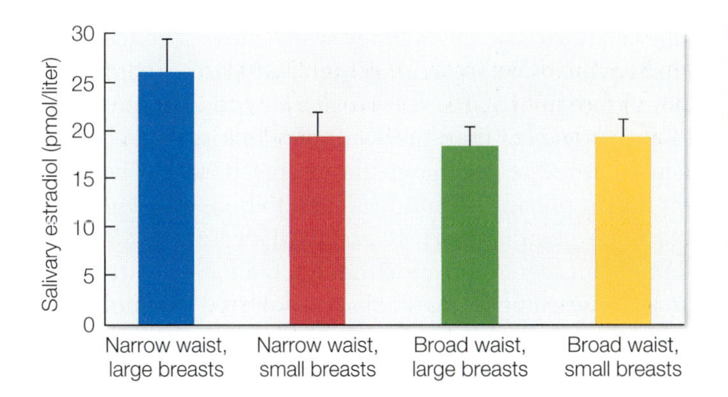

FIGURE 14.10 Body shape is correlated with fertility in women. Women with features that men often claim they prefer—namely a narrow waist and large breasts—are more fertile than women with other body shapes, based on a sample of healthy Polish women between 24 and 37 years of age who were not taking birth control pills. Bars depict mean +/– SE. (After Jasiénska et al. 2004.)

study (**FIGURE 14.10**) (Jasiénska et al. 2004). The fact that in different cultures (Brooks et al. 2010) men also prefer the waist-to-hip ratio associated with the hourglass figure, as do congenitally blind individuals who assess this ratio by touch, suggests that the Western visual media is not responsible for this aspect of male choice (Karremans et al. 2010); natural selection is.

Most or all of the physical attributes that men in Western societies tend to find sexually appealing may be linked to a woman's potential to become pregnant. But even highly fertile women are most likely to conceive only during the few days each month when they are ovulating. During this time, mating with high-quality men could provide genetic benefits for a woman's offspring, and if so, we can predict that ovulating women will be especially interested in competing for and securing high-quality sexual partners. Ovulation in female mammals is linked to a spike in circulating estrogen levels, and the higher these levels in humans, the more likely a woman is to become pregnant should she copulate at this time (Lipson and Ellison 1996). In other nonhuman primates, the fertile window is often widely and conspicuously advertised, such as in the bright red sexual swelling of the perineal skin in baboons.

Although it was once thought that neither men nor women could detect when a woman is fertile, there is growing evidence that both sexes can detect the changes associated with ovulation. For example, women with relatively high estrogen levels are considered to be more attractive by others (Durante and Li 2009). Ovulating women provide cues of this event with changes in facial appearance (**FIGURE 14.11**) (Roberts et al. 2004, Law-Smith et al. 2006), voice pitch (Bryant and Haselton

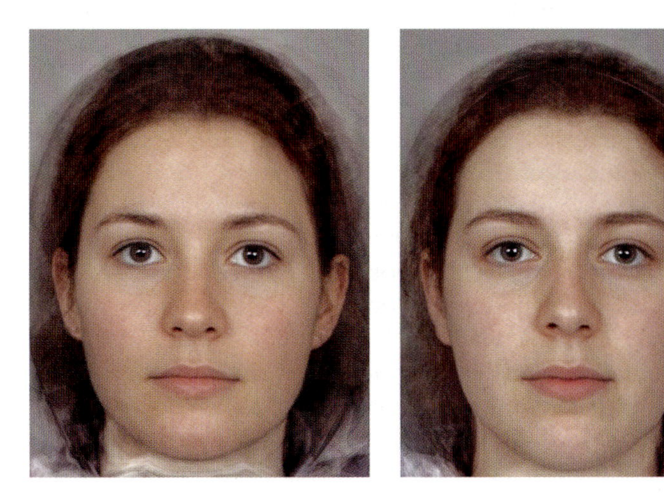

FIGURE 14.11 Facial appearance provides a cue of circulating estrogen levels. The faces shown are composite images generated by overlaying ten photographs of women with the highest urinary estrogen levels and ten photographs of women with the lowest urinary estrogen levels. You should be able to guess which composite image was derived from photographs of women at peak fertility. (From Law-Smith et al. 2006.)

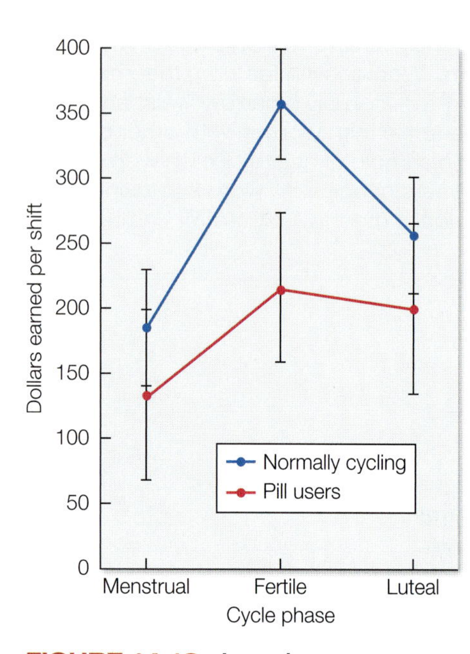

FIGURE 14.12 Lap dancers earn more money from tips when they are ovulating. During the fertile phase of their cycle when ovulating, lap dancers make more money than they make during the menstrual or luteal phases of their cycle. Additionally, dancers taking birth control pills make less in tips than dancers cycling normally during the fertile period. Values depict mean +/– SE. (After Miller et al. 2007.)

2009; but see Fischer et al. 2011), the way they walk (Provost et al. 2008), and their body odor. For example, Devendra Singh and Matthew Bronstad showed that men find the smell of a T-shirt worn by an ovulating woman to be "more pleasant and sexy" than the scent of a T-shirt worn by the same woman when she is not fertile (Singh and Bronstad 2001). The odor of women in their fertile phase also induces men to boost the levels of circulating testosterone in their bodies (Miller and Maner 2010).

Although male baboons can determine that a female is in estrous by looking at her red sexual swelling, since women do not have inflamed genitals when they are fertile, men must use some of the cues described above. Geoffrey Miller and colleagues used a novel method to demonstrate that men are quite capable of (unconsciously) assessing female fertility. The researchers recruited several lap dancers working in Albuquerque, some of whom were taking birth control pills and others who were not. These women were willing to cooperate with the researchers by telling them the amount they received as tips after their erotic performances. Because the research team considered it likely that men could detect cues associated with female fertility, and that men should find these cues attractive, the team predicted that lap dancers who were not taking birth control pills and were ovulating would receive more tips from their customers than would those women who were not ovulating, either because they were taking birth control pills or because they were in a nonovulatory phase of the menstrual cycle. Consistent with their prediction, the researchers found that the lap dancers took in about twice as much in tips when they were ovulating compared with when they were not ovulating or taking birth control pills (**FIGURE 14.12**) (Miller et al. 2007).

It is important to remember that to really understand correlational relationships such as this one, one must decipher the underlying mechanism. What was the proximate mechanism (factors such as shifts in body scent, facial attractiveness, soft-tissue symmetry, waist-to-hip ratio, or verbal creativity or fluency) underlying the result that lap dancers earned more in tips when they were fertile? Although the researchers did not directly test this in their study, they argued that the proximate mechanism was unrelated to shifts in stage-dance moves, clothing, or initial conversational content, because these cues did not vary much among dancers. Instead, the researchers speculated that because the pattern of tip earnings was similar to that of estradiol levels across the cycle, then perhaps estradiol levels might mediate tip earnings if males can somehow assess female hormone levels through smell or some other mechanism. This proximate hypothesis remains to be tested.

ovulatory shift hypothesis Women experience elevated sexual attraction to men with characteristics that reflect male genetic quality when the women are most fertile.

Contraceptive pills not only affect how women are perceived by men but may also influence female mate choice, for example by reducing the attractiveness of genetically dissimilar partners under some conditions (Roberts et al. 2008, Alvergne and Lummaa 2010). A meta-analysis of 50 studies examining more than 130 factors found robust support for the **ovulatory shift hypothesis**, which proposes that women experience elevated sexual attraction to men with characteristics that reflect male genetic quality when the women are most fertile (in other words, on high- versus low-fertility days of the women's menstrual cycle) (Gildersleeve et al. 2014). More generally, the sexual preferences of women, as well as men, change in relation to the menstrual cycle (Gangestad and Thornhill 2008, Jones et al. 2008). For example, when asked to evaluate potential partners for a brief sexual encounter, fertile women tended to favor the men with more masculine faces (Penton-Voak et al. 1999) and bodies (Little et al. 2007a), both traits that are likely to be related to genetic quality (**FIGURE 14.13**). Additionally, fertile women found men with symmetrical

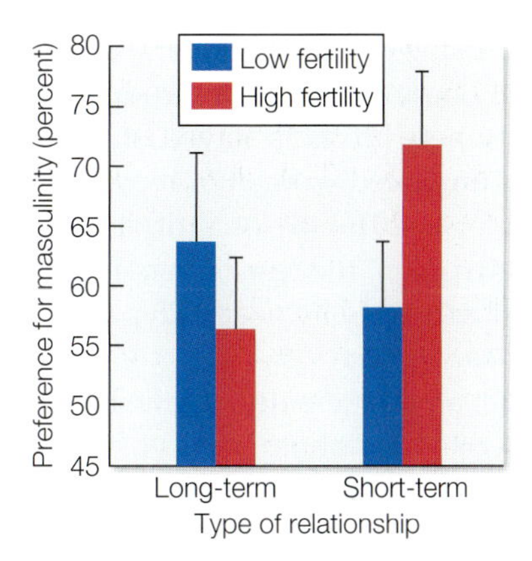

FIGURE 14.13 The menstrual cycle affects female mate choice. Women changed their evaluation of two male body images, one relatively feminized and the other masculinized, depending on the phase of their menstrual cycles. Fertile (ovulating) females (blue bars) who were asked to judge the two images for a prospective long-term relationship favored the feminized image, whereas they picked the masculinized image when asked about a short-term relationship preference. Bars depict mean +/– SE. (After Little et al. 2007a.)

faces more attractive (Little et al. 2007b) and preferred their smell (Thornhill and Gangestad 1999b). At the proximate level, changes in neural responses may underlie these changes in preferences, as the right medial orbitofrontal cortex of ovulating women is activated more strongly when these women are shown images of masculinized male faces as opposed to feminized ones (Rupp et al. 2009).

Hormones are surely also involved proximately in changes in the perception of male features that occur over both the ovulatory cycle and the lifetime of a woman. Before puberty and after menopause, women have low levels of estrogen. In keeping with this hormonal reality, preferences for masculine male faces are relatively low before puberty and after menopause (Little et al. 2010). In this way, changes in hormone levels (and in the analysis of men's faces) across the lifespan of a female mirror the changes that occur during the menstrual cycle of a postpubescent woman.

Why might women with social partners of average or low attractiveness attempt to copulate with men of above average quality? For the very same reasons we learned earlier that other female animals might mate outside of the pair-bond (see Chapter 9). These changes in brain activity and mate preference might occur because ovulating women with social partners of average or low quality are unconsciously attempting to secure "good genes" for their offspring, which could make their sons especially attractive or dominant. Similarly, selective women might be more likely to secure superior "complementary genes" from an extra-pair mate, which could generate better offspring genotypes, perhaps particularly with respect to immune system development. This idea of complimentary gene choice is supported by the finding that women usually prefer the smell of men with dissimilar major histocompatibility complex (MHC) genes, which play a critical role in the immune response to pathogens (Brown and Eklund 1994, Wedekind et al. 1995, Lie et al. 2010). According to this idea, the offspring of MHC-dissimilar couples should be more heterozygous and thus have better immune systems and greater resistance to infection. Interestingly, women in relationships with men whose MHC genes are similar to their own report lower satisfaction with their mates and a greater number of extra-pair partners than do women in relationships with men who are genetically dissimilar with respect to these genes (Garver-Apgar et al. 2006).

Reproductive and Sexual Conflict in Humans

Reproductive conflict among males for access to females is common in humans, as it is in other species. We need not look far into our own history to see examples where men started wars over a woman. But reproductive conflict among *females* for access to mates is also quite common in humans. As you learned in Box 13.4, female–female reproductive conflict is predicted to be common in social species

(Rubenstein and Lovette 2007), including humans. Using data from Finland going back to the seventeenth century, a team led by Virpi Lummaa studied reproductive conflict among females in so-called joint-families, where brothers stayed on their natal farms and sisters married out so that several unrelated women of reproductive age co-resided in the same households (Pettay et al. 2016). By quantifying the effects of simultaneous reproduction among these women, the researchers found that the risk of offspring mortality before adulthood increased by nearly 25 percent if co-resident women reproduced within 2 years of each other.

In another demonstration of reproductive conflict among females, Jaimie Krens and colleagues showed women who were partnered to men photographs of other women taken during either their ovulatory or nonovulatory menstrual cycle phases (Krems et al. 2016). The researchers found that the partnered women reported intentions to render their male partners inaccessible to other women, but only when their partners were highly desirable to other women. Moreover, exposure to ovulating women also increased the women's sexual desire for their own partners, if they were highly desirable ones. Together, these results indicate that women recognize the

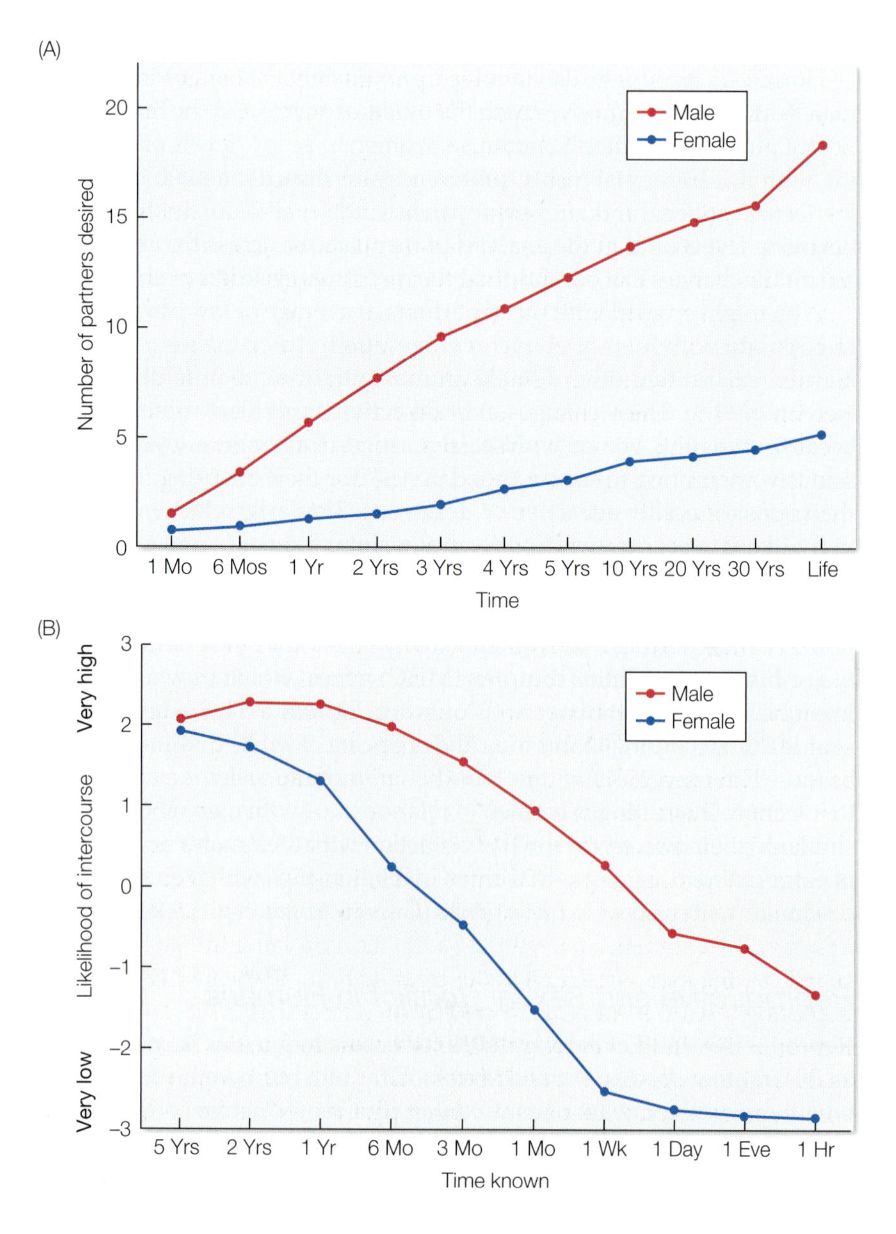

FIGURE 14.14 Sex differences in the desire for sexual variety. (A) Men and women differ in the number of sexual partners they say they would ideally like to have over different periods of time. (B) Men and women also differ in their estimates of the likelihood that they would agree to have sexual intercourse with an attractive member of the opposite sex after having known that individual for varying lengths of time. (After Buss and Schmitt 1993.)

potential for reproductive conflict and guard their mates in response to this perceived threat.

In addition to reproductive conflict, sexual conflict can be common in humans (Buss 2017). Indeed, because the mate preferences and genetic interests of men and women are not the same, we expect to find sexual conflict in our species, as we do in most other animals (see Chapter 9). Some major differences in reproductive behavior between men and women include their interest in the number or type of sexual partners. For example, men are much more likely to pay for sex with a prostitute or to view pornography than are women. Evolutionary psychologists have argued that the prostitution and pornography industries take advantage of the male (but not female) psyche (Symons 1979). David Buss and David Schmitt illuminated some of the differences in reproductive behavior between the sexes simply by asking a sample of college undergraduates how many sexual partners they would like to have over different periods of time. Unsurprisingly, the men in the study consistently wanted many more mates than did the women (**FIGURE 14.14A**). Moreover, when Buss and Schmitt asked individuals to evaluate the likelihood that they would be willing to have sex with a desirable potential mate after having known this person for periods ranging from 1 hour to 5 years, the differences between men and women were also dramatic (**FIGURE 14.14B**): "After knowing a potential mate for just 1 hour, men are slightly disinclined to consider having sex, but the disinclination is not strong. For most women, sex after just 1 hour is a virtual impossibility" (Buss and Schmitt 1993).

The typically greater enthusiasm of men compared with women for sexual variety is reflected in the results of another study conducted by Martie Haselton (Haselton 2003). Haselton asked roughly 100 undergraduate men and 100 undergraduate women whether they had had encounters with the opposite sex in which the other person evidently thought they were more (or less) interested sexually in this person than they actually were. Men reported about equal numbers of encounters during the preceding year in which women had "overperceived" and "underperceived" the males' romantic intentions. In contrast, women claimed that men were far more likely to think that they were sexually interested in them, when in fact the women were not, than to make the opposite mistake of underperception of sexual intent (**FIGURE 14.15**). This kind of bias was documented in another way by two social psychologists who sent an attractive young man and an attractive young woman on the following mission: they were to approach strangers of the opposite sex on a college campus, asking some of them, "Would you go to bed with me tonight?" Not one woman agreed to the proposition, but 75 percent of the men said yes, even though the men had known the woman in question for only a minute or less (Clark and Hatfield 1989). A study of French men and women produced similar results (Gueguen 2011), suggesting that this phenomenon is not unique to American college students.

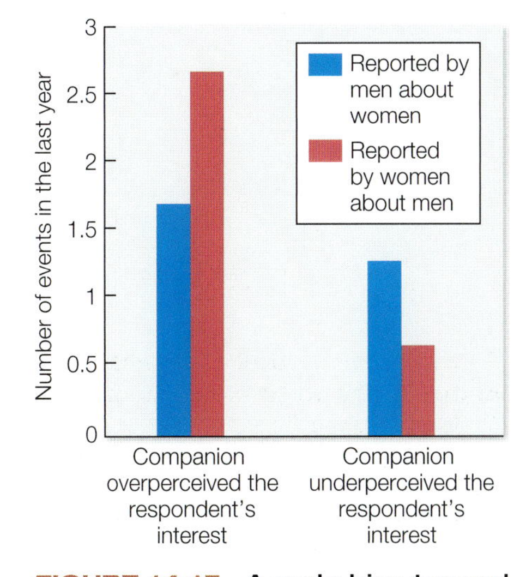

FIGURE 14.15 A male bias toward sexual overperception. Men tend to think that women have romantic inclinations toward them even when they do not. (After Haselton 2003.)

Clearly, men seem to be more keen to engage in sexual activity than women. When social psychologist Doug Kenrick surveyed a group of undergraduates about the minimum acceptable level of intelligence that they would require in a partner for interactions ranging from a first date to marriage, the men were far less demanding than women when it came to accepting a partner for a casual sexual encounter (**FIGURE 14.16**) (Kenrick et al. 1990). However, these results indicated something else about male and female sexual preferences. Although men were more willing than women to have casual sex with a partner with a low IQ, both men and women were similarly selective when considering requirements for a long-term partner. In other words, men and women diverge when it comes

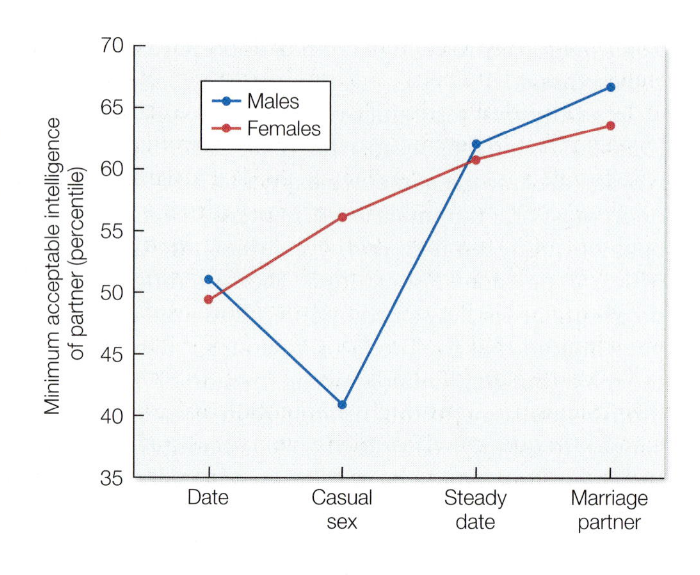

FIGURE 14.16 Sex differences in mate selectivity. College men differ from college women in the minimum intelligence they say they would require in a casual sexual partner. However, they have similar standards with respect to the minimum intelligence they say is essential for a marriage partner. "Intelligence" was scored on an IQ percentile scale such that a score of 50 meant that the acceptable individual had an IQ higher than half the population. (After Kenrick et al. 1990.)

to mate selectivity (casual sex), but converge when it comes to long-term partner selectivity (marriage).

Extreme Sexual Conflict in Humans: Polygamy and Extramarital Affairs

Another hallmark of sexual conflict in humans is deception, where men might exaggerate their feelings for a woman in order to have sex with her, or a woman might withhold sex to secure nonsexual resources from a man (Buss 2017). One significant source of sexual conflict arises if men are, on average, more interested in acquiring multiple sexual partners than are women. Although monogamous men may advance their reproductive success by helping one mate, polygamous men with several wives are likely to have more children than their monogamous counterparts. Polygamy has almost certainly been an option for men with substantial resources throughout our species' history, judging from the fact that the acquisition of several wives was culturally sanctioned in more than 80 percent of all preindustrial societies (Murdock 1967, Marlowe 2000). Yet despite being culturally sanctioned in many societies, polygamy has never been that common in human society. Even though the percentage of married men with more than one wife varied from 0 to 70 percent, on average only 12 percent (with a median of only 5 percent) of married men had more than one wife in these preindustrial societies (Marlowe 2000).

What are the fitness consequences for women paired with polygamous partners? Polygamy carries the risk for a female that her partner will divert resources from her and her children to another woman and her offspring. Throughout history, females of our species have probably tended to reproduce more successfully when they have had exclusive access to a husband's resources and parental care. In nineteenth-century Utah, for example, women who were married monogamously to relatively poor Mormon men had more surviving children on average (6.9 offspring) than women married to rich polygamous Mormons (5.5 offspring) (Heath and Hadley 1998). In fact, for every woman added to a polygamist's collection of wives, the reproductive success of the previous wives fell by about one child on average, further evidence that polygamy was probably not reproductively advantageous to Mormon women in the past (Moorad et al. 2011).

Even though polygamy is even rarer today than it once was, extramarital affairs also have the potential to increase the fitness of a male substantially, especially if his illegitimate children are cared for by his extramarital partners. In the United States, somewhere between 20 and 50 percent of married individuals engage in sexual infidelity (Buss 2016). Similarly, a survey of Australian students revealed that 28 percent of men and 22 percent of women engaged in extra-pair copulations (Simmons et al. 2004). However, extramarital activity has potential costs as well as benefits for males if a diversion of resources from existing children results from the men's pursuit of an additional partner (or two). Jeffrey Winking and his colleagues attempted to test whether this cost would moderate the polygamous tendencies of men in a traditional society, the Tsimane of Bolivia (Winking et al. 2007). The researchers predicted that if extramarital activity reduces the chances that existing offspring will achieve their maximum reproductive potential, due to loss of paternal investment, then the frequency with which men have sexual affairs will decline as a man and his primary mate have more children. Indeed, the predicted pattern does occur (**FIGURE 14.17**).

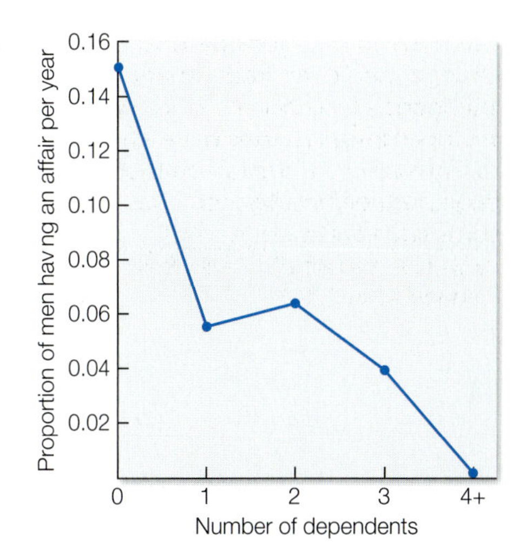

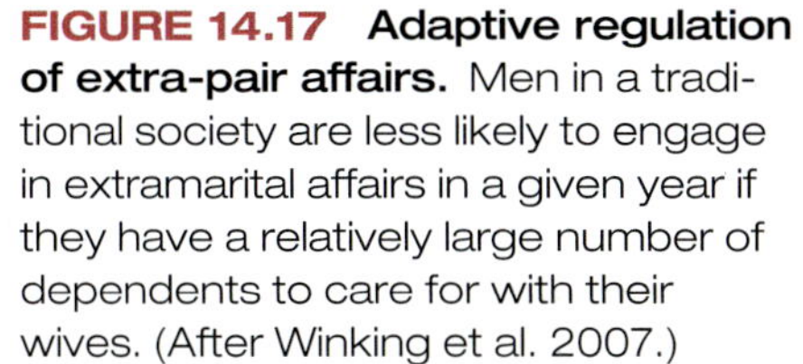
FIGURE 14.17 Adaptive regulation of extra-pair affairs. Men in a traditional society are less likely to engage in extramarital affairs in a given year if they have a relatively large number of dependents to care for with their wives. (After Winking et al. 2007.)

The potential benefits to men of polygamy and extramarital affairs increase the likelihood of conflict between husbands and wives. Yet, some women are receptive to extramarital affairs (in much the way females in many animal species are receptive to extra-pair copulations), which may enable them to acquire better genes for their offspring from their extra-pair partners. However, given that extra-pair paternity rates in humans are actually quite low, at less than 2 percent (Simmons et al. 2004, Anderson 2006, Boattini et al. 2015, Greeff and Erasmus 2015), there is little evidence to suggest that women are seeking good genes through the solicitation of extra-pair mates (Larmuseau et al. 2016). Instead, women may be seeking additional resources (things such as material goods or more protection), which requires only that they copulate with extra-pair males, not actually be fertilized by them. In addition to seeking resources from extramarital affairs, if a wealthy or powerful extra-pair partner becomes a woman's primary partner, she may be able to exchange a low-ranking partner for a socially superior one, with all the positive effects on her fitness that trading up affords (Schmitt et al. 2001), an idea referred to as the **mate switching hypothesis**. According to the mate switching hypothesis, partnered women are constantly assessing their partner's mate value, their own mate value, and the value (and interest) of potential mates (Buss et al. 2017). Extramarital affairs then serve as a way to assess potential alternative mates, as well as a form of mate insurance by securing a backup partner should a switch become necessary.

mate switching hypothesis Women may exchange a low-ranking partner for a superior one if it has positive effects on their reproduction or survival.

Given the potential for their mates to stray, men—just like many other male animals—have evolved ways of paternity assurance. Mate guarding, through vigilance and even violence, is one strategy that men use to reduce infidelity (Buss 2016). Additionally, because of the risk of lost paternity due to unfaithful wives, men are well known to pay special attention to the resemblance between themselves and their putative children. Wives are well aware of this interest, and they are often quick to suggest that their newborns look very much like their husbands (even when impartial judges detect a greater similarity between a baby's appearance and that of the mother) (Alvergne et al. 2007). Moreover, a father's similarity to his child (as judged by himself or by others) clearly influences his investment in that child (Apicella and Marlowe 2004, Alvergne et al. 2009).

Another result of the sexual conflicts between men and women may be a capacity for sexual jealousy, an emotional state that helps individuals prevent the fitness damage caused by an unfaithful mate. However, the nature of sexual jealousy should differ between the sexes because adulterous partners harm men and women in different ways. A wife whose husband acquires another partner loses some or all of her access to her husband's wealth, which she may need to support herself and her children. Therefore, from an ultimate perspective, a woman's sexual jealousy should be focused on the loss of a provider, which is more likely to occur when a man becomes emotionally involved with another woman. In contrast, a husband whose wife mates with another man may eventually care unknowingly for offspring fathered by that man. A man's sexual jealousy should therefore revolve around the potential loss of paternity and the misdirected parental investment arising from a wife's extramarital sexual activity rather than the potential loss of the other partner's resources (Daly et al. 1982).

If this view is correct, then if men and women were asked to imagine their responses to two scenarios—one in which a partner develops a deep friendship with another individual, and one in which a partner engages in sexual intercourse with another individual—women should find the first scenario more disturbing than the second, whereas men should be more upset at the thought of a mate copulating with another man. Data from several cultures confirm these predictions (Daly et al. 1982, Buss et al. 1992). For example, in a study involving Swedish university students, who live in a fairly sexually permissive culture, 63 percent of the women found the prospect of emotional infidelity more troubling, whereas almost exactly the same percentage of the men deemed sexual infidelity more upsetting (Wiederman and Kendall 1999).

Coercive Sex

In most modern societies, men aspire to monopolize or restrict sexual access to their mates, although they do not always succeed. Marriage institutionalizes these ambitions. Although one sometimes hears of societies in which complete sexual freedom is the norm, the notion that such cultures actually exist appears to have been a (wishful?) misinterpretation on the part of outside observers. Since sexual infidelity by a woman carries with it the risk of lost paternity (and fitness) for her mate, this risk appears to lie behind much sexual violence committed by men against their partners (Daly and Wilson 1992, Burch and Gallup 2004, Buss and Duntley 2011). In some societies, a woman known to have cuckolded her husband may be legally killed by her aggrieved mate (Daly et al. 1982). The fact that the murder of women suspected of adultery is still condoned in some parts of the world is one of the least attractive manifestations of sexual conflict in our species. Another component of male sexual behavior is the regular occurrence of forced copulation (rape) or coercive sex, a phenomenon that is not limited to human beings (**FIGURE 14.18**). Despite the fact that human rapists are often severely punished, coercive sex occurs in every culture studied to date (Thornhill and Palmer 2000).

Sociologists and psychologists have long been interested in the evolutionary basis of rape in human societies. There are generally three schools of thought on this controversial topic: (1) that rape is about the assertion of male power (**intimidation hypothesis**); (2) that it is an alternative male mating tactic for securing reproductive success (**alternative mating hypothesis**); and (3) that it is simply a maladaptive by-product of the male sexual psyche (**by-product hypothesis**) (**HYPOTHESES 14.1**). It is important to point out that the first hypothesis focuses on the proximate cause of rape, whereas the second is about the ultimate reproductive consequences. According to the levels-of-analysis approach we have employed throughout this book, these hypotheses are not mutually exclusive; it is possible for rapists to be motivated purely by a desire to hurt women and yet have children as a result of their aggression. Or as the third hypothesis suggests, the phenomenon may not even be adaptive at all.

HYPOTHESES 14.1

Non-mutually exclusive hypotheses for rape in humans

intimidation hypothesis
Rape allows men to intimidate and control women.

alternative mating hypothesis
Rape is an alternative male mating tactic for securing reproductive success.

by-product hypothesis
Rape is a maladaptive by-product of the male sexual psyche.

(A)

(B)

FIGURE 14.18 Rape occurs in animals other than humans. (A) In the beetle *Tegrodera aloga*, a male (right) can court a female (left) decorously by repeatedly drawing her antennae into grooves on his head; copulation ensues only if the female permits it. (B) Alternatively, a male (below) can force a female (above) to mate by running to her, grasping her, throwing her on her side, and inserting his everted genitalia as the female struggles to break free. (Photographs by John Alcock.)

In her book *Against Our Will*, Susan Brownmiller proposed the intimidation hypothesis and the idea that rapists act on behalf of all men to instill fear in all women to intimidate and control them (Brownmiller 1975). In its original formulation, the intimidation hypothesis implies that some males are willing to take the substantial punishment risks associated with rape to provide a benefit for many other men. This argument, however, suffers from all the logical problems inherent in "for the good of the group" hypotheses (with the added difficulty that groups composed of only one sex cannot be the focus of any realistic sort of group selection), but we can test it anyway (see Chapter 12 for a discussion of group selection). If the evolved function of the trait is to subjugate all women, then the rapist element in male society can be predicted to target older, dominant women (or young women who aspire to positions of power) to demonstrate the penalty that comes from stepping outside the traditional subordinate role. This prediction is not supported (Thornhill and Thornhill 1983). Although Brownmiller's version of the intimidation hypotheses was largely formulated around group selectionist arguments, the idea that rape is a means by which a man exerts dominance or power over a woman has merit. From the workplace to the White House and beyond, there is ample evidence in our society that sexual harassment, sexual assault, and rape are often perpetrated by men in positions of power.

As an alternative to the intimidation hypothesis, Randy and Nancy Thornhill proposed another hypothesis, which states that rape is an adaptive, alternative mating tactic for securing reproductive success (Thornhill and Thornhill 1983). As such, rape of strangers by men is analogous to forced copulation in *Panorpa* scorpionflies, in which males unable to offer nuptial gifts use the low-gain, last-chance tactic of trying to force females to copulate with them, or to forced extra-pair copulations in waterfowl (see Chapter 9). For this hypothesis to work, some victims of rape must become pregnant, which they do, even in modern societies in which many women take birth control pills (Thornhill and Palmer 2000). In fact, there is some evidence that copulatory rape may be more likely to result in pregnancy than consensual sex (Gottschall and Gottschall 2003). In the past, in the absence of reliable birth control technology and abortion procedures, rapists would have had a still higher probability of fathering children through forced copulation. Furthermore, if rape really is a product of an evolved reproductive mechanism, then rapists should more often target women of high fertility, just as bank swallows (*Riparia riparia*) and other birds identify fertile (egg-laying) females and try to force those individuals to copulate with them (Beecher and Beecher 1979, Sorenson 1994). An analysis of more than 300,000 sexual assaults in humans found that the modal age of victims was 15 years of age (Felson and Cundiff 2011). Whether this pattern was the result of greater access to adolescents (due to their degree of fertility) or because of some other factor remains open to debate. Finally, the notion that rape is about the assertion of male power or male aggression leads to the prediction that the age distribution of rape victims should be the same as that of women who are robbed (Felson and Cundiff 2011) or murdered (Thornhill and Thornhill 1983) by male assailants. However, crime data are at odds with this prediction: female victims of robbery who are also sexually assaulted are more likely to be between the ages of 15 and 29 than are female victims of robbery alone (**FIGURE 14.19**).

Although these findings suggest that rape could increase the fitness of some men, it is also entirely possible that rape is not adaptive at all. Instead, according to the by-product hypothesis, rape may be a maladaptive by-product of the male sexual psyche, which causes quick sexual arousal, a desire for variety in sexual partners, and an interest in impersonal sex, all attributes that generate many adaptive (or fitness-enhancing) consequences while also incidentally leading some men to rape some women (Thornhill and Palmer 2000). After all, men engage in many decidedly non-reproductive sexual activities, including masturbation, homosexual sex, and rape of postmenopausal women and prepubertal girls, just as males of many non-human species also exhibit sexual activity that cannot possibly result in offspring, such as the copulatory mounting of weaned pups by male northern elephant seals (*Mirounga angustirostris*) (Rose et al. 1991). Moreover, one attempt to estimate the reproductive consequences of rape for men in a traditional society yielded the conclusion that the fitness cost to a rapist exceeded the benefit by a factor of about ten (Smith et al. 2001), a result that strongly suggests rape is not an adaptation, at least not in that society. Even if coercive copulation usually reduces the fitness of its practitioners, the rape as by-product hypothesis would be tenable if the systems motivating male sexual behavior had a net positive effect on fitness. One prediction unique to the by-product hypothesis is that rapists will have unusually high levels of sexual activity with consenting as well as nonconsenting partners. Some evidence supports this prediction (Palmer 1991, Lalumiére et al. 1996), but as is true of many issues in human sexual behavior, more data are required. Nevertheless, in this case as in so many others, the adaptationist approach has generated novel hypotheses about a sensitive subject that are entirely testable in principle and practice. While no doubt controversial, the evolutionary perspective on rape is now available for skeptical scrutiny, and as a result, we may eventually gain a better understanding of the ultimate causes of the behavior. If and when we do, we will not in any way be obliged to

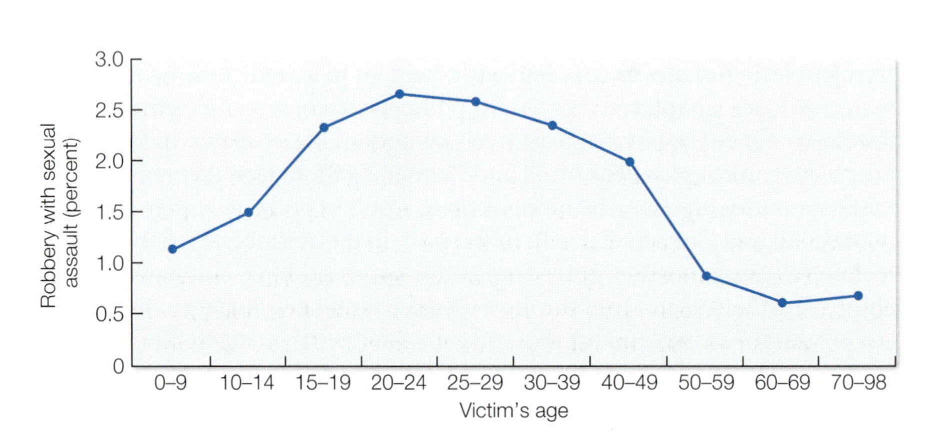

FIGURE 14.19 Testing alternative hypotheses for sexual assault. If rape were motivated purely by the intent to attack women and exert dominance, we would expect that the percentage of robbery victims who were sexually assaulted would not differ across age classes of the victims. Instead, women who were robbed *and* sexually assaulted are especially likely to be between 15 and 29 years of age. Crime statistics come from a database of 44,237 cases in which a lone male robbed a lone female. (After Felson and Cundiff 2011.)

be more understanding of the illegal and immoral activity of rapists (Thornhill and Palmer 2000), but we may generate new ideas about how to prevent rape.

Practical Applications of Behavioral Theory

Throughout this chapter, we have argued that most human behaviors can be viewed through the same lens as those of others animals. Indeed, numerous aspects of human behavior provide intriguing Darwinian puzzles that have attracted the attention of adaptationists. For example, in our society, some people adopt children who are unrelated to them, some restaurant-goers tip wait staff with whom they will never interact again, and some individuals go out of their way to punish social transgressors of various sorts, even though the punishers pay the price of doing the punishing and the rest of society derives the benefit without the cost. These cases would seem to be immune to the analysis of altruism based on kin selection theory because they occur exclusively among unrelated individuals, yet they occur frequently (but see West et al. 2011). In addition, there are cultural differences in human behavior that cannot be prescribed to genetic differences, such as the requirement in some places that a bride bring a dowry to her husband or his family upon the bride's marriage, while in other groups the husband-to-be must ante up a bridewealth payment to the family of his bride-to-be. Recall from our discussion of the Mukogodo of Kenya in Chapter 11 that these cultural differences in marriage practice have societal implications that are consistent with evolutionary thinking, including sex-ratio theory. Indeed, the view that these differences are the arbitrary effects of cultural norms has been debunked. Instead, much like genetically-influenced behaviors, culturally transmitted behaviors can be passed down family lines, spread quickly within the population, and lead to higher reproductive success (Hartung 1982, Gaulin and Boster 1990).

Although evolutionary biologists and psychologists are beginning to address and dissect many of these puzzles as they apply to human behavior, many students of the field might wonder if we can go beyond simply gaining an understanding of human behavior and actually apply an evolutionary approach to improve human livelihoods. Indeed, there are many areas where understanding the evolutionary principles that underlie human behaviors might help us improve society. For example, consider how examining resource use and competition in humans could influence the well-being of our planet. Not only do humans desire large quantities of goods to provision our offspring adequately, but many of us want to acquire wealth and use it ostentatiously, a key component of the reproductive strategy of men, or to marry someone who is rich, a common wish of unmarried women (Miller 2009). These factors and others have led the average resident of the United States to need all the resources that can be "harvested" from about 10 hectares of land. Since there are only about 12 billion hectares of productive land on the planet, and more than 300 million citizens in the United States, our single nation accounts for one-fourth of that entire amount. If all of the 8 billion people in the world were to achieve a standard of living equal to that of the average American, we would need 80 billion hectares of productive land to do the job. Clearly, that is not going to happen. Even now we are eating our planet alive in an unsustainable manner, destroying the capital that our descendants will need (Rees 2002).

Perhaps taking an evolutionary approach to consider how individuals compete for and optimally use resources to provide for their family versus use them to attract mates or signal dominance might someday help deal with overpopulation and resource use. An evolutionary approach provides a predictive approach with which to study human behavior. But can we actually use this approach to improve human livelihoods right now? That is, can thinking about adaptation or comparing our own behaviors to those of other species enhance our own well-being? Conservation and sustainable development might be topics where this approach could be useful in the long term, but one area of science that has already received a great deal of attention in this regard and is already paying off is medicine.

Evolutionary Medicine

Like any other trait, the health status of all individuals is the product of their evolutionary history as it interacts with their current environment. By considering why behaviors or responses to pathogens evolved, some researchers believe that we can better treat a myriad of diseases, pathologies, and symptoms (Gluckman et al. 2016). Indeed, adaptationist thinking has influenced how some physicians approach treating a range of disorders and symptoms such as asthma, allergy, fever, and coughing. Similarly, considering life history trade-offs in reproductive function versus somatic investment in health has been suggested to improve public health efforts to reduce the risk of hormone-sensitive cancers and other noncommunicable diseases (Jasieńska et al. 2017). In fact, "Darwinian," or "evolutionary," medicine as this approach has become known, has proven so illuminating that some medical schools are now requiring that students take a class in this area. Here we focus on just two topics related to evolutionary medicine: obesity and autism.

OBESITY Obesity has reached epidemic proportions in many developed nations. According to the U.S. Centers for Disease Control and Prevention, as of 2017 one in three adults and one in six children in America are obese, defined as having too much body fat. Obese individuals are more likely to suffer from diabetes, coronary heart disease, hypertension, and other maladies. The health costs attributed to obesity alone are in the hundreds of millions of dollars annually (Wolf and Colditz 1998). Since the increased occurrence of obesity in modern society is the result of both environmental change and hereditary predisposition from our evolutionary past, three evolutionary explanations have been proposed to explain this pattern (**HYPOTHESES 14.2**) (Speakman 2013). First, although overeating may have been positively selected for in the past because having additional calories may have increased our ancestor's fitness, it is no longer adaptive today (Lev-Ran 2001). According to the **thrifty gene hypothesis**, individuals carrying so-called thrifty genes that enabled the efficient storage of energy as fat between famines would have been at a selective advantage. This ability to store energy in the form of body fat may have prepared people for periods of energy shortfall in the same way that fat accumulation does for hibernating mammals and migratory birds (Speakman 2013). Yet recent genetic data fail to support the thrifty gene hypothesis. For example, molecular studies suggest that hundreds and even thousands of genes influence obesity, which would make it unlikely that selection occurs only during periods of food or energy shortage (Heberbrand et al. 2010). Moreover, some of the genetic variants related to obesity occur in nonhuman primates that do not exhibit obesity (Southam et al. 2010).

If obesity is not adaptive, could it simply be maladaptive but favored today as a by-product of selection on some other trait, such as the ability of brown adipose tissue (BAT) to burn off excess caloric intake? BAT is unique to mammals and is related to thermogenesis

HYPOTHESES 14.2

Alternative hypotheses for obesity in humans. The first two hypotheses suggest (mal) adaptive explanations at the ultimate level, whereas the third is a nonadaptive hypothesis resulting from neutral evolutionary processes.

thrifty gene hypothesis Ancestrally, individuals carrying genes that promoted efficient storage of energy as fat would have been at a selective advantage during times of food scarcity, but today such genes may make obesity more likely and are thus maladaptive.

indirect selection hypothesis Obesity is maladaptive but favored today as a by-product of selection on some other trait.

drifty gene hypothesis Mutations in the genes that predispose humans to obesity are neutral and have drifted over evolutionary time.

(heat production). According to this **indirect selection hypothesis**, BAT varies among individuals as a result of selection from living in different thermal environments, then only some individuals would be able to use it secondarily to burn off excess energy intake (Speakman 2013). However, the genes related to obesity are unrelated to BAT function (Speakman 2013), and knocking out a gene related to BAT function in mice that produce a protein called UCP1 did not lead to greater obesity in those individuals when fed a high-fat diet (Enerback et al. 1997).

If it is neither adaptive nor maladaptive, could obesity simply be due to genetic drift? According to the **drifty gene hypothesis**, most mutations in the genes that predispose us to obesity are neutral and have been drifting over evolutionary time, perhaps causing some people to be obesity prone and others to be obesity resistant (Speakman 2013). In fact, genetic studies find little evidence of strong selection on obesity, which is consistent with the drifty gene hypothesis. However, lack of evidence of positive selection on obesity-related genes does not necessarily mean that drift is the cause. It simply means that additional studies are needed to study the evolutionary basis of obesity. Importantly, failure to find strong support for any of these hypotheses does not negate the evolutionary approach to medicine. Instead, it just means we have more work to do before we can understand the evolutionary history of obesity. After all, our fondness for sugar and fat, rare substances in past diets, has an evolutionary foundation and in some way contributes to our enthusiasm for foods that are associated with obesity today.

AUTISM Another area of medicine that may benefit greatly from an evolutionary approach is the study of autism. Autism, or autism spectrum disorder, refers to a range of conditions characterized by challenges with social skills as well as speech and nonverbal communication. According to the U.S. Centers for Disease Control and Prevention, autism affects at least 1 in 45 children in the United States. At its core, autism is characterized by a lack of the capacity to engage in reciprocal social interactions. Despite millions of dollars spent annually on autism research, scientists still do not understand the disorder. Since autism has been a challenge for scientists to study in humans, some researchers have argued for studying similar conditions in other animal species. For example, a team led by Gene Robinson used a comparative genomics approach to determine whether superficial behavioral similarities between humans and the honey bee *Apis mellifera* in how they respond to social stimuli reflect shared molecular mechanisms (Shpigler et al. 2017). The researchers reasoned that since honey bees are highly social like humans, with a complex unspoken language, social behavior, and kin recognition, they might also share similar social deficits as humans. After demonstrating that bees that respond differently to social stimuli (resident intruders and nurturing interactions between adults and larvae) have different neurogenomic signatures, Robinson's team then probed the hereditary basis of social unresponsiveness. By comparing the list of genes they found to be related to social unresponsiveness in honey bees to those genes related to autism in humans, the researchers found significant overlap among the suites of genes. They concluded that there is deep conservation of genes implicated in autism in humans and those associated with social responsiveness in honey bees. Ultimately, this study demonstrates that social responsiveness may be an important phenotype related to autism, and suggests that to better understand how autism works in humans, we must study how more typical social behaviors grade smoothly into severe autism (Crespi 2017). Perhaps most important, this work illustrates how the comparative approach can help yield new insights into human medical disorders. This work does not end with honey bees, as other researchers are using the knowledge gained from studying the neurobiological basis of social attachment in voles (see Chapter 3) to help develop novel treatments for autism (Young and Barret 2015).

The Triumph of an Evolutionary Analysis of Human Behavior

Although we and other behavioral biologists see great utility in using an adaptive framework to study human behavior, the evolutionary explanations for elements of human behavior have not always been well received, especially by social scientists. The hostility to an evolutionary analysis of human behavior by a component of the biological community reached a peak soon after the 1975 publication of E. O. Wilson's *Sociobiology: The New Synthesis* (Wilson 1975). The controversy surrounding sociobiology, now generally known as behavioral ecology or evolutionary psychology, has largely faded, but certainly not entirely (Buss and Pinker 2009). Critiques include the idea that "we humans don't do things just because we want to raise our inclusive fitness." Indeed, although humans have a great many desires, the wish to maximize our inclusive fitness rarely, if ever, is at the top of our list (Rose 1998). While this may be true, if a baby cuckoo could talk, for example, it would not tell you that it rolled its host's eggs out of the nest so it could propagate as many copies of its own genes as possible. Neither cuckoos nor humans need be aware of the ultimate reasons for their activities in order to behave adaptively. If specific strategies are reproductively beneficial, then they will be favored by natural selection over time.

Still other critics of taking an evolutionary approach to understand human behavior might claim that "not all human behavior is biologically adaptive." Over the years, many people have claimed that an array of cultural practices, such as circumcision, prohibitions against eating perfectly edible foods, and the use of birth control devices to limit fertility, seem unlikely to advance individual fitness. If some humans do things that reduce their fitness, these people argue, then an adaptationist approach to human behavior must be invalid. This claim is based on the belief that natural selection theory requires that every aspect of every organism must be currently adaptive (Gould and Lewontin 1979). This assumption is incorrect, in part because some current environments of humans are different in many respects from the environments in which we evolved. Indeed, the goal of most behavioral biologists is not to assert that a behavior of interest is an adaptation but instead to identify evolutionary (or Darwinian) puzzles, produce plausible hypotheses, and test alternative ones by generating testable predictions taken from those possible explanations. Just as we showed earlier in this chapter in our discussions of rape and obesity, there is no guarantee that any selectionist hypothesis will withstand testing, which is as it should be. If evolutionary hypotheses about the adaptive value of a human behavior are incorrect, effective tests will tell us so, which enables us to remove some ideas from the table. After all, testing hypotheses simply entails using a possible explanation to generate predictions (or expectations) that can be tested, with the goal of eliminating hypotheses whose predictions do not cut the mustard. When one hypothesis fails to be supported, we can generate and test new hypotheses, an approach that is simply part of the scientific method.

As you have seen throughout this book, a hypothesis that a behavioral ability is adaptive does not mean that the characteristic is developmentally fixed. All biologists now understand that development is an interactive process involving both genes and environment. Change the environment, and you will change the gene–environment interactions underlying a behavioral phenotype, with the result that the phenotype may change. Our own environment has changed greatly over evolutionary time. From the forests and then savannas of East Africa, to the colder climates of Europe, to the modern cities in which many of us live today, the human environment has changed continually over the past few million years. And what about the social environment? If, for example, you change the cultural environment that a baby is exposed to so that, say, a child of Spanish-speaking parents is raised by adults who speak Urdu, you know what will happen. But if infants are to become language users, they need very special genes, especially those genes that code for proteins that promote the development and maintenance of the brain elements that underlie language learning. Natural selection has surely had a role to play in the evolution of these mechanisms, especially as our social and ecological environments have changed over the millennia. This is an explanatory hypothesis, not a claim about developmental inflexibility or a statement about the moral desirability of "natural" phenomena.

One thing should be clear to everyone by now, whether an opponent or proponent of an adaptationist approach to human behavior, namely that evolutionary analyses have demonstrated beyond doubt that just because something is "adaptive" or "natural" or "evolved" does not mean that it is "good" or "moral" or "desirable." Once you know that our genes have the capacity to make us work on their behalf without regard for our welfare or that of most other people, then it should be easier to fight against our evolved impulses (Williams 1996). Knowledge of evolutionary outcomes could, to name just one thing, make us (a vehicle) less susceptible to exploitation by unscrupulous individuals (Cialdini 2001, Colarelli and Dettmann 2003). Informed persons could also better avoid the moral certainty and self-righteousness that enable us to demonize and dehumanize our opponents prior to attacking and killing them. When Pogo declared "We have met the enemy and he is us," he was correct, and those of us who understand the role of natural selection in shaping our evolutionary history know why he was right. Perhaps some of you can use that understanding to help make the human species less an enemy to itself. At the very least, using this approach will allow us to understand the basic principles that guide the evolution of life all around us.

SUMMARY

1. Human beings are an animal species. Therefore, all four levels of analysis can be used to identify the causes of our behavior. All the neural and endocrine mechanisms in our bodies develop as a result of the continuous interplay between our genes and the environments in which they operate. The way in which our brains work provides the proximate foundation for our behavior. Some genetic differences lead to neurophysiological differences among people, with reproductive consequences for individuals. As a result of behavioral differences among people, some genes are passed on to the next generation more than others, leading to the evolution of our species and our behavior.
2. The adaptationist approach is the primary focus of animal behavior, behavioral ecology, sociobiology, evolutionary psychology, evolutionary anthropology, or whatever we want to label it. Researchers in these disciplines use natural selection theory to generate testable hypotheses about the possible adaptive value of our species' behavior. These hypotheses are designed to *explain* why people do what they do, *not* to justify these actions as moral or desirable.
3. The utility of the adaptationist approach for understanding the ultimate causes of human behavior is especially apparent in the studies that have been done on mate choice. Sexual selection theory leads us to suspect that men and women will differ in how they evaluate potential mates. Women probably focus on male traits ranging from skill in verbal courtship to indicators of wealth to evidence of parental potential, whereas men instead prefer the characteristics associated with high fertility in women, such as a youthful appearance, a low waist-to-hip ratio, and an intermediate body weight.
4. The hypothesis that our evolved psychological mechanisms increased the fitness of individuals living in the pre-contraceptive past is open to test. These tests have led some researchers to consider that we possess adaptive mechanisms that lead to behavior almost universally considered immoral and undesirable, such as violent conflict between the sexes and coercive sex, as well as conditions that we consider unhealthy, such as obesity.

(Continued)

SUMMARY *(continued)*

5. Applying an evolutionary approach that considers adaptive function or compares humans with other species has the potential to improve human well-being. One example of this is the burgeoning field of evolutionary medicine, which argues that considering why behaviors evolved might improve our ability to treat disease and other health disorders.
6. When a claim is made that something is adaptive or natural, this only means that the something tends to propagate our genes. Widespread acceptance of this point might enable us to understand that trusting our impulses is more likely to help us pass on our genes than to produce wise decisions that maximize either our personal happiness or the general good.

COMPANION WEBSITE

Go to **ab11e.com** for discussion questions and all of the audio and video clips.

Glossary

A

action potential A neural signal that transmits information via a self-regenerating change in membrane electrical charge that travels the length of a nerve cell, sometimes triggering further action potentials in adjacent nerve cells. [4]

activational effect An immediate but temporary regulatory effect of a hormone. Also see *organizational effect*. [5]

acute stressor Short-term event than can directly and rapidly affect an animal's behavior by temporarily elevating glucocorticoid levels. [5]

adaptation A characteristic that confers higher inclusive fitness to individuals than any existing alternative exhibited by other individuals in the population. An adaptation is a trait that has spread, is spreading, or is being maintained in a population as a result of natural selection. [1]

adaptive value The contribution that a trait or gene makes to inclusive fitness. [1]

allele A form of a gene. Different alleles typically code for distinctive variants of the same enzyme. [1]

alternative hypotheses When only one of a series of competing hypotheses could explain a given behavior. [1]

alternative mating strategy A type of behavioral polymorphism in which a mating behavior has a strong genetic component. Different forms of the behavior may be seen in different individuals (i.e., the behavior is polymorphic), but the form is fixed throughout a given individual's life. [9]

altricial Referring to young that reside in the nest for an extended period of time. [11]

altruism Cooperative behavior that lowers the donor's reproductive success while increasing the reproductive success of the recipient of the altruistic act. [12]

amplitude Intensity of a waveform. [2]

ancestral state reconstruction The recreation of traits of extinct species. Requires extrapolation from measured characteristics of extant species back in time to their common ancestors. [2]

animal personality Individual behaviors that are repeatable through time. [13]

anisogamy The fusion of two gametes that differ greatly in size. [9]

aposematism Warning coloration, in which an organism's highly visible or vivid coloration signals to potential predators that it is distasteful or dangerous. Also see *Batesian mimicry; Müllerian mimicry*. [6]

armaments Elaborate morphological traits that have been selected for because they can act as weapons in intrasexual battles. [9]

associated reproductive pattern The onset of reproductive behavior is tightly correlated with seasonal cues that trigger changes in circulating hormones and the gonads. Also see *dissociated reproductive pattern*. [5]

B

badges of status Signals that reveal information about an individual's size or dominance status. Although badge of status signals are typically not costly to produce, they may be costly to maintain due to social enforcement, a phenomenon referred to as a *maintenance cost*. [8]

Bateman's principle Because males achieve greater reproductive success they tend to have more mates than do females. As a result, males tend to have higher reproductive variance than females. [9]

Batesian mimicry When an edible species resembles a distasteful or dangerous one. [6]

behavioral polymorphism When behavioral phenotypes become relatively more influenced by genetic factors than by the environment. [3]

behavioral syndromes Behavioral consistency within individuals across contexts. [13]

bottom-up forces When food and other resources influence behavioral decisions. Also see *top-down forces*. [7]

C

carotenoid A type of dietary pigment that provides coloration to animal tissues and scavenges free radicals such as antioxidants that have been linked to disease and aging. [8]

central dogma of molecular biology Information encoded in DNA is transcribed into RNA, which in turn is translated into a protein. [3]

central pattern generator A group of cells in an organism's nervous system that produces a particular pattern of signals necessary for a functional behavioral response. [4]

chronic stressor Prolonged event that results in chronically elevated glucocorticoid levels, leading to negative physiological effects. [5]

circadian clock An internal oscillator modulated by external cues such as sunlight or temperature that regulates physiological processes. [4]

circadian rhythm A roughly 24-hour cycle of behavior that expresses itself independent of environmental changes. [5]

circannual rhythm An annual cycle of behavior that expresses itself independent of environmental changes. [5]

citizen science data Observations recorded by the public and collated by scientists to test hypotheses. [6]

coefficient of relatedness (r) The probability that an allele present in one individual will be present in a close relative; the proportion of the total genotype of one individual present in the other as a result of shared ancestry. [12]

coevolutionary arms race When two parties in conflict exert reciprocal selection pressure on each other, with an adaptive advance made by one often leading eventually to an adaptive counterresponse by the other. [11]

cognition The mental process of acquiring knowledge and understanding through thought, experience, and the senses. [4]

collective behavior The synchronized movements of individuals following a series of basic interaction rules. [13]

command center A neural cluster or integrated set of clusters that has primary responsibility for the control of a particular behavioral activity. [4]

communication The transfer of information from one individual (sender) to another (receiver) that affects current or future behavior and the fitness of one or both individuals. [8]

comparative approach An approach to behavioral biology that involves using comparisons among species that have evolved independently to study relationships among traits or historical and physical constraints on trait evolution. [1]

conditional strategy with alternative mating tactics The genetically based capacity of an individual to use different mating tactics under different environmental conditions; the inherited behavioral capacity to be flexible in response to certain cues or situations. More simply called *conditional mating tactics*. [9]

conspecific (intraspecific) brood parasitism When an animal exploits the parental care of individuals of their own species. [11]

convenience polyandry A form of polyandry in which a female will struggle with a male but acquiesces to his mating attempts in order to save time and energy. [10]

convergent evolution The independent acquisition over time through natural selection of similar characteristics in two or more unrelated species. Also see *divergent evolution*. [1]

cooperative breeding A social system in which more than two individuals care for young. [13]

cryptic female choice The ability of a female that receives sperm from more than one male to choose which sperm get to fertilize her eggs. [9]

cue An unintentional transfer of information between a sender and a receiver. [8]

D

Darwinian puzzle A trait that is maintained in a population even though it appears to reduce the fitness of individuals that possess it. Traits of this sort attract the attention of evolutionary biologists. [1]

deceitful signaling A form of communication in which a sender uses a specially evolved signal to manipulate the behavior of a receiver such that the sender receives a fitness benefit but the receiver pays a fitness cost. Sometimes referred to as *manipulation*. [8]

density-dependent habitat selection When settlement decisions are influenced by the intensity of intraspecific competition as reflected by the density of conspecifics in a location. [7]

descent with modification The foundational idea of evolution, that individuals pass varied genetic traits to their offspring, and that differences in reproductive success among individuals in a population cause a population or species to evolve over time. [1]

developmental homeostasis The capacity of developmental mechanisms within individuals to produce adaptive traits, despite potentially disruptive effects of mutant genes and suboptimal environmental conditions. [3]

developmental plasticity The ability to respond (often with changes in neural connections) to environmental cues through the adjustment of genotypic expression during early development. [3]

dilution effect When associating in groups makes it less likely that any one individual will be depredated. [6]

diplodiploidy Having two sets of the chromosomes, and therefore two copies of genes (one from a mother and from a father). Synonymous with *diploidy*. Also see *haploidy*. [12]

diploidy Having two sets of the chromosomes, and therefore two copies of genes (one from a mother and from a father). Synonymous with *diplodiploidy*. Also see *haploidy*. [12]

direct benefits Material benefits provided by a male that can increase a female's fitness. Examples include parental care, access to resources, safety from predators, and reduced harassment by other males. [9]

direct fitness A measure of the reproductive (genetic) success of an individual based on the number of its offspring that live to reproduce. Also see *inclusive fitness*, *indirect fitness*. [12]

dispersal The permanent movement from the birthplace to somewhere else. [7]

dissociated reproductive pattern The onset of reproductive behavior is apparently not triggered by a sharp change in circulating hormones. Also see *associated reproductive pattern*. [5]

divergent evolution The evolution by natural selection of differences between closely related species that live in different environments and are therefore subject to different selection pressures. Also see *convergent evolution.* [1]

DNA methylation Chemical modifications (typically the addition of methyl groups) to DNA molecules that influence gene expression. [3]

dominance hierarchy Social ranking within a group, in which some individuals give way to others, often conceding useful resources without a fight. [9]

E

eavesdropping The detection of signals from a legitimate signaler by an illegitimate receiver, to the detriment of the signaler and the benefit of the receiver. [8]

economic defensibility The trade-off in costs versus benefits for maintaining a territory. [7]

entrain To reset a biological clock so that an organism's activities are scheduled in keeping with local conditions. [5]

epigenetic modification Alterations to the genome that do not change the DNA sequence. Examples include *DNA methylation.* [3]

ethology The study of the proximate mechanisms and adaptive value of animal behavior is a discipline founded by Niko Tingerben and Konrad Lorenz. [4]

eusociality Describes species with overlapping generations, cooperative care of young, and reproductive division of labor where many individuals in a group are permanently sterile. [12]

evolutionarily stable strategy (ESS) That set of rules of behavior that when adopted by a certain proportion of the population cannot be replaced by any alternative strategy. [6]

evolutionary constraint A limitation or restriction on adaptive evolution. [1]

evolutionary game theory An evolutionary approach to the study of adaptive value in which the payoffs to individuals associated with one behavioral tactic are dependent on what the other members in the group are doing. [6]

evolutionary genetics A field of study that examines how DNA leads to evolutionary change by comparing sequence variation among species or individuals that exhibit different traits. [3]

evolutionary history The sequence of changes that have occurred over time as an ancestral trait becomes modified and takes on a new form (and sometimes, a new function). Also see *phylogeny.* [1]

experimental approach An approach to behavioral biology that involves manipulating features of the animal or its environment to more directly establish a causal relationship among traits. [1]

explosive breeding assemblage The temporary formation of a large group of mating individuals. [10]

extra-pair copulation A mating by a male or female with an individual other than his or her partner in a socially monogamous species. [9]

F

facultative siblicide Behavior that occasionally results in an individual killing a sibling or siblings. [11]

fitness A measure of the genes contributed to the next generation by an individual. Often stated in terms of the number of surviving offspring produced by the individual, technically called *direct fitness*. Also see *inclusive fitness; indirect fitness.* [1]

fitness benefit The positive effect of a trait on an individual's reproductive (and genetic) success. [1]

fitness cost The negative effect of a trait on an individual's reproductive (and genetic) success. [1]

fitness payoff The fitness gain or loss from a social interaction or other behavior. [13]

fixed action pattern (FAP) An innate, highly stereotyped response that is triggered by a well-defined, simple stimulus. Once the pattern is activated, the response is performed in its entirety. [4]

forward genetics The identification of the genes responsible for a given phenotype. [3]

free-running cycle The cycle of activity of an individual that is expressed in a constant environment. [5]

frequency The rate at which amplitude of a stimulus, such as sound, increases and decreases. [2]

frequency-dependent selection A form of natural selection in which those individuals that happen to belong to the less common of two types in the population are the ones that are more fit because of their lower frequency in the population. Can be negative (the fitness of a phenotype decreases as it becomes more common) or positive (the fitness of a phenotype increases as it becomes more common) in nature. [6]

functional genomics A field of study that examines the relationship between genotype and phenotype by comparing gene or protein expression in individuals or species that exhibit different traits. [3]

fundamental asymmetry of sex Males produce small sperm (and many of them) and females produce large eggs (and relatively few of them), based on the premise that sperm are energetically "cheaper" to produce than eggs. [9]

G

gene A segment of DNA, typically one that encodes information about the sequence of amino acids that make up a protein. [1]

genetic monogamy A form of monogamy in which a male and female form a pair-bond and mate only with each other. [10]

genotype The genetic constitution of an individual. Can refer either to the specific alleles of one gene possessed by the individual or to the individual's complete set of genes. [1]

group augmentation Individuals survive or reproduce better by living in larger groups. [13]

group selection The process that occurs when groups differ in their collective attributes and these differences affect the survival chances of the group. [12]

H

habitat saturation Occurs when territories or breeding sites are limiting to a population because most of the best sites are already occupied by other members of that species. [13]

Hamilton's rule The argument made by William D. Hamilton that altruism can spread through a population when $rB > C$ (where r is the coefficient of relatedness between the altruist and the individual helped, B is the fitness benefit received by the helped individual, and C is the cost of altruism in terms of the direct fitness lost by the altruist due to its actions). [12]

haplodiploidy A sex-determination system in which males develop from an unfertilized egg and are haploid (have one set of chromosomes), whereas females develop from a fertilized egg and are diploid (have two sets of chromosomes). [12]

haploidy Having one set of the chromosomes, and therefore one copy of each gene. Also see *diploidy* and *diplodiploidy*. [12]

histone modification A type of epigenetic modification to histone proteins that affects gene expression by altering chromatin structure or recruiting histone modifiers. [3]

honest signaling A form of communication in which both the sender and the receiver obtain a fitness benefit. [8]

honest signal Signal that indicates the quality of the sender, facilitates decision making by a receiver, and cannot be easily produced by the sender. [8]

hypothalamic–pituitary–adrenal (HPA) axis A cascade in which corticotropin-releasing hormone (CRH) secreted from the hypothalamus stimulates the release of adrenocorticotropic hormone (ACTH) from the anterior pituitary gland, which in turn leads to the production of glucocorticoids in the adrenal cortex. [5]

hypothalamic–pituitary–gonadal (HPG) axis A cascade of reproductive hormones in which gonadotropin-releasing hormone (GnRH) produced in the hypothalamus triggers the release of luteinizing hormone from the pituitary, and eventually the production of testosterone (in males) and estrogen (in females) in the gonads. [5]

hypothesis A possible, even speculative, explanation that is used as a starting point for further examination via testing predictions taken from the hypothesis. [1]

I

ideal free distribution theory A theoretical framework that enables behavioral biologists to predict what animals should do when choosing between alternative habitats of different quality in the face of competition for space, food, or other critical resources. [7]

imprinting A form of learning in which individuals exposed to certain key stimuli early in life form an association with an object or individual and may later attempt to mate with similar objects. [3]

inclusive fitness A total measure of an individual's contribution of genes to the next generation generated by both direct fitness (derived from reproduction) and indirect fitness (which depends on social interactions with relatives). [12]

indirect (genetic) benefit Genetic benefit that does not benefit a female directly but does increase the fitness of her offspring. [9]

indirect fitness A measure of the genetic success of an altruistic individual based on the number of relatives (or genetically similar individuals) that the altruist helps reproduce that would not otherwise have survived to do so. Also see *direct fitness, inclusive fitness*. [12]

indirect reciprocity A form of reciprocity in which a helpful action is repaid at a later date by individuals other than the recipient of assistance. [13]

infanticide The intentional killing of offspring, usually by a parent. [11]

innate releasing mechanism A conceptual neural mechanism thought to control an innate response to a sign stimulus. [4]

instinct A behavioral pattern that reliably develops in most individuals, promoting a functional response to a releaser stimulus the first time the behavior is performed. [4]

Institutional Animal Care and Use Committee (IACUC) An institutional ethics committee whose permission is required for research studies on nonhuman vertebrates. [14]

Institutional Review Board (IRB) An institutional ethics committee whose permission is required for research studies on humans. [14]

interactive theory of development The development of behavioral traits requires both genetic information and environmental inputs. [3]

interneuron A nerve cell- that relays signals either from sensory receptor neurons (e.g., touch receptors, odor receptors, light receptors) to the central nervous system (a sensory interneuron) or from the central nervous system to neurons commanding muscle cells (a motor interneuron). [4]

intersexual selection Sexual selection usually involving male courtship behavior or appearance that influences a female's choice of mate. [9]

interspecific brood parasitism When an animal exploits the parental care of individuals of another species. [11]

intrasexual selection Sexual selection in which members of the same sex compete for access to mates. [9]

K

kin recognition The ability to distinguish between close genetic kin and non-kin. [13]

kin selection A type of natural selection that favors the reproductive success of an organism's relatives, even at a cost to the organism's own fitness. [12]

kleptoparasitism Stealing food from an individual that has caught or collected it. [13]

L

landscape of fear The spatially explicit elicitation of fear in prey when cues in the environment (such as odors, alarm calls) lead to the perceived risk of predation. [6]

lek A traditional display site that females visit to select a mate from among the males displaying at their small, resource-free territories. [10]

levels of analysis The proximate (developmental and physiological) and ultimate (historical and adaptive evolutionary) causes of a behavior. [1]

load-lightening Helpers reduce the workload of parents in offspring care. [13]

M

maintenance cost A cost associated with the socially enforced maintenance of a signal. [8]

major histocompatibility complex (MHC) A group of cell surface proteins critical for the immune system to recognize foreign molecules. Allelic diversity in the MHC leads to a more robust immune system and thus may have a selective advantage. [9]

marginal value theorem A type of optimality model that predicts that an animal should leave a foraging patch when its rate of food intake in that patch drops below the average rate for the habitat, and that this marginal capture rate should be equalized over all patches within a habitat. [6]

maternal effect Where an individual's phenotype is determined not only by the environment it experiences and its genotype, but also by the environment and genotype of its mother. [3]

melanin A type of color-producing pigment that animals produce in tissues such as skin and hair. The two main forms of melanin are eumelanin, which produces brown and black coloration, and pheomelanin, which produces red and yellow coloration. [8]

migration The regular movement back and forth between two relatively distant locations by animals that use resources concentrated in these different sites. [7]

migratory connectivity The movement of individuals between summer and winter populations, including the stopover sites between the breeding and wintering grounds. [7]

monogamy A mating system in which one male mates with just one female, and one female mates with just one male in a breeding season. Also see *genetic monogamy; social monogamy*. [10]

monogynous Among social insects, refers to a colony that has only a single reproductive queen. [3]

Müllerian mimicry When two or more distasteful or dangerous species resemble one another. [6]

multilevel selection Selection that acts at both the group and the individual levels, originating with genes and progressing through cells, then organisms, and finally groups of organisms. According to the advocates of this form of selection, at higher levels, populations, multispecies communities, and even whole ecosystems can be subject to selection. [12]

multimodal signaling The use of multiple traits to signal to the same or different individuals. [8]

mutual benefit A type of social behavior when both interacting individuals receive a fitness benefit. [13]

N

natural selection The process that occurs when individuals differ in their hereditary traits and the differences are correlated with differences in reproductive success. Natural selection can produce evolutionary change. Also see *group selection; kin selection; sexual selection*. [1]

neural circuit A group of interconnected neurons that are able to regulate their own activity using a feedback loop. [4]

neuron A nerve cell. [4]

neurotransmitter A chemical signal that diffuses from one nerve cell to another across a synapse. [4]

non-mutually exclusive hypotheses When multiple hypotheses could apply to a given behavior. [1]

nuclei Dense clusters of neurons within central nervous systems. [2]

nuptial gift A food item transferred by a male to a female just prior to or during copulation. [9]

O

obligate siblicide Behavior that always results in an individual killing a sibling or siblings. [11]

observational approach An approach to behavioral biology that involves watching animals behaving in nature or in the lab. [1]

operant conditioning A kind of learning based on trial and error, in which an action, or operant, becomes more frequently performed if it is rewarded. [3]

operational sex ratio The ratio of sexually active males to sexually receptive females in a population. [9]

optimal foraging theory A model that predicts how an animal should behave when searching for food. [6]

optimality theory An evolutionary theory based on the assumption that the attributes of organisms are optimal; that is, the attributes present in the organism are better than other alternatives in terms of the ratio of fitness benefits to costs. The theory is used to generate hypotheses about the possible adaptive value of traits in terms of the net fitness gained by individuals that exhibit these attributes. [6]

organizational effect A relatively permanent effect of a hormone during development that causes changes to physiology and behavior. Also see *activational effect*. [5]

ornament An elaborate morphological trait that has apparently been selected for because it attracts mates. [9]

oscillogram A graph of amplitude of an acoustical stimulus as a function of time. [2]

P

parent–offspring conflict Evolutionary conflict arising from differences in optimal parental investment in an offspring from the standpoint of the parent versus that of the offspring. [11]

parental investment Costly parental activities that increase the likelihood of survival for some existing offspring but that may reduce the parent's chances of producing offspring in the future. [9]

parsimony The principle that the simplest explanation that fits the evidence is likely to be the correct one. [1]

parthenogenesis Development from an unfertilized egg. [12]

phenotype Any measurable aspect of an individual's body or behavior that arises from an interaction of the individual's genes with its environment. [1]

photoperiod The number of hours of light in a 24-hour period. [5]

phylogeny An evolutionary genealogy of the relationships among species or clusters of species; a representation of the evolutionary history of taxa. [1]

pigment A type of molecule that differentially absorbs and emits wavelengths of visible light. [8]

plural breeding A form of cooperative breeding system in which social groups contain more than one breeder of at least one sex. [13]

polyandry A mating system in which a female has several partners in a breeding season. Also see *convenience polyandry*. [10]

polygamy The human practice of having more than one wife or husband simultaneously. [11]

polygynandry A mating system in which both males and females have several partners with whom they form a pair-bonds. [10]

polygynous Among social insects, refers to a colony that has multiple reproductive queens. [3]

polygyny A mating system in which a male fertilizes the eggs of several partners in a breeding season. [10]

polyphenism The occurrence within a species of two or more alternative phenotypes whose differences are induced by key differences in the environments experienced by individual members of the species. [3]

precocial Referring to young that are mobile soon after hatching. [11]

predator swamping The movement of migratory individuals together in high densities to confuse predators or reduce the predation risk to the migratory individuals. Analagous to the dilution effect, where associating in groups makes it less likely that any one individual will be depredated. [7]

prediction An expectation that should follow if a hypothesis is true. [1]

preexisting bias An existing bias in an animal's sensory system that detects some features of an organism's world better than others. [8]

preexisting trait An existing behavioral, physiological, or morphological characteristic that already provides an informative cue to receivers. If the sender benefits from the receiver's response, the cue can be modified into a signal via a process called ritualization. [8]

prisoner's dilemma A game theory construct in which the fitness payoffs to individuals are set such that mutual cooperation between the players generates a lower return than defection (which occurs when one individual accepts assistance from the other but does not return the favor). [13]

production cost A cost associated with the production of an energetically expensive signal. [8]

promiscuity A mating system in which both males and females have several partners without forming a pair-bonds. [10]

proximate cause An immediate, underlying reason for why a behavior is the way it is based on the operation of internal mechanisms possessed by an individual. [1]

R

reciprocal altruism A helpful action that is repaid at a later date by the recipient of the assistance. Also known as *reciprocity*. Also see *indirect reciprocity*. [13]

releaser A sign stimulus given by an individual as a social signal to another individual. [4]

reproductive conflict Conflict among members within a social group over which individuals get to breed, as well as conflict over the resources that individuals need in order to breed.

reproductive skew The unequal partitioning of reproductive success within a population or social group. [9]

reproductive success The number of surviving offspring produced by an individual; direct fitness. [1]

reproductive value A measure of the probability that a given offspring will reach the age of reproduction, or the potential of an individual to leave surviving descendants in the future. [11]

resource-holding potential The inherent capacity of an individual to defeat others when competing for useful resources. [7]

resource selection A method for characterizing the distribution of a species using the known spatial distribution of its resources. [7]

reverse genetics The determination of which phenotypes will arise as a result of particular genetic sequences.

Often achieved through experimental manipulation of the genetic code. [3]

S

scientific method Observation, measurement, and experimentation to test hypotheses by seeing if the predictions produced by those hypotheses are correct. If not, hypotheses can be refined to undergo repeated testing. [1]

selection pressure An agent of differential reproduction or survival that causes a population to change genetically. [1]

selfish herd A group of individuals whose members use others as living shields against predators. [6]

selfishness When the donor benefits from a social interaction but the recipient does not. Sometimes called *deceit* or *manipulation*. [13]

sensory drive The fine-tuning of signals to work effectively in a particular environment. [8]

sensory exploitation The evolution of signals that activate established sensory systems of signal receivers in ways that elicit responses favorable to the signal sender. [8]

sex role reversal A change in the typical behavior patterns of males and females, as when, for example, females compete for access to males or when males choose selectively among potential mates. [9]

sex-biased dispersal When individuals of one sex disperse farther than those of the other. [7]

sexual arms race Conflict between males and females over mating as males evolve traits used in manipulating or forcing females to copute, and females evolve traits that help circumvent this (often violent) process. [9]

sexual conflict Conflict between males and females over mate choice. [9]

sexual dimorphism Difference in appearance between males and females of the same species. [9]

sexual selection A form of natural selection that acts on traits used to compete for mates with others of the same sex, or to attract members of the opposite sex in order to mate with them. [9]

siblicidal behavior Behavior that results in an individual killing a sibling or siblings. Also see *obligate siblicide; facultative siblicide*. [11]

sibling conflict Evolutionary conflict resulting from divergent interests among current and/or future siblings. [11]

sign stimulus The effective component of an action or object that triggers a fixed action pattern in an animal. [4]

signal A specially evolved message that contains information. [8]

singular breeding A form of cooperative breeding system in which social groups contain only one breeder of each sex. [13]

social monogamy A form of monogamy in which a male and female form a pair-bond but one or both sexes also may mate outside of the pair-bond. [10]

spectrogram A visual representation of sound frequencies over time. In a spectrogram, intensity is shown using color or grayscale. Sometimes referred to as a *sonogram*. [2]

sperm competition Competition among the sperm of different males that determines whose gametes will fertilize a female's eggs. [9]

spermatophore Often a type of nuptial gift given by a male to a female that contains both nutrients and sperm. [8]

spite When neither the donor nor the recipient benefits from a social interaction, but instead both pay a cost. [13]

stimulus Generally a sensory signal capable of triggering a complex behavior. [4]

stimulus filtering The capacity of nerve cells and neural networks to ignore stimuli that could potentially elicit a response from them. [4]

structural colors Colors produced by light interacting physically with the nanometer-scale arrangement of tissues and air. [8]

supergene A region of DNA containing many linked genes that can influence the development of a behavioral phenotype. [3]

syllable A distinct element of a vocalization, including birdsong. [2]

synapse The point of near-contact between one nerve cell and another. [4]

T

territory An area in which individuals exhibit a readiness to defend against intruders. [7]

Tinbergen's four questions A scheme developed by Nikolaas Tinbergen to address the proximate and ultimate causes of a behavior. Within the proximate level of analysis fall questions relating to development and to internal mechanism, and within the ultimate level of analysis fall questions relating to evolutionary history and adaptive function. [1]

top-down forces When predation risk influences behavioral decisions. Also see *bottom-up forces*. [7]

trait A character or feature of an organism. Also see *phenotype*. [1]

transcriptome All of the mRNA being expressed in a given tissue at given time point. [3]

U

ultimate cause An evolutionary, historical reason for why a behavior is the way it is. [1]

W

waveform Graphical representation of the change in pressure over time that composes a sound. Also see *amplitude; frequency*. [2]

worker policing Among eusocial insects, when a worker female eats or removes eggs that have been laid by other workers rather than those laid by a queen. [12]

References

Chapter 1

Alcock, J., Jones, C. E., and Buchmann, S. L. 1976. Location before emergence of female bee, *Centris pallida*, by its male (Hymenoptera: Anthophoridae). *Journal of Zoology* 179: 189–199.

Alcock, J. and Sherman, P. W. 1994. On the utility of the proximate-ultimate dichotomy in biology. *Ethology* 96: 58–62.

Beehner, J. C. and Bergman, T. J. 2008. Infant mortality following male takeovers in wild geladas. *American Journal of Primatology* 70: 1152–1159.

Borries, C., Launhardt, K., Epplen, C., Epplen, J. T., and Winkler, P. 1999. DNA analyses support the hypothesis that infanticide is adaptive in langur monkeys. *Proceedings of the Royal Society of London B* 266: 901–904.

Cullen, E. 1957. Adaptations in the kittiwake to cliff nesting. *Ibis* 99: 275–302.

Curtin, R. and Dolhinow, P. 1978. Primate social behavior in a changing world. *American Scientist* 66: 468–475.

Darwin, C. 1859. *On the Origin of Species*. London: Murray.

Dawkins, R. 1989. *The Selfish Gene*. Oxford: Oxford University Press.

Dobzhansky, T. 1973. Nothing makes sense except in the light of evolution. *American Biology Teacher* 35: 125–129.

Hofmann, H. A., Renn, S. C. P., and Rubenstein, D. R. 2016. Introduction to symposium. New frontiers in the integrative study of animal behavior: Nothing in neuroscience makes sense except in the light of behavior. *Integrative and Comparative Biology* 56: 1192–1196.

Holekamp, K. E. and Sherman, P. W. 1989. Why male ground squirrels disperse. *American Scientist* 77: 232–239.

Hoogland, J. L. and Sherman, P. W. 1976. Advantages and disadvantages of bank swallow (*Riparia riparia*) coloniality. *Ecological Monographs* 46: 33–58.

Hrdy, S. B. 1977. Infanticide as a primate reproductive strategy. *American Scientist* 65: 40–49.

Knörnschild, M., Ueberschaer, K., Helbig, M., and Kalko, E. K. V. 2011. Sexually selected infanticide in a polygynous bat. *PLoS One* 6: e25001.

Kruuk, H. 1964. Predators and anti-predator behaviour of the black-headed gull *Larus ridibundus*. *Behaviour Supplements* 11: 1–129.

Levitis, D. A., Lidicker, W. Z., and Freund, G. 2009. Behavioural biologists do not agree on what constitutes behaviour. *Animal Behaviour* 78: 103–110.

Lyon, J. E., Pandit, S. A., van Schalk, C. P., and Pradhan, G. R. 2011. Mating strategies in primates: A game theoretical approach to infanticide. *Journal of Theoretical Biology* 274: 103–108.

Massaro, M., Chardine, J. W., and Jones, I. L. 2001. Relationships between black-legged kittiwake nest site characteristics and susceptibility to predation by large gulls. *Condor* 103: 793–801.

Owings, D. H. and Coss, R. G. 1977. Snake mobbing by California ground squirrels: Adaptive variation and ontogeny. *Behaviour* 62: 50–69.

Phillips, R. A., Furness, R. W., and Stewart, F. M. 1998. The influence of territory density on the vulnerability of Arctic skuas *Stercorarius parasiticus* to predation. *Biological Conservation* 86: 21–31.

Reeve, H. K. and Sherman, P. W. 2001. Optimality and phylogeny. In S. H. Orzack and E. Sober (eds.), *Adaptationism and Optimality*, pp. 45–63. Cambridge: Cambridge University Press.

Rubenstein, D. R. and Hofmann, H. A. 2015. Editorial overview: The integrative study of animal behavior. *Current Opinion in Behavioral Sciences* 6: v–viii.

Rundus, A. S., Owings, D. S., Joshi, S. S., Chinn, E. and Giannini, N. 2007. Ground squirrels use an infrared signal to deter rattlesnake predation. *Proceedings of the National Academy of Sciences USA* 104: 14372–14374.

Sommer, V. 1987. Infanticide among free-ranging langurs (*Presbytis entellus*) of Jodhpur (Rajasthan/India): Recent observations and a reconsideration of hypotheses. *Primates* 28: 163–197.

Sordahl, T. A. 2004. Field evidence of predator discrimination abilities in American Avocets and Black-necked Stilts. *Journal of Field Ornithology* 75: 376–386.

Tinbergen, N. 1959. Comparative studies of the behaviour of gulls (Laridae): A progress report. *Behaviour* 15: 1–70.

Tinbergen, N. 1963. On the aims and methods of ethology. *Zeitschrift für Tierpsychologie* 20: 410–433.

Chapter 2

Anderson, R. C., Peters, S., and Nowicki, S. 2014. Effects of early auditory experience on the development of local song preference in female swamp sparrows. *Behavioral Ecology and Sociobiology* 68: 437–447.

Araya-Salas, M. and Wright, T. 2013. Open-ended song learning in a hummingbird. *Biology Letters* 9: 20130625.

Baker, M. C., Bottjer, S. W., and Arnold, A. P. 1984. Sexual dimorphism and lack of seasonal changes in vocal control regions of the white-crowned sparrow brain. *Brain Research* 295: 85–89.

Baker, M. C. and Cunningham, M. A. 1985. The biology of bird-song dialects. *Behavioral and Brain Sciences* 8: 85–133.

Ballentine, B., Hyman, J., and Nowicki, S. 2004. Vocal performance influences female response to male bird song: An experimental test. *Behavioral Ecology* 15: 163–168.

Baptista, L. F. and Morton, M. L. 1988. Song learning in montane white-crowned sparrows: From whom and when. *Animal Behaviour* 36: 1753–1764.

Baptista, L. F. and Petrinovich, L. 1984. Social interaction, sensitive phases, and the song template hypothesis in the white-crowned sparrow. *Animal Behaviour* 32: 172–181.

Beecher, M. D. and Brenowitz, E. A. 2005. Functional aspects of song learning in songbirds. *Trends in Ecology & Evolution* 20: 143–149.

Beecher, M. D. and Campbell, E. 2005. The role of unshared song in the singing interactions between neighbouring song sparrows. *Animal Behaviour* 70: 1297–1304.

Beecher, M. D., Campbell, E., and Nordby, J. C. 2000a. Territory tenure in song sparrows is related to song sharing with neighbours, but not to repertoire size. *Animal Behaviour* 59: 29–37.

Beecher, M. D., Campbell, S. E., Burt, J. M., Hill, C. E., and Nordby, J. C. 2000b. Song-type matching between neighbouring song sparrows. *Animal Behaviour* 59: 21–27.

Beecher, M. D., Stoddard, P. K., Campbell, S. E., and Horning, C. L. 1996. Repertoire matching between neighbouring song sparrows. *Animal Behaviour* 51: 917–923.

Bonaparte, K. M., Riffle-Yokoi, C., and Burley, N. T. 2011. Getting a head start: Diet, sub-adult growth, and associative learning in a seed-eating passerine. *PLoS One* 6: e23775.

Bottjer, S. W., Miesner, E. A., and Arnold, A. P. 1984. Forebrain lesions disrupt development but not maintenance of song in passerine birds. *Science* 224: 901–903.

Bradbury, J. W. and Vehrencamp, S. K. 2011. *Principles of Animal Communication*, 2nd ed. Sunderland, MA: Sinauer Associates.

Braindard, M. S. and Douple, A. J. 2000. Interruption of a basal ganglia–forebrain circuit prevents plasticity of learned vocalizations. *Nature* 404: 762–766.

Brenowitz, E. A. 1991. Evolution of the vocal control system in the avian brain. *Seminars in the Neurosciences* 3: 399–407.

Brenowitz, E. A., Lent, K., and Kroodsma, D. E. 1995. Brain space for learned song in birds develops independently of song learning. *Journal of Neuroscience* 15: 6281–6286.

Brenowitz, E. A., Margoliash, D., and Nordeen, K. W. 1997. An introduction to birdsong and the avian song system. *Journal of Neurobiology* 33: 495–500.

Buchanan, K. L. and Catchpole, C. K. 2000. Song as an indicator of male parental effort in the sedge warbler. *Proceedings of the Royal Society of London B* 267: 321–326.

Burt, J. M., Campbell, S. E., and Beecher, M. D. 2001. Song type matching as threat: A test using interactive playback. *Animal Behaviour* 62: 1163–1170.

Cardoso, G. C., Mota, P. G., and Depraz, V. 2007. Female and male serins (*Serinus serinus*) respond differently to derived song traits. *Behavioral Ecology and Sociobiology* 61: 1425–1436.

Catchpole, C. K. and Slater, P. J. B. 2008. *Bird Song: Biological Themes and Variations*. Cambridge, England: Cambridge University Press.

Chen, Y., Clark, O., and Woolley, S. C. 2017. Courtship song preferences in female zebra finches are shaped by developmental auditory experience. *Proceedings of the Royal Society of London B* 284: 20170054.

Chen, Y., Matheson, L. E., and Sakata, J. T. 2016. Mechanisms underlying the social enhancement of vocal learning in songbirds. *Proceedings of the National Academy of Sciences USA* 113: 6641–6646.

Chilton, G., Lein, M. R., and Baptista, L. F. 1990. Mate choice by female white-crowned sparrows in a mixed-dialect population. *Behavioral Ecology and Sociobiology* 27: 223–227.

Chilton, G. M., Baker, M. C., Barrentine, C. D., and Cunningham, M. A. 1995. White-crowned sparrow (*Zonotrichia leucophrys*), version 2.0. In P. G. Rodewald (ed.), *The Birds of North America*. Ithaca, NY: Cornell Lab of Ornithology.

Derryberry, E. P. 2007. Evolution of bird song affects signal efficacy: An experimental test using historical and current signals. *Evolution* 61: 1938–1945.

DeWolfe, B. B., Baptista, L. F., and Petrinovich, L. 1989. Song development and territory establishment in Nuttall's white-crowned sparrow. *Condor* 91: 297–407.

Engesser, S., Crane, J. M. S., Savage, J. L., Russell, A. F., and Townsend, S. W. 2015. Experimental evidence for phonemic contrasts in a nonhuman vocal system. *PLoS Biology* 13: e1002171.

Farrell, T. M., Weaver, K., An, Y. S., and MacDougall-Shackleton, S. A.
2012. Song bout length is indicative of spatial learning in European starlings. *Behavioral Ecology* 23: 101–111.

Farries, M. A. 2001. The oscine song system considered in the context of the avian brain: Lessons learned from comparative neurobiology. *Brain, Behavior and Evolution* 58: 80–100.

Gentner, T. Q. and Hulse, S. H. 2000. European starling preference and choice for variation in conspecific male song. *Animal Behaviour* 59: 443–458.

Gobes, S. M. H. and Bolhuis, J. J. 2007. Birdsong memory: A neural dissociation between song recognition and production. *Current Biology* 17: 789–793.

Hackett, S. J., Kimball, R. T., Reddy, S., Bowie, R. C. K., Braun, E. L., et al. 2008. A phylogenomic study of birds reveals their evolutionary history. *Science* 320: 1763–1768.

Harbison, H., Nelson, D. A., and Hahn, T. P. 1999. Long-term persistence of song dialects in the mountain white-crowned sparrow. *Condor* 101: 133–148.

Hofmann, H. A., Renn, S. C. P., and Rubenstein, D. R. 2016. Introduction to symposium. New frontiers in the integrative study of animal behavior: Nothing in neuroscience makes sense except in the light of behavior. *Integrative and Comparative Biology* 56: 1192–1196.

Hunter, M. L. and Krebs, J. R. 1979. Geographic variation in the song of the great tit (*Parus major*) in relation to ecological factors. *Journal of Animal Ecology* 48: 759–785.

Jackendoff, R. 1999. Possible stages in the evolution of the language capacity. *Trends in Cognitive Science* 3: 272–279.

Jarvis, E. D., Mirarab, S., Aberer, A. J., Houde, P., Li, C., et al. 2014. Whole-genome analyses resolve early branches in the tree of life of modern birds. *Science* 346: 1320–1331.

Jarvis, E. D., Ribeiro, S., da Silva, M. L., Ventura, D., Vielliard, J., and Mello, C. V. 2000. Behaviourally driven gene expression reveals song nuclei in hummingbird brain. *Nature* 406: 628–632.

Keen, S. C., Meliza, C. D., and Rubenstein, D. R. 2013. Flight calls signal group and individual identity but not kinship in a cooperatively breeding bird. *Behavioral Ecology* 24: 1279–1285.

Konishi, M. 1965. The role of auditory feedback in the control of vocalization in the white-crowned sparrow. *Zeitschrift für Tierpsychologie* 22: 770–783.

Konishi, M. 1985. Birdsong: From behavior to neurons. *Annual Review of Neuroscience* 8: 125–170.

Kroodsma, D., Hamilton, D., Sánchez, J. E., Byers, B. E., Fandiño-Mariño, H., et al. 2013. Behavioral evidence for song learning in the suboscine bellbirds (*Procnias* spp.; Cotingidae). *Wilson Journal of Ornithology* 125: 1–14.

Kroodsma, D. E. and Konishi, M. 1991. A suboscine bird (eastern phoebe, *Sayornis phoebe*) develops normal song without auditory feedback. *Animal Behaviour* 42: 477–487.

Kroodsma, D. E., Liu, W. C., Goodwin, E., and Bedell, P. A. 1999a. The ecology of song improvisation as illustrated by North American sedge wrens. *The Auk* 116: 373–386.

Kroodsma, D. E., Sánchez, J., Stemple, D. W., Goodwin, E., da Silva, M. L., and Vielliard, J. M. E. 1999b. Sedentary life style of neotropical sedge wrens promotes song imitation. *Animal Behaviour* 57: 855–863.

Leighton, G. M. 2017. Cooperative breeding influences the number and type of vocalizations in avian lineages. *Proceedings of the Royal Society of London B* 284: 20171508.

Leitner, S., Nicholson, J., Leisler, B., DeVoogd, T. J., and Catchpole, C. K. 2002. Song and the song control pathway in the brain can develop independently of exposure to song in the sedge warbler. *Proceedings of the Royal Society of London B* 269: 2519–2524.

Lipshutz, S. E., Overcast, I. A., Hickerson, M. J., Brumfield, R. T., and Derryberry, E. P. 2017. Behavioural response to song and genetic divergence in two subspecies of white-crowned sparrows (*Zonotrichia leucophrys*). *Molecular Ecology* 26: 3011–3027.

Luther, D. and Baptista, L. 2010. Urban noise and the cultural evolution of bird songs. *Proceedings of the Royal Society of London B* 277: 469–473.

MacDonald, I. F., Kempster, B., Zanette, L., and MacDougall-Shackleton, S. A. 2006. Early nutritional stress impairs development of a song-control brain region in both male and female juvenile song sparrows (*Melospiza melodia*) at the onset of song learning. *Proceedings of the Royal Society of London B* 273: 2559–2564.

MacDougall-Shackleton, S. A. 2011. The levels of analysis revisited. *Philosophical Transactions of the Royal Society B* 366: 2076–2085.

MacDougall-Shackleton, E. A., Derryberry, E. P., and Hahn, T. P. 2002. Nonlocal male mountain white-crowned sparrows have lower paternity and higher parasite loads than males singing local dialect. *Behavioral Ecology* 13: 682–689.

MacDougall-Shackleton, E. A., Hulse, S. H., and Ball, G. F. 1998. Neural bases of song preferences in female zebra finches (*Taeniopygia guttata*). *Neuroreport* 9: 3047–3052.

Marler, P. 1970. A comparative approach to vocal learning: Song development in white-crowned sparrows. *Journal of Comparative and Physiological Psychology Monograph*, 71: 1–25.

Marler, P. and Tamura, M. 1964. Culturally transmitted patterns of vocal behavior in sparrows. *Science* 146: 1483–1486.

McDonald, P. G. 2012. Cooperative bird differentiates between the calls of different individuals, even when vocalizations were from completely unfamiliar individuals. *Biology Letters* 8: 365–368.

Mello, C. V. and Ribeiro, S. 1998. Zenk protein regulation by song in the brain of songbirds. *Journal of Comparative Neurology* 383: 426–438.

Mooney, R., Hoese, W., and Nowicki, S. 2001. Auditory representation of the vocal repertoire in a songbird with multiple song types. *Proceedings of the National Academy of Sciences USA* 98: 12778–12783.

Moorman, S., Mello, C. V., and Bolhuis, J. J. 2011. From songs to synapses: Molecular mechanisms of birdsong memory. *Bioessays* 33: 377–385.

Nealen, P. M. and Perkel, D. J. 2000. Sexual dimorphism in the song system of the Carolina wren *thryothorus ludovicianus*. *Journal of Comparative Neurology* 418: 346–360.

Nelson, D. A., Hallberg, K. I., and Soha, J. A. 2004. Cultural evolution of Puget Sound white-crowned sparrow song dialects. *Ethology* 110: 879–908.

Nicholls, J. A. and Goldizen, A. W. 2006. Habitat type and density influence vocal signal design in satin bowerbirds. *Journal of Animal Ecology* 75: 549–558.

Nordby, J. C., Campbell, S. E., and Beecher, M. D. 1999. Ecological correlates of song learning in song sparrows. *Behavioral Ecology* 10: 287–297.

Nottebohm, F. and Arnold, A. P. 1976. Sexual dimorphism in vocal control areas of songbird brain. *Science* 194: 211–213.

Nowicki, S. and Searcy, W. A. 2004. Song function and the evolution of female preferences: Why birds sing, why brains matter. *Annals of the New York Academy of Sciences* 1016: 704–723.

Nowicki, S. and Searcy, W. A. 2014. The evolution of vocal learning. *Current Opinion in Neurobiology* 28: 48–53.

Nowicki, S., Hasselquist, D., Bensch, S., and Peters, S. 2000. Nestling growth and song repertoire size in great reed warblers: Evidence for song learning as an indicator mechanism in mate choice. *Proceedings of the Royal Society of London B* 267: 2419–2424.

Nowicki, S., Peters, S., and Podos, J. 1998. Song learning, early nutrition, and sexual selection in birds. *American Zoologist* 38: 179–190.

Nowicki, S., Searcy, W. A., and Peters, S. 2002a. Brain development, song learning, and mate choice in birds: A review and experimental test of the "nutritional stress hypothesis." *Journal of Comparative Physiology A* 188: 1003–1114.

Nowicki, S., Searcy, W. A., and Peters, S. 2002b. Quality of song learning affects female response to male birdsong. *Proceedings of the Royal Society of London B* 269: 1949–1954.

Odom, K. J., Hall, M. L., Riebel, K., Omland, K. E., and Langmore, N. E. 2014. Female song is widespread and ancestral in songbirds. *Nature Communications* 5: 3379.

Olkowicz, S., Kocourek, M., Lucan, R. K., Portes, M., Fitch, W. T., et al. 2016. Birds have primate-like numbers of neurons in the forebrain. *Proceedings of the National Academy of Sciences USA* 113: 7255–7260.

Ölveczky, B. P., Andalman, A. S., and Fee, M. S. 2005. Vocal experimentation in the juvenile songbird requires a basal ganglia circuit. *PLoS Biology* 3: e153.

Pfaff, J. A., Zanetter, L., MacDougall-Shackleton, S. A., and MacDougall-Shackleton, E. A. 2007. Song repertoire size varies with hvc volume and is indicative of male quality in song sparrows (*Melospiza melodia*). *Proceedings of the Royal Society of London B* 274: 2035–2040.

Pfenning, A. R., Hara, E., Whitney, O., Rivas, M. V., Wang, R., et al. 2014. Convergent transcriptional specializations in the brains of humans and song-learning birds. *Science* 346: 1256846.

Poesel, A., Fries, A., Miller, L., Gibbs, H. L., Soha, J. A., and Nelson, D. A. 2017. High levels of gene flow among song dialect populations of the Puget sound white-crowned sparrow. *Ethology* 123: 581–592.

Poesel, A. and Nelson, D. A. 2012. Delayed song maturation and territorial aggression in a songbird. *Biology Letters* 8: 369–371.

Prum, R. O., Berv, J. S., Dornburg, A., Field, D. J., Townsend, J. P., et al. 2015. A comprehensive phylogeny of birds (Aves) using targeted next-generation DNA sequencing. *Nature* 526: 569–573.

Rowley, I. and Chapman, G. 1986. Cross-fostering, imprinting, and learning in two sympatric species of cockatoos. *Behaviour* 96: 1–16.

Scharff, C. and Nottebohm, F. 1991. A comparative study of the behavioral deficits following lesions of various parts of the zebra finch song system: Implications for vocal learning. *Journal of Neuroscience* 11: 2896–2913.

Schmidt, K. L., Moore, S. D., MacDougall-Shackleton, E. A., and MacDougall-Shackleton, S. A. 2013. Early-life stress affects song complexity, song learning, and volume of the brain nucleus RA in adult male song sparrows. *Animal Behaviour* 86: 23–35.

Sibley, C. G., Ahlquist, J. E., and Monroe, Jr., B. L. 1988. A classification of the living birds of the world based on DNA-DNA hybridization studies. *The Auk* 105: 409–423.

Sockman, K. W., Salvante, K. G., Racke, D. M., Campbell, C. R., and Whitman, B. A. 2009. Song competition changes the brain and behavior of a male songbird. *Journal of Experimental Biology* 212: 2411–2418.

Sockman, K. W., Sewall, K. B., Ball, G. F., and Hahn, T. P. 2005. Economy of mate attraction in the Cassin's finch. *Biology Letters* 1: 34–37.

Soha, J. A., Nelson, D. A., and Parker, P. G. 2004. Genetic analysis of song dialect populations in Puget Sound white-crowned sparrows. *Behavioral Ecology* 15: 636–646.

Spencer, K. A., Wimpenny, J. H., Buchanan, K. L., Lovell, P. G., Goldsmith, A. R., and Catchpole, C. K. 2005. Developmental stress affects the attractiveness of male song and female choice in the zebra finch (*Taeniopygia guttata*). *Behavioral Ecology and Sociobiology* 58: 423–428.

Suh, A., Paus, M., Kiefmann, M., Churakov, G., Franke, F. A., et al. 2011. Mesozoic retroposons reveal parrots as the closest living relatives of passerine birds. *Nature Communications* 2: 443.

Suzuki, T. N., Wheatcroft, D., and Griesser, M. 2017. Wild birds use an ordering rule to decode novel call sequences. *Current Biology* 27: 2331–23336.

Toews, D. P. L. 2017. From song dialects to speciation in white-crowned sparrows. *Molecular Ecology* 26: 2842–2844.

Touchton, J. M., Seddon, N., and Tobias, J. A. 2014. Captive rearing experiments confirm song development without learning in a tracheophone suboscine bird. *PLoS One* 9: e95746.

Wang, N., Braun, E. L., and Kimball, R. T. 2012. Testing hypotheses about the sister group of the Passeriformes using an independent 30-locus data set. *Molecular Biology and Evolution* 29: 737–750.

Warrington, M. H., McDonald, P. G., and Griffith, S. 2015. Within-group vocal differentiation of individuals in the cooperatively breeding apostlebird. *Behavioral Ecology* 26: 493–501.

Weckstein, J. and Zink, R. M. 2001. Anomalous variation in mitochondrial genomes of white-crowned (*Zonotrichia leucophrys*) and golden-crowned (*Z. atricapilla*) sparrows: Pseudogenes, hybridization, or incomplete lineage sorting? *The Auk* 118

Wilbrecht, L., Crionas, A., and Nottebohm, F. 2002. Experience affects recruitment of new neurons but not adult neuron number. *Journal of Neuroscience* 22: 825–831.

Williams, H., Levin, I. I., Norris, D. R., Newman, A. E. M., and Wheelwright, N. T. 2013. Three decades of cultural evolution in savannah sparrow songs. *Animal Behaviour* 85: 213–223.

Woodgate, J. L., Leitner, S., Catchpole, C. K., Berg, M. L., Bennett, A. T., and Buchanan, K. L. 2011. Developmental stressors that impair song learning in males do not appear to affect female preferences for song complexity in the zebra finch. *Behavioral Ecology* 22: 566–573.

Woolley, S. C. and Doupe, A. J. 2008. Social context-induced song variation affects female behavior and gene expression. *PLoS Biology* 6: e62.

Chapter 3

Anstey, M. L., Rogers, S. M., Ott, S. R., Burrows M., and Simpson, S. J. 2009. Serotonin mediates behavioral gregarization underlying swarm formation in desert locusts. *Science* 323: 627–630.

Arnold, S. J. 1981. The microevolution of feeding behavior. In A. Kamil and T. Sargent (eds.), *Foraging Behavior: Ecological, Ethological, and Psychological Approaches*, pp. 409–453. New York: Garland STPM Press.

Balda, R. P. 1980. Recovery of cached seeds by a captive *Nucifraga caryocatactes. Zeitschrift für Tierpsychologie* 52: 331–346.

Balda, R. P. and Kamil, A. C. 1992. Long-term spatial memory in Clark's nutcracker, *Nucifraga columbiana. Animal Behaviour* 44: 761–769.

Ben-Shahar, Y., Robichon, A., Sokolowski, M. B., and Robinson, G. E.
2002. Influence of gene action across different time scales on behavior. *Science* 296: 741–744.

Bendesky, A., Kwon, Y.-M., Lassance, J.-M., Lewarch, C. L., Yao, S.,
et al. 2017. The genetic basis of parental care evolution in monogamous mice. *Nature* 544: 434–439.

Bentley, D. and Hoy, R. R. 1974. The neurobiology of cricket song. *Scientific American* 231: 34–44.

Berthold, P. 1991. Genetic control of migratory behaviour in birds. *Trends in Ecology & Evolution* 6: 254–257.

Berthold, P. and Pulido, F. 1994. Heritability of migratory activity in a natural bird population. *Proceedings of the Royal Society of London B* 257: 311–315.

Berthold, P. and Querner, U. 1995. Microevolutionary aspects of bird migration based on experimental results. *Israel Journal of Zoology* 41: 377–385.

Boonstra, R., Hik, D., Singelton, G. R., and Tinnikov A. 1998. The impact of predator-induced stress on the snowshoe hare cycle. *Ecological Monographs* 68: 371–394.

Boque, N., de la Iglesia, R., de la Garza, A., Milagro, F. I., Olivares, M., et al. 2013. Prevention of diet-induced obesity by apple polyphenols in Wistar rats through regulation of adipocyte gene expression and DNA methylation patterns. *Molecular Nutrition and Food Research* 57: 1473–1478.

Brakefield, P. M. 2011. Evo-devo and accounting for Darwin's endless forms. *Philosophical Transactions of the Royal Society B* 366: 2069–2075.

Burmeister, S. S., Jarvis, E. D., and Fernald, R. D. 2005. Rapid behavioral and genomic responses to social opportunity. *PLoS Biology* 3: e363.

Carroll, S. B. 2005. Evolution at two levels: On genes and form. *PLoS Biology* 3: e245.

Carroll, S. B., Grenier, J. K., and Weatherbee, S. D. 2005. *From DNA to Diversity: Molecular Genetics and the Evolution of Animal Design*. Malden, MA: Blackwell Publishing

Chandrasekaran, S., Ament, S. A., Eddy, J. A., Rodriguez-Zas, S. L., Schatz, B. R., et al. 2011. Behavior-specific changes in transcriptional modules lead to distinct and predictable neurogenomic states. *Proceedings of the National Academy of Sciences USA* 108: 18020–18025.

Collins, J. P. and Cheek, J. E. 1983. Effect of food and density on development of typical and cannibalistic salamander larvae in *Ambystoma tigrinum nebulosum. American Zoologist* 23: 77–84.

Colombelli-Negrel, D., Hauber, M. E., Robertson, J., Sulloway, F. J., Hoi, H., et al. 2012. Embryonic learning of vocal passwords in superb fairy-wrens reveals intruder cuckoo nestlings. *Current Biology* 22: 2155–2160.

Croston, R., Branch, C. L., Pitera, A. M., Kozlovsky, D. Y., Bridge, E. S., et al. 2017. Predictably harsh environment is associated with reduced cognitive flexibility in wild food-caching mountain chickadees. *Animal Behaviour* 123: 139–149.

Dantzer, B., Newman A. E. M., Boonstra, R., Palme, R., Boutin, S., et al. 2013. Density triggers maternal hormones that increase adaptive offspring growth in a wild mammal. *Science* 340: 1215–1217.

Dawkins, R. 1982. *The Extended Phenotype*. San Francisco: W.H. Freeman.

de Kort, S. R. and Clayton, N. S. 2006. An evolutionary perspective on caching by corvids. *Proceedings of the Royal Society of London B* 273: 417–423.

Donaldson, Z. R. and Young, L. J. 2008. Oxytocin, vasopressin, and the neurogenetics of sociality. *Science* 322: 900–904.

Dunlap, A. S., Chen, B. B., Bednekoff, P. A., Greene, T. M., and Balda, R. P. 2006. A state-dependent sex difference in spatial memory in pinyon jays, *Gymnorhinus cyanocephalus*: Mated females forget as predicted by natural history. *Animal Behaviour* 72: 401–411.

Emlen, D. J. 2000. Integrating development with evolution: A case study with beetle horns. *BioScience* 50: 403–418.

Emlen, D. J., Szafran, Q., Corley, L., and Dworkin I. 2006. Insulin signaling and limb-patterning: Candidate pathways for the origin and evolutionary diversification of beetle "horns." *Heredity* 97: 179–191.

Emlen, D. J., Warren, I. A., Johns, A., Dworkin, I., and Lavine, L. C. 2012. A mechanism of extreme growth and reliable signaling in sexually selected ornaments and weapons. *Science* 337: 860–864.

Feldman, C. R., Brodie Jr, E. D., Brodie III, E. D., and Pfrender, M. E. 2010. Genetic architecture of a feeding adaptation: Garter snake (*Thamnophis*) resistance to tetrodotoxin-bearing prey. *Proceedings of the Royal Society of London B* 277: 3317–3325.

Fernald, R. D. 1993. Cichlids in love. *The Sciences* 33: 27–31.

Fernald, R. D. 2011. Systems biology meets behavior. *Proceedings of the National Academy of Sciences USA* 108: 17861–17862.

Francis, R. C., Soma, K. K., and Fernald, R. D. 1993. Social regulation of the brain-pituitary-gonadal axis. *Proceedings of the National Academy of Sciences USA* 90: 7794–7798.

Gamboa, G. J. 2004. Kin recognition in eusocial wasps. *Annales Zoologici Fennici* 41: 789–808.

Garcia, J. and Ervin, F. R. 1968. Gustatory-visceral and telereceptor-cutaneous conditioning: Adaptation in internal and external milieus. *Communications in Behavioral Biology (A)* 1: 389–415.

Garcia, J., Hankins, W. G., and Rusiniak, K. W. 1974. Behavioral regulation of the *milieu interne* in man and rat. *Science* 185: 824–831.

Gaulin, S. J. C. and FitzGerald, R. W. 1989. Sexual selection for spatial-learning ability. *Animal Behaviour* 37: 322–331.

Geffeney, S., Brodie Jr., E. D., Ruben, P. C., and Brodie III, E. D. 2002. Mechanisms of adaptation in a predator-prey arms race: TTX-resistant sodium channels. *Science* 297: 1336–1339.

Geffeney, S. L., Fujimoto, E., Brodie III, E. D., Brodie Jr, E. D., and Ruben, P. C. 2005. Evolutionary diversification of TTX-resistant sodium channels in a predator–prey interaction. *Nature* 434: 759–763.

Getz, L. L. and Carter, C. S. 1996. Prairie vole partnerships. *American Scientist* 84: 56–62.

Göth, A. and Evans, C. S. 2004. Social responses without early experience: Australian brush-turkey chicks use specific visual cues to aggregate with conspecifics. *Journal of Experimental Biology* 207: 2199–2208.

Hammock, E. A., D. and Young, L. J. 2005. Microsatellite instability generates diversity in brain and sociobehavioral traits. *Science* 308: 1630–1634.

Hampton, R. R. and Shettleworth, S. J. 1996. Hippocampal lesions impair memory for location but not color in passerine birds. *Behavioral Neuroscience* 110: 831–835.

Harlow, H. F. and Harlow, M. K. 1962. Social deprivation in monkeys. *Scientific American* 207: 136–146.

Harlow, H. F., Harlow, M. K., and Suomi, S. J. 1971. From thought to therapy: Lessons from a primate laboratory. *American Scientist* 59: 538–549.

Hauber, M. E., Russo, S. A., and Sherman, P. W. 2001. A password for species recognition in a brood-parasitic bird. *Proceedings of the Royal Society of London B* 268: 1041–1048.

Hauber, M. E. and Sherman, P. W. 2001. Self-referent phenotype matching: theoretical considerations and empirical evidence. *Trends in Neurosciences* 24: 609–616.

Herb, B. R., Wolschin, F., Hansen, K. D., Aryee, M. J., Langmead, B., et al. 2012. Reversible switching between epigenetic states in honeybee behavioral subcastes. *Nature Neuroscience* 15: 1371–1373.

Hitchcock, C. L. and Sherry, D. F. 1990. Long-term memory for cache sites in the black-capped chickadee. *Animal Behaviour* 40: 701–712.

Holmes, W. G. 1986. Identification of paternal half-siblings by captive Belding's ground squirrels. *Animal Behaviour* 34: 321–327.

Holmes, W. G. and Sherman, P. W. 1982. The ontogeny of kin recognition in two species of ground squirrels. *American Zoologist* 22: 491–517.

Holmes, W. G. and Sherman, P. W. 1983. Kin recognition in animals. *American Scientist* 71: 46–55.

Huang, Z.-Y. and Robinson, G. E. 1992. Honeybee colony integration: Worker-worker interactions mediate hormonally regulated plasticity in division of labor. *Proceedings of the National Academy of Sciences USA* 89: 11726–11729.

Kamakura, M. 2011. Royalactin induces queen differentiation in honeybees. *Nature* 473: 478–483.

Kannisto, V., Christensen, K., and Vaupel, J. W. 1997. No increased mortality in later life for cohorts born during famine. *American Journal of Epidemiology* 145: 987–994.

Kasumovic, M. M. and Andrade, M. C. B. 2006. Male development tracks rapidly shifting sexual versus natural selection pressures. *Current Biology* 16: R242–R243.

Kent, C. F., Daskalchuk, T., Cook, L., Sokolowski, M. B., and Greenspan, R. J. 2009. The *Drosophila* foraging gene mediates adult plasticity and gene–environment interactions in behaviour, metabolites, and gene expression in response to food deprivation. *PLoS Genetics* 5: e1000609.

Kerverne, E. B. 1997. An evaluation of what the mouse knockout experiments are telling us about mammalian behaviour. *Bioessays* 19: 1091–1098.

Kijimoto, T. and Moczek, A. P. 2016. Hedgehog signaling enables nutrition-responsive inhibition of an alternative morph in a polyphenic beetle. *Proceedings of the National Academy of Sciences USA* 113: 5982–5987.

King, A. P. and West, M. J. 1983. Epigenesis of cowbird song: A joint endeavour of males and females. *Nature* 305: 704–706.

Krebs, C. J., Boutin, S., Booonstra, R., Sinclair, A. R. E., Smith, J. N. M., et al. 1995. Impact of food and predation on the snowshoe hare cycle. *Science* 269: 1112–1115.

Kroodsma, D. E. and Canady, R. A. 1985. Differences in repertoire size, singing behavior, and associated neuroanatomy among marsh wren populations have a genetic basis. *Auk* 102: 439–446.

Kucharski, R., Maleszka, J., Foret, S., and Maleszka, R. 2008. Nutritional control of reproductive status in honeybees via DNA methylation. *Science* 319: 1827–1830.

Küpper, C., Stocks, M., Risse, J. E., dos Remedios, N., Farrell, F. L., et al. 2015. A supergene determines highly divergent male reproductive morphs in the ruff. *Nature Genetics* 48: 79–83.

Lamichhaney, S., Fan, G., Widemo, F., Gunnarsson, U., Thalmann, D. S., et al. 2015. Structural genomic changes underlie alternative reproductive strategies in the ruff (*Philomachus pugnax*). *Nature Genetics* 48: 84–88.

Lea, A. J., Altmann, J., Alberts, S. C., and Tung J. 2015. Developmental constraints in a wild primate. *American Naturalist* 185: 809–821.

Lea, A. J., altmann, J., Alberts, S. C., and Tung J. 2016. Resource base influences genome-wide DNA methylation levels in wild baboons (*Papio cynocephalus*). *Molecular Ecology* 25: 1681–1696.

Leoncini, I., Le Conte, Y., Costagliola, G., Plettner, E., Toth, A. L., et al. 2004. Regulation of behavioral maturation by a primer pheromone produced by adult worker honey bees. *Proceedings of the National Academy of Sciences USA* 101: 17559–17564.

Levine, S. and Mullins, R. F. 1966. Hormonal influences on brain organization in infant rats. *Science* 152: 1585–1592.

Lim, M. M., Murphy, A. Z., and Young, L. J. 2004a. Ventral striatopallidal oxytocin and vasopressin V1a receptors in the monogamous prairie vole (*Microtus ochrogaster*). *Journal of Comparative Neurology* 468: 555–570.

Lim, M. M., Wang, X., Olazábal, D. E., Ren, X., Terwilliger, E. F., and Young, L. J. 2004b. Enhanced partner preference in a promiscuous species by manipulating the expression of a single gene. *Nature* 429: 754–757.

Lindauer, M. 1961. *Communication among Social Bees*. Cambridge, MA: Harvard University Press.

Lore, R. and Flannelly, K. 1977. Rat societies. *Scientific American* 236: 106–116.

Lorenz, K. Z. 1952. *King Solomon's Ring*. New York: Crowell.

Lyko, F., Foret, S., Kucharski, R., Wolf, S., Falckenhayn, C., and Maleszka R. 2010. The honey bee epigenomes: Differential methylation of brain DNA in queens and workers. *PLoS Biology* 8: e1000506.

Mabry, K. E., Streatfeild, C. A., Keane, B., and Solomon, N. G. 2011. *AVPR1A* length polymorphism is not associated with either social or genetic monogamy in free-living prairie voles. *Animal Behaviour* 81: 11–18.

MacDougall-Shackleton, S. A., Sherry, D. F., Clark, A. P., Pinkus, R., and Hernandez, A. M. 2003. Photoperiodic regulation of food-storing and hippocampus volume in black-capped chickadees *Poecile atricapilla*. *Animal Behaviour* 65: 805–812.

Maret, T. J. and Collins, J. P. 1994. Individual responses to population size structure: The role of size variation in controlling expression of a trophic polyphenism. *Oecologia* 100: 279–285.

Mateo, J. M. 2002. Kin-recognition abilities and nepotism as a function of sociality. *Proceedings of the Royal Society of London B* 269: 721–727.

Mateo, J. M. 2006. The nature and representation of individual recognition odours in Belding's ground squirrels. *Animal Behaviour* 71: 141–154.

Mateo, J. M. and Holmes, W. G. 1997. Development of alarm-call responses in Belding's ground squirrels: The role of dams. *Animal Behaviour* 54: 509–524.

Mateo, J. M. and Johnston, R. E. 2000. Kin recognition and the "armpit effect": Evidence of self-reference phenotype matching. *Proceedings of the Royal Society of London B* 267: 695–700.

McGlothlin, J. W., Kobiela, M. E., Feldman, C. R., Brodie Jr., E. D., Pfrender, M. E., and Brodie III, E. D. 2016. Historical contingency in a multigene family facilitates adaptive evolution of toxin resistance. *Current Biology* 26: 1616–1621.

Milius, S. 2004. Where'd I put that? *Science News* 165: 103–105.

Möller, A., Pavlick, B., Hile, A. G., and Balda, R. P. 2001. Clark's nutcrackers *Nucifraga columbiana* remember the size of their cached seeds. *Ethology* 107: 451–461.

Morley, R. and Lucas A. 1997. Nutrition and cognitive development. *British Medical Bulletin* 53: 123–124.

Mueller, J., Pulido, F., and Kempenaers B. 2011. Identification of a gene associated with avian migratory behaviour. *Proceedings of the Royal Society of London B* 278: 2848–2856.

Okhovat, M., Berrio, A., Wallace, G., Ophir, A. G., and Phelps, S. M. 2015. Sexual fidelity trade-offs promote regulatory variation in the prairie vole brain. *Science* 350: 1371–1374.

Olson, D. J., Kamil, A. C., Balda, R. P., and Nims, P. J. 1995. Performance of four seed-caching corvid species in operant tests of nonspatial and spatial memory. *Journal of Comparative Psychology* 109: 173–181.

Ophir, A. G., Wolff, J. O., and Phelps S. 2008. Variation in neural V1aR predicts sexual fidelity and space use among male prairie voles in semi-natural settings. *Proceedings of the National Academy of Sciences USA* 105: 1249–1254.

Parker, K. J., Kinney, L. F., Phillips, K. M., and Lee, T. M. 2001. Paternal behavior is associated with central neurohormone receptor binding patterns in meadow voles (*Microtus pennsylvanicus*). *Behavioral Neuroscience* 115: 1341–1348.

Parker, K. J. and Lee, T. M. 2001. Central vasopressin administration regulates the onset of facultative paternal behavior in *Microtus pennsylvanicus* (meadow voles). *Hormones and Behavior* 39: 285–294.

Pfennig, D. W. and Collins, J. P. 1993. Kinship affects morphogenesis in cannibalistic salamanders. *Nature* 362: 836–838.

Pfennig, D. W., Rice, A. M., and Martin, R. A. 2007. Field and experimental evidence for competition's role in phenotypic divergence. *Evolution* 61: 257–271.

Pfennig, D. W., Sherman, P. W., and Collins, J. P. 1994. Kin recognition and cannibalism in polyphenic salamanders. *Behavioral Ecology* 5: 225–232.

Picciotto, M. R. 1999. Knockout mouse models used to study neurobiological systems. *Critical Reviews in Neurobiology* 13: 103–149.

Pitkow, L. J., Sharer, C. A., Ren, X. L., Insel, T. R., Terwilliger, E. F., and Young, L. J. 2001. Facilitation of affiliation and pair-bond formation by vasopressin receptor gene transfer into the ventral forebrain of a monogamous vole. *Journal of Neuroscience* 21: 7392–7396.

Pravosudov, V. V. and de Kort, S. R. 2006. Is the western scrub-jay (*Aphelocoma californica*) really an underdog among food-caching corvids when it comes to hippocampal volume and food caching propensity? *Brain, Behavior and Evolution* 67: 1–9.

Purcell, J., Brelsford, A., Wurm, Y., Perrin, N., and Chapuisat M. 2014. Convergent genetic architecture underlies social organization in ants. *Current Biology* 24: 2728–2732.

Rasmussen, K. M. 2001. The "fetal origins" hypothesis: Challenges and opportunities for maternal and child nutrition. *Annual Review of Nutrition* 21: 73–95.

Ratcliffe, J. M., Fenton, M. B., and Galef, B. G. 2003. An exception to the rule: Common vampire bats do not learn taste aversions. *Animal Behaviour* 65: 385–389.

Robinson, G. E. 1998. From society to genes with the honey bee. *American Scientist* 86: 456–462.

Robinson, G. E. 2004. Beyond nature and nurture. *Science* 304: 397–399.

Ross, K. G. and Keller L. 1998. Genetic control of social organization in an ant. *Proceedings of the National Academy of Sciences USA* 95: 14232–14237.

Ross, K. G. and Keller L. 2002. Experimental conversion of colony social organization by manipulation of worker genotype composition in fire ants (*Solenopsis invicta*). *Behavioral Ecology and Sociobiology* 51: 287–295.

Rubenstein, D. R. and Hofmann, H. A. 2015. Proximate pathways underlying social behavior. *Current Opinion in Behavioral Sciences* 6: 154–159.

Rubenstein, D. R., Skolnik, H. E., Berrio, A., Champagne, F., Phelps, S., and Solomon, J. 2016. Sex-specific fitness effects of unpredictable early life conditions are associated with DNA methylation in the avian glucocorticoid receptor. *Molecular Ecology* 25: 1714–1728.

Sayol, F., Maspons, J., Lapiedra, O., Iwaniuk, A. N., Szekely, T., and Sol, D. 2016. Environmental variation and the evolution of large brains in birds. *Nature Communications* 7: 13971.

Seeley, T. D. 1982. Adaptive significance of the age polyethism schedule in honeybee colonies. *Behavioral Ecology and Sociobiology* 11: 287–293.

Sen Sarma, M., Rodriguez-Zas, S. L., Gernat, T., Nguyen, T., Newman, T., and Robinson, G. E. 2010. Distance-responsive genes found in dancing honey bees. *Genes, Brain and Behavior* 9: 825–830.

Sheriff, M. J., Krebs, C. J., and Boonstra R. 2009. The sensitive hare: Sublethal effects of predator stress on reproduction in snowshoe hares. *Journal of Animal Ecology* 78: 1249–1258.

Sheriff, M. J. and Love, O. P. 2012. Determining the adaptive potential of maternal stress. *Ecology Letters* 16: 271–280.

Sherry, D. F. 1984. Food storage by black-capped chickadees: Memory of the location and contents of caches. *Animal Behaviour* 32: 451–464.

Sherry, D. F., Forbes, M. R. L., Kjurgel, M., and Ivy, G. O. 1993. Females have a larger hippocampus than males in the brood-parasitic brown-headed cowbird. *Proceedings of the National Academy of Sciences USA* 90: 7839–7843.

Sherry, D. F. and Hoshooley, J. S. 2010. Seasonal hippocampal plasticity in food-storing birds. *Philosophical Transactions of the Royal Society B* 365: 933–943.

Shettleworth, S. J., Hampton, R. R., and Westwood, R. P. 1995. Effects of season and photoperiod on food storing by black-capped chickadees, *Parus atricapillus*. *Animal Behaviour* 49: 989–998.

Simpson, S. J., Despland, E., Hagele, B. F., and Dodgson T. 2001. Gregarious behavior in desert locusts is evoked by touching their back legs. *Proceedings of the National Academy of Sciences USA* 98: 3895–3897.

Simpson, S. J., Sword, G. A., and Lo, N. 2011. Polyphenism in insects. *Current Biology* 21: R738–R749.

Skinner, B. F. 1966. Operant behavior. In W. Honig (ed.), *Operant Behavior*. New York: Appleton-Century-Crofts.

Slagsvold, T. 1998. On the origin and rarity of interspecific nest parasitism in birds. *American Naturalist* 152: 264–272.

Slagsvold, T., Hansen, B. T., Johannessen, L. E., and Lifjeld, J. T. 2002. Mate choice and imprinting in birds studied by cross-fostering in the wild. *Proceedings of the Royal Society of London B* 269: 1449–1455.

Solomon, N. G., Richmond, A. R., Harding, P. A., Fries, A., Acquemin, S., et al. 2009. Polymorphism at the *avpr1a* locus in male prairie voles correlated with genetic but not social monogamy in field populations. *Molecular Ecology* 18: 4680–4695.

Stein, Z., Susser, M., Saenger, G., and Marolla, F. 1972. Nutrition and mental performance. *Science* 178: 708–713.

Streatfeild, C. A., Mabry, K. E., Keane, B., Crist, T. O., and Solomon, N. G. 2011. Intraspecific variability in the social and genetic mating systems of prairie voles, *Microtus ochrogaster*. *Animal Behaviour* 82: 1387–1398.

Sullivan, J. P., Jassim, O., Fahrbach, S. E., and Robinson, G. E. 2000. Juvenile hormone paces behavioral development in the adult worker honey bee. *Hormones and Behavior* 37: 1–14.

Susser, M. and Stein Z. 1994. Timing in prenatal nutrition: A reprise of the Dutch famine study. *Nutrition Reviews* 52: 84–94.

Taylor, S. and Campagna, L. 2016. Avian supergenes. *Science* 351: 446–447.

Tobi, E. W., Goeman, J. J., Monajemi, R., Gu, H., Putter, H., et al. 2014. DNA methylation signatures link prenatal famine exposure to growth and metabolism. *Nature Communications* 5: 5592–5614.

Toth, A. L. and Robinson, G. E. 2007. Evo-devo and the evolution of social behavior. *Trends in Genetics* 23: 334–341.

Toth, A. L., Varala, K., Henshaw, M. T., Rodriguez-Zas, S. L., Hudson, M. E., and Robinson, G. E. 2010. Brain transcriptomic analysis in paper wasps identifies genes associated with behaviour across social insect lineages. *Proceedings of the Royal Society of London B* 277: 2139–2148.

Toth, A. L., Varala, K., Newman, T. C., Miguez, F. E., Hutchison, S. K., et al. 2007. Wasp gene expression supports an evolutionary link between maternal behavior and eusociality. *Science* 318: 441–444.

Walum, H., Westberg, L., Henningsson, S., Neiderhiser, J. M., Reiss, D., et al. 2008. Genetic variation in the vasopressin receptor 1a gene (*AVPR1A*) associates with pair-bonding behavior in humans. *Proceedings of the National Academy of Sciences USA* 105: 14153–14156.

Wang, J., Wurm, Y., Nipitwattanaphon, M., Riba-Grognuz, O., Huang U.-C., et al. 2013. A Y-like social chromosome causes alternative colony organization in fire ants. *Nature* 493: 664–668.

Weaver, I. C. G., Cervoni, N., Champagne, F. A., D'Alessio, A. C., Sharma, S., et al. 2004. Epigenetic programming by maternal behavior. *Nature Neuroscience* 7: 847–854.

West-Eberhard, M. J. 2003. *Developmental Plasticity and Evolution*. New York: Oxford University Press.

White, S. A., Nguyen, T., and Fernald, R. D. 2002. Social regulation of gonadotropin-releasing hormone. *Journal of Experimental Biology* 205: 2567–2581.

Whitfield, C. W., Cziko, A. M., and Robinson, G. E. 2003. Gene expression profiles in the brain predict behavior in individual honey bees. *Science* 302: 296–299.

Young, L. J., Nilsen, R., Waymire, K. G., Grant R. MacGregor, G. R., and Insel, T. R. 1999. Increased affliative response to vasopressin in mice expressing the V1a receptor from a monogamous vole. *Nature* 400: 766–768.

Young, L. J. and Wang Z. 2004. The neurobiology of pair bonding. *Nature Neuroscience* 7: 1048–1054.

Chapter 4

Acharya, L. and McNeil, J. N. 1998. Predation risk and mating behavior: The responses of moths to bat-like ultrasound. *Behavioral Ecology* 9: 552–558.

Aiello, L. C. and Wheeler, P. 1995. The expensive-tissue hypothesis: The brain and the digestive system in human and primate evolution. *Current Anthropology* 36: 1999–1221.

Barber, J. R. and Conner, W. E. 2007. Acoustic mimicry in a predator–prey interaction. *Proceedings of the National Academy of Sciences USA* 104: 9331–9334.

Barton, D. and Dunbar, R. I. M. 1997. Evolution of the social brain. In A. Whiten and R. W. Byrne (eds.), *Machiavellian Intelligence II: Extensions and Evaluations*, 2nd ed., pp. 240–263. Cambridge: Cambridge University Press.

Bass, A. H. 1996. Shaping brain sexuality. *American Scientist* 84: 352–363.

Benson-Amrama, S., Dantzer, B., Strickere, G., Swanson, E. M., and Holekamp, K. E. 2016. Brain size predicts problem-solving ability in mammalian carnivores. *Proceedings of the National Academy of Sciences USA* 113: 2532–2537.

Boulcott, P. D., Walton, K., and Braithwaite, V. A. 2005. The role of ultraviolet wavelengths in the mate-choice decisions of female three-spined sticklebacks. *Journal of Experimental Biology* 208: 1453–1458.

Brodfuehrer, P. D. and Hoy, R. R. 1990. Ultrasound-sensitive neurons in the cricket brain. *Journal of Comparative Physiology A* 166: 651–662.

Brower, L. P. 1995. Understanding and misunderstanding the migration of the monarch butterfly (Nymphalidae) in North America, 1857–1995. *Journal of the Lepidopteran Society* 49: 304–385.

Burkhardt, R. W. 2004. *Patterns of Behavior: Konrad Lorenz, Niko Tinbergen, and the Founding of Ethology*. Chicago: University of Chicago Press.

Catania, K. C. 1995. Structure and innervation of the sensory organs on the snout of the star-nosed mole. Journal of Comparative Neurology 351: 536–548.

Catania, K. C. 2000. Cortical organization in Insectivora: The parallel evolution of the sensory periphery and the brain. *Brain Behavior and Evolution* 55: 311–321.

Catania, K. C. 2008. Worm grunting, fiddling, and charming: Humans unknowingly mimic a predator to harvest bait. *PLoS One* 3: e3472.

Catania, K. C. and Kaas, J. H. 1996. The unusual nose and brain of the star-nosed mole. *BioScience* 46: 578–586.

Catania, K. C. and Kaas, J. H. 1997. Somatosensory fovea in the star-nosed mole: Behavioral use of the star in relation to innervation patterns and cortical representation. *Journal of Comparative Neurology* 387: 215–233.

Catania, K. C. and Remple, F. E. 2004. Tactile foveation in the star-nosed mole. *Brain, Behavior and Evolution* 63: 1–12.

Catania, K. C. and Remple, F. E. 2005. Asymptotic prey profitability drives star-nosed moles to the foraging speed limit. *Nature* 433: 519–522.

Catania, K. C. and Remple, M. S. 2002. Somatosensory cortex dominated by the representation of teeth in the naked mole-rat brain. *Proceedings of the National Academy of Sciences USA* 99: 5692–5697.

Clarac, F. and Pearlstein, E. 2007. Invertebrate preparations and their contribution to neurobiology in the second half of the 20th century. *Brain Research Reviews* 54: 113–161.

Conner, W. E. and Corcoran, A. J. 2012. Sound strategies: The 65-million-year-old battle between bats and insects. *Annual Review of Entomology* 57: 21–39.

Corcoran, A. J., Barber, J. R., Hristov, N. I., and Conner, W. E. 2011. How do tiger moths jam bat sonar? *Journal of Experimental Biology* 214: 2416–2425.

Davies, N. B. 2000. *Cuckoos, Cowbirds, and Other Cheats.* London: T. and A. D. Poyser

DeCasien, A. R., Williams, S. A., and Higham, J. P. 2017. Primate brain size is predicted by diet but not sociality. *Nature Ecology & Evolution.* doi: 10.1038/s41559-017-0112.

Dowdy, N. J. and Conner, W. E. 2016. Acoustic aposematism and evasive action in select chemically defended arctiine (Lepidoptera: Erebidae) species: Nonchalant or not? *PLoS One* 11: e0152981.

Drea, C. M. and Carter, A. N. 2009. Cooperative problem solving in a social carnivore. *Animal Behaviour* 78: 967–977.

Dunbar, R. I. M. 2003. The social brain: Mind, language, and society in evolutionary perspective. *Annual Review of Anthropology* 32: 163–181.

Dunbar, R. I. M. and Shultz, S. 2007. Evolution in the social brain. *Science* 317: 1344–1347.

Feng, N. Y. and Bass, A. H. 2017. Neural, hormonal, and genetic mechanisms of alternative reproductive tactics: Vocal fish as model systems. In D. W. Pfaff and M. Joëls (eds.), *Hormones, Brain and Behavior*, pp. 47–68. Oxford: Academic Press.

Finarelli, J. A. and Flynn, J. J. 2009. Brain-size evolution and sociality in Carnivora. *Proceedings of the National Academy of Sciencs of USA*106: 9345–9349.

Frost, W. N., Hoppe, T. A., Wang, J., and Tian, L. M. 2001. Swim initiation neurons in *Tritonia diomedea. American Zoologist* 41: 952–961.

Froy, O., Gotter, A. L., Casselman, A. L., and Reppert, S. M. 2003. Illuminating the circadian clock in monarch butterfly migration. *Science* 300: 1303–1305.

Fullard, J. H. 1982. Echolocation assemblages and their effects on moth auditory systems. *Canadian Journal of Zoology* 60: 2572–2576.

Fullard, J. H. 1998. The sensory coevolution of moths and bats. In R. R. Hoy, A. N. Popper and R. R. Fay (eds.), *Comparative Hearing: Insects*, pp. 279–326. New York: Springer.

Fullard, J. H., Dawson, J. W., and Jacobs, D. S. 2003. Auditory encoding during the last moment of a moth's life. *Journal of Experimental Biology* 206: 281–294.

Getting, P. A. 1983. Mechanisms of pattern generation underlying swimming in *Tritonia.* II. Network reconstruction. *Journal of Neurophysiology* 49: 1017–1035.

Getting, P. A. 1989. A network oscillator underlying swimming in *Tritonia*. In J. W. Jacklet (ed.), *Neuronal and Cellular Oscillators*, pp. 215–236. New York: Dekker.

Geva-Sagiv, M., Las, L., Yovel, Y., and Ulanovsky, N. 2015. Spatial cognition in bats and rats: From sensory acquisition to multiscale maps and navigation. *Nature Reviews Neuroscience* 16: 94–108.

Griffin, D. R. 1958. *Listening in the Dark.* New Haven, CT: Yale University Press.

Hare, B., Brown, M., Williamson, C., and Tomasello, M. 2002. The domestication of social cognition in dogs. *Science* 298: 1634–1636.

Hawryshyn, C. W., Ramsden, S. D., Betke, K. M., and Sabbah, S. 2010. Spectral and polarization sensitivity of juvenile Atlantic salmon (*Salmo salar*): Phylogenetic considerations. *Journal of Experimental Biology* 213: 3187–3197.

Heinrich, B. 1999. *Mind of the Raven: Investigations and Adventures with Wolf-Birds.* New York: HarperCollins.

Holzhaider, J. C., Sibley, M. D., Taylor, A. H., Singh, P. J., Gray, R. D., and Hunt, G. R. 2011. The social structure of New Caledonian crows. *Animal Behaviour* 81: 83–92.

Immelmann, K., Piltz, A., and Sossinka, R. 1977. Experimental studies on the significance of mouth markings in zebra finch nestlings. *Zeitschrift für Tierpsychologie* 45: 210–218.

Kalko, E. K. V. 1995. Insect pursuit, prey capture and echolocation in pipistrelle bats (Microchiroptera). *Animal Behaviour* 50: 861–880.

Kell, C. A., von Kriegsterin, K., Rosler, R., Kleinschmidt, A., and Laufs, H. 2005. The sensory cortical representation of the human penis: Revisiting somatotopy in the male homunculus. *Journal of Neuroscience* 25: 5984–5987.

Kemp, D. J. 2008. Female mating biases for bright ultraviolet iridescence in the butterfly *Eurema hecabe* (Pieridae). *Behavioral Ecology* 19: 1–8.

Kruuk, H. 2004. *Niko's Nature: The Life of Niko Tinbergen and His Science of Animal Behavior*. Oxford: Oxford University Press.

Lappin, A. K., Brandt, Y., Husak, J. F., Macedonia, J. M., and Kemp, D. J. 2006. Gaping displays reveal and amplify a mechanically based index of weapon performance. *American Naturalist* 168: 100–113.

Lehmann, G. U. C. and Heller, K.-G. 1998. Bushcricket song structure and predation by the acoustically orienting parasitoid fly *Therobia leonidei* (Diptera: Tachinidae: Ormiini). *Behavioral Ecology and Sociobiology* 43: 239–245.

Madden, J. 2001. Sex, bowers, and brains. *Proceedings of the Royal Society of London B* 268: 833–838.

Marasco, P. D. and Catania, K. C. 2007. Response properties of primary afferents supplying Eimer's organ. *Journal of Experimental Biology* 210: 765–780.

Mason, A. C., Oshinsky, M. L., and Hoy, R. R. 2001. Hyperacute directional hearing in a microscale auditory system. *Nature* 410: 686–690.

May, M. 1991. Aerial defense tactics of flying insects. *American Scientist* 79: 316–329.

Menzel, R. and Giurfa, M. 2001. Cognitive architecture of a mini-brain: The honeybee. *Trends in Cognitive Sciences* 5: 62–71.

Miller, L. A. and Surlykke, A. 2001. How some insects detect and avoid being eaten by bats: Tactics and countertactics of prey and predator. *BioScience* 51: 570–581.

Moiseff, A., Pollack, G. S., and Hoy, R. R. 1978. Steering responses of flying crickets to sound and ultrasound: Mate attraction and predator avoidance. *Proceedings of the National Academy of Sciences USA* 75: 4052–4056.

Navarrete, A., van Schaik, C. P., and Isler, K. 2011. Energetics and the evolution of human brain size. *Nature* 480: 91–94.

Nolen, T. G. and Hoy, R. R. 1984. Phonotaxis in flying crickets: Neural correlates. *Science* 226: 992–994.

Payne, R. B. 2005. Nestling mouth markings and colors of old world finches, Estrildidae: Mimicry and coevolution of nesting finches and their *Vidua* brood parasites. Ann Arbor, MI: Miscellaneous Publication 194 of the University of Michigan Museum of Zoology.

Pérez-Barberia, F. J., Shultz, S., and Dunbar, R. I. M. 2007. Evidence for coevolution of sociality and relative brain size in three orders of mammals. *Evolution* 61: 2811–2821.

Pike, T. W., Bjerkeng, B., Blount, J. D., Lindstrom, J., and Metcalfe, N. B. 2011. How integument colour reflects its carotenoid content: A stickleback's perspective. *Functional Ecology* 25: 297–304.

Pike, T. W., Blount, J. D., Bjerkeng, B., Lindstrom, J., and Metcalfe, N. B. 2007. Carotenoids, oxidative stress, and female mating preference for longer-lived males. *Proceedings of the Royal Society of London B* 274: 1591–1596.

Pontzer, H., Brown, M. H., Raichlen, D. A., Dunsworth, H., Hare, B., et al. 2016. Metabolic acceleration and the evolution of human brain size and life history. *Nature* 533: 390–392.

Ratcliffe, J. M. 2009. Predator–prey interaction in an auditory world. In R. Dukas and J. M. Ratcliffe (eds.), *Cognitive Ecology*, 2nd ed., pp. 201–225. Chicago: University of Chicago Press.

Reppert, S. M., Zhu, H. S., and White, R. H. 2004. Polarized light helps monarch butterflies navigate. *Current Biology* 14: 155–158.

Robert, D., Amoroso, J., and Hoy, R. R. 1992. The evolutionary convergence of hearing in a parasitoid fly and its cricket host. *Science* 258: 1135–1137.

Roeder, K. D. 1963. *Nerve Cells and Insect Behavior*. Cambridge, MA: Harvard University Press.

Roeder, K. D. 1970. Episodes in insect brains. *American Scientist* 58: 378–389.

Roeder, K. D. and Treat, A. E. 1961. The detection and evasion of bats by moths. *American Scientist* 49: 135–148.

Roth, G. and Dicke, U. 2005. Evolution of the brain and intelligence. *Trends in Cognitive Sciences* 9: 250–257.

Rutowski, R. L. 1998. Mating strategies in butterflies. *Scientific American* 279: 64–69.

Sakai, S. T., Arsznov, B. M., Lundrigan, B. L., and Holekamp, K. E. 2011. Brain size and social complexity: A computed tomography study in Hyaenidae. *Brain Behavior and Evolution* 77: 91–104.

Saul-Gershenz, L. and Millar, J. G. 2006. Phoretic nest parasites use sexual deception to obtain transport to their host's nest. *Proceedings of the National Academy of Sciences USA* 103: 14039–14044.

Schuetz, J. G. 2005. Low survival of parasite chicks may result from their imperfect adaptation to hosts rather than expression of defenses against parasitism. *Evolution* 59: 2017–2224.

Sheehan, M. J. and Tibbetts, E. A. 2010. Selection for individual recognition and the evolution of polymorphic identity signals in *Polistes* paper wasps. *Journal of Evolutionary Biology* 23: 570–577.

Sheehan, M. J. and Tibbetts, E. A. 2011. Specialized face learning is associated with individual recognition in paper wasps. *Science* 334: 1272–1275.

Shlizerman, E., Phillips-Portillo, J., Forger, D. B., and Reppert, S. M. 2016. Neural integration underlying a time-compensated sun compass in the migratory monarch butterfly. *Cell Reports* 15: 683–691.

Shultz, S., Dunbar, R. I. M. 2010. Social bonds in birds are associated with brain size and contingent on the correlated evolution of life-history and increased parental investment. *Biological Journal of the Linnean Society* 100: 111–123.

Sisneros, J. A. 2007. Saccular potentials of the vocal plainfin midshipman fish, *Porichthys notatus*. *Journal of Comparative Physiology A* 193: 413–424.

Sisneros, J. A. and Bass, A. H. 2003. Seasonal plasticity of peripheral auditory frequency sensitivity. *Journal of Neuroscience* 23: 1049–1058.

Skals, N., Anderson, P., Kanneworff, M., Löfstedt, C., and Surlykke, A. 2005. Her odours make him deaf: Crossmodal modulation of olfaction and hearing in a male moth. *Journal of Experimental Biology* 208: 595–601.

Stapley, J. and Whiting, M. J. 2005. Ultraviolet signals fighting ability in a lizard. *Biology Letters* 2: 169–172.

Stumpner, A. and Lakes-Harlan, R. 1996. Auditory interneurons in a hearing fly (*Therobia leonidei*, Ormiini, Tachinidae, Diptera). *Journal of Comparative Physiology A* 178: 227–233.

Surlykke, A. 1984. Hearing in notodontid moths: A tympanic organ with a single auditory neuron. *Journal of Experimental Biology* 113: 323–334.

Taylor, A. H., Elliffe, D., Hunt, G. R., and Gray, R. D. 2010. Complex cognition and behavioural innovation in New Caledonian crows. *Proceedings of the Royal Society of London B* 277: 2637–2643.

ter Hofstede, H. M., Goerlitz, H. R., Ratcliffe, J. M., Holderied, M. W., and Surlykke, A. 2013. The simple ears of noctuoid moths are tuned to the calls of their sympatric bat community. *Journal of Experimental Biology* 216: 3954–3962.

ter Hofstede, H. M. and Ratcliffe, J. M. 2016. Evolutionary escalation: The bat–moth arms race. *Journal of Experimental Biology* 219: 1589–1602.

Tinbergen, N. 1951. *The Study of Instinct*. New York: Oxford University Press.

Tinbergen, N. and Perdeck, A. C. 1950. On the stimulus situations releasing the begging response in the newly hatched herring gull (*Larus argentatus* Pont.). *Behaviour* 3: 1–39.

Topál, J., Gergely, G., Erd´o´hegyi, A., Csibra, G., and Miklósi, A. 2009. Differential sensitivity to human communication in dogs, wolves, and human infants. *Science* 325: 1269–1272.

Udell, M. A. R., Dorey, N. R., and Wynne, C. D. L. 2008. Wolves outperform dogs in following human social cues. *Animal Behaviour* 76: 1767–1773.

Udell, M. A. R., Dorey, N. R., andWynne, C. D. L. 2010. What did domestication do to dogs? A new account of dogs' sensitivity to human actions. *Biological Reviews* 85: 327–345.

Vereecken, N. J. and Mahé, G. 2007. Larval aggregations of the blister beetle *Stenoria analis* (Schaum) (Coleoptera: Meloidae) sexually deceive patrolling males of their host, the solitary bee *Colletes hederae* Schmidt & Westrich (Hymenoptera: Colletidae). *Annales de la Société Entomologique de France* 43: 493–496.

Wickler, W. 1968. *Mimicry in Plants and Animals*. London: World University Library.

Willows, A. O. D. 1971. Giant brain cells in mollusks. *Scientific American* 224: 68–75.

Yack, J. E., Kalko, J. E. V., and Surlykke, A. 2007. Neuroethology of ultrasonic hearing in nocturnal butterflies (Hedyloidea). *Journal of Comparative Physiology A* 193: 577–590.

Yager, D. D. 2012. Predator detection and evasion by flying insects. *Current Opinion in Neurobiology* 22: 201–207.

Yager, D. D. and May, M. L. 1990. Ultrasound-triggered, flight-gated evasive maneuvers in the flying praying mantis, *Parasphendale agrionina*. II. Tethered flight. *Journal of Experimental Biology* 152: 41–58.

Yager, D. D. and Svenson, G. J. 2008. Patterns of praying mantis auditory system evolution based on morphological, molecular, neurophysiological, and behavioural data. *Biological Journal of the Linnean Society* 94: 541–568.

Zhu, H., Casselman, A., and Reppert, S. M. 2008. Chasing migration genes: A brain expressed sequence tag resource for summer and migratory monarch butterflies (*Danaus plexippus*). *PLoS One* 3: e1345.

Chapter 5

Adkins-Regan, E. 2005. *Hormones and Animal Social Behavior*. Princeton, NJ: Princeton University Press

Balthazart, J., Baillien, M., Charlier, T. D., Cornil, C. A., and Ball, G. F. 2003. The neuroendocrinology of reproductive behavior in Japanese quail. *Domestic Animal Endocrinology* 25: 69–82.

Baylies, M. K., Bargiello, T. A., Jackson, F. R., and Young, M. W. 1987. Changes in abundance or structure of the *Per* gene product can alter periodicity of the *Drosophila* clock. *Nature* 326: 390–392.

Benkman, C. W. 1990. Intake rates and the timing of crossbill reproduction. *Auk* 107: 376–386.

Bennett, M. F., Shriner, J., and Brown, R. A. 1957. Persistent tidal cycles of spontaneous motor activity in the fiddler crab, *Uca pugnax*. *Biological Bulletin* 112: 267–275.

Chapman, T., Bangham, J., Vinti, G., Lung, O., Wolfner, M. F., et al. 2003. The sex peptide of *Drosophila melanogaster*: Female post-mating responses analyzed by using RNA interference. *Proceedings of the National Academy of Sciences USA* 100: 9923–9928.

Cheng, M. Y., Bullock, C. M., Li, C. Y., Lee, A. G., Bermak, J. C., et al. 2002. Prokineticin 2 transmits the behavioural circadian rhythm of the suprachiasmatic nucleus. *Nature* 417: 405–410.

Colwell, C. S. 2012. Linking neural activity and molecular oscillations in the SCN. *Nature Reviews Neuroscience* 12: 553–569.

Creel, S., Dantzer, B., Goymann, W., and Rubenstein, D. R. 2012. The ecology of stress: Effects of the social environment. *Functional Ecology* 27: 66–80.

Crews, D. 1974. Castration and androgen replacement on male facilitation of ovarian activity in the lizard *Anolis carolinensis*. *Journal of Comparative and Physiological Psychology* 87: 963–969.

Crews, D. 1984. Gamete production, sex hormone secretion, and mating behavior uncoupled. *Hormones and Behavior* 18: 22–28.

Crews, D. 1991. Trans-seasonal action of androgen in the control of spring courtship behavior in male red-sided garter snakes. *Proceedings of the National Academy of Sciences USA* 88: 3545–3548.

Crews, D. and Greenberg, N. 1981. Function and causation of social signals in lizards. *American Zoologist* 21: 273–294.

Davis-Walton, J. and Sherman, P. W. 1994. Sleep arrhythmia in the eusocial naked mole-rat. *Naturwissenschaften* 81: 272–275.

DeCoursey, P. J. and Buggy, J. 1989. Circadian rhythmicity after neural transplant to hamster third ventricle: Specificity of suprachiasmatic nuclei. *Brain Research* 500: 263–275.

Deviche, P. and Sharp, P. J. 2001. Reproductive endocrinology of a free-living, opportunistically breeding passerine (white-winged crossbill, *Loxia leucoptera*). *General and Comparative Endocrinology* 123: 268–279.

Dittami, J. P. and Gwinner, E. 1985. Annual cycles in the African stonechat *Saxicola torquata axillaris* and their relationship to environmental factors. *Journal of Zoology* 207: 357–370.

Farner, D. S. 1964. Time measurement in vertebrate photoperiodism. *American Naturalist* 95: 375–386.

Farner, D. S. and Lewis, R. A. 1971. Photoperiodism and reproductive cycles in birds. *Photophysiology* 6: 325–370.

Follett, B. K., Mattocks, P. W., Jr, and Farner, D. S. 1974. Circadian function in the photoperiodic induction of gonadotropin secretion in the white-crowned sparrow, *Zonotrichia leucophrys gambelli*. *Proceedings of the National Academy of Sciences USA* 71: 1666–1669.

Froy, O., Gotter, A. L., Casselman, A. L., and Reppert, S. M. 2003. Illuminating the circadian clock in monarch butterfly migration. *Science* 300: 1303–1305.

Gettler, L. T., McDade, T. W., Feranil, A. B., and Kuzawa, C. W. 2012. Longitudinal evidence that fatherhood decreases testosterone in males. *Proceedings of the National Academy of Sciences USA* 108: 16194–16199.

Goymann, W. and Helm, B. 2014. Seasonality of life histories in tropical birds: Circannual rhythms and *Zeitgeber*. In H. Numata and B. Helm (eds.), *Annual, Lunar and Tidal Clocks*, pp. 247–275. New York: Springer.

Goymann, W. and Wingfield, J. C. 2004. Allostatic load, social status and stress hormones: The costs of social status matter. *Animal Behaviour* 67: 591–602.

Granados-Fuentes, D., Tseng, A., and Herzog, E. D. 2006. A circadian clock in the olfactory bulb controls olfactory responsivity. *Journal of Neuroscience* 26: 12219–12225.

Greives, T. J., McGlothlin, J. W., Jawor, J. M., Demas, G. E., and Ketterson, E. D. 2006. Testosterone and innate immune function inversely covary in a wild population of breeding dark-eyed juncos (*Junco hyemalis*). *Functional Ecology* 20: 812–818.

Gwinner, E. 1996. Circannual clocks in avian reproduction and migration. *Ibis* 138: 47–63.

Gwinner, E. and Dittami, J. 1990. Endogenous reproductive rhythms in a tropical bird. *Science* 249: 906–908.

Hahn, T. P. 1995. Integration of photoperiodic and food cues to time changes in reproductive physiology by an opportunistic breeder, the red crossbill, *Loxia curvirostra* (Aves: Carduelinae). *Journal of Experimental Zoology* 272: 213–226.

Hahn, T. P. 1998. Reproductive seasonality in an opportunistic breeder, the red crossbill, *Loxia curvirostra*. *Ecology* 79: 2365–2375.

Hahn, T. P., Cornelius, J. M., Sewall, K. B., Kelsey, T. R., Hau, M., and Perfito, N. 2008. Environmental regulation of annual schedules in opportunistically-breeding songbirds: Adaptive specializations or variations on a theme of white-crowned sparrow? *General and Comparative Endocrinology* 157: 217–226.

Hahn, T. P., Pereyra, M. E., Sharbaugh, S. M., and Bentley, G. E. 2004. Physiological responses to photoperiod in three cardueline finch species. *General and Comparative Endocrinology* 137: 99–108.

Hahn, T. P., Wingfield, J. C., Mullen, R., and Deviche, P. J. 1995. Endocrine bases of spatial and temporal opportunism in arctic-breeding birds. *American Zoologist* 35: 259–273.

Hamner, W. M. 1964. Circadian control of photoperiodism in the house finch demonstrated by interrupted-night experiments. *Nature* 203: 1400–1401.

Hau, M., Ricklefs, R. E., Wikelski, M., Lee, K. A., and Brawn, J. D. 2010. Corticosterone, testosterone and life-history strategies of birds. *Proceedings of the Royal Society of London B* 277: 3203–3212.

Hau, M., Stoddard, S. T., and Soma, K. A. 2004a. Territorial aggression and hormones during the non-breeding season in a tropical bird. *Hormones and Behavior* 45: 40–49.

Hau, M., Wikelski, M., Gwinner, H., and Gwinner, E. 2004b. Timing of reproduction in a Darwin's finch: Temporal opportunism under spatial constraints. *Oikos* 106: 489–500.

Hau, M., Wikelski, M., and Wingfield, J. C. 1998. A neotropical forest bird can measure the slight changes in tropical photoperiod. *Proceedings of the Royal Society of London B* 265: 89–95.

Hau, M., Wikelski, M., Soma, K. K., and Wingfield, J. C. 2000. Testosterone and year-round territorial aggression in a tropical bird. *General and Comparative Endocrinology* 117: 20–33.

Helm, B., Schwabl, I., and Gwinner, E. 2009. Circannual basis of geographically distinct bird schedules. *Journal of Experimental Biology* 212: 1259–1269.

Husak, J. F., Irschick, D. J., Meyers, J. J., Lailvaux, S. P., and Moore, I. T. 2007. Hormones, sexual signals, and performance of green anole lizards (*Anolis carolinensis*). *Hormones and Behavior* 52: 360–367.

Jenssen, T. A., Lovern, M. B., and Congdon, J. D. 2001. Field-testing the protandry-based mating system for the lizard, *Anolis carolinensis*: Does the model organism have the right model? *Behavioral Ecology and Sociobiology* 50: 162–171.

Johnson, C. H. and Hasting, J. W. 1986. The elusive mechanisms of the circadian clock. *American Scientist* 74: 29–37.

Johnston, J. D. and Skene, D. J. 2015. Regulation of mammalian neuroendocrine physiology and rhythms by melatonin. *Journal of Endocrinology* 226: T187–T189.

Ketterson, E. D. and Nolan, V., Jr. 1999. Adaptation, exaptation, and constraint: A hormonal perspective. *American Naturalist* 154 (Supplement): S4–S25.

Klein, S. L. 2000. The effects of hormones on sex differences in infection: From genes to behavior. *Neuroscience and Biobehavioral Reviews* 24: 627–638.

Krohmer, R. W., Bieganskia, G. J., Baleckaitisa, D. D., Haradab, N., and Balthazartc, J. 2002. Distribution of aromatase immunoreactivity in the forebrain of red-sided garter snakes at the beginning of the winter dormancy. *Journal of Chemical Neuroanatomy* 23: 59–71.

Krohmer, R. W., Boyle, M. H., Lutterschmidt, D. I., and Mason, R. T. 2010. Seasonal aromatase activity in the brain of the male red-sided garter snake. *Hormones and Behavior* 58: 485–492.

Langmore, N. E., Cockrem, J. F., and Candy, E. J. 2002. Competition for male reproductive investment elevates testosterone levels in female dunnocks, *Prunella modularis*. *Proceedings of the Royal Society of London B* 269: 2473–2478.

Lincoln, G. A., Guinness, F., and Short, R. V. 1972. The way in which testosterone controls the social and sexual behavior of the red deer stag (*Cervus elaphus*). *Hormones and Behavior* 3: 375–396.

Lockard, R. B. 1978. Seasonal change in the activity pattern of *Dipodomys spectabilis*. *Journal of Mammalogy* 59: 563–568.

Lockard, R. B. and Owings, D. H. 1974. Seasonal variation in moonlight avoidance by bannertail kangaroo rats. *Journal of Mammalogy* 55: 189–193.

Loher, W. 1972. Circadian control of stridulation in the cricket *Teleogryllus commodus* Walker. *Journal of Comparative Physiology* 79: 173–190.

Loher, W. 1979. Circadian rhythmicity of locomotor behavior and oviposition in female *Teleogryllus commodus*. *Behavioral Ecology and Sociobiology* 5: 383–390.

Mak, G. K., Enwere, E. K., Gregg, C., Pakarainen, T., Poutanen, M., et al. 2007. Male pheromone-stimulated neurogenesis in the adult female brain: Possible role in mating behavior. *Nature Neuroscience* 10: 1003–1011.

Mansukhani, V., Adkins-Regan, E., and Yang, S. 1996. Sexual partner preference in female zebra finches: The role of early hormones and social environment. *Hormones and Behavior* 30: 506–513.

Marler, C. A., Walsberg, G., White, M. L., and Moore, M. C. 1995. Increased energy expenditure due to increased territorial defense in male lizards after phenotypic manipulation. *Behavioral Ecology and Sociobiology* 37: 225–231.

McGlothlin, J. W., Jawor, J. M., and Ketterson, E. D. 2007. Natural variation in a testosterone-mediated trade-off between mating effort and parental effort. *American Naturalist* 170: 864–875.

Moore, M. C. and Kranz, B. 1983. Evidence for androgen independence of male mounting behavior in white-crowned sparrows (*Zonotrichia leucophrys gambelii*). *Hormones and Behavior* 17: 414–423.

Muehlenbein, M. P. and Watts, D. P. 2010. The costs of dominance: Testosterone, cortisol and intestinal parasites in wild male chimpanzees. *BioPsychoSocial Medicine* 4: 1184–1121.

Neal, J. K. and Wade, J. 2007. Courtship and copulation in the adult male green anole: Effects of season, hormone and female contact on reproductive behavior and morphology. *Behavioural Brain Research* 177: 177–185.

Nelson, R. J., Badura, L. L., and Goldman, B. D. 1990. Mechanisms of seasonal cycles of behavior. *Annual Reviews of Psychology* 41: 81–108.

O'Donnell, R. P., Shine, R., and Mason, R. T. 2004. Seasonal anorexia in the male red-sided garter snake, *Thamnophis sirtalis parietalis*. *Behavioral Ecology and Sociobiology* 56: 413–419.

Packer, C. 1994. *Into Africa*. Chicago: University of Chicago Press.

Packer, C., Swanson, A., Ikanda, D., and Kushnir, H. 2011. Fear of darkness, the full moon, and the nocturnal ecology of African lions. *PLoS One* 6: e22285.

Page, T. L. 1985. Clocks and circadian rhythms. In G. A. Kerkut and L. I. Gilbert (eds.), *Comprehensive Insect Physiology, Biochemistry, and Pharmacology*, pp. 577–652. New York: Pergamon Press.

Pengelley, E. T. and Asmundson, S. J. 1974. Circannual rhythmicity in hibernating animals. In E. T. Pengelley (ed.), *Circannual Clocks*, pp. 95–160. New York: Academic Press.

Pereyra, M. E., Sharbaugh, S. M., and Hahn, T. P. 2005. Interspecific variation in photo-induced GnRH plasticity among nomadic cardueline finches. *Brain, Behavior and Evolution* 66: 35–49.

Perfito, N., Zann, R. A., Bentley, G. E., and Hau, M. 2007. Opportunism at work: Habitat predictability affects reproductive readiness in free-living zebra finches. *Functional Ecology* 21: 291–301.

Perrigo, G., Bryant, W. C., and vom Saal, F. S. 1990. A unique neural timing system prevents male mice from harming their own offspring. *Animal Behaviour* 39: 535–539.

Ralph, M. R., Foster, R. G., Davis, F. C., and Menaker, M. 1990. Transplanted suprachiasmatic nucleus determines circadian rhythm. *Science* 247: 975–978.

Ralph, M. R. and Menaker, M. 1988. A mutation of the circadian system in golden hamsters. *Science* 241: 1225–1227.

Reed, W. L., Clark, M. E., Parker, P. G., Raouf, S. A., Arguedas, N., et al. 2006. Physiological effects on demography: A long-term experimental study of testosterone's effects on fitness. *American Naturalist* 167: 667–683.

Rodriguez-Zas, S. L., Southey, B. R., Shemsh, Y., Rubin, E. B., Cohen, M., et al. 2012. Microarray analysis of natural socially regulated plasticity in circadian rhythms of honey bees. *Journal of Biological Rhythms* 27: 12–24.

Roff, D. A. and Fairbairn, D. J. 2007. The evolution and genetics of migration in insects. *BioScience* 57: 155–164.

Rubenstein, D. R. and Wikelski, M. 2003. Seasonal changes in food quality: A proximate cue for reproductive timing in marine iguanas. *Ecology Letters* 84: 3013–3023.

Runfeldt, S. and Wingfield, J. C. 1985. Experimentally prolonged sexual-activity in female sparrows delays termination of reproductive activity in their untreated mates. *Animal Behaviour* 33: 403–410.

Safran, R. J., Adelman, J. S., McGraw, K. J., and Hau, M. 2008. Sexual signal exaggeration affects male physiological state in barn swallows. *Current Biology* 18: R461–R462.

Sapolsky, R. M., Romero, L. M., and Munck, A. U. 2000. How do glucocorticoids influence stress responses? Integrating permissive, suppressive, stimulatory, and preparative actions. *Endocrine Reviews* 21: 55–89.

Schneider, J. S., Stone, M. K., Wynne-Edwards, K. E., Horton, T. H., Lydon, J., et al. 2003. Progesterone receptors mediate male aggression toward infants. *Proceedings of the National Academy of Sciences USA* 100: 2951–2956.

Sinervo, B., Miles, D. B., Frankino, W. A., Klukowski, M., and DeNardo, D. F. 2000. Testosterone, endurance, and darwinian fitness: Natural and sexual selection on the physiological bases of alternative male behaviors in side-blotched lizards. *Hormones and Behavior* 38: 222–233.

Small, T. W., Sharp, P. J., and Deviche, P. 2007. Environmental regulation of the reproductive system in a flexibly breeding Sonoran Desert bird, the rufous-winged sparrow, *Aimophila carpalis*. *Hormones and Behavior* 51: 483–495.

Soma, K. K., Tramontin, A. D., and Wingfield, J. C. 2000. Oestrogen regulates male aggression in the non-breeding season. *Proceedings of the Royal Society of London B* 267: 1089–1092.

Spoelstra, K., Wikelski, M., Daand, S., Loudone, A. S. I., and Hau, M. 2016. Natural selection against a circadian clock gene mutation in mice. *Proceedings of the National Academy of Sciences USA* 113: 686–691.

Stillman, J. H. and Barnwell, F. H. 2004. Relationship of daily and circatidal activity rhythms of the fiddler crab, *Uca princeps*, to the harmonic structure of semidiurnal and mixed tides. *Marine Biology* 144: 473–482.

Strand, C. R., Small, T. W., and Deviche, P. 2007. Plasticity of the rufous-winged sparrow, *Aimophila carpalis*, song control regions during the monsoon-associated summer breeding period. *Hormones and Behavior* 52: 401–408.

Toh, K. L., Jones, C. R., He, Y., Eide, E. J., Hinz, W. A., et al. 2001. An h*Per2* phosphorylation site mutation in familial advanced sleep phase syndrome. *Science* 291: 1040–1043.

Tökölyi, J., McNamara, J. M., Houston, A. I., and Barta, Z. 2012. Timing of avian reproduction in unpredictable environments. *Evolutionary Ecology* 26: 25–42.

Toma, D. P., Bloch, G., Moore, D., and Robinson, G. E. 2000. Changes in *period* mRNA levels in the brain and division of labor in honey bee colonies. *Proceedings of the National Academy of Sciences USA* 97: 6914–6919.

Tomioka, K. and Matsumoto, A. 2015. Circadian molecular clockworks in non-model insects. *Current Opinion in Insect Science* 7: 58–64.

Trainor, B. C., Bird, I. M., Alday, N. A., Schlinger, B. A., and Marler, C. A 2003. Variation in aromatase activity in the medial preoptic area and plasma progesterone is associated with the onset of paternal behavior. *Neuroendocrinology* 78: 36–44.

Turek, F. W., McMillan, J. P., and Menaker, M. 1976. Melatonin: Effects of the circadian rhythms of sparrows. *Science* 194: 1441–1443.

Vitousek, M. N., Rubenstein, D. R., and Wikelski, M. 2007. The evolution of foraging behavior in the Galápagos marine iguana: Natural and sexual selection on body size drives ecological, morphological, and behavioral specialization. In S. M. Reilly, D. B. Miles, and L. D. McBrayer (eds.), *Foraging Behavior in Lizards*, pp. 491–507. Cambridge: Cambridge University Press.

Wingfield, J. C., Lynn, S. E., and Soma, K. K. 2001. Avoiding the "costs" of testosterone: Ecological bases of hormone–behavior interactions. *Brain, Behavior and Evolution* 57: 239–251.

Wingfield, J. C. and Moore, M. C. 1987. Hormonal, social and environmental factors in the reproductive biology of free-living male birds. In D. Crews (ed.), *Psychobiology of Reproductive Behavior: An Evolutionary Perspective*, pp. 149–175. Englewood Cliffs, NJ: Prentice-Hall.

Wingfield, J. C. and Ramenofsky, M. 1997. Corticosterone and facultative dispersal in response to unpredictable events. *Ardea* 85: 155–166.

Xu, X., Coats, J. K., Yang, C. F., Wang, A., Ahmed, O. M., et al. 2012. Modular genetic control of sexually dimorphic behaviors. *Cell* 148: 596–607.

Yapici, N., Kim, Y.-J., Ribiero, C., and Dickson, B. J. 2008. A receptor that mediates the post-mating switch in *Drosophila* reproductive behaviour. *Nature* 451: 33–38.

Young, A. J., Carlson, A. A., Monfort, S. L., Russell, A. F., Bennett, N. C., and Clutton-Brock, T. 2006. Stress and the suppression of subordinate reproduction in cooperatively breeding meerkats. *Proceedings of the National Academy of Sciences USA* 103: 12005–12010.

Young, M. W. 2000. Marking time for a kingdom. *Science* 288: 451–453.

Young, C., Majoloc, B., Heistermannd, M., Schülkea, O., and Ostner, J. 2014. Responses to social and environmental stress are attenuated by strong male bonds in wild macaques. *Proceedings of the National Academy of Sciences USA* 111: 18195–18200.

Zera, A. J., Zhao, Z., and Kaliseck, K. 2007. Hormones in the field: Evolutionary endocrinology of juvenile hormone and ecdysteroids in field populations of the wing-dimorphic cricket *Gryllus firmus*. *Physiological and Biochemical Zoology* 80: 592–606.

Zhu, H., Sauman, I., Yuan, A., Emery-Le, M., Emery, P., and Reppert, S. M. 2008. Cryptochromes define a novel circadian clock mechanism in monarch butterflies that may underlie sun compass navigation. *PLoS Biology* 6: e4.

Zucker, I. 1983. Motivation, biological clocks, and temporal organization of behavior. In E. Satinoff and P. Teitelbaum (eds.), *Handbook of Behavioral Neurobiology: Motivation*, pp. 3–21.New York: Plenum Press.

Zuk, M., Johnsen, T. S., and MacLarty, T. 1995. Endocrine-immune interactions, ornaments, and mate choice in red jungle fowl. *Proceedings of the Royal Society of London B* 260: 205–210.

Chapter 6

Beckmann, C. and Shine, R. 2011. Toad's tongue for breakfast: Exploitation of a novel prey type, the invasive cane toad, by scavenging raptors in tropical Australia. *Biological Invasions* 13: 1447–1455.

Brady, P. C., Gilerson, A. A., Kattawar, G. W., Sullivan, J. M., Twardowski, M. S., et al. 2015. Open-ocean fish reveal an omnidirectional solution to camouflage in polarized environments. *Science* 350: 965–969.

Brower, J. V. Z. 1958. Experimental studies of mimicry in some North American butterflies. 1. The monarch, *Danaus plexippus*, and viceroy, *Limenitis archippus*. *Evolution* 12: 3–47.

Brower, L. P. and Calvert, W. H. 1984. Chemical defence in butterflies. In R. I. Vane-Wright and P. R. Ackery (eds.), *The Biology of Butterflies*. London: Academic Press.

Bura, V. L., Rohwer, V. G., Martin, P. R., and Yack, J. E. 2011. Whistling in caterpillars (*Amorpha juglandis*, Bombycoidea): Sound-producing mechanism and function. *Journal of Experimental Biology* 214: 30–37.

Burger, J. and Gochfeld, M. 2001. Smooth-billed ani (*Crotophaga ani*) predation on butterflies in Mato Grosso, Brazil: Risk decreases with increased group size. *Behavioral Ecology and Sociobiology* 49: 482–492.

Caro, T. M. 1986a. The functions of stotting in Thomson's gazelles: Some tests of the predictions. *Animal Behaviour* 34: 663–684.

Caro, T. M. 1986b. The functions of stotting: A review of the hypotheses. *Animal Behaviour* 34: 649–662.

Christianson, D. and Creel, S. 2010. A nutritionally mediated risk effect of wolves on elk. *Ecology* 91: 1184–1191.

Clutton-Brock, T. H., O'Riain, M. J., Brotherton, P. N. M., Gaynor, D., Kansky, R., et al. 1999. Selfish sentinels in cooperative mammals. *Science* 284: 1640–1644.

Cook, L. M., Grant, B. S., Saccheri, I. J., and Mallet, J. 2012. Selective bird predation on the peppered moth: The last experiment of Michael Majerus. *Biology Letters* 8: 609–612.

Coyne, J. 1998. Not black and white. *Nature* 396: 35–36.

Creel, S. and Christianson, D. 2008. Relationships between direct predation and risk effects. *Trends in Ecology & Evolution* 23: 194–201.

Creel, S., Winnie, J., Jr., Maxwell, B., Hamlin, K., and Creel, M. 2005. Elk alter habitat selection as an antipredator response to wolves. *Ecology* 86: 3387–3397.

Davis Rabosky, A. R., Cox, C. L., Rabosky, D. L., Title, P. O., Holmes, I. A., et al. 2016. Coral snakes predict the evolution of mimicry across New World snakes. *Nature Communications* 7: 11484.

Dawkins, R. 1980. Good strategy or evolutionarily stable strategy? In G. W. Barlow and J. Silverberg (eds.), *Sociobiology: Beyond Nature/Nurture?* pp. 331–367. Boulder, CO: Westview Press.

Dawkins, R. 1989. *The Selfish Gene*. Oxford: Oxford University Press.

de Belle, J. S. and Sokolowski, M. B. 1987. Heredity of *rover/sitter*: Alternative foraging strategies of *Drosophila melanogaster* larvae. *Heredity* 59: 73–83.

Droge, E., Creel, S., Becker, M. S., and M'soka, J. 2017. Risky times and risky places interact to affect prey behaviour. *Nature Ecology & Evolution* 1: 1123–1128.

Eberhard, W. G. 1982. Beetle horn dimorphism: Making the best of a bad lot. *American Naturalist* 119: 420–426.

Fischhoff, I. R., Sundaresan, S. R., Cordingley, J., and Rubenstein, D. I. 2007. Habitat use and movements of plains zebra (*Equus burchelli*) in response to predation danger from lions. *Behavioral Ecology* 18: 725–729.

Fitzpatrick, M. J., Feder, E., Rowe, L., and Sokolowski, M. B. 2007. Maintaining a behaviour polymorphism by frequency-dependent selection on a single gene. *Nature* 447: 210–212.

Flower, T. P., Gribble, M., and Ridley, A. R. 2014. Deception by flexible alarm mimicry in an African bird. *Science* 344: 513–516.

Gallagher, A. J., Creel, S., Wilson, R. P., and Cooke, S. J. 2017. Energy landscapes and the landscape of fear. *Trends in Ecology & Evolution* 32: 88–96.

Gill, F. B. and Wolf, L. L. 1975a. Economics of feeding territoriality in the golden-winged sunbird. *Ecology* 56: 333–345.

Gill, F. B. and Wolf, L. L. 1975b. Foraging strategies and energetics of East African sunbirds at mistletoe flowers. *American Naturalist* 109: 491–510.

Goldbogen, J. A., Calambokidis, J., Oleson, E., Potvin, J., Pyenson, N. D., et al. 2011. Mechanics, hydrodynamics and energetics of blue whale lunge feeding: Efficiency dependence on krill density. *Journal of Experimental Biology* 214: 131–146.

Goodenough, A. E., Little, N., Carpenter, W. S., and Hart, A. G. 2017. Birds of a feather flock together: Insights into starling murmuration behaviour revealed using citizen science. *PLoS One* 12: e0179277.

Grant, B. S. 1999. Fine tuning the peppered moth paradigm. *Evolution* 53: 980–984.

Grant, B. S., Owen, D. F. and Clarke, C. A. 1996. Parallel rise and fall of melanic peppered moths in America and Britain. *Journal of Heredity* 87: 351–357.

Greene, E., Orsak, L. T., and Whitman, D. W. 1987. A tephritid fly mimics the territorial displays of its jumping spider predators. *Science* 236: 310–312.

Gross, M. R. 1996. Alternative reproductive strategies and tactics: Diversity within species. *Trends in Ecology & Evolution* 11: 92–98.

Gross, M. R. and MacMillan, A. M. 1981. Predation and the evolution of colonial nesting in bluegill sunfish (*Lepomis macrochirus*). *Behavioral Ecology and Sociobiology* 8: 163–174.

Hamilton, W. D. 1971. Geometry for the selfish herd. *Journal of Theoretical Biology* 31: 295–311.

Hogan, B. G., Hildrenbrandt, H., Scott-Samuel, N. E., Cuthill, I. C., and Hemelrijk, C. K. 2017. The confusion effect when attacking simulated three-dimensional starling flocks. *Royal Society Open Science* 4: 160564.

Hori, M. 1993. Frequency-dependent natural selection in the handedness of scale-eating cichlid fish. *Science* 260: 216–219.

Howlett, R. J. and Majerus, M. E. N. 1987. The understanding of industrial melanism in the peppered moth (*Biston betularia*) (Lepidoptera: Geometridae). *Biological Journal of the Linnean Society* 30: 31–44.

Ioannou, C. C., Bartumeus, F., Krause, J., and Ruxton, G. D. 2011. Unified effects of aggregation reveal larger prey groups take longer to find. *Proceedings of the Royal Society of London B* 278: 2985–2990.

Johansson, J., Turesson, H., and Persson, A. 2004. Active selection for large guppies, *Poecilia reticulata*, by the pike cichlid, *Crenicichla saxatilis*. *Oikos* 105: 595–605.

Johnsson, J. I. and Sundström, F. 2007. Social transfer of predation risk information reduces food locating ability in European minnows (*Phoxinus phoxinus*). *Ethology* 113: 166–173.

Kettlewell, H. B. D. 1955. Selection experiments on industrial melanism in the Lepidoptera. *Heredity* 9: 323–343.

Krama, T. and Krama, I. 2005. Cost of mobbing call to breeding pied flycatcher, *Ficedula hypoleuca*. *Behavioral Ecology* 16: 37–40.

Krause, J. and Ruxton, G. D. 2002. *Living in Groups*. Oxford: Oxford University Press.

Krebs, J. R., Ryan, J., and Charnov, E. L. 1974. Hunting by expectation or optimal foraging? A study of patch use by chickadees. *Animal Behaviour* 22: 953–964.

Kruuk, H. 1964. Predators and anti-predator behaviour of the black-headed gull (*Larus ridibundus*). *Behaviour Supplements* 11: 1–129.

Laundre, J. W., Hernandez, L., and Altendorf, K. B. 2001. Wolves, elk, and bison: reestablishing the "landscape of fear" in Yellowstone National Park, U.S.A. *Canadian Journal of Zoology* 79: 1401–1409.

Leal, M. 1999. Honest signalling during prey-predator interactions in the lizard *Anolis cristatellus*. *Animal Behaviour* 58: 521–526.

Lemon, W. C. and Barth, R. H. 1992. The effects of feeding rate on reproductive success in the zebra finch, *Taeniopyga guttata*. *Animal Behaviour* 44: 851–857.

Mather, M. H. and Roitberg, B. D. 1987. A sheep in wolf's clothing: Tephritid flies mimic spider predators. *Science* 236: 308–310.

Maynard Smith, J. and Price, G. R. 1973. The logic of animal conflict. *Nature* 246: 15–18.

Meehan, C. J., Olson, E. J., Reudink, M. W., Kyser, T. K., and Curry, R. L. 2009. Herbivory in a spider through exploitation of an ant-plant mutualism. *Current Biology* 19: R892–R893.

Meire, P. M. and Ervynck, A. 1986. Are oystercatchers (*Haemoptopus ostralegus*) selecting the most profitable mussels (*Mytilus edulis*)? *Animal Behaviour* 34: 1427–1435.

Molleman, F. 2011. Puddling: From natural history to understanding how it affects fitness. *Entomologia Experimentalis et Applicata* 134: 107–113.

Morrell, L. J., Ruxton, G. D., and James, R. S. 2011. Spatial positioning in the selfish herd. *Behavioral Ecology* 22: 16–22.

Olofsson, M., Vallin, A., Jakobsson, S., and Wiklund, C. 2011. Winter predation on two species of hibernating butterflies: Monitoring rodent attacks with infrared cameras. *Animal Behaviour* 81: 529–534.

Oosthuizen, J. H., and Davies, R. W. 1994. The biology and adaptations of the hippopotamus leech *Placobdelloides jaegerskioeldi* (Glossiphoniidae) to its host. *Canadian Journal of Zoology* 72: 418–422.

Palmer, A. R. 2010. Scale-eating cichlids: From hand(ed) to mouth. *Journal of Biology* 9: 11.

Pfennig, D. W., Harcombe, W. R., and Pfennig, K. S. 2001. Frequency-dependent Batesian mimicry. *Nature* 410: 323.

Pietrewicz, A. T. and Kamil, A. C. 1977. Visual detection of cryptic prey by blue jays (*Cyanocitta cristata*). *Science* 195: 580–582.

Pitcher, T. 1979. He who hesitates lives: Is stotting anti-ambush behavior? *American Naturalist* 113: 453–456.

Pitman, R. L., Deecke, V. B., Gabriele, C. M., Srinivasan, M., Black, N., et al. 2017. Humpback whales interfering when mammal-eating killer whales attack other species: Mobbing behavior and interspecific altruism? *Marine Mammal Science* 33: 7–58.

Plowright, R. C., Fuller, G. A., and Paloheimo, J. E. 1989. Shell-dropping by northwestern crows: A reexamination of an optimal foraging study. *Canadian Journal of Zoology* 67: 770–771.

Pyke, G. H., Pulliam, H. R., and Charnov, E. L. 1977. Optimal foraging: A selective review of theory and tests. *Quarterly Review of Biology* 52: 137–154.

Quinn, J. L. and Cresswell, W. 2006. Testing domains of danger in the selfish herd: Sparrowhawks target widely spaced redshanks in flocks. *Proceedings of the Royal Society of London B* 273: 2521–2526.

Rudge, D. W. 2006. Myths about moths: A study in contrasts. *Endeavour* 30: 19–23.

Sargent, T. D. 1976. *Legion of Night: The Underwing Moths.* Amherst, MA: University of Massachusetts Press.

Sherratt, T. N. 2001. The coevolution of warning signals. *Proceedings of the Royal Society of London B* 269: 741–746.

Snowberg, L. K. and Benkman, C. W. 2009. Mate choice based on a key ecological performance trait. *Journal of Evolutionary Biology* 22: 762–769.

Stankowich, T. and Campbell, L. A. 2016. Living in the danger zone: Exposure to predators and the evolution of spines and body armor in mammals. *Evolution* 70: 1501–1511.

Stankowich, T., Caro, T., and Cox, M. 2011. Bold coloration and the evolution of aposematism in terrestrial carnivores. *Evolution* 65: 3090–3099.

Stankowich, T. and Romero, A. N. 2016. The correlated evolution of antipredator defences and brain size in mammals. *Proceedings of the Royal Society of London B* 284: 201661857.

Stuart-Fox, D. M., Moussalli, A., Marshall, N. J., and Owens, I. P. F. 2003. Conspicuous males suffer higher predation risk: Visual modelling and experimental evidence from lizards. *Animal Behaviour* 66: 541–550.

Stucker, A. M. M., Venegas, P. J., and Summers, K. 2014. Experimental evidence for predator learning and Müllerian mimicry in Peruvian poison frogs (*Ranitomeya*, Dendrobatidae). *Evolutionary Ecology* 28: 413–426.

Sweeney, B. W. and Vannote, R. L. 1982. Population synchrony in mayflies: A predator satiation hypothesis. *Evolution* 36: 810–821.

Turner, G. F. and Pitcher, T. J. 1986. Attack abatement: A model for group protection by combined avoidance and dilution. *American Naturalist* 128: 228–240.

Urban, M. C. 2007. Risky prey behavior evolves in risky habitats. *Proceedings of the National Academy of Sciences USA* 104: 14377–14382.

Vetter, R. S., Visscher, P. K., and Camazine, S. 1999. Mass envenomations by honey bees and wasps. *The Western Journal of Medicine* 170: 223–227.

Walther, B. A. and Gosler, A. G. 2001. The effects of food availability and distance to protective cover on the winter foraging behaviour of tits (Aves: *Parus*). *Oecologia* 129: 312–320.

Watt, P. J. and Chapman, R. 1998. Whirligig beetle aggregations: What are the costs and the benefits? *Behavioral Ecology and Sociobiology* 42: 179–184.

Whitfield, D. P. 1990. Individual feeding specializations of wintering turnstone *Arenaria interpres*. *Journal of Animal Ecology* 59: 193–211.

Wiersma, P. and Verhulst, S. 2005. Effects of intake rate on energy expenditure, somatic repair, and reproduction of zebra finches. *Journal of Experimental Biology* 208: 4091–4098.

Williams, C. K., Lutz, R. S., and Applegate, R. D. 2003. Optimal group size and northern bobwhite coveys. *Animal Behaviour* 66: 377–387.

Wirsing, A. J., Heithaus, M. R., and Dill, L. M. 2007. Can you dig it? Use of excavation, a risky foraging tactic, by dugongs is sensitive to predation danger. *Animal Behaviour* 74: 1085–1091.

Zach, R. 1979. Shell-dropping: Decision-making and optimal foraging in northwestern crows. *Behaviour* 68: 106–117.

Chapter 7

Aires, R. F., Olivejra, G. A., Oliveira, T. F., Ros, A. F. H., and Oliveria, R. F. 2015. Dear enemies elicit lower androgen responses to territorial challenges than unfamiliar intruders in a cichlid fish. *PLoS One* 10: e0137705.

Alcock, J. and Bailey, W. J. 1997. Success in territorial defence by male tarantula hawk wasps *Hemipepsis ustulata*: The role of residency. *Ecological Entomology* 22: 377–383.

Alerstam, T., Hedenstrom, A., and Akesson, S. 2003. Long-distance migration: Evolution and determinants. *Oikos* 103: 247–260.

Anderson, J. B. and Brower, L. P. 1996. Freeze-protection of overwintering monarch butterflies in Mexico: Critical role of the forest as a blanket and an umbrella. *Ecological Entomology* 21: 107–116.

Anderson, R. C. 2009. Do dragonflies migrate across the western Indian Ocean? *Journal of Tropical Ecology* 25: 347–358.

Angelier, F., Vleck, C. M., Holberton, R. L., and Marra, P. P. 2013. Telomere length, non-breeding habitat and return rate in male American redstarts. *Functional Ecology* 27: 342–350.

Baird, T. A. and Curtis, J. L. 2010. Context-dependent acquisition of territories by male collared lizards: The role of mortality. *Behavioral Ecology* 21: 753–758.

Baker, A. J., Gonzalez, P. M., Piersma, T., Niles, L. J., do Nascimento, I. D. S., et al. 2004. Rapid population decline in red knots: Fitness consequences of decreased refuelling rates and late arrival in Delaware Bay. *Proceedings of the Royal Society of London B* 275: 875–882.

Beletsky, L. D. and Orians, G. H. 1989. Territoriality among male red-winged blackbirds. III. Testing hypotheses of territorial dominance. *Behavioral Ecology and Sociobiology* 24: 333–339.

Bell, C. P. 2000. Process in the evolution of bird migration and pattern in avian ecogeography. *Journal of Avian Biology* 31: 258–265.

Bergman, M., Gotthard, K., Berger, D., Olofsson, M., Kemp, D. J., and Wiklund, C. 2007. Mating success of resident versus non-resident males in a territorial butterfly. *Proceedings of the Royal Society of London B* 274: 1659–1665.

Berthold, P., Helbig, A. J., Mohr, G., and Querner, U. 1992. Rapid microevolution of migratory behaviour in a wild bird species. *Nature* 360: 668–670.

Bonte, D., Van Dyck, H., Bullock, J. M., Coulon, A., Delgado, M., et al. 2012. Costs of dispersal. *Biological Reviews* 87: 290–312.

Boyce, M. S. and McDonald, L. L. 1999. Relating populations to habitats using resource selection functions. *Trends in Ecology & Evolution* 14: 268–272.

Brower, L. P. 1996. Monarch butterfly orientation: Missing pieces of a magnificent puzzle. *Journal of Experimental Biology* 199: 93–103.

Brower, L. P., Fink, L. S., and Walford, P. 2006. Fueling the fall migration of the monarch butterfly. *Integrative and Comparative Biology* 46: 1123–1142.

Brown, J. L. 1964. The evolution of diversity in avian territorial systems. *Wilson Bulletin* 76: 160–169.

Brown, J. L. 1975. *The Evolution of Behavior*. New York: W. W. Norton.

Bull, C. M. and Freake, M. J. 1999. Home-range fidelity in the Australian sleepy lizard, *Tiliqua rugosa*. *Australian Journal of Zoology* 47: 125–132.

Buston, P. M. 2004. Territory inheritance in clownfish. *Proceedings of the Royal Society of London B* 271: S252–S254.

Calvert, W. H. and Brower, L. P. 1986. The location of monarch butterfly (*Danaus plexippus* L.) overwintering colonies in Mexico in relation to topography and climate. *Journal of the Lepidopterists' Society* 40: 164–187.

Caro, T. (ed.). 1998. *Behavioral Ecology and Conservation Biology*. New York: Oxford University Press.

Carpenter, S. J., Erickson, J. M., and Holland, F. D. 2003. Migration of a Late Cretaceous fish. *Nature* 423: 70–74.

Clark, T. D., Furey, N. B., Rechisky, E. L., Gale, M. K., Jeffries, K. M., et al. 2016. Tracking wild sockeye salmon smolts to the ocean reveals distinct regions of nocturnal movement and high mortality. *Ecological Applications* 26: 959–978.

Coombs, W. P., Jr. 1990. Behavior patterns of dinosaurs. In D. B. Weishampel, P. Dodson, and H. Osmólska (eds.), *The Dinosauria*, pp. 32–42. Berkeley: University of California Press.

Cox, G. W. 1985. The evolution of avian migration systems between temperate and tropical regions of the New World. *American Naturalist* 126: 452–474.

Davies, N. B. 1978. Territorial defence in the speckled wood butterfly (*Pararge aegeria*): The resident always wins. *Animal Behaviour* 26: 138–147.

Fischer, K., Perlick, J., and Galetz, T. 2008. Residual reproductive value and male mating success: Older males do better. *Proceedings of the Royal Society of London B* 275: 1517–1524.

Fisher, J. 1954. Evolution and bird sociality. In J. Huxley, A. C. Hardy, and E. B. Ford (eds.), *Evolution as a Process*, pp. 71–83. London: Allen & Unwin.

Franchini, P., Irisarri, I., Fudickar, A., Schmidt, A., Meyer, A., et al. 2017. Animal tracking meets migration genomics: Transcriptomic analysis of a partially migratory bird species. *Molecular Ecology* 26: 3204–3216.

Fretwell, S. D. and Lucas, H. K., Jr. 1969. On territorial behavior and other factors influencing habitat distribution in birds. I. Theoretical development. *Acta Biotheoretica* 19: 16–36.

Fricke, H. C., Hencecroth, J., and Hoerner, M. E. 2011. Lowland-upland migration of sauropod dinosaurs during the Late Jurassic epoch. *Nature* 480: 513–515.

Furey, N. B., Hinch, S. G., Bass, A. L., Middleton, C. T., Minke-Martin, V., and Lotto, A. G. 2016. Predator swamping reduces predation risk during nocturnal migration of juvenile salmon in a high-mortality landscape. *Journal of Animal Ecology* 85: 948–959.

Greenwood, P. J. 1980. Mating systems, philopatry, and dispersal in birds and mammals. *Animal Behaviour* 28: 1140–1162.

Hasselquist, D. 1998. Polygyny in great reed warblers: A long-term study of factors contributing to fitness. *Ecology* 79: 2376–2350.

Hawkes, L. A., Balachandran, S., Batnayar, N., Butler, P. J., Frappell, P. B., and Milsom, W. K. 2011. The trans-Himalayan flights of bar-headed geese (*Anser indicus*). *Proceedings of the National Academy of Sciences USA* 108: 9516–9519.

Hedenström, A. 2010. Extreme endurance migration: What is the limit to non-stop flight? *PLoS Biology* 8: e1000362.

Hinsch, M. and Komdeur, J. 2017. What do territory owners defend against? *Proceedings of the Royal Society of London B* 284: 20162356.

Holekamp, K. E. 1984. Natal dispersal in Belding's ground squirrels (*Spermophilus beldingi*). *Behavioral Ecology and Sociobiology* 16: 21–30.

Höner, O. P., Wachter, B., East, M. L., Streich, W. J., Wilhelm, K., et al. 2007. Female mate-choice drives the evolution of male-biased dispersal in a social mammal. *Nature* 448: 798–801.

Hopcraft, J. G., Morales, J. M., Beyer, H. L., Borner, M., Mwangomo, E., et al. 2014. Competition, predation, and migration: Individual choice patterns of Serengeti migrants captured by hierarchical models. *Ecological Monographs* 84: 355–372.

Hyman, J., Hughes, M., Searcy, W. A., and Nowicki, S. 2004. Individual variation in the strength of territory defense in male song sparrows: Correlates of age, territory tenure, and neighbor aggressiveness. *Behaviour* 141: 15–27.

Jiménez, J. A., Hughes, K. A., Alaks, G., Graham, L., and Lacy, R. C. 1994. An experimental study of inbreeding depression in a natural habitat. *Science* 266: 271–273.

Kays, R., Crofoot, M. C., Jetz, W., and Wikelski, M. 2015. Terrestrial animal tracking as an eye on life and planet. *Science* 348: aaa2478.

Kemp, D. J. 2002. Sexual selection constrained by life history in a butterfly. *Proceedings of the Royal Society of London B* 269: 1341–1345.

Kemp, D. J. and Wiklund, C. 2003. Residency effects in animal contests. *Proceedings of the Royal Society of London B* 271: 1707–1711.

Klaassen, R. H. G., Hake, M., Strandberg, R., Koks, B. J., Trierweiler, C., et al. 2014. When and where does mortality occur in migratory birds? Direct evidence from long-term satellite tracking of raptors. *Journal of Animal Ecology* 83: 176–184.

Kramer, G. R., Andersen, D. E., Buehler, D. A., Wood, P. B., Peterson, S. M., et al. 2018. Population trends in *Vermivora* warblers are linked to strong migratory connectivity. *Proceedings of the National Academy of Sciences USA* 115: E3192–E3200.

Krebs, J. R. 1982. Territorial defence in the great tit (*Parus major*): Do residents always win? *Behavioral Ecology and Sociobiology* 11: 185–194.

Lacey, E. A. and Wieczorek, J. R. 2001. Territoriality and male reproductive success in arctic ground squirrels. *Behavioral Ecology* 12: 626–631.

Lank, D. B., Butler, R. W., Ireland, J., and Ydenberg, R. C. 2003. Effects of predation danger on migration strategies of sandpipers. *Oikos* 103: 303–319.

Latta, S. C. and Brown, C. 1999. Autumn stopover ecology of the blackpoll warbler (*Dendroica striata*) in thorn scrub forest of the Dominican Republic. *Canadian Journal of Zoology* 77: 1147–1156.

Levey, D. J. and Stiles, F. G. 1992. Evolutionary precursors of long-distance migration: Resource availability and movement patterns in Neotropical landbirds. *American Naturalist* 140: 447–476.

Lucia, K. E. and Keane, B. 2012. A field test of the effects of familiarity and relatedness on social associations and reproduction in prairie voles. *Behavioral Ecology and Sociobiology* 66: 13–27.

Lundberg, P. 1985. Dominance behavior, body-weight and fat variations, and partial migration in European blackbirds *Turdus merula*. *Behavioral Ecology and Sociobiology* 17: 185–189.

Lundberg, P. 1988. The evolution of partial migration in birds. *Trends in Ecology & Evolution* 3: 172–176.

Mandel, J. T., Bildstein, K. L., Bohrer, G., and Winkler, D. W. 2008. Movement ecology of migration in turkey vultures. *Proceedings of the National Academy of Sciences USA* 105: 19102–129107.

Manley, B. F. J., McDonald, L. L., Thomas, D. L., McDonald, T. L., and Erickson, W. P. 2004. *Resource Selection by Animals: Statistical Design and Analysis for Field Studies*. New York: Kluwer Academic Publishers.

Marden, J. H. and Waage, J. K. 1990. Escalated damselfly territorial contests and energetic wars of attrition. *Animal Behaviour* 39: 954–959.

Margulis, S. W. and Altmann, J. 1997. Behavioural risk factors in the reproduction of inbred and outbred oldfield mice. *Animal Behaviour* 54: 397–408.

Marler, C. A. and Moore, M. C. 1989. Time and energy costs of aggression in testosterone-implanted free-living male mountain spiny lizards (*Sceloporus jarrovi*). *Physiological Zoology* 62: 1334–1350.

Marler, C. A. and Moore, M. C. 1991. Supplementary feeding compensates for testosterone-induced costs of aggression in male mountain spiny lizards, *Sceloporus jarrovi*. *Animal Behaviour* 42: 209–219.

Marra, P. P., Hobson, K. A., and Holmes, R. T. 1998. Linking winter and summer events in a migratory bird by using stable-carbon isotopes. *Science* 282: 1884–1886.

Marra, P. P. and Holmes, R. T. 2001. Consequences of dominance-mediated habitat segregation in American Redstarts during the nonbreeding season. *Auk* 118: 92–104.

Marvin, G. A. 2001. Age, growth, and long-term site fidelity in the terrestrial plethodontid salamander *Plethodon kentucki*. *Copeia* 2001: 108–117.

Maynard Smith, J. 1974. The theory of games and the evolution of animal conflicts. *Journal of Theoretical Biology* 47: 209–221.

Maynard Smith, J. and Parker, G. A. 1976. The logic of asymmetric contests. *Animal Behaviour* 24: 159–175.

McGowan, C. P., Smith, D. R., Sweka, J. A., Martin, J., Nichols, J. D., et al. 2011. Multispecies modeling for adaptive management of horseshoe crabs and red knots in the Delaware Bay. *Natural Resource Modeling* 24: 117–156.

McLoughlin, P. D., Morris, D. W., Fortin, D., Vander Wal, E., and Contasti, A. 2010. Considering ecological dynamics in resource selection functions. *Journal of Animal Ecology* 79: 4–12.

McNair, D. B., Massiah, E. B., and Frost, M. D. 2002. Ground-based autumn migration of blackpoll warblers at Harrison Point, Barbados. *Caribbean Journal of Science* 38: 239–248.

Mitani, J. C., Watts, D. P., and Amsler, S. J. 2010. Lethal intergroup aggression leads to territorial expansion in wild chimpanzees. *Current Biology* 20: R507–R508.

Musiega, D. E., Kazadi, S. N., and Fukuyama, K. 2006. A framework for predicting and visualizing the East African wildebeest migration-route patterns in variable climatic conditions using geographic information system and remote sensing. *Ecological Research* 21: 530–543.

Niles, L. J., Bart, J., Sitters, H. P., Dey, A. D., Clark, K. E., et al. 2009. Effects of horseshoe crab harvest in Delaware Bay on red knots: Are harvest restrictions working? *BioScience* 59: 153–164.

Norris, D. R., Marra, P. P., Kyser, T. K., Sherry, T. W., and Ratcliffe, L. M. 2004. Tropical winter habitat limits reproductive success on the temperate breeding grounds in a migratory bird. *Proceedings of the Royal Society of London B* 271: 59–64.

O'Neill, K. M. 1983. Territoriality, body size, and spacing in males of the bee wolf *Philanthus basilaris* (Hymenoptera; Sphecidae). *Behaviour* 86: 295–321.

Orians, G. H. 1969. On the evolution of mating systems in birds and mammals. *American Naturalist* 103: 589–603.

Outlaw, D. C., Voelker, G., Mila, B., and Girman, D. J. 2003. Evolution of long-distance migration in and historical biogeography of *Catharus* thrushes: A molecular phylogenetic approach. *Auk* 120: 299–310.

Papaj, D. R. and Messing, R. H. 1998. Asymmetries in physiological state as a possible cause of resident advantage in contests. *Behaviour* 135: 1013–1030.

Piper, W. H. 2011. Making habitat selection more "familiar": A review. *Behavioral Ecology and Sociobiology* 65: 1329–1351.

Plaistow, S. and Siva-Jothy, M. T. 1996. Energetic constraints and male mate securing tactics in the damselfly *Calopteryx splendens xanthosoma* (Charpentier). *Proceedings of the Royal Society of London B* 263: 1233–1238.

Powell, G. V. N. and Bjork, R. D. 2004. Habitat linkages and the conservation of tropical biodiversity as indicated by seasonal migrations of three-wattled bellbirds. *Conservation Biology* 18: 500–509.

Pryke, S. R. and Andersson, S. 2003. Carotenoid-based epaulettes reveal male competitive ability: Experiments with resident and floater red-shouldered widowbirds. *Animal Behaviour* 66: 217–224.

Pusey, A. E. and Wolf, M. 1996. Inbreeding avoidance in animals. *Trends in Ecology & Evolution* 11: 201–206.

Quaintenne, G., van Gils, J. A., Bocher, P., Dekinga, A., and Piersma, T. 2011. Scaling up ideals to freedom: Are densities of red knots across western Europe consistent with ideal free distribution? *Proceedings of the Royal Society of London B* 278: 2728–2736.

Rachlow, J. L., Berkeley, E. V., and Berger, J. 1998. Correlates of male mating strategies in white rhinos (*Ceratotherium simum*). *Journal of Mammalogy* 79: 1317–1324.

Ralls, K., Brugger, K., and Ballou, J. 1979. Inbreeding and juvenile mortality in small populations of ungulates. *Science* 206: 1101–1103.

Ramirez, M. I., Azcarate, J. G., and Luna, L. 2003. Effects of human activities on monarch butterfly habitat in protected mountain forests, Mexico. *Forestry Chronicle* 79: 242–246.

Rubenstein, D. R., Chamberlain, C. P., Holmes, R. T., Ayres, M. P., Waldbauer, J. R., et al. 2002. Linking breeding and wintering ranges of a migratory songbird using stable isotopes. *Science* 295: 1062–1065.

Ruegg, K. C. and Smith, T. B. 2002. Not as the crow flies: A historical explanation for circuitous migration in Swainson's thrush (*Catharus ustulatus*). *Proceedings of the Royal Society of London B* 269: 1375–1381.

Sandberg, R. and Moore, F. R. 1996. Migratory orientation of red-eyed vireos, *Vireo olivaceus*, in relation to energetic condition and ecological context. *Behavioral Ecology and Sociobiology* 39: 1–10.

Schwabl, H. 1983. Auspragung und Bedeutung des Teilzugverhaltnes einer sudwestdeutschen Population der Amsel *Turdus merula*. *Journal für Ornithologie* 124: 101–116.

Shaffer, S. A., Tremblay, Y., Weimerskirch, H., Scott, D., Thompson, D. R., et al. 2006. Migratory shearwaters integrate oceanic resources across the Pacific Ocean in an endless summer. *Proceedings of the National Academy of Sciences USA* 103: 12799–12802.

Sherry, T. W. 2018. Identifying migratory birds' population bottlenecks in time and space. *Proceedings of the National Academy of Sciences USA* 115: 3515–3517

Shier, D. M. and Swaisgood, R. R. 2011. Fitness costs of neighborhood disruption in translocations of a solitary mammal. *Conservation Biology* 26: 116–123.

Tobias, J. 1997. Asymmetric territorial contests in the European robin: The role of settlement costs. *Animal Behaviour* 54: 9–21.

Trochet, A., Courtois, E. A., Stevens, V. M., Baguette, M. 2016. Evolution of sex-biased dispersal. *Quarterly Review of Biology* 91: 297–321.

Urquhart, F. A. 1960. *The Monarch Butterfly*. Toronto: University of Toronto Press.

Weber, J. M. 2009. The physiology of long-distance migration: Extending the limits of endurance metabolism. *Journal of Experimental Biology* 212: 593–597.

Webster, M. S., Marra, P. P., Haig, S. M., Bensch, S., and Holmes, R. T. 2002. Links between worlds: Unraveling migratory connectivity. *Trends in Ecology & Evolution* 17: 76–83.

Weidinger, K. 2000. The breeding performance of blackcap *Sylvia atricapilla* in two types of forest habitat. *Ardea* 88: 225–233.

Weimerskirch, H., Martin, J., Clerquin, Y., Alexandre, P., and Jiraskova, S. 2001. Energy saving in flight formation. *Nature* 413: 697–698.

Welty, J. 1982. *The Life of Birds*. Philadelphia, PA: Saunders College Publishing.

Whiting, M. J. 1999. When to be neighbourly: Differential agonistic responses in the lizard *Platysaurus broadleyi*. *Behavioral Ecology and Sociobiology* 46: 210–214.

Wikelski, M., Kays, R. W., Kasdin, N. J., Thorup, K., Smith, J. A., and Swenson, G. W. J. 2007. Going wild: What a global small-animal tracking system could do for experimental biologists. *Journal of Experimental Biology* 210: 181–186.

Williams, T. C. and Williams, J. M. 1978. An oceanic mass migration of land birds. *Scientific American* 239: 166–176.

Winger, B. M., Barker, F. K., and Ree, R. H. 2014. Temperate origins of long-distance seasonal migration in New World songbirds. *Proceedings of the National Academy of Sciences USA* 111: 12115–12120.

Winger, B. M., Lovette, I. J., and Winkler, D. W. 2011. Ancestry and evolution of seasonal migration in the Parulidae. *Proceedings of the Royal Society of London B* 279: 610–618.

Winker, K. and Pruett, C. L. 2006. Seasonal migration, speciation, and morphological convergence in the genus *Catharus* (Turdidae). *Auk* 123: 1052–1068.

Winkler, D. W., Gandoy, F. A., Areta, J. I., Iliff, M. J., Rakhimberdiev, E., et al. 2017. Long-distance range expansion and rapid adjustment of migration in a newly established population of barn swallows breeding in Argentina. *Current Biology* 27: 10801084.

Wolanski, E., Gereta, E., Borner, M., and Mduma, S. 1999. Water, migration, and the Serengeti ecosystem. *American Scientist* 87: 526–533.

Yoder, J. M., Marschall, E. A., and Swanson, D. A. 2004. The cost of dispersal: Predation as a function of movement and site familiarity in ruffed grouse. *Behavioral Ecology* 15: 469–476.

Zeh, D. W. and Zeh, J. A. 1992. Dispersal-generated sexual selection in a beetle-riding pseudoscorpion. *Behavioral Ecology and Sociobiology* 30: 135–142.

Zeh, J. A. 1997. Polyandry and enhanced reproductive success in the harlequin beetle-riding pseudoscorpion. *Behavioral Ecology and Sociobiology* 40: 111–118.

Zuniga, D., Gager, Y., Kokko, H., Fudickar, A. M., Schmidt, A., et al. 2017. Migration confers winter survival benefits in a partially migratory songbird. *eLife* 6: e28123.

Chapter 8

Alcock, J. 2000. Interactions between the sexually deceptive orchid *Spiculaea ciliata* and its wasp pollinator *Thynnoturneria* sp. (Hymenoptera: Thynninae). *Journal of Natural History* 34: 629–636.

Alcock, J. and Bailey, W. J. 1995. Acoustical communication and the mating system of the Australian whistling moth *Hecatesia exultans* (Noctuidae: Agaristidae). *Journal of Zoology* 237, 337–352.

Andersson, M. 1982. Female choice selects for extreme tail length in a widowbird. *Nature* 299: 818–820.

Andersson, S. and Prager, M. 2006. Quantifying colors. In G. E. Hill and K. J. McGraw (eds.), *Bird Coloration: Mechanisms and Measurements*, pp. 41–89. Cambridge, MA: Harvard University Press.

Andersson, S., Pryke, S. R. S., Ornborg, J., Lawes, M. J., and Andersson, M. 2002. Multiple receivers, multiple ornaments, and a trade-off between agonistic and epigamic signaling in a widowbird. *American Naturalist* 160: 683–691.

Bailey, N. W. and Zuk, M. 2008. Acoustic experience shapes female mate choice in field crickets. *Proceedings of the Royal Society of London B* 275: 2645–2650.

Barbero, F., Bonelli, S., Thomas, J. A., Balletto, E., and Schonrogge, K. 2009. Acoustical mimicry in a predatory social parasite of ants. *Journal of Experimental Biology* 212: 4084–4090.

Basolo, A. L. 1990. Female preference predates the evolution of the sword in swordtail fish. *Science* 250: 808–810.

Basolo, A. L. 1995. Phylogenetic evidence for the role of a pre-existing bias in sexual selection. *Proceedings of the Royal Society of London B* 259: 307–311.

Baugh, A. T. and Ryan, M. J. 2011. The relative value of call embellishment in túngara frogs. *Behavioral Ecology and Sociobiology* 65: 359–367.

Beekman, M., Gloag, R. S., Even, N., Wattanachaiyingchareon, W., and Oldroyd, B. P. 2008. Dance precision of *Apis florea*: Clues to the evolution of the honeybee dance language? *Behavioral Ecology and Sociobiology* 62: 1259–1265.

Bernal, X. E., Page, R. A., Rand, A. S., and Ryan, M. J. 2007. Cues for eavesdroppers: Do frog calls indicate prey density and quality? *American Naturalist* 169: 409–415.

Biesmeijer, J. C. and Seeley, T. D. 2005. The use of waggle dance information by honey bees throughout their foraging careers. *Behavioral Ecology and Sociobiology* 59: 133–142.

Borgia, G. 2006. Preexisting traits are important in the evolution of elaborated male sexual display. *Advances in the Study of Behavior* 36: 249–303.

Bradbury, J. W. and Vehrencamp, S. L. 2011. *Principles of Animal Communication* (2nd ed.). Sunderland, MA: Sinauer Associates

Brandt, Y. 2003. Lizard threat display handicaps endurance. *Proceedings of the Royal Society of London B* 270: 1061–1068.

Burgener, N., Dehnhard, M., Hofer, H., and East, M. L. 2009. Does anal gland scent signal identity in the spotted hyaena? *Animal Behaviour* 77: 707–715.

Burley, N. T. and Symanski, R. 1998. "A taste for the beautiful": Latent aesthetic mate preferences for white crests in two species of Australian grassfinches. *American Naturalist* 152: 792–802.

Calleia, F. D. O., Rohe, F., and Gordo, M. 2009. Hunting strategy of the margay (*Leopardus wiedii*) to attract the wild tamarin (*Saguinus bicolor*). *Neotropical Primates* 16: 31–34.

Cardinal, S. and Danforth, B. N. 2011. The antiquity and evolutionary history of social behavior in bees. *PLoS ONE* 6: e21086.

Carlsen, S. M., Jacobsen, G., and Romundstad, P. 2006. Maternal testosterone levels during pregnancy are associated with offspring size at birth. *European Journal of Endocrinology* 155: 365–370.

Christy, J. H. 1995. Mimicry, mate choice, and the sensory trap hypothesis. *American Naturalist* 146: 171–181.

Chuang, C.-Y., Yang, E.-C., and Tso, I.-M. 2008. Deceptive color signaling in the night: A nocturnal predator attracts prey with visual lures. *Behavioral Ecology* 19: 237–244.

Conley, A. J., Corbin, C. J., Browne, P., Mapes, S. M., Place, N. J., et al. 2006. Placental expression and molecular characterization of aromatase cytochrome P450 in the spotted hyena (*Crocuta crocuta*). *Placenta* 28: 668–675.

Crespi, B. J. 2000. The evolution of maladaptation. *Heredity* 84: 623–629.

Crews, D. (ed.). 1987. *Psychobiology of Reproductive Behavior: An Evolutionary Perspective*. Prentice-Hall, Englewood Cliffs, NJ.

Crews, D. and Moore, M. C. 1986. Evolution of mechanisms controlling mating behavior. *Science* 231: 121–125.

Crothers, L. R. and Cummings, M. E. 2015. A multifunctional warning signal behaves as an agonistic status signal in a poison frog. *Behavioral Ecology* 26: 560–568.

Dall, S. R. X., Giraldeau, L.-A., Olsson, O., McNamara, J. M., and Stephens, D. W. 2005. Information and its use by animals in evolutionary ecology. *Trends in Ecology & Evolution* 20: 187–193.

Danforth, B. N., Conway, L., and Ji, S. Q. 2003. Phylogeny of eusocial *Lasioglossum* reveals multiple losses of eusociality within a primitively eusocial clade of bees (Hymenoptera : Halictidae). *Systematic Biology* 52: 23–36.

Davies, N. B. and Halliday, T. R. 1978. Deep croaks and fighting assessment in toads *Bufo bufo*. *Nature* 275: 683–685.

Dawkins, R. 1982. *The Extended Phenotype*. San Francisco: W.H. Freeman.

Deere, K. A., Grether, G. F., Sun, A., and Sinsheimer, J. S. 2012. Female mate preference explains countergradient variation in the sexual coloration of guppies (*Poecilia reticulata*). *Proceedings of the Royal Society of London B* 279: 1684–1690.

Dloniak, S. M., French, J. A., Place, N. J., Weldele, M. L., Glickman, S. E., and Holekamp, K. E. 2004. Non-invasive monitoring of fecal androgens in spotted hyenas (*Crocuta crocuta*). *General and Comparative Endocrinology* 135: 51–61.

Dodson, G. N. 1997. Resource defense mating system in antlered flies, *Phytalmia* spp. (Diptera: Tephritidae). *Annals of the Entomological Society of America* 90: 496–504.

Donaldson, M. C., Lachmann, M., and Bergstrom, C. T. 2007. The evolution of functionally referential meaning in a structured world. *Journal of Theoretical Biology* 246: 225–233.

Dornhaus, A. and Chittka, L. 2001. Food alert in bumblebees (*Bombus terrestris*): Possible mechanisms and evolutionary implications. *Behavioral Ecology and Sociobiology* 50: 570–576.

Drea, C. M. 2011. Endocrine correlates of pregnancy in the ring-tailed lemur (*Lemur catta*): Implications for the masculinization of daughters. *Hormones and Behavior* 59: 417–427.

Drea, C. M., Place, N. J., Weldele, M. L., Coscia, E. M., Licht, P., Glickman, S. E. 2002. Exposure to naturally circulating androgens during foetal life incurs direct reproductive costs in female spotted hyenas, but is prerequisite for male mating. *Proceedings of the Royal Society of London B* 269: 1981–1987.

Drea, C. M., Weldele, M. L., Forger, N. G., Coscia, E. M., Frank, L. G., et al. 1998. Androgens and masculinization of genitalia in the spotted hyaena (*Crocuta crocuta*). 2. Effects of prenatal anti-androgens. *Journal of Reproduction and Fertility* 113: 117–127.

East, M. L., Burke, T., Wilhelm, K., Greig, C., and Hofer, H. 2003. Sexual conflicts in spotted hyenas: Male and female mating tactics and their reproductive outcome with respect to age, social status and tenure. *Proceedings of the Royal Society of London B* 270: 1247–1254.

East, M. L., Hofer, H., and Wickler, W. 1993. The erect "penis" is a flag of submission in a female-dominated society: Greetings in Serengeti spotted hyenas. *Behavioral Ecology and Sociobiology* 33: 355–370.

East, M. L., Höner, O. P., Wachter, B., Wilhelm, K., Burke, T., and Hofer, H. 2009. Maternal effects on offspring social status in spotted hyenas. *Behavioral Ecology* 20: 478–483.

Endler, J. A. 1992. Signals, signal conditions, and the direction of evolution. *American Naturalist* 139: 125–153.

Endler, J. A. 1993. Some general comments on the evolution and design of animal communication systems. *Philosophical Transactions of the Royal Society B* 340: 215–225.

Endler, J. A. and Basolo, A. L. 1998. Sensory ecology, receiver biases, and sexual selection. *Trends in Ecology & Evolution* 13: 415–420.

Estes, R. D. 2012. *The Behavior Guide to African Mammals.* Berkeley, California: University of California Press.

Ewer, R. F. 1973. *The Carnivores*. Ithaca, NY: Cornell University Press.

Firebaugh, A. and Haynes, K. J. 2016. Experimental tests of light-pollution impacts on nocturnal insect courtship and dispersal. *Oecologia* 182: 1203–1211.

Frank, L. G. 1986. Social organization of the spotted hyaena (*Crocuta crocuta*). II. Dominance and reproduction. *Animal Behaviour* 34: 1510–1527.

Frank, L. G., Holekamp, H. E., and Smale, L. 1995a. Dominance, demographics, and reproductive success in female spotted hyenas: A long-term study. In A. R. E. Sinclair and P. Arcese (eds.), *Serengeti II: Research, Management, and Conservation of an Ecosystem*, pp. 364–384. Chicago: University of Chicago Press.

Frank, L. G., Weldele, M. L., and Glickman, S. E. 1995b. Masculinization costs in hyaenas. *Nature* 377: 584–585.

Fullard J. H. 1998. The sensory coevolution of moths and bats. In R. R. Hoy, A. N. Popper, and R. R. Fay (eds.), *Comparative Hearing: Insects*, pp. 279–326. New York: Springer.

Gaskett, A. C. and Herberstein, M. E. 2008. Orchid sexual deceit provokes pollinator ejaculation. *American Naturalist* 171: E206–E212.

Girard, M. B., Elias, D. O. and Kasumovic, M. M. 2015. Female preference for multi-modal courtship: Multiple signals are important for male mating success in peacock spiders. *Proceedings of the Royal Society of London B* 282: 20152222.

Girard, M. B., Kasumovic, M. M., and Elias, D. O. 2011. Multi-modal courtship in the peacock spider, *Maratus volans* (O. P. Cambridge, 1874). *PLoS One* 6: e25390.

Glickman, S. E., Cunha, G. R., Drea, C. M., Conley, A. J., and Place, N. J. 2006. Mammalian sexual differentiation: Lessons from the spotted hyena. *Trends in Endocrinology and Metabolism* 17: 349–356.

Glickman, S. E., Frank, L. G., Licht, P., Yalckinkaya, T., Siiteri, P. K., and Davidson, J. 1993. Sexual differentiation of the female spotted hyena: One of nature's experiments. *Annals of the New York Academy of Sciences USA* 662: 135–159.

Golla, W., Hofer, H., and East, M. L. 1999. Within-litter sibling aggression in spotted hyaenas: Effect of maternal nursing, sex, and age. *Animal Behaviour* 58: 715–726.

Gould, S. J. 1981. Hyena myths and realities. *Natural History* 90: 16–24.

Gould, S. J. 1986. Evolution and the triumph of homology, or why history matters. *American Scientist* 74: 60–69.

Goymann, W., East, M. L., and Hofer, H. 2001. Androgens and the role of female "hyperaggressiveness" in spotted hyenas (*Crocuta crocuta*). *Hormones and Behavior* 39: 83–92.

Gray, D. A. and Cade, W. H. 1999. Sex, death, and genetic variation: Natural and sexual selection on cricket song. *Proceedings of the Royal Society of London B* 266: 707–709.

Grether, G. F. 2000. Carotenoid limitation and mate preference evolution: A test of the indicator hypothesis in guppies (*Poecilia reticulata*). *Evolution* 54: 1712–1714.

Grether, G. F. 2010. The evolution of mate preferences, sensory biases, and indicator traits. *Advances in the Study of Behavior* 41: 35–76.

Grether, G. F., Cummings, M. E., and Hudon, J. 2005. Countergradient variation in the sexual coloration of guppies (*Poecilia reticulata*): Drosopterin synthesis balances carotenoid availability. *Evolution* 59: 175–188.

Grether, G. F., Hudon, J., and Endler, J. A. 2001. Carotenoid scarcity, synthetic pteridine pigments, and the evolution of sexual coloration in guppies (*Poecilia reticulata*). *Proceedings of the Royal Society of London B* 268: 1245–1253.

Gruter, C. and Farina, W. M. 2009. The honeybee waggle dance: Can we follow the steps? *Trends in Ecology & Evolution* 5: 242–247.

Hagelin, J. C. 2002. The kinds of traits involved in male–male competition: A comparison of plumage, behavior, and body size in quail. *Behavioral Ecology* 13: 32–41.

Haskell, D. G. 1999. The effect of predation on begging-call evolution in nestling wood warblers. *Animal Behaviour* 57: 893–901.

Hebets, E. A., Barron, A. B., Balakrishnan, C. N., Hauber, M. E., Mason, P. H., and Hoke, K. L. 2016. A systems approach to animal communication. *Proceedings of the Royal Society of London B* 283: 20152889.

Hibbitts, T. J., Whiting, M. J., and Stuart-Fox, D. M. 2007. Shouting the odds: Vocalization signals status in a lizard. *Behavioral Ecology and Sociobiology* 61: 1169–1176.

Hofer, H. and East, M. L. 2000. Conflict management in female-dominated spotted hyenas. In F. Aureli (ed.), *Natural Conflict Resolution*, pp. 232–234. Berkeley: University of California Press.

Hofer, H. and East, M. L. 2003. Behavioral processes and costs of co-existence in female spotted hyenas: A life history perspective. *Evolutionary Ecology* 17: 315–331.

Holekamp, K. E. and Dloniak, S. M. 2010. Intraspecific variation in the behavioral ecology of a tropical carnivore, the spotted hyena. In R. Macedo (ed.), *Advances in the Study of Behavior*, pp. 189–229. New York: Elsevier/Academic Press.

Höner, O. P., Wachter, B., Hofer, H., Wilhelm, K., Thierer, D., et al. 2010. The fitness of dispersing spotted hyaena sons is influenced by maternal social status. *Nature Communications* 1: 7.

Huffard, C. L., Saarman, N., Hamilton, H., and Simison, W. B. 2010. The evolution of conspicuous facultative mimicry in octopuses: An example of secondary adaptation? *Biological Journal of the Linnean Society* 101: 68–77.

Jacob, F. 1977. Evolution and tinkering. *Science* 196: 1161–1166.

Jones, I. L. and Hunter, F. M. 1998. Heterospecific mating preferences for a feather ornament in least auklets. *Behavioral Ecology* 9: 187–192.

Klump, G. M., Kretzschmar, E., and Curio, E. 1986. The hearing of an avian predator and its avian prey. *Behavioral Ecology and Sociobiology* 18: 317–324.

Kruuk, H. 1972. *The Spotted Hyena*. Chicago: University of Chicago Press.

Lappin, A. K., Brandt, Y., Husak, J. F., Macedonia, J. M., and Kemp, D. J. 2006. Gaping displays reveal and amplify a mechanically based index of weapon performance. *American Naturalist* 168: 100–113.

Leech, S. M. and Leonard, M. L. 1997. Begging and the risk of predation in nestling birds. *Behavioral Ecology* 8: 644–646.

Lewis, S. M. 2016. *Silent Sparks: The Wondrous World of Fireflies*. Princeton, New Jersey: Princeton University Press.

Lewis, S. M. and Cratsley, C. K. 2008. Flash signal evolution, mate choice, and predation in fireflies. *Annual Review of Entomology* 53: 293–321.

Lindauer, M. 1961. *Communication among Social Bees*. Cambridge, MA: Harvard University Press.

Lloyd, J. E. 1965. Aggressive mimicry in *Photuris*: Firefly femmes fatales. *Science* 149: 653–654.

Lloyd, J. E. 1966. Studies on the flash communication systems of *Photinus* fireflies. *University of Michigan* 130: 1–95.

Lloyd, J. E. 1975. Aggressive mimicry in *Photuris* fireflies: Signal repertoires by *femmes fatales*. *Science* 197: 452–453.

Logue, D. M., Abiola, I. O., Rains, D., Bailey, N. W., Zuk, M., and Cade, W. H. 2010. Does signalling mitigate the cost of agonistic interactions? A test in a cricket that has lost its song. *Proceedings of the Royal Society of London B* 277: 2571–2575.

Marler, P. 1955. Characteristics of some animal calls. *Nature* 176: 6–8.

Matthews, L. H. 1939. Reproduction in the spotted hyena *Crocuta crocuta* (Erxleben). *Philosophical Transactions of the Royal Society of London B* 230: 1–78.

Maynard Smith, J. 1974. The theory of games and the evolution of animal conflicts. *Journal of Theoretical Biology* 47: 209–221.

McGraw, K. J. 2006a. Mechanics of carotenoid-based coloration. In G. E. Hill and K. J. McGraw (eds.), *Bird Coloration: Mechanisms and Measurement*, pp. 177–242. Cambridge, MA: Harvard University Press.

McGraw, K. J. 2006b. Mechanics of melanin-based coloration. In G. E. Hill and K. J. McGraw (eds.), *Bird Coloration: Mechanisms and Measurement*, pp. 243–294. Cambridge, MA: Harvard University Press.

Meyer, A. 1997. The evolution of sexually selected traits in male swordtail fishes (*Xiphophorus*: Poecilidae). *Heredity* 79: 329–337.

Meyer, A., Morrisey, J. M., and Schartl, M. 1994. Recurrent origin of a sexually selected trait in *Xiphophorus* fishes inferred from a molecular phylogeny. *Nature* 368: 539–542.

Money, J. and Ehrhardt, A. A. 1972. *Man and Woman, Boy and Girl*. Baltimore, MD: Johns Hopkins University Press.

Muller, M. N. and Wrangham, R. 2002. Sexual mimicry in hyenas. *Quarterly Review of Biology* 77: 3–16.

Nakano, R., Skals, N., Takanashi, T., Surlykke, A., Koike, T., et al. 2008. Moths produce extremely quiet ultrasonic courtship songs by rubbing specialized scales. *Proceedings of the National Academy of Sciences USA* 105: 11812–11817.

Nakano, R., Takanashi, T., Surlykke, A., Skals, N., and Ishikawa, Y. 2003. Evolution of deceptive and true courtship songs in moths. *Scientific Reports* 3: 10.1038/srep02003.

Nakano, R., Takanashi, T., Skals, N., Surlykke, A., and Ishikawa, Y. 2010. To females of a noctuid moth, male courtship songs are nothing more than bat echolocation calls. *Biology Letters* 6: 582–584.

Nakano, R., Takanashi, T., and Surlykke, A. 2015. Moth hearing and sound communication. *Journal of Comparative Physiology A* 201: 111–121.

Nieh, J. C. 2004. Recruitment communication in stingless bees (Hymenoptera, Apidae, Meliponini). *Apidologie* 35: 159–182.

Östlund-Nilsson, S. and Holmlund, M. 2003. The artistic three-spined stickleback (*Gasterosteus aculeatus*). *Behavioral Ecology and Sociobiology* 53: 214–220.

Otte, D. 1974. Effects and functions in the evolution of signaling systems. *Annual Review of Ecology, Evolution and Systematics* 5: 385–471.

Partan, S. and Marler, P. 1999. Communication goes multimodal. *Science* 283: 1272–1273.

Place, N. J., Coscia, E. M., Dahl, N. J., Drea, C. M., Holekamp, K. E., et al. 2011. The anti-androgen combination, flutamide plus finasteride, paradoxically suppressed LH and androgen concentrations in pregnant spotted hyenas, but not in males. *General and Comparative Endocrinology* 170: 455–459.

Place, N. J. and Glickman, S. E. 2004. Masculinization of female mammals: Lessons from nature. In L. Baskin (ed.), *Hypospadias and Genital Development*, pp. 243–353. New York: Springer.

Proctor, H. C. 1991. Courtship in the water mite *Neumania papillator*: Males capitalize on female adaptations for predation. *Animal Behaviour* 42: 589–598.

Proctor, H. C. 1992. Sensory exploitation and the evolution of male mating behaviour: A cladistic test. *Animal Behaviour* 44: 745–752.

Prum, R. O. 2006. Anatomy, physics, and evolution of structural colors. In G. E. Hill and K. J. McGraw (eds.), *Bird Coloration: Mechanisms and Measurement*, pp. 295–353. Cambridge, MA: Harvard University Press.

Pryke, S. R. S., Andersson, S., and Lawes, M. J. 2001. Sexual selection of multiple handicaps in red-collared widow-birds: Female choice of tail length but not carotenoid display. *Evolution* 55: 1452–1463.

Quinn, V. S. and Hews, D. K. 2000. Signals and behavioural responses are not coupled in males: Aggression affected by replacement of an evolutionarily lost colour signal. *Proceedings of the Royal Society of London B* 267: 755–758.

Racey, P. A. and Skinner, J. D. 1979. Endocrine aspects of sexual mimicry in spotted hyenas *Crocuta crocuta*. *Journal of Zoology* 187: 315–326.

Rendall, D., Owren, M. J., and Ryan, M. J. 2009. What do animal signals mean? *Animal Behaviour* 78: 233–240.

Ridsdill-Smith, T. J. 1970. The biology of *Hemithynnus hyalinatus* (Hymenoptera, Tiphiidae), a parasite of scarabaeid larvae. *Journal of the Australian Entomological Society* 9: 183–195.

Rodd, F. H., Hughes, K. A., Grether, G. F., and Baril, C. T. 2002. A possible non-sexual origin of mate preference: Are male guppies mimicking fruit? *Proceedings of the Royal Society of London B* 269: 475–481.

Rodríguez, R. L. and Snedden, W. A. 2004. On the functional design of mate preferences and receiver biases. *Animal Behaviour* 68: 427–432.

Rosenthal, G. G. and Evans, C. S. 1998. Female preference for swords in *Xiphophorus helleri* reflects a bias for large apparent size. *Proceedings of the National Academy of Sciences USA* 95: 4431–4436.

Ryan, M. J. 1983. Frequency modulated calls and species recognition in a neotropical frog, *Physaleumus pustulosus*. *Journal of Comparative Physiology* 150: 217–221.

Ryan, M. J. 1985. *The Túngara Frog*. Chicago: University of Chicago Press.

Ryan, M. J., Bernal, X. E. and Rand, A. S. 2010. Female mate choice and the potential for ornament evolution in túngara frogs *Physalaemus pustulosus*. *Current Zoology* 56: 343–357.

Ryan, M. J. and Cummings, M. E. 2013. Perceptual biases and mate choice. *Annual Review of Ecology and Systematics* 44: 437–459.

Ryan, M. J., Fox, J. H., Wilczynski, W., and Rand, A. S. 1990. Sexual selection for sensory exploitation in the frog *Physalaemus pustulosus*. *Nature* 343: 66–67.

Ryan, M. J., Tuttle, M. D., and Taft, L. K. 1981. The costs and benefits of frog chorusing behavior. *Behavioral Ecology and Sociobiology* 8: 273–278.

Ryan, M. J. and Wagner, W. E. Jr. 1987. Asymmetries in mating behavior between species: Female swordtails prefer heterospecific males. *Science* 236: 595–597.

Ryan, M. T. 1998. Sexual selection, receiver biases, and the evolution of sex differences. *Science* 281: 1999–2003.

Schiestl, F. P. 2010. Pollination: Sexual mimicry abounds. *Current Biology* 20: R1020–R1022.

Schiestl, F. P. and Dötteri, S. 2012. The evolution of floral scent and olfactory preferences in pollinators: Coevolution or pre-existing bias? *Evolution* 66: 2042–2055.

Schlaepfer, M. A., Runge, M. C., and Sherman, P. W. 2002. Ecological and evolutionary traps. *Trends in Ecology & Evolution* 17: 474–480.

Seeley, T. D. 2010. *Honeybee Democracy*. Princeton, NJ: Princeton University Press.

Seyfarth, R. M., Cheney, D. L., and Marler, P. 1980. Monkey responses to three different alarm calls: evidence of predator classification and semantic communication. *Science* 210: 801–803.

Smith, J. E., Powning, K. S., Dawes, S. E., Estrada, J. R., Hopper, A. L., et al. 2011. Greetings promote cooperation and reinforce social bonds among spotted hyaenas. *Animal Behaviour* 81: 401–415.

Smith, J. E., Van Horn, R. C., Powning, K. S., Cole, A. R., Graham, K. E., et al. 2010. Evolutionary forces favoring intragroup coalitions among spotted hyenas and other animals. *Behavioral Ecology* 21: 284–303.

Suzuki, T. N. 2011. Parental alarm calls warn nestlings about different predatory threats. *Current Biology* 21: R15–R16.

Tibbetts, E. A. and Dale, J. 2004. A socially enforced signal of quality in a paper wasp. *Nature* 432: 218–222.

Tibbetts, E. A., Forrest, T., Vernier, C., Jinn, J., and Madagame, A. 2015. Socially selected ornaments and fitness: Signals of fighting ability in paper wasps are positively associated with survival, reproductive success, and rank. *Evolution* 69: 2917–2926.

Tibbetts, E. A. and Izzo, A. 2010. Social punishment of dishonest signalers caused by mismatch between signal and behavior. *Current Biology* 20: 1637–1640.

Vanpé, C., Gaillard, J. M., Kjellander, P., Mysterud, A., Magnien, P., et al. 2007. Antler size provides an honest signal of male phenotypic quality in roe deer. *American Naturalist* 169: 481–493.

von Frisch, K. 1965. *The Dance Language and Orientation of Bees*. Heidelberg, Berlin: Springer.

Watts, H. E. and Holekamp, K. E. 2007. Hyena societies. *Current Biology* 17: R657–R660.

Watts, H. E., Tanner, J. B., Lundrigan, B. L., and Holekamp, K. E. 2009. Post-weaning maternal effects and the evolution of female dominance in the spotted hyena. *Proceedings of the Royal Society of London B* 276: 2291–2298.

West-Eberhard, M. J. 1979. Sexual selection, social competition, and evolution. *Proceedings of the American Philosophical Society* 123: 222–234.

Wiens, J. J. 2001. Widespread loss of sexually selected traits: How the peacock lost its spots. *Trends in Ecology & Evolution* 19: 517–523.

Wilkins, M. R., Shizuka, D., Joseph, M. B., Hubbard, J. K., and Safran, R. J. 2015. Multimodal signalling in the North American barn swallow: A phenotype network approach. *Proceedings of the Royal Society of London B* 282: 20151574.

Wilkinson, G. S. and Dodson, G. N. 1997. Function and evolution of antlers and eye stalks in flies. In J. C. Choe and B. J. Crespi (eds.), *The Evolution of Mating Systems in Insects and Arachnids*, pp. 310–328. Cambridge: Cambridge University Press.

Wilson, E. O. 1971. *The Insect Societies*. Cambridge, MA: Harvard University Press.

Wilson, R. S., Angelitta, M. J. Jr, James, R. S., Navas, C., and Seebacher, F. 2007. Dishonest signals of strength in male slender crayfish (*Cherax dispar*) during agonistic encounters. *American Naturalist* 170: 284–291.

Wyatt, T. D. 2009. Pheromones and other chemical communication in animals. *Encyclopedia of Neuroscience* 7: 611–616.

Zuk, M., Rotenberry, J. T., and Tinghitella, R. M. 2006. Silent night: Adaptive disappearance of a sexual signal in a parasitized population of field crickets. *Biology Letters* 2: 521–524.

Chapter 9

Aisenberg, A. and Barrantes, G. 2011. Sexual behavior, cannibalism, and mating plugs as sticky traps in the orb weaver spider *Leucauge argyra* (Tetragnathidae). *Naturwissenschaften* 98: 605–613.

Aisenberg, A. and Eberhard, W. G. 2009. Female cooperation in plug formation in a spider: Effects of male copulatory courtship. *Behavioral Ecology* 20: 1236–1241.

Alberts, S. C., Buchan, J. C., and Altmann, J. 2006. Sexual selection in wild baboons: From mating opportunities to paternity success. *Animal Behaviour* 72: 1177–1196.

Alberts, S. C., Watts, H. E., and Altmann, J. 2003. Queuing and queue-jumping: Long-term patterns of reproductive skew in male savannah baboons, *Papio cynocephalus*. *Animal Behaviour* 65: 821–840.

Alcock, J., Jones, C. E., and Buchmann, S. L. 1977. Male mating strategies in the bee *Centris pallida* Fox (Hymenoptera: Anthophoridae). *American Naturalist* 111: 145–155.

Alonzo, S. H., Stiver, K. A., and Marsh-Rollo, S. E. 2016. Ovarian fluid allows directional cryptic female choice despite external fertilization. *Nature Communications* 7: 12452.

Ancona, S., Drummond, H., and Zaldívar-Rae, J. 2010. Male whiptail lizards adjust energetically costly mate guarding to male-male competition and female reproductive value. *Animal Behaviour* 79: 75–82.

Andersson, M. 1982. Female choice selects for extreme tail length in a widowbird. *Nature* 299: 818–820.

Andersson, M. 1994. *Sexual Selection*. Princeton, NJ: Princeton University Press.

Andrade, M. C. B. 1996. Sexual selection for male sacrifice in the Australian redback spider. *Science* 271: 70–72.

Andrade, M. C. B. 2003. Risky mate search and male self-sacrifice in redback spiders. *Behavioral Ecology* 14: 531–538.

Arnold, S. J. 1983. Sexual selection: The interface of theory and empiricism. In P. P. G. Bateson (ed.), *Mate Choice*, pp. 67–107. Cambridge: Cambridge University Press.

Arnqvist, G. and Kirkpatrick, M. 2005. The evolution of infidelity in socially monogamous passerines: The strength of direct and indirect selection on extrapair copulation behavior in females. *American Naturalist* 165: S26–S37.

Bakst, M. R. 1998. Structure of the avian oviduct with emphasis on sperm storage in poultry. *Journal of Experimental Zoology* 282: 618–626.

Barske, J., Schlinger, B. A., Wikelski, M., and Fusani, L. 2011. Female choice for male motor skills. *Proceedings of the Royal Society of London B* 278: 3523–3528.

Bateman, A. J. 1948. Intra-sexual selection in *Drosophila*. *Heredity* 2: 349–368.

Berglund, A. and Rosenqvist, G. 2001. Male pipefish prefer ornamented females. *Animal Behaviour* 61: 345–350.

Berglund, A., Rosenqvist, G., and Robinson-Wolrath, S. 2006. Food or sex? Males and females in a sex role-reversed pipefish have different interests. *Behavioral Ecology and Sociobiology* 60: 281–287.

Berglund, A., Rosenqvist, G., and Svensson, I. 1986. Mate choice, fecundity, and sexual dimorphism in two pipefish species (Syngnathidae). *Behavioral Ecology and Sociobiology* 19: 301–307.

Birkhead, T. R. 1998. Cryptic female choice: Criteria for establishing female sperm choice. *Evolution* 52: 1212–1218.

Birkhead, T. R. and Møller, A. P. 1992. *Sperm Competition in Birds: Evolutionary Causes and Consequences*. London: Academic Press.

Bjork, A., Dallai, I., and Pitnick, S. 2007. Adaptive modulation of sperm production rate in *Drosophila bifurca*, a species with giant sperm. *Biology Letters* 3: 517–519.

Blanckenhorn, W. U. 2005. Behavioral causes and consequences of sexual size dimorphism. *Ethology* 11: 977–1016.

Blount, J. D., Metcalfe, N. B., Birkhead, T. R., and Surai, P. F. 2003. Carotenoid modulation of immune function and sexual attractiveness in zebra finches. *Science* 300: 125–127.

Bodony, D. J., Day, L. B., Friscia, A. R., Fusani, L., Kharon, A., et al. 2016. Determination of the wingsnap sonation mechanism of the Golden-collared manakin (*Manacus vitellinus*). *Journal of Experimental Biology* 219: 1524–1534.

Borgia, G. 1985. Bower quality, number of decorations and mating success of male satin bowerbirds (*Ptilonorhynchus violaceus*). *Animal Behaviour* 33: 266–271.

Borgia, G. 1986. Sexual selection in bowerbirds. *Scientific American* 254: 92–100.

Borgia, G. 2006. Preexisting traits are important in the evolution of elaborated male sexual display. *Advances in the Study of Behavior* 36: 249–303.

Borgia, G., Egeth, M., Uy, J. A. C., and Patricelli, G. L. 2004. Juvenile infection and male display: Testing the bright male hypothesis across individual life histories. *Behavioral Ecology* 15: 722–728.

Brennan, P. L. R. 2010. Sexual selection. *Nature Education Knowledge* 3: 79.

Brennan, P. L. R., Clark, C., and Prum, R. O. 2010. Explosive eversion and functional morphology of waterfowl penis supports sexual conflict in genitalia. *Proceedings of the Royal Society of London B* 277: 1309–1314.

Brennan, P. L. R. and Prum, R. O. 2012. The limits of sexual conflict in the narrow sense: New insights from waterfowl biology. *Philosophical Transactions of the Royal Society B* 367: 2324–2338.

Brennan, P. L. R., Prum, R. O., McCracken, K. G., Sorenson, M. D., Wilson, R. E., and Birkhead, T. R. 2007. Coevolution of male and female genital morphology in waterfowl. *PLoS One* 2: e418.

Briskie, J. V. and Montgomerie, R. 1997. Sexual selection and the intromittent organ of birds. *Journal of Avian Biology* 28: 73–86.

Brownmiller, S. 1975. *Against Our Will*. New York: Simon and Schuster.

Bruning, B., Phillips, B. L., and Shine, R. 2010. Turgid female toads give males the slip: A new mechanism of female mate choice in the Anura. *Biology Letters* 6: 322–324.

Buskirk, R. E., Frolich, C., and Ross, K. G. 1984. The natural selection of sexual cannibalism. *American Naturalist* 123: 612–625.

Buzatto, B. A. and Machado, G. 2008. Resource defense polygyny shifts to female defense polygyny over the course of the reproductive season of a Neotropical harvestman. *Behavioral Ecology and Sociobiology* 63: 85–94.

Byers, J., Hebets, E., and Podos, J. 2010. Female mate choice based upon male motor performance. *Animal Behaviour* 79: 771–778.

Casselman, S. J. and Montgomerie, R. 2004. Sperm traits in relation to male quality in colonial spawning bluegill. *Journal of Fish Biology* 64: 1700–1711.

Clark, A. G., Begun, D. J. and Prout, T. 1999. Female x male interactions in *Drosophila* sperm competition. *Science* 283: 217–220.

Coker, C. R., McKinney, F., Hays, H., Briggs, S., and Cheng, K. 2002. Intromittent organ morphology and testis size in relation to mating system in waterfowl. *Auk* 119: 403–413.

Cordero, C. and Eberhard, W. G. 2003. Female choice of sexually antagonistic male adaptations: A critical review of some current research. *Journal of Evolutionary Biology* 16: 1–6.

Córdoba-Aguilar, A., Uhía, E., and Rivera, A. C. 2003. Sperm competition in Odonata (Insecta): The evolution of female sperm storage and rivals' sperm displacement. *Journal of Zoology* 261: 381–398.

Cowlishaw, G. and Dunbar, R. I. M. 1991. Dominance rank and mating success in male primates. *Animal Behaviour* 41: 1045–1056.

Cunningham, E. J. A. and Russell, A. F. 2000. Egg investment is influenced by male attractiveness in the mallard. *Nature* 404: 74–77.

Dakin, R. and Montgomerie, R. 2011. Peahens prefer peacocks displaying more eyespots, but rarely. *Animal Behaviour* 82: 21–28.

Daly, M. and Wilson, M. 1983. *Sex, Evolution and Behavior*. Boston: Willard Grant Press.

Darwin, C. 1871. *The Descent of Man and Selection in Relation to Sex*. London: Murray.

Davies, N. B. 1983. Polyandry, cloaca-pecking and sperm competition in dunnocks. *Nature* 302: 334–336.

Dawkins, R. 1980. Good strategy or evolutionarily stable strategy? In G. W. Barlow and J. Silverberg (eds.), *Sociobiology: Beyond Nature/Nurture?* pp. 331–367. Boulder, CO: Westview Press.

Dawkins, R. 1986. *The Blind Watchmaker*. New York: W. W. Norton.

Dean, R., Nakagawa, S., and Pizzari, T. 2011. The risk and intensity of sperm ejection in female birds. *American Naturalist* 178: 343–354.

Dickinson, J. L. 1995. Trade-offs between postcopulatory riding and mate location in the blue milkweed beetle. *Behavioral Ecology* 6: 280–286.

Dickinson, J. L. and Rutowski, R. L. 1989. The function of the mating plug in the chalcedon checkerspot butterfly. *Animal Behaviour* 38: 154–162.

Doucet, S. M. and Montgomerie, R. 2003. Multiple sexual ornaments in satin bowerbirds: ultraviolet plumage and bowers signal different aspects of male quality. *Behavioral Ecology* 14: 503–509.

Drăgăniou, T. I., Nagle, L., and Kreutzer, M. 2002. Directional female preference for an exaggerated male trait in canary (*Serinus canaria*) song. *Proceedings of the Royal Society of London B* 269: 2525–2531.

Drews, C. 1996. Contests and patterns of injuries in free-ranging male baboons (*Papio cynocephalus*). *Behaviour* 133: 443–474.

Eberhard, W. G. 1982. Beetle horn dimorphism: Making the best of a bad lot. *American Naturalist* 119: 420–426.

Eberhard, W. G. 1996. *Female Control: Sexual Selection by Cryptic Female Choice*. Princeton, NJ: Princeton University Press.

Eberhard, W. G. 2005. Evolutionary conflicts of interest: Are female conflicts of interest different? *American Naturalist* 165: S19–S25.

Emlen, D. J. 2008. The evolution of animal weapons. *Annual Review of Ecology, Evolution and Systematics* 39: 387–413.

Emlen, S. T. and Oring, L. W. 1977. Ecology, sexual selection and the evolution of mating systems. *Science* 197: 215–223.

Field, S. A. and Keller, M. A. 1993. Alternative mating tactics and female mimicry as post-copulatory mate-guarding behaviour in the parasitic wasp *Cotesia rubecula*. *Animal Behaviour* 46: 1183–1189.

Fisher, H. S., Giomi, L., Hoekstra, H. E., and Mahadevan, L. 2014. The dynamics of sperm cooperation in a competitive environment. *Proceedings of the Royal Society of London B* 281: 201440296.

Fisher, H. S. and Hoekstra, H. E. 2010. Competition drives cooperation among closely related sperm of deer mice. *Nature* 463: 801–803.

Fisher, R. A. 1930. *The Genetical Theory of Natural Selection*. Oxford: Clarendon Press.

Flanagan, S. P., Johnson, J. B., Rose, E., and Jones, A. G. 2014. Sexual selection on female ornaments in the sex-role-reversed Gulf pipefish (*Syngnathus scovelli*). *Journal of Evolutionary Biology* 27: 2457–2467.

Foellmer, M. W. and Fairbairn, D. J. 2003. Spontaneous male death during copulation in an orb-weaving spider. *Proceedings of the Royal Society of London B* 270: S183–S185.

Folstad, I. and Karter, A. J. 1992. Parasites, bright males, and the immunocompetence handicap. *American Naturalist* 139: 603–622.

Fox, E. A. 2002. Female tactics to reduce sexual harassment in the Sumatran orangutan (*Pongo pygmaeus abelii*). *Behavioral Ecology and Sociobiology* 52: 93–101.

Friberg, U. and Arnqvist, G. 2003. Fitness effects of female mate choice: Preferred males are detrimental for *Drosophila melanogaster* females. *Journal of Evolutionary Biology* 16: 797–811.

Fu, P., Neff, B. D., and Gross, M. R. 2001. Tactic-specific success in sperm competition. *Proceedings of the Royal Society of London B* 268: 1105–1112.

Funk, D. H. and Tallamy, D. W. 2000. Courtship role reversal and deceptive signals in the long-tailed dance fly, *Rhamphomyia longicauda*. *Animal Behaviour* 59: 411–421.

Fusani, L., Day, L. B., Canoine, V., Reinemann, D., Hernandez, E., and Schlinger, B. A. 2007. Androgen and the elaborate courtship behavior of a tropical lekking bird. *Hormones and Behavior* 51: 62–68.

Fuxjager, M. J., Eaton, J., Lindsay, W. R., Salwiczek, L. H., Rensel, M. A., et al. 2015. Evolutionary patterns of adaptive acrobatics and physical performance predict expression profiles of androgen receptor—but not oestrogen receptor—in the forelimb musculature. *Functional Ecology* 29: 1197–1208.

Fuxjager, M. J. and Schlinger, B. A. 2015. Perspectives on the evolution of animal dancing: A case study of manakins. *Current Opinion in Behavioral Sciences* 6: 7–12.

García-Navas, V., Ferrer, E. S. and Sanz, J. J. 2012. Plumage yellowness predicts foraging ability in the blue tit *Cyanistes caeruleus*. *Biological Journal of the Linnean Society* 106: 418–429.

Gesquiere, L. R., Wango, E. O., Alberts, S. C. and Altmann, J. 2007. Mechanisms of sexual selection: Sexual swellings and estrogen concentrations as fertility indicators and cues for male consort decisions in wild baboons. *Hormones and Behavior* 51: 114–125.

Gil, D., Leboucher, G., Lacroix, A., Cue, R. and Kreutzer, M. 2004. Female canaries produce eggs with greater amounts of testosterone when exposed to preferred male song. *Hormones and Behavior* 45: 64–70.

Gilbert, L., Williamson, K. A. and Graves, J. A. 2012. Male attractiveness regulates daughter fecundity non-genetically via maternal investment. *Proceedings of the Royal Society of London B* 279: 523–528.

Gomendio, M. and Roldan, E. R. S. 1991. Sperm competition influences sperm size in mammals. *Proceedings of the Royal Society of London B* 243: 181–185.

Gomes, C. M. and Boesch, C. 2009. Wild chimpanzees exchange meat for sex on a long-term basis. *PLoS One* 4: e5116.

Gowaty, P. A., Kim, and Y.-K., Anderson, W. W. 2012. No evidence of sexual selection in a repetition of Bateman's classic study of *Drosophila melanogaster*. *Proceedings of the National Academy of Sciences USA* 109: 11740–11745.

Gray, E. M. 1997. Female red-winged blackbirds accrue material benefits from copulating with extra-pair males. *Animal Behaviour* 53: 625–639.

Grether, G. F. 2000. Carotenoid limitation and mate preference evolution: A test of the indicator hypothesis in guppies (*Poecilia reticulata*). *Evolution* 54: 1712–1714.

Griffith, S. C., Owens, I. P. F. and Thuman, K. A. 2002. Extra pair paternity in birds: a review of interspecific variation and adaptive function. *Molecular Ecology* 11: 2195–2212.

Griggio, M., Biard, C., Penn, D. J., and Hoi, H. 2011. Female house sparrows "count on" male genes: Experimental evidence for MHC-dependent mate preference in birds. *BMC Evolutionary Biology* 11: doi:10.1186/1471-2148-1111-1144.

Gross, M. R. 1996. Alternative reproductive strategies and tactics: Diversity within species. *Trends in Ecology & Evolution* 11: 92–98.

Gwynne, D. T. 1981. Sexual difference theory: Mormon crickets show role reversal in mate choice. *Science* 213: 779–780.

Gwynne, D. T. 1984. Sexual selection and sexual differences in Mormon crickets (Orthoptera, Tettigoniidae, *Anabrus simplex*). *Evolution* 38: 1011–1022.

Gwynne, D. T. 1993. Food quality controls sexual selection in Mormon crickets by altering male mating investment. *Ecology* 74: 1406–1413.

Gwynne, D. T., Bussiere, L. F., and Ivy, T. M. 2007. Female ornaments hinder escape from spider webs in a role-reversed swarming dance fly. *Animal Behaviour* 73: 1077–1082.

Hamilton, W. D. and Zuk, M. 1982. Heritable true fitness and bright birds: A role for parasites? *Science* 218: 384–387.

Hausfater, G. 1975. Dominance and reproduction in baboons (*Papio cynocephalus*): A quantitative analysis. *Contributions in Primatology* 7: 1–150.

Healy, S. D. and Rowe, C. 2007. A critique of comparative studies of brain size. *Proceedings of the Royal Society of London B* 274: 453–464.

Hidalgo-García, S. 2006. The carotenoid-based plumage coloration of adult blue tits *Cyanistes caeruleus* correlates with the health status of their brood. *Ibis* 148: 727–734.

Hill, G. E. and Montgomerie, R. 1994. Plumage color signals nutritional condition in the house finch. *Proceedings of the Royal Society of London B* 258: 47–52.

Himuro, C. and Fujisaki, K. 2008. Males of the seed bug *Togo hemipterus* (Heteroptera: Lygaeidae) use accessory gland substances to inhibit remating by females. *Journal of Insect Physiology* 54: 1538–1542.

Holland, B. and Rice, W. R. 1998. Chase-away sexual selection: Antagonistic seduction versus resistance. *Evolution* 52: 1–7.

Holland, B. and Rice, W. R. 1999. Experimental removal of sexual selection reverses intersexual antagonistic coevolution and removes a reproductive load. *Proceedings of the National Academy of Sciences USA* 96: 5083–5088.

Hoving, H. J. T., Lipinski, M. R., Videler, J. J., and Bolstad, K. S. R. 2010. Sperm storage and mating in the deep-sea squid *Taningia danae* Joubin, 1931 (Oegopsida: Octopoteuthidae). *Marine Biology* 157: 393–400.

Hunt, J. and Simmons, L. W. 2002. Confidence of paternity and paternal care: Covariation revealed through the experimental manipulation of the mating system in the beetle *Onthophagus taurus*. *Journal of Evolutionary Biology* 15: 784–795.

Johnsen, A., Pärn, H., Fossøy, F., Kleven, O., Laskemoen, T., and Lifjeld, T. J. 2008. Is female promiscuity constrained by the presence of her social mate? An experiment with bluethroats *Luscinia svecica*. *Behavioral Ecology and Sociobiology* 62: 1761–1767.

Jordan, W. C. and Bruford, M. W. 1998. New perspectives on mate choice and the MHC. *Heredity* 81: 239–245.

Jukema, J. and Piersma, T. 2005. Permanent female mimics in a lekking shorebird. *Biology Letters* 2: 161–164.

Keagy, J., Savard, J. F., and Borgia, G. 2011. Complex relationship between multiple measures of cognitive ability and male mating success in satin bowerbirds, *Ptilonorhynchus violaceus*. *Animal Behaviour* 81: 1063–1070.

Kim, Y. J., Bartalska, K., Audsley, N., Yamanaka, N., Yapici, N., et al. 2009. MIPs are ancestral ligands for the sex peptide receptor. *Proceedings of the National Academy of Sciences USA* 107: 6520–6525.

Kirkpatrick, M. 1982. Sexual selection and the evolution of female choice. *Evolution* 36: 1–12.

Kodric-Brown, A. 1993. Female choice of multiple male criteria in guppies: Interacting effects of dominance, coloration and courtship. *Behavioral Ecology and Sociobiology* 32: 415–420.

Kodric-Brown, A. and Brown, J. H. 1984. Truth in advertising: The kinds of traits favored by sexual selection. *American Naturalist* 124: 309–323.

Kokko, H., Jennions, M. D. and Brooks, D. R. 2006. Unifying and testing models of sexual selection. *Annual Review of Ecology, Evolution and Systematics* 37: 43–46.

Komdeur, J., Burke, T. and Richardson, D. S. 2007. Explicit experimental evidence for the effectiveness of proximity as mate-guarding behaviour in reducing extra-pair fertilization in the Seychelles warbler. *Molecular Ecology* 16: 3679–3688.

Komdeur, J., Kraaijeveld-Smit, F., Kraaijeveld, K., and Edelaar, P. 1999. Explicit experimental evidence for the role of mate guarding in minimizing loss of paternity in the Seychelles warbler. *Proceedings of the Royal Society of London B* 266: 2075–2081.

Kraaijeveld, K., Gregurke, J., Hall, C., Komdeur, J., and Mulder, R. A. 2004. Mutual ornamentation, sexual selection, and social dominance in the black swan. *Behavioral Ecology* 15: 380–389.

Küpper, C., Stocks, M., Risse, J. E., dos Remedios, N., Farrell, F. L., et al. 2015. A supergene determines highly divergent male reproductive morphs in the ruff. *Nature Genetics* 48: 79–83.

Lack, D. 1968. *Ecological Adaptations for Breeding in Birds*. London: Methuen.

Lamichhaney, S., Fan, G., Widemo, F., Gunnarsson, U., Thalmann, D. S., et al. 2015. Structural genomic changes underlie alternative reproductive strategies in the ruff (*Philomachus pugnax*). *Nature Genetics* 48: 84–88.

Lande, R. 1981. Models of speciation by sexual selection of polygenic traits. *Proceedings of the National Academy of Sciences USA* 78: 3721–3725.

Lelito, J. P. and Brown, W. D. 2008. Mate attraction by females in a sexually cannibalistic praying mantis. *Behavioral Ecology and Sociobiology* 63: 313–320.

Lewis, S. M. and Cratsley, C. K. 2008. Flash signal evolution, mate choice, and predation in fireflies. *Annual Review of Entomology* 53: 293–321.

Li, D., Oh, J., Kralj-Fisher, S., and Kuntner, M. 2012. Remote copulation: Male adaptation to female cannibalism. *Biology Letters* 8: 512–515.

Low, M. 2005. Female resistance and male force: Context and patterns of copulation in the New Zealand stitchbird *Notiomystis cincta*. *Journal of Avian Biology* 36: 436–448.

Low, M. 2006. The energetic cost of mate guarding is correlated with territorial intrusions in the New Zealand stitchbird. *Behavioral Ecology* 17: 270–276.

Loyau, A., Jalme, M. S., Cagniant, C., and Sorci, G. 2005a. Multiple sexual advertisements honestly reflect health status in peacocks (*Pavo cristatus*). *Behavioral Ecology and Sociobiology* 58: 552–557.

Loyau, A., Jalme, M. S., and Sorci, G. 2005b. Intra- and intersexual selection for multiple traits in the peacock (*Pavo cristatus*). *Ethology* 111: 810–820.

Loyau, A., Saint Jalme, M., Mauget, R., and Sorci, G. 2007. Male sexual attractiveness affects the investment of maternal resources into the eggs in peafowl (*Pavo cristatus*). *Behavioral Ecology and Sociobiology* 61: 1043–1052.

Lung, O., Tram, U., Finnerty, C. M., Eipper-Mains, M. A., Kalb, J. M., and Wolfner, M. F. 2002. The *Drosophila melanogaster* seminal fluid protein Acp62F is a protease inhibitor that is toxic upon ectopic expression. *Genetics* 160: 211–224.

Lupold, S., Pitnick, S., Berben, K. S., Blengini, C. S., Belote, J. M., and Mainer, M. K. 2013. Female mediation of competitive fertilization success in *Drosophila melanogaster*. *Proceedings of the National Academy of Sciences USA* 110: 10693–10698.

Madden, J. R. 2003a. Bower decorations are good predictors of mating success in the spotted bowerbird. *Behavioral Ecology and Sociobiology* 53: 269–277.

Madden, J. R. 2003b. Male spotted bowerbirds preferentially choose, arrange, and proffer objects that are good predictors of mating success. *Behavioral Ecology and Sociobiology* 53: 263–268.

Manier, M. K., Belote, J. M., Berben, K. S., Lupold, S., Ala-Honkola, O., et al. 2013. Rapid diversification of sperm precedence traits and processes among three sibling *Drosophila* species. *Evolution* 67: 2348–2362.

Marshall, R. C., Buchanan, K. L., and Catchpole, C. K. 2003. Sexual selection and individual genetic diversity in a songbird. *Proceedings of the Royal Society of London B* 270: S248–S250.

Matsumoto-Oda, A., Hamai, M., Hayaki, H., Hosaka, K., Hunt, K. D., et al. 2007. Estrus cycle asynchrony in wild female chimpanzees, *Pan troglodytes schweinfurthii*. *Behavioral Ecology and Sociobiology* 61: 661–668.

Maxwell, M. R., Gallego, K. M., and Barry, K. L. 2010. Effects of female feeding regime in a sexually cannibalistic mantid: Fecundity, cannibalism, and male response in *Stagmomantis limbata* (Mantodea). *Ecological Entomology* 35: 775–787.

McGraw, K. J. and Ardia, D. R. 2003. Carotenoids, immunocompetence, and the information content of sexual colors: An experimental test. *American Naturalist* 162: 704–712.

McGraw, K. J. and Hill, G. E. 2000. Differential effects of endoparasitism on the expression of carotenoid- and melanin-based ornamental coloration. *Proceedings of the Royal Society of London B* 267: 1525–1531.

Meslin, C., Plakke, M. S., Deutsch, A. B., Small, B. S., Morehouse, N. I., and Clark, N. L. 2015. Digestive organ in the female reproductive tract borrows genes from multiple organ systems to adopt critical functions. *Molecular Biology and Evolution* 32: 1567–1580.

Michl, G., Török, J., Griffith, S. C., and Sheldon, B. C. 2002. Experimental analysis of sperm competition mechanisms in a wild bird population. *Proceedings of the National Academy of Sciences USA* 99: 5466–5470.

Milinski, M. and Bakker, T. C. 1990. Female sticklebacks use male coloration in mate choice and hence avoid parasitized males. *Nature* 344: 330–333.

Miller, J. A. 2007. Repeated evolution of male sacrifice behavior in spiders correlated with genital mutilation. *Evolution* 61: 1301–1315.

Moore, H., Dvorakova, K., Jenkins, N., and Breed, W. 2002. Exceptional sperm cooperation in the wood mouse. *Nature* 418: 174–177.

Morehouse, N. I. and Rutowksi, R. L. 2010a. Developmental responses to variable diet composition in a butterfly: The role of nitrogen, carbohydrates, and genotype. *Oikos* 119: 636–645.

Morehouse, N. I. and Rutowksi, R. L. 2010b. In the eyes of the beholders: Female choice and avian predation risk associated with an exaggerated male butterfly color. *American Naturalist* 176: 768–784.

Muller, M. N., Kahlenberg, S. M., Thompson, M. E., and Wrangham, R. W. 2007. Male coercion and the costs of promiscuous mating for female chimpanzees. *Proceedings of the Royal Society of London B* 274: 1009–1014.

Noë, R. and Sluijter, A. A. 1990. Reproductive tactics of male savanna baboons. *Behaviour* 113: 117–170.

Obara, Y., Fukano, Y., Watanabe, K., Ozawa, G., and Sasaki, K. 2011. Serotonin-induced mate rejection in the female cabbage butterfly, *Pieris rapae crucivora*. *Naturwissenschaften* 98: 989–993.

Östlund, S. and Ahnesjö, I. 1998. Female fifteen-spined sticklebacks prefer better fathers. *Animal Behaviour* 56: 1177–1183.

Paczolt, K. A. and Jones, A. G. 2010. Post-copulatory sexual selection and sexual conflict in the evolution of male pregnancy. *Nature* 464: 401–405.

Palombit, R. A., Seyfarth, R. M., and Cheney, D. L. 1997. The adaptive value of "friendships" to female baboons: Experimental and observational evidence. *Animal Behaviour* 54: 599–614.

Parker, G. A. 1970. Sperm competition and its evolutionary consequences in the insects. *Biological Reviews* 45: 526–567.

Parker, G. A. 2006. Sexual conflict over mating and fertilization: An overview. *Philosophical Transactions of the Royal Society B* 361: 235–259.

Partridge, C., Boettcher, A., and Jones, A. G. 2013. The role of courtship behavior and size in mate preference in the sex-role-reversed gulf pipefish, *Syngnathus scovelli*. *Ethology* 119: 692–701.

Pasch, B., George, A. S., Campbell, P., and Phelps, S. M. 2011. Androgen-dependent male vocal performance influences female preference in Neotropical singing mice. *Animal Behaviour* 82: 177–183.

Patricelli, G. L., Uy, J. A. C., and Borgia, G. 2003. Multiple male traits interact: Attractive bower decorations facilitate attractive behavioural displays in satin bowerbirds. *Proceedings of the Royal Society of London B* 270: 2389–2395.

Patricelli, G. L., Uy, J. A. C., and Borgia, G. 2004. Female signals enhance the efficiency of mate assessment in satin bowerbirds (*Ptilonorhynchus violaceus*). *Behavioral Ecology* 15: 297–304.

Petrie, M. 1992. Peacocks with low mating success are more likely to suffer predation. *Animal Behaviour* 44: 585–586.

Petrie, M. 1994. Improved growth and survival of offspring of peacocks with more elaborate trains. *Nature* 371: 585–586.

Petrie, M., Cotgreave, P., and Pike, T. W. 2009. Variation in the peacock's train shows a genetic component. *Genetica* 135: 7–11.

Petrie, M. and Halliday, T. 1994. Experimental and natural changes in the peacock's (*Pavo cristatus*) train can affect mating success. *Behavioral Ecology and Sociobiology* 35: 213–217.

Pike, T. W., Blount, J. D., Lindstrom, J., and Metcalfe, N. B. 2007. Dietary carotenoid availability influences a male's ability to provide parental care. *Behavioral Ecology* 18: 1100–1105.

Pischedda, A. and Chippindale, A. K. 2006. Intralocus sexual conflict diminishes the benefits of sexual selection. *PLoS Biology* 4: e356.

Pitnick, S. and García-González, F. 2002. Harm to females increases with male body size in *Drosophila melanogaster*. *Proceedings of the Royal Society of London B* 269: 1821–1828.

Pitnick, S., Jones, K. E. and Wilkinson, G. S. 2006. Mating system and brain size in bats. *Proceedings of the Royal Society of London B* 273: 719–724.

Pizzari, T. and Birkhead, T. R. 2000. Female feral fowl eject sperm of subdominant males. *Nature* 405: 787–789.

Plakke, M. S., Deutsch, A. B., Meslin, C., Clark, N. L., and Morehouse, N. I. 2015. Dynamic digestive physiology of a female reproductive organ in a polyandrous butterfly. *Journal of Experimental Biology* 218: 1548–1555.

Polak, M., Wolf, L. L., Starmer, W. T., and Barker, J. S. F. 2001. Function of the mating plug in *Drosophila hibisc*i Bock. *Behavioral Ecology and Sociobiology* 49: 196–205.

Pratt, D. M. and Anderson, V. H. 1985. Giraffe social behavior. *Journal of Natural History* 19: 771–781.

Preston, B. T., Jalme, M. S., Hingrat, Y., Lacroix, F., and Sorci, G. 2015. The sperm of aging male bustards retards their offspring's development. *Nature Communications* 6: 6146.

Preston, B. T., Stevenson, I. R., Pemberton, J. M., and Wilson, K. 2001. Dominant rams lose out by sperm depletion. *Nature* 409: 681–682.

Prete, F. R. 1995. Designing behavior: A case study. *Perspectives in Ethology* 11: 255–277.

Pruett-Jones, S. and Pruett-Jones, M. 1994. Sexual competition and courtship disruptions: Why do male bowerbirds destroy each other's bowers? *Animal Behaviour* 47: 607–620.

Prum, R. O. 2010. The Lande-Kirkpatrick mechanism is the null model of evolution by intersexual selection: Implications for meaning, honesty and design in intersexual signals. *Evolution* 64: 3085–3100.

Queller, D. C. 1997. Why do females care more than males? *Proceedings of the Royal Society of London B* 264: 1555–1557.

Rantala, M. J., Moore, F. R., Skrinda, I., Krama, T., Kivleniece, I., et al. 2012. Evidence for the stress-linked immunocompetence handicap hypothesis in humans. *Nature Communications* 3: 694.

Reinhardt, K., Naylor, R., and Siva-Jothy, M. T. 2003. Reducing a cost of traumatic insemination: female bedbugs evolve a unique organ. *Proceedings of the Royal Society of London B* 270: 2371–2375.

Reinhardt, K., Naylor, R. A., and Siva-Jothy, M. T. 2009. Ejaculate components delay reproductive senescence while elevating female reproductive rate in an insect. *Proceedings of the National Academy of Sciences USA* 106: 21743–21747.

Reynolds, S. M., Dryer, K., Bollback, J., Uy, J. A. C., Patricelli, G. L., et al. 2007. Behavioral paternity predicts genetic paternity in Satin Bowerbirds (*Ptilonorhynchus violaceus*), a species with a non-resource-based mating system. *Auk* 124: 857–867.

Rezac, M. 2009. The spider *Harpactea sadistica*: Co-evolution of traumatic insemination and complex female genital morphology in spiders. *Proceedings of the Royal Society of London B* 276: 2697–2701.

Rintamäki, P. T., Lundberg, A., Alatalo, R. V., and Höglund, J. 1998. Assortative mating and female clutch investment in black grouse. *Animal Behaviour* 56: 1399–1403.

Rosenqvist, G. 1990. Male mate choice and female-female competition for mates in the pipefish *Nerophis ophidion*. *Animal Behaviour* 39: 1110–1116.

Rowland, W. J. 1994. Proximate determinants of stickleback behavior: An evolutionary perspective. In M. Bell and S. Foster (eds.), *The Evolutionary Biology of the Threespine Stickleback*. pp 297–344. Oxford: Oxford University Press.

Rubenstein, D. R. 2007. Female extrapair mate choice in a cooperative breeder: Trading sex for help and increasing offspring heterozygosity. *Proceedings of the Royal Society of London B* 274: 1895–1903.

Rubenstein, D. R. 2012. Sexual and social competition: broadening perspectives by defining female roles. *Philosophical Transactions of the Royal Society B* 367: 2248–2252.

Rubenstein, D. R. and Lovette, I. J. 2009. Reproductive skew and selection on female ornamentation in social species. *Nature* 462: 786–789.

Rubenstein, D. R. and Wikelski, M. 2003. Seasonal changes in food quality: A proximate cue for reproductive timing in marine iguanas. *Ecology Letters* 84: 3013–3023.

Ryan, M. J., Fox, J. H., Wilczynski, W., and Rand, A. S. 1990. Sexual selection for sensory exploitation in the frog *Physalaemus pustulosus*. *Nature* 343: 66–67.

Sato, H. 1998. Male participation in nest building in the dung beetle *Scarabaeus catenatus* (Coleoptera: Scarabaeidae): Mating effort versus paternal effort. *Journal of Insect Behavior* 11: 833–843.

Schärer, L., Rowe, L., and Arnqvist, G. 2012. Anisogamy, chance and the evolution of sex roles. *Trends in Ecology & Evolution* 27: 260–264.

Schlinger, B. A., Day, L. B., and Fusani, L. 2008. Behavior, natural history and neuroendocrinology of a tropical bird. *General and Comparative Endocrinology* 157: 254–258.

Schwabl, H., Mock, D. W., and Gieg, J. A. 1997. A hormonal mechanism for parental favouritism. *Nature* 386: 231.

Schwartz, S. K., Wagner Jr., W. E., and Hebets, E. A. 2014. Obligate male death and sexual cannibalism in dark fishing spiders. *Animal Behaviour* 93: 151–156.

Schwartz, S. K., Wagner Jr., W. E., and Hebets, E. A. 2016. Males can benefit from sexual canabilism facilitated by self-sacrifice. *Current Biology* 26: 2794–2799.

Senar, J. C., Figuerola, J., and Pascual, J. 2002. Brighter yellow blue tits make better parents. *Proceedings of the Royal Society of London B* 269: 257–261.

Shuster, S. M. 1989. Male alternative reproductive strategies in a marine isopod crustacean (*Paracerceis sculpta*): The use of genetic markers to measure differences in the fertilization success among alpha, beta, and gamma-males. *Evolution* 43: 1683–1689.

Shuster, S. M. 1992. The reproductive behaviour of alpha, beta, and gamma morphs in *Paracerceis sculpta*, a marine isopod crustacean. *Behaviour* 121: 231–258.

Shuster, S. M. and Wade, M. J. 1991. Equal mating success among male reproductive strategies in a marine isopod. *Nature* 350: 608–610.

Simmons, L. W. 2001. *Sperm Competition and Its Evolutionary Consequences in the Insects*. Princeton, NJ: Princeton University Press.

Simmons, L. W. amd Emlen, D. J. 2006. Evolutionary trade-off between weapons and testes. *Proceedings of the National Academy of Sciences USA* 103: 16346–16351.

Simmons, R. E. and Scheepers, L. 1996. Winning by a neck: Sexual selection in the evolution of giraffe. *American Naturalist* 148: 771–786.

Simpson, S. J., Sword, G. A., Lorch, P. D., and Couzin, I. D. 2006. Cannibal crickets on a forced march for protein and salt. *Proceedings of the National Academy of Sciences USA* 103: 4152–4156.

Singer, A. G., Beauchamp, G. K. and Yamazaki, K. 1997. Volatile signals of the major histocompatibility complex in male mouse urine. *Proceedings of the National Academy of Sciences USA* 94: 2210–2214.

Stockley, P., Gage, M. J. G., Parker, G. A. and Moller, A. P. 1997. Sperm competition in fishes: The evolution of testis size and ejaculate characteristics. *American Naturalist* 149: 933–954.

Stoltz, J. A., Elias, D. O., and Andrade, M. C. B. 2008. Females reward courtship by competing males in a cannibalistic spider. *Behavioral Ecology and Sociobiology* 62: 689–697.

Strum, S. C. 1987. *Almost Human*. New York: W.W. Norton.

Stutt, A. D. and Siva-Jothy, M. T. 2001. Traumatic insemination and sexual conflict in the bed bug *Cimex lectularius*. *Proceedings of the National Academy of Sciences USA* 98: 5683–5687.

Svensson, B. G. 1997. Swarming behavior, sexual dimorphism, and female reproductive status in the sex role-reversed dance fly species *Rhamphomyia marginata*. *Journal of Insect Behavior* 10: 783–804.

Taborsky, M., Hudde, B., and Wirtz, P. 1987. Reproductive behavior and ecology of *Symphodus* (*Crenilabrus*) *Ocellatus*, a European wrasse with four types of male behavior. *Behaviour* 102: 82–118.

Takahashi, M., Arita, H., Hiraiwa-Hasegawa, M., and Hasegawa, T. 2008. Peahens do not prefer peacocks with more elaborate trains. *Animal Behaviour* 75: 1209–1219.

Thornhill, R. 1976. Sexual selection and nuptial feeding behavior in *Bittacus apicalis* (Insecta: Mecoptera). *American Naturalist* 119: 529–548.

Thornhill, R. 1981. *Panorpa* (Mecoptera: Panorpidae) scorpionflies: Systems for understanding resource-defense polygyny and alternative male reproductive efforts. *Annual Review of Ecology and Systematics* 12: 355–386.

Tregenza, T., Simmons, L. W., Wedell, N., and Zuk, M. 2006. Female preference for male courtship song and its role as a signal of immune function and condition. *Animal Behaviour* 72: 809–818.

Trivers, R. L. 1972. Parental investment and sexual selection. In B. Campbell (ed.), *Sexual Selection and the Descent of Man*, pp. 136–179. Chicago: Aldine.

Tuttle, E. M., Pruett-Jones, S., and Webster, M. S. 1996. Cloacal protuberances and extreme sperm production in Australian fairy-wrens. *Proceedings of the Royal Society of London B* 263: 1359–1364.

Uhl, G., Nessler, S., and Schneider, J. 2011. Securing paternity in spiders? A review on occurrence and effects of mating plugs and male genital mutilation. *Genetica* 138: 75–104.

Uy, J. A. C., Patricelli, G. L., and Borgia, G. 2001. Complex mate searching in the satin bowerbird *Ptilonorhynchus violaceus*. *American Naturalist* 158: 530–542.

Vahed, K. 1998. The function of nuptial feeding in insects: Review of empirical studies. *Biological Reviews* 73: 43–78.

Vallet, E., Beme, I., and Kreutzer, M. 1998. Two-note syllables in canary songs elicit high levels of sexual display. *Animal Behaviour* 55: 291–297.

Van Voorhies, W. A. 1992. Production of sperm reduces nematode lifespan. *Nature* 360: 456–458.

Vitousek, M. N., Rubenstein, D. R., Nelson, K., and Wikelski, M. 2008. Are hotshots always hot? A longitudinal study of hormones, behavior, and reproductive success in male marine iguanas. *General and Comparative Endocrinology* 157: 227–232.

Waage, J. K. 1979. Dual function of the damselfly penis: Sperm removal and transfer. *Science* 203: 916–918.

Waage, J. K. 1997. Parental investment: Minding the kids or keeping control? In P. A. Gowaty (ed.), *Feminism and Evolutionary Biology: Boundaries, Interactions, and Frontiers*, pp. 527–553. New York: Chapman and Hall.

Wedell, N., Gage, J. G., and Parker, G. A. 2002. Sperm competition, male prudence, and sperm-limited females. *Trends in Ecology & Evolution* 17: 313–320.

Widdig, A., Bercovitch, F. B., Streich, W. J., Sauermann, U., Nürnberg, P., and Krawczak, M. 2004. A longitudinal analysis of reproductive skew in male rhesus macaques. *Proceedings of the Royal Society of London B* 271: 819–826.

Widemo, F. 1998. Alternative reproductive strategies in the ruff, *Philomachus pugnax*: A mixed ESS? *Animal Behaviour* 56: 329–336.

Wikelski, M. and Baurle, S. 1996. Pre-copulatory ejaculation solves time constraints during copulations in marine iguanas. *Proceedings of the Royal Society of London B* 263: 439–444.

Wikelski, M., Carbone, C., and Trillmich, F. 1996. Lekking in marine iguanas: Female grouping and male reproductive strategies. *Animal Behaviour* 52: 581–596.

Wojcieszek, J. M., Nicholls, J. A., and Goldizen, A. W. 2007. Stealing behavior and the maintenance of a visual display in the satin bowerbird. *Behavioral Ecology* 18: 689–695.

Wolff, J. O., Mech, S. G., Dunlap, A. S., and Hodges, K. E. 2002. Multi-male mating by paired and unpaired female prairie voles (*Microtus ochrogaster*). *Behaviour* 139: 1147–1160.

Zahavi, A. 1975. Mate selection: A selection for a handicap. *Journal of Theoretical Biology* 53: 205–214.

Zeh, J. A. and Zeh, D. W. 2003. Toward a new sexual selection paradigm: Polyandry, conflict, and incompatibility. *Ethology* 109: 929–950.

Ziegler, A., Kentenich, H., and Uchanska-Ziegler, B. 2005. Female choice and the MHC. *Trends in Ecology & Evolution* 26: 496–502.

Chapter 10

Akçay, E. and Roughgarden, J. 2007. Extra-pair paternity in birds: Review of the genetic benefits. *Evolutionary Ecology Research* 9: 855–868.

Alcock, J., Eickwort, G. C., and Eickwort, K. R. 1977. The reproductive behavior of *Anthidium maculosum* and the evolutionary significance of multiple copulations by females. *Behavioral Ecology and Sociobiology* 2: 385–396.

Andrade, M. C. B. 1996. Sexual selection for male sacrifice in the Australian redback spider. *Science* 271: 70–72.

Andrade, M. C. B. 2003. Risky mate search and male self-sacrifice in redback spiders. *Behavioral Ecology* 14: 531–538.

Apollonio, M., Festa-Bianchet, M., Mari, F., Bruno, E., and Locati, M. 1998. Habitat manipulation modifies lek use in fallow deer. *Ethology* 104: 603–612.

Arct, A., Drobniak, S. M., and Cichoń, M. 2015. Genetic similarity between mates predicts extra-pair paternity: A meta-analysis of bird studies. *Behavioral Ecology* 26: 959–968.

Arct, A., Drobniak, S. M., Podmokła, E., Gustafson, L., and Cichoń, M. 2013. Benefits of extra-pair mating may depend on environmental conditions: An experimental study in the blue tit (*Cyanistes caeruleus*). *Behavioral Ecology and Sociobiology* 67: 1809–1815.

Arnold, T. W. 1999. What limits clutch size in waders? *Journal of Avian Biology* 30: 216–220.

Balmford, A., Deutsch, J. C., Nefdt, R. J. C., and Clutton-Brock, T. 1993. Testing hotspot models of lek evolution: Data from three species of ungulates. *Behavioral Ecology and Sociobiology* 33: 57–65.

Bissoondath, C. J. and Wiklund, C. 1995. Protein content of spermatophores in relation to monandry/polyandry in butterflies. *Behavioral Ecology and Sociobiology* 37: 365–372.

Blyth, J. E. and Gilburn, A. 2006. Extreme promiscuity in a mating system dominated by sexual conflict. *Journal of Insect Behavior* 19: 447–455.

Borgia, G. 1985. Bower quality, number of decorations and mating success of male satin bowerbirds (*Ptilonorhynchus violaceus*). *Animal Behaviour* 33: 266–271.

Borries, C., Launhardt, K., Epplen, C., Epplen, J. T., and Winkler, P. 1999. Males as infant protectors in Hanuman langurs (*Presbytis entellus*) living in multimale groups: Defence pattern, paternity, and sexual behavior. *Behavioral Ecology and Sociobiology* 46: 350–356.

Borries, C., Savini, T., and Koenig, A. 2011. Social monogamy and the threat of infanticide in larger mammals. *Behavioral Ecology and Sociobiology* 65: 685–693.

Bradbury, J. W. 1977. Lek mating behavior in the hammer-headed bat. *Zeitschrift für Tierpsychologie* 45: 225–255.

Bradbury, J. W. 1981. The evolution of leks. In R. D. Alexander and D. W. Tinkle (eds.), *Natural Selection and Social Behavior*, pp. 138–169. New York: Chiron Press.

Bradbury, J. W. and Gibson, R. M. 1983. Leks and mate choice. In P. Bateson (ed.), *Mate Choice*, pp. 109–138. Cambridge: Cambridge University Press.

Bradbury, J. W., Vehrencamp, S. L., and Gibson, R. M. 1989. Dispersion of displaying male sage grouse. I. Patterns of temporal variation. *Behavioral Ecology and Sociobiology* 24: 1–14.

Brennan, P. L. R. 2009. Incubation in great tinamou (*Tinamus major*). *Wilson Journal of Ornithology* 121: 506–511.

Breven, K. A. 1981. Mate choice in the wood frog, *Rana sylvatica*. *Evolution* 35: 707–722.

Bro-Jørgensen, J. and Durant, S. M. 2003. Mating strategies of topi bulls: Getting in the centre of attention. *Animal Behaviour* 65: 585–594.

Brockmann, H. J. and Penn, D. 1992. Male mating tactics in the horseshoe crab, *Limulus polyphemus*. *Animal Behaviour* 44: 653–665.

Brouwer, L., Barr, I., van de Pol, M., Burke, T., Komdeur, J., and Richardson, D. S. 2010. MHC-dependent survival in a wild population: Evidence for hidden genetic benefits gained through extra-pair fertilizations. *Molecular Ecology* 19: 3444–3455.

Byrne, P. G. and Whiting, M. J. 2011. Effects of simultaneous polyandry on offspring fitness in an African tree frog. *Behavioral Ecology* 22: 385–391.

Cantoni, D. and Brown, R. 1997. Paternal investment and reproductive success in the California mouse, *Peromyscus californicus*. *Animal Behaviour* 54: 377–386.

Cockburn, A., Dalziell, A. H., Blackmore, C. J., Double, M. C., Kokko, H., et al. 2009. Superb fairy-wren males aggregate into hidden leks to solicit extragroup fertilizations before dawn. *Behavioral Ecology* 20: 501–510.

Concannon, M. R., Stein, A. C., and Uy, A. L. 2012. Kin selection may contribute to lek evolution and trait introgression across an avian hybrid zone. *Molecular Ecology* 21: 1477–1486.

Dale, J., Montgomerie, R., Michaud, D., and Boag, P. 1999. Frequency and timing of extra-pair fertilisation in the polyandrous red phalarope (*Phalaropus fulicarius*). *Behavioral Ecology and Sociobiology* 46: 50–56.

Dale, S., Rinden, H., and Slagsvold, T. 1992. Competition for a mate restricts mate search of female pied flycatchers. *Behavioral Ecology and Sociobiology* 30: 165–176.

Davies, N. B. 1985. Cooperation and conflict among dunnocks, *Prunella modularis*, in a variable mating system. *Animal Behaviour* 33: 628–648.

Davies, N. B. 1986. Reproductive success of dunnocks, *Prunella modularis*, in a variable mating system. I. Factors influencing provisioning rate, nestling weight, and fledging success. *Journal of Animal Ecology* 55: 123–138.

Davies, N. B., Hartley, I. R., Hatchwell, B. J., Desrochers, A., Skeer, J., and Nebels, D. 1995. The polygynandrous mating system of the alpine accentor, *Prunella collaris*. I. Ecological causes and reproductive conflicts. *Animal Behaviour* 49: 769–788.

Davies, N. B. and Houston, A. I. 1986. Reproductive success of dunnocks, *Prunella modularis*, in a variable mating system. II. Conflicts of interest among breeding adults. *Journal of Animal Ecology* 55: 139–154.

Davies, N. B. and Lundberg, A. 1984. Food distribution and a variable mating system in the dunnock, *Prunella modularis*. *Journal of Animal Ecology* 53: 895–912.

Deutsch, J. C. 1994. Uganda kob mating success does not increase on larger leks. *Behavioral Ecology and Sociobiology* 34: 451–459.

Dickinson, J. L. 2001. Extra-pair copulations in western bluebirds (*Sialia mexicana*): female receptivity favors older males. *Behavioral Ecology and Sociobiology* 50: 423–429.

Dodson, G. N. 1997. Resource defense mating system in antlered flies, *Phytalmia* spp. (Diptera: Tephritidae). *Annals of the Entomological Society of America* 90: 496–504.

Double, M. C. and Cockburn, A. 2003. Subordinate superb fairy-wrens (*Malurus cyaneus*) parasitize the reproductive success of attractive dominant males. *Proceedings of the Royal Society of London B* 270: 379–384.

Edvardsson, M. 2007. Female *Callosobruchus maculatus* mate when they are thirsty: Resource-rich ejaculates as mating effort in a beetle. *Animal Behaviour* 74: 183–188.

Eggert, A.-K. and Sakaluk, S. K. 1995. Female-coerced monogamy in burying beetles. *Behavioral Ecology and Sociobiology* 37: 147–154.

Emlen, S. T. and Oring, L. W. 1977. Ecology, sexual selection, and the evolution of mating systems. *Science* 197: 215–223.

Emlen, S. T., Wrege, P. H., and Webster, M. S. 1998. Cuckoldry as a cost of polyandry in the sex-role-reversed wattled jacana, *Jacana jacana*. *Proceedings of the Royal Society of London B* 265: 2359–2364.

Faaborg, J., Parker, P. G., DeLay, L., de Vries, T. J., Bednarz, J. C., et al. 1995. Confirmation of cooperative polyandry in the Galapagos hawk (*Buteo galapagoensis*). *Behavioral Ecology and Sociobiology* 36: 83–90.

Fiske, P., Rintamäki, P. T., and Karvonen, E. 1998. Mating success in lekking males: A meta-analysis. *Behavioral Ecology* 9: 328–338.

Foltz, D. W. and Schwagmeyer, P. L. 1989. Sperm competition in the 13-lined ground squirrel: Differential fertilization success under field conditions. *American Naturalist* 133: 257–265.

Forstmeier, W., Nakagawa, S., Griffith, S., and Kempenaers, B. 2014. Female extra-pair mating: adaptation or genetic constraint? *Trends in Ecology & Evolution* 29: 456–464.

Fossøy, F., Johnsen, A., and Lifjeld, J. T. 2008. Multiple genetic benefits of female promiscuity in a socially monogamous passerine. *Evolution* 62: 145–156.

Francis, C. M., Elp, A., Brunton, J. A., and Kunz, T. H. 1994. Lactation in male fruit bats. *Nature* 367: 691–692.

Frith, C. B. and Frith, D. W. 2001. Nesting biology of the spotted catbird, *Ailuroedus melanotis*, a monogamous bowerbird (Ptilonorhynchidae), in Australian Wet Tropics upland rainforests. *Australian Journal of Zoology* 49: 279–310.

Gerlach, N. M., McGlothlin, J. W., Parker, P. G., and Ketterson, E. D. 2012. Promiscuous mating produces offspring with higher lifetime fitness. *Proceedings of the Royal Society of London B* 279: 860–866.

Getz, L. L. and Carter, C. S. 1996. Prairie-vole partnerships. *American Scientist* 84: 56–62.

Gibson, R. M. 1996. A re-evaluation of hotspot settlement in lekking sage grouse. *Animal Behaviour* 52: 993–1005.

Gibson, R. M., Aspbury, A. S., and McDaniel, L. L. 2002. Active formation of mixed-species grouse leks: A role for predation in lek evolution? *Proceedings of the Royal Society of London B* 269: 2503–2507.

Gilburn, A. and Day, T. H. 1994. Sexual dimorphism, sexual selection and the αβ chromosomal inversion polymorphism in the seaweed fly, *Coelopa frigida*. *Proceedings of the Royal Society of London B* 257: 303–309.

Gray, E. M. 1997. Female red-winged blackbirds accrue material benefits from copulating with extra-pair males. *Animal Behaviour* 53: 625–639.

Griffith, S. C., Owens, I. P. F., and Thuman, K. A. 2002. Extra pair paternity in birds: a review of interspecific variation and adaptive function. *Molecular Ecology* 11: 2195–2212.

Gubernick, D. J. and Teferi, T. 2000. Adaptive significance of male parental care in a monogamous mammal. *Proceedings of the Royal Society of London B* 267: 147–150.

Gwinner, H. and Schwabl, H. 2005. Evidence for sexy sons in European starlings (*Sturnus vulgaris*). *Behavioral Ecology and Sociobiology* 58: 375–382.

Harcourt, A. H., Harvey, P. H., Larson, S. G., and Short, R. V. 1981. Testis weight, body weight and breeding system in primates. *Nature* 293: 55–57.

Hartley, I. R., Davies, N. B., Hatchwell, B. J., Desrochers, A., Nebels, D., and Burke, T. 1995. The polygynandrous mating system of the alpine accentor, *Prunella collaris*. II. Multiple paternity and parental effort. *Animal Behaviour* 49: 789–803.

Heinze, J., Hölldobler, B., and Yamauchi, K. 1998. Male competition in *Cardiocondyla* ants. *Behavioral Ecology and Sociobiology* 42: 239–246.

Heistermann, M., Ziegler, T., van Schaik, C. P., Launhardt, K., Winkler, P., and Hodges, J. K. 2001. Loss of oestrus, concealed ovulation and paternity confusion in free-ranging Hanuman langurs. *Proceedings of the Royal Society of London B* 268: 2445–2451.

Höglund, J., Alatalo, R. V., Lundberg, A., Rintamlki, P. T., and Lindell, J. 1999. Microsatellite markers reveal the potential for kin selection on black grouse leks. *Proceedings of the Royal Society of London B* 266: 813–816.

Höglund, J. and Lundberg, A. 1987. Sexual selection in a monomorphic lek-breeding bird: Correlates of male mating success in the great snipe *Gallinago media*. *Behavioral Ecology and Sociobiology* 21: 211–216.

Hohoff, C., Franzen, K., and Sachser, N. 2003. Female choice in a promiscuous wild guinea pig, the yellow-toothed cavy (*Galea musteloides*). *Behavioral Ecology and Sociobiology* 53: 341–349.

Jamieson, I. G. 1997. Testing reproductive skew models in a communally breeding bird, the pukeko, *Porphyrio porphyrio*. *Proceedings of the Royal Society of London B* 264: 335–340.

Jiguet, F. and Bretagnolle, V. 2006. Manipulating lek size and composition using decoys: An experimental investigation of lek evolution models. *American Naturalist* 168: 758–768.

Johnsen, A., Andersen, V., Sunding, C., and Lifjeld, J. T. 2000. Female bluethroats enhance offspring immunocompetence through extra-pair copulations. *Nature* 406: 296–299.

Johnsen, A., Andersson, S., Ornberg, J., and Lifjeld, J. T. 1998. Ultraviolet plumage ornamentation affects social mate choice and sperm competition in bluethroats (Aves: *Luscinia s. svecica*): A field experiment. *Proceedings of the Royal Society of London B* 265: 1313–1318.

Jones, J. S. and Wynne-Edwards, K. E. 2000. Paternal hamsters mechanically assist the delivery, consume amniotic fluid and placenta, remove fetal membranes, and provide parental care during the birth process. *Hormones and Behavior* 37: 116–125.

Just, J. 1988. Siphonoecetinae (Corophiidae). 6: A survey of phylogeny, distribution, and biology. *Crustaceana, Supplement* 13: 193–208.

Keil, A. and Sachser, N. 1998. Reproductive benefits from female promiscuous mating in a small mammal. *Ethology* 104: 897–903.

Kempenaers, B., Verheyen, G. R., and Dhondt, A. A. 1997. Extra-pair paternity in the blue tit (*Parus caeruleus*): Female choice, male characteristics, and offspring quality. *Behavioral Ecology* 8: 481–492.

Kempenaers, B., Verheyen, G. R., van der Broeck, M., Burke, T., van Broeckhoven, C., and Dhondt, A. A. 1992. Extra-pair paternity results from female preference for high-quality males in the blue tit. *Nature* 357: 494–496.

Kingma, S. A., Hall, M. L., and Peters, A. 2014. Breeding synchronization facilitates extra-pair mating for inbreeding avoidance. *Behavioral Ecology* 24: 1390–1397.

Komers, P. E. and Brotherton, P. N. M. 1997. Female space use is the best predictor of monogamy in mammals. *Proceedings of the Royal Society of London B* 264: 1261–1270.

Lack, D. 1968. *Ecological Adaptations for Breeding in Birds*. London: Methuen.

Lank, D. B., Oring, L. W., and Maxson, S. J. 1985. Mate and nutrient limitation of egg-laying in a polyandrous shorebird. *Ecology* 66: 1513–1524.

Lebigre, C., Timmermans, C., and Soulsbury, C. D. 2016. No behavioural response to kin competition in a lekking species. *Behavioral Ecology and Sociobiology* 70: 1457–1465.

Lill, A. 1974. Sexual behavior of the lek-forming white-bearded manakin (*Manacus manacus trinitatis* Hartert). *Zeitschrift für Tierpsychologie* 36: 1–36.

Lloyd, J. E. 1980. Insect behavioral ecology: Coming of age in bionomics, or, compleat biologists have revolutions too. *Florida Entomologist* 63: 1–4.

Loiselle, B. A., Ryder, T. B., Duraes, R., Tori, W., Blake, J. G., and Parker, P. G. 2006. Kin selection does not explain male aggregation at leks of four manakin species. *Behavioral Ecology* 18: 287–291.

Loyau, A., Saint Jalme, M., and Sorci, G. 2007. Non-defendable resources affect peafowl lek organization: A male removal experiment. *Behavioural Processes* 74: 64–70.

Lukas, D. and Clutton-Brock, T. H. 2013. The evolution of social monogamy in mammals. *Science* 341: 526–530.

Lyon, B. E., Montgomerie, R. D., and Hamilton, L. D. 1987. Male parental care and monogamy in snow buntings. *Behavioral Ecology and Sociobiology* 20: 377–382.

Mattila, H. R. and Seeley, T. D. 2007. Genetic diversity in honey bee colonies enhances productivity and fitness. *Science* 317: 362–364.

McCracken, G. F. and Bradbury, J. W. 1981. Social organization and kinship in the polygynous bat *Phyllostomus hastatus*. *Behavioral Ecology and Sociobiology* 8: 11–34.

Miller, J. A. 2007. Repeated evolution of male sacrifice behavior in spiders correlated with genital mutilation. *Evolution* 61: 1301–1315.

Monteiro, N. M., Carneiro, D., Antunes, A., Queiroz, N., Vieira, M. N., and Jones, A. G. 2017. The lek mating system of the worm pipefish (*Nerophis lumbriciformis*): A molecular maternity analysis and test of the phenotype-linked fertility hypothesis. *Molecular Ecology* 26: 1371–1385.

Moreno, J., Veiga, J. P., Cordero, P. J., and Mínguez, E. 1999. Effects of paternal care on reproductive success in the polygynous spotless starling *Sturnus unicolor*. *Behavioral Ecology and Sociobiology* 47: 47–53.

Mulder, R. A., Dunn, P. O., Cockburn, A., Lazenby-Cohen, K. A., and Howell, M. J. 1994. Helpers liberate female fairy-wrens from constraints on extra-pair mate choice. *Proceedings of the Royal Society of London B* 255: 223–229.

Murphy, C. G. 2003. The cause of correlations between nightly numbers of male and female barking treefrogs (*Hyla gratiosa*) attending choruses. *Behavioral Ecology* 14: 274–281.

Nazareth, T. M. and Machado, G. 2010. Mating system and exclusive postzygotic paternal care in a Neotropical harvestman (Arachnida: Opiliones). *Animal Behaviour* 79: 547–554.

Newcomer, S. D., Zeh, J. A., and Zeh, D. W. 1999. Genetic benefits enhance the reproductive success of polyandrous females. *Proceedings of the National Academy of Sciences USA* 96: 10236–10241.

Nunn, C. L., Gittleman, J. L., and Antonovics, J. 2000. Promiscuity and the primate immune system. *Science* 290: 1168–1170.

Oldroyd, B. P. and Fewell, J. H. 2007. Genetic diversity promotes homeostasis in insect colonies. *Trends in Ecology & Evolution* 22: 408–413.

Opie, C., Atkinson, Q. D., Dunbar, R. I. M., and Shultz, S. 2013. Male infanticide leads to social monogamy in primates. *Proceedings of the National Academy of Sciences USA* 110: 13328–13332.

Orians, G. H. 1969. On the evolution of mating systems in birds and mammals. *American Naturalist* 103: 589–603.

Oring, L. W. 1985. Avian polyandry. *Current Ornithology* 3: 309–351.

Oring, L. W., Colwell, M. A., and Reed., J. M. 1991. Lifetime reproductive success in the spotted sandpiper (*Actitis macularia*): Sex differences and variance components. *Behavioral Ecology and Sociobiology* 28: 425–432.

Oring, L. W., Fleischer, R. C., Reed, J. M., and Marsden, K. E. 1992. Cuckoldry through stored sperm in the sequentially polyandrous spotted sandpiper. *Nature* 359: 631–633.

Oring, L. W. and Knudson, M. L. 1973. Monogamy and polyandry in the spotted sandpiper. *The Living Bird* 11: 59–73.

Packer, C., Scheel, D., and Pusey, A. E. 1990. Why lions form groups: Food is not enough. *American Naturalist* 136: 1–19.

Panova, M., Bostrom, J., Hofving, T., Areskoug, T., Eriksson, A., et al. 2010. Extreme female promiscuity in a non-social invertebrate species. *PLoS One* 5: e9640.

Partecke, J., von Haeseler, A., and Wikelski, M. 2002. Territory establishment in lekking marine iguanas, *Amblyrhynchus cristatus*: Support for the hotshot mechanism. *Behavioral Ecology and Sociobiology* 51: 579–587.

Pietsch, T. W. 1975. Precocious sexual parasitism in the deep sea ceratioid anglerfish, *Cryptopsaras couesi* Gill. *Nature* 256: 38–40.

Pietsch, T. W. 2005. Dimorphism, parasitism, and sex revisited: modes of reproduction among deep-sea ceratioid anglerfishes (Teleostei: Lophiiformes). *Icthyological Research* 52: 207–236.

Pitafi, K. D., Simpson, R., Stephen, J. J., and Day, T. H. 1990. Adult size and mate choice in seaweed flies (*Coelopa frigida*). *Heredity* 65: 91–97.

Podmokta, E., Dubiec, A., Arct, A., Drobniak, S. M., Gustafsson, L., and Cichoń, M. 2015. Malaria infection status predicts extra-pair paternity in the blue tit. *Journal of Avian Biology* 46: 303–306.

Poiani, A. and Wilks, C. 2000. Sexually transmitted diseases: A possible cost of promiscuity in birds? *Auk* 117: 1061–1065.

Pribil, S. and Searcy, W. A. 2001. Experimental confirmation of the polygyny threshold model for red-winged blackbirds. *Proceedings of the Royal Society of London B* 268: 1643–1646.

Quinn, J. S., Woolfenden, G. E., Fitzpatrick, J. W., and White, B. N. 1999. Multi-locus DNA fingerprinting supports genetic monogamy in Florida scrub-jays. *Behavioral Ecology and Sociobiology* 45: 1–10.

Reid, J. M., Monaghan, P., and Ruxton, G. D. 2002. Males matter: The occurrence and consequences of male incubation in starlings (*Sturnus vulgaris*). *Behavioral Ecology and Sociobiology* 51: 255–261.

Reynolds, S. M., Uy, J. A. C., Patricelli, G. L., Coleman, S. W., Braun, M. J., and Borgia, G. 2014. Tests of the kin selection model of mate choice and inbreeding avoidance in satin bowerbirds. *Behavioral Ecology* 25: 1005–1014.

Richardson, D. S, Komdeur, J., Burke, T., and von Schantz, T. 2005. MHC-based patterns of social and extra-pair mate choice in the Seychelles warbler. *Proceedings of the Royal Society of London B* 272, 759–767.

Rintamäki, P. T., Alatalo, R. V., Höglund, J., and Lundberg., A. 1995. Male territoriality and female choice on black grouse leks. *Animal Behaviour* 49: 759–767.

Rintamäki, P. T., Höglund, J., Alatalo, R. V., and Lundberg, A. 2001. Correlates of male mating success on black grouse (*Tetrao tetrix* L.) leks. *Annales Zoologici Fennici* 38: 99–109.

Rittenhouse, T. A. G., Semlitsch, R. D., and Thompson, F. R. 2009. Survival costs associated with wood frog breeding migrations: Effects of timber harvest and drought. *Ecology* 90: 1620–1630.

Rodríguez-Muñoz, R., Bretman, A., and Tregenza, T. 2011. Guarding males protect females from predation in a wild insect. *Current Biology* 21: 1716–1719.

Rubenstein, D. I. 1994. The ecology of female social behavior in horses, zebras, and asses. In P. Jarman and A. Rossiger (eds.), *Animal Societies: Individuals, Interactions, and Organization*, pp. 13–28. Kyoto: Kyoto University Press.

Rubenstein, D. R. 2007. Female extra-pair mate choice in a cooperative breeder: Trading sex for help and increasing offspring heterozygosity. *Proceedings of the Royal Society of London B* 274: 1895–1903.

Runcie, M. J. 2000. Biparental care and obligate monogamy in the rock-haunting possum, *Petropseudes dahli*, from tropical Australia. *Animal Behaviour* 59: 1001–1008.

Ryder, T. B., Blake, J. G., and Loiselle, B. R. A. 2006. A test of the environmental hotspot hypothesis for lek placement in three species of manakins (Pipridae) in Ecuador. *Auk* 123: 247–258.

Santos, E. S. A. and Nakagawa, S. 2013. Breeding biology and variable mating system of a population of introduced dunnocks (*Prunella modularis*) in New Zealand. *PLoS One* 8: e69329.

Santos, E. S. A., Santos, L. L. S., Lagisz, M., and Nakagawa, S. 2015. Conflict and cooperation over sex: The consequences of social and genetic polyandry for reproductive success in dunnocks. *Journal of Animal Ecology* 84: 1509–1519.

Sardell, R. J., Arcese, P., Keller, L. F., and Reid, J. M. 2011. Sex-specific differential survival of extra-pair and within-pair offspring in song sparrows, *Melospiza melodia*. *Proceedings of the Royal Society of London B* 278: 3251–3259.

Sato, T. 1994. Active accumulation of spawning substrate: A determinant of extreme polygyny in a shell-brooding cichlid fish. *Animal Behaviour* 48: 669–678.

Schamel, D., Tracy, D. M., and Lank, D. B. 2004. Male mate choice, male availability and egg production as limitations on polyandry in the red-necked phalarope. *Animal Behaviour* 67: 847–853.

Schwagmeyer, P. L. 1994. Competitive mate searching in the 13-lined ground squirrel: Potential roles of spatial memory? *Ethology* 98: 265–276.

Schwagmeyer, P. L. 1995. Searching today for tomorrow's mates. *Animal Behaviour* 50: 759–767.

Seeley, T. D. and Tarpy, D. R. 2007. Queen promiscuity lowers disease within honeybee colonies. *Proceedings of the Royal Society of London B* 274: 67–72.

Setchell, J. M. 2005. Do female mandrills prefer brightly colored males? *International Journal of Primatology* 26: 715–735.

Setchell, J. M., Charpentier, M. J. E., Abbott, K. M., Wickings, E. J., and Knapp, L. A. 2010. Opposites attract: MHC-associated mate choice in a polygynous primate. *Journal of Evolutionary Biology* 23: 136–148.

Shorey, L. 2002. Mating success on white-bearded manakin (*Manacus manacus*) leks: Male characteristics and relatedness. *Behavioral Ecology and Sociobiology* 52: 451–457.

Shorey, L., Piertney, S., Stone, J., and Hoglund, J. 2000. Fine-scale genetic structuring on *Manacus manacus* leks. *Nature* 408: 352–353.

Sih, A., Montiglio, P.-O., Wey, T. W., and Fogarty, S. 2017. Altered physical and social conditions produce rapidly reversible mating systems in water striders. *Behavioral Ecology* 28: 632–639.

Simone-Finstrom, M., Walz, M., and Tarpy, D. R. 2016. Genetic diversity confers colony-level benefits due to individual immunity. *Biology Letters* 12: 20151007.

Slatyer, R. A., Mautz, R. B., Blackwell, P. R. Y., and Jennions, M. D. 2011. Estimating genetic benefits of polyandry from experimental studies: A meta-analysis. *Biological Reviews* 87: 1–33.

Snow, D. W. 1956. Courtship ritual: The dance of the manakins. *Animal Kingdom* 59: 86–91.

Sogabe, A., Matsumoto, K., and Yanagisawa, Y. 2007. Mate change reduces the reproductive rate of males in a monogamous pipefish *Corythoichthys haematopterus*: The benefit of long-term pair bonding. *Ethology* 113: 764–771.

Stenmark, G., Slagsvold, T., and Lifjeld, J. T. 1988. Polygyny in the pied flycatcher, *Ficedula hypoleuca*: A test of the deception hypothesis. *Animal Behaviour* 36: 1646–1657.

Strassmann, J. E. 2001. The rarity of multiple mating by females in the social Hymenoptera. *Insectes Sociaux* 48: 1–13.

Tarpy, D. R. and Nielsen, D. I. 2002. Sampling error, effective paternity, and estimating the genetic structure of honey bee colonies (Hymenoptera: Apidae). *Annals of the Entomological Society of America* 95: 513–528.

Taylor, M. L., Price, T. A. R., and Wedell, N. 2015. Polyandry in nature: A global analysis. *Trends in Ecology & Evolution* 29: 376–383.

Taylor, M. L., Wedell, N., and Hosken, D. J. 2007. The heritability of attractiveness. *Current Biology* 17: R959–R960.

Temrin, H. and Arak, A. 1989. Polyterritoriality and deception in passerine birds. *Trends in Ecology & Evolution* 4: 106–108.

Townsend, A. K., Bowman, R., Fitzpatrick, J. W., Dent, M., and Lovette, I. J. 2011. Genetic monogamy across variable demographic landscapes in cooperatively breeding Florida scrub-jays. *Behavioral Ecology* 22: 464–470.

Tregenza, T. and Wedell, N. 2000. Genetic compatibility, mate choice, and patterns of parentage: Invited review. *Molecular Ecology* 9: 1013–1027.

Tschirren, B., Postma, E., Rutstein, A. N., and Griffith, S. C. 2012. When mothers make sons sexy: Maternal effects contribute to the increased sexual attractiveness of extra-pair offspring. *Proceedings of the Royal Society of London B* 279: 1233–1240.

Varian-Ramos, C. W. and Webster, M. S. 2012. Extra-pair copulations reduce inbreeding for female red-backed fairy-wrens, *Malurus melanocephalus*. *Animal Behaviour* 83: 857–864.

Vincent, A. C. J., Marsden, A. D., Evans, K. L., and Sadler, L. M. 2004. Temporal and spatial opportunities for polygamy in a monogamous seahorse, *Hippocampus whitei*. *Behaviour* 141: 141–156.

Vincent, A. C. J. and Sadler, L. M. 1995. Faithful pair bonds in wild seahorses, *Hippocampus whitei*. *Animal Behaviour* 50: 1557–1569.

Vitousek, M. N., Rubenstein, D. R., Nelson, K., and Wikelski, M. 2008. Are hotshots always hot? A longitudinal study of hormones, behavior, and reproductive success in male marine iguanas. *General and Comparative Endocrinology* 157: 227–232.

Vollrath, F. 1998. Dwarf males. *Trends in Ecology & Evolution* 13: 159–163.

Waage, J. K. 1973. Reproductive behavior and its relation to territoriality in *Calopteryx maculata* (Beauvois) (Odonata: Calopterygidae). *Behaviour* 47: 240–256.

Weatherhead, P. J. and Robertson, R. J. 1979. Offspring quality and the polygyny threshold: The "sexy son" hypothesis. *American Naturalist* 113: 201–208.

Weathers, W. W. and Sullivan, K. A. 1989. Juvenile foraging proficiency, parental effort, and avian reproductive success. *Ecological Monographs* 59: 223–246.

Webster, M. S. 1994. Female-defence polygyny in a Neotropical bird, the Montezuma oropendula. *Animal Behaviour* 48: 779–794.

Webster, M. S. and Robinson, S. K. 1999. Courtship disruptions and male mating strategies: Examples from female-defense mating systems. *American Naturalist* 154: 717–729.

Webster, M. S., Tarvin, K. A., Tuttle, E. M., and Pruett-Jones, S. 2007. Promiscuity drives sexual selection in a socially monogamous bird. *Evolution* 61: 2205–2211.

Wedell, N. and Tregenza, T. 1999. Successful fathers sire successful sons. *Evolution* 53: 620–625.

Westneat, D. F., Sherman, P. W., and Morton, M. L. 1990. The ecology and evolution of extra-pair copulations in birds. *Current Ornithology* 7: 330–369.

Wiklund, C., Karlsson, B., and Leimar, O. 2001. Sexual conflict and cooperation in butterfly reproduction: A comparative study of polyandry and female fitness. *Proceedings of the Royal Society of London B* 268: 1661–1667.

Wilson, A. B. and Martin-Smith, K. M. 2007. Genetic monogamy despite social promiscuity in the pot-bellied seahorse (*Hippocampus abdominalis*). *Molecular Ecology* 16: 2345–2352.

Woodroffe, R. and Vincent, A. 1994. Mother's little helpers: Patterns of male care in mammals. *Trends in Ecology & Evolution* 9: 294–297.

Woyciechowski, M., Kabat, L., and Król., E. 1994. The function of the mating sign in honey bees, *Apis mellifera* L.: New evidence. *Animal Behaviour* 47: 733–735.

Yoshizawa, K., Ferreira, R. L., Kamimura, Y., and Liehard, C. 2014. Female penis, male vagina, and their correlated evolution in a cave insect. *Current Biology* 24: 1006–1010.

Zeh, J. A. 1997. Polyandry and enhanced reproductive success in the harlequin beetle-riding pseudoscorpion. *Behavioral Ecology and Sociobiology* 40: 111–118.

Zeh, J. A., Newcomer, S. D., and Zeh, D. W. 1998. Polyandrous females discriminate against previous mates. *Proceedings of the National Academy of Sciences USA* 95: 13273–13736.

Zeh, J. A. and Zeh, D. W. 1996. The evolution of polyandry I: Intragenomic conflict and genetic incompatibility. *Proceedings of the Royal Society of London B* 263: 1711–1717.

Zeh, J. A. and Zeh, D. W. 1997. The evolution of polyandry II: Post-copulatory defenses against genetic incompatibility. *Proceedings of the Royal Society of London B* 264: 69–75.

Chapter 11

Anderson, D. J. 1990. Evolution of obligate siblicide in boobies. 1. A test of the insurance-egg hypothesis. *American Naturalist* 135: 334–350.

Aragón, S., Møller, A. P., Soler, J. J., and Soler, M. 1999. Molecular phylogeny of cuckoos supports a polyphyletic origin of brood parasitism. *Journal of Evolutionary Biology* 12: 495–506.

Balcombe, J. P. 1990. Vocal recognition of pups by mother Mexican free-tailed bats, *Tadarida brasiliensis mexicana*. *Animal Behaviour* 39: 960–966.

Balshine-Earn, S. 1995. The costs of parental care in Galilee St. Peter's fish, *Sarotherodon galilaeus*. *Animal Behaviour* 50: 1–7.

Beecher, M. D., Beecher, I. M., and Hahn, S. 1981. Parent-offspring recognition in bank swallows, *Riparia riparia*: II. Development and acoustic basis. *Animal Behaviour* 29: 95–101.

Bleeker, M., Kingma, S. A., Szentirmai, I., Szekely, T., and Komdeur, J. 2005. Body condition and clutch desertion in the penduline tit *Remiz pendulinus*. *Behaviour* 142: 1465–1478.

Brown, K. M. 1998. Proximate and ultimate causes of adoption in ring-billed gulls. *Animal Behaviour* 56: 1529–1543.

Buzatto, B. A., Requena, G. S., Martins, E. G., and Machado, G. 2007. Effects of maternal care on the lifetime reproductive success of females in a neotropical harvestman. *Journal of Animal Ecology* 76: 937–945.

Cameron, E. Z. 2004. Facultative adjustment of mammalian sex ratios in support of the Trivers–Willard hypothesis: evidence for a mechanism. *Proceedings of the Royal Society of London B* 271: 1723–1728.

Caro, S. M., Griffin, A. S., Hinde, C. A., and West, S. A. 2016a. Unpredictable environments lead to the evolution of parental neglect in birds. *Nature Communications* 7: 10985.

Caro, S. M., West, S. A., and Griffin, A. S. 2016b. Sibling conflict and dishonest signaling in birds. *Proceedings of the National Academy of Sciences USA* 113: 13803-13808.

Charrier, I., Mathevon, N., and Jouventin, P. 2003. Vocal signature recognition of mothers by fur seal pups. *Animal Behaviour* 65: 543–550.

Clark, A. B. 1978. Sex ratio and local resource competition in a prosimian primate. *Science* 201: 163–165.

Clifford, L. D. and Anderson, D. J. 2001a. Experimental demonstration of the insurance value of extra eggs in an obligately siblicidal seabird. *Behavioral Ecology* 12: 340–347.

Clifford, L. D. and Anderson, D. J. 2001b. Food limitation explains most clutch size variation in the Nazca booby. *Journal of Animal Ecology* 70: 539–545.

Clotfelter, E. D., Schubert, K. A., Nolan, V., and Ketterson, E. D. 2003. Mouth color signals thermal state of nestling dark-eyed juncos (*Junco hyemalis*). *Ethology* 109: 171–182.

Clutton-Brock, T. H. 1991. *The Evolution of Parental Care*. Princeton, NJ: Princeton University Press.

Clutton-Brock, T. H. and Parker, G. A. 1992. Potential reproductive rates and the operation of sexual selection. *Quarterly Review of Biology* 67: 437–456.

Colombelli-Negrel, D., Hauber, M. E., Robertson, J., Sulloway, F. J., Hoi, H., et al. 2012. Embryonic learning of vocal passwords in superb fairy-wrens reveals intruder cuckoo nestlings. *Current Biology* 22: 2155–2160.

Cronk, L. 1993. Parental favoritism to daughters. *American Scientist* 81: 272–279.

Cronk, L. 2000. Female-biased parental investment and growth performance among the Mukogodo. In L. Cronk, N. Chagnon and W. Irons (eds.), *Adaptation and Human Behavior: An Anthropological Perspective*, pp. 203-221. New York: Transaction Publishers.

Davies, N. B. and de L. Brooke, M. 1988. Cuckoos versus reed warblers: Adaptations and counteradaptations. *Animal Behaviour* 36: 262–284.

Davies, N. B., de L. Brooke, M., and Kacelnik, A. 1996. Recognition errors and probability of parasitism determine whether reed warblers should accept or reject mimetic cuckoo eggs. *Proceedings of the Royal Society of London B* 263: 925–931.

Davies, N. B., Kilner, R. M., and Noble, D. G. 1998. Nestling cuckoos *Cuculus canorus* exploit hosts with begging calls that mimic a brood. *Proceedings of the Royal Society of London B* 265: 673-678.

de Ayala, R. M., Saino, N., Møller, A. P., and Anselmi, C. 2007. Mouth coloration of nestlings covaries with offspring quality and influences parental feeding behavior. *Behavioral Ecology* 18: 526–534.

Del-Claro, K. and Tizo-Pedroso, E. 2009. Ecological and evolutionary pathways of social behavior in pseudoscorpions (Arachnida: Pseudoscorpiones). *Acta Ethologica* 12: 13–22.

Drummond, H. and Chavelas, C. G. 1989. Food shortage influences sibling aggression in the blue-footed booby. *Animal Behaviour* 37: 806–819.

Eadie, J. M. and Fryxell, J. M. 1992. Density dependence, frequency-dependence, and alternative nesting strategies in goldeneyes. *American Naturalist* 140: 621–641.

Emlen, S. T. and Oring, L. W. 1977. Ecology, sexual selection, and the evolution of mating systems. *Science* 197: 215–223.

Feeney, W. E., Stoddard, M. C., Kilner, R. M. and Langmore, N. E. 2014a. "Jack-of-all-trades" egg mimicry in the brood parasitic Horsfield's bronze-cuckoo? *Behavioral Ecology* 25: 1365–1373.

Feeney, W. E., Welbergen, J. A. and Langmore, N. E. 2014b. Advances in the study of coevolution between avian brood parasites and their hosts. *Annual Review of Ecology and Systematics* 45: 227–246.

Fisher, R. A. 1934. *The Genetical Theory of Natural Selection*. Oxford: Clarendon Press.

Ghalambor, C. K. and Martin, T. E. 2001. Fecundity–survival trade-offs and parental risk-taking in birds. *Science* 292: 494–497.

Gibbs, H. L., Sorenson, M. D., Marchetti, K., Brooke, M. d. L., Davies, N. B., and Nakamura, H. 2000. Genetic evidence for female host-specific races of the common cuckoo. *Nature* 407: 183-186.

Godfray, H. C. J. 1995. Signaling of need between parents and young: Parent-offspring conflict and sibling rivalry. *American Naturalist* 146: 1–24.

Gowaty, P. A. and Lennartz, M. R. 1985. Sex ratios of nestling and fledging red-cockaded woodpeckers (*Picoides borealis*) favor males. *American Naturalist* 126: 347–353.

Grafen, A. 1990. Biological signals as handicaps. *Journal of Theoretical Biology* 144: 517–546.

Graves, J. A., Whiten, A. 1980. Adoption of strange chicks by herring gulls, *Larus argentatus. Zeitschrift für Tierpsychologie* 54: 267–278.

Gross, M. R. and Sargent, R. C. 1985. The evolution of male and female parental care in fishes. *American Zoologist* 25: 807–822.

Hamilton, W. D. 1967. Extraordinary sex ratios. *Science* 156: 477–488.

Hartung, J. 1982. Polygyny and inheritance of wealth. *Current Anthropology* 23: 1–12.

Hauber, M. E. 2003. Hatching asynchrony, nestling competition, and the cost of interspecific brood. *Behavioral Ecology* 14: 227–235.

Heeb, P., Schwander, T., and Faoro, S. 2003. Nestling detectability affects parental feeding preferences in a cavity-nesting bird. *Animal Behaviour* 66: 637–642.

Holley, A. J. F. 1984. Adoption, parent–chick recognition, and maladaptation in the herring gull *Larus argentatus. Zeitschrift für Tierpsychologie* 64: 9–14.

Hoover, J. P. and Robinson, S. K. 2007. Retaliatory mafia behavior by a parasitic cowbird favors host acceptance of parasitic eggs. *Proceedings of the National Academy of Sciences USA* 104: 4479–4483.

Jesseau, S. A., Holmes, W. G., and Lee, T. M. 2008. Mother–offspring recognition in communally nesting degus, *Octodon degus*. *Animal Behaviour* 75: 573–582.

Kilner, R. M. and Langmore, N. E. 2011. Cuckoos versus hosts in insects and birds: Adaptations, counter-adaptations and outcomes. *Biological Reviews* 86: 836–852.

Kilner, R. M., Madden, J. R., and Hauber, M. E. 2004. Brood parasitic cowbird nestlings use host young to procure resources. *Science* 305: 877–879.

Kleindorfer, S., Hoi, H., Evans, C., Mahr, K., Robertson, J., et al. 2014. The cost of teaching embryos in superb fairy-wrens *Behavioral Ecology* 25: 1131–1135.

Knudsen, B. and Evans, R. M. 1986. Parent–young recognition in herring gulls (*Larus argentatus*). *Animal Behaviour* 34: 77–80.

Kölliker, M. 2007. Benefits and costs of earwig (*Forficula auricularia*) family life. *Behavioral Ecology and Sociobiology* 61: 1489–1497.

Komdeur, J. 1996. Facultative sex ratio bias in the offspring of Seychelles warblers. *Proceedings of the Royal Society of London B* 263: 661–666.

Krüger, O., Davies, N. B. 2002. The evolution of cuckoo parasitism: A comparative analysis. *Proceedings of the Royal Society of London B* 269: 375–381.

Langmore, N. E., Hunt, S., and Kilner, R. M. 2003. Escalation of a coevolutionary arms race through host rejection of brood parasitic young. *Nature* 422: 157–160.

Langmore, N. E. and Kilner, R. M. 2007. Breeding site and host selection by Horsfield's bronze-cuckoos, *Chalcites basalis*. *Animal Behaviour* 74: 995–1004.

Lin, C. P., Danforth, B. N., and Wood, T. K. 2004. Molecular phylogenetics and evolution of maternal care in Membracine treehoppers. *Systematic Biology* 53: 400–421.

Lotem, A., Nakamura, H., and Zahavi, A. 1995. Constraints on egg discrimination and cuckoo–host co-evolution. *Animal Behaviour* 49: 1185–1209.

Lougheed, L. W. and Anderson, D. J. 1999. Parent blue-footed boobies suppress siblicidal behavior of offspring. *Behavioral Ecology and Sociobiology* 45: 11–18.

Lyon, B. E. 1993. Conspecific brood parasitism as a flexible female reproductive tactic in American coots. *Animal Behaviour* 46: 911–928.

Lyon, B. E. 2003. Ecological and social constraints on conspecific brood parasitism by nesting female American coots (*Fulica americana*). *Journal of Animal Ecology* 72: 47–60.

Lyon, B. E. 2007. Mechanism of egg recognition in defenses against conspecific brood parasitism: American coots (*Fulica americana*) know their own eggs. *Behavioral Ecology and Sociobiology* 61: 455–463.

Lyon, B. E. and Eadie, J. M. 2008. Conspecific brood parasitism in birds: A life-history perspective. *Annual Review of Ecology, Evolution, and Systematics* 39: 343–363.

Lyon, B. E., Eadie, J. M., and Hamilton, L. D. 1994. Parental choice selects for ornamental plumage in American coot chicks. *Nature* 371: 240–243.

Machado, G., Requena, G. S., Buzatto, B. A., Osses, F., and Rossetto, L. M. 2004. Five new cases of paternal care in harvestmen (Arachnida: Opiliones): Implications for the evolution of male guarding in the neotropical family Gonyleptidae. *Sociobiology* 44: 577–598.

Mahler, B., Confalonieri, V. A., Lovette, I. J., and Reboreda, J. C. 2007. Partial host fidelity in nest selection by the shiny cowbird (*Molothrus bonariensis*), a highly generalist avian brood parasite. *Journal of Evolutionary Biology* 20: 1918–1923.

Martin, T. E. and Schwabl, H. 2008. Variation in maternal effects and embryonic development rates among passerine species. *Philosophical Transactions of the Royal Society B* 363: 1663–1674.

McCracken, G. F. 1984. Communal nursing in Mexican free-tailed bat maternity colonies. *Science* 223: 1090–1091.

McCracken, G. F. and Gustin, M. K. 1991. Nursing behavior in Mexican free-tailed bat maternity colonies. *Ethology* 89: 305–321.

McLaren, C. M. and Sealey, S. G. 2003. Factors influencing susceptibility of host nests to brood parasitism. *Ethology* 15: 343-353.

Medvin, M. B. and Beecher, M. D. 1986. Parent-offspring recognition in the barn swallow (*Hirundo rustica*). *Animal Behaviour* 34: 1627–1639.

Medvin, M. B., Stoddard, P. K., and Beecher, M. D. 1993. Signals for parent-offspring recognition: A comparative analysis of the begging calls of cliff swallows and barn swallows. *Animal Behaviour* 45: 841–850.

Merrill, L., Chiavacci, S. J., Paitz, R. T., and Benson, T. J. 2017. Rates of parasitism, but not allocation of egg resources, vary among and within hosts of a generalist avian brood parasite. *Oecologia* 184: 399–410.

Miller, D. E. and Emlen, J. T., Jr. 1975. Individual chick recognition and family integrity in the ring-billed gull. *Behaviour* 52: 124–144.

Mock, D. W. 1984. Siblicidal aggression and resource monopolization in birds. *Science* 225: 731–733.

Mock, D. W., Drummond, H., and Stinson, C. H. 1990. Avian siblicide. *American Scientist* 78: 438–449.

Mock, D. W. and Parker, G. A. 1997. *The Evolution of Sibling Rivalry*. Oxford: Oxford University Press.

Mock, D. W. and Ploger, B. J. 1987. Parental manipulation of optimal hatch asynchrony in cattle egrets: An experimental study. *Animal Behaviour* 35: 150–160.

Nazareth, T. M. and Machado, G. 2010. Mating system and exclusive postzygotic paternal care in a neotropical harvestman (Arachnida: Opiliones). *Animal Behaviour* 79: 547–554.

Neff, B. D. 2003. Decisions about parental care in response to perceived paternity. *Nature* 422: 716–719.

Osorno, J. L. and Drummond, H. 1995. The function of hatching asynchrony in the blue-footed booby. *Behavioral Ecology and Sociobiology* 37: 265–274.

Paquet, M. and Smiseth, P. T. 2017. Females manipulate behavior of caring males via prenatal maternal effects. *Proceedings of the National Academy of Sciences USA* 114: 6800–6805.

Pierotti, R. and Murphy, E. C. 1987. Intergenerational conflicts in gulls. *Animal Behaviour* 35: 435–444.

Pollet, T. V., Fawcett, T. W., Buunk, A. P., and Nettle, D. 2009. Sex-ratio biasing towards daughters among lower-ranking co-wives in Rwanda. *Biology Letters* 5: 765–768.

Queller, D. C. 1997. Why do females care more than males? *Proceedings of the Royal Society of London B* 264: 1555–1557.

Requena, G. S., Buzatto, B. A., Munguía-Steyer, R., and Machado, G. 2009. Efficiency of uniparental male and female care against egg predators in two closely related syntopic harvestmen. *Animal Behaviour* 78: 1169–1176.

Rohwer, S. and Spaw, C. D. 1988. Evolutionary lag versus bill-size constraints: A comparative study of the acceptance of cowbird eggs by old hosts. *Evolutionary Ecology* 2: 27–36.

Saino, N., Ninni, P., Calza, S., Martinelli, R., de Bernardi, F., and Møller, A. P. 2000. Better red than dead: Carotenoid-based mouth coloration reveals infection in barn swallow nestlings. *Proceedings of the Royal Society of London B* 267: 57–61.

Salomon, M., Schneider, J., and Lubin, Y. 2005. Maternal investment in a spider with suicidal maternal care, *Stegodyphus lineatus* (Araneae, Eresidae). *Oikos* 109: 614–622.

Santos, E. S. A. and Nakagawa, S. 2017. Facultative adjustment of paternal care in the face of female infidelity in dunnocks. *bioRxiv*: http://dx.doi.org/10.1101/158816.

Schielzeth, H. and Bolund, E. 2010. Patterns of conspecific brood parasitism in zebra finches. *Animal Behaviour* 79: 1329–1337.

Schwabl, H., Mock, D. W., and Gieg, J. A. 1997. A hormonal mechanism for parental favouritism. *Nature* 386: 231.

Sealy, S. G. 1995. Burial of cowbird eggs by parasitized yellow warblers: An empirical and experimental study. *Animal Behaviour* 49: 877–889.

Seidelmann, K. 2006. Open-cell parasitism shapes maternal investment patterns in the red mason bee *Osmia rufa*. *Behavioral Ecology* 17: 839–846.

Seidelmann, K., Ulbrich, K., and Mielenz, N. 2010. Conditional sex allocation in the Red Mason bee, *Osmia rufa*. *Behavioral Ecology and Sociobiology* 64: 337–347.

Semel, B. and Sherman, P. W. 2001. Intraspecific parasitism and nest-site competition in wood ducks. *Animal Behaviour* 61: 787–803.

Shizuka, D. and Lyon, B. E. 2010. Coots use hatch order to learn to recognize and reject conspecic brood parasitic chicks. *Nature* 463: 223–226.

Shizuka, D. and Lyon, B. E. 2011. Hosts improve the reliability of chick recognition by delaying the hatching of brood parasitic eggs. *Current Biology* 21: 515–519.

Sieff, D. F. 1990. Explaining biased sex ratios in human populations: A critique of recent studies. *Current Anthropology* 31: 25–48.

Slagsvold, T. 1998. On the origin and rarity of interspecific nest parasitism in birds. *American Naturalist* 152: 264–272.

Smiseth, P. T., Ward, R. J. S., and Moore, A. J. 2007. Parents influence asymmetric sibling competition: Experimental evidence with partially dependent young. *Ecology* 88: 3174–3182.

Smith, M. S., Kish, B. J., and Crawford, C. B. 1987. Inheritance of wealth as human kin investment. *Ethology and Sociobiology* 8: 171–182.

Smith, R. L. 1997. Evolution of paternal care in giant water bugs (Heteroptera: Belostomatidae). In J. C. Choe and B. Crespi (eds.), *Social Competition and Cooperation among Insects and Arachnids, II. Evolution of Sociality*, pp. 116–149. Cambridge: Cambridge University Press.

Smith, R. L. and Larsen, E. 1993. Egg attendance and brooding by males of the giant water bug *Lethocerus medius* (Guerin) in the field (Heteroptera, Belostomatidae). *Journal of Insect Behavior* 6: 93–106.

Soler, M., Soler, J. J., Martinez, J. G., and Møller, A. P. 1995. Magpie host manipulation by great spotted cuckoos: Evidence for an avian Mafia? *Evolution* 49: 770–775.

Sorenson, M. D. and Payne, R. B. 2001. A single ancient origin of brood parasitism in African finches: Implications for host–parasite coevolution. *Evolution* 55: 2550–2567.

Sorenson, M. D. and Payne, R. B. 2005. Molecular systematics: Cuckoo phylogeny inferred from mitochondrial DNA sequences. In R. B. Payne (ed.), *Bird Familes of the World: Cuckoos*, pp. 68–94. Oxford: Oxford University Press.

Spottiswoode, C. N., Stryjewski, K. F., Quader, S., Colebrook-Robjent, J. F. R., and Sorenson, M. D. 2011. Ancient host specificity within a single species of brood parasitic bird. *Proceedings of the National Academy of Sciences USA* 108: 17738–17742.

Stevens, M., Troscianko, J., and Spottiswoode, C. N. 2013. Repeated targeting of the same hosts by a brood parasite compromises host egg rejection. *Nature Communications* 4: 2475.

Stoddard, M. C., Kilner, R. M., and Town, C. 2014. Pattern recognition algorithm reveals how birds evolve individual egg pattern signatures. *Nature Communications* 5: 4117.

Summers, K. and Tumulty, J. 2014. Parental care, sexual selection, and mating systems in neotropical poison frogs. In R. Maceo and G. Maceo (eds.), *Sexual Selection: Perspectives and Models from the Neotropics*, pp. 289–320. New York: Academic Press.

Tallamy, D. W. 2001. Evolution of exclusive paternal care in arthropods. *Annual Review of Entomology* 46: 139–165.

Thogerson, C. M., Brady, C. M., Howard, R. D., Mason, G. J., Pajor, E. A., et al. 2013. Winning the genetic lottery: Biasing birth sex ratio results in more grandchildren. *PLoS One* 8: e67867.

Trivers, R. L. 1972. Parental investment and sexual selection. In B. Campbell (ed.), *Sexual Selection and the Descent of Man*, pp. 136–179. Chicago: Aldine.

Trivers, R. L. 1974. Parent–offspring conflict. *American Zoologist* 14: 249–264.

Trivers, R. L. and Willard, D. E. 1973. Natural selection of parental ability to vary the sex ratio of offspring. *Science* 179: 90–92.

van Dijk, R. E., Szekely, T., Komdeur, J., Pogany, A., Fawcett, T. W., and Weissing, F. J. 2011. Individual variation and the resolution of conflict over parental care in penduline tits. *Proceedings of the Royal Society of London B* 279: 1927–1936.

van Dijk, R. E., Szentirmai, I., Komdeur, J., and Szekely, T. 2007. Sexual conflict over parental care in penduline tits *Remiz pendulinus*: The process of clutch desertion. *Ibis* 149: 530–534.

Vincze, O., Kosztolanyi, A., Barta, Z., Kupper, C., Alrashidi, M., et al. 2016. Parental cooperation in a changing climate: Fluctuating environments predict shifts in care division. *Global Ecology and Biogeography* 26: 347–358.

Wilkinson, M., Kupfer, A., Marques-Porto, R., Jeffkins, H., Antoniazzi, M. M., and Jared, C. 2008. One hundred million years of skin feeding? Extended parental care in a neotropical caecilian (Amphibia: Gymnophiona). *Biology Letters* 4: 358–361.

Winfree, R. 1999. Cuckoos, cowbirds, and the persistence of brood parasitism. *Trends in Ecology & Evolution* 14: 338–343.

Yom-Tov, Y., Geffen, E. 2006. On the origin of brood parasitism in altricial birds. *Behavioral Ecology* 17: 196–205.

Zahavi, A. 1979. Parasitism and nest predation in parasitic cuckoos. *American Naturalist* 113: 157–159.

Chapter 12

Abbot, P., Abe, J., Alcock, J., Alizon, S., Alphedrinha, J. A. C., et al. 2011. Inclusive fitness theory and eusociality. *Nature* 471: E1–E4.

Alpendrinha, J., West, S. A., and Gardner, A. 2013. Haplodiploidy and the evolution of eusociality: Worker reproduction. *American Naturalist* 182: 421–438.

Boomsma, J. J. 2007. Kin selection versus sexual selection: Why the ends do not meet. *Current Biology* 17: R673–R683.

Boomsma, J. J. 2009. Lifetime monogamy and the evolution of eusociality. *Philosophical Transactions of the Royal Society B* 364: 3191–3207.

Bourke, A. F. G. 2008. Social evolution: Daily self-sacrifice by worker ants. *Current Biology* 18: R1100–R1101.

Bourke, A. F. G. 2011a. *Principles of Social Evolution*. Oxford: Oxford University Press.

Bourke, A. F. G. 2011b. The validity and value of inclusive fitness theory. *Proceedings of the Royal Society of London B* 278: 3313–3320.

Brock, D. A., Douglas, T. E., Queller, D. C., and Strassman, J. E. 2011. Primitive agriculture in a social amoeba. *Nature* 469: 393–398.

Dawkins, R. 1982. *The Extended Phenotype*. San Francisco: W.H. Freeman.

Dawkins, R. 1989. *The Selfish Gene*. Oxford: Oxford University Press.

Dunn, C. 2009. Siphonophores. *Current Biology* 19: R233–R234.

Eldakar, O. T. and Wilson, D. S. 2011. Eight criticisms not to make about group selection. *Evolution* 65: 1523–1526.

Elgar, M. A. 2015. Integrating insights across diverse taxa: Challenges for understanding social evolution. *Frontiers in Ecology and Evolution* 3: 124.

Fortunato, A., Strassman, J. E., Santorelli, L., and Queller, D. C. 2003. Co-occurrence in nature of different clones of the social amoeba, *Dictyostelium discoideum*. *Molecular Ecology* 12: 1031–1038.

Galaktionov, K. V., Podvyaznaya, I. M., Nikolaev, K. E., and Levakin, I. A. 2015. Self-sustaining infrapopulation or colony? Redial clonal groups of *Himasthla elongata* (Mehlis, 1831) (Trematoda: Echinostomatidae) in *Littorina littorea* (Linnaeus) (Gastropoda: Littorinidae) do not support the concept of eusocial colonies in trematodes. *Folia Parasitologica* 62: 067.

Garcia-Vedrenne, A. E., Quintana, A. C. E., DeRogatis, A. M., Martyn, K., Kuris, A. M., and Hechinger, R. F. 2016. Social organization in parasitic flatworms: Four additional echinostomoid trematodes have a soldier caste and one does not. *Journal of Parasitology* 102: 11–20.

Gardner, A., Alpedrinha, J., and West, S. A. 2012. Haplodiploidy and the evolution of eusocialty: Split sex ratios. *American Naturalist* 179: 240–256.

Gilbert, O. M., Foster, K. R., Mehdiabadi, N. J., Strassman, J. E., and Queller, D. C. 2007. High relatedness maintains multicellular cooperation in a social amoeba by controlling cheater mutants. *Proceedings of the National Academy of Sciences USA* 104: 8913–8917.

Hamilton, W. D. 1964. The genetical theory of social behaviour, I, II. *Journal of Theoretical Biology* 7: 1–52.

Hamilton, W. D. 1996. Foreword. In S. Turillazzi and M. J. West-Eberhard (eds.), *Natural History and the Evolution of Paper Wasps*, pp. v–vi. Oxford: Oxford University Press.

Hechinger, R. F., Wood, A. C., and Kuris, A. M. 2010. Social organization in a flatworm: Trematode parasites form soldier and reproductive castes. *Proceedings of the Royal Society of London B* 278: 656–665.

Hölldobler, B. and Wilson, E. O. 2008. *The Superorganism: The Beauty, Elegance, and Strangeness of Insect Societies*. New York: W. W. Norton & Company.

Hughes, W. O., Oldroyd, B. P., Beekman, M., and Ratnieks, F. L. 2008. Ancestral monogamy shows kin selection is key to evolution of sociality. *Science* 320: 1213–1216.

Kapheim, K. M., Pan, H., Salzberg, S. L., Puiu, D., Magoc, T., et al. 2015. Genomic signatures of evolutionary transitions from solitary to group living. *Science* 348: 1139–1143.

Kronauer, D. J. C. and Boomsma, J. J. 2007. Multiple queens means fewer mates. *Current Biology* 17: R753–R755.

Kuzdzal, J. J., Foster, K. R., Queller, D. C., and Strassman, J. E. 2007. Exploiting new terrain: An advantage to sociality in the slime mold *Dictyostelium discoideum*. *Behavioral Ecology* 18: 433–437.

Liao, X., Rong, S., and Queller, D. C. 2015. Relatedness, conflict, and the evolution of eusociality. *PLoS Biology* 13: e1002098.

Liebig, J., Peeters, C., and Hölldobler, B. 1999. Worker policing limits the number of reproductives in a ponerine ant. *Proceedings of the Royal Society of London B* 266: 1865–1870.

Marshall, J. A. R. 2011. Group selection and kin selection: Formally equivalent approaches. *Trends in Ecology & Evolution* 26: 325–332.

Maschwitz, U. and Maschwitz, E. 1974. Platzende Arbeiterinnen: Eine neue Art der Feindabwehr bei sozialen Hautflüglern. *Oecologia* 14: 289–294.

Maynard Smith, J. 1964. Group selection and kin selection. *Nature* 201: 1145–1147.

Maynard Smith, J. and Szathmáry, E. 1995. *The Major Transitions in Evolution*. Oxford: Oxford University Press.

McKenzie, S. K., Fetter-Pruneda, I., Ruta, V., and Kronauer, D. J. C. 2017. Transcriptomics and neuroanatomy of the clonal raider ant implicate an expanded clade of odorant receptors in chemical communication. *Proceedings of the National Academy of Sciences USA* 113: 14091–14096.

Meunier, J., West, S. A., and Chapuisat, M. 2008. Split sex ratios in the social Hymenoptera: A meta-analysis. *Behavioral Ecology* 19: 382–390.

Mikheyev, A. S. and Linksvayer, T. A. 2015. Genes associated with ant social behavior show distinct transcriptional and evolutionary patterns. *eLife* 4: e04775.

Monnin, T., Ratnieks, F. L. W., Jones, G. R., and Beard, R. 2002. Pretender punishment induced by chemical signalling in a queenless ant. *Nature* 419: 61–65.

Mouritsen, K. and Halvorsen, F. J. 2015. Social flatworms: The minor caste is adapted for attacking competing parasites. *Marine Biology* 162: 1503–1509.

Nowak, M. A., Tarnita, C. E., and Wilson, E. O. 2011. The evolution of eusociality. *Nature* 466: 1057–1062.

Okasha, S. 2010. Levels of selection. *Current Biology* 20: R306–R307.

Page, R. E., Robinson, G. E., Fondrk, M. K., and Nasr, M. E. 1995. Effects of worker genotypic diversity on honey-bee colony development and behavior (*Apis mellifera* L.). *Behavioral Ecology and Sociobiology* 36: 387–396.

Queller, D. C. 1996. The measurement and meaning of inclusive fitness. *Animal Behaviour* 51: 229–232.

Queller, D. C., Strassman, J. E., and Hughes, C. R. 1993. Microsatellites and kinship. *Trends in Ecology & Evolution* 8: 285–288.

Ratnieks, F. L. and Helanterä, H. 2009. The evolution of extreme altruism and inequality in insect societies. *Philosophical Transactions of the Royal Society B* 364: 3169–3179.

Ratnieks, F. L. W., Foster, K. R., and Wenseleers, T. 2011. Darwin's special difficulty: The evolution of "neuter insects" and current theory. *Behavioral Ecology and Sociobiology* 65: 481–492.

Reeve, H. K. 2000. Review of *Unto Others: The Evolution and Psychology of Unselfish Behavior*. *Evolution and Human Behavior* 21: 65–72.

Reeve, H. K. and Hölldobler, B. 2007. The emergence of a superorganism through intergroup competition. *Proceedings of the National Academy of Sciences USA* 104: 9736–9740.

Rittschof, C. C., Bukhari, S. A., Sloofman, L. G., Troy, J. M., Caetano-Anolles, D., et al. 2014. Neuromolecular responses to social challenge: Common mechanisms across mouse, stickleback fish, and honey bee. *Proceedings of the National Academy of Sciences USA* 111: 17929–17934.

Robinson, G. E. 1999. Integrative animal behaviour and sociogenomics. *Trends in Ecology & Evolution* 14: 202–205.

Robinson, G. E. 2005. Sociogenomics: Social life in molecular terms. *Nature Reviews Genetics* 6: 257–270.

Rubenstein, D. R. 2012. Family feuds: Social competition and sexual conflict in complex societies. *Philosophical Transactions of the Royal Society B* 367: 2304–2313.

Rubenstein, D. R. and Abbot, P. 2017a. *Comparative Social Evolution*. Cambridge: Cambridge University Press.

Rubenstein, D. R. and Abbot, P. 2017b. The evolution of social evolution. In D. R. Rubenstein and P. Abbott (eds.), *Comparative Social Evolution*, pp. 1–18. Cambridge: Cambridge University Press.

Schlotterer, C. 2004. The evolution of molecular markers—just a matter of fashion? *Nature Reviews Genetics* 5: 63–69.

Strassman, J. E., Zhu, Y., and Queller, D. C. 2000. Altruism and social cheating in the social amoeba *Dictyostelium discoideum*. *Nature* 408: 965–967.

Strassmann, J. E. and Queller, D. C. 2007. Insect societies as divided organisms: The complexities of purpose and cross-purpose. *Proceedings of the National Academy of Sciences USA* 104: 8619–8626.

Szathmáry, E. 2015. Toward major evolutionary transitions theory 2.0. *Proceedings of the National Academy of Sciences USA* 112: 10104–10111.

Szathmáry, E. and Maynard Smith, J. 1995. The major evolutionary transitions. *Nature* 374: 227–232.

Tarpy, D. R. 2003. Genetic diversity within honeybee colonies prevents severe infections and promotes colony growth. *Proceedings of the Royal Society of London B* 270: 99–103.

Tofilski, A., Couvillon, M. J., Evison, S. E. F., Helanterä, H., Robinson, E. J. H., and Ratnieks, F. L. W. 2008. Preemptive defensive self-sacrifice by ant workers. *American Naturalist* 172: E239–E243.

Toth, A. L. and Robinson, G. E. 2007. Evo-devo and the evolution of social behavior. *Trends in Genetics* 23: 334–341.

Toth, A. L. and Robinson, G. E. 2009. Evo-devo and the evolution of social behavior: Brain gene expression analyses in social insects. *Cold Spring Harbor Symposia on Quantitative Biology* 74: 026.

Trivers, R. L. and Hare, H. 1976. Haplodiploidy and the evolution of the social insects. *Science* 191: 249–263.

Wenseleers, T. and Ratnieks, F. L. W. 2004. Tragedy of the commons in *Melipona* bees. *Proceedings of the Royal Society of London B* 271

Wenseleers, T. and Ratnieks, F. L. W. 2006a. Comparative analysis of worker reproduction and policing in eusocial Hymenoptera supports relatedness theory. *American Naturalist* 168: E163–E179.

Wenseleers, T. and Ratnieks, F. L. W. 2006b. Enforced altruism in insect societies. *Nature* 444: 50.

West, S. A., El Mouden, C., and Gardner, A. 2011. Sixteen common misconceptions about the evolution of cooperation in humans. *Evolution and Human Behavior* 32: 231–262.

Williams, G. C. 1966. *Adaptation and Natural Selection*. Princeton, N J: Princeton University Press.

Wilson, D. S. 1975. Theory of group selection. *Proceedings of the National Academy of Sciences USA* 72: 143–146.

Wilson, D. S. and Sober, E. 1994. Reintroducing group selection to the human behavioral sciences. *Behavioral and Brain Sciences* 17: 585–608.

Wilson, D. S. and Wilson, E. O. 2007. Rethinking the theoretical foundation of sociobiology. *Quarterly Review of Biology* 82: 327–348.

Wilson, D. S. and Wilson, E. O. 2008. Evolution "for the good of the group." *American Scientist* 96: 380–389.

Wilson, E. O. 1971. *The Insect Societies*. Cambridge, MA: Harvard University Press.

Woodard, S. H., Fischman, B. J., Venkat, A., Hudson, M. E., Varala, K., et al. 2011. Genes involved in convergent evolution of eusociality in bees. *Proceedings of the National Academy of Sciences USA* 108: 7472–7477.

Wynne-Edwards, V. C. 1962. *Animal Dispersion in Relation to Social Behaviour*. Edinburgh: Oliver & Boyd.

Chapter 13

Alexander, R. D. 1974. The evolution of social behavior. *Annual Review of Ecology and Systematics* 5: 325–383.

Alpin, L. M., Farine, D. R., Morand-Ferron, J., Cockburn, A., Thornton, A., and Sheldon, B. C. 2015. Experimentally induced innovations lead to persistent culture via conformity in wild birds. *Nature* 518: 538–541.

Arnold, K. E. and Owens, I. P. F. 1998. Cooperative breeding in birds: A comparative test of the life history hypothesis. *Proceedings of the Royal Society of London B* 265: 739–745.

Axelrod, R. and Hamilton, W. D. 1981. The evolution of cooperation. *Science* 211: 1390–1396.

Balshine-Earn, S., Neat, F. C., Reid, H. and Taborsky, M. 1998. Paying to stay or paying to breed? Field evidence for direct benefits of helping behavior in a cooperatively breeding fish. *Behavioral Ecology* 9: 432–438.

Beekman, M. and Jordan, L. A. 2017. Does the field of animal personality provide any new insights for behavioral ecology? *Behavioral Ecology* 28: 617–623.

Bell, M. B. V., Cant, M. A., Borgeaud, C., Thavarajah, N., Samson, J., and Clutton-Brock, T. H. 2014. Suppressing subordinate reproduction provides benefits to dominants in cooperative societies of meerkats. *Nature Communications* 5: 4499.

Bergmuller, R., Heg, D., and Taborsky, M. 2005. Helpers in a cooperatively breeding cichlid stay and pay or disperse and breed, depending on ecological constraints. *Proceedings of the Royal Society of London B* 272: 325–331.

Boomsma, J. J. 2007. Kin selection versus sexual selection: Why the ends do not meet. *Current Biology* 17: R673–R683.

Braude, S. 2000. Dispersal and new colony formation in wild naked mole-rats: Evidence against inbreeding as the system of mating. *Behavioral Ecology* 11: 7–12.

Brown, C. R. 1988. Social foraging in cliff swallows: Local enhancement, risk sensitivity, competition, and the avoidance of predators. *Animal Behaviour* 36: 780-792.

Brown, C. R. and Brown, M. B. 1986. Ecto-parasitism as a cost of coloniality in cliff swallows (*Hirundo pyrrhonota*). *Ecology* 67: 1206–1218.

Brown, C. R. and Brown, M. B. 2004. Empirical measurement of parasite transmission between groups in a colonial bird. *Ecology* 85: 1619–1626.

Brown, J. L. 1987. *Helping and Communal Breeding in Birds: Ecology and Evolution*. Princeton, NJ: Princeton University Press.

Burda, H., Honeycutt, R. L., Begall, S., Locker-Grutjen, O., and Scharff, A. 2000. Are naked and common mole-rats eusocial and if so, why? *Behavioral Ecology and Sociobiology* 47: 293–303.

Buston, P. 2003. Size and growth modification in clownfish. *Nature* 424: 145–146.

Cant, M. A., Hodge, S. J., Bell, M. B., Gilchrist, J. S., and Nichols, H. J. 2010. Reproductive control via eviction (but not the threat of eviction) in banded mongooses. *Proceedings of the Royal Society of London B* 277: 2219–2226.

Cant, M. A., Nichols, H. J., Johnstone, R. A., and Hodge, S. J. 2014. Policing of reproduction by hidden threats in a cooperative mammal. *Proceedings of the National Academy of Sciences USA* 111: 326–330.

Carter, A. J., English, S., and Clutton-Brock, T. H. 2014. Cooperative personalities and social niche specialization in female meerkats. *Journal of Evolutionary Biology* 27: 815–825.

Carter, G. G., Farine, D. R., and Wilkinson, G. S. 2017. Social bet-hedging in vampire bats. *Biology Letters* 13: 20170112.

Carter, G. G. and Wilkinson, G. S. 2015. Social benefits of non-kin food sharing by female vampire bats. *Proceedings of the Royal Society of London B* 282: 20152524.

Clutton-Brock, T. H. 2002. Breeding together: Kin selection and mutualism in cooperative vertebrates. *Science* 296: 69-72.

Clutton-Brock, T. H. 2009a. Cooperation between non-kin in animal societies. *Science* 462: 51–57.

Clutton-Brock, T. H. 2009b. Structure and function in mammalian societies. *Philosophical Transactions of the Royal Society B* 364: 3229–3242.

Clutton-Brock, T. H., Hodge, S. J., Spong, G., Russell, A. F., Jordan, N. R., et al. 2006. Intrasexual competition and sexual selection in cooperative mammals. *Nature* 444: 1065–1068.

Cockburn, A., Sims, R. A., Osmond, H. L., Green, D. J., Double, M. C., and Mulder, R. A. 2008. Can we measure the benefits of help in cooperatively breeding birds: The case of superb fairy-wrens *Malurus cyaneus*? *Journal of Animal Ecology* 77: 430–438.

Connor, R. C. 2010. Cooperation beyond the dyad: On simple models and a complex society. *Philosophical Transactions of the Royal Society B* 365: 2687–2697.

Cornwallis, C. K., West, S. A., Davis, K. E., and Griffin, A. S. 2010. Promiscuity and the evolutionary transition to complex societies. *Nature* 466: 969–972.

Dale, J., Dey, C. J., Delhey, K., Kempenaers, B., and Valcu, M. 2015. The effects of life history and sexual selection on male and female plumage colouration. *Nature* 527: 367–370.

Day, C. J., O'Connor, C. M., Wilkinson, H., Schultz, S., Balshine, S., and Fitzpatrick, J. L. 2017. Direct benefits and evolutionary transitions to complex societies. *Nature Ecology & Evolution* 1: 0137.

DuVal, E. H. 2007. Adaptive advantages of cooperative courtship for subordinate male lance-tailed manakins. *American Naturalist* 169: 423–432.

Edelman, A. J. and McDonald, D. B. 2014. Structure of male cooperation networks at long-tailed manakin leks. *Animal Behaviour* 97: 125–133.

Edwards, H. A., Burke, T., and Dugdale, H. L. 2017a. Repeatable and heritable behavioural variation in a wild cooperative breeder. *Behavioral Ecology* 28: 668-676.

Edwards, H. A., Dugdale, H. L., Richardson, D. S., Komdeur, J., and Burke, T. 2017b. Extra-pair parentage and personality in a cooperatively breeding bird. *Behavioral Ecology and Sociobiology* 72: 37.

Eikenaar, C., Richardson, D. S., Brouwer, L., and Komdeur, J. 2007. Parent presence, delayed dispersal, and territory acquisition in the Seychelles warbler. *Behavioral Ecology* 18: 874–879.

Emlen, S. T. 1982. The evolution of helping. I. An ecological constraints model. *American Naturalist* 119: 29-39.

Emlen, S. T., Wrege, P. H., and Demong, N. J. 1995. Making decisions in the family: An evolutionary perspective. *American Scientist* 83: 148-157.

Foster, M. S. 1977. Odd couples in manakins: A study of social organization and cooperative breeding in *Chiroxiphia linearis*. *American Naturalist* 111: 845–853.

Fricke, H. W. 1979. Mating system, resource defense and sex change in the anemonefish *Amphiprion akallopisos*. *Zeitschrift für Tierpsychologie* 50: 313–326.

Gardner, A. and West, S. A. 2004. Spite among siblings. *Science* 305: 1413–1414.

Gaston, A. J. 1978. The evolution of group territorial behavior and cooperative breeding. *American Naturalist* 112: 1091–1100.

Giron, D., Dunn, D. W., Hardy, I. C. W., and Strand, M. R. 2004. Aggression by polyembryonic wasp soldiers correlates with kinship but not resource competition. *Nature* 430: 676–679.

Giron, D., Harvey, J. A., Johnson, J. A., and Strand, M. R. 2007. Male soldier caste larvae are non-aggressive in the polyembryonic wasp *Copidosoma floridanum*. *Biology Letters* 3: 431–434.

Greene, E., Lyon, B. E., Muehter, V. R., Ratcliffe, L., Oliver, S. J., and Boag, P. T. 2000. Disruptive sexual selection for plumage colouration in a passerine bird. *Nature* 407: 1000-1003.

Greenwood, P. J. and Harvey, P. H. 1982. The natal and breeding dispersal of birds. *Annual Review of Ecology and Systematics* 13: 1–21.

Griesser, M. and Ekman, J. 2004. Nepotistic alarm calling in the Siberian jay, *Perisoreus infaustus*. *Animal Behaviour* 67: 933–939.

Griesser, M. and Ekman, J. 2005. Nepotistic mobbing behaviour in the Siberian jay, *Perisoreus infaustus*. *Animal Behaviour* 69: 345–352.

Griffin, A. S. and West, S. A. 2003. Kin discrimination and the benefit of helping in cooperatively breeding vertebrates. *Science* 302: 634–636.

Groenewoud, F., Frommen, J. G., Josi, D., Tanaka, H., Jungwirth, A., and Taborsky, M. 2016. Predation risk drives social complexity in cooperative breeders. *Proceedings of the National Academy of Sciences USA* 113: 4104–4109.

Grutter, A. S. 1999. Cleaner fish really do clean. *Nature* 398: 672–673.

Hofmann, H. A., Beery, A. K., Blumstein, D. T., Couzin, I. D., Earley, R. L., et al. 2014. An evolutionary framework for studying mechanisms of social behavior. *Trends in Ecology & Evolution* 29: 581–589.

Hughes, W. O., Oldroyd, B. P., Beekman, M., and Ratnieks, F. L. 2008. Ancestral monogamy shows kin selection is key to evolution of sociality. *Science* 320: 1213–1216.

Ito, F., Bernard, E., and Torres, R. A. 2016. What is for dinner? First report of human blood in the diet of the hairy-legged vampire bat *Diphylla ecaudata*. *Acta Chiropterologica* 18: 509-515.

Jennions, M. D. and Macdonald, D. W. 1994. Cooperative breeding in mammals. *Trends in Ecology & Evolution* 9: 89-93.

Jetz, W. and Rubenstein, D. R. 2011. Environmental uncertainty and the global biogeography of cooperative breeding in birds. *Current Biology* 21: 72–78.

Kennedy, P., Higginson, A. D., Radford, A. N., and Summer, S. 2018. Altruism in a volatile world. *Nature* 555: 359–362.

Kingma, S. A. 2017. Direct benefits explain interspecific variation in helping behaviour among cooperatively breeding birds. *Nature Communications* 8: 1094.

Kingma, S. A., Hall, M. L., Arriero, E., and Peters, A. 2010. Multiple benefits of cooperative breeding in purple-crowned fairy-wrens: A consequence of fidelity? *Journal of Animal Ecology* 79: 757–768.

Koenig, W. D., Mumme, R. L., Stanback, M. T., and Pitelka, F. A. 1995. Patterns and consequences of egg destruction among joint-nesting acorn woodpeckers. *Animal Behaviour* 50: 607–621.

Kokko, H., Johnstone, R. A., and Clutton-Brock, T. H. 2001. The evolution of cooperative breeding through group augmentation. *Proceedings of the Royal Society of London B* 268: 187–196.

Komdeur, J. 1992. Importance of habitat saturation and territory quality for evolution of cooperative breeding in the Seychelles warbler. *Nature* 358: 493–495.

Komdeur, J., Huffstadt, A., Prast, W., Castle, G., Mileto, R., and Wattel, J. 1995. Transfer experiments of Seychelles warblers to new islands: Changes in dispersal and helping behavior. *Animal Behaviour* 49: 695–708.

Krakauer, A. H. 2005. Kin selection and cooperative courship in wild turkeys. *Nature* 434: 69–72.

Lacey, E. A. and Sherman, P. 1991. Social organization of naked mole-rat colonies: Evidence for divisions of labor. In P. W. Sherman, J. U. M. Jarvis and R. D. Alexander (eds.), *The Biology of the Naked Mole-Rat*, pp. 275–336. Princeton, NJ: Princeton University Press.

Leadbeater, E., Carruthers, J. M., Green, J. P., Rosser, N. S., and Field, J. 2011. Nest inheritance is the missing source of direct fitness in a primitively eusocial insect. *Science* 333: 874–876.

Leimgruber, K. L., Rosati, A. G., and Santos, L. R. 2016. Capuchin monkeys punish those who have more. *Evolution and Human Behavior* 37: 236–244.

Lukas, D. and Clutton-Brock, T. H. 2012. Cooperative breeding and monogamy in mammalian societies. *Proceedings of the Royal Society of London B* 279: 2151–2156.

Lukas, D. and Clutton-Brock, T. H. 2017. Climate and the distribution of cooperative breeding in mammals. *Royal Society Open Science* 4: 160897.

Maia, R., Rubenstein, D. R., and Shawkey, M. D. 2013. Key ornamental innovations facilitate diversification in an avian radiation. *Proceedings of the National Academy of Sciences USA* 110: 10687–10692.

Maia, R., Rubenstein, D. R., and Shawkey, M. D. 2016. Selection, constraint, and the evolution of coloration in African starlings. *Evolution* 70: 1064–1079.

McDonald, D. B. 2010. A spatial dance to the music of time in the leks of long-tailed manakins. *Advances in the Study of Behavior* 42: 55–81.

McDonald, D. B. and Potts, W. K. 1994. Cooperative display and relatedness among males in a lek-mating bird. *Science* 266: 1030-1032.

Mumme, R. L., Koenig, W. D., and Pitelka, F. A. 1983. Reproductive competition in the communal acorn woodpecker: Sisters destroy each other's eggs. *Nature* 306: 583–684.

Nowak, M. A. and Sigmund, K. 2005. Evolution of indirect reciprocity. *Nature* 437: 1291–1298.

Reale, D., Reader, S. M., Sol, D., McDougall, P. T., and Dingemanse, N. J. 2007. Integrating animal temperament within ecology and evolution. *Biological Reviews* 82: 291–318.

Reyer, H.-U. 1984. Investment and relatedness: A cost/benefit analysis of breeding and helping in the pied kingfisher. *Animal Behaviour* 32: 1163–1178.

Riehl, C. 2013. Evolutionary routes to non-kin cooperative breeding in birds. *Proceedings of the Royal Society of London B* 2801: 20132245.

Rubenstein, D. R. 2011. Spatiotemporal environmental variation, risk aversion, and the evolution of cooperative breeding as a bet-hedging strategy. *Proceedings of the National Academy of Sciences USA* 108: 10816–10822.

Rubenstein, D. R. 2012. Family feuds: social competition and sexual conflict in complex societies. *Philosophical Transactions of the Royal Society B* 367: 2304–2313.

Rubenstein, D. R and Abbot, P. 2017a. *Comparative Social Evolution*. Cambridge: Cambridge University Press.

Rubenstein, D. R. and Abbot, P. 2017b. The evolution of social evolution. In D. R. Rubenstein and P. Abbot (eds.), *Comparative Social Evolution*, pp. 1–18. Cambridge: Cambridge University Press.

Rubenstein, D. R., Botero, C. A., and Lacey, E. A. 2016. Discrete but variable structure of animal societies leads to the false perception of a social continuum. *Royal Society Open Science* 3: 160147.

Rubenstein, D. R. and Lovette, I. J. 2007. Temporal environmental variability drives the evolution of cooperative breeding in birds. *Current Biology* 17: 1414–1419.

Rubenstein, D. R. and Lovette, I. J. 2009. Reproductive skew and selection on female ornamentation in social species. *Nature* 462: 786–789.

Russell, A. F., Langmore, N. E., Cockburn, A., Astheimer, L. B., and Kilner, R. M. 2007a. Reduced egg investment can conceal helper effects in cooperatively breeding birds. *Science* 317: 941–944.

Russell, A. F., Young, A. J., Spong, G., Jordan, N. R., and Clutton-Brock, T. H. 2007b. Helpers increase the reproductive potential of offspring in cooperative meerkats. *Proceedings of the Royal Society of London B* 274: 513–520.

Sachs, J. L., Mueller, U. G., Wilcox, T. P., and Bull, J. J. 2004. The evolution of cooperation. *Quarterly Review of Biology* 79: 135–160.

Schino, G. and Aureli, F. 2010. The relative roles of kinship and reciprocity in explaining primate altruism. *Ecology Letters* 13: 45–50.

Shen, S.-F., Emlen, S. T., Koenig, W. D., and Rubenstein, D. R. 2017. The ecology of cooperative breeding behaviour. *Ecology Letters* 20: 708–720.

Sherman, P. W. 1977. Nepotism and the evolution of alarm calls. *Science* 197: 1246–1253.

Sherman, P. W., Jarvis, J. U. M., and Alexander, R. D. (eds.). 1991. *The Biology of the Naked Mole-Rat*. Princeton, NJ: Princeton University Press.

Sih, A. and Bell, A. M. 2008. Insights for behavioral ecology from behavioral syndromes. *Advances in the Study of Behavior* 32: 227–281.

Sih, A., Bell, A. M., and Johnson, J. C. 2004. Behavioral syndromes: An ecological and evolutionary overview. *Trends in Ecology & Evolution* 19: 372–378.

Sih, A., Mathot, K. J., Moiron, M., Montiglio, P.-O., Wolf, M., and Dingemanse, N. J. 2015. Animal personality and state–behaviour feedbacks: A review and guide for empiricists. *Trends in Ecology & Evolution* 30: 50–60.

Smith, B. R. and Blumstein, D. T. 2008. Fitness consequences of personality: A meta-analysis. *Behavioral Ecology* 19: 448–455.

Spottiswoode, C. N., Begg, K. S., and Begg, C. M. 2016. Reciprocal signaling in honeyguide–human mutualism. *Science* 353: 387–389.

St-Pierre, A., Larose, K., and Dubois, F. 2009. Long-term social bonds promote cooperation in the iterated Prisoner's Dilemma. *Proceedings of the Royal Society of London B* 276: 4223–4228.

Strandburg-Peshkin, A., Farine, D. R., Couzin, I. D., and Crofoot, M. C. 2015. Shared decision-making drives collective movement in wild baboons. *Science* 348: 1358-1361.

Strandburg-Peshkin, A., Farine, D. R., Crofoot, M. C., and Couzin, I. D. 2017. Habitat and social factors shape individual decisions and emergent group structure during baboon collective movement. *eLife* 6: e19505.

Taborsky, M. 1994. Sneakers, satellites, and helpers: Parasitic and cooperative behavior in fish reproduction. *Advances in the Study of Behavior* 23: 1–100.

Trivers, R. L. 1971. The evolution of reciprocal altruism. *Quarterly Review of Biology* 46: 35–57.

Webster, M. S., Varian, C. W., and Karubian, J. 2008. Plumage color and reproduction in the red-backed fairy-wren: Why be a dull breeder? *Behavioral Ecology* 19: 517–524.

Wedekind, C. and Milinski, M. 1996. Human cooperation in the simultaneous and the alternating Prisoner's Dilemma: Pavlov versus Generous Tit-for-Tat. *Proceedings of the National Academy of Sciences USA* 93: 2686–2689.

West, S. A., El Mouden, C., and Gardner, A. 2011. Sixteen common misconceptions about the evolution of cooperation in humans. *Evolution and Human Behavior* 32: 231–262.

West, S. A., Griffin, A. S., and Gardner, A. 2007. Social semantics: Altruism, cooperation, mutualism, strong reciprocity, and group selection. *Journal of Evolutionary Biology* 20: 415–432.

Wilkinson, G. S. 1984. Reciprocal food sharing in the vampire bat. *Nature* 308: 181–184.

Young, A. J., Carlson, A. A., Monfort, S. L., Russell, A. F., Bennett, N. C., and Clutton-Brock, T. H. 2006. Stress and the suppression of subordinate reproduction in cooperatively breeding meerkats. *Proceedings of the National Academy of Sciences USA* 103: 12005–12010.

Chapter 14

Allison, T., Ginter, H., McCarthy, G., Nobre, A. C., Puce, A., et al. 1994. Face recognition in human extrastriate cortex. *Journal of Neurophysiology* 71: 821–825.

Allison, T., Puce, A., and McCarthy, G. 2000. Social perception from visual cues: The role of the STS region. *Trends in Cognitive Sciences* 4: 267–278.

Alvergne, A., Faurie, C., and Raymond, M. 2007. Differential facial resemblance of young children to their parents: Who do children look like more? *Evolution and Human Behavior* 28: 135–144.

Alvergne, A., Faurie, C., and Raymond, M. 2009. Father-offspring resemblance predicts paternal investment in humans. *Animal Behaviour* 78: 61–69.

Alvergne, A. and Lummaa, V. 2010. Does the contraceptive pill alter mate choice in humans? *Trends in Ecology & Evolution* 25: 171–179.

Anderson, K. G. 2006. How well does paternity confidence match actual paternity? *Current Anthropology* 47: 513–520.

Antfolk, J., Salo, B., Alanko, K., Bergen, E., Corander, J., et al. 2015. Women's and men's sexual preferences and activities with respect to the partner's age: Evidence for female choice. *Evolution and Human Behavior* 36: 73–79.

Apicella, C. L. and Marlowe, F. W. 2004. Perceived mate fidelity and paternal resemblance predict men's investment in children. *Evolution and Human Behavior* 25: 371–378.

Arslan, R. C., Willfuhr, K. P., Frans, E. M., Verweij, K. J. H., Burkner, P.-C., et al. 2017. Older fathers' children have lower evolutionary fitness across four centuries and in four populations. *Proceedings of the Royal Society of London B* 284: 2017.1562.

Bateman, A. J. 1948. Intra-sexual selection in *Drosophila*. *Heredity* 2: 349–368.

Beecher, M. D. and Beecher, I. M. 1979. Sociobiology of bank swallows: Reproductive strategy of the male. *Science* 205: 1282–1285.

Boattini, A., Sarno, S., Pedrini, P., Medoro, C., Carta, M., et al. 2015. Traces of medieval migrations in a socially stratified population from Northern Italy. Evidence from uniparental markers and deep-rooted pedigrees. *Heredity* 114: 155–162.

Boone, J. L., III. 1986. Parental investment and elite family structure in preindustrial states: A case study of late medieval-early modern Portuguese genealogies. *American Anthropologist* 88: 859–878.

Borgerhoff Mulder, M. 1987. Resources and reproductive success in women with an example from the Kipsigis of Kenya. *Journal of Zoology* 213: 489–505.

Bressler, E. R., Martin, R. A. and Balshine, S. 2006. Production and appreciation of humor as sexually selected traits. *Evolution and Human Behavior* 27: 121–130.

Brooks, R., Shelly, J. P., Fan, J., Zhai, L., and Chau, D. K. P. 2010. Much more than a ratio: Multivariate selection on female bodies. *Journal of Evolutionary Biology* 23: 2238–2248.

Brown, J. L. and Eklund, A. 1994. Kin recognition and the major histocompatibility complex: An integrative review. *American Naturalist* 143: 435–461.

Brownmiller, S. 1975. *Against Our Will*. New York: Simon and Schuster

Bryant, G. A. and Haselton, M. G. 2009. Vocal cues of ovulation in human females. *Biology Letters* 5: 12–15.

Bryant, J. I. 2010. Dialect. In R. L. Jackson II (ed.), *Encyclopedia of Identity*, pp. 219–220. Urbana, IL.: Sage Publications.

Burch, R. L. and Gallup, G. G. 2004. Pregnancy as a stimulus for domestic violence. *Journal of Family Violence* 19: 243–247.

Buss, D. M. 1989. Sex differences in human mate preferences: Evolutionary hypothesis tested in 37 cultures. *Behavioral and Brain Sciences* 12: 1–149.

Buss, D. M. 2003. *The Evolution of Desire: Strategies of Human Mating, 3rd Edition*. New York: Free Press.

Buss, D. M. 2012. *Evolutionary Psychology: The New Science of the Mind*. Upper Saddle River, NJ: Pearson.

Buss, D. M. 2016. *The Evolution of Desire: Stategies of Human Mating, 4th Edition*. New York: Basic Books.

Buss, D. M. 2017. Sexual conflict in human mating. *Current Directions in Psychological Science* 26: 307–313.

Buss, D. M. and Duntley, J. D. 2011. The evolution of intimate partner violence. *Aggression and Violent Behavior* 16: 411–419.

Buss, D. M., Goetz, C., Duntley, J. D., Asaso, K., and Conroy-Beam, D. 2017. The mate switching hypothesis. *Personality and Individual Differences* 104: 143–149.

Buss, D. M., Larsen, R. J., Westen, D., and Semmelroth, J. 1992. Sex differences in jealousy: Evolution, physiology, and psychology. *Psychological Science* 3: 251–255.

Buss, D. M. and Pinker, S. 2009. Pop psychology probe. *Scientific American* 300: 10–11.

Buss, D. M. and Schmitt, D. P. 1993. Sexual strategies theory: An evolutionary perspective on human mating. *Psychological Review* 100: 204–232.

Buss, D. M. and Shackelford, T. K. 2008. Attractive women want it all: Good genes, economic investment, parenting proclivities, and emotional commitment. *Evolutionary Psychology* 6: 134–146.

Buston, P. M. and Emlen, S. T. 2003. Cognitive processes underlying human mate choice: The relationship between self-perception and mate preference in Western society. *Proceedings of the National Academy of Sciences USA* 100: 8805–8810.

Carroll, S. 2005. *Endless Forms Most Beautiful*. New York: W. W. Norton.

Case, A., Lubotsky, D., and Paxson, C. 2002. Economic status and health in childhood: The origins of the gradient. *American Economic Review* 92: 1308–1334.

Chabout, J., Sarkar, A., Parel, S. R., Radden, T., Dunson, D. B., et al. 2016. A *Foxp2* mutation implicated in human speech deficits alters sequencing of ultrasonic vocalizations in adult male mice. *Frontiers in Behavioral Neuroscience* 10: 197.

Cialdini, R. B. 2001. The science of persuasion. *Scientific American* 284: 76–81.

Clark, R. D. and Hatfield, E. 1989. Gender differences in receptivity to sexual offers. *Journal of Psychology and Human Sexuality* 2: 39–55.

Colarelli, S. M. and Dettmann, J. R. 2003. Intuitive evolutionary perspectives in marketing practices. *Psychology & Marketing* 20: 837–865.

Crespi, B. J. 2017. Shared sociogenetic basis of honey bee behavior and human risk for autism. *Proceedings of the National Academy of Sciences USA* 114: 9502–9504.

Daly, M., Wilson, M. 1983. *Sex, Evolution and Behavior*. Boston: Willard Grant Press

Daly, M., Wilson, M. 1992. The man who mistook his wife for a chattel. In , *The Adapted Mind*, Barkow, J., Cosmides, L., and Tooby, J. (eds.), pp. 289–322. New York: Oxford University Press.

Daly, M., Wilson, M., and Weghorst, S. T. 1982. Male sexual jealousy. *Ethology and Sociobiology* 3: 11–27.

Dunbar, R. I. M. 1996. *Grooming, Gossip, and the Evolution of Language*. Cambridge, MA: Harvard University Press

Durante, K. M. and Li, N. P. 2009. Oestradiol level and opportunistic mating in women. *Biology Letters* 5: 179–182.

Easton, J. A., Goetz, C. D., and Buss, D. M. 2015. Human mate choice, evolution of. *International Encyclopedia of the Social & Behavioral Sciences*, 2nd Edition 11: 340–347.

Enerback, S., Jacobsson, A., Simpson, E. M., Guerra, C., and Yamashita, H. 1997. Mice lacking mitochondrial uncoupling protein are cold-sensitive but not obese. *Nature* 387: 90–94.

Feinberg, D. R. 2008. Are human faces and voices ornaments signaling common underlying cues to mate value? *Evolutionary Anthropology* 17: 112–118.

Felson, R. B. and Cundiff, P. R. 2011. Age and sexual assault during robberies. *Evolution and Human Behavior* 33: 10–16.

Fink, B., Neave, N., and Seydel, H. 2007. Male facial appearance signals physical strength to women. *American Journal of Human Biology* 19: 82–87.

Fischer, J., Semple, S., Fickenscher, G., Jürgens, R., Kruse, E., et al. 2011. Do women's voices provide cues of the likelihood of ovulation? The importance of sampling regime. *PLoS One* 6: e24490.

Folstad, I. and Karter, A. J. 1992. Parasites, bright males, and the immunocompetence handicap. *American Naturalist* 139: 603–622.

Foo, Y. Z., Rhodes, G., and Simmons, L. W. 2017. The carotenoid beta-carotene enhances facial color, attractiveness and perceived health, but not actual health, in humans. *Behavioral Ecology* 28: 570–578.

Gallup, A. C., White, D. D., and Gallup, G. G. 2007. Handgrip strength predicts sexual behavior, body morphology, and aggression in male college students. *Evolution and Human Behavior* 28: 423–429.

Gangestad, S. W. and Thornhill, R. 2008. Human oestrus. *Proceedings of the Royal Society of London B* 275: 991–1000.

Garver-Apgar, C. E., Gangestad, S. W., Thornhill, R., Miller, R. D., and Olp, J. J. 2006. Major histocompatibility complex alleles, sexual responsivity, and unfaithfulness in romantic couples. *Psychological Science* 17: 830–835.

Gaulin, S. J. C. and Boster, J. S. 1990. Dowry as female competition. *American Anthropologist* 92: 994–1005.

Geary, D. C. 2010. *Male, Female: The Evolution of Human Sex Differences.* Washington, DC: American Psychological Association

Gildersleeve, K., Haselton, M. G., and Fales, M. R. 2014. Do women's mate preferences change across the ovulatory cycle? A meta-analytic review. *Psycholoigcal Bulletin* 140: 1205–1259.

Gluckman, P., Beedle, A., Buklijas, T., Low, F., and Hanson, M. 2016. *Principles of Evolutionary Medicine*: Oxford University Press.

Gottschall, J. A. and Gottschall, T. A. 2003. Are per-incident rape-pregnancy rates higher than per-incident consensual pregnancy rates? *Human Nature* 14: 1–20.

Gould, S. J. and Lewontin, R. C. 1979. The spandrels of San Marco and the Panglossian paradigm: A critique of the adaptationist programme. *Proceedings of the Royal Society of London B* 205: 581–598.

Grammar, K., Fink, B., Møller, A. P., and Thornhill, R. 2003. Darwinian aesthetics: Sexual selection and the biology of beauty. *Biological Reviews* 78: 385–407.

Greeff, J. M. and Erasmus, J. C. 2015. Three hundred years of low non-paternity in human populations. *Heredity* 115: 396–404.

Greengross, G. and Miller, G. 2011. Humor ability reveals intelligence, predicts mating success, and is higher in males. *Intelligence* 39: 188–192.

Groszer, M., Keays, D. A., Deacon, R. M. J., de Bono, J. P., Prasad-Mulcare, S., et al. 2008. Impaired synaptic plasticity and motor learning in mice with a point mutation implicated in human speech deficits. *Current Biology* 18: 354–362.

Gueguen, N. 2011. Effects of solicitor sex and attractiveness on receptivity to sexual offers: A field study. *Archives of Sexual Behavior* 40: 915–919.

Haesler, S., Rochefort, C., Licznerksi, P., Osten, P., and Scharff, C. 2007. Incomplete and inaccurate vocal imitation after knockdown of *FoxP2* in songbird basal ganglia nucleus area X. *PLoS Biology* 5: e321.

Hartung, J. 1982. Polygyny and inheritance of wealth. *Current Anthropology* 23: 1–12.

Haselton, M. G. 2003. The sexual overperception bias: Evidence of a systematic bias in men from a survey of naturally occurring events. *Journal of Research in Personality* 37: 34–47.

Heath, K. M. and Hadley, C. 1998. Dichotomous male reproductive strategies in a polygynous human society: Mating versus parental effort. *Current Anthropology* 39: 369–374.

Heberbrand, J., Volckmar, A. L., Knoll, H., and Hinney, A. 2010. Chipping away the "missing heritability": GIANT steps forward in the molecular elucidation of obesity—but still lots to go. *Obesity Facts* 3: 294–303.

Hitsch, G. J., Hotacsu, A., and Ariely, D. 2010. What makes you click? Mate preference in online dating. *Quantitative Marketing and Economics* 8: 393–427.

Hobaiter, C. and Byrne, R. W. 2011. The gestural repertoire of the wild chimpanzee. *Animal Cognition* 14: 745–767.

Holveck, M. J. and Riebel, K. 2011. Low-quality females prefer low-quality males when choosing a mate. *Proceedings of the Royal Society of London B* 277: 153–160.

Hurst, J. A., Baraitser, M., Auger, E., Graham, F., and Norell, S. 1990. An extended family with a dominantly inherited speech disorder. *Developmental Medicine & Child Neurology* 32: 352–355.

Irons, W. 1979. Cultural and biological success. In N. A. Chagnon and W. Irons (eds.), *Evolutionary Biology and Human Social Behavior: An Anthropological Perspective*, pp. 257–272. North Scituate, MA: Duxbury Press.

Jasiénska, G., Bribiescas, R. G., Furberg, A.-S., Helle, S., and Nuñez-de la Mora, A. 2017. Human reproduction and health: An evolutionary perspective. *The Lancet* 390: 510–520.

Jasiénska, G., Ziomkiewicz, A., Ellison, P. T., Lipson, S. F., and Thune, I. 2004. Large breasts and narrow waists indicate high reproductive potential in women. *Proceedings of the Royal Society of London B* 271: 1213–1217.

Jones, B. C., DeBruine, L. M., Perrett, D. I., Little, A. C., Feinberg, D. R., and Smith, M. J. L. 2008. Effects of menstrual cycle phase on face preferences. *Archives of Sexual Behavior* 37: 78–84.

Kanazawa, S. 2003. Can evolutionary psychology explain reproductive behavior in the contemporary United States? *Sociological Quarterly* 44: 291–302.

Kaplan, H. and Hill, K. 1985. Hunting ability and reproductive success among male Ache foragers: Preliminary results. *Current Anthropology* 26: 131–133.

Karremans, J. C., Frankenhuis, W. E., and Arons, S. 2010. Blind men prefer a low waist-to-hip ratio. *Evolution and Human Behavior* 31: 182–186.

Kellogg, W. N. and Kellogg, L. A. 1933. *The Ape and the Child*. New York: Hafner Publishing.

Kenrick, D. T. and Keefe., R. C. 1992. Age preferences in mates reflect sex differences in reproductive strategies. *Behavioral and Brain Sciences* 15: 75–133.

Kenrick, D. T., Sadalla, E. K., Groth, G., and Trost, M. R. 1990. Evolution, traits, and the stages of human courtship: Qualifying the parental investment model. *Journal of Personality* 58: 97–116.

Krause, J., Lalueza-Fox, C., Orlando, L., Enard, W., Green, R. E., et al. 2007. The derived *FOXP2* variant of modern humans was shared with neandertals. *Current Biology* 17: 1908–1912.

Krems, J. A., Neuberg, S. L., Neel, R., Puts, D. A., and Kenrick, D. T. 2016. Women selectively guard their (desirable) mates from ovulating women. *Journal of Personality and Social Psychology* 110: 551–573.

Kuhl, P. K. 2010. Brain mechanisms in early language acquisition. *Neuron* 67: 713–727.

Lalumiére, M. L., Chalmers, L. J., Quinsey, V. L., and Seto, M. C. 1996. A test of the mate deprivation hypothesis of social coercion. *Ethology and Sociobiology* 17: 299–318.

Larmuseau, M. H. D., Matthijs, K., and Wenseleers, T. 2016. Cuckolded fathers rare in human populations. *Trends in Ecology & Evolution* 31: 327–329.

Law-Smith, M. J., Perrett, D. I., Jones, B. C., Cornwell, R. E., Moore, F. R., et al. 2006. Facial appearance is a cue to oestrogen levels in women. *Proceedings of the Royal Society of London B* 273: 135–140.

Lea, A. M. and Ryan, M. J. 2015. Irrationality in mate choice revealed by túngara frogs. *Science* 349: 964–966.

Leonard, W. R., Robertson, M. L., Snodgrass, J. J., and Kuzawa, C. W. 2003. Metabolic correlates of hominid brain evolution. *Comparative Biochemistry and Physiology A* 136: 5–15.

Lev-Ran, A. 2001. Human obesity: An evolutionary approach to understanding our bulging waistline. *Diabetes/Metabolism Research and Reviews* 17: 347–362.

Lewkowicz, D. J. and Hansen-Tift, A. M. 2012. Infants deploy selective attention to the mouth of a talking face when learning speech. *Proceedings of the National Academy of Sciences USA* 109: 1431–1436.

Li, N. P., Bailey, J. M., Kenrick, D. T., and Linsenmeier, J. A. W. 2002. The necessities and luxuries of mate preferences: Testing the tradeoffs. *Journal of Personality and Social Psychology* 82: 947–955.

Li, N. P., Valentine, K. A., and Patel, L. 2011. Mate preferences in the US and Singapore: A cross-cultural test of the mate preference priority model. *Personality and Individual Differences* 50: 291–294.

Lie, H. C., Simmons, L. W., and Rhodes, G. 2010. Genetic dissimilarity, genetic diversity, and mate preferences in humans. *Evolution and Human Behavior* 31: 48–58.

Lieberman, P. 2007. The evolution of human speech: Its anatomical and neural bases. *Current Anthropology* 48: 39–66.

Lieberman, P. and McCarthy, R. 2001. Tracking the evolution of language and speech: Comparing vocal tracts to indentify speech capabilities. *Expedition Magazine* 49: 15–20.

Lipson, S. F. and Ellison, P. T. 1996. Comparison of salivary steroid profiles in naturally occurring conception and non-conception cycles. *Human Reproduction* 11: 2090–2096.

Little, A. C., Burt, D. M., Penton-Voak, I. S., and Perrett, D. I. 2001. Self-perceived attractiveness influences human female preferences for sexual dimorphism and symmetry in male faces. *Proceedings of the Royal Society of London B* 268: 39–44.

Little, A. C., Jones, B. C., and Burriss, R. P. 2007a. Preferences for masculinity in male bodies change across the menstrual cycle. *Hormones and Behavior* 51: 633–639.

Little, A. C., Jones, B. C., and Burt, D. M. 2007b. Preferences for symmetry in faces change across the menstrual cycle. *Biological Psychology* 76: 209–216.

Little, A. C., Saxton, T. K., Roberts, S. C., Jones, B. C., DeBruine, L. M., et al. 2010. Women's preferences for masculinity in male faces are highest during reproductive age-range and lower around puberty and post-menopause. *Psychoneuroendocrinology* 35: 912–920.

Mace, R. 1998. The coevolution of human fertility and wealth inheritance strategies. *Philosophical Transactions of the Royal Society of London B* 353: 389–397.

Maricic, T., Guntner, V., Georgiev, O., Gehre, S., Curlin, M., et al. 2012. A recent evolutionary change affects a regulatory element in the human *FOXP2* gene. *Molecular Biology and Evolution* 30: 844–852.

Marlowe, F. W. 2000. Paternal investment and the human mating system. *Behavioral Processes* 51: 45–61.

Massaro, D. W. and Stork, D. G. 1998. Speech recognition and sensory integration. *American Scientist* 86: 236–244.

McGurk, H. and Macdonald, J. 1976. Hearing lips and seeing voices. *Nature* 264: 746–748.

Miller, G. F. 2000. *The Mating Mind*. New York: Doubleday.

Miller, G. F. 2009. *Spent: Sex, Evolution and Consumerism*. New York: Penguin Group.

Miller, G. F., Tybur, J., and Jordan, B. 2007. Ovulatory cycle effects on tip earnings by lap-dancers: Economic evidence for human estrus? *Evolution and Human Behavior* 28: 375–381.

Miller, R. M., Sanchez, K., and Rosenblum, L. D. 2010. Alignment to visual speech information. *Attention, Perception and Psychophysics* 72: 1614–1625.

Miller, S. L. and Maner, J. K. 2010. Scent of a woman: Men's testosterone responses to olfactory ovulation cues. *Psychological Science* 21: 276–283.

Moorad, J. A., Promislow, D. E. L., Smith, K. R., and Wade, M. J. 2011. Mating system change reduces the strength of sexual selection in an American frontier population of the 19th century. *Evolution and Human Behavior* 32: 147–155.

Morgan, T. J. H., Uomini, N. T., Rendell, L. E., Chouinard-Thuly, L., Street, S. E., et al. 2015. Experimental evidence for the co-evolution of hominin tool-making teaching and language. *Nature Communications* 6: 6029.

Mueller, U. and Mazur, A. 2001. Evidence of unconstrained directional selection for male tallness. *Behavioral Ecology and Sociobiology* 50: 302–311.

Murdock, G. P. 1967. *Ethnographic Atlas*. Pittsburgh, PA: University of Pittsburgh Press.

Murugan, M., Harward, S., Scharff, C., and Mooney, R. 2013. Diminished *FoxP2* levels affect dopaminergic modulation of corticostriatal signaling important to song variability. *Neuron* 80: 1464–1476.

Nedelec, J. L. and Beaver, K. M. 2014. Physical attractiveness as a phenotypic marker of health: An assessment using a nationally representative sample of American adults. *Evolution and Human Behavior* 35: 456–463.

Palmer, C. T. 1991. Human rape: Adaptation or by-product? *Journal of Sex Research* 28: 365–386.

Pawłowski, B. and Dunbar, R. I. M. 1999. Impact of market value on human mate choice decisions. *Proceedings of the Royal Society of London B* 266: 281–285.

Penton-Voak, I. S., Perrett, D. I., Castles, D. L., Kobayashi, T., Burt, D. M., et al. 1999. Menstrual cycle alters face preference. *Nature* 399: 741–742.

Perusse, D. 1993. Cultural and reproductive success in industrial societies: Testing the relationship at the proximate and ultimate levels. *Behavioral and Brain Sciences* 16: 267–283.

Pettay, J. E., Lahdenpera, M., Rotkirch, A., and Lummaa, V. 2016. Costly reproductive competition between co-resident females in humans. *Behavioral Ecology* 27: 1601–1608.

Pilowsky, J. A. and Rubenstein, D. R. 2013. Social context and the lack of sexual dimorphism in song in an avian cooperative breeder. *Animal Behaviour* 85: 709–714.

Pinker, S. 1994. *The Language Instinct*. New York: W. Morrow & Co.

Pinker, S. 2003. Language as an adaptation to the cognitive niche. In S. Kirby and M. Christiansen (eds.), *Language Evolution: States of the Art*. New York: Oxford University Press.

Prokop, P. and Fedor, P. 2011. Physical attractiveness influences reproductive success of modern men. *Journal of Ethology* 29: 453–458.

Prokosch, M. D., Coss, R. G., Scheib, J. E., and Blozis, S. A. 2009. Intelligence and mate choice: Intelligent men are always appealing. *Evolution and Human Behavior* 30: 11–20.

Provost, M. P., Quinsey, V. L., and Troje, N. F. 2008. Differences in gait across the menstrual cycle and their attractiveness to men. *Archives of Sexual Behavior* 37: 598–604.

Rees, W. E. 2002. An ecological economics perspective on sustainability and prospects for ending poverty. *Population and Environment* 24: 15–46.

Reid, R. L. and van Vugt, D. A. 1987. Weight related change in reproductive function. *Fertility and Sterility* 48: 905–913.

Rivas, E. 2005. Recent use of signs by chimpanzees (*Pan troglodytes*) in interactions with humans. *Journal of Comparative Psychology* 119: 404–417.

Roberts, S. C., Gosling, L. M., Carter, V., and Petrie, M. 2008. MHC-correlated odour preferences in humans and the use of oral contraceptives. *Proceedings of the Royal Society of London B* 275: 2715–2722.

Roberts, S. C., Havlicek, J., Flegr, J., Hruskova, M., Little, A. C., et al. 2004. Female facial attractiveness increases during the fertile phase of the menstrual cycle. *Proceedings of the Royal Society of London B* 271: S270–S272.

Rose, M. 1998. *Darwin's Spectre*. Princeton, NJ: Princeton University Press.

Rose, N. A., Deutsch, C. J., and Le Boeuf, B. J. 1991. Sexual behavior of male northern elephant seals: III. The mounting of weaned pups. *Behaviour* 119: 171–192.

Rubenstein, D. R. and Lovette, I. J. 2007. Temporal environmental variability drives the evolution of cooperative breeding in birds. *Current Biology* 17: 1414–1419.

Rupp, H. A., James, T. W., Ketterson, E. D., Sengelaub, D. R., Janssen, E., and Heiman, J. R. 2009. Neural activation in the orbitofrontal cortex in response to male faces increases during the follicular phase. *Hormones and Behavior* 56: 66–72.

Scharff, C. and Petri, J. 2011. Evo-devo, deep homology and *FoxP2*: Implications for the evolution of speech and language. *Philosophical Transactions of the Royal Society B* 366: 2124–2140.

Schmitt, D. P., Shackelford, T. K., Duntley, J., Tooke, W., and Buss, D. M. 2001. The desire for sexual variety as a key to understanding basic human mating strategies. *Personal Relationships* 8: 425–455.

Scott, I. M., Clark, A. P., Josephson, S. C., Boyette, A. H., Cuthill, I. C., et al. 2014. Human preferences for sexually dimorphic faces may be evolutionarily novel. *Proceedings of the National Academy of Sciences USA* 111: 14388–14393.

Sedgewick, J. R., Flath, M. E., and Elias, L. J. 2017. Presenting your best self(ie): The influence of gender on vertical orientation of selfies on Tinder. *Frontiers in Psychology* 8: 604.

Sevi, B., Aral, T., and Eskenazi, T. 2017. Exploring the hook-up app: Low sexual disgust and high sociosexuality predict motivation to use Tinder for casual sex. *Personality and Individual Differences*, in press.

Shettleworth, S. J. 2010. *Cognition, Evolution and Behavior*. New York: Oxford University Press.

Shpigler, H. Y., Saul, M. C., Corona, F., Block, L., Ahmed, A. C., et al. 2017. Deep evolutionary conservation of autism-related genes. *Proceedings of the National Academy of Sciences USA* 114: 9654–9658.

Simmons, L. W., Firman, R. C., Rhodes, G., and Peters, M. 2004. Human sperm competition: testis size, sperm production and rates of extrapair copulations. *Animal Behaviour* 68: 297–302.

Singh, D. 1993. Adaptive significance of female physical attractiveness: Role of the waist-to-hip ratio. *Journal of Personality and Social Psychology* 65: 293–307.

Singh, D. and Bronstad, P. M. 2001. Female body odour is a potential cue to ovulation. *Proceedings of the Royal Society of London B* 268: 797–801.

Smith, E. A., Borgerhoff Mulder, M., and Hill, K. 2001. Controversies in the evolutionary social sciences: A guide for the perplexed. *Trends in Ecology & Evolution* 16: 128–135.

Sorenson, L. G. 1994. Forced extra-pair copulation in the white-cheeked pintail: Male tactics and female responses. *Condor* 96: 400–410.

Southam, L., Soranzo, N., Montgomerie, S. B., Frayling, T. M., and McCarthy, M. I. 2010. Is the thrifty genotype hypothesis supported by evidence based on confirmed type 2 diabetes and obesity susceptibility variants? *Diabetologia* 52: 846–851.

Speakman, J. R. 2013. Evolutionary perspectives on the obesity epidemic: Adaptive, maladaptive, and neutral view points. *Annual Review of Nutrition* 33: 289–317.

Swaddle, J. P. and Reierson, G. W. 2002. Testosterone increases perceived dominance but not attractiveness in human males. *Proceedings of the Royal Society of London B* 269: 2285–2289.

Symons, D. 1979. *The Evolution of Human Sexuality*. New York: Oxford University Press.

Thornhill, R. and Gangestad, S. W. 1999a. Facial attractiveness. *Trends in Cognitive Sciences* 3: 452–460.

Thornhill, R. and Gangestad, S. W. 1999b. The scent of symmetry: A human sex pheromone that signals fitness? *Evolution and Human Behavior* 20: 175–201.

Thornhill, R. and Palmer, C. T. 2000. *A Natural History of Rape: The Biological Bases of Sexual Coercion*. Cambridge, MA: MIT Press.

Thornhill, R. and Thornhill, N. W. 1983. Human rape: An evolutionary analysis. *Ethology and Sociobiology* 4: 137–173.

Tovée, M. J., Maisey, D. S., Emery, J. L., and Cornelissen, P. L. 1999. Visual cues to female physical attractiveness. *Proceedings of the Royal Society of London B* 266: 211–218.

Townsend, J. M. 1989. Mate selection criteria: A pilot study. *Ethology and Sociobiology* 10: 241–253.

Trivers, R. L. 2011. *The Folly of Fools: Deceit and Self-Deception in Human Affairs*. New York: Basic Books.

Tyson, G., Perta, V. C., Haddadi, H., and Seto, M. C. 2016. A first look at user activity on Tinder. https://arxiv.org/pdf/1607.01952.pdf

Ujhelyi, M. 1996. Is there any intermediate stage between animal communication and language? *Journal of Theoretical Biology* 18: 71–76.

Uomini, N. T. and Meyer, G. F. 2013. Shared brain lateralization patterns in language and Acheulean stone tool production: A functional transcranial doppler ultrasound study. *PLoS One* 8: e72693.

Vargha-Khadem, F., Gadian, D. G., Copp, A., and Mishkin, M. 2005. *FOXP2* and the neuroanatomy of speech and language. *Nature Reviews Neuroscience* 6: 131–138.

Waynforth, D. and Dunbar, R. I. M. 1995. Conditional mate choice strategies in humans: Evidence from lonely hearts advertisements. *Behaviour* 132: 755–779.

Wedekind, C., Seebeck, T., Bettens, F., and Paepke, A. J. 1995. MHC-dependent mate preferences in humans. *Proceedings of the Royal Society of London B* 260: 245–249.

West, S. A., El Mouden, C., and Gardner, A. 2011. Sixteen common misconceptions about the evolution of cooperation in humans. *Evolution and Human Behavior* 32: 231–262.

Wiederman, M. W. and Allgeier., E. R. 1992. Gender differences in mate selection criteria: Sociobiological or socioeconomic explanation? *Ethology and Sociobiology* 13: 115–124.

Wiederman, M. W. and Kendall, E. 1999. Evolution, sex, and jealousy: Investigation with a sample from Sweden. *Evolution and Human Behavior* 20: 121–128.

Williams, G. C. 1996. *The Pony Fish's Glow*. New York: Basic Books.

Wilson, E. O. 1975. *Sociobiology: The New Synthesis*. Cambridge, MA: Harvard University Press.

Wilson, M. L., Miller, C. M., and Crouse, K. N. 2017. Humans as a model species for sexual selection research. *Proceedings of the Royal Society of London B* 284: 20171320.

Winking, J., Kaplan, H., Gurven, M., and Rucas, S. 2007. Why do men marry and why do they stray? *Proceedings of the Royal Society of London B* 274: 1643–1649.

Wolf, A. M. and Colditz, G. A. 1998. Current estimates of the economic cost of obesity in the United States. *Obesity Research* 6: 97–106.

Wood, B. and Harrison, T. 2011. The evolutionary context of the first hominins. *Nature* 470: 347–352.

Young, L. J. and Barret, C. E. 2015. Can oxytocin treat autism? *Science* 347: 825–826.

Index

Page numbers in *italic* denote entries that are included in a figure.

A

B

E

F

H

N

O

S

T

U

V

About the Book

Publisher: Andrew D. Sinauer
Acquisitions Editor: Rachel Meyers
Production Editor: Martha Lorantos
Production Manager: Chris Small
Production Specialist: Beth Roberge Friedrichs
Book Designer: Joanne Delphia
Photo Researcher: Mark Siddall
Copy Editor: Elizabeth Pierson